The Genesis of General Relativity

BOSTON STUDIES IN THE PHILOSOPHY OF SCIENCE

VOLUME 250

The Genesis of General Relativity

Edited by Jürgen Renn

Volume 3

GRAVITATION IN THE TWILIGHT OF CLASSICAL PHYSICS: BETWEEN MECHANICS, FIELD THEORY, AND ASTRONOMY

Editors

Jürgen Renn and Matthias Schemmel
Max Planck Institute for the History of Science, Germany

Associate Editors

Christopher Smeenk
UCLA, U.S.A.

Christopher Martin
Indiana University, U.S.A.

Assistant Editor

Lindy Divarci
Max Planck Institute for the History of Science, Germany

 Springer

A C.I.P. Catalogue record for this book is available from the Library of Congress.

ISBN-10 1-4020-3999-9 (HB)
ISBN-13 978-1-4020-3999-7 (HB)
ISBN-10 1-4020-4000-8 (e-book)
ISBN-13 978-1-4020-4000-9 (e-book)

As a complete set for the 4 volumes

Published by Springer,
P.O. Box 17, 3300 AA Dordrecht, The Netherlands.

www.springer.com

Printed on acid-free paper

TABLE OF CONTENTS

Volume 3

Gravitation in the Twilight of Classical Physics: An Introduction 1
Jürgen Renn and Matthias Schemmel

The Gravitational Force between Mechanics and Electrodynamics

The Third Way to General Relativity: Einstein and Mach in Context 21
Jürgen Renn

Source text 1901: Gravitation 77
Jonathan Zenneck

Source text 1900: Considerations on Gravitation 113
Hendrik A. Lorentz

Source text 1896: Absolute or Relative Motion? 127
Benedict and Immanuel Friedlaender

Source text 1904: On Absolute and Relative Motion 145
August Föppl

An Astronomical Road to a New Theory of Gravitation

The Continuity between Classical and Relativistic Cosmology
in the Work of Karl Schwarzschild 155
Matthias Schemmel

Source text 1897: Things at Rest in the Universe 183
Karl Schwarzschild

A New Law of Gravitation Enforced by Special Relativity

Breaking in the 4-Vectors: the Four-Dimensional Movement
in Gravitation, 1905–1910 193
Scott Walter

Source text 1906: On the Dynamics of the Electron (Excerpts) 253
Henri Poincaré

Source text 1908: Mechanics and the Relativity Postulate 273
Hermann Minkowski

Source text 1910: Old and New Questions in Physics (Excerpt) 287
Hendrik A. Lorentz

The Problem of Gravitation as a Challenge
for the Minkowski Formalism

The Summit Almost Scaled: Max Abraham as a Pioneer
of a Relativistic Theory of Gravitation 305
Jürgen Renn

Source text 1912: On the Theory of Gravitation 331
Max Abraham

Source text 1912: The Free Fall 341
Max Abraham

Source text 1913: A New Theory of Gravitation 347
Max Abraham

Source text 1915: Recent Theories of Gravitation 363
Max Abraham

A Field Theory of Gravitation in the
Framework of Special Relativity

Einstein, Nordström, and the Early Demise of Scalar,
Lorentz Covariant Theories of Gravitation 413
John D. Norton

Source text 1912: The Principle of Relativity and Gravitation 489
Gunnar Nordström

Source text 1913: Inertial and Gravitational Mass
in Relativistic Mechanics 499
Gunnar Nordström

Source text 1913: On the Theory of Gravitation from the Standpoint
of the Principle of Relativity 523
Gunnar Nordström

Source text 1913: On the Present State of the Problem of Gravitation . . . 543
Albert Einstein

From Heretical Mechanics to a New Theory of Relativity

Einstein and Mach's Principle. 569
Julian B. Barbour

Source text 1914: On the Relativity Problem 605
Albert Einstein

Source text 1920: Ether and the Theory of Relativity. 613
Albert Einstein

Volume 4
(parallel volume)

From an Electromagnetic Theory of Matter to a New Theory of Gravitation

Mie's Theories of Matter and Gravitation. 623
Christopher Smeenk and Christopher Martin

Source text 1912–1913: Foundations of a Theory of Matter (Excerpts). . 633
Gustav Mie

Source text 1914: Remarks Concerning Einstein's
Theory of Gravitation 699
Gustav Mie

Source text 1915: The Principle of the Relativity of the
Gravitational Potential. 729
Gustav Mie

Source text 1913: The Momentum-Energy Law in the
Electrodynamics of Gustav Mie 745
Max Born

Including Gravitation in a Unified Theory of Physics

The Origin of Hilbert's Axiomatic Method 759
Leo Corry

Hilbert's Foundation of Physics: From a Theory of Everything
to a Constituent of General Relativity. 857
Jürgen Renn and John Stachel

Einstein Equations and Hilbert Action: What is Missing
on Page 8 of the Proofs for Hilbert's First Communication
on the Foundations of Physics? 975
Tilman Sauer

Source text 1915: The Foundations of Physics
(Proofs of First Communication) 989
David Hilbert

Source text 1916: The Foundations of Physics
(First Communication) 1003
David Hilbert

Source text 1917: The Foundations of Physics
(Second Communication) 1017
David Hilbert

From Peripheral Mathematics to a New Theory of Gravitation

The Story of Newstein or: Is Gravity just another Pretty Force? 1041
John Stachel

Source text 1877: On the Relation of Non-Euclidean
Geometry to Extension Theory 1079
Hermann Grassmann

Source text 1916: Notion of Parallelism on a General Manifold
and Consequent Geometrical Specification of the Riemannian
Curvature (Excerpts) 1081
Tullio Levi-Civita

Source text 1918: Purely Infinitesimal Geometry (Excerpt) 1089
Hermann Weyl

Source text 1923: The Dynamics of Continuous Media
and the Notion of an Affine Connection on Space-Time 1107
Elie Cartan

Index: Volumes 3 and 4 1131

JÜRGEN RENN AND MATTHIAS SCHEMMEL

GRAVITATION IN THE TWILIGHT OF CLASSICAL PHYSICS: AN INTRODUCTION

AIM AND STRUCTURE OF THIS BOOK

More than is the case for any other theory of modern physics, general relativity is usually seen as the work of one man, Albert Einstein. In taking this point of view, however, one tends to overlook the fact that gravitation has been the subject of controversial discussion since the time of Newton. That Newton's theory of gravitation assumes action at a distance, i.e., action without an intervening mechanism or medium, was perceived from its earliest days as being problematical. Around the turn of the last century, in the twilight of classical physics, the problems of Newtonian gravitation theory had become more acute. Consequently, there was a proliferation of alternative theories of gravitation which were quickly forgotten after the triumph of general relativity. In order to understand this triumph, it is necessary to compare general relativity to its contemporary competitors. As we shall see, general relativity owes much to this competition. A historical analysis of the struggle between alternative theories of gravitation and the different approaches to the problem of gravitation thus complements the analysis of Einstein's efforts. An account of the genesis of general relativity that does not discuss these competitors remains incomplete and biased. At the same time, this wider perspective on the emergence of general relativity provides an exemplary case of alternatives in the history of science, presenting a whole array of alternative theories of gravitation and the eventual emergence of a clear winner. It is thus an ideal topic for addressing long-standing questions in the philosophy of science on the basis of detailed historical evidence.

The present book, comprising volumes 3 and 4 of the series, discusses alternative theories of gravitation that were relevant to the genesis of general relativity and thus constitute its immediate scientific context. Many of these theories figured in the discussions of Einstein and his contemporaries. The set of theories covered here is not complete as far as gravitation theories in the late nineteenth and early twentieth centuries are concerned. But even a comprehensive treatment of this narrower set of theories represents a considerable challenge for the history of science. Unlike Einstein's work, many of the theories dealt with here are known only to a few specialists. This situation has only just begun to improve through recent work on the genesis of general relativity and much remains to be done in the future.

Jürgen Renn (ed.). *The Genesis of General Relativity*, Vol. 3
Gravitation in the Twilight of Classical Physics: Between Mechanics, Field Theory, and Astronomy.
© 2007 Springer.

What is presented in these volumes are two types of texts, sources and interpretations. The sources are key documents relevant to the history of general relativity. Many of these texts were originally written in German and are presented here for the first time in English translation. The interpretative texts are essays, most of them specially written for these volumes. They provide historical context and analysis of the theories presented in the sources. The book is divided into sections reflecting a classification of approaches to the problem of gravitation. Different subdisciplines of classical physics generated different ways of approaching the problem of gravitation. The emergence of special relativity further raised the number of possible approaches while creating new requirements that all approaches had to come to terms with. Each section of this book is dedicated to one of these approaches and, as a rule, consists of an historical essay and several sources.

THE UNFOLDING OF ALTERNATIVE THEORIES OF GRAVITATION

From the perspective of an epistemologically oriented history of science, the unfolding of alternative theories of gravitation in the twilight of classical physics can be interpreted as the realization of the potential embodied in the knowledge system of classical physics to address the problem of gravitation, this knowledge system eventually being transformed by the special-relativistic revolution. The dynamics of this unfolding was largely governed by internal tensions of the knowledge system rather than by new empirical knowledge, which at best played only a minor role. A central problem of the Newtonian theory of gravitation was, as already mentioned, that it assumed the action between two attracting bodies to be instantaneous and that it did not provide any explanation for the instantaneous transport of action along arbitrary distances. This characteristic feature of the gravitational force, called action at a distance, became even more dubious after the mid-nineteenth century when it was recognized that electromagnetic forces did not comply with the idea of action at a distance. This internal tension of the knowledge system of classical physics was intensified, but not created, by the advent of the theory of special relativity, according to which the notion of an instantaneous distance between two bodies as it appears in Newton's force law can no longer be accepted.

The attempts to resolve these kinds of tensions typically crystallized around mental models representing the gravitational interaction on the basis of other familiar physical processes and phenomena. A mental model is conceived here as an internal knowledge representation structure serving to simulate or anticipate the behavior of objects or processes, like imagining electricity as a fluid. Mental models are flexible structures of thinking that are suitable for grasping situations about which no complete information is available. They do so by relying on default assumptions that result from prior experiences and can be changed if additional knowledge becomes available without having to give up the model itself.

Thus, in what may be called the *gas model*, gravitation could be conceived as resulting from pressure differences in a gaseous aether. Or, in what may be called the *umbrella model*, the attraction of two bodies could be imagined to result from the mutual shielding of the two bodies immersed in an aether whose particles rush in random directions and, in collisions with matter atoms, push them in the direction of the particles' motions. Or one could think of gravitation in analogy to the successful description of electromagnetism by the *Lorentz model*, accepting a dichotomy of gravitational field on the one hand and charged particles—masses—that act as sources of the field on the other. The elaboration of these approaches, with the help of mathematical formalism, led typically to a further proliferation of alternative approaches and, at the same time, provided the tools to explore these alternatives to a depth that allowed new tensions to be revealed. The history of these alternative approaches can thus be read, in a way similar to Einstein's work, as an interaction between the physical meaning embodied in various models and the mathematical formalism used to articulate them.

THE POTENTIAL OF CLASSICAL PHYSICS

The history of treatments of gravitation in the nineteenth century reflects the transition from an era in which mechanics constituted the undisputed fundamental discipline of physics to an era in which mechanics became a subdiscipline alongside electrodynamics and thermodynamics.

From the time of its inception, the action-at-a-distance conception of Newtonian gravitation theory was alien to the rest of mechanics, according to which interaction always involved contact. This explains the early occurrence of attempts to interpret the gravitational force by means of collisions, for instance, by invoking the umbrella model described above. During these early days the comparison of the gravitational force to electric and magnetic forces had already been suggested as well. However, the analogy with electricity and magnetism became viable only after theories on these subjects had been sufficiently elaborated. There were even attempts at thermal theories of gravitation after thermodynamics had developed into an independent subdiscipline of physics. Besides providing new foundational resources for approaching the problem of gravitation, the establishment of independent subdisciplines and the questioning of the primacy of mechanics that resulted from it affected the development of the theoretical treatment of gravitation in yet another way, namely through the emergence of revisionist formulations of mechanics. This *heretical mechanics*, as we shall call it, consisted in attempts to revise the traditional formulation given to mechanics by Newton, Euler and others, and often amounted to questioning its very foundations.

Approaches to the problem of gravitation in the context of these developments of classical physics are covered by the first and last sections of the first volume of this book, *The Gravitational Force between Mechanics and Electrodynamics*, and *From Heretical Mechanics to a New Theory of Relativity*. Further stimuli for rethinking gravitation came from the development of astronomy and mathematics. This approach is addressed by the second section, *An Astronomical Road to a New Theory of Gravitation*.

The Mechanization of Gravitation

Before the advent of the special theory of relativity, the validity of Newton's law of gravitation was essentially undisputed in mainstream physics. Alternative laws of gravitation were, of course, conceivable but Newton's law proved to be valid to a high degree of precision. While the minute discrepancies between the observed celestial motions and those predicted by Newtonian theory, most prominently the advance of Mercury's perihelion, could be resolved by one of these alternatives, they could also be resolved by adjusting lower-level hypotheses such as those regarding the distribution of matter in the solar system. In any case, the empirical knowledge at that time did not force a revision of Newtonian gravitation theory. The more pressing problem of this theory was that it did not provide a convincing model for the propagation of the gravitational force.

The most elaborate theories to address this problem made use of the umbrella model. These theories start from the idea of an impact of aether particles on matter, as formulated by Le Sage in the late eighteenth century. The gravitational aether is imagined to consist of particles that move randomly in all directions. Whenever such an aether atom hits a material body it pushes the body in the direction of its movement. A single body remains at rest since the net impact of aether particles from all sides adds up to zero. However, if two bodies are present, they partly shield each other from the stream of aether particles. As a result, the impact of aether particles on their far sides outweighs that on their near sides and the two bodies are driven towards each other.

Caspar Isenkrahe, Sir William Thomson (Lord Kelvin), and others developed different theories based on this idea in the late nineteenth century. But regardless of the details, this approach suffers from a fundamental problem related to the empirical knowledge about the proportionality of the force of gravity with mass. In order to take this into account one needs to allow the aether particles to penetrate a material body in such a way that they can interact equally with all of its parts. This requirement is better fulfilled the more transparent matter is to the aether particles. But, the more transparent matter is, the less shielding it provides from the aether particles on which the very mechanism for explaining gravity is based. Hence, without shielding there is no gravitational effect; without penetration there is no proportionality of the gravitational effect to the total mass. Furthermore, in theories explaining gravitation by the mechanical action of a medium, the problem of heat exchange between the medium and ordinary matter arises (in analogy to electromagnetic heat radiation), in most approaches leading to an extreme heating of matter.

From a broader perspective, such attempts at providing a mechanical explanation of gravity had lost their appeal by the end of the nineteenth century after the successful establishment of branches of physics that could not be reduced to mechanics, such as Maxwell's electrodynamics and Clausius' thermodynamics. Nevertheless, this development led indirectly to a contribution of the mechanical tradition to solving the problem of gravitation by provoking the emergence of revised formulations of mechanics, referred to here as heretical mechanics.

Heretical Mechanics

A critical revision of mechanics, pursued in different ways by Carl Neumann, Ludwig Lange, and Ernst Mach among others, had raised the question of the definition and origin of inertial systems and inertial forces, as well as their possible relations to the distribution of masses in the universe. Through the latter issue, this revision of mechanics was also important for the problem of gravitation. It also gave rise to attempts at formulating mechanics in purely relational terms, that is, exclusively in terms of the mutual distances of the particles and derivatives of these distances. Such attempts are documented, for instance, in the texts presented in this book of Immanuel and Benedict Friedlaender and of August Föppl. As becomes clear from these texts, heretical mechanics contributed to understanding the relation between gravitational and inertial forces as both are due to the interaction of masses. According to Föppl there must be velocity-dependent forces between masses although he did not think of these forces as being gravitational. The Friedlaender brothers also conceived of inertia as resulting from an interaction between masses and did speculate on its possible relation to gravitation. In spite of such promising hints, heretical mechanics remained marginal within classical physics, in part because it lacked a framework with which one could explore the relation between gravitation and inertia. This relation was established by Einstein within the framework of field theory, first in 1907 through his principle of equivalence, and more fully with the formulation of general relativity.

Einstein's successful heuristic use of Machian ideas in his relativistic theory of gravitation encouraged the mechanical tradition to continue working toward a purely relational mechanics in the spirit of Mach. Attempts in this direction were made by Hans Reissner, Erwin Schrödinger, and, more recently, Julian Barbour and Bruno Bertotti. The success of general relativity provided a touchstone for the viability of these endeavors. At the same time, the question to which extent the issues raised by heretical mechanics, such as a relational understanding of inertia, have been settled by general relativity is still being discussed today.

From Peripheral Mathematics to a New Theory of Gravitation

The success or failure of a physical idea hinges to a large extent on the mathematical tools available for expressing it. In view of the crucial role of the mathematical concept of affine connection at a later state in the development of the general theory of relativity, it is interesting to consider the impact this tool might have had on the formulation of physical theories had it been part of mathematics by the latter half of the nineteenth century. That this counter-factual assumption is actually not that far-fetched can be seen from the work of Hermann Grassmann, Heinrich Hertz, Tullio Levi-Civita, and Elie Cartan, in part reproduced in this book. Such a fictive development might have given rise to a kind of heretical gravitation theory driven by peripheral mathematics and formulated by some "Newstein" long before the advent of special relativity. Perhaps the search for a different conceptualization of mechanics in which gravitation and inertia are treated alike, as is the case according to Einstein's equivalence principle, could have provided a physical motivation for such an alternative formulation of classical mechanics with the help of affine connections. Perhaps Heinrich Hertz's attempt to exclude forces from mechanics, replacing them by geometrical constraints, might have served as a starting point for such a development, triggering a geometrization of physics, had it not been so marginal to the mainstream of late nineteenth-century physics.

As with ordinary classical mechanics, Newstein's theory would have eventually conflicted with the tradition of electrodynamics and its implication of a finite propagation speed for physical interactions, which ultimately leads to the metrical structure of special relativity with its constraints on physical interactions. Then the problem that arose from this conflict could be—in contrast to the actual course of history—formulated directly in terms of the compatibility of two well-defined mathematical structures, the affine connection expressing the equality in essence of gravitation and inertia, and the metric tensor expressing the causal structure of spacetime. This formulation of the problem would have smoothed the pathway to general relativity considerably since the heretical aspect of Einstein's work—the incorporation of the equality in essence of gravitation and inertia—would have already been implemented in Newstein's predecessor theory. General relativity might thus have been the outcome of mainstream research.

The Potential of Astronomy

Another field of classical science that might have contributed more than it actually did to the emergence of general relativity is astronomy. This is made evident by the sporadic interventions by astronomers such as Hugo von Seeliger, who questioned the seemingly self-evident foundations of the understanding of the universe in classical science. Their work was stimulated by new mathematical developments such as the emergence of non-Euclidean geometries or by heretical mechanics insofar as it raised questions relevant to astronomy, for instance, concerning the definition of inertial systems. It was further stimulated by the recognition of astronomical deviations from the predictions of Newton's law (such as the perihelion advance of Mercury), or by the paradoxes resulting from applying classical physics to the universe-at-large when this is assumed to be infinite (such as the lack of definiteness in the expression of the gravitational force, or Olbers' paradox of the failure of the night sky to be as bright as the Sun).

Although the full extent to which these problems were connected became clear only after the establishment of general relativity, the astronomer Karl Schwarzschild, who was exceptional in his interdisciplinary outlook, addressed many of them and was even able to relate them to one another. He explored, for instance, the cosmological implications of non-Euclidean geometry and considered the possibility of an anisotropic large-scale structure of the universe in which interial frames can only be defined locally. With less entrenched disciplinary boundaries of late nineteenth-century classical science, such considerations could have had wider repercussions on the foundations of physics, perhaps giving rise to the emergence of a non-classical cosmology.

A Thermodynamic Analogy

In rejecting the assumption of an instantaneous propagation of gravitational interactions, it makes sense to modify classical gravitation theory by drawing upon analogies with other physical processes that have a finite propagation speed, such as the propagation of electromagnetic effects or the transport of heat in matter. Such analogies obviously come with additional conceptual baggage. A gravitational theory built according to the model of electrodynamic field theory, for instance, was confronted with the question of whether the gravitational analogue of electromagnetic waves really exist, or the question of why there is only one kind of charge (gravitational mass) in gravitation theory as opposed to two in electromagnetism (positive and negative charge). To avoid such complications, one could also consider amending Newtonian theory by extending the classical Poisson equation for the gravitational potential into a diffusion equation by adding a term with a first-order time derivative term, exploiting the analogy with heat transport in thermodynamics. In 1911, such a theory was proposed by Gustav Jaumann without, however, taking into account the spacetime framework of special relativity. As a consequence, it had little impact.

Electromagnetism as a Paradigm for Gravitation

Since early modern times magnetism served as a model for action at a distance as it apparently occurs between the constituents of the solar system. However, as long as there was no mathematical formulation describing magnetic forces, no quantitative description of gravitation could be obtained from this analogy. After Newton had established a quantitative description of gravitation, this could now conversely be used as a model for describing magnetic and electric forces, as realized in the laws of Coulomb, Ampère, and Biot-Savart. With the further development of electromagnetic theory as represented by velocity-dependent force laws and Maxwellian field theory, it regained its paradigmatic potential for understanding gravitation. After the striking success of Einstein's field theory of gravitation, which describes the gravitational force in terms of the geometry of spacetime, gravitation took the lead again as attempts were made that aimed at a geometrical description of electromagnetism and the other fundamental interactions with a view toward the unification of all natural forces. Such attempts are still being made today.

The motive of unification also underlay nineteenth-century attempts to reduce gravitation to electricity, such as those of Ottaviano Fabrizio Mossotti and Karl Friedrich Zöllner, who interpreted gravity as a residual effect of electric forces. They assumed that the attractive electric force slightly outweighs the repulsive one, resulting in a universal attraction of all masses built up from charged particles. Ultimately, however, this interpretation amounts to little more than the statement that there is a close analogy between the fundamental force laws of electrostatics and Newtonian gravitation.

The paradigmatic role of electromagnetism for gravitation theory was boosted dramatically when electrodynamics emerged as the first field theory of physics. A field-theoretic reformulation of Newtonian gravity modelled on electrostatics was provided by the Poisson equation for the Newtonian gravitational potential. Even though the Poisson equation was merely a mathematical reformulation of Newton's law, it had profound implications for the physical interpretation of gravitation and introduced new possibilities for the modification of Newtonian gravitation theory. The analogy with electromagnetism raised the question of whether gravitational effects propagate with a finite speed like electromagnetic effects. A finite speed of propagation further suggested the existence of velocity-dependent forces among gravitating bodies, amounting to a gravitational analogue to magnetic forces. It also suggested the possibility of gravitational waves. In short, a field theory of gravitation opened up a whole new world of phenomena that might or might not be realized in nature.

The uncertainty of the existence of such phenomena was in any case not the most severe problem that a field theory of gravitation was confronted with. If gravitation is conceived of as a field with energy content, the fact that like "charges" always attract has a number of problematic consequences. First and foremost, ascribing energy to the gravitational field itself leads to a dilemma that does not occur in the electromagnetic case. In the latter case, the work performed by two attracting charges as they approach each other can be understood to be extracted from the field, and the field energy disappears when they meet at one point. In contrast, while work can also be performed by two approaching gravitating masses, the field energy is enhanced, rather than diminished, as they come together at one point. (Accordingly no equivalent of a black hole is known in electrodynamics.) As Gustav Mie explains in his paper on the gravitational potential presented in this book, the gravitational field is peculiar in that it becomes stronger when work is released. While a similar effect occurs with the magnetic field of two current-bearing conductors, the source of the energy is obvious in this case. The energy comes from an external energy supply such as a battery. Such an external supply is missing in the case of gravitation. A plausible escape strategy was to assume that the energy of the gravitational field is negative so that, when the field becomes stronger, positive energy is released, which can be exploited as work. For the plausible option of formulating a theory of gravitation in strict analogy to electrodynamics by simply postulating Maxwell's equations with appropriately changed signs for the gravitational field, this negative energy assumption has dramatic consequences when considering dynamic gravitational fields. A minute deviation of a gravitating system from equilibrium will cause the field to release more and more energy, while the system deviates further and further from its original state of equilibrium. In fact, due to the reversed sign, gravitational induction, if conceived in analogy to electromagnetic induction, becomes a self-accelerating process. This will be referred to here as the *negative energy problem*.

Despite this problem, Hendrik Antoon Lorentz took up the thread of Mossotti and others and proposed to treat gravitation as a residual force resulting from electromagnetism. While the electromagnetic approach to gravitation offered, in principle, the possibility to account for observed deviations from Newtonian gravitation theory, the field theories actually elaborated by Lorentz and others failed to yield the correct value for the perihelion advance of Mercury, a commonly used touchstone.

All in all, the analogy of gravitation with electromagnetism, promising as it must have appeared, could not be as complete as advocated by its proponents. The considerable potential of the tradition of field theory for formulating a new theory of gravitation still needed to be explored and the key to disclosing its riches had yet to be discovered.

The attempts to subsume gravitation under the familiar framework of electromagnetism were later followed by approaches that aimed at a unification of physics on a more fundamental level, still focusing, however, on gravitation and electromagnetism. The most prominent attempts along these lines, contemporary to the genesis of general relativity, were those of Gustav Mie and David Hilbert. Their works are covered in the sections *From an Electromagnetic Theory of Matter to a New Theory of Gravitation* and *Including Gravitation in a Unified Theory of Physics*. These attempts, however, only led to a formal integration of the two forces without offering any new insights into the nature of gravity.

The key to successfully exploiting the resources of field theory for a new theory of gravitation was only found when the challenge of formulating a gravitational field theory was combined with insights from heretical mechanics. Instead of attempting a formal unification of two physical laws, Einstein combined the field theoretic approach with the idea of an equality in essence of gravitation and inertia, and eventually achieved an integration of two knowledge traditions hitherto separated due to the high degree of specialization of nineteenth-century physics.

THE CHALLENGE OF SPECIAL RELATIVITY FOR GRAVITATION

The advent of special relativity in 1905 made the need for a revision of Newtonian gravitation theory more urgent since an instantaneous propagation of gravitation was incompatible with the new spacetime framework in which no physical effect can propagate faster than the speed of light. A revision of this kind could be achieved in various ways. One could formulate an action-at-a-distance law involving a finite time of propagation as had been developed in electromagnetism, e.g. by Wilhelm Weber. Or one could formulate a genuine field theory of gravitation. The four-dimensional formulation of special relativity emerging from the work of Henri Poincaré, Hermann Minkowski, and Arnold Sommerfeld brought about a set of clearly distinguished alternative approaches for realizing such a field theory of gravitation. Eventually, however, due to the implications of special relativity not only for the kinematic concepts of space and time but also for the dynamic concept of mass, gravitation was bursting out of the framework of special relativity.

A New Law of Gravitation Enforced by Special Relativity

The simplest way to make gravitation theory consistent with special relativity was to formulate a new direct particle interaction law of gravitation in accordance with the conditions imposed by special relativity, e.g., that the speed of propagation of the gravitational force be limited by the speed of light. This kind of approach, which was pursued by Poincaré in 1906 and by Minkowski in 1909, and which is presented here in the section *A New Law of Gravitation Enforced by Special Relativity*, could rely on the earlier attempts to introduce laws of gravitation with a finite speed of propagation. However, the stricter condition of Lorentz invariance now had to be satisfied.

While the formulation of a relativistic law of gravitation could solve the particular problem of consolidating gravitation theory with the new theory of special relativity, it disregarded older concerns about Newtonian gravitation, such as those relating to action at a distance. Furthermore, questions concerning fundamental principles of physics such as that of the equality of action and reaction emerged in these formulations. It remained, in any case, unclear to which extent the modified laws of gravitation could be integrated into the larger body of physical knowledge.

Towards a Field Theory of Gravitation

More important and more ambitious than the attempts at a new direct-particle interaction law of gravitation was the program of formulating a new field theory of gravitation. As pointed out above, if gravitation—in analogy to electromagnetism—is transmitted by a field with energy content, the fact that in the gravitational case like "charges" (masses) attract has problematic consequences, such as the negative energy problem. A promising approach to the negative energy problem was the assumption that masses also have energy content defined in such a way that the energy content of two attracting masses decreases when the masses approach each other. This effect can in turn be ascribed to a direct contribution of the gravitational potential to the energy content of the masses. Hence, there is a way to infer a relation between mass and energy content by considering the negative energy problem of a gravitational field theory.

The above considerations on the negative energy problem of gravitational field theory suggest that the potential plays a greater role in such a theory than it does in classical electromagnetic field theory. How to represent the gravitational potential is further directly connected with the question of how to represent the gravitational mass, or, more generally, the source of the gravitational field, since both are related through the field equation. The following three mathematical types of potentials were considered before the establishment of general relativity with the corresponding implications on the field strengths and the sources.

- *Scalar theories*. Potential and source are Lorentz scalars and the field strength is a (Lorentz) four-vector.
- *Vector theories*. Potential and source are four-vectors and the field is what was then called a "six-vector" (an antisymmetric second-rank tensor).
- *Tensor theories*. Potential and source are symmetric second-rank tensors and the field is represented by some combination of derivatives of the potential.

From what has been said above about a theory of gravitation construed in analogy with electrodynamics, the problems of a vector theory become apparent. In contrast to the electromagnetic case, where the charge density is one component of the four-current, the gravitational mass density is not one component of a four-vector. From this it follows in particular that no expression involving the mass is available to solve the negative energy problem by forming a scalar product of source and potential in order to adjust the energy expression.

Having thus ruled out vector theories, only scalar theories and tensor theories remain. Einstein's theories, in particular the *Entwurf* theory and his final theory of general relativity, belong to the latter class. Further alternative tensor theories of gravitation were proposed, but only after the success of general relativity, which is why they are not discussed here. As concerns scalar theories, a further branching of alternatives occurs as shall be explained in the following.

Every attempt to embed the classical theory of gravitation into the framework of special relativity had to cope not only with its kinematic implications, that is, the new spacetime structure which required physical laws be formulated in a Lorentz covariant manner, but also with its dynamical implications, in particular, the equivalence of energy and mass expressed by the formula $E = mc^2$. Since, in a gravitational field, the energy of a particle depends on the value of the gravitational potential at the position of the particle, the equivalence of energy and mass suggests that either the particle's mass or the speed of light (or both) must also be a function of the potential. Choosing the speed of light as a function of the potential immediately exits the framework of special relativity, which demands a constant speed of light. It thus may seem that choosing the inertial mass to vary with the gravitational potential is preferable since it allows one to stay within that framework.

According to contemporary evidence and later recollections, Einstein in 1907 explored both possibilities, a variable speed of light and a variable mass. He quickly came to the conclusion that the attempt to treat gravitation within the framework of special relativity leads to the violation of a fundamental tenet of classical physics, which may be called *Galileo's principle*. It states that in a gravitational field all bodies fall with the same acceleration and that hence two bodies dropped from the same height with the same initial vertical velocity reach the ground simultaneously. The latter formulation generalizes easily to special relativity. If the inertial mass increases with the energy content of a physical system, as is implied by special relativity, a body with a horizontal component of motion will have a greater inertial mass than the same body without such a motion, and hence fall more slowly than the latter.

The same conclusion can be drawn by purely kinematic reasoning in the framework of special relativity. Consider two observers, one at rest, the other in uniform horizontal motion. When the two observers meet, they both drop identical bodies and watch them fall to the ground. From the viewpoint of the stationary observer, the body he has dropped will fall vertically, while the body the moving observer has dropped will fall along a parabolic trajectory. From the viewpoint of the moving observer, the roles of the two bodies are interchanged: the first body will fall along a parabolic trajectory while the second will fall vertically.

If one now assumes that, in the reference frame of the stationary observer, the bodies will touch the ground simultaneously, as is required by Galileo's principle in the above formulation, the same cannot hold true in the moving system due to the relativity of simultaneity. In other words, Galileo's principle cannot hold for both observers. Thus, the assumption of Galileo's principle leads to a violation of the principle of relativity. On the other hand, if one assumes, in accordance with the principle of relativity, that the two observers both measure the same time of fall for the body falling vertically in their respective frame of reference, the time needed for the body to fall along a parabolic path can be determined from this time by taking time dilation into account. It thus follows that the time needed for the fall along a parabolic path is longer than the time needed for the vertical fall, in accordance with the conclusion drawn from the dynamical assumption of a growth of inertial mass with energy content.

Both of the possibilities considered by Einstein, a dependence on the gravitational potential either of the speed of light or of the inertial mass, were later explored by Max Abraham and Gunnar Nordström respectively. These theories, which represented the main competitors of Einstein's theories of gravitation, are discussed in the sections *The Problem of Gravitation as a Challenge for the Minkowski Formalism* and *A Field Theory of Gravitation in the Framework of Special Relativity*, to which the remainder of this introduction is devoted.

The Problem of Gravitation as a Challenge for the Minkowski Formalism

The assumption of a dependence of the speed of light on the gravitational potential made it necessary to generalize the Minkowski formalism, although the full consequences of this generalization became clear only gradually. It was Max Abraham who took the first steps in this direction by implementing Einstein's 1907 suggestion of a variable speed of light related to the gravitational potential within this formalism. Questioned by Einstein about the consistency of the modified formalism with

Minkowski's framework, he introduced the variable line element of a non-flat four-dimensional geometry.

Abraham's theory stimulated Einstein in 1912 to resume work on a theory of gravitation. Apart from developing his own theory, Abraham also made perceptive observations on alternative options for developing a relativistic theory of gravity, on internal difficulties as well as on physical and astronomical consequences such as energy conservation in radioactive decay or the stability of the solar system.

A Field Theory of Gravitation in the Framework of Special Relativity?

While Abraham explored the implications of a variable speed of light, Nordström pursued the alternative option of a variable mass. Nordström thus remained within the kinematic framework of special relativity. As in all such approaches, however, he did so at the price of violating to some extent Galileo's principle.

More importantly, Nordström also faced the problem that in a special relativistic theory of gravitation the dynamical implications of special relativity need to be taken into account as well. These dynamical consequences suggested, for example, ascribing to energy not only an inertial but also a gravitational mass, which immediately implies that light rays are curved in a gravitational field. This conclusion, however, is incompatible with special relativistic electrodynamics in which the speed of light is constant.

Another implication of the dynamic aspects of special relativity concerns the source of the gravitational field. If any quantity other than the energy-momentum tensor of matter is chosen as a source-term in the gravitational field equation, as is the case in all scalar theories including Nordström's, gravitational mass cannot be fully equivalent to inertial mass, whose role has been taken in special relativistic physics by the energy-momentum tensor. However, while such conceptual considerations cast doubt on the viability of special relativistic theories of gravitation, they were not insurmountable hurdles for such theories. In fact, Nordström's final version of his theory remained physically viable as long as no counter-evidence was known. Einstein's successful calculation of Mercury's perihelion advance on the basis of general relativity in late 1915 undermined Nordström's theory, which did not yield the correct value. This, however, did not constitute a fatal blow as long as other astrophysical explanations of Mercury's anomalous motion remained conceivable. The fatal blow only came when the bending of light in a gravitational field was observed in 1919. Nordström's theory did not predict such an effect. For the final version of his theory this can easily be seen by observing that the trace of the energy-momentum tensor, which acts as the source of the gravitational field in that theory, vanishes for electromagnetic fields. Another way of seeing this makes use of the work of Einstein and Adriaan Fokker, who showed that Nordström's theory can be viewed as a special case of a metric theory of gravitation with the additional condition that the speed of light is a constant, thus excluding a dispersion of light waves that gives rise to the bending of light rays.

Before Nordström's theory matured to its final version, which constitutes a fairly satisfactory special relativistic theory of gravitation, several steps were necessary in which the original idea was elaborated, in particular regarding the choice of an appropriate source expression. The most obvious choice and the first considered by Nordström is the rest mass density. The problem with this quantity is, however, that it is not a Lorentz scalar. Nordström's second choice was the Lagrangian of a particle. This, however, leads to a violation of the equality of gravitational and inertial mass. While according to special relativity, kinetic energy, (e.g. the thermal motion of the particles composing a body), adds to the body's inertial mass, it is subtracted from the potential energy in the Lagrangian. If that Lagrangian hence describes the gravitational mass, the difference between the two masses increases as more kinetic energy is involved. In his final theory Nordström chose, at Einstein's suggestion, the trace of the energy-momentum tensor, the Laue scalar, thus extending the validity of the equivalence principle from mass points at rest to "complete static systems." A complete static system is a system for which there exists a reference frame in which it is in static equilibrium. In such a frame, the mechanical behavior of the system is essentially determined by a single scalar quantity. In fact, since in special relativity the inertial behavior of matter is determined by the energy-momentum tensor, the requirement of equality of inertial and gravitational mass implies that a scalar responsible for the coupling of matter to the gravitational field must be derived from the energy-momentum tensor.

The problem in choosing the Laue-scalar as a source expression is how deal with the transport of stresses in a gravitational field while maintaining energy conservation. Einstein argued that such stresses may be used—unless appropriate provisions are taken—to construct a *perpetuum mobile*, since—by creating or removing stresses—one can, so to speak, apparently switch gravitational mass on and off. In other words, while the work required for creating a stress can simply be recovered by removing it, the gravitational mass created by the stress can meanwhile be used to perform work in the presence of a gravitational field. Given that stresses depend on the geometry of the falling object under consideration, a solution can be found by appropriately adjusting the geometry, as Nordström showed. In other words, the assumption that gravitational mass can be generated by stresses led, in conjunction with the requirement of energy-momentum conservation, to the conclusion that the geometry has to vary with the gravitational potential.

According to Einstein's assessment of Nordström's final theory in his Vienna lecture, the theory satisfies all one can require from a theory of gravitation based on contemporary knowledge, which did not yet include the observation of light deflection in a gravitational field. At that time no known gravitation theory was able to explain Mercury's perihelion advance. Einstein's only remaining objection concerned the fact that what he considered to be Mach's principle—the assumption that inertia is caused by the interaction of masses—appears not to be satisfied in Nordström's theory.

But as we have seen, because of the role of stresses for gravitational mass, Nordström had to assume that the behavior of rods and clocks also depends on the gravitational potential. Indeed, as becomes clear from the hindsight of general relativity, it is arguable whether his theory really fits the special relativistic framework, corresponding as it does to a spacetime theory that is only conformally flat, i.e., based on a metric that is flat besides a scalar factor. The way that Nordström's theory stands to general relativity in that it attributes transformations to material bodies, which in the later theory are understood as transformations of spacetime, is reminiscent of the way that Lorentz's theory of the aether stands to special relativity.

ACKNOWLEDGEMENTS

For their careful reading of this text, extensive discussions, and helpful commentaries, we would like to thank Michel Janssen, Domenico Giulini, Christopher Smeenk, and John Stachel. We would also like to thank Lindy Divarci, Christoph Lehner, Carmen Hammer, Miriam Gabriel, Yoonsuhn Chung, and Dieter Hoffmann for their considerable effort and stamina during the preparation of the index.

NOTES ON THE TRANSLATIONS

The translation of sources in these volumes, if not indicated otherwise, were produced through a collective effort. The majority of the texts were first translated by Dieter Brill. Werner Heinrich also contributed considerably to the initial translations. Other texts were translated by June Inderthal and Barbara Stepansky. The translations were further reworked by Dieter Brill, Lindy Divarci, Christopher Martin, Matthias Schemmel, and Christopher Smeenk. We would also like to thank Don Salisbury for helping us to improve the translation of Max Abraham's *Eine neue Gravitationstheorie* and Erhard Scholz for helping to improve the translation of Weyl's text.

For clarification of terminology, certain words are given in the original language. The occurrence of page breaks in the original texts are marked and page numbers are given in the margin.

SOURCES

The following list contains the literature referred to in this introduction. It further contains references to a selection of the numerous contributions to the understanding of gravitation that were published in the period between the end of the nineteenth century and the advent of general relativity, without claiming to be complete. Further references as well as historical commentaries on the development of gravitational theories are found, e.g., in N. T. Roseveare *Mercury's Perihelion from Le Verrier to Einstein* (Clarendon Press, 1982); Matthew R. Edward (ed.) *Pushing Gravity: New Perspectives on Le Sage's Theory of Gravitation* (Apeiron, 2002); G. J. Whitrow and G. E. Mroduch "Relativistic Theories of Gravitation: A comparative analysis with particular reference to astronomical tests" *Vistas in Astronomy* 1: 1–67; D. Giulini "Some remarks on the notions of general covariance and background independence" arXiv:gr-qc/0603087—a contribution to *An assessment of current paradigms in the physics of fundamental interactions,* I.O. Stamatescu (ed.), (Springer, to appear); as well as the *Einstein Studies* series, edited by Don Howard and John Stachel (Birkhäuser, 1989–). Sources which are reproduced (in part or in whole) in these volumes are marked in the following list in boldface.

Abraham, Max. 1912. "Zur Theorie der Gravitation." *Physikalische Zeitschrift* **13: 1–4.**
——. 1912. "Das Elementargesetz der Gravitation." *Physikalische Zeitschrift* 13: 4–5.
——. **1912. "Der freie Fall."** *Physikalische Zeitschrift* **13: 310–311.**
——. 1912. "Das Gravitationsfeld." *Physikalische Zeitschrift* 13: 793–797.
——. 1912. "Die Erhaltung der Energie und der Materie im Schwerkraftfelde." *Physikalische Zeitschrift* 13: 311–314.
——. 1912. "Relativität und Gravitation. Erwiderung auf eine Bemerkung des Herrn A. Einstein." *Annalen der Physik* 38: 1056–1058.
——. 1912. "Nochmals Relativität und Gravitation. Bemerkungen zu A. Einsteins Erwiderung." *Annalen der Physik* 39: 444–448.
——. **1913. "Eine neue Gravitationstheorie."** *Archiv der Mathematik und Physik* **20: 193–209.**
——. **1915. "Neuere Gravitationstheorien."** *Jahrbuch der Radioaktivität und Elektronik* **(11) 4: 470–520.**
Barbour, Julian and Bruno Bertotti. 1977. "Gravity and Inertia in a Machian Framework." *Nuovo Cimento* 38B: 1–27.
——. 1982. "Mach's Principle and the Structure of Dynamical Theories." *Proceedings of the Royal Society London* 382: 295–306.
Behacker, Max. 1913. "Der freie Fall und die Planetenbewegung in Nordströms Gravitationstheorie." *Physikalische Zeitschrift* (14) 20: 989–992.
Born, Max. 1914. "Der Impuls-Energie-Satz in der Elektrodynamik von Gustav Mie." *Nachrichten der Königlichen Gesellschaft der Wissenschaften zu Göttingen* **1914, 23–36.**
Cartan, Elie. 1986. *On Manifolds With An Affine Connection And The Theory Of General Relativity.* **Naples: Bibliopolis, chap. 1, 31–55.**
Cunningham, E. 1914. *The Principles of Relativity.* Cambridge: Cambridge University Press.
Einstein, Albert. 1907. "Über das Relativitätsprinzip und die aus demselben gezogenen Folgerungen." *Jahrbuch der Radioaktivität und Elektronik* 4, 411–462.
——. **1913. "Zum Gegenwärtigen Stande des Gravitationsproblems."** *Physikalische Zeitschrift* **14: 1249–1269.**
——. **1914. "Zum Relativitätsproblem."** *Scientia* **15: 337–348.**
——. **1920.** *Äther und Relativitätstheorie.* **Berlin: Springer.**
Einstein, Albert, and Adriaan D. Fokker. 1914. "Die Nordströmsche Gravitationstheorie vom Standpunkt des absoluten Differentialkalküls." *Annalen der Physik* 44: 321–328.
Fokker, Adriaan D. 1914. "A Summary of Einstein and Grossmann's Theory of Gravitation." *The London, Edinburgh, and Dublin Philosophical Magazine and Journal of Science* 29: 77–96.
Föppl, August. 1904. "Über absolute und relative Bewegung." *Sitzungsberichte der Bayerischen Akademie der Wissenschaften, mathematisch-physikalische Klasse* **34: 383–395.**
——. 1905. "Über einen Kreiselversuch zur Messung der Umdrehungsgeschwindigkeit der Erde." *Königlich Bayerische Akademie der Wissenschaften, München, mathematisch-physikalische Klasse, Sitzungsberichte* 34: 5–28.
Friedlaender, Benedict and Immanuel. 1896. *Absolute oder relative Bewegung?* **Berlin: Leonard Simion.**
Gans, Richard. 1905. "Gravitation und Elektromagnetismus." *Physikalische Zeitschrift* 6: 803–805.
——. 1912. "Ist die Gravitation elektromagnetischen Ursprungs?" In *Festschrift Heinrich Weber zu seinem siebzigsten Geburtstag am 5. März gewidmet von Freunden und Schülern.* Leipzig, Berlin: Teubner, 75–94.
Gerber, Paul. 1917. "Die Fortpflanzungsgeschwindigkeit der Gravitation." *Annalen der Physik* 52: 415–441.
Gramatzki, H. J. 1905. *Elektrizität und Gravitation im Lichte einer mathematischen Verwandtschaft. Versuch zur Grundlage einer einheitlichen Mechanik der elektrischen gravitierenden und trägen Massen mit Hilfe der phänomenologischen Interpretation gewisser mathematischer Begriffsvorgänge.* München: Lindauer.
Grassmann, Hermann. 1844. *Die lineale Ausdehnungslehre, ein neuer Zweig der Mathematik.* Leipzig: Otto Wigand.

————. 1995. *A New Branch of Mathematics: The Ausdehnungslehre of 1844 and Other Works*. **Translated by Lloyd C. Kannenberg. Chicago and LaSalle: Open Court.**

Hertz, Heinrich. 1894. *Die Prinzipien der Mechanik in neuem Zusammenhange dargestellt*. Leipzig: Barth.

Hilbert, David. 1916. "Die Grundlagen der Physik. (Erste Mitteilung.)" *Nachrichten von der Königlichen Gesellschaft der Wissenschaften zu Göttingen, Mathematisch-physikalische Klasse* 1915, 395–407.

————. **1917. "Die Grundlagen der Physik. (Zweite Mitteilung.)"** *Nachrichten von der Königlichen Gesellschaft der Wissenschaften zu Göttingen, Mathematisch-physikalische Klasse* 1916, 53–76.

Hofmann, Wenzel. 1904. *Kritische Beleuchtung der beiden Grundbegriffe der Mechanik: Bewegung und Trägheit und daraus gezogene Folgerungen betreffs der Achsendrehung der Erde und des Foucault'schen Pendelversuchs*. Wien, Leipzig: M. Kuppitsch Witwe.

Isenkrahe, Caspar. 1879. *Das Räthsel von der Schwerkraft: Kritik der bisherigen Lösungen des Gravitationsproblems und Versuch einer neuen auf rein mechanischer Grundlage*. Braunschweig: Vieweg.

Ishiwara, Jun. 1912. "Zur Theorie der Gravitation." *Physikalische Zeitschrift* 13: 1189–1193.

————. 1914. "Grundlagen einer relativistischen elektromagnetischen Gravitationstheorie." Parts I and II. *Physikalische Zeitschrift 15*: 294–298, 506–510.

Jaumann, G. 1911. "Geschlossenes System physikalischer und chemischer Differentialgesetze." *Sitzungsberichte der math.-nw. Klasse der Kaiserl. Akademie der Wissenschaften, Wien* 120: 385–530.

————. 1912. "Theorie der Gravitation." *Sitzungsberichte der math.-nw. Klasse der Kaiserl. Akademie der Wissenschaften, Wien* 121: 95–182.

Lange, Ludwig. 1886. *Die geschichtliche Entwicklung des Bewegungsbegriffes und ihr voraussichtliches Endergebnis: ein Beitrag zur historischen Kritik der mechanischen Principien*. Leipzig: Engelmann.

Laue, Max von. 1917. "Die Nordströmsche Gravitationstheorie." *Jahrbuch der Radioaktivität und Elektronik* 14: 263–313.

LeSage, Georges-Louis. 1784. "Lucrèce Newtonien." *Mémoires de l'Académie royale des Sciences et Belles Lettres de Berlin, pour 1782*.

Levi-Civita, Tullio. 1916. "Nozione di parallelismo in una varietà qualunque e conseguente specificazione geometrica della curvatura riemanniana." *Circolo Matematico di Palermo. Rendiconti*, **42: 173–204.**

Lorentz, Hendrik A. 1900. "Considerations on Gravitation." *Proceedings Royal Academy Amsterdam* **2: 559–574.**

————. **1910. "Alte und neue Fragen der Physik."** *Physikalische Zeitschrift* **11: 1234–1257.**

————. 1914. "La Gravitation." *Scientia* (16) 36: 28–59.

Mach, Ernst. 1883. *Die Mechanik in ihrer Entwickelung: historisch-kritisch dargestellt*. Leipzig: Brockhaus.

Mie, Gustav. 1912. "Grundlagen einer Theorie der Materie I, II und III." *Annalen der Physik* **37: 511–534; 39 (1912), pp. 1–40; 40 (1913): 1–66.**

————. **1914. "Bemerkungen zu der Einsteinschen Gravitationstheorie."** *Physikalische Zeitschrift* **15: 115–122.**

————. **1915. "Das Prinzip von der Relativität des Gravitationspotentials."** In *Arbeiten aus den Gebieten der Physik, Mathematik, Chemie: Festschrift Julius Elster und Hans Geitel zum sechzigsten Geburtstag*. **Braunschweig: Vieweg, 251–268.**

Minkowski, Hermann. 1908. "Die Grundgleichungen für elektromagnetische Vorgänge in bewegten Körpern." *Nachrichten der Königlichen Gesellschaft der Wissenschaften zu Göttingen, Mathematisch-Physikalische Klasse*, **53–111.**

Neumann, Carl. 1870. *Ueber die Principien der Galilei-Newton'schen Theorie*. Leipzig: Teubner.

Nordström, Gunnar. 1912. "Relativitätsprinzip und Gravitation." *Physikalische Zeitschrift* **13: 1126–1129.**

————. **1913. "Träge und schwere Masse in der Relativitätsmechanik."** *Annalen der Physik* **40: 856–878.**

————. **1913. "Zur Theorie der Gravitation vom Standpunkt des Relativitätsprinzips."** *Annalen der Physik* **42: 533–554.**

————. 1914. "Über den Energiesatz in der Gravitationstheorie." *Physikalische Zeitschrift* 14: 375–380.

————. 1914. "Über die Möglichkeit, das elektromagnetische Feld und das Gravitationsfeld zu vereinigen." *Physikalische Zeitschrift* 15: 504–506.

————. 1914. "Zur Elektrizitäts- und Gravitationstheorie." *Finska Vetenskaps-Societetens Förhandlingar (LVII. A)* 4: 1–15.

Poincaré, Henri. 1906. "Sur la dynamique de l'électron." *Rendiconti del Circolo Matematico di Palermo 21: 129–175.*

Reißner, Hans. 1914. "Über die Relativität der Beschleunigungen in der Mechanik." *Physikalische Zeitschrift* 15: 371–375.

————. 1915. "Über eine Möglichkeit die Gravitation als unmittelbare Folge der Relativität der Trägheit abzuleiten." *Physikalische Zeitschrift* (16) 9/10: 179–185.

Ritz, Walter. 1909. "Die Gravitation." *Scientia* 5: 241–255.

Schrödinger, Erwin. 1925. "Die Erfüllbarkeit der Relativitätsforderung in der klassischen Mechanik." *Annalen der Physik* 77, pp. 325–336.

Schwarzschild, Karl. 1897. "Was in der Welt ruht." *Die Zeit, Vienna* Vol. 11, No. 142, 19 June 1897, 181–183.

————. 1900. "Über das zulässige Krümmungsmaass des Raumes." *Vierteljahresschrift der Astronomischen Gesellschaft* 35: 337–347.

Seeliger, Hugo von. 1895. "Über das Newton'sche Gravitationsgesetz." *Astronomische Nachrichten* 137: 129–136.

————. 1909. "Über die Anwendung der Naturgesetze auf das Universum." *Sitzungsberichte der Königlichen Bayerischen Akademie der Wissenschaften, Mathematische-physikalische Klasse*, 3–25.

Sommerfeld, Arnold. 1910. "Zur Relativitätstheorie, II: Vierdimensionale Vektoranalysis." *Annalen der Physik* 33: 649–689.

Thomson, William (Lord Kelvin). 1873. "On the ultramundane corpuscles of Le Sage." *Phil. Mag.* 4th ser. 45: 321–332.

Wacker, Fritz. 1909. *Über Gravitation und Elektromagnetismus*. (Ph. D. thesis, Eberhard Karls Universität Tübingen). Leipzig: Noske.

Webster, D. L. 1912. "On an Electromagnetic Theory of Gravitation." *Proceedings of the American Academy of Sciences* 47: 559–581.

Weyl, Hermann. 1918. "Reine Infinitesimalgeometrie." *Mathematische Zeitschrift* 2: 384–411.

Zenneck, Jonathan. 1903. "Gravitation." In Arnold Sommerfeld (ed.) *Encyklopädie der mathematischen Wissenschaften*, Vol. 5 (Physics). Leipzig: Teubner, 25–67.

Zöllner, Friedrich, Wilhelm Weber, and Ottaviano F. Mossotti. 1882. *Erklärung der universellen Gravitation aus den statischen Wirkungen der Elektricität und die allgemeine Bedeutung des Weber'schen Gesetzes*. Leipzig: Staackmann.

————. 1898. "Die räumliche und zeitliche Ausbreitung der Gravitation." *Zeitschrift für Mathematik und Physik* 43: 93–104.

THE GRAVITATIONAL FORCE
BETWEEN MECHANICS
AND ELECTRODYNAMICS

JÜRGEN RENN

THE THIRD WAY TO GENERAL RELATIVITY: EINSTEIN AND MACH IN CONTEXT

1. INTRODUCTION

The relationship between Einstein and Mach is often discussed as a prototypical case of the influence of philosophy on physics.[1] It is, on the other hand, notoriously difficult to accurately pinpoint such influences of philosophy on science, in particular with regard to modern physics. To a working scientist, such influences must seem to belong to a past era. There seems to be little room left for philosophy in the practice of today's physics. It plays virtually no part in the physics curriculum, and scholars who are both active physicists and philosophers are rare exceptions. In view of this situation it may be appropriate to reexamine the mythical role that philosophy played for one of the founding heroes of modern physics, Albert Einstein. It is conceivable that the disjointed remarks on philosophy that are dispersed throughout his œuvre can be integrated into a coherent image of what may then rightly be called "his philosophy." But even if such a reconstruction should be successful and yield more than an eclectic collection of occasional reflections, the more decisive question of the utility of philosophy for his science would be left unanswered. In fact, Einstein as a philosopher may have been a rather different *persona* from Einstein the physicist, and having two souls in one breast would not be an atypical state of affairs for a German intellectual. This contribution will therefore not undertake a systematic attempt at reconstructing his philosophy, but rather be limited to a case study of the interaction between philosophy and physics, reexamining the impact of Mach's philosophical critique of classical mechanics on Einstein's discovery of general relativity.[2] This reexamination is made possible by newly discovered documentary evidence concerning Einstein's research as well as by the achievements of recent studies in the history of general relativity.[3] Both factors contribute to an historical understanding of the

---

1   The literature on this subject is considerable; for more or less comprehensive accounts, see among others, (Blackmore 1992; Boniolo 1988; Borzeszkowski and Wahsner 1989, in particular, pp. 49–64; Goenner 1981; Hoefer 1994; Holton 1986, chap. 7; Norton 1993; Pais 1982, 282–288; Reichenbach 1958; Sciama 1959; Sewell 1975; Stein 1977; Torretti 1978, 1983, 194–202; Wolters 1987), as well as other literature quoted below. An earlier version of the present paper has been widely circulated in preprint form since 1994, an Italian version of this can be found in (Pisent and Renn 1994). Its themes have been taken up in various subsequent publications, see, e.g., (Barbour and Pfister 1995; CPAE 8).

2   For Mach's critique, see (Mach 1883: translated in Mach 1960).

Jürgen Renn (ed.). *The Genesis of General Relativity*, Vol. 3
Gravitation in the Twilight of Classical Physics: Between Mechanics, Field Theory, and Astronomy.
© 2007 Springer.

relation between Mach's philosophy and Einstein's physics that is not only richer in detail, but also in context, and hence able to reveal the alternatives available to the historical actors in the search for a new theory of gravitation.

The main result of the analysis presented below is that the theory of general relativity can be seen to have emerged as the result of one among several possible strategies dealing with conceptual problems of classical physics, strategies which were worked out to different degrees in the course of the historical development. Since this development was, in other words, not completely determined by the intrinsic features of the scientific problems which the historical actors confronted, it is now possible to evaluate more clearly the external factors affecting the choice between different strategies.[4] The approach pursued by Einstein can be characterized as a combination of field theoretical and mechanistic approaches shaped by his philosophical outlook on foundational problems of physics. In the following, two conclusions are drawn in particular:

i) The heuristics under the guidance of which Einstein elaborated general relativity was rooted in the heterogeneous conceptual traditions of classical physics. At least in its intermediate stages of development, the conceptual framework of Einstein's theory resembled the peculiar combination of field theoretic and mechanistic elements in Lorentz's electron theory, rather than the coherent and self-contained conceptual framework of special relativity, which had superseded the conceptual patchwork of Lorentz's theory.[5] Mach's ideas were one element in this mixture of traditional conceptual frameworks; their interpretation by Einstein depended on the context provided by the other elements. In particular, the heuristic role of Mach's ideas has to be seen in the wider context of the role that classical mechanics played for the emergence of general relativity. As with the other heuristic elements, Mach's ideas were eventually superseded by the conceptual consequences of general relativity, as Einstein saw them. In particular, Mach's concept of inertia as a property not of space but of the interaction between physical masses played a role comparable to that of the aether in Lorentz's theory of electrodynamics: it introduced a helpful heuristics that led to its own elimination, since the conceptual preconditions of the development of general relativity turned out to be incompatible with its outcome.

ii) What distinguished Einstein's early approach to the problem of gravitation from that of his contemporaries was his refusal to accept that a mechanistic and a field theoretic outlook on physics were mutually exclusive alternatives. It was his philosophical perspective on foundational problems of physics that allowed him to conceive of field theory and mechanics as complementary resources for the formulation of a new theory of gravitation. Contrary to most contemporary physicists dealing

---

3   For new evidence, see in particular, the various volumes of *The Collected Papers of Albert Einstein*. For recent historical studies of the development of general relativity, see the contributions to vols. 1 and 2 of this series and the references given therein.

4   For a similar kind of argumentation, see (Freudenthal 1986).

5   See the extensive discussion in "Classical Physics in Disarray ..." and "Pathways out of Classical Physics ..." (both in vol. 1 of this series).

with the problem of gravitation, he attempted to incorporate in his new theory both foundational assumptions of classical mechanics and their critical revision by Mach; and contrary to most physicists searching for a physical implementation of Mach's analysis of the foundations of mechanics, he took into account the antimechanistic philosophical intentions of this critique. Einstein's philosophical perspective is, however, not only characterized by his interest in and understanding of such philosophical intentions, but even more by his integrative outlook on the conceptual foundations of physics. His peculiar approach to the specific problem of gravitation can only be understood if one acknowledges that for him the problem of a new theory of gravitation was simultaneously the problem of developing new conceptual foundations for the entire body of physics. Although it may not be common to label such an integrative perspective as "philosophical"—in view of the predominantly metatheoretical concerns of the philosophy of science—it was also no longer a self-evident preoccupation of science at the beginning of this century, let alone of science today. Nevertheless, the fruitfulness of Einstein's approach argues for its reconsideration by both philosophy and science.

In the following, it will first be discussed how Einstein's project of generalizing the principle of relativity emerged in the context of his own research as well as in that of other contemporary approaches to the problem of gravitation (section 2). Some of the historical presuppositions of the conceptual innovation represented by general relativity will then be examined, paying particular attention to the contributions of mechanics and field theory to its development. The aim is to describe the horizon of possibilities open to the historical actors (section 3).[6] Next, the influence of Mach's critique of classical mechanics on the creation and interpretation of general relativity by Einstein (section 4) will be traced in some detail. Finally, the question of Einstein's philosophical perspective on the foundational problems of physics and its role in the emergence of general relativity (section 5) will be addressed once again.

2. A NEW THEORY OF GRAVITATION IN THE CONTEXT OF COMPETING WORLDVIEWS

2.1 A Relativistic Theory of Gravitation as a Problem of "Normal Science"

In 1907, when Einstein first dealt with the problem of a relativistic theory of gravitation, philosophical interests seemed to be far from his main concerns. Although he was employed by the Swiss patent office at that time, he was no longer an outsider to academic physics. By way of his publications, correspondence, and personal relationships, he was already becoming a well-respected member of the physics establishment. The times had passed when philosophical readings in the mock "Olympia" academy, which Einstein had founded some years earlier together with other bohemian friends, formed one of the centers of his intellectual life. Einstein was first con-

---

6   For the concept of horizon, see (Damerow and Lefèvre 1981).

fronted with the task of revising Newton's theory of gravitation in light of the relativity theory of 1905 when he was asked to write a review on relativity theory that would also cover its implications for various areas of physics not directly related to the field from which it originated, namely the electrodynamics of moving bodies.[7] Hence, the revision of Newton's theory of gravitation entered Einstein's intellectual horizon, not as the consequence of a philosophically minded ambition to go beyond the original special theory towards a more general theory of relativity, but as a necessary part of the usual "mopping up operation" whereby new results are integrated with the traditional body of knowledge. The necessity to modify the classical theory of gravitation appeared to Einstein and his contemporaries all the more pressing, as within the conceptual framework of classical physics an asymmetry could already be observed between the instantaneous propagation of the gravitational force and the propagation of the electromagnetic field with the finite speed of light. It therefore comes as no surprise that not only Einstein but also several of his contemporaries addressed the problem of formulating a field theory of gravitation that was to be in agreement with the principles suggested by the theory of the electromagnetic field and, most importantly, with the new kinematics of relativity theory.[8]

2.2 The Proliferation of Alternative Approaches to the Problem of Gravitation

It appears to be a phenomenon characteristic of the development of science that in a situation of conceptual conflict of this kind, alternative approaches to the solution of the conflict begin to proliferate. Among the factors accounting for this proliferation are the diverse resources upon which the alternative approaches can draw. Even after the establishment of special relativity, the instruments available for a revision of Newton's theory of gravitation had to be taken essentially from the arsenal of classical physics, in particular from classical mechanics and classical electrodynamics. As these two branches of classical physics were founded on different conceptual structures—on the one hand the direct interaction between point particles, and on the other hand, the propagation of continuous fields in time—the use of resources from one or the other branch to solve the same problem could present itself as a choice between conceptual alternatives. In this way, the problem of a new theory of gravitation assumed right from the beginning the character of a borderline problem of classical physics.[9] The choice among alternative approaches to the problem of gravitation was

---

7   See (Einstein 1907b, sec. V). See also Einstein's later recollections, e.g. those reported in (Wheeler 1979, 188). For a historical discussion of this paper, see (Miller 1992).

8   See, among others, the source papers in this and in vol. 4 of this series (Lorentz 1910; Minkowski 1908; Poincaré 1906), as well as the various papers by Abraham, Nordström, and Mie. See also (Abraham 1912a, 1912b; Lorentz 1910; Mie 1914; Minkowski 1911a [1908], in particular, pp. 401–404; Minkowski 1911b [1909], in particular, pp. 443–444; Nordström 1912; Poincaré 1905, in particular, pp. 1507–1508; 1906, in particular, pp. 166–175; Ritz 1909).

9   For the concept of borderline problem, see "Classical Physics in Disarray ..." (in vol. 1 of this series).

therefore also related to the way in which such borderline problems were handled at that time.

Even before the turn of the century, that is, long before the great conceptual revolutions of early twentieth century physics, many physicists saw themselves at a crossroads, forced to decide between alternative conceptual foundations for their field.[10] Mechanics had long played the dual role of a subdiscipline and of an ontological foundation of physics, and at the threshold to the twentieth century, there were still physicists who adhered to the ontological primacy of mechanics, and who were therefore convinced that the entire body of physics should be built on conceptual foundations drawn from mechanics. With the formulation of classical electrodynamics by Maxwell, Hertz, and Lorentz, the difficulty of achieving such a reduction of physics to the conceptual apparatus of mechanics became increasingly evident. Although field theory itself was initially formulated in a mechanical language, towards the end of the century it came to represent an autonomous conceptual framework largely independent of that of mechanics. To some physicists, such as Wien and Lorentz, field theory even appeared to offer an alternative conceptual foundation for all of physics; they speculated about an electrodynamic worldview in which mechanics would have to be reformulated as a field theory rather than the other way around. Finally, with the development of classical thermodynamics in the mid-nineteenth century, including the formulation of the principle of conservation of energy, a third alternative conceptual foundation of physics (discussed under the name of "energetics") seemed to volunteer itself.[11] Hence, the mechanistic conception of physics, the electromagnetic worldview, and energetics distinguished themselves by the choice of the subdiscipline of classical physics to which they granted a foundational role for the entire field.

The formulation of a field theory of gravitation in analogy with, or even on the basis of, the Maxwell-Lorentz theory of the electromagnetic field was thus not a farfetched thought in the context of the electrodynamic world picture and had been approached by several authors.[12] In such a theory, gravitation, like electrodynamical interactions, would have to propagate with a finite speed. The establishment of the theory of relativity in 1905 did not make attempts in this direction obsolete; on the contrary, the issue of formulating a theory of gravity became even more urgent, since Newton's theory clearly violated one of the fundamental principles of the relativity theory—the requirement that no physical action propagates with a velocity greater than that of light. The primary task was to reformulate the experimentally well-con-

---

10   For further discussion of the conceptual foundations of classical physics at the turn of the century, see "Classical Physics in Disarray ..." (in vol. 1 of this series). For a brief account of the different approaches prevalent at the turn of the century, see also (Jungnickel and McCormach 1986, chap. 24).

11   For a comprehensive historical analysis of energetics, see (Deltete 2000).

12   For contemporary reviews, see "Gravitation" (Zenneck 1903); "Recent Theories of Gravitation" (Abraham 1915), (both in this volume). For the heuristic role of electrodynamics for Einstein's formulation of a field theory of gravitation, see "Pathways out of Classical Physics ..." (in vol. 1 of this series).

firmed Newtonian law of gravitation in accordance with the principles of the new kinematics, in particular with the Lorentz transformations of space and time coordinates, under which the classical law does not remain invariant. It is in fact not difficult to formulate a Lorentz covariant field equation which can be interpreted as a direct generalization of Newton's law. Around 1907 Einstein apparently pursued this line of research without, however, achieving satisfying results. Indeed, if such a Lorentz covariant generalization of Newton's theory could have been formulated without problems, there would have been no reason for Einstein to look beyond the special theory of relativity of 1905 and choose the thorny path that was to lead him to the formulation of the general theory of relativity in 1915.

One of the difficulties encountered by Einstein concerns the concept of mass, or more precisely the relation between the *two* concepts of mass in classical mechanics: gravitational and inertial. According to the special theory of relativity the inertial mass of a body depends on its energy content.[13] On the other hand, it was empirically known in the context of classical mechanics that the inertial mass is always exactly equal to the gravitational mass. In a relativistic theory of gravitation, the gravitational mass of a physical system should therefore also depend on its total energy. In a later recollection, Einstein summarized his view of this implication of classical mechanics and the special theory of relativity for a relativistic theory of gravitation:

> If the theory did not accomplish this or could not do it naturally, it was to be rejected.
> The condition is most naturally expressed as follows: the acceleration of a system falling freely in a given gravitational field is independent of the nature of the falling system (specially therefore also of its energy content).[14]

It was precisely this requirement, however, which turned out not to be fulfilled in the early attempts at a special relativistic theory of gravitation.[15]

In other words, a straightforward relativistic generalization of Newton's gravitational law seemed to be in conflict with "Galileo's principle," i.e., with the principle that the accelerations of bodies falling in a gravitational field are equal.[16] Quantitatively, however, the failure of the Galileo's principle may have been negligibly small, as Mie, for instance, claimed for his later special relativistic theory of gravitation.[17] Researchers such as Mie, whose outlook on this issue was shaped by the electrody-

---

13   See (Einstein 1907a), in which this conclusion is rederived in a general way, possibly with the problems of a relativistic theory of gravitation already in mind.

14   "Wenn die Theorie dies nicht oder nicht in natürlicher Weise leistete, so war sie zu verwerfen. Die Bedingung lässt sich am natürlichsten so aussprechen: die Fall-Beschleunigung eines Systems in einem gegebenen Schwerefelde ist von der Natur des fallenden Systems (speziell also auch von seinem Energie-Inhalte) unabhängig." (Einstein 1992, 64, 65)

15   For a reconstruction of Einstein's failed attempt to incorporate gravitation within the relativity theory of 1905, see "Classical Physics in Disarray ..." sec. 2.9 (in vol. 1 of this series), see also "Einstein, Nordström, and the Early Demise of Scalar, Lorentz Covariant Theories of Gravitation" (in this volume). For Einstein's later recollections, see (Einstein 1992, 58–63).

16   Galileo's name is usually (but incorrectly) associated with the introduction of the principle of inertia, while the principle which is named after him here can indeed be found in his work; for historical discussion, see (Damerow et al. 2004, chap. 3).

namic worldview, were all the more willing to give up the Galileo's principle as they did not feel obliged to consider the implications of classical mechanics as foundational for physics, unless they perceived an unavoidable conflict with experimental evidence. Einstein, however, somewhat prematurely gave up this line of research. In the years 1912 to 1914, Nordström, with the help of contributions from von Laue and Einstein himself, attempted to formulate a consistent special relativistic field theory of gravitation and eventually succeeded to some extent in including the equality of gravitational and inertial mass.[18] This theory even triggered insights—e.g., that clocks and rods are affected by the gravitational field—upon which its further development in the direction of general relativity could have been based. Hence, it constituted at least the beginning of an independent road towards a theory similar to general relativity, "the route of field theory."

2.3 Mach's Critique of Mechanics and the Three Routes to General Relativity

From the conflict between classical mechanics and the special theory of relativity, which Einstein perceived in 1907, he drew a conclusion that was diametrically opposed to that of the followers of an electromagnetic worldview. For him the equality of inertial and gravitational mass was not just an empirically confirmed but otherwise marginal result of classical mechanics; he held onto it as a principle upon which a new theory of gravitation was to be based. He was therefore ready to accept that this theory would no longer fit into the framework of special relativity.[19] Hence Einstein's

---

17 See (Mie 1913, 50). Similar views are also found in other authors pursuing a special relativistic field theory of gravitation, see e.g. (Nordström 1912, 1129): "From a letter from Herr Prof. Dr. A. Einstein I learn that earlier he had already concerned himself with the possibility I used above for treating gravitational phenomena in a simple way. However, he became convinced that the consequences of such a theory cannot correspond with reality. In a simple example he shows that, according to this theory, a rotating system in a gravitational field will acquire a smaller acceleration than a non-rotating system. I do not find this result dubious in itself, for the difference is too small to yield a contradiction with experience." ("Aus einer brieflichen Mitteilung von Herrn Prof. Dr. A. Einstein erfahre ich, daß er sich bereits früher mit der von mir oben benutzten Möglichkeit befaßt hat, die Gravitationserscheinungen in einfacher Weise zu behandeln, daß er aber zu der Überzeugung gekommen ist, daß die Konsequenzen einer solchen Theorie der Wirklichkeit nicht entsprechen können. Er zeigt an einem einfachen Beispiel, daß nach dieser Theorie ein rotierendes System im Schwerkraftfelde eine kleinere Beschleunigung erhalten wird als ein nichtrotierendes. Diese Folgerung finde ich an sich nicht bedenklich, da der Unterschied zu klein ist, um einen Widerspruch mit der Erfahrung zu geben.")

18 For Nordström's work, see the section "A Field Theory of Gravitation in the Framework of Special Relativity," in particular "Einstein, Nordström and the Early Demise of Scalar, Lorentz Covariant Theories of Gravitation" (in this volume).

19 Einstein remarked with regard to the violation of Galileo's principle in Abraham's and Mie's theories of gravitation (Einstein 1914, 343): "Due to their smallness, these effects are certainly not accessible to experiments. But it seems to me that there is much to be said for taking the connection between inertial and gravitational mass to be warranted *in principle*, regardless of what forms of energy are taken into account." ("Diese Wirkungen wären zwar wegen ihrer Kleinheit dem Experiment nicht zugänglich. Aber es scheint mir viel dafür zu sprechen, dass der Zusammenhang zwischen der trägen und schweren Masse *prinzipiell* gewahrt ist, abgesehen von der Art der auftretenden Energieformen.")

further considerations did not lead him away from mechanics, but rather brought him into contact with its foundational questions, in particular with the role of inertial systems in classical mechanics.

Mach's philosophical critique of the foundations of classical mechanics suggested to Einstein that the problem of a new theory of gravitation had to be resolved in connection with a generalization of the relativity principle of classical mechanics and special relativity. Quite apart from the specific problem of gravitation, some of Mach's contemporary readers, as well as researchers who had independently arrived at similar views, had drawn the conclusion that one should look for a new, generally relativistic formulation of mechanics.[20] Their conceptual and technical resources were mostly confined to those of classical mechanics, and their chances of making contact with the more advanced results of physics at the turn of the century, which to a large extent were based on field theory (in particular, classical electrodynamics), were, at least at that time, slender. Nevertheless, the line of research that extends from the work of these early followers of Mach (discussed in more detail in the next section) to the recent work of Julian Barbour and Bruno Bertotti, Fred Hoyle and Jayant Narlikar, André Assis and others demonstrates that the project of formulating a generally relativistic theory of mechanics, including a treatment of gravitation, could be as successfully pursued as the project of a purely field theoretic approach to the problem of gravitation, as represented in particular by the work of Nordström.[21] In the following, this approach will be called "the mechanistic generalization of the relativity principle."

In view of this historical context, the heuristics that guided Einstein's formulation of the general theory of relativity can now be identified as a "third way," a peculiar mixture of field theoretical and mechanical elements. This affirmation suggests several questions, which are addressed in the following: What are the advantages and the disadvantages of the different strategies? What exactly are the contributions of the field theoretical and of the mechanical tradition to Einstein's heuristic strategy? What is the relation between the conceptual structures guiding Einstein's research and those that were newly established by it? As the development of the general theory of relativity was apparently not uniquely determined by the intrinsic nature of the problem to be solved, what then were the external factors that shaped Einstein's perspective and what role did philosophical positions play among them?

---

20   See, for example, the source texts in the first part of this volume "The Gravitational Force between Mechanics and Electrodynamics." For a survey of the interpretation of Mach's critique by contemporary readers, see also (Norton 1995).

21   For historical overviews of attempts to incorporate Mach's critique in physical theories, see (Assis 1995; Barbour 1993, Barbour and Pfister 1995; Goenner 1970, 1981).

3. ROOTS OF GENERAL RELATIVITY IN CLASSICAL PHYSICS

3.1 Resources and Stumbling Blocks Presented by the Tradition of Field Theory

The conceptual roots of general relativity in the tradition of field theory are more familiar than those in the tradition of mechanics. As mentioned above, not only special relativity but even the classical theory of the electromagnetic field made it plausible to conceive of gravitation as a field propagated with finite velocity. But there were also other contributions from this tradition which sooner or later found their way into the development of general relativity. Notably, field theory endows space with physical properties and thus contributes to blurring the distinction between matter and space. That this tendency (even taken by itself) could suggest the introduction of non-Euclidean geometry as a physical property of space is illustrated by the work of Georg Friedrich Bernhard Riemann and William Clifford in the nineteenth century.[22] In any case, field theory enriched the limited ontology of classical mechanics by introducing the field as a reality in its own right, an apparently trivial consequence, which, as we shall see, took considerable time to achieve a firm standing even within the development of general relativity. Field theory also suggested the existence of forces more general than the two-particle interactions usually considered in point mechanics, as is illustrated, for instance, by the transition from Coulomb forces between point charges to electrodynamic interactions such as induction; and it offered a mechanism for unifying separate forces as aspects of one more general field, as can again be illustrated by the example of electrodynamics conceived of as a unification of electric and magnetic interactions. It was therefore natural for those who pursued the program of formulating a field theory of gravitation either on the basis of, or in analogy to electrodynamics to search for the dynamic aspects of the gravitational field, considering Newton's law (in analogy to Coulomb's) as a description of the field's static aspects only. But the knowledge of the Newtonian special case could, and also did serve as a touchstone for any attempt at a more general theory—including Einstein's general theory of relativity, in whose development the question of the "Newtonian limit" was to play a crucial role.[23] The mature formulation of electrodynamic field theory by H. A. Lorentz also suggested a model for the essential elements of a field theory of gravitation and for their interplay: a field equation was needed to describe the effect of sources on the field, and an equation of motion was needed to describe the motion of bodies in the field.[24] Finally, those who looked for an "electromagnetic" theory of gravitation were also very clear about the

---

22 See (Clifford 1976 [1889]; Riemann 1868). On p. 149 of his paper, Riemann claims that non-Euclidean geometry could be important in physics if the concept of body should turn out not to be independent of that of space. He expected this consideration to be of relevance for a future microphysics.

23 See (Norton 1989b) and "Pathways out of Classical Physics ..." (in vol. 1 of this series).

24 For a discussion of the historical continuity between Lorentz's electron theory and Einstein's theory of general relativity, see (McCormmach 1970) and "Pathways out of Classical Physics..." (in vol. 1 of this series).

experimental evidence to be accounted for by the new theory: the explanation of the perihelion shift of Mercury was in fact mentioned as an empirical check in almost all discussions of electromagnetic theories of gravitation, which, in this sense, can be said to have left a very tangible patrimony to general relativity in pointing to one of its classical tests.[25]

But as much as the tradition of field theory was able to contribute to the conceptual development of general relativity, it did not determine a heuristic strategy that clearly outlined the way to a satisfactory solution of the problem of gravitation. What is more, in hindsight, from the perspective of the completed theory of general relativity, it becomes evident that the tradition of classical field theory also included conceptual components that must be considered as stumbling blocks on the way to such a solution. In first turning to the problem of the heuristic ambiguity of field theory, as mentioned above, there were indeed several different lines to follow in formulating a field theory of gravitation within this tradition.[26] One of the factors accounting for this proliferation of alternatives lay in the uncertainty as to which principles of mechanics were to be maintained in a field theory of gravitation, given the necessity of revising at least some of them. The electromagnetic approach to the problem of gravitation tended, in any case, to ignore the foundational problems of mechanics, as long as this seemed experimentally acceptable. An early example of this tendency, characteristic of the electromagnetic world picture, is provided by the stepmotherly way in which, before the advent of special relativity, the principle of relativity and the principle of the equality of action and reaction was treated in Lorentz's electron theory. The same attitude characterized his later attempts to integrate gravitation into the conceptual framework of field theory. For instance, in a 1910 review paper (Lorentz 1910) Lorentz seemed unperturbed by the fact that the relativistic law of gravitation he proposed violated the principle of the equality of action and reaction. This difficulty is just one representative example of the problems associated with the task of reconstructing the shared knowledge accumulated in mechanics on the basis of purely field theoretic foundations. In addition to these problems, there was little experimental guidance in how to proceed in building the new theory of gravitation—apart from the speculations about the perihelion shift of Mercury mentioned above. To use a metaphor employed by Einstein (1913, 1250): the task of constructing a field theory of gravitation was similar to finding Maxwell's equations exclusively on the basis of Coulomb's law of electrostatic forces, that is, without any empirical knowledge of non-static gravitational phenomena.

Let us now address the problem of the conceptual stumbling blocks. Their evaluation naturally depends on the point of view one takes. In view of the conceptual framework of the finished general theory of relativity, classical field theory must have been misleading in several respects. One obvious aspect is the linearity of the classi-

---

25   See (Zenneck 1903). For a contemporary survey of the problem of gravitation and the role of the peri-
     helion shift, see (Roseveare 1982).
26   See note 8 above.

cal theory in contrast to the non-linearity of the field equations of general relativity. A related aspect is the independence of the field equation and the equation of motion from each other in the classical theory, as opposed to their interdependence in general relativity. Closely associated with these more structural aspects—and perhaps even more important—are the conceptual changes with respect to classical physics brought about by general relativity. These changes include the introduction of new concepts of space and time, the new role of the gravitational field acting as its own source, and the changes of the concepts of energy and force manifested, for instance, by the absence of a gravitational stress-energy tensor in general relativity, in contrast to the existence of such a stress-energy tensor for the electromagnetic field in classical field theory. These changes could not have been anticipated on the basis of classical field theory; furthermore, in the search for a new theory of gravitation, classical field theory necessarily engendered expectations which were flatly contradicted by the outcome of that search.

3.2 The Foundational Critique of Mechanics and the Mechanistic Generalization of the Relativity Principle

The heuristic contributions of classical physics to the development of general relativity as well as the conceptual stumbling blocks it presented for this development obviously require a more detailed treatment and should be discussed in particular in the context of the concrete theories which are subsumed here under the rather general heading of "classical physics." For the purposes of the present contribution, an examination of this kind will be attempted only for the tradition of mechanics, for which one particular strand was of primary influence on the development of general relativity—both directly and as an alternative to Einstein's theory. This strand was represented by a reevaluation of mechanics, which was the outcome of a debate on its foundations in the second half of the nineteenth century. In this period some basic concepts of classical mechanics had ceased to be as self-evident as they had once appeared in the Newtonian tradition.

A central example is Newton's claim that even a single body in an otherwise empty universe possesses inertia, a claim which—in spite of its metaphysical character—played a crucial role in his argument in favor of the existence of absolute space.[27] This argument involves a bucket filled with water, which is considered first in a state in which the bucket rotates but the water is at rest and its surface flat, and second in a state in which both the bucket and the water rotate, producing a curved surface. According to Newton's interpretation of this experiment, the second case represents an absolute rotation, whereas the first case represents only a relative motion between water and bucket that does not cause physical effects. The conclusion that this argument provides evidence for the existence of absolute space is, however, only legitimate if other physical causes of the curvature of the water in the

---

27   This has been shown in detail in (Freudenthal 1986), on which also the following remarks are based.

second case can be excluded; in other words, the argument is convincing only under the physically unrealizable assumption that a rotational motion of the water in an otherwise empty universe would also give rise to the same effect. This assumption in turn is based on the metaphysical premise that a system is composed of parts which carry their essential properties (such as inertia in the case of a material system) even when they exist in isolation in empty space. It was also on this premise that Newton considered gravitation—in contrast to inertia—to be a universal but not an essential property of a material body.[28]

In the middle of the nineteenth century, a motivation for revisiting such metaphysical foundations of mechanics was provided by the establishment of non-mechanical theories such as electrodynamics and thermodynamics as mature subdisciplines of classical physics.[29] As a consequence of this development, mechanics not only lost its privileged status as the only conceivable candidate providing a conceptual basis for the entire building of physics, a status which was often associated with a claim to *a priori* truth, but also the conceptual foundation of mechanics itself could now be critically reexamined, including, for instance, the concept of absolute space and its justification by Newton. This revision of the status of the fundamental concepts of mechanics alone helped to prepare the conditions for a change of these concepts, should such a change become necessary in view of the growing body of knowledge.[30]

In any case, the critical reevaluation of the conceptual presuppositions of mechanics led to a proliferation of alternative approaches to the problem of gravitation, much as with the proliferation of alternative approaches within the framework of field theory. It was possible to elaborate more clearly the presuppositions upon which classical Newtonian mechanics was built, to revise the theory by attempting to eliminate those assumptions, which now appeared to be no longer acceptable (without any other substantial changes), or to formulate a new theory altogether. Carl Neumann's paper "On the Principles of the Galilean-Newtonian Theory" of 1869 provides an example of the first alternative: in order to replace Newton's concept of absolute space, he introduced the "body alpha" as the material embodiment of an absolute reference frame, comparing it with the luminiferous aether of electrodynamics as a like-

28   See the explanation of *Regula* III in (Newton 1972 [1726], 389).

29   Compare also the sequence in which Einstein, in his *Autobiographical Notes* (Einstein 1992), treats the *external* criticism of mechanics (the critique of mechanics as the basis of physics, pp. 22–23) and the "internal," conceptual criticism (pp. 24–31).

30   Compare e.g. the remark by Carl Neumann in 1869 (Neumann 1993 [1870], 367): "Finally, just as the present theory of electrical phenomena may perhaps one day be replaced by *another* theory, and the notion of an electric fluid could be removed, it is also the case that it is not an absolute impossibility that the Galilean-Newtonian theory will be supplanted one day by another theory, by some other picture, and the body alpha be made superfluous." ("Ebenso endlich, wie die gegenwärtige Theorie der elektrischen Erscheinungen vielleicht dereinst durch eine *andere* Theorie ersetzt, und die Vorstellung des elektrischen Fluidums beseitigt werden könnte; ebenso ist es wohl auch kein Ding der absoluten Unmöglichkeit, dass die Galilei-Newton'sche Theorie dereinst durch eine andere Theorie, durch ein anderes Bild verdrängt, und jener Körper Alpha überflüssig gemacht werde.") For the "body alpha" see below.

wise hypothetical, yet legitimate, conceptual element of the theory.[31] Nevertheless, by this reformulation Neumann did not intend to change the substance of Newton's theory, in particular not with respect to the question of relative and absolute motion, as the following passage illustrates:

> This seems to be the right place for an observation which forces itself upon us and from which it clearly follows how unbearable are the contradictions that arise when motion is conceived as something relative rather than something absolute.
>
> Let us assume that among the stars there is one which is composed of fluid matter and is somewhat similar to our terrestrial globe and that it is rotating around an axis that passes through its center. As a result of such a motion, and due to the resulting centrifugal forces, this star would take on the shape of a flattened ellipsoid. We now ask: *What shape will this star assume if all remaining heavenly bodies are suddenly annihilated (turned into nothing)?*
>
> These centrifugal forces are dependent only on the state of the star itself; they are totally independent of the remaining heavenly bodies. Consequently, this is our answer: These centrifugal forces and the spherical ellipsoidal form dependent on them will persist regardless of whether the remaining heavenly bodies continue to exist or suddenly disappear.[32]

The critical examinations of the foundations of classical mechanics in (Lange 1886) and (Mach 1960) correspond to the second alternative mentioned above, since both authors were guided by the intention to revise mechanics by eliminating problematic assumptions. They may be considered as attempts to provide a conceptual reinterpretation of the existing formalism of classical mechanics (possibly even including

---

31 "But a further question arises, whether this body exists—really, concretely, as the earth, the sun, and the remaining heavenly bodies do. We may answer this question, as I see it, by saying that its existence can be stated with the same right, with the same certainty, as the existence of the luminiferous ether or the electrical fluid." ("Aber es erhebt sich die weitere Frage, ob jener Körper denn eine wirkliche, concrete Existenz besitze gleich der Erde, der Sonne und den übrigen Himmelskörpern. Wir könnten, wie mir scheint, hierauf antworten, dass seine Existenz mit demselben Recht, mit derselben Sicherheit behauptet werden kann wie etwa die Existenz des Licht-Aethers oder die des elektrischen Fluidums.") (Neumann 1993 [1870], 365).

32 "Es mag hier eine Betrachtung ihre Stelle finden, welche sich leicht aufdrängt, und aus welcher deutlich hervorgeht, wie unerträglich die Widersprüche sind, welche sich einstellen, sobald man die Bewegung nicht als etwas Absolutes, sondern nur als etwas Relatives auffasst.

Nehmen wir an, dass unter den Sternen sich einer befinde, der aus *flüssiger* Materie besteht, und der— ebenso etwa wie unsere Erdkugel—in rotirender Bewegung begriffen ist um eine durch seinen Mittelpunkt gehende Axe. In Folge einer solchen Bewegung, infolge der durch sie entstehenden Centrifugalkräfte wird alsdann jener Stern die Form eines abgeplateten Ellipsoids besitzen. *Welche Form wird*—fragen wir nun—*der Stern annehmen, falls plötzlich alle übrigen Himmelskörper vernichtet (in Nichts verwandelt) würden?*

Jene Centrifugalkräfte hängen nur ab von dem Zustande des Sternes selber; sie sind völlig unabhängig von den übrigen Himmelskörpern. Folglich werden—so lautet unsere Antwort—jene Centrifugalkräfte und die durch sie bedingte ellipsoidische Gestalt ungeändert *fortbestehen*, völlig gleichgültig ob die übrigen Himmelskörper fortexistiren oder plötzlich verschwinden.") (Neumann 1993 [1870], 366, n. 8)

minor adjustments of this formalism), with no ambition to formulate a new theory or to cover new empirical ground. Lange's approach is today the less well known, precisely because his contribution was the introduction of the concept of an inertial system, a contribution that was successful in becoming part of the generally accepted conceptual interpretation of classical mechanics. Mach's widely discussed critique of the foundations of classical mechanics, on the other hand, is characterized by vacillation between more or less successful attempts to reformulate Newtonian mechanics on a clearer and leaner conceptual basis and suggestions to create a new theory. It seems plausible to assume that this ambiguity was actually not in conflict with Mach's intentions, as the principal aim of his reformulation of elements of classical mechanics was to stress and clarify the dependence of this theory on experience, and hence to open up the possibility of revising the theory if required by new empirical evidence.[33]

One of the principal targets of Mach's critique was Newton's interpretation of the bucket experiment as evidence in favor of the existence of absolute space.[34] To Newton's argument, according to which the curvature of the surface of the rotating water is a physical effect of the rotation with respect to absolute space, he objected that in our actual experience this rotation can be considered as a relative rotation, namely with respect to the fixed stars:

> Try to fix Newton's bucket and rotate the heaven of fixed stars and then prove the absence of centrifugal forces. (Mach 1960, 279)

Thus, Mach questioned the fundamental metaphysical presupposition of Newton's conclusion that physical effects of absolute space would also occur if the rotation took place in an otherwise empty universe, i.e. the presupposition that all elements of a system retain their essential properties independently of their relation to the composite system:

> Nature does not begin with elements, as we are obliged to begin with them. (Mach 1960, 287–288)

On the grounds of his different philosophical view, Mach demanded that the entire corpus of mechanics should be reformulated in terms of the motion of material bodies relative to each other. For instance, he introduced a new definition of the concept of mass based on the mutual accelerations of bodies with respect to each other. He also suggested that inertial frames of reference should be determined on the basis of the observable relative motions of bodies in the universe, e.g. by determining a frame of reference in which the average acceleration of a mass with respect to other—ide-

---

33 This seems to be the most natural explanation for Mach's rather indifferent reaction to the controversy about the purpose of his critique as observed in (Norton 1995). Compare Mach's remarks on his revised principle of inertia: "It is impossible to say whether the new expression would still represent the true condition of things if the stars were to perform rapid movements among one another. The general experience cannot be constructed from the particular case given us. We must, on the contrary, *wait* until such an experience presents itself." (Mach 1960, 289)

34 See (Mach 1960, chap. 2, sec. 6), in particular, pp. 279–284.

ally all—bodies in the universe vanishes. On the one hand, Mach's proposals for a
reformulation of classical mechanics clearly presuppose the validity of classical
mechanics: both his new definition of mass by mutual accelerations and his idea of
introducing increasingly improved inertial frames of reference by taking into account
more and more bodies, over whose relative motion an average can be taken, assume
that the concept of an inertial frame makes sense exactly as understood in classical
mechanics. In other words, his proposal presupposes that there is indeed such a privileged class of reference frames and that they can be realized physically with sufficient
approximation.[35] However, Mach's analysis also indicated the limits of the validity
of classical mechanics, in particular by explicitly relating the concept of inertial
frame to the motion of cosmic masses. Thus, without changing the substance of classical mechanics, he succeeded nevertheless in making clear—by proposing an alternative formulation based on different philosophical presuppositions—that the range
of application of classical mechanics may be more limited than hitherto assumed, and
that the theory might have to be changed eventually, for instance, in light of growing
astronomical knowledge. Only on the basis of such an increased knowledge could it
then be decided whether Mach's suggestion to reformulate classical mechanics in
terms of relative motions would actually amount to proposing a new theory, substantially different from Newton's.

Attempts to formulate such a new theory, even in the absence of new empirical evidence, form a third alternative reaction to the critical reevaluation of the foundations
of mechanics in the second half of the nineteenth century. Whether these attempts
were stimulated by Mach or not, their common starting point was the rejection of
Newton's philosophical presupposition that the properties of the elements of a physical system could be ascribed to each one of them even if they existed alone in empty
space. It was this presupposition that enabled Newton to infer from the nature of the
inertial effects present in the bucket experiment to that of the inertial behavior of a single particle in empty space, and from there, to the physical reality of empty space.
Only by introducing an entity such as "absolute space" did Newton succeed in distinguishing between the kinematical and dynamical aspects of motion. Hence, if now this
presupposition had become questionable, so had the entire relation between dynamics
and kinematics. In particular, the distinction between force-free motions and those
explained by the action of forces had to be given a new grounding in terms of relative
motions between ponderable bodies. While Mach had essentially presupposed the
validity of classical mechanics and attempted to reconstruct its achievements on this
new basis, it was also conceivable to start from first principles and reformulate dynamics from the beginning in terms of relative motions between ponderable bodies, possibly even without using the concept of an inertial frame in the sense of classical
mechanics. Attempts in this direction of a *mechanistic generalization of the relativity
principle* were first undertaken around the turn of the century by Benedict and Imman-

---

35   See the penetrating analysis in (Wahsner and von Borzeszkowski 1992, 324–328).

uel Friedlaender, August Föppl, and Wenzel Hofmann, then decades later by Reissner and Schrödinger, and in our days by Barbour, Bertotti and others.[36]

Physicists of at least the first generation in this genealogy were confronted with the difficulty of taking up once again many of the foundational questions of mechanics discussed centuries earlier by Galileo Galilei, René Descartes, Isaac Newton, Gottfried Wilhelm Leibniz, and Christiaan Huyghens, and they attempted to recreate mechanics essentially from scratch. Indeed, apart from the foundational role given to the concept of relative motion even in dynamics and the known laws of classical mechanics, this approach of a mechanistic generalization of the relativity principle had few heuristic clues to go on. One of these clues was directly related to the criticism of Newton's interpretation of the bucket experiment: if it is indeed true that the curvature of the rotating water in the bucket is due to an interaction between the water and the distant cosmic masses, then a similar but smaller effect should be observable if large but still manipulable terrestrial masses are brought into rotation with respect to a test body. Experiments along these lines were suggested by several of these researchers and conducted by, among others, the Friedlaender brothers and Föppl — all with a negative result.[37] Nevertheless, the theoretical efforts continued — even though they remained marginal — and eventually found additional resources and inspiration in the theory of general relativity formulated in 1915 by Einstein.

3.3 Resources and Stumbling Blocks Presented by the Tradition of Mechanics

After this discussion of the historical roots of the mechanistic generalization of the relativity principle, we can now summarize some of the principal heuristic contributions and obstacles which the tradition of mechanics presented to the development of general relativity, as in the beginning of this section for field theory. First and foremost it was the idea of abolishing the privileged status of the inertial frame, which emerged from the foundational critique of mechanics in the nineteenth century, that proved to be an essential component of both Einstein's early research concerning generalized relativity, and of the considerations surrounding the competing traditions to provide a mechanistic generalization of the relativity principle. In fact, if separable material bodies are to be the ultimate basis of reality, as they are in the approach of a mechanistic generalization of the relativity principle, each material body should be equally well suited to defining a reference frame, and therefore should enter into the laws of physics on the same level with all other bodies. The idea of abolishing the privileged status of inertial frames was associated with the interpretation of the so-called inertial forces — such as those acting on the rotating water in Newton's bucket — as aspects of a new, yet to be discovered, velocity-dependent physical interaction between masses in relative motion with respect to each other. Under the desig-

---

36   See, e.g., (Friedlaender and Friedlaender 1896; Föppl 1905a, 1905b; Hofmann 1904; Reissner 1914, 1915; Schrödinger 1925, Barbour and Bertotti 1977).
37   See (Friedlaender and Friedlaender 1896; Föppl 1905a).

nation of "dragging effects," such interactions became an important theme of the later general theory of relativity; there they can be understood as a new aspect of the gravitational interaction between masses, which was unknown in Newtonian mechanics. Finally, Mach's definition of inertial mass in terms of the accelerations that two bodies induce in each other brought the concept of inertial mass even closer to the concept of gravitational mass than their quantitative identity in classical mechanics had so far. It follows from this definition that, contrary to Newtonian mechanics, inertial mass, in contrast to gravitational mass, can no longer be considered as a property that bodies possess independently of their interaction with each other. The search for effects of the presence of other bodies on the inertial mass of a test body became a component of the heuristics guiding Einstein's research on a generalized theory of relativity.

While these were the specific and crucial contributions of the foundational critique of mechanics, other aspects of mechanics in the nineteenth century also contained important heuristic hints and conceptual resources for the development of general relativity; these resources, however, cannot be dealt with here systematically. In particular, the introduction of laws of motion expressed in generalized coordinates, the formulation of mechanics for non-Euclidean geometry, and the attempts at an elimination of the concept of force all represent resources which could be, and in part were, exploited in the development of general relativity.[38] The study of motion constrained to curved surfaces in classical mechanics provided, for instance, the blueprints for the formulation of the geodesic law of motion as a generalization of the principle of inertia in general relativity: in both cases the essential assertion is that motion not subject to external forces follows a geodesic line.

But unlike the foundational critique of mechanics, these other aspects of the development of classical mechanics did not themselves constitute another independent research program for formulating a substantially new mechanics that could lead to a theory comparable to general relativity. Rather their heuristic contribution to formulating such a new theory became relevant only in the context of Einstein's later attempt to solve the problem of gravitation, and only on the basis of results that lay outside their scope. For instance, Hertz's mechanics (Hertz 1894) is a reformulation of classical mechanics in which the elimination of the concept of force requires the introduction of hypothetical invisible masses acting as constraints for the visible motions. Not only does its formalism and, in particular, its generalized geodesic law of motion bear a number of similarities with the formalism of general relativity, but also the general approach of replacing the concept of force by geometrical concepts is common to both theories.[39] Although, even in the context of classical mechanics, the concept of force can be eliminated in the specific case of the gravitational interaction without introducing Hertz's speculative entities—merely on the basis of Galileo's principle, that is, by realizing that all bodies move with equal speeds in a gravitational

---

38   For a historical account of these developments, see (Lützen 1993).
39   The geometrical interpretation of general relativity is, however, a largely post-1915 development.

field—a systematic exploitation of formal results such as Hertz's required not only a restriction of mechanics to the special case of gravitational interaction, but also the introduction of Minkowski's reformulation of special relativity uniting the time with the space-coordinates into one spacetime continuum. Only under these presuppositions did the formal achievements of nineteenth-century mechanics become a resource for the insight that force-free motion in a gravitational field can also be understood as geodesic motion in a non-Euclidean spacetime continuum.

Considered in hindsight, however, these contributions to the development of general relativity that were rooted in the tradition of classical mechanics also presented conceptual obstacles to its development. First of all, as in the case of field theory discussed above, there was much ambiguity in the research program of a mechanistic generalization of the relativity principle. It is impossible to assess the direction that this program would have taken by itself without the guidance of Einstein's achievement, since the general theory of relativity was formulated in 1915—long before an elaborate and more or less successful realization of this program emerged. Around 1915 it was far less advanced than the attempts to solve the problem of gravitation in the context of field theory. The papers proposing a mechanistic generalization of the relativity principle are mostly in the form of programmatic treatises. They contain few technical details, and show, even by their style, that they deal with foundational problems of mechanics as commonly discussed by Galileo and his contemporaries in early modern times. In particular, in exploring the postulated velocity-dependent interaction, the mechanistic approach had few tools comparable to the powerful methods developed in the field-theoretic context: of particular interest were the tools for coping with the interaction of electrically charged masses in motion with respect to each other. Also on the experimental level, the mechanistic generalization of the relativity principle failed to identify evidence in favor of this new interaction between moving masses. It is therefore not surprising that the followers of a mechanistic generalization of the relativity principle were few and played only a marginal role in contemporary discussions. In addition to its weaknesses as an independent program of research, the idea of a mechanistic generalization of the relativity principle included aspects that, if judged from the perspective of the accomplished theory of general relativity, were both stimulating and misleading: while the ideal of a theory in which all physical aspects of space are derived from the relations between separable material bodies was an essential motivation for the search for a general theory of relativity, it turned out to be incompatible with its outcome since in general relativity the gravitational field has an existence in its own right, which cannot be reduced to the effects of matter in motion.

3.4 The Example of Benedict and Immanuel Friedlaender

The opportunities and difficulties presented by a mechanistic generalization of the relativity principle are best illustrated by the contribution of the Friedlaender brothers. Their philosophical starting point is the critique of the concept of motion of a single body in an otherwise empty space, on which, as we have seen, Newton's argument for absolute space was founded:

> Now think (if you can) of a progressive motion of a *single* body in a universal space that is otherwise imagined to be totally empty; how can the motion be detected, i.e. distinguished from rest? By *nothing* we should think; indeed the whole idea of such an absolute, progressive motion is meaningless.[40]

Like other critics of Newtonian mechanics, Immanuel and Benedict Friedlaender question the meaning of inertial frames and postulate a new velocity-dependent interaction between moving masses. But unlike other representatives of a mechanistic generalization of the relativity principle, they explicitly link this new interaction to gravitation:

> If this phenomenon was verifiable, this would be the incentive for a reformulation of mechanics and at the same time a further insight would have been gained into the nature of gravity, since these phenomena must be due to the actions at a distance of masses, and here in particular to the dependence of these actions at a distance on relative rotations.[41]

How far they went in anticipating the relation between gravitation and inertia as understood in general relativity becomes clear from speculations presented towards the end of their paper:

> It is also apparent that according to our conception the motions of the bodies of the solar system could be seen as pure *inertial motions*, whereas according to the usual view the inertial motion, or rather its permanent gravitationally modified tendency, would strive to produce a rectilinear-tangential motion.[42]

And, in another suggestive passage:

> But it seems to me that the correct formulation of the law of inertia will be found only when *relative inertia* as an effect of masses on each other, and gravity, which is after all also an effect of masses on each other, are reduced to a *unified law*.[43]

At a first glance, the insight formulated by the Friedlaenders into the relation between velocity-dependent inertial forces and gravitation seems to contradict the claim that a

---

40 "Nun denke man sich aber, (wenn man kann,) eine fortschreitende Bewegung eines *einzigen* Körpers in dem als sonst völlig leer gedachten Weltenraume; woran wäre die Bewegung bemerklich, d.h. von Ruhe unterscheidbar? An *Nichts* sollten wir meine; ja, die ganze Vorstellung einer solchen absoluten, fortschreitenden Bewegung ist sinnleer." The first part of their jointly published booklet, pp. 5–17, is by Immanuel Friedlaender and the second part, pp. 18–35, by Benedict Friedlaender. (Friedlaender and Friedlaender 1896, 20)

41 "War diese Erscheinung nachzuweisen, so war der Anstoß zu einer Umformung der Mechanik gegeben und zugleich ein weiterer Ausblick in das Wesen der Gravitation gewonnen, da es sich ja dabei nur um Fernwirkungen von Massen und zwar hier der Abhängigkeit dieser Fernwirkungen von relativen Rotationen handeln kann." (Friedlaender and Friedlaender 1896, 15)

42 "Es ist auch leicht ersichtlich, daß nach unsrer Auffassung die Bewegungen der Körper des Sonnensystems als reine *Beharrungsbewegungen* angesehen werden könnten; während nach der üblichen Anschauung die Beharrungsbewegung, oder vielmehr deren fortwährend durch die Gravitation abgeänderte Tendenz eine geradlinig-tangentiale Bewegung hervorzurufen bestrebt wäre." (Friedlaender and Friedlaender 1896, 33)

mechanistic generalization of the relativity principle did not possess tools comparable to those used in the electromagnetic tradition to treat the interaction of charged
masses in motion with respect to one another. However, a footnote to the same passage makes it clear that the source of this insight into a possible relation between
gravity and inertia actually is the *combination* of the introduction of velocity-dependent forces by the mechanistic generalization of the relativity principle and of the
treatment of velocity-dependent forces in the electromagnetic tradition:

> For this it would be very desirable to resolve the question whether Weber's law applies to
> gravity, as well as the question concerning gravity's speed of propagation.[44]

The reference is to Wilhelm Weber's fundamental law for the force between electric
point charges, which is a generalization of Coulomb's law for the electrostatic force,
taking into account also the motion of the charges. By including velocity-dependent
terms, Weber's law represents an attempt to cover electrodynamic interactions too,
while maintaining the form of an action-at-a-distance law, that is, of a direct interaction between the point charges without an intervening medium. In other words, the
Friedlaenders established a connection between their foundational critique of
mechanics and the contemporary discussions about an electromagnetic theory of
gravitation.[45]

By the time their paper was published, however, action-at-a-distance laws such as
Weber's had been largely superseded by the field-theoretic approach to electromagnetism taken by Maxwell, Hertz, and others, who assumed a propagation of the electromagnetic force by an intervening medium, the aether.[46] The Friedlaenders
themselves seem to have entertained considerations along these lines, without, however, drawing any technical consequences from them:

> No mind thinking scientifically could ever have permanently and seriously believed in
> unmediated action at a distance; the apparent action at a distance can be nothing other
> than the result of the action of forces that are transmitted in some way by the medium
> being situated between the two gravitating bodies.[47]

---

43   "Mir will aber scheinen, daß die richtige Fassung des Gesetzes der Trägheit erst dann gefunden ist,
     wenn die *relative Trägheit* als eine Wirkung von Massen auf einander und die Gravitation, die ja auch
     eine Wirkung von Massen auf einander ist, auf ein *einheitliches Gesetz* zurückgeführt sein werden."
     (Friedlaender and Friedlaender 1896, 17)
44   "Es wäre dazu sehr zu wünschen, daß die Frage, ob das Webersche Gesetz auf die Gravitation anzu-
     wenden ist, sowie die nach der Fortpflanzungsgeschwindigkeit der Schwerkraft gelöst würden."
     (Friedlaender and Friedlaender 1896, 17)
45   Hints to such a connection are also found in other authors, even if they are less explicit; see, e.g.,
     (Föppl 1905b, 386–394; Mach 1960, 296), with reference to the Friedlaender brothers and Föppl. For
     a discussion of Mach's position, see (Wolters 1987), in particular, pp. 37–70.
46   For the role of Weber's law in the later tradition of generally relativistic mechanics, see (Assis 1989,
     1995; see also Barbour 1992, 145).
47   "An die unvermittelte *Fernwirkung* kann kein naturwissenschaftlich denkender Kopf jemals andau-
     ernd und ernstlich geglaubt haben; die scheinbare *Fernkraft* kann nichts anderes sein, als das Resultat
     von Kraftwirkungen, die durch das zwischen beiden gravitirenden Körpern befindliche Medium in
     irgend einer Art vermittelt werden." (Friedlaender and Friedlaender 1896, 19)

But whether in the field-theoretic or in the action-at-a-distance form, it was the tools of the electromagnetic tradition of classical physics which allowed the Friedlaenders to establish the link between the new understanding of inertia and gravitation. It is therefore not surprising that they treat the dragging effects of masses in relative motion to each other in analogy with electromagnetic induction:

> ... only in order to indicate the extent to which the problem of motion that we have raised and hypothetically solved here is related to that of the nature of gravity, but at the same time that comes rather close to the known effects of electric forces, will the following parallel be pointed out: a body that approaches a second one or moves away from it would be without influence on the latter as long as the velocity of approach (to be taken either with a positive or a negative sign) remains unchanged; any change of this velocity would entail the above-demonstrated [dragging] effect.
>
> As is well known, the presence of a current in a conductor is not sufficient for the generation of induction effects, either the magnitude of the current or the distance must vary; in our case the change of distance, i.e. the motion, would not suffice for the generation of the attractive or repulsive effects, but rather the velocity itself has to change.[48]

3.5 The Historical Horizon Before Einstein's Contribution

In summary, we have identified and discussed two entirely different strategies—both pursued at the time when Einstein began to work seriously on a relativistic theory of gravitation—for dealing with important conceptual issues at the foundations of mechanics and gravitation theory. The field theoretic approach to the problem of gravitation was, around this time, mainly stimulated by the incompatibility between Newton's theory of gravitation and the special theory of relativity, while the starting point of the mechanistic generalization of the relativity principle was a philosophical critique of the foundations of Newtonian mechanics based on newly established branches of classical physics. Their relation can be understood in the context of the two principal competing worldviews of classical physics around the turn of the century, the electromagnetic worldview and the mechanical worldview. In particular, these worldviews determined the different conceptual resources from which the two strategies drew rather exclusively, those of field theory and of classical mechanics respectively. Whereas the mechanistic generalization of the relativity principle

---

48 "... und nur, um anzudeuten, in wie fern das hier angeregte und hypothetisch gelöste Bewegungsproblem mit demjenigen des Wesens der Gravitation zusammenhängt, sich zu gleicher Zeit aber den bekannten Wirkungsweisen elektrischer Kräfte einigermaßen nähert, sei auf folgende Parallele hingewiesen: Ein Körper, der sich einem zweiten nähert oder von ihm entfernt, würde ohne Einfluß auf diesen sein, solange die positiv oder negativ zu nehmende Annäherungsgeschwindigkeit unverändert bleibt; jede Aenderung der Geschwindigkeit hingegen würde die vorher gezeigte Wirkung ausüben.
Das Vorhandensein eines Stromes in einem Leiter genügt bekanntlich zur Erzeugung von Induktionswirkungen nicht, es muß entweder die Stromstärke oder die Entfernung wechseln; in unserem Falle würde nun zur Erzeugung der anziehenden oder abstoßenden Wirkungen auch die Entfernungsänderung, d.h. die Bewegung nicht ausreichen, es muß vielmehr die Geschwindigkeit selbst sich ändern." (Friedlaender and Friedlaender 1896, 30)

remained in the margin of contemporary physics, the field theoretic approach to grav-itation, at least for a while, played a larger part in contemporary discussions, and both strategies were pursued in ignorance of one another.

The two strategies encountered problems which, in hindsight, can be recognized as being closely related to each other. On a general level, the difficulties of the two strategies were in an inverse relation to each other: those following the field theoretic approach were confronted with the problem of reconstructing on a new conceptual basis the shared knowledge accumulated in classical mechanics, e.g. the insight into the equality of gravitational and inertial mass. The followers of a mechanistic gener-alization of the relativity principle, on the other hand, had to face the task of keeping up in their terms with the immense contribution of field theory to the progress of physics in the nineteenth century, a formidable challenge even for current attempts to pursue the tradition of the mechanistic generalization of the relativity principle. But on the specific level of the gravitational and inertial interactions of masses, the prob-lems faced by the two approaches are better characterized as being complementary to each other: on the basis of concise theoretical considerations, the electromagnetic approach to the problem of gravitation required the existence of a velocity-dependent gravitational interaction in analogy to electromagnetic induction, for which there was, however, little, if any, experimental evidence; the mechanistic generalization of the relativity principle, on the other hand, postulated a new velocity-dependent inter-action between inertial masses in order to explain well-known observations such as the curvature of the water's surface in Newton's bucket experiment, but failed to develop a theoretical framework for its systematic treatment. Since the two traditions remained isolated from each other—with the remarkable but inconsequential excep-tion of the Friedlaender brothers—their complementary strengths were not exploited before Einstein's contribution.

4. MACH'S PRINCIPLE:
BETWEEN A MECHANISTIC GENERALIZATION OF THE
RELATIVITY PRINCIPLE AND A FIELD THEORY OF GRAVITATION

4.1 The Emergence of a Link Between Einstein's Research on Gravitation and Mach's Critique of Mechanics in 1907

The problems of a field theory of gravitation, from which Einstein had started in 1907, pointed in two ways to Mach's critique of Newton's mechanics, namely, to his redefinition of the concept of mass and to his rejection of absolute space as a founda-tion for the understanding of inertial motion. As discussed in the previous section, the concept of inertial mass and the concept of absolute space were in fact connected through Newton's assumption that the essential properties of the elements of a system are independent of these elements' part in the larger (composite) system. The rejec-tion of this assumption shattered both Newton's distinction between inertial and grav-itational mass as essential and non-essential properties of a body respectively, and his demonstration of absolute space. Einstein had been familiar with Mach's critique of Newton's mechanics since his student days[49] and probably reread the corresponding chapters of the *Mechanik* after his first attack on the problem of gravitation in 1907.[50]

The physical asymmetry between inertial and gravitational mass, which, as perceived by Einstein in 1907, was at the heart of the conflict between a special relativistic theory of gravitation and classical mechanics, may have directed his attention to their more general asymmetry in Newtonian mechanics. According to Newtonian mechanics, inertial mass is a property that can also be ascribed to a single body in an otherwise empty universe, whereas gravitational mass can only be conceived as a property of a system of bodies. Mach's analysis of the concept of inertial mass can be considered as an attempt to remove just this asymmetry, at least on the level of an operational definition of inertial mass. According to this definition, inertial mass is determined, as we have seen, on the basis of the mutual accelerations within a system of bodies, i.e. not as the independent property of a single body. Although Mach's intention was probably only to give a more concise account of classical mechanics without changing its content, nevertheless his definition makes it clear that, in principle, the interaction between two masses, and hence their magnitude, may depend on the presence of other masses in the world (recall that the inertial frame within which the accelerations are measured is, according to Mach's critique of absolute space, determined by the distribution of masses in the universe). In any case, according to Mach, inertial mass and gravitational mass both depend upon interactions between bodies. This lends additional strength to Einstein's conclusion that the equivalence of inertial and gravitational mass in classical mechanics points to a deeper conceptual unity that is to be preserved also in a new theory of gravitation.

Einstein's introduction of the principle of equivalence, which expresses the equality of inertial and gravitational mass independent of the specific laws of motion of classical mechanics, indicated a connection to Mach's critical discussion of Newton's purported demonstration of absolute space. The successful use of a uniformly accelerated frame of reference to describe the behavior of bodies falling in a constant gravitational field must naturally have raised questions about the relation between arbitrarily accelerated reference frames and more general gravitational fields. For Einstein, such questions pointed in particular to the problem of the privileged role of inertial frames in classical mechanics, as he confirms in the recollection already quoted in the first section:

---

49 For an early reference to Mach, see Einstein to Mileva Marić, 10 September 1899 (CPAE 1, Doc. 54; Renn and Schulmann 1992, 14, 85). For later recollections mentioning Mach, see (Einstein 1933, 1954b, 1992).

50 For contemporary evidence of Einstein's rereading, see p. 58 of Einstein's Scratch Notebook 1910–1914? (CPAE 3, 592, app. A), where Einstein wrote the title of the crucial sec. 6 of chap. 2 of Mach's *Mechanics* (Mach 1960); pp. 7–8 of Einstein's *Lecture Notes for an Introductory Course on Mechanics* at the University of Zurich, Winter semester 1909/1910, (CPAE 3, 15–16, discussed in more detail below); and the discussion of Mach's ideas in a notebook on Einstein's Course on Analytical Mechanics, Winter semester 1912/13, by Walter Dällenbach, (for a brief description, see (CPAE 4, app. A).

> So, if one considers pervasive gravitational fields, not *a priori* restricted by spatial
> boundary conditions, physically possible, then the concept of 'inertial system' becomes
> completely empty. The concept of 'acceleration relative to space' then loses all meaning
> and with it the principle of inertia along with the paradox of Mach.[51]

In other words, the appearance of accelerated frames of reference in an argument
concerning gravitation made it possible to relate to each other two theoretical traditions which had until then led essentially separate existences, the tradition of a field
theory of gravitation in the sense of electrodynamics and the tradition of foundational
critique of mechanics in the sense of what is called here "mechanistic generalization
of the relativity principle." In the previous section, we have seen that the idea of
including accelerated frames of reference on an equal footing with inertial systems
was as alien to the tradition of field theory as the idea of a field theory of gravitation
was to the tradition of the mechanistic generalization of the relativity principle.

Now, however, Mach's critical examination of the privileged role of inertial
frames in classical mechanics provided Einstein with the context for considering his
introduction of an accelerated frame of reference in the equivalence principle argument, not only as a technical trick to deal with a specific aspect of the problem of formulating a field theory of gravitation, but as a hint to the solution of a foundational
problem of classical mechanics. But while Mach's critique justified the consideration
of arbitrary frames of reference as a basis for the description of physical processes,
and hence the extension of the equivalence principle argument to include more general accelerated frames, such as the rotating frame of Newton's bucket,[52] it did not
provide Einstein with the conceptual tools for dealing with the strange effects
encountered in such frames. The tradition of field theory, in the context of which he
had first approached the problem of gravitation, offered him, on the other hand, just
the conceptual tools that allowed him to interpret the inertial forces in accelerated
frames of reference as aspects of a more general notion of a gravito-inertial field, in
the same sense that electromagnetic field theory makes it possible to conceive induction as an aspect of a more general notion of an electric field.

In other words, instead of attempting to resolve Mach's paradox of the privileged
role of inertial frames in the context of a revised version of classical mechanics, as
did the adherents of a mechanistic generalization of the relativity principle, Einstein
was now able to address this foundational problem of mechanics in the context of a
field theory of gravitation in which inertial forces could be understood as an aspect of

---

51 "Wenn man also das Verhalten der Körper inbezug auf das letztere Bezugssystem als durch ein "wirk-
    liches" (nicht nur scheinbares) Gravitationsfelder als möglich betrachtet, so wird der Begriff des Iner-
    tialsystems völlig leer. Der Begriff "Beschleunigung gegenüber dem Raume" verliert dann jede
    Bedeutung und damit auch das Trägheitsprinzip samt dem Mach'schen Paradoxon." (Einstein 1992,
    62–63)

52 For the particular role of rotating frames in motivating this generalization, compare Einstein's later
    remark concerning an objection against the privileged role of inertial frames in classical mechanics
    and in special relativity: "The objection is of importance more especially when the state of motion of
    the reference-body is of such a nature that it does not require any external agency for its maintenance,
    e.g. in the case when the reference body is rotating uniformly." (Einstein 1961, 72)

a unified gravito-inertial field. By establishing a "missing link" between the traditions of a mechanistic generalization of the relativity principle and field theory, he had found the key to the problems which appeared to be unsolvable within each of the two traditions taken separately. Where the followers of a field theory of gravitation searched in vain for an empirical clue which could have guided them beyond "Coulomb's law" of static gravitation (i.e. Newton's law) to a gravitational dynamics, Einstein succeeded with the help of Mach's critique in recognizing in the inertial effects of a rotating system, such as Newton's bucket, the case of a stationary gravitational field caused by moving masses. He interpreted this case as a gravitational analogue to a magnetostatic field in electrodynamics which can also be conceived as being caused by moving (in this case: electrical) masses.[53] And vice versa, where the adherents of a mechanistic generalization of the relativity principle searched in vain for new effects that could reveal more about the mysterious interaction between distant masses in relative motion with respect to each other, which in the only case known to them was responsible for the curvature of the water's surface in Newton's bucket, Einstein had no qualms about identifying this force as a dynamical aspect of universal gravitation and thus relating the unknown force to a well-explored domain of classical physics. In summary, Einstein's experiences with a field theory of gravitation and his familiarity with the foundational problems of mechanics had set the stage for his reception of whatever these two traditions had to offer for his program to build a relativistic theory of gravity that was also to be a theory of general relativity. What had previously seemed to be mutually exclusive approaches, to some extent now became, from his perspective, complementary.

4.2 Hints at a Machian Theory of Mechanics in Einstein's Research on Gravitation Between 1907 and 1912

The following is a brief account of those features of Einstein's heuristics that reflect the complementary influence of the two traditions in the sense outlined above. While there is no direct contemporary evidence for the role of Mach's critique of mechanics on Einstein's 1907 formulation of what later became known as the equivalence principle, such an influence very likely forms the background for Einstein's reaction to the problems of a relativistic theory of gravitation.[54] Beyond shaping this reaction and opening the perspective towards a generalization of relativity theory, Mach's influence on the further development of this theory remained secondary, even when Einstein began to elaborate his original insight into the equivalence principle in papers published in 1911 and 1912.[55] The principal reason for this secondary status is that, in

---

53   See Einstein to Paul Ehrenfest, 20 June 1912 (CPAE 5, Doc. 409), discussed below.

54   See, in particular, (Einstein 1954b) for evidence that Einstein's perspective was indeed shaped at a very early stage by Mach's critique of mechanics. For a discussion of the relation between equivalence principle and Mach's interpretation of the bucket experiment, see "Classical Physics in Disarray …" (in vol. 1 of this series).

55   See, in particular, (Einstein 1911, 1912a, 1912b, 1912c).

this period, he drew mainly on the resources of field theory with the aim of construct-ing a field equation—analogous to the classical field equation for Newton's gravita-tional field—for the static gravitational field of his elevator thought experiment.[56]

Nevertheless, between 1907 and 1912 Einstein seems also to have collected hints pointing at a Machian theory of mechanics. For instance, he made use of Mach's analysis of the conceptual foundations of mechanics in preparing a course on classi-cal mechanics at the University of Zurich for the winter semester 1909/1910[57] and referred to it in connection with his research on gravitation in correspondence to Ernst Mach of the same period.[58] At about the same time, he wrote the following in a letter to a friend:

> I am just now lecturing on the foundations of that poor, dead mechanics, which is so beautiful. What will its successor look like? With that question I torment myself cease-lessly.[59]

In the notes Einstein prepared for his lecture course he introduces Mach's definition of mass.[60] He emphasized the close relation between gravitational and inertial mass, following from the independence of gravity from material properties:

> The fact that the force of gravity is independent of the material demonstrates a close kin-ship between inertial mass on the one hand and gravitational action on the other hand.[61]

The dependence of inertial mass on the entire system of bodies in the universe implicit in Mach's definition of mass made it conceivable for Einstein that the inertial mass of a given body may also be a function of the system of other bodies, which var-ies with their distribution around the given body.[62] In (Einstein 1912c), he partially confirmed this conclusion by calculating the effect on the inertial mass of a body due to the presence of a massive spherical shell around it; the paper also deals with the effect on this body by an accelerated motion of the spherical shell. This paper, dedi-cated to Einstein's theory of the static gravitational field, is not only the first paper in

---

56 For an extensive discussion, see "Classical Physics in Disarray ..." (in vol. 1 of this series).
57 See Einstein's *Lecture Notes for an Introductory Course on Mechanics* at the University of Zurich, Winter semester 1909/1910 in (CPAE 3).
58 See Einstein to Ernst Mach, 9 August 1909 (CPAE 5, Doc. 174, 204) and Einstein to Ernst Mach, 17 August 1909 (CPAE 5, Doc. 175, 205).
59 "Ich lese gerade die Fundamente der armen gestorbenen Mechanik, die so schön ist. Wie wird ihre Nachfolgerin aussehen? Damit plage ich mich unaufhörlich." Einstein to Heinrich Zangger, 15 November 1911 (CPAE 5, Doc. 305, 349).
60 See pp. 7–8 of Einstein's *Lecture Notes for an Introductory Course on Mechanics* at the University of Zurich, Winter semester 1909/1910 (CPAE 3, 15–16).
61 "Die Thatsache, dass die Kraft der Schwere vom Material unabhängig ist, zeigt eine nahe Verwand-schaft zwischen träger Masse einerseits und Gravitationswirkung andererseits." See p. 15 of Ein-stein's *Lecture Notes for an Introductory Course on Mechanics* at the University of Zurich, Winter semester 1909/1910 (CPAE 3, 21; my translation)
62 This is in disagreement with the claim expressed in (Barbour 1992, 135), that Einstein was not justi-fied in maintaining that he was a following a stimulation by Mach in considering a dependence of inertial mass on the presence of other masses in the universe.

which he publicly mentions Mach's critique as a heuristic motivation behind his search for a generalized theory of relativity, but it also carries a title expressing the translation of this heuristics into the language of field theory: "Is there a gravitational effect which is analogous to electrodynamic induction?"

In 1912 Mach's critique gained a new importance for Einstein's work on gravitation for yet another reason. After convincing himself that he had found a more or less satisfactory theory of the static gravitational field, he turned to what he considered to be the next simple case, the stationary field represented by the inertial forces in a rotating frame. In other words, after exhausting, at least for the time being, the heuristic potential of the "elevator," he now turned to that of the "bucket." His contemporary correspondence confirms that he considered this case both from the perspective of field theory and from that of the mechanistic generalization of the relativity principle. In a letter to Ehrenfest from 20 June 1912, with reference to his theory of the static gravitational field and to the generalization necessary to cope with situations such as that of a rotating ring, he wrote:

> In the theory of electricity my case corresponds to the electrostatic field, while the more general static case would further include the analogue of the static magnetic field. I am not yet that far.[63]

In a letter to Besso dated 26 March 1912, Einstein remarked—probably referring to the same topic, i.e., the treatment of the inertial forces in a rotating frame as generalized gravitational effects in a frame at rest—in the spirit of Mach's remark on Newton's bucket: "You see that I am still far from being able to conceive rotation as rest!".[64] Not only Einstein's publications and correspondence but also his private research notes document the influence of both traditions—electrodynamics and mechanics—on the terminology in which he expressed the heuristics of his theory. Thus, we can exclude the possibility that his choice of words was merely a matter of making himself understood by his audience.[65]

4.3 Einstein's Machian Heuristics in his Research on a Relativistic Theory of Gravitation between 1912 and 1915

Einstein found it difficult to accomplish the transition from his treatment of the static special case to a more general theory that included the dynamical aspects of the gravitational field. In the summer of 1912, however, he attained the insight into the crucial role of non-Euclidean geometry for formulating the gravitational field theory he

---

63 "Mein Fall entspricht in der Elektrizitätstheorie dem elektrostatischen Felde, wogegen der allgemeine[r]e statische Fall noch das Analogon des statischen Magnetfeldes mit einschliessen würde." Einstein to Paul Ehrenfest, 20 June 1912, (CPAE 5, Doc. 409, 486).

64 "Du siehst, dass ich noch weit davon entfernt bin, die Drehung als Ruhe auffassen zu können." Einstein to Michele Besso, 26 March 1912, (CPAE 5, Doc. 377, 436).

65 See, in particular, Einstein's comments on his calculation of the effect of rotation and linear acceleration of a massive shell on a test particle in his and Michele Besso's May 1913 "Manuscript on the Motion of the Perihelion of Mercury" in (CPAE 4, Doc. 14).

searched for, an insight which, in spite of the many difficulties still to be resolved, paved the way for the final theory of general relativity published in 1915. This insight provides an important example of the fruitfulness of the combined heuristics of "elevator" (i.e., Einstein's equivalence principle) and "bucket" (i.e., Newton and Mach's bucket in Einstein's interpretation).[66] The heuristics of the bucket, i.e. the Machian idea to consider water in a bucket as constituting a frame at rest, first provided the qualitative insight into a possible role of non-Euclidean geometry in a rotating frame of reference.[67] The heuristics of the elevator, i.e. the elaboration of the theory of the static gravitational field, then prepared, in combination with Minkowski's four-dimensional formalism, the technical environment for the concrete application of this insight to the problem of gravitation. The crucial link between the general idea and this technical environment was provided by Gaussian surface theory, which made it possible to interpret the equation of motion suggested by the formalism of the static theory as the geodesic equation of a non-Euclidean geometry. It was only possible, however, to exploit the formal similarity between the two equations because of the deeper conceptual similarity between the problem of motion in a gravitational field and the problem of inertial motion in an accelerated frame of reference, as suggested by Einstein's Machian interpretation of inertia. This conceptual similarity may have helped Einstein to think of Gaussian surface theory in the first place, as he had been familiar since his student days with the relation in classical mechanics between motion constrained to a surface without external forces—which also can be conceived of as generalized inertial motion—and the geodesic equation in Gaussian surface theory.[68]

But even after Einstein recognized that the gravitational potential of his static theory could be interpreted as a component of the metric tensor in four-dimensional geometry, he would nevertheless have been, at least in principle, in the same situation as those searching for a dynamic theory of the gravitational field starting from Newton's theory as the only known special case. It was his "Machian" insight that the inertial effects in accelerated frames can be considered as an aspect of a more general gravito-inertial field that provided him with an entire class of examples supporting the relation between the equation of motion, metric tensor, and gravito-inertial field, which emerged from the generalization of the static theory. In fact, Einstein could

---

66   Compare also Einstein's Kyoto Lecture (Ishiwara 1971, 78–88).

67   This was first stressed in (Stachel 1989). For a more extensive reconstruction, see "Classical Physics in Disarray ..." and "The First Two Acts" (both in vol. 1 of this series).

68   This is suggested by the similarity between a page in the Zurich Notebook by Einstein (p. 41R of "Research Notes on a Generalized Theory of Relativity," dated ca. August 1912, in (CPAE 4, Doc. 10) and p. 88 of the student notes on Geiser's lecture course on infinitesimal geometry, taken by Einstein's friend Marcel Grossmann in 1898 (Eidgenössische Technische Hochschule, Zürich, Bibliothek, Hs 421: 16); for Einstein's attendance of this course in the summer semester 1898, see (CPAE 1, 366); for his later recollections on the significance of this course for his work on general relativity, see (Ishiwara 1971, 78–88). The connection between Einstein's research notes and Grossmann's student notes was identified by Tilman Sauer; see also (Castagnetti et al. 1994) and "Commentary" (in vol. 2 of this series).

easily show that the inertial motion of a particle in an arbitrarily accelerated frame of reference can be described by the same type of equation as that published in May 1912 for a static gravitational field,[69] involving not just one variable but indeed a 4-by-4 metric tensor.

The introduction of the metric tensor provided Einstein with the framework for capturing the resources of the two traditions, field theory and the mechanistic generalization of the relativity principle, as well as those of the mathematical tradition established, among others, by Riemann and Christoffel. The tradition of field theory suggested, for instance, that—following the model of Poisson's equation for the gravitational potential in classical physics—some second-order differential operator was to be applied to the metric tensor in order to yield the left-hand-side of a gravitational field equation.[70] It therefore comes as no surprise to find that the first entries in the Zurich Notebook, in which Einstein tackled the problem of gravitation, reflect his attempt to translate the field equation of the theory for the static field into a second order differential equation for the metric tensor.[71] However, the construction of a satisfactory field equation for the gravitational field was an incredibly difficult task that would demand Einstein's attention for the following three years. In his search, he could rely on the tradition of the mechanistic generalization of the relativity principle which offered him concrete examples for metric tensors to be covered by the new theory, such as the metric tensor for Minkowski space described from the perspective of a rotating frame of coordinates. The inertial forces arising in such a rotating frame are well known from classical physics and could hence serve as criteria for the theory to be constructed.

In the course of Einstein's long search for a gravitational field equation, he continued to exploit the heuristics of the "elevator" and "bucket" in particular, and the traditions of field theory and mechanics in general, in order to build up a considerable "machinery" of formalisms, mathematical techniques, and conceptual insights. This machinery eventually developed a dynamics of its own and led to a "conceptual drift"; i.e., to results that were not always compatible with Einstein's heuristic starting points, whether they were rooted in field theory or in the mechanistic generalization of the relativity principle.[72]

One of the first indications of such a conceptual drift was a revision published in 1912 of the theory of the static gravitational field, which conflicted with the "heuristics of the elevator," and also with an expectation raised by traditional field theory.[73]

---

69 See (Einstein 1912b).

70 For more extensive discussion, see "Pathways out of Classical Physics ..." (in vol. 1 of this series).

71 See p. 39L of "Research Notes on a Generalized Theory of Relativity" (dated ca. August 1912) in (CPAE 4, Doc. 10). See also "Facsimile and Transcription of the Zurich Notebook" (in vol. 1 of this series).

72 See (Elkana 1970).

73 For Einstein's first theory, see (Einstein 1912a), for his second, revised theory, see (Einstein 1912b). For historical discussion, see "The First Two Acts" and "Pathways out of Classical Physics ..." (both in vol. 1 of this series).

The revision of Einstein's static theory became necessary after he found out that his theory violated the principle of the equality of action and reaction. The non-linearity of the revised field equation turned out to be incompatible with the equivalence principle as formulated by Einstein in 1907. The homogeneous static gravitational field which he replaced by a uniformly accelerated frame of reference was simply no longer a solution of the revised, non-linear field equation.[74] In other words, after the revision, the theory of the static gravitational field contradicted its own heuristic starting point. Consequently, Einstein had to restrict the principle of equivalence. But from the perspective of our present discussion, the most significant implication of this episode was that the gravitational field had entered the scene in its own right, on a par with the material bodies acting as its source. Hence it became, at least in principle, conceivable that non-trivial gravito-inertial fields could exist without being caused by material bodies. Einstein, however, remained hesitant to accept this conclusion—which is in obvious contradiction with the Machian requirement that all inertial effects are due to ponderable masses—even after he had formulated the final theory of general relativity.

During Einstein's work on his generalized theory of relativity in the years 1912 and 1913, the "heuristics of the bucket" did not fare much better. In Einstein's research notes from this period, one encounters again and again the metric tensor representing the Minkowski space as seen from a rotating frame of reference.[75] However, it remained unclear for some time whether or not the field equation of the preliminary theory of gravitation, which Einstein published in 1913 with his mathematician friend Marcel Grossmann (Einstein and Grossmann 1913), satisfied this requirement of incorporating the Machian bucket. Einstein's eventual discovery that the "*Entwurf*" theory conflicted with this expectation was an important motive for discarding this theory and for beginning anew the search for a theory that promised to fulfill his original goals.[76] In this way, the "heuristics of the bucket" once more played a crucial role in the genesis of the general theory of relativity.

4.4 Attempts at a Machian Interpretation of General Relativity in the Period 1915–1917

After Einstein had formulated his theory in 1915, the tension between his original heuristics and the implications of the new theory remained unresolved; this tension continued to characterize the further development of the theory until at least 1930. Initially, one motive behind Einstein's emphasis on epistemological arguments based

74 For an extensive evaluation of Einstein's principle of equivalence, see (Norton 1989a), and, in particular, for the present discussion, p. 18.
75 See, e.g., pp. 42R, 43L, 11L, 12L, 12R, 24R, and 25R of "Research Notes on a Generalized Theory of Relativity" (dated ca. August 1912) in (CPAE 4, Doc. 10). See also "Facsimile and Transcription of the Zurich Notebook" (in vol. 1 of this series).
76 See, e.g., Einstein to Arnold Sommerfeld, 28 November 1915, (CPAE 8, Doc. 153). For historical discussion, see (Janssen 1999) and "What Did Einstein Know ..." (in vol. 2 of this series).

on the relation between the new theory and its Machian heuristics may have been his desire to make his achievement acceptable to the scientific community. In fact, an important element of the empirical confirmation of the theory was only supplied when the eclipse expedition of 1919 confirmed the bending of light in a gravitational field. In 1913 Einstein had written to Mach that the agreement which he had found between the consequences of his then preliminary theory of gravitation and Mach's critique of Newtonian mechanics was practically the only argument he had in its favor.[77] Also in his early publications on the final theory he continued to insist on its epistemological advantages, which provided additional support for its claim of superiority with regard to competing theories.[78]

But Einstein's insistent pursuit of the Machian aspects of general relativity in the early years after its formulation was determined less by tactical motives than by the perceived need for a physical interpretation of the technical features of the new theory in light of his original heuristics. For instance, Einstein soon realized that, as a rule, the field equation determines the gravitational field only if, in addition to the matter distribution, boundary conditions are specified. This technical feature of the theory had to be brought together with his intention to realize a generally relativistic theory and his Machian hopes of explaining inertial behavior by material bodies only.[79] For some time in 1916 and early 1917, he attempted to formulate boundary conditions that would somehow comply with his original intentions.[80] He searched, for example, for boundary conditions in which the components of the metric tensor take on degenerate values since he assumed that a singular metric tensor would remain invariant under general coordinate transformations, and thus make it possible to maintain the requirement of general covariance even at the boundary region of spacetime. He also searched for a way to define a region outside the system of masses that constituted the physical universe in which a test body would possess no inertia, so that he might then be able to claim that inertia is indeed created by the physical system circumscribed by this empty boundary region.[81] In the course of these attempts, the expectation that general relativity was to provide a Machian explanation of inertia began to be silently transformed from a requirement concerning the nature of the theory to a criterion to be applied to specific solutions of the theory. Since Minkowski's flat spacetime, with inertial properties familiar from classical mechanics and special relativity, was a solution to the vacuum field equations of general rela-

---

77 See Einstein to Ernst Mach, second half of December 1913 (CPAE 5, 583–584).

78 See, e.g., (Einstein 1916a, 771–772).

79 See Einstein to Lorentz, 23 January 1915 (CPAE 8, Doc. 47), and the extensive historical discussion in (Kerszberg 1989a, 1989b), as well as in (Hoefer 1994), on which the following account is based. See also the introduction to (CPAE 8); "The Einstein-de Sitter-Weyl-Klein Debate," (CPAE 8, 351–357).

80 See, e.g., Einstein to Michele Besso, 14 May 1916, (CPAE 8, Doc. 219).

81 See Einstein to de Sitter, 4 November 1916 (CPAE 8, Doc. 273) and Einstein to Gustav Mie, 8 February 1918 (Doc. 460).

tivity, it simply could not be true in general that in this theory inertial effects are explained by the presence of matter.

After Einstein's failure to find a satisfactory treatment of the supposed Machian properties of general relativity along the road of singular boundary conditions, in (Einstein 1917) he advanced a completely different proposal for dealing with the cosmological aspects of the theory. He introduced a spacetime that satisfied all his expectations concerning the constitution of the universe, including the explanation of its inertial properties by the masses acting as sources of the gravitational field, but at the price of modifying the field equations to which this spacetime was a solution. As Einstein's cosmological paper of 1917 has been discussed a number of times, it may suffice to briefly emphasize its place in the development of the tensions between Einstein's Machian heuristics and the implications of the new theory.[82] The solution to the field equations—modified by the introduction of a "cosmological constant"—which Einstein considered in 1917 describes a spatially closed, static universe with a uniform matter distribution. It therefore entirely avoided the problem of specifying appropriate boundary conditions and, at the same time, was believed by him to correspond to a more or less realistic picture of the universe as known at that time. In general, though, Einstein tended to neglect the relation between the new theory and astronomy, as well as the exploration of the properties of the solutions to its field equations. In contrast, Willem de Sitter—at the time Einstein's principal opponent in the discussion of the allegedly Machian features of general relativity— repeatedly emphasized the astronomical consequences of the various solutions to the field equations.[83] In any case, Einstein not only hoped that his radical step of modifying the field equations of general relativity allowed him to find at least one acceptable solution to the field equations, but he also assumed that he would succeed in excluding altogether empty space solutions in which inertial properties are present in spite of the absence of matter.[84] It was therefore an unpleasant surprise—which he found difficult to digest and at first attempted to refute—when de Sitter demonstrated shortly after the publication of Einstein's paper that even the modified field equations allow such an empty space solution.[85] In 1918 Einstein published a critical note on de Sitter's solution in which he wrote:

> If de Sitter's solution were valid everywhere, then it would be thereby shown that the purpose which I pursued with the introduction of the λ-term [the cosmological con-

82 See, in particular, (Hoefer 1994) for a detailed discussion of this paper from the point of view of Mach's influence on Einstein.

83 See, e.g., Einstein to Willem de Sitter, before 12 March 1917, (CPAE 8, Doc. 311), where he referred to his solution as a "Luftschloss," (castle in the air) having the principal purpose of showing that his theory is free of contradictions. See also Einstein to Michele Besso, 14 May 1916, for the Machian motivations of Einstein's construction. For a historical account of the controversy between Einstein and de Sitter on the implementation of Machian ideas and cosmological considerations in general relativity, see (Kerszberg 1989a, 1989b).

84 See Einstein to de Sitter, 24 March 1917, (CPAE 8, Doc. 317).

85 See de Sitter to Einstein, 20 March 1917, (CPAE 8, Doc. 313).

stant–J R.] has not been reached. In my opinion the general theory of relativity only forms a satisfactory system if according to it the physical qualities of space are *completely* determined by matter alone. Hence no $g_{\mu\nu}$ -field must be possible, i.e., no space-time-continuum, without matter that generates it.[86]

4.5 The Introduction of "Mach's principle" in 1918

The increasing tension between Einstein's original intentions and the ongoing exploration of the consequences of the new theory was accompanied by attempts to rephrase the criteria of what it meant to satisfy the philosophical requirements corresponding to the heuristics that had guided the discovery of the theory. Characteristically, Einstein (Einstein 1918a, 241–242) introduced and defined the very term "Mach's principle" in the context of a controversy over whether or not the general theory of relativity in fact represented a realization of his intention to implement a generalization of the relativity principle of classical mechanics and special relativity. His paper of 1918 was a response to the argument by Kretschmann that the general covariance of the field equations of general relativity does not imply such a generalization of the relativity principle, but can be considered as a mathematical property only. Einstein argued that he had so far not sufficiently distinguished between two principles, which he now introduced as the principle of relativity and Mach's principle.[87]

The first principle, as defined by Einstein, states that the only physically meaningful content of a relativistic theory are coincidences of physical events at points of space and time. Since the occurrence of these point coincidences is independent of whether they are described in one or the other coordinate frame, their most appropriate description is by a generally covariant theory. This principle had, of course, not been the starting point of Einstein's search for a generally relativistic theory of gravitation, but rather constitutes a result of his reflection on complications encountered in a long, but eventually successful, search for such a theory.[88] For our purposes here, it is particularly remarkable that this formulation of the principle of relativity no longer appeals to the intuition of a world of isolated bodies distributed in an otherwise empty space whose physical interactions should depend only on their relative distances, velocities, etc. As we have seen, this intuition was characteristic of the mechanistic generalization of the relativity principle, and was at the root of Einstein's search for a generalized theory of relativity.

---

86   "Bestände die De Sittersche Lösung überall zu Recht, so würde damit gezeigt sein, daß der durch die Einführung des "λ- Gleides" von mir beabsichtigte Zweck nicht erreicht wäre. Nach meiner Meinung bildet die allgemeine Relativitätstheorie nämlich nur dann ein befriedigendes System, wenn nach ihr die physikalischen Qualitäten des Raumes allein durch die Materie vollständig bestimmt werden. Es darf also kein $g_{\mu\nu}$ -Feld, d. h. dein Raum-Zeit-Kontinuum, möglich sein ohne Materie, welche es erzeugt." (Einstein 1918b, 271)

87   For historical discussions of this paper and its context, on which the following account is based, see (Norton 1992a, in particular, pp. 299–301, 1993, 806–809; Rynasiewicz 1999).

88   See the various discussions of Einstein's "hole argument" in the recent literature, e.g. in (Norton 1989b, sec. 5). See also the discussion in "Untying the Knot ..." (in vol. 2 of this series).

This original intuition in fact included Mach's suggestion to conceive of inertial effects as the result of physical interactions between the bodies of such a world. Now, however, the causal link between inertial effects and matter suggested by Mach's critical analysis of the foundations of classical mechanics needed to be reinterpreted in light of the newly developed formalism of general relativity. According to this formalism, inertial effects are described by the metric tensor representing the gravito-inertial field, while matter is described by the energy-momentum tensor representing the source term of the field equations for the gravitational field. It was therefore natural for Einstein to translate the supposed causal nexus between inertial forces and matter into the requirement that the gravitational field be entirely determined by the energy-momentum tensor. It is this requirement which he chose in 1918 to call "Mach's principle."[89] Certainly, this was not a mathematically concise criterion to determine whether general relativity as a theory, or, as particular solutions of the theory, do or do not satisfy Mach's principle. Two aspects of this principle are nevertheless clear. First, the translation of Mach's original suggestion into the language of general relativity transferred it from the conceptual context of mechanics into that of field theory, as both terms in Einstein's 1918 definition of Mach's principle—the gravitational field and the energy-momentum tensor—are basically field theoretical concepts. Second, it is obvious from the context of this definition, discussed above, that whatever was precisely intended, Einstein considered empty space solutions of the gravitational field equations as a violation of this principle.

4.6 The Conceptual Drift from Mach's Principle to "Mach's Aether" (1918–1920)

Ironically, both these aspects of Einstein's first explicit definition of Mach's principle in his writings contributed to the preparation for its eventual rejection. As a first step towards this rejection, de Sitter established that not only Einstein's gravitational field equations of 1915, but even the equations modified by the introduction of the cosmological constant, admit empty space solutions. As a consequence, Mach's principle now definitely took on the role of a selection principle for solutions to the field equations. It seems that one interpretative reaction by Einstein to this serious defeat of his principle was to extend the field theoretical interpretation of general relativity at the expense of the emphasis on the mechanical roots of his original heuristics. By 1920, the 1918 attempt to define Mach's principle in terms of the conceptual building blocks of his theory had been complemented by the introduction of a "Machian aether" as a means of capturing its conceptual implications.[90] In a lecture given in Leiden, Einstein (1920) exploited the time-honored concept of an aether, to which

---

89 "Mach's principle: The G-field is *completely* determined by the masses of bodies. Since mass and energy are identical in accordance with the results of the special theory of relativity and the energy is described formally by means of the symmetric energy tensor ($T_{\mu\nu}$), the G-field is conditioned and determined (*bedingt und bestimmt*) by the energy tensor of the matter." See (Einstein 1918a, 241–242), quoted from (Barbour 1992, 138).

Lorentz had given the definitive form in the realm of electrodynamics, in order to explain the new concept of space which had emerged with general relativity. He now directly turned against Mach's interpretation of inertial effects as caused by cosmic masses, because this interpretation presupposed an action at a distance, a notion incompatible with both field theory and relativity theory. Instead, contrary to his original heuristics, Einstein (1920, 11–12) associated these inertial effects with the nature of space, which he now conceived as being equipped with physical qualities, and which he hence appropriately called aether. Contrary to Lorentz's aether, however, Mach's aether, which Einstein thought of as being represented by the metric tensor, was supposed not only to condition but also to be conditioned, at least in part, by matter. This capacity of being influenced by the presence of matter was, apparently, the last resort for the Machian idea of the generation of inertial effects by the interaction of material bodies in Einstein's conceptual framework.

For the time being, however, two aspects of the relation between matter and space remained open problems: with space—under the name Machian aether—taking on the role of an independent physical reality, the question presented itself of whether matter had not lost all claims to primacy in a causal nexus between space and matter. In his Leiden lecture, Einstein (1920, 14) noted that it was possible to imagine a space without an electromagnetic field, but not one without a gravitational field, as space is only constituted by the latter; he concluded that matter, which for him was represented by the electromagnetic field, appears to be only a secondary phenomenon of space. In (Einstein 1919), he had made an attempt at a derivation of the properties of matter from the gravitational and the electrodynamic field, an attempt which he considered as still being unsatisfactory but which, for him, constituted the beginning of a new line of research in the tradition of the electrodynamic—or field theoretical—worldview. This kind of research program held out the possibility not only of reintroducing the concept of an aether in order to represent the physical qualities of space, but also of providing a theoretical construction of matter as an aspect of this aether. The other outstanding question concerning the relation between matter and space, which was left unclarified even after Einstein's introduction of a Machian aether, was the astronomical problem of the distribution of masses and of the large-scale spatial structure of the universe. Both questions, the theoretical as well as the empirical, turned out to be significant, not only for Einstein's further exploration of general relativity, but indirectly for the fate of Mach's principle as well.

---

90 For historical discussions, see (Illy 1989; Kox 1989; Kostro 1992, 2000; Renn 2003). Probably under the influence of Lorentz, Einstein had begun to reconsider the concept of aether already in 1916. On 17 June of this year he had written to H. A. Lorentz: "I admit that the general theory of relativity is closer to the aether hypothesis than the special theory." ("Ich gebe Ihnen zu, dass die allgemeine Relativitätstheorie der Aetherhypothese näher liegt als die spezielle Relativitätstheorie." (CPAE 8, Doc 226; English translation in Kostro 1992, 262.) At that time, however, as the same letter suggests, Einstein took it for granted that the aether is entirely determined by material processes. The transition to the aether concept as explained in the following seems to be complete by the end of 1919, see Einstein to H. A. Lorentz, 15 November 1919, (CPAE 9, Doc. 165).

4.7 Mach's Principle: From the Back Burner to Lost in Space (1920–1932)

The program of interpreting general relativity along the lines of Mach's philosophical critique of classical mechanics ceased to play a significant role in Einstein's research after 1920. In addition to the difficulty of implementing Machian criteria in the elaboration of the theory, his exploration during the twenties of the heuristic potential that general relativity offered for the formulation of a unified theory of gravitation and electrodynamics was probably responsible for this shift of interest.[91] As this heuristic potential for a further unification of physics was associated with the field theoretic aspects of general relativity, the relation of the theory to the foundational problems of mechanics naturally faded into the background. Nevertheless, on several occasions during his ongoing research on a unified theory of gravitation and electromagnetism, Einstein hoped that he could link the program of a unified field theory with a satisfactory solution of the cosmological problem in the sense of his Machian heuristics. In 1919, for example, he emphasized that his new theory had the advantage that the cosmological constant appears in the fundamental equations as a constant of integration, and no longer as a universal constant peculiar to the fundamental law; he made a point of showing that again a spherical world results from his new equations (Einstein 1919, 353; 1923b, 36). An additional reason for not completely rejecting Mach's principle may have been Einstein's awareness, in a period which saw the triumph of quantum mechanics, that, after all, the corpuscular foundation of physics and not the field theoretic might prevail in the end; fields would then indeed have to be conceived as epiphenomena of matter, like the gravitational field according to Mach's principle.[92]

There was also a rather mundane reason why Mach's principle did not figure prominently in Einstein's publications of this period, while not being entirely dismissed by him. More than its definition in 1918, its association with the cosmological model of 1917 had brought the principle to an end point of its theoretical development, to a point where the question of whether or not Mach's principle could be implemented in general relativity had become a question of its confirmation or refutation by astronomical data. In 1921 Einstein remarked, with reference to the possibility of explaining inertia in the context of his cosmological model:

> Experience alone can finally decide which of the two possibilities is realized in nature. (Einstein 1922a, 42)[93]

In any case, for the time being, he remained convinced that astronomical research on the large systems of fixed stars would bear a model of the universe compatible with his Machian expectations. In 1921 he also wrote:

---

91   See (Pais 1982, 287–288); see also the extensive discussion in (Vizgin 1994).

92   See, in particular, Einstein's views expressed in connection with theoretical and experimental studies of radiation in this period, for example: "It is thus proven with certainty that the wave field has no real existence, and that the Bohr emission is an instantaneous process in the true sense." Einstein to Max Born, 30 December 1921, my translation; see also the discussion in (Vizgin 1994, 176).

93   The German original is (Einstein 1921a).

> A final question has reference to the cosmological problem. Is inertia to be traced to
> mutual action with distant masses? And connected with the latter: Is the spatial extent of
> the universe finite? It is here that my opinion differs from that of Eddington. With Mach,
> I feel that an affirmative answer is imperative, but for the time being nothing can be
> proved. (Einstein 1921b, 784)[94]

In other words, although in the period between 1920 and 1930 Einstein invested his hopes and his research efforts mainly in the creation of a unified field theory, he nevertheless kept Mach's principle on the back burner as long as it was not contradicted by astronomical data.

Einstein's firm conviction made him sceptical with respect to the possibility of alternative cosmological models. In 1922 he criticized, among other proposals, Friedmann's paper on solutions to the original field equations which correspond to a dynamical universe.[95] He mistakenly identified a calculational error in Friedmann's solution, which he had viewed with suspicion from the beginning. In another paper of the same year (Einstein 1922b, 437), he explicitly criticized a cosmological model for its incompatibility with "Mach's postulate." In 1923, however, Einstein recognized that he had committed an error in rejecting Friedmann's dynamical solutions. He published a retraction (Einstein 1923c) of his earlier criticism and henceforth no longer expected an astronomical confirmation of his Machian cosmology with the same certainty as before. The change of Einstein's attitude is apparent from a comparison between the published retraction of his criticism with a manuscript version that has been preserved. In the manuscript version Einstein wrote:

> It follows that the field equations, besides the static solution, permit dynamic (that is,
> varying with the time coordinate) spherically symmetric solutions for the spatial struc-
> ture, to which a physical significance can hardly be ascribed.

In the published paper, on the other hand, Einstein omitted the last half-sentence.[96] In another paper of the same year, Einstein referred with scepticism to "Mach's postulate" and to the modification of the field equations that it required, because the introduction of the cosmological constant was not founded on experience. He concluded:

> For this reason the suggested solution of the 'cosmological problem' can, for the time
> being, not be entirely satisfactory.[97]

Nevertheless, until the end of the twenties Einstein did not give up hope that Mach's principle could be maintained as a feature of a cosmologically plausible solution of the field equations of general relativity. When he discussed the "aether" of general

---

94 Einstein's astronomical views in this period were strongly under the influence of his Machian belief,
    see, e.g., (Einstein 1922b, 436).
95 See (Einstein 1922d); for Einstein's criticism of other proposals, see (Einstein 1922b, 1922c).
96 This has been noted by John Stachel. For the translation of the passage, see also (Stachel 1986, 244).
97 "Aus diesem Grunde kann die angedeutete Lösung des kosmologischen Problems einstweilen nicht
    völlig befriedigen." (Einstein 1923a, 8; my translation.) He also modified an earlier version of an
    attempt to formulate a unified field theory by omitting the cosmological constant, see (Vizgin 1994,
    192–193).

relativity in (Einstein 1924, 90), he added that it is determined by ponderable masses and that this determination is complete if the world is spatially finite and closed in itself. In the same paper, he dealt both with the possibility that a unification of gravitation and electrodynamics can be achieved by field theory and with the possibility that an understanding of the quantum problem can be achieved without field theoretical components.[98] As suggested above, it is conceivable that this ambivalence as to which of the foundational concepts—field or corpuscle—would eventually prevail may have reinforced the role of Mach's principle in Einstein's thinking. In 1926, he discussed the cosmological implications of general relativity in line with his earlier arguments in favor of a finite static universe.[99] In 1929 he wrote:

> Nothing certain is known of what the properties of the space-time continuum may be as a whole. Through the general theory of relativity, however, the view that the continuum is infinite in its time-like extent but finite in its space-like extent has gained in probability. (Einstein 1929, 107)

Around 1930, however, things began to change. Primarily driven by his strong intellectual engagement in the program to formulate a unified field theory, Einstein expressed himself even more definitely than earlier in favor of a causal primacy of space in relation to matter—in sharp contrast to his original Machian heuristics. He would still ask the question:

> If I imagine all bodies completely removed, does empty space still remain?

and suggest a negative answer. But now this question was not so much intended to refer to the constitution of the universe, but was rather an epistemological inquiry regarding the construction of the concept of space:

> But how is the concept of space itself constructed? If I imagine all bodies completely removed, does empty space still remain? Or is even this concept to be made dependent on the concept of body? Yes, certainly, I reply.[100]

While in the sequel of the paper, Einstein develops at length his reasons for suggesting a *cognitive* primacy of the concept of physical object with respect to the concept of space, he concludes his discussion of the state of research on the foundations of physics with the following remark:

> Space, brought to light by the corporeal object, made a physical reality by Newton, has in the last few decades swallowed ether and time and seems about to swallow also the field and the corpuscles, so that it remains as the sole medium of reality.[101]

---

98 See (Einstein 1924), in particular, pp. 92–93.
99 See (Einstein 1926–1927) and, for historical discussion, (Vizgin 1994, 212–213).
100 "Wie kommt aber der Raumbegriff selbst zustande? Wenn ich die Körper allesamt weggenommen denke, bleibt doch wohl der leere Raum über? Soll etwa auch dieser vom Körperbegriff abhängig gemacht werden? Nach meiner Überzeugung ganz gewiß!" (Einstein 1930a, 180)
101 "Der Raum, ans Licht gebracht durch das körperliche Objekt, zur physikalischen Realität erhoben durch Newton, hat in den letzten Jahrzehnten den Äther und die Zeit verschlungen und scheint im Begriffe zu sein, auch das Feld und die Korpuskeln zu verschlingen, so daß er als alleiniger Träger der Realität übrig bleibt." (Einstein 1930a, 184)

In a lecture given in 1930 Einstein formulated his view even more drastically:

> The strange conclusion to which we have come is this—that now it appears that space will have to be regarded as a primary thing and that matter is derived from it, so to speak, as a secondary result. Space is now turning around and eating up matter. We have always regarded matter as a primary thing and space as a secondary result. Space is now having its revenge, so to speak, and is eating up matter. (Einstein 1930b, 610)

In the course of his work on unified field theory, assisted by his epistemological reflections, Einstein had come a long way from believing that a successful implementation of Mach's principle would entail a synthesis of physics in which the concept of matter would play a primary and the concept of space a secondary role. Nevertheless, as the development of Mach's principle in his thinking had become so closely associated with his cosmological ideas, the question of Mach's principle remained open precisely to the extent that the decision about Einstein's static universe was left open by observational cosmology. In the period between 1917 and 1930, a prevailing problem debated by researchers in this field was whether de Sitter's or Einstein's static universe is a better model of reality, while the question of expanding universes, raised by Friedmann in 1922 and by Lemaître in 1927, largely remained outside the horizon of observational cosmology.[102] The range of theoretical alternatives taken into account by contemporary researchers testifies to the persistent role of Einstein's Machian interpretation of general relativity for cosmology, even if this interpretation gradually became a mere connotation of one of the cosmological alternatives rather than being the primary issue.

With the stage thus set for an observational decision on Mach's principle, a definitive blow to Einstein's belief in it came with the accumulation of astronomical evidence in favor of an expanding universe, the decisive contribution being Hubble's work published in 1929.[103] Einstein became familiar with these results early in 1931 during a stay at the California Institute of Technology. As is suggested by an entry in his travel diary from 3 January 1931, Richard Tolman convinced Einstein that his doubts about the correctness of Tolman's arguments in favor of the role of nonstatic models for a solution of the cosmological problem were not justified.[104] In March of the same year Einstein wrote to his friend Michele Besso:

> The Mount Wilson Observatory people are excellent. They have recently found that the spiral nebulae are spatially approximately uniformly distributed and show a strong Doppler effect proportional to their distance, which follows without constraint from the theory of relativity (without cosmological constant).[105]

---

102 See (Ellis 1989, 379–380).
103 For historical discussion, see (Ellis 1989, 376–378).
104 "Doubts about correctness of Tolman's work on cosmological problem. Tolman, however, was in the right." Quoted from (Stachel 1986, n. 53, 249); for a discussion of Tolman's contribution, see (Ellis 1989, 379–380).
105 Einstein to Michele Besso, 1 March 1931, quoted from (Stachel 1986, 245).

Almost immediately after his return to Berlin, Einstein published a paper (Einstein 1931b) on the cosmological problem in which he stated that the results of Hubble had made his assumption of a static universe untenable. As it was even easier for general relativity to account for Hubble's results than for a static universe—because no modification of the field equations by the introduction of a cosmological constant was required—his earlier static solution now appeared unlikely to Einstein, given the empirical evidence (Einstein 1931b, 5).

In a lecture given in October of 1931, he still mentioned his static solution in connection with the implementation of Mach's ideas in general relativity, but, in spite of the numerous remaining difficulties of the dynamical conception of the universe, he now had definitely given up his belief in a Machian world (Einstein 1932). In 1932, in a joint paper with de Sitter—his main antagonist in the earlier controversy about a Machian explanation of inertia—Einstein himself presented an expanding universe solution to the unmodified field equations. In this paper, the original Machian motivation for Einstein's static universe solution is no longer even mentioned:

> Historically the term containing the "cosmological constant" λ was introduced into the field equations in order to enable us to account theoretically for the existence of a finite mean density in a static universe. It now appears that in the dynamical case this end can be reached without the introduction of λ. (Einstein and de Sitter 1932, 213)

In other words, in the course of the evolution of Einstein's cosmological views, from his adherence to a static world to his acceptance of an expanding universe, Mach's principle had simply disappeared.

4.8 Reflections in the Aftermath of Mach's Principle

Although Einstein continued to acknowledge the role of Mach's critique of classical mechanics for the emergence of general relativity even after 1930, one can nevertheless notice a tendency to reinterpret even the heuristics which had originally guided his formulation of the theory. In his later accounts of the conceptual foundations of general relativity, he appealed to the field concepts in order to point out those weaknesses of classical physics that he had discussed earlier in the spirit of Mach's critique of mechanics. He emphasized, for instance, that it was due to the introduction of the field concept that the standpoint of considering space and time as independent realities had been surmounted.[106] Or he argued (Einstein 1961, app. V, 153) that the principle of equivalence, which had originally motivated the extension of the relativity principle beyond the special theory of relativity, already demonstrated the existence of the field as a reality in its own right, that is, independent of matter, since for the field experienced by an observer in an accelerated frame of reference the question of sources does not arise.

When the occasion presented itself, Einstein also became quite explicit about his rejection of his earlier Machian heuristics. In a letter to Felix Pirani, for instance, he

---

106 See, e.g., (Einstein 1961, app. V, 144).

explains with reference to Mach's principle, as he himself had earlier defined it, that he no longer finds it plausible that matter represented by the energy-momentum tensor could completely determine the gravitational field, since the specification of the energy-momentum tensor itself already presupposes knowledge of the metric field. In the same letter Einstein explicitly revokes Mach's principle:

> In my view one should no longer speak of Mach's principle at all. It dates back to the time in which one thought that the "ponderable bodies" are the only physically real entities and that all elements of the theory which are not completely determined by them should be avoided. (I am well aware of the fact that I myself was long influenced by this *idée fixe*.)[107]

He similarly explains in his *Autobiographical Notes*:

> Mach conjectures that in a truly reasonable theory inertia would have to depend upon the interaction of the masses, precisely as was true for Newton's other forces, a conception that for a long time I considered in principle the correct one. It presupposes implicitly, however, that the basic theory should be of the general type of Newton's mechanics: masses and their interaction as the original concepts. Such an attempt at a resolution does not fit into a consistent field theory, as will be immediately recognized. (Einstein 1992, 27)

In summary, this section has shown that Mach's critique of classical mechanics was a crucial element in the heuristics guiding Einstein's way to the formulation of the general theory of relativity. It played this role as one among several aspects of the tradition of classical physics and was, just as many of these other elements, eventually superseded by the development of general relativity. At the outset, it opened up Einstein's perspective towards a generalization of the relativity principle and towards an explanation of inertial effects, and hence of the physical properties of space, by material bodies. By conceptualizing inertial forces as an interaction of bodies in motion, it provided a decisive complement to the prospect of a dynamical theory of gravitation, which was suggested by the conceptual tradition of field theory, but lacked an empirical substantiation that could offer orientation among a variety of possible research directions. The results which Einstein accumulated in the course of his search for a general theory of relativity enforced several adjustments and reformulations of his original heuristics. Eventually, it became impossible for him to bring the progress of general relativity into agreement with these heuristics.[108] Here we have seen that this is the case for those aspects of his heuristics which were founded on the stimulation received from Mach's critique of mechanics. It turns out, however, that the incompatibility between the conceptual framework that shaped Einstein's original heuristics and that which emerged from the final theory can be demonstrated more generally.[109]

---

107 Einstein to Felix Pirani, 2 February 1954 (my translation). (Einstein Archives: call number 17 - 447.00.)

108 See also the systematic discussions of the relation between Mach's principle and the progress of general relativity in (Goenner 1970, 1981; Torretti 1983, 199–201).

109 For extensive discussion, see "Pathways out of Classical Physics ..." (in vol. 1 of this series)

5. EINSTEIN'S PHILOSOPHICAL PERSPECTIVE ON THE
FOUNDATIONAL PROBLEMS OF PHYSICS

5.1 Einstein's Route to General Relativity: Between Physics and Philosophy

The account given in the previous section of the impact of Mach's critique on the development of general relativity seems to provide a strong case in point for an influence of philosophy on physics. Einstein himself confirms in many contemporary comments as well as in later recollections that he conceived the emergence of general relativity at least in part as a response to Mach's analysis of the foundations of classical mechanics.[110] He indeed continued his search for such a response even when more simple alternative approaches to the problem of gravitation seemed to be available and when only epistemological arguments could motivate the continuation of his search for a generalization of the relativity principle.[111] The fact that also the followers of a mechanistic generalization of the relativity principle could refer to Mach's analysis as the philosophical background of their enterprise, however, raises some doubts as to how significant the contribution of philosophy to Einstein's particular approach actually was. The starting point of Einstein's revision of the foundations of mechanics was in fact, as we have seen, in contrast to that of these "Machians," not a general philosophical concern but a concrete problem which he encountered in the course of his research. It was not that the principle of equivalence had been formulated as a consequence of Einstein's search for a generalization of the principle of relativity, but vice versa, that the introduction of the equivalence principle in the context of a problem of "normal science" had opened up the perspective towards the foundational questions of mechanics. In a recollection from 1919 Einstein laconically states with reference to the emergence of general relativity:

> The epistemological urge begins only in 1907.[112]

There is, however, a crucial distinction between the reaction of Einstein and that of the adherents of a mechanistic generalization of the relativity principle to Mach's critique of the foundations of mechanics. In Einstein's view, the primary philosophical force of Mach's critique was directed against precisely what seemed to be for the "Machian relativists"—at least within the context of this particular research problem—an undisputed presupposition of their thinking, namely the mechanistic ontol-

---

110 For contemporary evidence, see, e.g., Einstein's correspondence with Mach quoted above, for a later recollection, see, e.g., (Einstein 1954a, 133–134). The significance of Mach's philosophical critique of mechanics for Einstein is exhaustively treated in (Wolters 1987, chap. 1).

111 See (Einstein 1914, 344), where Einstein comments on Nordström's competing theory.

112 "Das erkenntnistheoretische Bedürfnis beginnt erst 1907." Einstein to Paul Ehrenfest, 4 December 1919 (CPAE 9, Doc. 189 - my translation). See also (Wheeler 1979, 188), for a later recollection by Einstein, according to which he recognized the significance of the equality of inertial and gravitational mass only as a consequence of his failure to formulate a special relativistic theory of gravitation. For a different interpretation, see (Barbour 1992, 130, 133).

ogy on the basis of which they attempted a generalization of the relativity principle. Einstein himself later remembered that questioning the self-evident character of the concepts of mechanics was one of the principal effects that Mach's philosophy had upon him:

> We must not be surprised, therefore, that, so to speak, all physicists of the last century saw in classical mechanics a firm and definitive foundation for all physics, yes, indeed, for all natural science, and that they never grew tired in their attempts to base Maxwell's theory of electromagnetism, which, in the meantime, was slowly beginning to win out, upon mechanics as well. ... It was Ernst Mach who, in his *History of Mechanics*, shook this dogmatic faith; this book exercised a profound influence upon me in this regard while I was a student.[113]

In other words, in contrast to those physicists whose reception of Mach's critique of mechanics was shaped only by the perspective of this one subdiscipline of physics, Einstein read Mach as a philosopher and understood the central philosophical intention behind Mach's historical and critical account of mechanics, which was directed against the special status which mechanics had had for a long time among the subdisciplines of physics.

We may therefore ask whether it was this philosophical sensibility with regard to the epistemological character of some of the foundational problems of classical physics which protected Einstein from the temptation to seek a solution to these problems within one of the subdisciplines of classical physics, as for instance, the adherents of a mechanistic generalization of the relativity principle. Indeed, there can be little doubt that Einstein's thinking was characterized by such a sensibility, which was heightened by his reading of philosophical authors such as Kant, Hume, Helmholtz, Mach, and Poincaré.[114] But it seems doubtful, on the other hand, whether philosophical scepticism with regard to false pretensions of a conceptual system is sufficient to overcome its limitations. At the turn of the century, philosophical critics of the privileged status of classical mechanics, often associated as it was with the pretension of an *a priori* character, may themselves serve as counter examples. Neither Mach nor Poincaré built the foundations of a new mechanics upon the basis of their respective epistemological critiques, let alone the foundations of a new conceptual framework for all of physics. As late as 1910, Poincaré—who had emphasized the conventional character of scientific concepts—was nevertheless of the opinion that the principles of mechanics may turn out to be victorious in their struggle with the new theory of relativity, and

---

113 "Wir dürfen uns daher nicht wundern, dass sozusagen alle Physiker des letzten Jahrhunderts in der klassischen Mechanik eine feste und endgültige Grundlage der ganzen Physik, ja der ganzen Naturwissenschaft sahen, und dass sie nicht müde wurden zu versuchen, auch die indessen langsam sich durchsetzende Maxwell'sche Theorie des Elektromagnetismus auf die Mechanik zu gründen. [...] Ernst Mach war es, der in seiner *Geschichte der Mechanik* an diesem dogmatischen Glauben rüttelte; dies Buch hat gerade in dieser Beziehung einen tiefen Einfluss auf mich als Student ausgeübt." (Einstein 1992, 20–21) See also (Holton 1986, chap. 7, 237–277, in particular, p. 241; 1988, chap. 4, 77–104; Wolters 1987, chap. 1, 20–36).

114 For a list of some of Einstein's philosophical readings, see the introduction to (CPAE 2).

that it was hence unjustified to prematurely abandon these principles.[115] Mach (1960, 295–296) had left it open, as we have seen, that new empirical evidence may require a modification of the principles of mechanics. Contrary to Einstein, he speculated that an electromagnetic worldview may provide a new universal conceptual framework for the entire body of physics, while his own contributions to such a unity remained rather on the level of a metatheoretical reflection on science.[116] Einstein, in any case, was convinced that one should not attempt to identify Mach's crucial contribution in what can also be found in the works of Bacon, Hume, Mill, Kirchhoff, Hertz, or Helmholtz, but rather in his concrete analysis of scientific content.[117]

In addition, it can be historically documented that Einstein's scepticism, with respect to the competing worldviews based on mechanics, electrodynamics, or thermodynamics, was rooted in his precise knowledge of their respective scientific failings and not only in his epistemological awareness.[118] Shortly after the turn of the century, when the electromagnetic worldview still appealed to many physicists as the most promising starting point for a new conceptual foundation of physics, Einstein had already recognized the devastating consequences which the discovery by Planck of the law of heat radiation had for classical electrodynamics and hence for the conceptual backbone of a worldview based on traditional field theory. But does this observation not imply that Einstein's philosophical perspective on the foundational problems of physics simply dissolves, in the end, into technical competence in physics? This conclusion would only be justified if one accepted the conceptual distinction between philosophy of physics and physics as accepted today, that is, as a distinction between a methodological, epistemological, or metaphysical—in any case, a metatheoretical—study of physics and the concrete occupation with its scientific problems. In order to respond to the question of the philosophical character of Einstein's perspective, we therefore have to examine briefly the historical situation of the relation between physics and philosophy in Einstein's time.

5.2 The Historical Context of Einstein's Philosophical Perspective on Physics

At the turn of the century, the separation between philosophy of science and science in the sense accepted today had long been complete. The more recent history of this separation can be understood as a consequence of the failure of traditional philosophy to integrate the natural sciences into its reflective enterprise. This failure is partly due to the explosive growth of the shared knowledge of the various disciplines, and partly

---

115 See (Poincaré 1911); see also (Cuvaj 1970, 108) for a historical discussion.
116 For an extensive discussion of Mach's attitude with respect to the electromagnetic worldview, see (Wolters 1987, 29–36). For Mach's attempt to integrate mechanics into the body of physics on the level of methodological reflections, see (Mach 1960, chap. 5).
117 See his remarks to this effect in his obituary for Mach, (Einstein 1916b, 154–155).
118 See, in particular, Einstein's own account in his *Autobiographical Notes* (Einstein 1992), in particular, pp. 42–45, which is confirmed by contemporary evidence such as Einstein's letters to Mileva Marić, see (Renn and Schulmann 1992; CPAE 1).

to the change of the cultural and political role which philosophy, and philosophy of science in particular, underwent in the nineteenth century. In German academic philosophy of the second half of the nineteenth century, for instance, neo-Kantianism, which saw itself as a critical reaction to the philosophy of German idealism, played a weighty role.[119] Its stance was that of a politically neutral epistemology which—in contrast to the natural philosophy of German idealism—often anything but politically neutral—no longer issued any prescriptions for science but just attempted to capture the epistemological and methodological structures that made scientific progress possible. Although neo-Kantianism and the tradition in philosophy of science that pursued its metatheoretical concerns took the natural sciences as their guidepost, they did not, however, offer a theoretical framework that enabled scientists to reflect upon the body of scientific knowledge in its totality, let alone to discuss the social and cultural conditions and implications of science.

On the other hand, since the middle of the nineteenth century, the intrinsic necessity of dealing with science as a social and cultural phenomenon had been approached primarily on a pragmatic level, as is witnessed by the increasing role in the development of the large-scale structure of science played by state science and education policy and the creation of funding agencies and scientific organizations (such as the Kaiser-Wilhelm-Gesellschaft in Germany). Attempts to achieve an intellectual integration of scientific knowledge, for instance in the form of a scientific worldview, remained in the shadow of this development towards a practical control of the sciences as a social system, and was only later supplemented by theoretical studies of science policy and the sociology of science.[120] As a consequence of this complex dynamics of the social and the intellectual development of science, the transfer of knowledge beyond disciplinary boundaries, and the establishment of connections between disparate branches of the body of knowledge, remained a process largely left to chance and to the initiative of the individual researcher. Only to a small degree was this process systematically furthered by the requirements of the intellectual integration of science for the purposes of education, to mention one extreme, and in the context of a few, themselves highly specialized interdisciplinary research projects, to mention the other extreme. The lack of a global intellectual synthesis of scientific knowledge was, on the other hand, only poorly compensated for by a popular scientific literature whose aim was often less the distribution and mediation of scientific knowledge than its mystification.

The lack of a systematic place in the social system of the sciences and of academic philosophy for reflection on the contents of science beyond the narrow requirements of disciplinary specialization lent a particular importance to the philosophical efforts by scientists themselves. For Einstein's intellectual development it is in fact clear that the writings of scientists such as Mach, Duhem, Poincaré, and Helmholtz

---

119 For this and the following, see the detailed study, (Köhnke 1986).
120 For an attempt to assess this historical situation from the point of view of a systematic historical epistemology, see (Damerow and Lefèvre 1994).

had a greater impact on his philosophical reflection on science than the works of contemporary academic philosophers, precisely because they often dealt with the philosophical implications of concrete problems at the forefront of research. Nevertheless, it would be misleading to consider Einstein's own philosophical contribution only as a continuation of the tradition of epistemological and methodological reflections by nineteenth-century philosopher-scientists. Although this view is naturally suggested by the separation of physics and philosophy as understood today, it is too restrictive to capture the peculiar way in which research in physics and philosophical reflection are intertwined in Einstein's work. In fact, Einstein's scientific contributions to many branches of physics, from thermodynamics to statistical mechanics, from the theory of relativity to quantum physics, cannot be understood without assuming the background of a scientific world picture holding together otherwise disparate chunks of knowledge. As student, Einstein already possessed an extraordinary overview of the state of physics of his time. This enabled him to recognize foundational questions of physics in problems which others preferred to see only from the point of view of their area of specialization.[121] In comparison to Einstein's perception of the entire body of physics and its conceptual incongruences, the claim of those who undertook the construction of, say, an electromagnetic world picture almost appears to be an attempt to conceal the limitations of a specialists' outlook. In any case, Einstein's perspective distinguished itself profoundly—and with significant consequences—from the mutual ignorance that characterized the field theoretical approach to the problem of gravitation and the approach of a mechanistic generalization of the relativity principle, as we have seen earlier.

5.3 Einstein and the Culture of Science

From this sketch of the historical relation between physics and philosophy, it should be clear that the roots of the scientific worldview, which shaped Einstein's perception of physics at the beginning of his career, could only have been of a highly eclectic and backward character. From what is known of his early biography, it is clear that his reading of popular scientific books, together with his exposure to the technical culture associated with the business activities of his family, played a crucial role in the early development of his scientific worldview.[122] The popular scientific books that he devoured as an adolescent combined an easily accessible and conceptually organized overview of scientific knowledge with the claim that the enterprise of science also serves as a model for the development of moral and political standards.[123] These works represented an attempt to transmit the values of democracy and of political and technological progress (which had been defeated on the political scene with

---

121 For a reconstruction of Einstein's discoveries of 1905 along these lines, see (Renn 1993). See also (Holton 1988, chap. 4).
122 For evidence, see (Einstein 1992), as well as the documents collected in (CPAE 1); for historical discussion, see (Damerow and Lefèvre 1994; Gregory 2000; Pyenson 1985; Renn 1993).
123 See, in particular, (Bernstein 1867–1869).

the failure of the revolution of 1848) in the medium of popular science.[124] Einstein's scientific worldview, which apparently had some of its roots in his early fascination with these popular scientific books, has indeed much in common with their image of science as a substitute for religion, with their appeal to the moral and also political ideals of science, and with their effort to achieve a conceptual unification of scientific knowledge beyond its disciplinary boundaries.[125]

The conceptual framework that formed the basis of this effort was a rather primitive combination of remnants of the old natural philosophy from the beginning of the nineteenth century, and of scientific results roughly on the level of the state of knowledge at the middle of the century. It was, however, apparently sufficient to provide the young Einstein with a global perspective on science to which he could then assimilate a broad array of detailed knowledge without committing himself to a premature specialization. In any case, during his entire scientific career he pursued the idea of a conceptual unity of physics, whose first primitive image he may have encountered in his early reading of popular scientific literature. The history of Einstein's formulation of the special theory of relativity, for instance, illustrates not only that he saw, even at the start of his career, in the conceptual diversity of mechanics and field theory a challenge to this unity of physics, but also that he was aware that neither of the two subdisciplines alone could provide the basis for a solution of this conflict. The foundation of the special theory of relativity—the principle of relativity being rooted in classical mechanics, and the principle of constancy of the speed of light in the tradition of field theory—makes it clear that the conceptual innovation represented by this theory presupposed an integration of the knowledge accumulated in these two branches of classical physics.[126]

In Einstein's reaction to the clash between classical mechanics and field theory in the case of gravitation it is now possible to recognize an intellectual attitude that was deeply rooted in his scientific worldview and shaped by his experience with the creation of the special theory of relativity.[127] The approach of a mechanistic generalization of the relativity principle had a function for the emergence of general relativity which is similar to the function mechanics had for the development of special relativity: it supported the principle of relativity with a network of arguments that went beyond the narrow scope of the specific questions under examination, whether these concerned the electrodynamics of moving bodies or the integration of Newton's theory of gravitation into a relativistic field theory. Although Einstein's perspective on the foundational problems of physics encompassed the entire range of classical phys-

---

124 The biographical background of Bernstein, the author of the book which apparently played a key role for Einstein's early intellectual development, is discussed in (Gregory 2000). For more on the relation between popular scientific literature and political developments in the nineteenth century, see (Gregory 1977); see also (Lefèvre 1990).

125 For a systematic analysis of the role of "images of science" as a mediator between science and its external influences, see (Elkana 1981). For a discussion of the religious dimensions of Einstein's scientific worldview, see (Renn 2005).

126 For further discussion, see "Classical Physics in Disarray ..." (in vol. 1 of this series).

ics, there can be no doubt that it was dominated by the tension between its two major conceptual strands, field theory and mechanics. In 1931, for instance, he wrote:

> In a special branch of theoretical physics the continuous field appeared side by side with the material particle as the representative of physical reality. This dualism, though disturbing to any systematic mind, has today not yet disappeared.

He then added with specific reference to Lorentz's theory of electrons, as well as with respect to the special and general theories of relativity:

> The successful physical systems that have been set up since then represent rather a compromise between these two programmes, and it is precisely this character of compromise that stamps them as temporary and logically incomplete, even though in their separate domains they have led to great advances. (Einstein 1931a, 69–70, 72)

For Einstein, the insight into the need to overcome the dualism of matter and field was not just paying lip service to the conceptual unity of physics, but one of the principal determinants of his research program. While his perspective was broader than that of many contemporary physicists, it was also limited by this same program. The extent to which Einstein's intellectual horizon was actually circumscribed by the problem of reconciling the fundamental, conceptual conflict he perceived at the heart of classical physics can be seen in his role, up to the twenties, in the exploration of the consequences of the theory of general relativity. Contrary to other researchers who took part in this research, Einstein's interest focused almost exclusively on what might be called the "philosophical closure" of the new theory. Whether concerning boundary conditions for the gravitational field, or the exact solutions to the field equations, his interest in these emerging research topics was not guided by a program of exploring new features of the theoretical structures he had created, nor by comparing these structures with the empirical results of astronomy, but rather by the question of whether or not a deeper understanding of general relativity would reveal its agreement with the heuristics that had guided its discovery. This interest merely reflects the perspective which had accompanied Einstein's work on general relativity since its inception: he had not searched for a theoretical foundation of cosmology, but rather for a contribution to the conceptual unification of classical physics and, in particular, a synthesis of the field theoretical and mechanical aspects of gravitation.

---

127 Einstein himself compared the heuristics which motivated his search for a general theory of relativity with that guiding his formulation of special relativity: "The theory has to account for the equality of the inertial and the gravitational mass of bodies. This is only achieved if a similar relation is established between inertia and gravitation as that [which is established] by the original theory of relativity between Lorentz's electromotive force and the action of electrical field strength on an electrical mass. (Depending on the choice of the frame of reference, one is dealing with one or the other.)" ("Die Theorie muss Rechenschaft geben von der Gleichheit der trägen und schweren Masse der Körper. Dies wird nur erzielt, wenn zwischen Trägheit und Schwerkraft eine ähnliche Beziehung hergestellt wird, wie durch die urprüngl. Relativitätstheorie zwischen Lorentz'scher elektromotorischer Kraft und Wirkung elektrischer Feldstärke auf eine elektrische Masse. (Je nach der Wahl des Bezugssystems liegt das eine oder das andere vor.)") Einstein to H. A. Lorentz, 23 January 1915 (CPAE 8, Doc. 47 - my translation).

In spite of these limitations of Einstein's perspective, and in spite of the conflict between his heuristic expectations and the conceptual implications of what he had found, it is remarkable that in the course of his work on general relativity he was nevertheless gradually able to overcome his own preconceived expectations and to adapt the interpretation of his theory to new results. This contrasts with many other cases of conceptual innovation in science, in which the crucial step of conceptual innovation takes place at a generational transition, in the transmission of knowledge from "master" to "disciple" so to speak, as was actually the case in the transformation of Lorentz's electrodynamics into Einstein's special theory of relativity.[128] Einstein's own significant contribution to the conceptual understanding of general relativity is related to the fact that, from his earliest efforts to formulate such a theory until the end of his life, he expounded unceasingly the conceptual presuppositions and consequences of his research in accounts accessible also to the non-specialist. Einstein was himself one of the great authors of popular scientific literature. With only minimal technical content, his writings made the intellectual core of his scientific problems accessible to readers. That Einstein's general accounts of the theory of relativity functioned not only to disseminate expert knowledge to the layman, and that they also formed a medium for his own reflection on the conceptual aspects of scientific problems, are facts often overlooked by philosophers of science. But the gradual adaptation of Einstein's Machian heuristics to the implications of general relativity, and its eventual definitive abandonment in the light of these implications, provide a vivid illustration of the impact these reflective accounts had on Einstein's own understanding of general relativity.

In general, it is hardly possible to overlook the significance that the effort to explain scientific knowledge to laymen had for Einstein's intellectual biography, in particular his capacity to address foundational questions beyond the limits imposed by disciplinary specialization. In Bern, as well as in Zurich, he shared his ideas with a group of friends, most of whom were not physicists. We know with certainty that Einstein was indebted to Michele Besso for a decisive inspiration which made possible the breakthrough in the formulation of the special theory of relativity.[129] Einstein also belonged to amateur science societies in Bern and in Zurich that offered an institutional framework for an exchange of ideas which transgressed the usual academic and social boundaries. Even before studying physics in Zurich, he attended an unusual high school in Aarau whose intellectual atmosphere presented no sharp demarcation between research and education, and in which he could experience the spirit of a *res publica litterarum*. Teachers, who were also scientists, such as the physicists Conrad Wüest and August Tuchschmid, or the linguist Jost Winteler, must have confirmed Einstein's conviction that science could offer the foundation for making a life, and not just an intellectual life.[130]

---

128 See (Damerow et al. 2004; Renn 1993).
129 See the acknowledgement in (Einstein 1905) as well as the recollection in (Ishiwara1971).

To conclude: a culture of science which includes the effort of explanation as well as the search for conceptual unity in the diversity of scientific knowledge, that is, a "culture of scientific mediation," forms an essential background for Einstein's philosophical perspective on the foundational problems of physics. The historical preconditions that made this perspective possible were already fragile at the time: evidently, neither popular scientific literature nor amateur science societies could halt the disciplinary fragmentation of scientific knowledge and the loss of possibilities for a single individual to achieve a comprehensive overview. Despite the claim by many physicists of Einstein's generation of the proximity of their field to philosophy, Einstein was in fact already part of a small minority who continually attempted to reflect upon the whole of physics and to search for its conceptual unity. The isolation in which he worked on his later attempts to create a unified field theory testify to his failure to achieve a unity of physics along these lines. But considering how much a single individual could accomplish, even on the basis of inadequate presuppositions, we can read the history of Einstein's achievements as the challenge and the encouragement to work on a culture of science that responds to the needs of today.

ACKNOWLEDGEMENTS

For permission to quote from unpublished Einstein documents I am grateful to Ze'ev Rosenkranz, former curator of the Albert Einstein Archives of the Hebrew University, Jerusalem. I would like to thank my colleagues from the Eidgenössische Technische Hochschule in Zurich—where part of the research for this paper was done—for their hospitality and friendly support, in particular Elmar Holenstein. For their careful reading of earlier versions of this text, as well as for helpful suggestions I am grateful to Met Bothner, Leo Corry, Peter Damerow, Yehuda Elkana, Gideon Freudenthal, Hubert Goenner, John Norton, Wolfgang Lefèvre, Peter McLaughlin, Fiorenza Renn, Ted Richards, Tilman Sauer, and Gereon Wolters. To Frederick Gregory, Michel Janssen, and Tilman Sauer I am particularly indebted for making preliminary versions of their papers accessible to me.

REFERENCES

Abraham, Max. 1912a. "Zur Theorie der Gravitation." *Physikalische Zeitschrift* 13: 1–4. (English translation in this volume.)
———. 1912b. "Das Elementargesetz der Gravitation." *Physikalische Zeitschrift* 13: 4–5.
———. 1915. "Neuere Gravitationstheorien." *Jahrbuch der Radioaktivität und Elektronik* 11: 470–520. (English translation in this volume.)
Assis, A. K. T. 1989. "On Mach's Principle." *Foundation of Physics Letters* 2: 301–318.
———. 1995. "Weber's Law and Mach's Principle." In (Barbour and Pfister 1995).
Barbour, Julian B. 1990. "The Part Played by Mach's Principle in the Genesis of Relativistic Cosmology." In R. Balbinot, B. Bertotti, S. Bergia, and A. Messina (eds.), *Modern Cosmology in Retrospect*. Cambridge, New York, Port Chester, Melbourne, Sydney: Cambridge University Press, 47–66.

---

130 See the documents collected in (CPAE 1; CPAE 5). For historical discussion, see (Pyenson 1985) and the introduction to (Renn and Schulmann 1992).

————. 1992. "Einstein and Mach's Principle." In J. Eisenstaedt and A. J. Kox (eds.), *Studies in the History of General Relativity,* (*Einstein Studies,* vol. 3). Boston: Birkhäuser, 125–153.

————. 1993. "The Search for True Alternatives and its Unexpected Outcome: General Relativity is Perfectly Machian." In *International Conference "Mach's Principle: From Newton's Bucket to Quantum Gravity."* Tübingen: Preprint.

Barbour, Julian B. and Bruno Bertotti. 1977. "Gravity and Inertia in a Machian Framework." *Il Nuovo Cimento* B 38: 1–27.

Barbour, Julian B. and Herbert Pfister (eds.). 1995. *Mach's Principle: From Newton's Bucket to Quantum Gravity.* (*Einstein Studies,* vol. 6.) Boston: Birkhäuser.

Bernstein, Aaron. 1867–1869. *Naturwissenschaftliche Volksbücher.* 20 vols. Berlin: Duncker.

Blackmore, John (ed.). 1992. *Ernst Mach: A Deeper Look. Documents and New Perspectives. Boston Studies in the Philosophy of Science*, vol 143. Dordrecht: Kluwer Academic Publishers.

Boniolo, Giovanni. 1988. *Mach e Einstein.* Rome: Armando.

Borzeszkowski, Horst-Heino von and Wahsner, Renate. 1989. *Physikalischer Dualismus und dialektischer Widerspruch: Studien zum physikalischen Bewegungsbegriff.* Darmstadt: Wissenschaftliche Buchgesellschaft.

Castagnetti, Giuseppe, Peter Damerow, Werner Heinrich, Jürgen Renn, and Tilman Sauer. 1994. *Wissenschaft zwischen Grundlagenkrise und Politik. Einstein in Berlin.* Berlin: Max-Planck-Institut für Bildungsforschung. Forschungsbereich Entwicklung und Sozialisation. Arbeitsstelle Albert Einstein.

Clifford, William K. 1976. "On the Space Theory of Matter." In M. Capek (ed.), *The Concepts of Space and Time.* Dordrecht/Boston: Reidel Publishing Company, 295–296.

CPAE 1. 1987. *The Collected Papers of Albert Einstein,* vol. 1: *The Early Years 1879–1902,* ed. J. Stachel, R. Schulmann, D. Cassidy, and J. Renn. Princeton: Princeton University Press.

CPAE 2. 1989. *The Collected Papers of Albert Einstein,* vol. 2: *The Swiss Years: Writings, 1900–1909,* ed. J. Stachel, R. Schulmann, D. Cassidy, J. Renn, D. Howard, and A. J. Kox. Princeton: Princeton University Press.

CPAE 3. 1993. *The Collected Papers of Albert Einstein,* vol. 3: The Swiss Years: Writings, 1909–1911, ed. M. Klein, A. J. Kox, J. Renn, and R. Schulmann. Princeton: Princeton University Press.

CPAE 4. 1995. *The Collected Papers of Albert Einstein,* vol. 4: The Swiss Years: Writings, 1912–1914, ed. M. J. Klein, A. J. Kox, J. Renn, and R. Schulmann. Princeton: Princeton University Press.

CPAE 5. 1993. *The Collected Papers of Albert Einstein,* vol. 5: The Swiss Years: Correspondence, 1902–1914, ed. M. Klein, A. J. Kox, J. Renn, R. Schulmann, et al. Princeton: Princeton University Press.

CPAE 6. 1996. A. J. Kox, Martin J. Klein, and Robert Schulmann (eds.), *The Collected Papers of Albert Einstein.* Vol. 6. *The Berlin Years: Writings, 1914–1917.* Princeton: Princeton University Press.

CPAE 8. 1998. Robert Schulmann, A. J. Kox, Michel Janssen, and József Illy (eds.), *The Collected Papers of Albert Einstein,* vol. 8. *The Berlin Years: Correspondence, 1914–1918.* Princeton: Princeton University Press.

CPAE 9. 2004. Diana Kormos Buchwald, Robert Schulmann, Jósef Illy, Daniel J. Kennefick, and Tilman Sauer (eds.). *The Collected Papers of Albert Einstein,* vol. 9: *The Berlin Years: Correspondence, January 1919–April 1920.* Princeton: Princeton University Press.

Cuvaj, Camillo. 1970. "A History of Relativity: The Role of Henri Poincaré and Paul Langevin." Ph. D., Yeshiva University.

Damerow, Peter, Gideon Freudenthal, Peter McLaughlin, and Jürgen Renn. 2004. *Exploring the Limits of Preclassical Mechanics,* 2nd ed. New York: Springer Verlag.

Damerow, Peter, and Wolfgang Lefèvre. 1981. *Rechenstein, Experiment, Sprache: historische Fallstudien zur Entstehung der exakten Wissenschaften.* Stuttgart: Klett-Cotta.

————. 1994. "Wissenssysteme im geschichtlichen Wandel." In F. Klix and H. Spada (eds.), *Enzyklopädie der Psychologie.* Band G, Themenbereich C: Theorie und Forschung - Serie II: Kognition.

Deltete, Robert J. (ed.). 2000. *The Historical Development of Energetics by Georg Helm.* (Translated from the German and with an introductory essay by Robert J. Deltete.) Dordrecht: Kluwer.

Ehrenfest, Paul. 1909. "Gleichförmige Rotation starrer Körper und Relativitätstheorie." *Physikalische Zeitschrift* 10: 918.

Einstein, Albert. 1905. "Zur Elektrodynamik bewegter Körper." *Annalen der Physik* 17: 891–921, (CPAE 2, Doc. 23).

————. 1907a. "Über die vom Relativitätsprinzip geforderte Trägheit der Energie." *Annalen der Physik* 23: 371–384, (CPAE 2, Doc. 45).

————. 1907b. "Über das Relativitätsprinzip und die aus demselben gezogenen Folgerungen." *Jahrbuch für Radioaktivität und Elektronik* 4: 411–462, (CPAE 2, Doc. 47).

————. 1911. "Über den Einfluß der Schwerkraft auf die Ausbreitung des Lichtes." *Annalen der Physik* 35: 898–908, (CPAE 3, Doc. 23).

————. 1912a. "Lichtgeschwindigkeit und Statik des Gravitationsfeldes." *Annalen der Physik* 38: 355–369, (CPAE 4, Doc. 3).

————. 1912b. "Zur Theorie des statischen Gravitationsfeldes" and "Nachtrag zur Korrektur." *Annalen der Physik* 38: 443–458, (CPAE 4, Doc. 4).

————. 1912c. "Gibt es eine Gravitationswirkung, die der elektrodynamischen Induktionswirkung analog ist?" *Vierteljahrsschrift für gerichtliche Medizin und öffentliches Sanitätswesen* 44: 37–40, (CPAE 4, Doc. 7).

————. 1913. "Zum gegenwärtigen Stande des Gravitationsproblems." *Physikalische Zeitschrift* 14: 1249–1262, (CPAE 4, Doc. 17). (English translation in this volume.)

————. 1914. "Zum Relativitätsproblem." *Scientia* 15: 337–348, (CPAE 4, Doc. 31). (English translation in this volume.)

————. 1916a. "Die Grundlagen der allgemeinen Relativitätstheorie." *Annalen der Physik* 49: 769–822, (CPAE 6, Doc. 30).

————. 1916b. "Ernst Mach." *Physikalische Zeitschrift* 17: 101–104, (CPAE 6, Doc. 29).

————. 1917. "Kosmologische Betrachtungen zur allgemeinen Relativitätstheorie." *Preussische Akademie der Wissenschaften. Sitzungsberichte* 142–152, (CPAE 6, Doc. 43).

————. 1918a. "Prinzipielles zur allgemeinen Relativitätstheorie." *Annalen der Physik* 55: 241–244, (CPAE 7, Doc. 4).

————. 1918b. "Kritisches zu einer von Hrn. De Sitter gegebenen Lösung der Gravitationsgleichungen." *Preussische Akademie der Wissenschaften, Sitzungsberichte* 270–272, (CPAE 7, Doc. 5).

————. 1919. "Spielen Gravitationsfelder im Aufbau der materiellen Elementarteilchen eine wesentliche Rolle?" *Preussische Akademie der Wissenschaften, Sitzungsberichte* 349–356, (CPAE 7, Doc. 17).

————. 1920. *Äther und Relativitätstheorie. Rede. Gehalten am 5. Mai 1920 an der Reichs-Universität zu Leiden.* Berlin: Julius Springer, (CPAE 7, Doc. 38; Meyenn 1990, 111–123). (English translation in this volume.)

————. 1921a. "Geometrie und Erfahrung." *Preussische Akademie der Wissenschaften, Sitzungsberichte* 123–130, (CPAE 7, Doc. 52).

————. 1921b. "A Brief Outline of the Development of the Theory of Relativity." *Nature* 106: 782–784, (CPAE 7, Doc. 53).

————. 1922a. "Geometry and Experience." In A. Einstein (ed.), *Sidelights of Relativity.* London: Methuen, 27–56.

————. 1922b. "Bemerkungen zu der Franz Seletyschen Arbeit 'Beiträge zum kosmologischen System'." *Annalen der Physik* 69: 436–438.

————. 1922c. "Bemerkungen zu der Abhandlung von E. Trefftz 'Das statische Gravitationsfeld zweier Massenpunkte in der Einsteinschen Theorie'." *Preussische Akademie der Wissenschaften, Sitzungsberichte* 448–449.

————. 1922d. "Bemerkung zu der Arbeit von A. Friedman 'Über die Krümmung des Raumes'." *Zeitschrift für Physik* 11: 326.

————. 1923a. "Grundgedanken und Probleme der Relativitätstheorie." In *Les Prix Nobel en 1921–1922.* Stockholm: Imprimerie Royale. (Lecture held in Gotenburg on 11 July 1923.)

————. 1923b. "Zur allgemeinen Relativitätstheorie." *Preussische Akademie der Wissenschaften, Sitzungsberichte* 32–38.

————. 1923c. "Notiz zu der Arbeit von A. Friedman 'Über die Krümmung des Raumes'." *Zeitschrift für Physik* 12: 228.

————. 1924. "Über den Äther." *Verhandlungen der Schweizerischen Naturforschenden Gesellschaft* 105: 85–93.

————. 1926–27. "Über die formale Beziehung des Riemann'schen Krümmungstensors zu den Feldgleichungen der Gravitation." *Mathematische Annalen* 97: 99–103.

————. 1929. "Space-time." In *Encyclopedia Britannica*, vol. 21. 105–108.

————. 1930a. "Space, Ether and the Field in Physics." In *Forum Philosophicum*, vol. 1. 180–184. (Translated from "Raum, Äther und Feld in der Physik" by Edgar S. Brightman.)

————. 1930b. "Address at the University of Nottingham." *Science* 71: 608–610. (Address given in German and translated by Dr. I. H. Brose.)

————. 1931a. "Maxwell's Influence on the Conception of Physical Reality." In *James Clerk Maxwell: A Commemoration Volume.* Cambridge: Cambridge University Press, 66–73.

————. 1931b. "Zum kosmologischen Problem der allgemeinen Relativitätstheorie." *Preussische Akademie der Wissenschaften. Sitzungsberichte* 235–237.

————. 1932. "Der gegenwärtige Stand der Relativitätstheorie." *Die Quelle* 82: 440–442.

————. 1933. *The Origins of the General Theory of Relativity. Being the First Lecture of the Georg A. Gibson Foundation in the University of Glasgow Delivered on June 20th, 1933.* Glasgow: Jackson, Wylie and Co.

————. 1954a. "On the Theory of Relativity." In *Ideas and Opinions.* New York: Crown, 246–249.

————. 1954b. "Notes on the Origin of the General Theory of Relativity." In *Ideas and Opinions*. New York: Crown, 285–290.

————. 1961. *Relativity. The Special and General Theory. A Popular Exposition*. New York: Bonanza Books.

————. 1992. *Autobiographical Notes*. La Salle, Illinois: Open Court.

Einstein, Albert, and Willem de Sitter. 1932. "On the Relation between the Expansion and the Mean Density of the Universe." *Proceedings of the National Academy of Sciences* 18: 213–214.

Einstein, Albert, and Marcel Grossmann. 1913. *Entwurf einer verallgemeinerten Relativitätstheorie und einer Theorie der Gravitation*. Leipzig: B. G. Teubner, (CPAE 4, Doc. 13).

Elkana, Yehuda. 1970. "Helmholtz' 'Kraft': An Illustration of Concepts in Flux." *Historical Studies in the Physical Sciences* 2: 263–298.

————. 1981. "A Programmatic Attempt at an Anthropology of Knowledge." In E. Mendelsohn and Y. Elkana (eds.), *Sciences and Cultures. Sociology of the Sciences*, vol. 5. Dordrecht: Reidel, 1–76.

Ellis, George F. R. 1989. "The Expanding Universe: A History of Cosmology from 1917 to 1960." In (Howard and Stachel 1989, 367–431).

Föppl, August. 1905a. "Über einen Kreiselversuch zur Messung der Umdrehungsgeschwindigkeit der Erde." *Königlich Bayerische Akademie der Wissenschaften, München, mathematisch-physikalische Klasse, Sitzungsberichte* (1904) 34: 5–28.

————. 1905b. "Über absolute und relative Bewegung." *Königlich Bayerische Akademie der Wissenschaften, München, mathematisch-physikalische Klasse, Sitzungsberichte* (1904) 34: 383–395. (English translation in this volume.)

Freudenthal, Gideon. 1986. *Atom and Individual in the Age of Newton*. Dordrecht: Reidel.

Friedlaender, Benedict and Immanuel Friedlaender. 1896. *Absolute oder Relative Bewegung?* Berlin: Leonhard Simion. (English translation in this volume.)

Goenner, Hubert. 1970. "Mach's Principle and Einstein's Theory of Gravitation." In R. S. Cohen and R. J. Seeger (eds.), *Ernst Mach. Physicist and Philosopher. Boston Studies in the Philosophy of Science*, vol. 6. Dordrecht: Reidel, 200–215.

————. 1981. "Machsches Prinzip und Theorien der Gravitation." In N. Jürgen, J. Pfarr and E.-W. Stachow (eds.), *Grundlagenprobleme der modernen Physik*. Mannheim/Wien/ Zürich: Bibliographisches Institut, 83–101.

Gregory, Frederick. 1977. *Scientific Materialism in 19th-Century Germany. Studies in the History of Modern Science*, vol. 1. Dordrecht: Reidel.

————. 2000. "The Mysteries and Wonders of Natural Science: Aaron Bernstein's 'Naturwissenschaftliche Volksbücher' and the Adolescent Einstein." In D. Howard and J. Stachel (eds.), *Einstein's Formative Years*. (*Einstein Studies*, vol. 8.) Boston: Birkhäuser,

Hertz, Heinrich. 1894. *Die Prinzipien der Mechanik in neuem Zusammenhange dargestellt*. Leipzig: Johann Ambrosius Barth.

Hoefer, Carl. 1994. "Einstein and Mach's Principle." *Studies in History and Philosophy of Science* 25: 287–335.

Hofmann, Wenzel. 1904. *Kritische Beleuchtung der beiden Grundbegriffe der Mechanik: Bewegung und Trägheit und daraus gezogene Folgerungen betreffs der Achsendrehung der Erde und des Foucault'schen Pendelversuchs*. Wien, Leipzig: M. Kuppitsch Witwe.

Holton, Gerald. 1986. *The Advancement of Science, and its Burdens: The Jefferson Lecture and Other Essays*. Cambridge/New York/Sydney: Cambridge University Press.

————. 1988. *Thematic Origins of Scientific Thought. Kepler to Einstein* (rev. ed.). Cambridge/London: Harvard University Press.

Howard, Don and John Stachel (eds.). 1989. *Einstein and the History of General Relativity*. (*Einstein Studies*, vol. 1.) Boston: Birkhäuser.

Illy, József. 1989. "Einstein Teaches Lorentz, Lorentz Teaches Einstein. Their Collaboration in General Relativity, 1913–1920." *Archive for History of Exact Sciences* 39: 247–289.

Ishiwara, Jun. 1971. *Einstein Kyozyo-Koen-roku*. Tokyo: Kabushika Kaisha.

Janssen, Michel. 1999. "Rotation as the Nemesis of Einstein's '*Entwurf*' Theory." In H. Goenner, J. Renn, J. Ritter, and T. Sauer (eds.), *The Expanding Worlds of General Relativity*. (*Einstein Studies*, vol. 7.) Boston: Birkhäuser: Birkhäuser, 127–157.

Jungnickel, Christa, and Russel McCormmach. 1986. *The Now Mighty Theoretical Physics. Intellectual Mastery of Nature*, vol. 2. Chicago and London: The University of Chicago Press.

Kerszberg, Pierre. 1989a. "The Einstein-de Sitter Controversy of 1916–1917 and the Rise of Relativistic Cosmology." In (Howard and Stachel 1989, 325–366).

————. 1989b. *The Invented Universe. The Einstein-De Sitter Controversy (1916–17) and the Rise of Relativistic Cosmology*. Oxford: Clarendon Press.

Köhnke, Klaus Christian. 1986. *Entstehung und Aufstieg des Neukantianismus. Die deutsche Universitätsphilosophie zwischen Idealismus und Positivismus*. Frankfurt am Main: Suhrkamp.

Kostro, Ludwik. 1992. "An Outline of the History of Einstein's Relativistic Ether Conception." In J. Eisenstaedt and A. J. Kox (eds.), *Studies in the History of General Relativity, (Einstein Studies*, vol. 3). Boston: Birkhäuser, 260–280.

———. 2000. *Einstein and the Ether.* Montreal: Apeiron.

Kox, A. J. 1989. "Hendrik Antoon Lorentz, the Ether, and the General Theory of Relativity." In (Howard and Stachel 1989, 201–212).

Lange, Ludwig. 1886. *Die geschichtliche Entwicklung des Bewegungsbegriffes und ihr voraussichtliches Endergebnis. Ein Beitrag zur historischen Kritik der mechanischen Principien.* Leipzig: Wilhelm Engelmann.

Lefèvre, Wolfgang. 1990. "Die wissenschaftshistorische Problemlage für Engels 'Dialektik der Natur'." In H. Kimmerle, W. Lefèvre, and R. Meyer (eds.), *Hegel-Jahrbuch* 1989. Bochum: Germinal, 455–464.

———. 1994. "La raccomandazione di Max Talmey - L'esperienza formativa del giovane Einstein." In (Pisent and Renn 1994).

Lorentz, H. A. 1910. "Alte und neue Fragen der Physik." *Physikalische Zeitschrift* 11: 1234–1257. (English translation in this volume.)

Lützen, Jesper. 1993. *Interactions between Mechanics and Differential Geometry in the 19th Century.* Preprint Series, vol. 25. København: Københavns Universitet. Matematisk Institut.

Mach, Ernst. 1883. *Die Mechanik in ihrer Entwickelung.* Leipzig: Brockhaus.

———. 1960. *The Science of Mechanics.* La Salle, Illinois: Open Court.

McCormmach, Russel. 1970. "Einstein, Lorentz and the Electron Theory." *Historical Studies in the Physical Sciences* 2: 41–87.

Meyenn, Karl von (ed.). 1990. *Albert Einsteins Relativitätstheorie: die grundlegenden Arbeiten.* Braunschweig: Vieweg.

Mie, Gustav. 1913. "Grundlagen einer Theorie der Materie. Dritte Mitteilung." *Greifswald, Physical Institute* 40: 1–65. (English translation in vol. 4 of this series.)

———. 1914. "Bemerkungen zu der Einsteinschen Gravitationstheorie. I und II." *Physikalische Zeitschrift* 14: 115–122, 169–176. (English translation in vol. 4 of this series.)

Miller, Arthur I. 1992. "Albert Einstein's 1907 *Jahrbuch* Paper: The First Step from SRT to GRT." In J. Eisenstaedt and A. J. Kox (eds.), *Studies in the History of General Relativity, (Einstein Studies,* vol. 3). Boston: Birkhäuser, 319–335.

Minkowski, Hermann. 1908. "Mechanik und Relativitätspostulat," appendix to "Die Grundgleichungen für elektromagnetischen Vorgänge in bewegten Körpern." *Nachrichten der Königlichen Gesellschaft der Wissenschaften zu Göttingen, Mathematisch-Physikalische Klasse,* 53–111. (English translation in this volume.)

———. 1911a. "Die Grundgleichungen für die elektromagnetischen Vorgänge in bewegten Körpern." In *Gesammelte Abhandlungen,* vol. 2. Leipzig: Teubner, 352–404.

———. 1911b. "Raum und Zeit." In *Gesammelte Abhandlungen,* vol. 2. Leipzig, 431–444.

Neumann, Carl. 1993. "The Principles of the Galilean-Newtonian Theory." *Science in Context* 6: 355–368. (Taken from a lecture held at the University of Leipzig on 3 November 1869 [1870] *Ueber die Principien der Galilei-Newton'schen Theorie*. Leipzig: Teubner.)

Newton, Isaac. 1972. *Philosophiae Naturalis Principia Mathematica* (1726 ed.), A. Koyré and I. B. Cohen (eds.). Cambridge: Harvard University Press.

Nordström, Gunnar. 1912. "Relativitätsprinzip und Gravitation." *Physikalische Zeitschrift* 13: 1126–1129. (English translation in this volume.)

Norton, John. 1989a. "What Was Einstein's Principle of Equivalence?" In (Howard and Stachel 1989, 5–47).

———. 1989b. "How Einstein found his Field Equations: 1912–1915." In (Howard and Stachel 1989, 101–159).

———. 1992a. "The Physical Content of General Covariance." In J. Eisenstaedt and A. J. Kox (eds.), *Studies in the History of General Relativity, (Einstein Studies,* vol. 3). Boston: Birkhäuser, 281–315.

———. 1993. "General Covariance and the Foundations of General Relativity: Eight Decades of Dispute." *Reports on Progress in Physics* 56: 791–858.

———. 1995. *Mach's Principle before Einstein.* In (Barbour and Pfister1995, 9–57).

Pais, Abraham. 1982. *Subtle is the Lord. The Science and Life of Albert Einstein.* Oxford/New York/Toronto/Melbourne: Oxford University Press.

Pisent, Gualtiero, and Jürgen Renn (eds.). 1994. *L'eredità di Einstein. Percorsi della scienza storia testi problemi,* vol. 4. Padova: il poligrafo.

Poincaré, Henri. 1905. "Sur la dynamique de l'électron." *Comptes rendus des séances de l'académie des sciences* 140: 1504–1508.

———. 1906. "Sur la dynamique de l'électron." *Rendiconto del Circolo Matematico di Palermo* 21: 129–175. (English translation in this volume.)

————. 1911. "Die neue Mechanik." *Himmel und Erde* 23: 97–116.

Pyenson, Lewis. 1985. *The Young Einstein: The Advent of Relativity*. Bristol: Hilger.

Reichenbach, Hans. 1958. *The Philosophy of Space and Time*. Translated by Marie Reichenbach and John Freund. New York: Dover.

Reissner, Hans. 1914. "Über die Relativität der Beschleunigungen in der Mechanik." *Physikalische Zeitschrift* 15: 371–375.

————. 1915. "Über eine Möglichkeit die Gravitation als unmittelbare Folge der Relativität der Trägheit abzuleiten." *Physikalische Zeitschrift* 16: 179–185.

Renn, Jürgen. 1993. "Einstein as a Disciple of Galileo: A Comparative Study of Conceptual Development in Physics." In M. Beller, R. S. Cohen and J. Renn (eds.), *Einstein in Context*. A special issue of *Science in Context*, 311–341.

————. 2003. "Book Review: Einstein and the Ether by Ludwik Kostro." *General Relativity and Gravitation* (35) 6: 1127–1130.

————. 2005. "Wissenschaft als Lebensorientierung: Eine Erfolgsgeschichte?" In E. Herms (ed.), *Leben: Verständnis. Wissenschaft. Technik*. (*Veröffentlichungen der Wissenschaftlichen Gesellschaft für Theologie*, vol. 24.) Tübingen: Gütersloher Verlagshaus, 15–31.

Renn, Jürgen and Robert Schulmann (eds.). 1992. *Albert Einstein - Mileva Marić: The Love Letters*. Princeton: Princeton University Press.

Riemann, Bernard. 1868. "Ueber die Hypothesen, welche der Geometrie zu Grunde liegen." *Abhandlungen der Königlichen Gesellschaft der Wissenschaft zu Göttingen* 13: 133–150.

Ritz, Walter. 1909. "Die Gravitation." *Scientia* 5: 241–255.

Roseveare, N. T. 1982. *Mercury's Perihelion from Le Verrier to Einstein*. Oxford: Clarendon Press.

Rynasiewicz, Robert. 1999. "Kretschmann's Analysis of Covariance and Relativity Principles." In H. Goenner, J. Renn, J. Ritter, and T. Sauer (ed.), *The Expanding Worlds of General Relativity*. (*Einstein Studies*, vol. 7.) Boston: Birkhäuser, 431–462.

Schrödinger, Erwin. 1925. "Die Erfüllbarkeit der Relativitätsforderung in der klassischen Mechanik." *Annalen der Physik* 77: 325–336.

Sciama, Dennis W. 1959. *The Unity of the Universe*. Garden City, New York: Doubleday & Co.

Sewell, William Clyde. 1975. "Einstein, Mach, and the General Theory of Relativity." Ph. D., Case Western Reserve University.

Sommerfeld, Arnold. (ed). 1903–1926. *Physik*, 3 vols. (*Encyklopädie der mathematischen Wissenschaften mit Einschluss ihrer Anwendungen*, vol. 5.) Leipzig: Teubner.

Stachel, John. 1986. "Eddington and Einstein." In E. Ullmann-Margalit (ed.), *The Prism of Science*. Dordrecht and Boston: Reidel, 225–250.

————. 1989. "The Rigidly Rotating Disk as the 'Missing Link' in the History of General Relativity." In (Howard and Stachel 1989, 48–62).

Stein, Howard. 1977. "Some Philosophical Prehistory of General Relativity." In C. Glymour, J. Earman and J. Stachel (eds.), *Foundations of Space-Time Theories*. Minneapolis: University of Minnesota Press, 3–49.

Torretti, Roberto. 1978. *Philosophy of Geometry from Riemann to Poincaré*. Dordrecht/Boston/London: Reidel.

————. 1983. *Relativity and Geometry*. Oxford: Pergamon Press.

Vizgin, Vladimir P. 1994. *Unified Field Theories in the First Third of the 20th Century. Science Networks, Historical Studies*, E. Hiebert and H. Wussing (eds.), vol. 13. Basel/Boston/Berlin: Birkhäuser.

Wahsner, Renate, and Horst-Heino von Borzeszkowski. 1992. *Die Wirklichkeit der Physik: Studien zu Idealität und Realität in einer messenden Wissenschaft*. Europäische Hochschulschriften, Frankfurt am Main/Berlin/Bern/New York/Paris/Wien: Lang.

Wheeler, John A. 1979. "Einstein's Last Lecture." In G. E. Tauber (ed.), *Albert Einstein's Theory of General Relativity*. New York: Crown.

Wolters, Gereon. 1987. *Mach I, Mach II, Einstein und die Relativitätstheorie: Eine Fälschung und ihre Folgen*. Berlin/New York: de Gruyter.

Zenneck, Jonathan. 1903. "Gravitation." In (Sommerfeld 1903–1926), 1: 25–67. (Printed in this volume.)

JONATHAN ZENNECK

GRAVITATION

Originally published as the entry "Gravitation" in the Encyklopädie der mathematischen Wissenschaften, 5th Volume: Physics, A. Sommerfeld, ed. B. G. Teubner, Leipzig 1903–1921, pp. 25–67. Finished in August, 1901, as noted by the author.

TABLE OF CONTENTS

1. Newton's law

1. Determination of the Gravitational Constant

2. Significance of these Determinations
3. Survey of Various Methods
4. Determinations with the Torsion Balance
   a. Static Method
   b. Dynamical Method
5. Determinations with the Double Pendulum
6. Determinations with an Ordinary Balance
7. Determinations with a Plumbline and Pendulum
   a. Static Method: Plumbline Deflection
   b. Dynamical Method: Pendulum Observation
8. Calculating the Gravitational Constant
9. The Result of the Determinations

2. Astronomical and Experimental Examination of Newton's Law

10. General
11. Dependence on Mass: Astronomical Test
12. Dependence on Mass: Experimental Test for Masses of the Same Material
13. Dependence on Mass: Experimental Test for Masses with Various Chemical Compositions
14. Dependence on Mass: Experimental Test for Masses with Various Structures
15. Dependence on Distance: Astronomical Test
16. Dependence on Distance: Experimental Test
17. Influence of the Medium on Gravitation

Jürgen Renn (ed.). *The Genesis of General Relativity*, Vol. 3
Gravitation in the Twilight of Classical Physics: Between Mechanics, Field Theory, and Astronomy.
© 2007 Springer.

18. Influence of Temperature
19. Dependence on Time: Constancy of Action of Forces
20. Dependence on Time: Finite Speed of Propagation

3. Extension of Newton's Law to Moving Bodies

21. Transferring Fundamental Electrodynamic Laws to Gravitation
22. Transferring Lorentz's Fundamental Electromagnetic Equations to Gravitation |

[26]
23. Laplace's Assumption
24. Gerber's Assumption

4. Extension of Newton's Law to Infinitely Large Masses

25. Difficulty with Newton's Law for Infinitely Large Masses
26. Elimination of the Difficulty by Altering the Law of Attraction
27. Elimination of the Difficulty by Introducing Negative Masses

5. Attempts at a Mechanical Explanation of Gravitation

28. Pressure Differences and Currents in the Aether
29. Aether Vibrations
30. Aether Impacts: Original Ideas of Lesage
31. Aether Impacts: Further Development of Lesage's Theory
32. Aether Impacts: Difficulties of This Theory
33. Aether Impacts: Jarolimek's Objections and Theory

6. Reduction of Gravitation to Electromagnetic Phenomena

34. Gravitation as a Field Effect
35. Electromagnetic Vibrations
36. Mossotti's Assumption and its Further Development

Comprehensive Literature

can be found at the beginning of each section in notes 1, 2, 36, 47, 48, 77, 82, 107.

Introductory Note

In this paper astronomical questions, which are not dealt with in detail until Volume VI, must be mentioned several times. The present paper does not aspire to completeness in this regard, but only draws upon as much astronomical material as is unavoidable for treating the subject.

1. Newton's Law

As is well known, the fundamental gravitational law was first[1] conceived clearly by Newton, and formulated in the third book of his *Philosophiae naturalis principia mathematica*, propositions I–VII. |

It states: If at a certain instant of time two mass elements with masses m_1 and m_2 [27] are at distance r from each other, then at the same instant a force acts on each of the two elements in the direction of the other with a magnitude

$$G\frac{m_1 m_2}{r^2}.$$

In this expression, G is a universal—i.e., only dependent on the system of units— constant, the so-called gravitational constant.

1. DETERMINATION OF THE GRAVITATIONAL CONSTANT[2]

2. Significance of this Determination

The inherent significance of the absolute determination of any physical constant is enhanced in the case of the gravitational constant for two additional reasons:

1. If the gravitational constant is known, the acceleration due to gravity g and the dimensions of the Earth yield the mass and the mean density of the Earth.[3] The latter

---

1   About Newton's forerunners cf. F. Rosenberger, *Isaac Newton und seine physikalischen Prinzipien*, Leipzig 1895. A compilation of nearly all papers (up to 1869) which are in some way related to the *mathematical* implementation of the law of attraction can be found in I. Todhunter, *History of the mathematical theories of attraction and the figure of the Earth*, 2 Vols., London 1873.

2   Principal review literature about absolute determinations: J.H. Poynting, *The Mean Density of the Earth*, London 1894; F. Richarz and O. Krigar-Menzel, *Berl. Abh.* 1898, Appendix; C.V. Boys, *Rapp. congrès internat. phys.* 3, Paris 1900, p. 306–349. Then there are Gehler's *Physikalisches Handwörterbuch*, Leipzig 1825, articles: *Anziehung, Drehwage, Erde, Materie*; S. Günther, *Lehrbuch der Geophysik* 1, 2nd ed., Stuttgart 1879; F. Richarz, Leipzig, *Vierteljahrschr. astr. Ges.* 24 (1887), p. 18–32 and 184–186.

3   If Δ is the mean density of the Earth and R its radius, to first approximation we get

$$g = \frac{4}{3} R \pi \Delta G.$$

Considering the corrections which are caused by flattening, centrifugal force and their differences within various latitudes, then one arrives at the relation explained in detail by F. Richarz and O. Krigar-Menzel[2]

$$9.7800\frac{m}{\sec^2} = \frac{4}{3} \cdot R_p \pi \Delta G \left(1 + \mathfrak{a} - \frac{3}{2}\mathfrak{c}\right),$$

where

    R_p = earth radius at the pole = 6356079m,

    $\mathfrak{a}$ = flattening = 0.0033416,

    $\mathfrak{c}$ = relation between centrifugal force and gravity at the equator = 0.0034672.

was the ultimate aim of most determinations; therefore, they are usually known as *determinations of the mean density of the Earth*.

[28]   2. If the Earth's mass is known, the masses of the | other planets and of the Sun follow, since the proportion of these masses to the Earth's mass is supplied by astronomical observation.[4]

3. Survey of Various Methods

The various methods chosen to gain the value for the gravitational constant *G* in absolute terms can be divided essentially into three main classes:

1. The force that masses of known magnitudes at known distances exert on each other, was determined directly: determinations with the torsion balance, the double pendulum, and the ordinary balance.[5]

2. Changes in the direction or magnitude of the acceleration due to gravity *g* caused by masses of known magnitudes were measured: deflection of the plumbline, pendulum observations.

3. It was attempted to calculate the Earth's mean density and thereby the gravitational constant from the density at the surface, based on a more or less hypothetical law about the increase of density towards the center of the Earth.

4. Determinations with the Torsion Balance

a. Static method. The weights attached to the balance beam are attracted by masses *next to* the beam. The resulting rotation of the beam is a measure of the attractive force's magnitude.

This method, which was probably first suggested by Reverend J. Michell,[6] was used by H. Cavendish,[7] F. Reich,[8] F. Baily,[9] A. Cornu and J. Baille,[10] C.V. Boys,[11] and finally by C. Braun.[12]

Reich's advance over Cavendish lies primarily in his use of a mirror arrangement to make measurements. Baily's measurements are particularly valuable because they were extended to a large number of materials, and were varied in other, manifold ways. Cornu and Baille have shown that the same | accuracy (the same deflection [29]   angle) can be achieved despite any reduction of scale, if only a suitable choice in sus-

---

4   But cf. section 11.
5   In the latter method, *g* enters into the result.
6   Quoted from Cavendish, *Lond. Trans.* 88 (1798).
7   See above note.
8   "Versuche über die mittlere Dichtigkeit der Erde mittelst der Drehwage," Freiberg 1838, and "Neue Versuche mit der Drehwage", Leipzig 1852.
9   *Lond. Astr. Soc. Mem.* 14 (1843).
10 Paris, *C. R.* 76 (1873), p. 954–58.
11 *Lond. Trans.* 186 (1889), p. 1–72.
12 *Wien. Denkschr.* 64 (1897), p. 187–285. Report on this: F. Richarz, *Leipzig Vierteljahrsschr. astr. Ges.* 33 (1898), p. 33–44.

pension provides equal oscillation periods of the torsion balance. Consequently, they used much smaller dimensions for their apparatus and so avoided a number of disturbances. Boys[13] extended this reduction to a smaller scale and made it possible to replace the metal suspension wires with much more favorable quartz fibers. Braun uses a torsion balance in a vacuum to avoid totally the worst disturbance while measuring with the torsion balance, namely the air currents.

Boys partly evaded the deficiency of the extremely small dimensions he used by skillful arrangement of his torsion balance; however, the disadvantage remains that with small dimensions, apart from the strong damping of the oscillations, errors in length determination and deficient homogeneity of the material can easily spoil the result's accuracy.[14] To avoid this deficiency of small dimensions and nevertheless reach high sensitivity, F. R. Burgess[15] suggested that arranging the weights to float on mercury would enable the use of large masses and thin suspension wires. In a pre-experiment with weights of 10×2 kg on both sides, he found a 12° deflection, but has not yet carried out his determinations.

b. Dynamical method. The attracting masses are placed in line with the balance beam. Their attraction serves to reinforce the restoring torque of the suspension. The resulting decrease of the oscillation period gives a measure of the attractive force's magnitude.

With this method, C. Braun obtained a value of G that agrees very well with results of his measurements via the static method. R. von Eötvös[16] suggested modifying this I method but has not yet published final results.   [30]

5. Determinations with the Double Pendulum

J. Wilsing's vertical double pendulum[17] — a vertical balance beam with a weight at each end, attracted by horizontally displaced masses — does not work by torsion of wires, but uses gravity as the restoring force. The torque is reduced to a minimum by placing the double pendulum's center of gravity only ca. $0,01$ mm beneath the edge. Such an arrangement combines high sensitivity[18] with significant stability, and moreover, in contrast to the torsion balance, has the advantage of being influenced to a lesser degree by air currents.

---

13 At length 2.3 cm of the balance beam, loaded on both sides with 1.3 to 3.98 g and deflected at each side by 7.4 kg, Boys received a deflection of ca. 370 scale sections. For Cavendish the quantities concerned were 196 cm, 730 g, 158 kg; he received a deflection of 6–14 scale sections.

14 Cf. F. Richarz's report cited in note 12.

15 Paris, C.R. 129 (1899), p. 407–409. Poynting[2] had already performed a similarly arranged experiment, but abandoned this arrangement due to interfering currents in the fluid.

16 Ann. Phys. Chem. 59 (1896), p. 354–400.

17 Potsdam. Astr.-physik. Obs. 6 (1887), No. 22 and 23.

18 At 325×0.54 kg, 1 to $10'$ deflection.

6. Determinations with the Ordinary Balance

The principle of this method, seemingly already presented by Descartes,[19] is as follows. Two equal weights m are placed on the scales of a balance. Underneath one of the two scales—possibly simultaneously above the other one—a mass M is brought in. The weight difference now observed provides a measure of the attraction M has on m.

For the purpose of absolute determination of the gravitational constant this method was probably first used by Ph. von Jolly,[20] later by J.H. Poynting,[2] and then by F. Richarz and O. Krigar-Menzel.[2]

Jolly's arrangement, which was already used in Newton's time by Hooke[21] in a quite similar way to determine a decrease of g with height, has the disadvantage that vertical air currents caused by temperature differences can disturb the weighing process by friction on the long suspending wires. Poynting avoided this shortcoming; furthermore, he saw to it that the angle by which the balance beam rotates can be read precisely, and that the attracting weights can be removed or brought closer without having to lock the balance or to expose it to vibration. Richarz's and Krigar-Menzel's method has the advantage of tolerating extraordinarily large attracting | masses (100,000 kg lead) without excessive difficulties, and moreover of allowing an effective four-fold attraction of this mass. However, the method suffers from the drawback that relieving and locking the balance becomes necessary in the course of determination.

7. Determinations with Plumbline and Pendulum

a. Static method (plumbline deflection). Deflecting masses were always mountains, and their deflection of the plumbline was determined by measuring the difference of geographical latitude between two points, if possible taken to the north and south of the deflecting mountain, once astronomically—where the direction of the plumbline enters—and then trigonometrically. The difference between the two determinations is twice the deflection caused by the mountain. The dimensions and the specific weight of the rocks determine the mass of the mountain.

Determinations of this kind were carried out by Bouguer[22] at Chimborazo, by N. Maskelyne and C. Hutton,[23] later by James[24] and Clarke at mountains in the Scottish highlands, by E. Pechmann[25] in the Alps and under particularly favorable conditions by E.D. Preston[26] on the Hawaiian islands.

[31]

19   Cited in *Observ. Sur la physique*, 2, Paris 1773.
20   *Münchn. Abh.* (2) 14 (1881); *Ann. Phys. Chem.* 14 (1881), p. 331–335.
21   Cited in Rosenberger, note 1.
22   *La figure de la terre*, Paris 1749, sec. VII, chap. IV.
23   *Lond. Trans.* 1775, p. 500–542; 1778, p. 689–788; 1821, p. 276–292.
24   *Phil. Mag.* (4) 12 (1856), p. 314–316; 13 (1856), p. 129–132 and *Lond. Trans.* 1856, p. 591–607.
25   *Wien. Denkschr. (math.-naturw. Kl.)* 22 (1864), p. 41–88.
26   Washington, *Bull. Phil. Soc.* 12 (1895), p. 369–395.

It has been suggested to use the sea at low and high tide[27] or a drainable lake[28] instead of a mountain, but a determination never seems to have been performed this way, though it would have significant advantages over using a mountain.

b. Dynamical method (pendulum observation). The scheme of such determinations is the following. Either at the foot and the top of a mountain or on the Earth's surface and in the depth of a mine, the oscillation period of the same pendulum is observed. The measured difference in oscillation period and hence in the acceleration due to gravity at the two points | yields the attraction of the mountain or the layer of Earth above the mine, respectively.[29]   [32]

Determinations of the first kind are due to Bouguer[22] (Cordills), Carlini[30] (and Plana) (Mont Cenis), under particularly favorable conditions from Mendenhall[31] (Fujiyama, Japan) and E.D. Preston[26] (Hawaiian islands).

Determinations of the second kind were first suggested by Drobisch,[32] later carried out by G.B. Airy[33] and in large number by R.v. Sterneck.[34]

A third method, in principle definitely more favorable, was attempted by A. Berget:[35] artificial alteration of g due to a difference in the water level of a drainable lake. However, his determination was spoiled by an unsuitable measurement of this change in g.

8. Calculation of the Gravitational Constant[36]

1. Laplace[37] based his calculations on the following conditions, as did Clairaut and Legendre:

a. The Earth consists of separate ellipsoidal layers. The density within each layer is constant.

b. The rotation is so slow that the deviation from the spherical shape becomes small, as well as the influence of the centrifugal force on g.

c. The Earth's substance shall be regarded as fluid.

---

27 By Robison 1804 (cited by Richarz and Krigar-Menzel, see note 2), Boscowich 1807 (cited in *Monatl. Korrespondenz z. Beförd. d. Erd- u. Himmelskunde* 21 (1810)), furthermore by von Struve (cited in *Astr. Nachr.* 22 (1845), p. 31 f.)

28 F. Keller, *Linc. Rend.* 3 (1887), p. 493.

29 Cf. for this and the following numbers vol. VI of the *Encyclop., Geophysik.*

30 *Milano Effem.* 1824. Cf. E. Sabine, *Quart. J.* 2 (1827), p. 153 and C.J. Guilio, *Torino Mem.* 2 (1840), p. 379

31 *Amer. J. of Science.* (3) 21 (1881), p. 99–103.

32 *De vera lunae figura* etc., Lipsiae 1826.

33 *Lond. Trans.* 1856, p. 297–342 and 343–352. For calculation cf. S. Haughton, *Phil. Mag.* (4) 12 (1856), p. 50–51 and F. Folie, *Bruxelles Bull.* (2) 33 (1872), p. 389–409.

34 Wien. *Mitteil. d. milit.-geogr. Inst.* 2–6, 1882–1886 and Wien. *Ber.* 108 (2a), p. 697–766.

35 Paris *C. R.* 116 (1893), p. 1501–1503. Cf. Gouy's objection (Paris *C. R.* 117 (1893), p. 96) that the temperature would have had to be constant at least to 0.2×10^{-6} degrees.

36 Cf. F. Tisserand, *Mécan. cél.* 2, Paris 1891, chap. XIV and XV.

37 *Méc. cél.* 5 (1824), Livr. 11, chap. 5.

[33] Under these circumstances Laplace calculates the conditions of equilibrium, into which, besides the Earth's elliptical property, the I law expressing the density of an Earth layer as a function of its distance from the center enters. Laplace makes two assumptions for this law:

$$\rho = \rho_0[1 + e(1 - a)], \tag{1}$$

$$\rho = \frac{A}{a}\sin(an), \tag{2}$$

where ρ stands for the density, a for the distance of a layer to the Earth's center (Earth radius = 1), and ρ_0, e as well as A, n are constants. These constants are determined on the one hand by the value of ρ on the Earth's surface, and on the other hand by the derived equilibrium conditions. One then obtains a relation between the mean density of the Earth (and hence of the gravitational constant) and the Earth's surface density ρ_0; that is, from the first assumption regarding increasing density towards the center of the Earth, it follows that

$$\Delta_1 = 1.587 \cdot \rho_0,$$

and from the second assumption, that

$$\Delta_2 = 2.4225 \cdot \rho_0,$$

if the Earth's ellipticity is assumed to equal 0,00326.
2. On essentially the same basis, using Laplace's second assumption regarding increasing density towards the center of the Earth, Clairant's formula for the equilibrium of the rotating Earth, taken to be a fluid, and the value 0.00346 for the ellipticity of the Earth, J. Ivory[38] arrives at the relation:[39]

$$\Delta = 1.901 \cdot \rho_0.$$

3. The recent extensive literature on this question (Lipschitz, Stieltjes, Tisserand, Roche, Maurice Lévy, Saigey, Callandreau, Radau, Poincaré, Tumlirz) can not be discussed here; therefore we refer to the previously cited chapters in Tisserand[40] or to Vol. VI of the Encyclopedia.

9. The Result of the Determinations

Regarding the question of what should be taken as the most probable value for the gravitational constant, the results of the methods discussed in sections 7 and 8 must immediately be excluded. I

---

38   *Phil. Mag.* 66 (1825), p. 321 f.
39   Taking for the mean density all over the Earth surface S. Haughton's[33] value ρ_0= 2.059, one would obtain according to Laplace: Δ_1 = 3.268, Δ_2 = 4.962, and according to Ivory: Δ= 3.914.
40   Cf. note 36.[1]

Indeed, the terrestrial methods (section 7) actually carried out have the advantage [34] over laboratory methods (sections 4–6) in that the attracting masses and hence the differences to be observed have a relatively significant magnitude. But this advantage is more than outweighed by the fact that dimensions and density of the attracting masses are known only incompletely, and that the inadequately observed mass distribution below the place of observation plays an essential, but entirely uncontrollable, part.[41]

Those terrestrial methods, however, that could have had prospects for success, because not only do they allow for using very large masses, but also because the magnitude of the attracting masses could be determined with sufficient accuracy, and because the influence of the environment—such as change of magnitude or direction of g by a lake or the sea at different levels—would drop out, have not been carried out at all or were carried out only in a flawed manner.

The attempts to calculate the gravitational constant (section 8) can not provide a reasonably reliable result, either. Apart from other uncertainties, the mean surface density of the Earth enters this calculation, and this is far from being known with as much accuracy as the gravitational constant itself when obtained by laboratory determination.

Thus only the results of laboratory determinations remain (sections 4–6). Considering the two most recent determinations from each method only, we get the following compilation:

		Δ	G
Torsion balance	*Boys*	5.527	$6.658 \cdot 10^{-8} \, \text{cm}^3 \, \text{s}^{-2} \, \text{g}^{-1}$
	Braun	5.5270[a]	
(Double pendulum	*Wilsing*	5.577	$6.596 \cdot 10^{-8} \, \text{cm}^3 \, \text{s}^{-2} \, \text{g}^{-1}$)
	Poynting	5.4934	$6.698 \cdot 10^{-8} \, \text{cm}^3 \, \text{s}^{-2} \, \text{g}^{-1}$
Balance	*Richarz* and *Krigar-Menzel*	5.5050	$6.685 \cdot 10^{-8} \, \text{cm}^3 \, \text{s}^{-2} \, \text{g}^{-1}$
Mean value of these determinations		5.513[b]	$6.675 \cdot 10^{-8} \, \text{cm}^3 \, \text{s}^{-2} \, \text{g}^{-1}$

a.　In copies issued later, Braun assumed the most probable result of his observations to be $\Delta = 5{,}52725$ (communicated by Prof. F. Richarz).

b.　As we know, Newton (*Principia lib. III, prop. X*) estimated the Earth's mean density to be 5–6. The mean 5.5 thus agrees with the mean value from the most recent measurements to $1/4 \, \%$.

41　Cf. W.S. Jacob, *Phil. Mag.* (4) 13 (1857), p. 525–528. Conversely, such determinations can be significant because they allow for a conclusion about mass distribution close to the place of observation. Cf. R. v. Sterneck's *Messungen*.[34]

[35] | The good agreement[42] between the values obtained by the same method on the
one hand, and the relatively significant disagreements among results of different
methods[43] on the other, show that these disagreements can only be due to deficiencies
in principle of the methods. As long as these have not been cleared up, none of the
results can be given more weight than another. It is a pity that Wilsing's method has
not yet been employed by a second observer to check Wilsing's result, and that the
influence of magnetic permeability of the double pendulum has not yet been examined.[44] Therefore, we did not take Wilsing's result into consideration in the calculation of the mean value above.

2. ASTRONOMICAL AND EXPERIMENTAL TESTS
OF NEWTON'S LAW

10. General

Two independent fields insure that Newton's law, even if not absolutely accurate, represents real conditions with a far-reaching accuracy unmatched by hardly any other
law.

In the *astronomical*[45] domain, this law yields planetary motions not only to the
first approximation (Kepler's laws); but even to the second approximation, the deviations of planetary motion due to pertubations by other planets follow from Newton's
law so accurately that the observed perturbations led to the prediction of the orbit and
relative mass of a hitherto unknown planet (Neptune).

[36]   On the other hand there are a number of astronomical | observations that show
deviations compared to calculations based on Newton's law. This deviation amounts
to[46]

  1. ca. $40''$ per century in the perihelion motion of Mercury;

  2. 5-fold probable error in the motion of the node of Venus' orbit;

  3. 3-fold probable error in the perihelion motion of Mars; and

  4. 2-fold probable error (uncertain!) in the eccentricity of Mercury's orbit.

In addition there are:

  5. significant anomalies in the motion of Encke's comet; and

---

42   Between torsion balance determinations there is a difference $\leq 0.012\,\%$, between balance determi-
     nations there is a difference of ca. 0.2%.

43   The largest difference between balance and double pendulum determinations is ca. 1.5%.

44   According to F. Richarz and O. Krigar-Menzel (*Bemerkungen zu dem ... von Herrn C.V. Boys über die
     Gravitationskonstante ... erstatteten Bericht*, Greifswald 1901) the deviation of Wilsing's result from
     others could be caused by such an influence.

45   Discussion of the validity of Newton's law in the astronomical domain in Tisserand, *Méc. cél.* 4
     (1896), chap. 29 and S. Newcomb.[48]

46   S. Newcomb, *The elements of the four inner planets* etc., Washington 1895. On page 109 ff. is a dis-
     cussion of possible explanations of these deviations.

6. small irregularities in the Moon's orbit.

Small corrections of Newton's law are therefore not excluded by the astronomical evidence,[47] even if it is not at all settled—particularly in the cases listed in 5) and 6) where conditions are more complicated and uncertain than for planetary orbits—that the above differences are due to an inaccuracy of the gravitational law.[46]

In the *experimental* area the best determinations of the gravitational constant, which all rest on the assumption of the validity of Newton's law, yield results in rather good agreement.[48] As these determinations were carried out with masses of a great variety of magnitudes, materials and distances, this agreement therefore excludes any considerable inaccuracy of | Newton's law, and allows at most for small corrections.   [37]

11. Dependence on Mass: Astronomical Test

Newton inferred that the force that two bodies exert on each other is proportional to the mass of each body as follows:

a. Observation shows Jupiter conferring *acceleration* to its satellites, the Sun to the planets, Earth to the Moon, and the Sun to Jupiter and its satellites, which is equal at equal distance. Hence, it follows that in these cases the *force* must be proportional to the mass of the *attracted* body.

---

47 Th. von Oppolzer (*Tagebl. d. 54. Vers. d. Naturf. u. Ärzte*, Salzburg, 1881) even draws quite an apodictic conclusion: "The theory of the Moon makes a conjecture quite probable, the theory of Mercury points at it firmly, Encke's comets lift it up to an irrefutable certainty that the theories built solely on Newton's law of attraction *in present form* are not sufficient for explaining the motion of heavenly bodies."

48 To compare the best terrestrial and laboratory methods:

	Observer	attracting mass	Δ
Laboratory methods	Boys	7.4 kg	5.527
	Braun	9.1 "	5.5270
	Poynting	154 "	5.4934
	Wilsing	325 "	5.577
	Richarz and Krigar-Menzel	100.000 "	5.5050

	Observer	attracting mass	Δ	
	Mendenhall	mountain of 3.800 m height	5.77	
	E. D. Preston	" 3.000 "	5.57	5.35
Terrestrial methods	"	" 4.000 "	5.13	
	von Sterneck	strata of various thickness	5.275	
	(Wien. Ber. 108)	"	5.56	
	"	"	5.3	
	"	"	5.35	

b. The principle of action and reaction then implies that the force must be proportional to the mass of the *attracting* body as well.

M.E. Vicaire[49] raised the following objection against this line of reasoning, which would however require discussion.[50] The examples presented represent a very special case: a very large body attracting a relatively very small body. But then the assumption that at equal distance the attraction can only be a function of the two masses already provides the result that the attractive force must be approximately proportional to the small body's mass.

This is because the function A_{Mm}, which expresses the attraction a large mass M has on small mass m, is certainly homogeneous in M and m. One can hence put: |

$$A_{Mm} = M^k \cdot f(\frac{m}{M})$$

[38]

$$= M^k\left[\frac{m}{M} \cdot f'(0) + \left(\frac{m}{M}\right)^2 \cdot \frac{f''(0)}{1 \cdot 2} + \dots\right]$$

$$= M^{k-1} \cdot m \cdot f'(0) \qquad \text{approx.,}$$

so the attraction is to first approximation proportional to m. Hence, from the fact that this proportionality is confirmed by observation, one must not conclude that the attraction is also proportional to the mass of the attracting large body. However, from this it would follow that calculations of planetary masses in relation to the Sun's mass based on Kepler's third law are in principle misguided.

Vicaire also objects to supporting these calculations by the usual calculations from the planetary perturbations. The *secular* perturbations of a planet m by another m', which are primarily observed and drawn upon in these calculations, do not at all result in the relative mass of planet m', but in the proportion $A_{mm'} : A_{Mm}$, which according to the above does not need to be identical to $m' : M$. Only the *periodic* perturbations could provide information about $m' : M$.

12. Dependence on Mass:
Experimental Test for Masses of the Same Material

The G-determinations of Poynting[2] and of Richarz and Krigar-Menzel[2] are of special value in relation to the question of how far the proportionality of the attractive force to the mass is guaranteed for masses of the same material. Both experimenters used unobjectionable laboratory methods carried out with the greatest care. Both

---

49   Paris, *C.R.* 78 (1874), p. 790–794.

50   This is opposed by the agreement within probable error between the mass of the planets determined from the perturbations which they exert on other planets, and the mass of the same planets determined from the motion of their moons, if they have any. For example, the mass of Mars from Jupiter's perturbations results in   = $1/2.812.526$,   from the elongation motions of its moons   = $1/3.093.500$.   Cf. as well F.W. Bessel, *Berl. Abh.* 1824 and *Ges. Werke* 1, p. 84.

GRAVITATION   89

determinations employed the same material (lead) and the same method of measurement, but masses of very different magnitudes (154 or 100.000 kg). Even though in one case the mass was 650 times greater than the other, the results agree to approximately 0.2%.

13. Dependence on Mass:
Experimental Test for Masses of Various Chemical Compositions

Three different methods have been used to examine whether the proportionality of the attractive force to mass is also strictly valid for masses of different chemical compositions.

a. The gravitational constant was determined for masses of different materials.

F. Baily[9] carried out a large number of measurements of this kind. If his results are arranged according to the specific weight I of the mass which was suspended from the torsion balance,[51] and if we take for each material the mean value from all measurements, the following is revealed. The values of Δ increase—the values of G decrease accordingly—as the specific weight of the mass is decreased.[52] However, there is reason to assume that these disagreements are a matter of a basic error in his arrangement or calculation.[53]   [39]

In any case, the fact that the results of Boys[11] and Braun[12] agree to 0.01%, although they refer to different materials, counts against the assumption that these different results are due to a different value of the gravitational constant for different substances. Likewise, with the help of a particularly sensitive torsion balance, v. Eötvös[16] claims to have found that the difference of attraction of glass, antimony, and corkwood from that of brass is less than $1/2 \cdot 10^{-7}$ and of air from that of brass less than $1 \cdot 10^{-5}$ of the total attraction.

b. Pendulums were produced out of various materials to compare their periods of oscillation.

---

51   Same attracting substance everywhere = lead.

52

Substance	specific weight	Δ	
Platinum	21	5.609	
Lead	11.4	5.622	
Brass	8.4	5.638	
Zinc	7	5.691	
Glass	ca. 2.6	5.748	exception
Ivory	1.8	5.745	

53   Cf. also F. Reich in the paper cited in note 8, "*Neue Versuche* etc.", p. 190.

This method, already employed by Newton,[54] has been refined by F.W. Bessel,[55] in particular. While Newton could only conclude from his experiments that the difference of attraction which the Earth exerts on bodies of very different composition is smaller than $1 \cdot 10^{-3}$ of the total attraction, Bessel managed to squeeze this limit down to $1/6 \cdot 10^{-4}$.

c. A sealed vessel which contains two different chemical substances is weighed, then the substances combine, and after completion of the chemical reaction the vessel is weighed again. |

[40] The first experiments of this kind by D. Kreichgauer[56] with mercury and bromine, and with mercury and iodine gives the result "that with the bodies employed, a change of attraction by the Earth due to chemical forces should stay below 1/20,000,000 of the total attraction." But H. Landolt[57] found under conditions as simple as possible—except for reactions, where a change of weight could not be determined with certainty—the following:

1. For reduction of silver sulphate by ferrosulphate in three series of experiments, a weight decrease by 0.167, 0.131 and 0.130 mg.

2. For iodic acid and hydrogen iodide weight decreases in six experimental series, varying between 0.01 and 0.177 mg.

Not only do these decreases in weight exceed probable measurement errors, but some of them also exceed the largest deviation among single measurements. A. Heydweiller[58] resumed these measurements after M. Hänsel[59] established that the deviations observed by Landolt in the first example can not be explained by the influence of magnetic forces. He also obtains decreasing weight in a series of cases and reaches the conclusion: "one may regard a change in weight as ascertained: in the effect iron has on copper sulphate in acid or basic solution ... , regarding the dissolution of acid copper sulphate ... , and in the effect potassium hydroxide has on copper sulphate ... ."

The cases presented above are therefore *well established but for the time being completely unexplained deviations from the proportionality of the action of gravity to mass.*

---

54 *Principia lib. III, propos. VI.*
55 *Astr. Nachr.* 10 (1833), p. 97.
56 *Berl. physik. Ges.* 10 (1891), p. 13–16.
57 *Zeitschr. physik. Chem.* 12 (1894), p. 11. He cites that in the synthesis of iodine and bromide silver J. S. Stas always obtained less than equivalent of the initial quantities. Indeed, the difference amounted on average in five experiments to $1/4 \cdot 10^{-4}$ of the total mass.
58 *Ann. Phys.* 5 (1901), p. 394–420.
59 Diss. Breslau 1899.

14. Dependence on Mass:
Experimental Test for Masses of Various Structures

The conjecture that attraction between two masses could depend on their structure is suggested by several theories explaining gravitation. This was examined experimentally in two directions. |

    a. Kreichgauer[56] examined whether a body (acetic sodium) changes weight while   [41] crystallizing. He found, however, that any change of weight is below $1/2 \cdot 10^{-7}$ of the total attraction.[60]

    b. A. S. Mackenzie[61] as well as J. H. Poynting and P. L. Grey[62] deal with the question of whether the gravitational effect of a crystalline substance varies with different directions. Mackenzie tested calcite against lead, and also calcite against calcite, but he found the difference to be smaller than 1/200 of the total attraction. Poynting and Grey arrive at the result that the attraction of quartz to quartz at parallel and crossed axes differs less than 1/16500 of the total attraction, and that at parallel axes, when one of the crystals is rotated by 180°, the attraction changes by less than 1/2850 of the total.

15. Dependence on Distance: Astronomical Test (cf. Vol. VI)[2]

S. Newcomb[63] discussed the question of the extent to which the $1/r^2$ in Newton's law is fixed by astronomical data. He reaches the following result:

    a. The agreement between the observed parallax of the Moon and that calculated from the magnitude of g on the Earth surface shows that for values of r that lie between Earth's radius and the radius of the Moon's orbit, the 2 in r^2 is guaranteed up to 1/5000 of its value.

    b. The agreement between the observed perturbation of the Moon by the Sun and the calculation based on Newton's law proves (with about the same accuracy) the validity of r^2 up to distances of the order of magnitude of the Earth's orbit, i.e. approximately up to 24000 times the Earth's radius.

    c. The validity of Newton's law up to the limits of the entire planetary system follows from the validity of Kepler's third law; that is, up to distances which amount to 20 times the Earth orbit's radius. Yet for this range the | accuracy with which $1/r^2$ [42] can be established from observation cannot be stated with certainty.

    One more touchstone of the same question, as Newton[64] already emphasized, is related to the fact that *perihelion motion* of the planets would result from a deviation in the exponent of the distance from 2. While on the one hand such a deviation can

---

60   Earlier Bessel,[57] and more recently von Eötvös,[16] found no difference between crystalline and amorphous bodies in their experiments.

61   *Phys. Rev.* 2 (1895), p. 321–343.

62   *Lond. Trans.* A 192 (1899), p. 245–256.

63   In the paper cited in note 46.

64   *Principia lib. I, sec. IX.*

not be large, because otherwise this would result in perihelion motions which contradict observation, on the other hand the observed anomalous perihelion motions could be rooted in a minor inaccuracy of the gravitational law. Indeed, M. Hall[65] proved that the law previously examined by G. Green,[66] which replaces $1/r^2$ with $1/r^{2+\lambda}$, where λ stands for a small number, is sufficient to explain the anomalous perihelion motion of Mercury, if $\lambda = 16 \cdot 10^{-8}$. This figure for λ would also give the right result for the observed anomalous perihelion motion of Mars, though for Venus and Earth the consequence would be somewhat too large a perihelion motion.[67] However, Newcomb, after discussing the respective conditions, states that Hall's assumption seems to him "provisionally not inadmissible."

16. Dependence on Distance: Experimental Test

This question was examined directly by Mackenzie,[61] by measuring at various distances the attraction of the same bodies with the torsion balance. He found that the discrepancy between the observed result and that calculated from Newton's law is in any case smaller than 1/500 of the total attraction.

From a theoretical perspective, our confidence in the 2 in the exponent of Newton's law stems essentially from the fact that from the standpoint of field theory (section 34) this law alone is compatible with the assumption of a general, source-free distribution of field strength; i.e., the concept of lines of force of the gravitational field is meaningful only if this law is valid precisely.

17. Influence of the Medium on Gravitation

[43]

The analogy of electric and magnetic charges, whose effect depends to a large degree on the medium in which they are contained, makes it seem altogether possible that such an influence is present in gravitation I as well, and that hence the gravitational constant is not universal, as Newton assumed, but rather depends on the medium. Just the relatively good agreement, in spite of the very different *form* of the employed masses, among G-determinations excludes a fairly considerable influence of bodies in the region between the attracting masses.[68] Furthermore, with a torsion balance L.W. Austin and C.B. Thwing[69] directly examined the question of whether a body with a different permeability for gravitation than air exists. Between two bodies attracting each other they inserted plates of various substances whose thickness was 1/3 the distance between the attracting masses. The result was that the difference would have to be smaller than 0.2% of the total attraction.

In another direction, Laplace[70] discussed the question of a possible influence of the medium. He assumes that bodies except air may possess a small absorption coef-

---

65   *Astr. Journ.* 14, p. 45.
66   *Cambr. Trans.* 1835, p. 403.
67   Cf. Newcomb in the paper cited in note 46, p. 109.

ficient α for gravitation so that the gravitational law for two mass elements m_1 and m_2 embedded in such a medium would be:

$$K = G\frac{m_1 \cdot m_2}{r^2} \cdot e^{-\alpha r}.$$

The application of this law to the Sun-Moon-Earth system, however, leads him to the conclusion that the value for Earth (radius R) would have to be:[71]

$$\alpha R < \frac{1}{10^6}.$$

18. Influence of Temperature

Some mechanical theories about the nature of gravitation[72] make it seem quite possible that the gravitational effect is modified by the temperature of I the medium. A [44] direct examination of this question has not yet been carried out; however, von Jolly points out that in his absolute determinations, the temperature difference was maximally 29.6°, without any difference in the results exceeding the magnitude of experimental error.

19. Dependence on Time: Constancy

The tacit assumption of the gravitational effect's independence on time in Newton's law has been challenged in two respects:

a. Is the gravitational constant also a constant with respect to time, or does it change over the course of time?

b. Does gravitation need time to take effect—does it have a finite speed of propagation, or is the gravitational effect instantaneous?

---

68   Wilsing uses long cylinders, Boys and Braun use spheres, Richarz and Krigar-Menzel use cubes, nevertheless good agreement, namely:

Wilsing	$\Delta = 5.577$	Difference 0.9%
Boys and Braun	$\Delta = 5.527$	
Richarz and Krigar-Menzel	$\Delta = 5.505$	" 0.4%.

Cf. in particular note 48.

69   Phys. Rev. 5 (1897), p. 294–300.
70   Méc. Cél. 5, book XVI, chap. IV, §6.
71   Poynting presents an indirect proof against the existence of a specific gravitational permeability: A deflection (refraction) of the gravitational effect has never been observed. However this question appears not to have been carefully examined to date.
72   Cf. part V of this article.

R. Pictet[73] discussed the first question based on the idea that gravitation is caused by impacts of aether particles.[72] His reasoning is the following: The total energy of the solar system consists of two parts: 1) the *vis viva* of planets and Sun; 2) the *vis viva* of aether particles. The *vis viva* of planets varies strongly depending on their momentary position with respect to the Sun. If the total energy of the solar system remains constant, then it follows that the *vis viva* of aether atoms and thereby the gravitational constant have to change with the course of time.

Experiments to prove such a temporal change of the gravitational constant would have a chance to succeed according to R. Pictet and P. Cellérier,[74] since the difference in the *vis viva* of the planets — the decisive ones are Jupiter and Saturn — e.g., between the minimum of year 1898–99 to the maximum of 1916–17, comes to about 18%.

20. Dependence on Time: Finite Speed of Propagation[75]

The second question, whether gravitation operates instantaneously or has a finite speed of propagation, was examined recently in terms of planetary motion by R. Lehmann-Filhès[76] and J. v. Hepperger.[77] |

[45]   Both works introduce a finite propagation velocity in the same way. At the moment when the planet (mass m) is at distance r from the Sun (mass M), the force $G \cdot M \cdot m / r^2$ as per Newton's law is propagated from the Sun with a finite velocity. This force then takes effect on the planet at a time when its distance from the Sun is different from r in direction as well as in magnitude. The same holds for the force that the planet exerts on the Sun.

There is some difference between Lehmann-Filhès and von Hepperger in their equations of motion, as the former takes the Sun's velocity, the latter the velocity of the Sun's and the planet's center of gravity, to be constant.

Both arrive at the result that the most influential change of planetary motion would be a secular change of the mean radius. From this it follows: first, that the introduction of a finite speed propagation *while retaining Newton's law* does not contribute anything to remove the difficulties regarding planetary orbits as presented on p. 86 [p. 36]; and second that the hypothetical propagation velocity would have to be much larger than the velocity of light, because otherwise a secular change of the mean radius would result, by an amount that would contradict observation. If the velocity of the Sun's proper motion lies between 1 and 5 km/sec,[78] then the propaga-

73   *Genève Bibl.* (6 sér., 3 période) 7 (1882), p. 513–521.
74   *Genève Bibl.* (6 sér., 3 période) 7 (1882), p. 522–535.
75   Lecture on this question: S. Oppenheim, *Jahresber. kais. kgl. akad. Gymn. Wien* 1894–1895, p. 3–28; F. Tisserand, *Méc. cél.* 4 (1896), chap. 28; F. Drude, *Ann. Phys. Chem.* 62 (1897).
76   *Astr. Nachr.* 110 (1885), p. 208.
77   *Wien. Ber.* 97 (1888), p. 337–362.
78   According to recent examinations, however, this should be appr. 15 km/sec. (cf. H. C. Vogel, *Astr. Nachr.* 132 (1893), p. 80 f.

tion velocity of gravitation would have to be at least 500 times larger than the velocity of light, according to v. Hepperger.

A stricter test of the assumption of a temporal propagation velocity is provided by its application to the Moon's motion, as carried out by R. Lehmann-Filhès.[79] He draws the conclusion that in order to keep the perturbations of the Moon's radius below an acceptable amount while retaining Newton's law, the propagation velocity of gravitation would have to be given an enormous value, perhaps a million times the velocity of light. Also the sign of the perturbation does not correspond to the discrepancy found between observation and theory for the Moon.

Th. v. Oppolzer[47] comes across similar difficulties when | applying the assumed finite propagation velocity to calculate orbits of comets.   [46]

3. EXTENSION OF NEWTON'S LAW TO MOVING BODIES[80]

21. Transferring Fundamental Electrodynamic Laws to Gravitation

The result of the attempts to introduce a finite propagation of gravitation while retaining Newton's law for moving bodies as well, and thereby to remove the existing disagreements between observation and calculation, must be characterized as rather unsatisfactory. It is therefore small wonder that attempts were made to question the validity of Newton's law for moving bodies, to regard it merely as a special case for bodies at rest, and to replace it with an extended law for moving bodies.

Above all it was examined whether the previously known electrodynamic fundamental laws were sufficient for this purpose.

C. Seegers[81] and G. Holzmüller[82] applied Weber's fundamental law, according to which the potential for two mass elements m_1 and m_2 at a distance r is

$$ P = \frac{G \cdot m_1 \cdot m_2}{r^2}\left[1 - \frac{1}{c^2}\cdot\left(\frac{dr}{dt}\right)^2\right] \qquad (c = \text{velocity of light}), $$

which as is well known Zöllner thought to be the fundamental law of all action-at-a-distance forces, to planetary motion in general, and the planetary motions were calculated numerically by F. Tisserand[83] and H. Servus.[84] For Mercury, the application of Weber's law results in an anomalous secular perihelion motion of ca. 14".

---

79 Münch. Ber. 25 (1896), p. 371.
80 Reviews of a part of the work in this area in S. Oppenheim,[77] P. Drude,[77] and F. Tisserand.[77]
81 Diss. Göttingen 1804.
82 Zeitschr. Math. Phys. 1870, p. 69–91.
83 Paris, C. R. 75 (1872), p. 760 and 110 (1890), p. 313.
84 Diss. Halle 1885. F. Zöllner cites (based on correspondence) that W. Schreibner calculated a secular perihelion motion of 6.7" for Mercury based on Weber's law. The reason for this figure's two-fold deviation from the figures cited in the text is that Schreibner equates the constant c in Weber's law with $\sqrt{2}$ times the velocity of light.

Transferring Gauss'[85] fundamental electrodynamic law to gravitation, in the sense that one introduces an attractive force K between two mass elements with co-ordinates x_1, y_1, z_1 and x_2, y_2, z_2, given by: |

[47]
$$K = \frac{G \cdot m_1 m_2}{r^2} \left\{ 1 + \frac{2}{c^2} \left[\left(\frac{d(x_1 - x_2)}{dt} \right)^2 + \left(\frac{d(y_1 - y_2)}{dt} \right)^2 + \left(\frac{d(z_1 - z_2)}{dt} \right)^2 - \frac{3}{2} \left(\frac{dr}{dt} \right)^2 \right] \right\},$$

gives a secular perihelion motion of Mercury of only 28", according to F. Tisserand's[86] calculation.

Riemann's[87] fundamental law

$$P = \frac{G \cdot m_1 \cdot m_2}{r} \left\{ 1 - \frac{1}{c^2} \left[\left(\frac{dx}{dt} \right)^2 + \left(\frac{dy}{dt} \right)^2 + \left(\frac{dz}{dt} \right)^2 \right] \right\} \qquad \begin{array}{l} (x, y, z \text{ are co-ordinates} \\ \text{of } m_1 \text{ relative to } m_2), \end{array}$$

would imply, according to M. Lévy,[88] twice the perihelion motion of Mercury that follows from Weber's law.

Therefore, Levy suggested a combination of Riemann's and Weber's laws in the form:

$$P = P_{Weber} + \alpha (P_{Riemann} - P_{Weber})$$

$$= \frac{G \cdot m_1 \cdot m_2}{r} \left\{ 1 - \frac{1}{c^2} \left[(1 - \alpha) \left(\frac{dr}{dt} \right)^2 + \alpha \left(\left(\frac{dx}{dt} \right)^2 + \left(\frac{dy}{dt} \right)^2 + \left(\frac{dz}{dt} \right)^2 \right) \right] \right\}$$

where α was then to be determined from the observed secular perihelion motion of Mercury. Assuming the perihelion motion of 38" as observed and 14.4" [89] as given by Weber's law, one finds $\alpha = 1.64 = $ approx. $5/3$.[88] On the basis of a perihelion motion of Mercury of 41.25", as given by other observers, and a motion given by Weber's law of 13.65",[89] α becomes $= 2.02$.

The law one obtains in this way has the decisive advantage of matching the achievement of Riemann's and Weber's laws in electrodynamics, and moreover it represents an extension of Newton's law to moving bodies that eliminates the worst disagreement between observation and calculation that has persisted until now.

---

85 *Ges. Werke* 5, p. 616 f., Nachlass.
86 Paris, *C.R.* 110 (1890), p. 313.
87 *Schwere, Elektrizität und Magnetismus*, ed. Hattendorf, Hannover 1896, p. 313 ff.
88 Paris, *C. R.* 110 (1890), p. 545–551. For motion of two masses the law was earlier discussed in general by O. Limann (Diss. Halle 1886).
89 Tisserand[85] (Paris, *C. R.* 75), and Servus.[86]

22. Transferring Lorentz's Fundamental Electromagnetic Equations to Gravitation

H.A. Lorentz[90] has attempted to use Maxwell's equations,[91] as extended by him to moving bodies, for gravitation. His conception | of the constitution of gravitating molecules is essentially in agreement with that of F. Zöllner, though in slightly modernized form. The foundation of Lorentz's approach is covered in section 36. [48]

The additional forces Lorentz obtains, apart from the ones given by Newton's law, have a factor of either $(p/c)^2$ or $(p \cdot w)/c^2$, where p is the velocity of the central body taken to be constant, w is the velocity of the planet relative to the central body, and c is the velocity of light. These additional forces are so small that they will probably be beyond observation in all cases; in the case of Mercury they are certainly below what is observable, as shown by Lorentz's calculation. It follows that Lorentz's equations, combined with Zöllner's conception of the nature of gravitating molecules, can be applied to gravitation,[92] but they do not contribute to removing existing disagreements between observation and calculation.

23. Laplace's Assumption

Previously Laplace[93] envisaged an extension of Newton's law for moving bodies in quite a different way. He seems to imagine the force coming from an attracting body m_1 as a sort of wave, which exerts an attractive force on each body m_2 it encounters of magnitude $G \cdot m_1 m_2 / r^2$ in the direction in which it propagates. The effect such a wave has on a moving body m_2 depends only on the relative motion of wave and body. One can thus imagine body m_2 at rest in space, if one ascribes to the wave another velocity component apart from its velocity in the r-direction, equal and opposite to the velocity of m_2. If $v =$ velocity of m_2 and c indicates the propagation velocity of gravitation, then the body m_2 receives a force component opposite to its orbit's direction and of value $(m_1 m_2 / r^2) \cdot (v/c)$,[94] rather than receiving only a force component in the r-direction. |

Following through with this point of view gives little satisfaction with respect to the planets: it does not result in a perihelion motion at all, but in a secular change of the mean radius; e.g., this change for the Moon has a value such that the lowest limit for c would have to be about 100,000,000 times the velocity of light. It is, however, not uninteresting that Laplace's conception achieves the same effect as a *resistance of the medium* proportional to the velocity of the planet. [49]

---

90 *Amsterdam Versl.*, April 1900.
91 Harlem, *Arch. Néerl.* 25 (1892), p. 363.
92 This includes the possibility that *the propagation velocity of gravitation is equal to the velocity of light*.
93 *Méc. cél.* 4, book X, chap. VII, §19 and 22.
94 These conditions would therefore correspond completely with those for the aberration of light.

According to Encke[95] and v. Oppolzer,[96] a resistance of the medium—however, proportional to the square of the velocity—could perhaps explain the irregularities of Encke's comet presented on p. 36. The anomalies of Winnecke's comet, presupposed by Oppolzer and explained the very same way, have since been shown to be non-existent by E. v. Haerdtl's[97] calculations.

24. Gerber's Assumption

P. Gerber's[98] two premises are:

a. The potential P transmitted from a mass μ to a second one m is $\frac{\mu}{r}$, where r is the distance from μ to m *at the moment of transmitting* the potential. This potential propagates with finite velocity c.

b. A certain time is necessary for the potential "to reach m, to impart itself to the mass; i.e., to evoke in m the state of motion corresponding to the potential." "If the masses are at rest, the motion of the potential passes m with its own velocity; then its value transmitted to m is in inverse proportion to the distance. If the masses speed towards each other, the time of transmission as well as the transmitted potential decrease proportionally to the ratio of the characteristic velocity of the potential to the sum of its and the masses' velocity, as the potential has this total velocity relative to m."

Gerber arrives at the value that the potential must have under these assumptions in the following manner:

[50] "The potential moves with the velocity of the attracting mass in addition to its own velocity c. The space | $r - \Delta r$ [99] traversed in time Δt by the two motions, one of the potential and the other one of the attracted mass, is thus

$$\Delta t\left(c - \frac{\Delta r}{\Delta t}\right),$$

while $r = c\Delta t$. So for the distance where the potential starts developing and to which it is in inverse proportion, one obtains

$$r - \Delta r = r\left(1 - \frac{1}{c}\frac{\Delta r}{\Delta t}\right).$$

Since, moreover, the velocity with which the motions pass each other has the value

$$c - \frac{\Delta r}{\Delta t},$$

---

95   Cited by von Oppolzer.
96   *Astr. Nachr.* 97, p. 150–154 and 228–235.
97   *Wien. Denkschr.* 56 (1889), p. 179 f.
98   *Zeitschr. Math. Phys.* 43 (1898), p. 93–104.
99   $\Delta r > 0$ for increasing r.

the potential turns out also to be proportional to

$$\frac{c}{c - \dfrac{\Delta r}{\Delta t}}.$$

due to the time consumed to impart itself to m. Thus one finds:

$$P = \frac{\mu}{r\left(1 - \dfrac{1}{c}\dfrac{\Delta r}{\Delta t}\right)^2}.$$

As long as the distance Δr is short and therefore $\Delta r / \Delta t$ small compared to c, one may replace the latter by dr / dt. So it becomes

$$P = \frac{\mu}{r\left(1 - \dfrac{1}{c}\dfrac{dr}{dt}\right)^2},$$

from which it follows with help of the binomial law to the second power:

$$P = \frac{\mu}{r}\left[1 + \frac{2}{c}\frac{dr}{dt} + \frac{3}{c^2}\left(\frac{dr}{dt}\right)^2\right].”$$

The application of this equation to planetary motions yields the following remarkable result: If the propagation velocity c is determined from Mercury's observed perihelion motion, then one finds $c = 305.500$ km/sec, which is the velocity of light with surprising accuracy. In other words, *if in Gerber's equation the velocity of light replaces the propagation velocity of gravitation, then this equation yields exactly the observed anomalous perihelion motion of Mercury.*

No difficulties for the other planets follow from Gerber's assumption, | except for Venus, where Gerber's approach gives a slightly too large secular perihelion motion of 8″. [51]

Gerber's assumption thus shows, as does Lévy's, that a propagation velocity of gravitation of the same magnitude as the velocity of light is not only possible, but can also serve to eliminate the worst disagreement that has existed between astronomical observation and calculation so far. To be sure, this was achieved only by confining the validity of Newton's Law to bodies at rest and postulating an extended law for moving bodies.

4. EXTENSION OF NEWTON'S LAW TO INFINITELY LARGE MASSES

25. Difficulty with Newton's Law for Infinitely Large Masses

Doubts have been expressed concerning the universal validity of Newton's law leading in quite a different direction, and the necessity of an extension has been considered.

In case the universe contains infinitely many masses, to obtain the force acting at any point one would strictly speaking have to solve the problem: to specify the effect of infinitely many masses of finite size at one particular point.

C. Neumann[100] was probably the first to point out that in this case the forces resulting from Newton's law may become indefinite. H. Seeliger[101] examined this question in a more general way, and showed that for infinite masses Newton's law can produce infinitely large forces as well as leaving them completely indefinite.

26. Elimination of the Difficulty by Altering the Law of Attraction

Seeliger suggests a slight modification of Newton's law in order to eliminate this difficulty, and he discusses various possibilities.

The form already discussed by Laplace

$$K = \frac{G \cdot m_1 \cdot m_2}{r^2} \cdot e^{-\alpha r}$$

[52] l is physically expected to suffice for the above purpose, as it corresponds to the assumption of absorption by the medium. In fact, it does suffice, and would moreover have the advantage of giving planetary perihelion motion. Yet the value of $\alpha = 0.00000038$ taken from Mercury's observed perihelion motion gives perihelion motions for the other planets which are difficult to reconcile with observations.[102]

The laws discussed by C. Neumann, according to which the potential P takes the form

$$P = G \cdot m_1 m_2 \left(\frac{A e^{-\alpha r}}{r} + \frac{B e^{-\beta r}}{r} + \ldots \right),$$

serve the same purpose, but the resulting perihelion motions of the planets stand in severe contradiction to observation.

In contrast, Green-Hall's law discussed earlier,

---

100 *Leipz. Abh.* 1874.
101 *Astr. Nachr.* 137 (1895), p. 129–136; *Münchn. Ber.* 26 (1896), p. 373–400. Controversy between
    J. Wilsing and H. Seeliger about this issue, *Astr. Nachr.* 137 and 138.
102 *Münchn. Ber.* 26 (1896), p. 388.

$$P = \frac{G \cdot m_1 m_2}{r^{1+\lambda}},$$

which would be suitable to account for the perihelion motions of the planets, retains the same problems as Newton's law with regards to infinite masses.

27. Elimination of the Difficulty by Introducing Negative Masses

A. Föppl[103] introduced the idea that the difficulty of Newton's law emphasized by Neumann and Seeliger is to be eliminated by introducing "negative masses" and maintaining the law, rather than by altering the form of the law. As with the gravitational force lines emitted by the familiar positive masses, there would be force lines flowing into negative masses. If the sum of the negative masses is taken to equal that of the positive ones, the total would be 0; as in the electric and the magnetic domain there would be the same number of sources and sinks.

With this assumption, the expression for field energy cannot be based upon the usual one,

$$\frac{1}{2} a |\Re|^2 dS,$$

where a is a constant of the medium, $\Re$ the vector of field strength defining the gravitational field, $|\Re|$ its I absolute value, and dS a volume element. Rather, as Maxwell already pointed out, one must replace this expression by   [53]

$$\left(C - \frac{1}{2} a |\Re|^2 \right) dS \qquad (C = \text{constant})$$

to obtain an attraction between masses of the same sign. Cf. section 34 regarding the significance of the constant C.

Prior to Föppl, C. Pearson[122] had already suggested the mere introduction of negative masses of the same magnitude as the familiar positive masses. This suggestion is actually a consequence of his theory, which attempts to derive electrical, optical, chemical and gravitational phenomena from suitably chosen aether motions.

The introduction of negative masses hardly causes any problems. For the fact that repulsion between two masses has never been observed, i.e. a negative mass has never been noticed, points to the possibility—though not to the necessity—that such masses were driven to spaces no longer accessible to observation due to the repulsion from positive masses in our system. On the other hand, according to A. Schuster,[104] who had the same thought (though merely in a "holiday dream"), the introduction of negative masses could perhaps serve to shed completely new light upon several phenomena, such as comet tails.

---

103 *Münchn. Ber.* 27 (1897), p. 93–99.
104 *Nature* 58 (1898), p. 367 and 618.

5. ATTEMPTS TO EXPLAIN GRAVITATION THROUGH MECHANICS[105]

28. Pressure Differences and Currents in the Aether[106]

The conjecture that gravitation could be caused by *pressure differences* in the supposedly homogeneous aether surrounding gravitating | masses stems from Newton[107] himself. According to him the aether would become denser the further it is from masses. Since each body has the tendency—later on he speaks of an elastic force of the medium—to go from the denser parts of the medium to the less dense ones, each of the two bodies must move in the direction of the other.

[54] Similar ideas have been worked out by Ph. Villemot,[108] L. Euler,[109] J. Herapath[110] and in a slightly different way by J. Odstrčil.[110]

A consequence of the assumption of pressure differences in the aether, combined with the idea that aether behaves like a fluid or a gas, is that *aether currents* must flow into the atoms.[3] According to J. Bernoulli,[108] B. Riemann,[111] and J. Yarkovski,[112] it is these aether currents which carry the body along and hence cause gravitation. G. Helm,[150] as well as C. Pearson,[113] arrived at a similar conception while trying "to explain gravitation with energy transfer in the aether."

Yarkovski pondered the question of the *cause* of the aether currents, but produced an explanation that is physically not tenable.

Among the many objections raised against these theories, there is also the question of what happens to the aether that flows into the atoms. There are only two possible answers: either the aether accumulates or it disappears inside them. Bernoulli, Helm, Yarkowski have decided for the former possibility; Riemann for the latter, who allows matter in ponderable bodies constantly to make a transition "from the physical world into the spiritual world."

---

105 Review articles: W.B. Taylor, *Smithson. Inst. Rep. for 1876* (1877), p. 205–282: Detailed discussion of papers up to 1873. C. Isenkrahe, a) *Isaac Newton und die Gegner seiner Gravitationstheorie* etc., *Progr. Gymn. Crefeld*, 1877–1878. b) *das Rätsel von der Schwerkraft*, Braunschweig 1879. c) *Zeitschr. Math. Phys.* 37, Suppl. (1892), p. 161–204; P. Drude[77]; partly also H. Gellenthin, "Bemerkungen über neuere Versuche, die Gravitation zu erklären etc.", *Progr. Realgymn. Stettin* 1884 and Gehler,[2] Articles: *Anziehung, Materie*.

106 The term "aether" is not always used with the same meaning in the following text; also in the original papers it is not always sufficiently defined. What is meant roughly in each case, is given by the context.

107 According to W. B. Taylor,[107] Newton expressed this view in a letter and repeated it in his *Optice*.

108 Cf. Taylor.[107]

109 Cf. Taylor[107] and especially Isenkrahe.[107]

110 *Wien. Ber.* 89 (1884), p. 485–491.

111 *Ges. Werke*, 2nd ed. 1853, p. 529.

112 *Hypothèse cinétique de la gravitation universelle* etc. Moscou 1888.

113 *Amer. J. of math.* 13 (1898), p. 419.

29. Aether Vibrations

The idea that aether vibrations in the form of longitudinal waves may cause not only the phenomena of light and heat, but also gravitation, has been developed in two directions. |

1. According to the first view the attracting body, or its atoms, are supposed to [55] vibrate themselves; these vibrations pass on to the aether, propagate to the attracted body and cause its approach.

Hooke,[114] Newton's inventive rival, already expressed this conception, which was taken up again by J. Guyot and F. Guthrie. The latter two seem to have arrived at this through the observation that light objects close to a vibrating body are pushed towards it. However, the fact that the approach takes place only under very particular conditions, and that under different conditions one observes an apparent repulsion— such was also included by F.A.E. and E. Keller[115] for explaining gravitation—proves that the assumption of an elastic aether and vibrating atoms does not suffice to explain gravitation. There must be at least one more assumption which produces conditions that guarantee an *attraction* under all circumstances.

To find these conditions, J. Callis[116] examined the following question analytically and in detail: What effect do longitudinal waves in a fluid whose pressure changes proportionally to changes in density, have on small inelastic, smooth spheres embedded in the elastic fluid medium? He reaches the conclusion that if the wave length is large compared to the spheres' radius, the spheres are then pushed towards the center of the spherical wave. For explaining gravitation, one would thus need to assume that there are vibrations whose wave lengths in the aether are large compared to the dimensions of the gravitating atoms.

A deficiency of this treatment is the requirement that only the attracting body emits waves. Such a principled differentiation between attracting and attracted body is incompatible with the nature of gravitation. The question must not be what effect do spherical waves have on bodies at rest, but rather what effect do they have on a body which is itself vibrating.

This complete problem was probably first handled mathematically by C.A. | Bjerknes,[117] for the case of an incompressible aether and pure pulsation of the [56] spheres (atoms). He proved that two pulsating spheres, whose radius is small compared to their separation, show an apparent attraction, and that this attraction is proportional to the intensity of pulsation and inversely proportional to the square of distance, if their pulsations agree in frequency and phase. If gravitation is to be attributed to pulsating atoms and molecules,[2] then at least the following additional assumptions are needed:

---

114 Cf. W.B. Taylor[105] and F. Rosenberger.|

115 Paris, *C. R.* 56 (1863), p. 530–533; also cf. Taylor.[105]

116 E.g. *Phil. Mag.* (4) 18 (1859), p. 321–334 and 442–451, cf. Taylor[105] about other work by Callis.

117 Cf. the compilation in V. Bjerknes, "Vorlesungen über hydrodynamische Fernkräfte nach C.A. Bjerknes' Theorie", Leipzig 1900.

a. The pulsations of all atoms or molecules must agree in frequency and phase.
b. The intensity of pulsation must be proportional to the mass.
There is one more thing. A.H. Lealy[118] pointed out that for the case of a *compressible* fluid, the effect of two spheres pulsating with equal phase and frequency reverses its sign if the distance between them exceeds half a wave length. So if one wants to use Bjerknes' results for gravitation, one would have to suppose either that the aether is *completely* incompressible (Bjerknes) or that it has such low compressibility that half the wave length of aether vibrations is larger than the distances for which observations have established the validity of Newton's law (A. Korn[119]). Only then is attraction always guaranteed, in agreement with observation.

Bjerknes' conception received further development by C. Pearson[120] and in the work by A. Korn just mentioned. The latter extended these ideas mainly to electromagnetic phenomena, the former to phenomena of optics and molecular physics, assuming complicated modes of vibration of the atoms. In his last paper, Pearson abandoned the assumption of oscillations for gravitation and only retained this [57] assumption for optics and molecular physics, while replacing the I pulsating atoms by places in the incompressible aether at which aether continuously flows in and out in an oscillatory manner ("aether squirts"). For gravitation, he assumes that there is a constant flow in addition to the oscillating one at the locations concerned. With this requirement, the assumed incompressibility of the aether leads directly to the conclusion that apart from places of emission (source points, ordinary masses) there must be just as many places of absorption (sink points, "negative masses").[121]

Using Bjerknes' results for explaining gravitation suffers from the obvious deficiency that assumptions are required which would first have to be explained themselves. Attempts to supply real reasons have been made for only *one* of these assumptions, the synchronous pulsation of atoms. J.H. Weber[122] points out that in the attempt to demonstrate Bjerknes' results the synchronization of the two spheres happens quickly "on its own"; i.e., due to the forces which are caused by the vibrations in the fluid, even if the pulsations were not synchronous at first. From this he concludes that if atoms pulsate at all, the pulsations should become synchronous "on their own" (as specified above).

According to Korn, the assumption of synchronous pulsation can be replaced by another one, which is that the whole solar system is exposed to a periodic pressure. This assumption may be preferred due to its simplicity, but this is the only advantage over Bjerknes's assumption.

---

118 *Cambr. Trans.* 14 (1) (1885), p. 45, 188.
119 "Eine Theorie der Gravitation und der elektrischen Erscheinungen auf Grundlage der Hydrodynamik", 2nd ed., Berlin 1898.
120 *Quart. J.* 20 (1883), p. 60, 184; *Cambr. Trans.* 14 (1889), p. 71 ff.; *Lond. math. Proc.* 20 (1888–1889), p. 38–63; *Amer. J. of math.* 13 (1898).
121 See section 27 of this article.
122 *Prometheus* 9 (1898), p. 241-244, 257–262.

2. The second class of attempts to base an explanation of gravitation on aether vibrations assumes that the atoms are not themselves vibrating, but that their activity consists only in a kind of shielding or absorption of aether vibrations.

Representatives of this view include F. and E. Keller,[115] Lecoq de Bois-baudran,[123] and, in a slightly different way, N. von Dellinghausen.[124]

30. Aether Impacts: The Original Ideas of Le Sage

The starting point of all aether impact theories is an idea which I Le Sage[125] devel-   [58]
oped in a particularly clear and skillful way. According to him the gravitational aether surrounding the atoms of a body consists of discrete particles—"corpuscules ultra-mondains"—which zoom about in all directions with the same extraordinarily high velocity. No continuous motion is imparted to a single atom embedded in this aether due to the impacts of these aether particles, since the effect of aether impacts from all directions cancels out. But if two atoms A_1 and A_2 are brought into this aether, the conditions change in two respects:

a. A_2 shields A_1 from a part of the aether atoms: The side of A_1 turned towards A_2 is hit by fewer aether particles than its side turned away from A_2. The consequence would have to be that A_1 is driven towards A_2 by the action of the aether impacts, and conversely A_2 is driven towards A_1.

If the atoms are assumed to be very large in comparison to the aether particles, it follows directly that this *shielding effect* of one atom of a body on another decreases with the square of distance. To make the shielding proportional to mass, Le Sage introduces the assumption that the gravitating masses are extraordinarily porous[126] to the aether particles so that the efficacy of the whole body becomes proportional to the number of atoms it contains.[127]

b. Due to the *reflection* of aether particles on A_2, a number of aether particles also hit the atom A_1 that would not have hit A_1 without the presence of A_2.[128] If these reflected aether atoms had the same velocity as those hitting A_1 directly, then they would cancel out the approach of A_1 towards A_2 caused by the shielding effect of A_2; thus, gravitation would not be produced.

---

123 Cf. Paris, *C. R.* 69 (1869), p. 703–705; cf. Taylor.[107]

124 "Die Schwere oder das Wirksamwerden der potentiellen Energie," *Kosmos* 1, Stuttgart 1884. Cf. C. Isenkrahe.[107]

125 *Berlin Mém.* 1782 and in P. Prévost, *Deux traîtés de Physique mécanique*, Paris 1818. In the last paper it is quoted that similar theories have been established before (by Nicolas Fatio and F. A. Redecker).

126 Strangely enough, Le Sage extends the assumption of very high porosity to every single atom of a body and therefore arrives at the conception of the peculiar "box atoms" [*Kastenatome*].

127 But cf. section 32, c).

128 In P. Drude[77] we find the note that Le Sage simply ignores reflection and thus his observation lacks rigor. This is probably a mistake: Le Sage devotes chapter IV to reflection in P. Prévost.

This is why Le Sage assumes furthermore that the aether particles are *absolutely*
[59] *inelastic*—"privé de toute élasticité"—and I states that under this assumption the
average velocity of the reflected atoms = $2/3$ of the non-reflected ones.[129]

Thus the difference of effects a) and b) still results in both atoms approaching
each other.

31. Aether Impacts: Further Development of Le Sage's Theory

Recently, Le Sage's theory was primarily defended by C. Isenkrahe, who emphasized
in particular the assumption that collisions between the aether particles and atoms are
subject to the laws of *inelastic* collision. Isenkrahe's progress beyond Le Sage consists
of the following points:

a. He ascribes to gravitational aether the properties of a gas in the sense of kinetic
gas theory. Hence he is giving up the assumption of *equal* velocity[130] of the aether
atoms.

b. He does not explain the porosity of the body to aether particles by porosity of
the atoms themselves, but by assuming that the distance between atoms[131] is large
compared to their dimensions.

c. To achieve the proportionality between attraction and mass, which was guaranteed by Le Sage's assumption only for bodies of the same composition, he assumes
that "the final components of matter are all of equal size; they may be the aether
atoms themselves."

A. Rysáneck's[132] assumptions are very similar. His achievement consists of the
precise implementation of the ideas of kinetic gas theory.[133] In his calculations he
actually takes into consideration that the velocities of aether atoms are distributed
according to Maxwell's law, whereas e.g. Isenkrahe does assume different velocities
of aether atoms, but replaces them by *one* average velocity in all his derivations.

[60] Prior to Isenkrahe S.T. Preston[134] I already pointed out that Le Sage's conceptions
could be suitably replaced by ideas of kinetic gas theory, if the mean free path of
aether atoms is assumed to be of the order of planetary distances. He developed this
idea in several papers, though without going into details as carefully as Isenkrahe and
Rysáneck.

---

129 About justification and validity of this information cf. C. Isenkrahe[107] in paper b., p. 155 ff.
130 Which Le Sage also chose merely for simplicity, as he explicitly points out the *different* velocities in
     the reflections of aether particles and atoms.
131 Which are assigned a spherical form for simplicity.
132 *Repert. Exp.-Phys.* 24 (1887), p. 90–115.
133 But cf. section 33.
134 *Phil. Mag.* (5) 4 (1877); *Wien Ber.* 87 (1882); *Phil. Mag.* (5) 11 (1894); *Diss. München* 1894.

32. Aether Impacts: Difficulties of these Theories

a. A necessary condition for a gravitational effect is that aether atoms lose translational velocity upon collision with atoms, which is achieved most easily by assuming inelastic collisions.

However, this assumption leads to the problem of where the energy lost at impact goes. P. Leray[135] and later P.A. Secchi,[136] W. Thomson,[137] S.T. Preston,[134] then A. Vaschy,[138] Isenkrahe himself and Rysáneck tried to avoid this problem in many different ways. None of these attempts, however, is itself unobjectionable.[139]

b. J. Croll[140] turns against the assumption made in most aether impact theories, which is that the distance between two molecules is very large compared to their dimensions, or rather compared to their spheres of action. He notes that this assumption grossly contradicts W. Thomson's estimates regarding the size of the molecules and their number per unit volume.

c. Objections can be raised from a different angle against the assumption of high porosity of the body for the aether atoms. If the porosity is presumed to be so large that the aether atoms that passed one layer of a body hit the next layer with completely undiminished velocity, then the proportionality between attraction and mass would be strictly preserved. At the same time, this requirement excludes any attraction at all. Hence, one must assume that the aether atoms forfeit a noticeable amount of their energy when passing a body layer. A.M. Bock[141] has shown that this assumption I is not incompatible with the required strict proportionality between attraction [61] and mass.

d. Bock pointed out one more problem. If a third mass comes between two masses, then the attraction of the two masses is considerably modified, such that the third mass seems to have a larger permeability, as shown by a mathematical examination of this case based on aether impact theories. Because this case is not rare, e.g., for the Moon, Earth and Sun, there would have to be observable perturbations over the course of time. But in fact no perturbations of this type have ever been observed.

e. Le Sage already discussed another objection against aether impact theories. If any body, e.g. a planet, moves in an aether with the assumed properties, then it must experience resistance. But none has been observed for the planets.

The last question was examined more precisely by Rysáneck, Bock and W. Browne[142] on the basis of astronomical data.[143] Since the secular changes of

135 Paris, *C.R.* 69 (1869), p. 615–621; also cf. Taylor.
136 Cited in Isenkrahe.[107b]
137 *Phil. Mag.* (4) 45 (1871), p. 321–332.
138 *J. de Phys.* (2) 5 (1886), p. 165–172.
139 Cf. C. Isenkrahe[107b]; Maxwell, *Encycl. Brit.*, 9th ed. Article: *Atom und Scient. Pap.* 2, p. 445, Cambridge 1890.
140 *Phil. Mag.* (5) 5 (1877), p. 45–46.
141 Diss. München 1891. Isenkrahe[107b] has already examined this question, though not fully.
142 *Phil. Mag.* (5) 10 (1894), p. 437–445.
143 Cf. also section 23.

planetary orbits give an upper limit on this hypothetical resistance, the aether impact theories yield a lower limit on the velocity of the aether atoms, if their density is assumed to be known. By using a density of the same order of magnitude as has been estimated for the optical aether, one obtains enormous numbers as the lower limit on the mean velocity. Rysáneck, e.g. based on calculations on Neptune's orbit, obtained the number $5 \cdot 10^{19}$ cm/sec.

f. Of all the objections that P. du Bois-Reymond[144] raised against the aether impact theories, one is particularly noteworthy.

Think of a ponderable truncated cone (cross-section $ABCD$) with a molecule α close to the top. According to the aether impact theories, the acceleration which α receives towards the cone is the difference between the effect the aether atoms with solid angle ω_1 and the effect the aether atoms with solid angle ω_2 have on the molecule. The first effect remains unaltered, but the second decreases if R, the distance between base CD and cone top O, increases. I Therefore, the total effect always remains smaller than the effect of the aether atoms of solid angle ω_1.

[62]

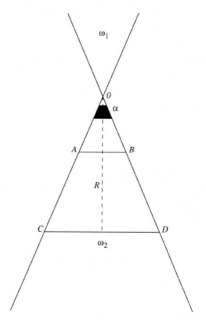

Figure 1

On the other hand, since according to Newton's law the attraction of the cone on α increases with R and exceeds any specifiable number, if the same is assumed about R, there are then only two possibilities: either to presume that the effect of

---

144 *Naturw. Rundschau* 3 (1888), p. 169–178.

aether atoms in space ω_1 on molecule α is infinitely large, or to assume that Newton's law no longer holds for infinitely extended masses.[145]

Isenkrahe[146] countered P. du Bois-Reymond's objection with the latter assumption. However, the difficulty persists, in that one must ascribe an enormous, if not infinite, magnitude to the aether atom's effect, which leads to shortcomings in other areas.[147]

33. Aether Impacts: Jarolimek's Objections and Theory

A. Jarolimek[148] emphasized a deficiency of all aether impact theories that suppose aether to be a gas in the sense of kinetic gas theory. The derivation of the law of gravity in these theories is based on a simple calculation with a certain mean free path of the aether atoms, which fails to take the variety of paths into account.

Regarding this point, Jarolimek notes that *in order to produce mutual attraction of two molecules, only those aether atoms whose actual path length is larger than the distance between the two molecules can be effective.* Thus it depends precisely upon the *absolute* |—*and not the average*—path length. But if one takes the variety of   [63] absolute path lengths into account, under the usual conditions of aether impact theory one does not obtain Newton's law at all.

Referring to Isenkrahe's[149] assumption that the atoms of a body may themselves be an aggregate of the extremely fine aether atoms, Jarolimek points out one more difficulty: this assumption contradicts a shielding effect of two body elements that decreases with the square of the distance. If these elements are identical to aether atoms, then an element could shield another only from those aether atoms whose center lies exactly on the line connecting the two elements; the shielding effect would therefore not depend on the distance at all, if the distance is so large compared to the element's radius that the element could be regarded as devoid of size.

Jarolimek establishes the following theory based on such considerations. He keeps Isenkrahe's assumption—the ultimate elements of atoms are identical to gravitational aether atoms. This practically fully frees him from a shielding effect. He arrives at a decrease of the gravitational effect with the square of the distance in the following way: "One has to think of the infinite number of *swarming* aether atoms as uniformly distributed in space at every instant, and one has to imagine that atoms *bouncing off* from one point fly off in straight lines in all directions. Considering a cone bundle, whose vertex is the point of origin and whose cross-section accordingly increases in quadratic proportion with distance to the vertex and which therefore contains at increasing distance *more uniformly distributed aether atoms* in quadratic proportion, one must realize that for atoms that bounced off (of which a *definite* number

145 Cf. section 4.
146 In the book: *Über die Fernkraft und das durch P. du Bois-Reymond aufgestellte* etc., Leipzig 1889.
147 The field representation leads to a similar problem (see section 34).
148 Wien. Ber. 88^2 (1883), p. 897–911.
149 Cf. section 31.

passes the observed cone bundle from the vertex), the probability to hit another atom in space must increase in quadratic proportion to the distance between the two.

From this it follows directly that the number of freely and linearly moving atoms decreases with increasing distance in quadratic proportion, or in other words: *that the aether contains n^2 times as many atoms with free path r as with free path nr.*" Con-

[64]   sequently, "*the simplest explanation for the law of gravity | is provided by the inequality of the path lengths of the aether molecules.*"

6. REDUCTION OF GRAVITATION TO ELECTROMAGNETIC PHENOMENA

34. Gravitation as a Field Effect

Before reporting on explanatory attempts based on electromagnetism, the empirical facts contained in Newton's law will be mathematically formulated by describing the "gravitational field" without reference to any particular conception of its nature.[150]

One is used to regarding Newton's law as the most distinguished example of an action at a distance. On the other hand, it must be emphasized that its content can be formulated by the following statement, which corresponds to the field theory standpoint: "*The field strength of gravitation is irrotational, and source-free in those regions of space where there are no masses. Where there are masses, the divergence of the field strength is proportional to the local mass density* ρ."

We understand by field strength the attractive force exerted on a *unit mass*; the force exerted on *mass* m_1 is m_1-times as large as the field strength. The proportionality factor for the divergence of the field strength is identical to $4\pi G$. In formulas,[151] the expression of our description of the gravitational field takes the following form, if $\Re$ stands for the vector of field strength:

$$\text{curl}\,\Re = 0, \qquad \text{div}\,\Re = 0 \quad \text{or} \quad = -4\pi G\rho.$$

This formulation and the classical formulation given in section 1 are mathematically exactly equivalent; in particular, it follows from the above differential equations according to the laws of potential theory that the field strength due to a single mass m_2 at distance r is calculated to be

$$\Re = \text{grad}\frac{m_2 G}{r}.$$

This yields the following value of the field strength (or its magnitude in the r-direction) in agreement with Newton's law:

---

150 Field representations of particular kinds are given by G. Helm, *Ann. Phys. Chem.* 14 (1881), p. 149; O. Heaviside, *Electrician* 31 (1893), p. 281 and 359.

151 Because of the significance of vector symbols rot, div, grad, cf. the beginning of the 2nd semi-volume V of the encyclopedia.

$$|\mathfrak{R}| = \frac{d}{dr}\frac{m_2 G}{r}.$$

| So far the field interpretation of gravitation offers neither advantages nor disad-   [65]
vantages over the action-at-a-distance interpretation. An advantage of the former
would arise if a finite propagation velocity of gravitational effects could be proven
with certainty, especially if it turned out to be equal to the velocity of light. Then the
above differential equations for stationary gravitational effects would have to be
extended to the case of a time-varying gravitational effect, which could easily be pat-
terned after the example of the electromagnetic equations. On the other hand, the
field interpretation involves a serious problem which Maxwell[152] pointed out.
Enquiring about the gravitational energy contained in a volume element dS of the
field, one must, in order to get an *attraction* of masses of the same sign, assume the
form

$$\left(C - \frac{1}{2}a|\mathfrak{R}|^2\right)dS$$

for this energy, where the constant a is identical to $1/4\pi G$. The constant C would
have to be larger than $(a/2)|\mathfrak{R}'|^2$, so that the gravitational energy has a positive
value throughout, where $|\mathfrak{R}'|$ stands for the largest value of field strength at any point
in space. However, it would follow from this that at points of vanishing field strength,
e.g. between the Earth and Sun at the point where the Sun's and the Earth's attraction
compensate each other, the energy content of space would have to have the enormous
quantity C per unit volume. Maxwell adds that he can not possibly imagine a
medium with such a property.

35. Electromagnetic Vibrations

The hypothesis that gravitation could be caused by aether vibrations, already discussed in section 29, was examined by H.A. Lorentz[92] under the following assumptions:

    a. The gravitating molecules consist of ions possessing an electric charge.

    b. The aether vibrations are electromagnetic vibrations whose wave length is
small compared to all those distances over which Newton's law is still valid.

    Lorentz arrives at this result: an attraction is possible under these conditions only
if | electromagnetic energy flows continuously into the volume elements that contain   [66]
gravitating molecules. If the assumptions are changed so that such a disappearance of
electromagnetic energy is avoided, then no attractive forces are obtained. This is why
Lorentz himself dismisses this theory and in the further course of his discussion joins
Mossotti-Zöllner's conception (see below).

---

152 *Lond. Trans.* 155 (1865), p. 492 = *Scient. Papers* 1, p. 570, Cambridge 1890.

36. Mossotti's Assumption and its Modern Development

In a completely different direction, O.F. Mossotti[153] tried to reduce gravitation to electrical forces, apparently following Aepinus. He assumes that a repulsion takes place between two molecules and likewise between two "aether atoms," but that there is an attractive force between a molecule and an aether atom which exceeds the repulsion between two molecules or two aether atoms. This assumption provides an attraction between two molecules embedded in the aether, as required by Newton's law.

This idea was simplified by F. Zöllner.[154] He imagines that each gravitating molecule or atom consists of one negatively and one positively charged particle, and assumes that the repulsion between two equal charges is smaller than the attraction between two unequal ones of the same size.

Zöllner's assumption was examined mathematically by W. Weber[155] based on his electrodynamic fundamental law. This was applied to moving bodies only recently by H.A. Lorentz,[90] using his generalized Maxwell equations (cf. section 22). Lorentz's approach is continued in a paper by W. Wien.[156] Cf. the end of article 14 of this volume concerning this most recent phase of the gravitational problem.

As attractive as the explanatory attempts based on electromagnetism, in particular, may appear today, it behooves one to wait, since little of the subject has been worked [67] out, to see if tangible advantages arise from it for understanding the | gravitational effect and for release from persistent problems. According to section 22, it seems that not much can be gained in this direction by the electromagnetic conception.

For the time being, one will have to summarize the above considerations as follows: all attempts to connect gravitation with other phenomena in a satisfying way are to be regarded as unsuccessful or as not yet adequately established. With this, however, one has, at the beginning of the 20th century, returned to the view of the 18th century, to the view that takes gravitation to be a fundamental property of all matter.

EDITORIAL NOTES

[1] In the original, Zenneck mistakenly refers to note 46 (44 according to our numbering), rather than 36.

[2] This reference is to volume 6 of the *Encyklopädie der mathematischen Wissenschaften*, which covers astronomy.

[3] The terms "Körperatom" and "Körpermolekül" have been translated as "atom" and "molecule" throughout this article, whereas "Ätheratom" and "Äthermolekül" have been translated as "aether atom" and "aether molecule."

---

153 Sur les forces qui régissent la constitution intérieure des corps, Turin 1836.
154 *Erklärung der universellen Gravitation aus den statischen Wirkungen der Elektrizität*, Leipzig 1882.
155 Cf. F. Zöllner.[156]
156 Über die Möglichkeit einer elektromagnetischen Begründung der Mechanik, *Arch. Néerl.* 1900.

HENDRIK A. LORENTZ

CONSIDERATIONS ON GRAVITATION

Originally published in Proceedings Royal Academy Amsterdam 2 (1900), pp. 559–574. A Dutch version appeared under the title "Beschouwingen over de zwaartekracht" in Verslag Koninklijke Akademie van Wetenschapen te Amsterdam 8 (1900), pp. 603–620. A French version appeared under the title "Considérations sur la Pesanteur" in Archives néerlandaises 7 (1902), pp. 325–338.

§1. After all we have learned in the last twenty or thirty years about the mechanism of electric and magnetic phenomena, it is natural to examine in how far it is possible to account for the force of gravitation by ascribing it to a certain state of the aether. A theory of universal attraction, founded on such an assumption, would take the simplest form if new hypotheses about the aether could be avoided, i. e. if the two states which exist in an electric and a magnetic field, and whose mutual connection is expressed by the well known electromagnetic equations were found sufficient for the purpose.

If further it be taken for granted that only electrically *charged* particles or ions, are directly acted on by the aether, one is led to the idea that every particle of ponderable matter might consist of two ions with equal opposite charges—or at least might contain two such ions—and that gravitation might be the result of the forces experienced by these ions. Now that so many phenomena have been explained by a theory of ions, this idea seems to be more admissible than it was ever before.

As to the electromagnetic disturbances in the aether which might possibly be the cause of gravitation, they must at all events be of such a nature, that they are capable of penetrating all ponderable bodies without appreciably diminishing in intensity. Now, electric vibrations of extremely small wave-length possess this property; hence the question arises what action there would be between two ions if the aether were traversed in all directions by trains of electric waves of small wave-length.

The above ideas are not new. Every physicist knows Le Sage's theory in which innumerable small corpuscula are supposed to move with great velocities, producing gravitation by their impact against the coarser particles of ordinary ponderable matter. I shall not here discuss this theory which is not in harmony with modern physical views. But, when it had been found that a pressure against a body may be produced as well by trains of electric waves, by rays of light e. g., as by moving projectiles and when the Röntgen-rays with their remarkable penetrating power had been discovered,

Jürgen Renn (ed.). *The Genesis of General Relativity*, Vol. 3
Gravitation in the Twilight of Classical Physics: Between Mechanics, Field Theory, and Astronomy.
© 2007 Springer.

it was natural to replace Le Sage's corpuscula by vibratory motions. Why should there not exist radiations, far more penetrating than even the X-rays, and which might therefore serve to account for a force which as far as we know, is independent of all intervening ponderable matter?

[560]   I have deemed it worthwhile to put this idea to the test. In I what follows, before passing to considerations of a different order (§5), I shall explain the reasons for which this theory of rapid vibrations as a cause of gravitation can not be accepted.

§2. Let an ion carrying a charge e and having a certain mass, be situated at the point $P(x,y,z)$; it may be subject or not to an elastic force, proportional to the displacement and driving it back to P, as soon as it has left this position. Next, let the aether be traversed by electromagnetic vibrations, the dielectric displacement being denoted by $\mathfrak{d}$, and the magnetic force by $\mathfrak{H}$, then the ion will be acted on by a force

$$4\pi V^2 e \mathfrak{d},$$

whose direction changes continually, and whose components are

$$X = 4\pi V^2 e \mathfrak{d}_x, \qquad Y = 4\pi V^2 e \mathfrak{d}_y, \qquad Z = 4\pi V^2 e \mathfrak{d}_z. \tag{1}$$

In these formulae V means the velocity of light.

By the action of the force (1) the ion will be made to vibrate about its original position P, the displacement (x, y, z) being determined by well known differential equations.

For the sake of simplicity we shall confine ourselves to simple harmonic vibrations with frequency n. All our formulae will then contain the factor, $\cos nt$ or $\sin nt$ and the forced vibrations of the ion may be represented by expressions of the form

$$x = a e \mathfrak{d}_x - b e \dot{\mathfrak{d}}_x,$$
$$x = a e \mathfrak{d}_y - b e \dot{\mathfrak{d}}_y, \tag{2}$$
$$x = a e \mathfrak{d}_z - b e \dot{\mathfrak{d}}_z,$$

with certain constant coefficients a and b. The terms with $\dot{\mathfrak{d}}_x$, $\dot{\mathfrak{d}}_y$ and $\dot{\mathfrak{d}}_z$, have been introduced in order to indicate that the phase of the forced vibration differs from that of the force (X, Y, Z); this will be the case as soon as there is a resistance, proportional to the velocity, and the coefficient b may then be shown to be positive. One cause of a resistance lies in the reaction of the aether, called forth by the radiation of which the vibrating ion itself becomes the center, a reaction which determines at the same time an apparent increase of the mass of the particle. We shall suppose however that we have kept in view this reaction in establishing the equations of motion, and in

[561]   assigning their values to the coefficients a and b. I Then, in what follows, we need only consider the forces due to the state of the aether, in so far as it not directly produced by the ion itself.

Since the formulae (2) contain e as a factor, the coefficients a and b will be independent of the charge; their sign will be the same for a negative ion and for a positive one.

Now, as soon as the ion has shifted from its position of equilibrium, new forces come into play. In the first place, the force $4\pi V^2 e\mathfrak{d}$ will have changed a little, because, for the new position, $\mathfrak{d}$ will be somewhat different from what it was at the point P. We may express this by saying that, in addition to the force (1), there will be a new one with the components

$$4\pi V^2 e\left(x\frac{\partial \mathfrak{d}_x}{\partial x} + y\frac{\partial \mathfrak{d}_x}{\partial y} + z\frac{\partial \mathfrak{d}_x}{\partial z}\right), \text{etc.} \tag{3}$$

In the second place, in consequence of the velocity of vibration, there will be an electromagnetic force with the components

$$e(\dot{y}\mathfrak{H}_z - \dot{z}\mathfrak{H}_y), \text{etc.} \tag{4}$$

If, as we shall suppose, the displacement of the ion be very small, compared with the wave-length, the forces (3) and (4) are much smaller than the force (1); since they are periodic—with the frequency $2n$,—they will give rise to now vibrations of the particle. We shall however omit the consideration of these slight vibrations, and examine only the mean values of the forces (3) and (4), calculated for a rather long lapse of time, or, what amounts to the same thing, for a full period $2\pi/n$.

§3. It is immediately clear that this mean force will be 0 if the ion is *alone* in a field in which the propagation of waves takes place equally in all directions. It will be otherwise, as soon as a second ion Q has been placed in the neighborhood of P; then, in consequence of the vibrations emitted by Q after it has been itself put in motion, there may be a force on P, of course in the direction of the line QP. In computing the value of this force, one finds a great number of terms, which depend in different ways on the distance r. We shall retain those which are inversely proportional to r or r^2, but we shall neglect all terms varying inversely as the higher powers of r; indeed, the influence of these, compared with that of the first mentioned terms will be of the order λ/r if λ is the | wave-length, and we shall suppose this to be a very small fraction. [562]

We shall also omit all terms containing such factors as $\cos 2\pi kr/\lambda$ or $\sin 2\pi kr/\lambda$ (k a moderate number). These reverse their signs by a very small change in r; they will therefore disappear from the resultant force, as soon as, instead of *single* particles P and Q, we come to consider systems of particles with dimensions many times greater than the wave-length.

From what has been said, we may deduce in the first place that, in applying the above formulae to the ion P, it is sufficient, to take for $\mathfrak{d}$ and $\mathfrak{H}$ the vectors that would exist if P were removed from the field. In each of these vectors two parts are to be distinguished. We shall denote by $\mathfrak{d}_1$, and $\mathfrak{H}_1$, the parts existing independently of Q and by $\mathfrak{d}_2$ and $\mathfrak{H}_2$ the parts due to the vibrations of this ion.

Let Q be taken as origin of coordinates, QP as axis of x, and let us begin with the terms in (2) having the coefficient a.

To these corresponds a force on P, whose first component is

$$4\pi V^2 e^2 a\left(\eth_x \frac{\partial \eth_x}{\partial x} + \eth_y \frac{\partial \eth_x}{\partial y} + \eth_z \frac{\partial \eth_x}{\partial z}\right) + e^2 a(\dot{\eth}_y \mathfrak{H}_z - \dot{\eth}_z \mathfrak{H}_y). \tag{5}$$

Since we have only to deal with the mean values for a full period, we may write for the last term

$$-e^2 a(\dot{\eth}_y \dot{\mathfrak{H}}_z - \dot{\eth}_z \dot{\mathfrak{H}}_y)$$

and if, in this expression, $\dot{\mathfrak{H}}_y$ and $\dot{\mathfrak{H}}_z$ be replaced by

$$4\pi V^2\left(\frac{\partial \eth_z}{\partial x} - \frac{\partial \eth_x}{\partial z}\right) \text{ and } 4\pi V^2\left(\frac{\partial \eth_x}{\partial y} - \frac{\partial \eth_y}{\partial x}\right)$$

becomes

$$2\pi V^2 e^2 a \frac{\partial(\eth^2)}{\partial x}, \tag{6}$$

where $\eth$ is the numerical value of the dielectric displacement.

Now, $\eth^2$ will consist of three parts, the first being $\eth_1^2$, the second $\eth_2^2$ and the third depending on the combination of $\eth_1$ and $\eth_2$.

Evidently, the value of (6), corresponding to the first part, will be 0.

As to the second part, it is to be remarked that the dielectric displacement, produced by Q, is a periodic function of the time. At distant points the amplitude takes [563] the form c/r where c is independent ‖ of r. The mean value of $\eth^2$ for a full period is $c^2/2r^2$ and by differentiating this with regard to x or to r, we should get r^3 in the denominator.

The terms in (6) which correspond to the part

$$2(\eth_{1x}\eth_{2x} + \eth_{1y}\eth_{2y} + \eth_{1z}\eth_{2z})$$

in $\eth^2$, may likewise be neglected. Indeed, if these terms contain no factors such as to $\cos 2\pi kr/\lambda$ or $\sin 2\pi kr/\lambda$ there must be between $\eth_1$ and $\eth_2$, either no phase-difference at all, or a difference which is independent of r. This condition can only be fulfilled, if a system of waves, proceeding in the direction of QP, is combined with the vibrations excited by Q, in so far as this ion is put in motion by that system itself. Then, the two vectors $\eth_1$ and $\eth_2$ will have a common direction perpendicular to QP, say that of the axis of y, and they will be of the form

$$\eth_{1y} = q\cos n\left(t - \frac{x}{V} + \varepsilon_1\right)$$

$$\delta_{2y} = \frac{c}{r}\cos n\left(t - \frac{x}{V} + \varepsilon_2\right).$$

The mean value of $\delta_{1y}\delta_{2y}$ is

$$\frac{1}{2}\frac{qc}{r}\cos n(\varepsilon_1 - \varepsilon_2),$$

and its differential coefficient with regard to x has r^2 in the denominator. It ought therefore to be retained, were it not for the extremely small intensity of the. systems of waves which give rise to such a result. In fact, by the restriction imposed on them as to their direction, these waves form no more then a very minute part of the whole motion.

§4. So, it is only the terms in (2), with the coefficient b, with which we are concerned. The corresponding forces are

$$-4\pi V^2 e^2 b\left(\delta_x\frac{\partial \delta_x}{\partial x} + \delta_y\frac{\partial \delta_x}{\partial y} + \delta_z\frac{\partial \delta_x}{\partial z}\right) \tag{7}$$

and

$$-e^2 b(\ddot{\delta}_y \mathfrak{H}_z - \ddot{\delta}_z \mathfrak{H}_y). \tag{8}$$

| If Q were removed, these forces together would be 0, as has already been [564] remarked. On the other hand, the force (8), taken by itself, would then likewise be 0. Indeed, its value is

$$n^2 e^2 b(\delta_y \mathfrak{H}_z - \delta_z \mathfrak{H}_y) \tag{9}$$

or, by Poynting's theorem $(n^2 e^2 b S_x / V^2)$, if S_x be the flow of energy in a direction parallel to the axis of x. Now, it is clear that, in the absence of Q, any plane must be traversed in the two directions by equal amounts of energy.

In this way we come to the conclusion that the force (7), in so far as it depends on the part (δ_1), is 0, and from this it follows that the total value of (7) will vanish, because the part arising from the combination of (δ_1) and (δ_2), as well as that which is solely due to the vibrations of Q, are 0. As to the first part, this may be shown by a reasoning similar to that used at the end of the preceding section. For the second part, the proof is as follows.

The vibrations excited by Q in any point A of the surrounding aether are represented by expressions of the form

$$\frac{1}{r}\vartheta\cos n\left(t - \frac{r}{V} + \varepsilon\right),$$

where ϑ depends on the direction of the line QA, and r denotes the length of this line. If, in differentiating such expressions, we wish to avoid in the denominator powers of r, higher than the first—and this is necessary, in order that (7) may remain free from powers higher than the second— $1/r$ and ϑ have to be treated as constants. Moreover, the factors ϑ are such, that the vibrations are perpendicular to the line QA. If, now, A coincides with P, and QA with the axis of x, in the expression for d_x we shall have $\vartheta = 0$, and since this factor is not to be differentiated, all terms in (7) will vanish.

Thus, the question reduces itself to (8) or (9). If, in this last expression, we take for d and H their real values, modified as they are by the motion of Q, we may again write for the force

$$\frac{n^2 e^2 b}{V^2} S_x;$$

this time, however wo have to understand by S_x the flow of energy as it is in the actual case. |

[565]   Now, it is clear that, by our assumptions, the flow of energy must be symmetrical all around Q; hence, if an amount E of energy traverses, in the outward direction, a spherical surface described around Q as center with radius r, we shall have

$$S_x = \frac{E}{4\pi r^2}$$

and the force on P will be

$$K = \frac{n^2 e^2 bE}{4\pi V^2 r^2}.$$

It will have the direction of QP prolonged.

In the space surrounding Q the state of the aether will be stationary; hence, two spherical surfaces enclosing this particle must be traversed by equal quantities of energy. The quantity E will be independent of r, and the force K inversely proportional to the square of the distance.

If the vibrations of Q were opposed by no other resistance but that which results from radiation, the total amount of electromagnetic energy enclosed by a surface surrounding Q would remain constant; E and K would then both be 0. If, on the contrary, in addition to the just mentioned resistance, there were a resistance of a different kind, the vibrations of Q would be accompanied by a continual loss of electromagnetic energy; less energy would leave the space within one of the spherical surfaces than would enter that space. E would be negative, and, since b is positive, there would be attraction. It would be independent of the signs of the charges of P and Q.

The circumstance however, that this attraction could only exist, if in some way or other electromagnetic energy were continually disappearing, is so serious a difficulty, that what has been said cannot be considered as furnishing an explanation of gravitation. Nor is this the only objection that can be raised. If the mechanism of gravitation consisted in vibrations which cross the aether with the velocity of light, the attraction ought to be modified by the motion of the celestial bodies to a much larger extent than astronomical observations make it possible to admit.

§5. Though the states of the aether, the existence and the laws of which have been deduced from electromagnetic phenomena, are found insufficient to account for universal attraction, yet one may try to establish a theory which is not wholly different from that of I electricity, but has some features in common with it. In order to obtain a [566] theory of this kind, I shall start from an idea that has been suggested long ago by Mossotti and has been afterwards accepted by Wilhelm Weber and Zöllner.

According to these physicists, every particle of ponderable matter consists of two oppositely electrified particles. Thus, between two particles of matter, there will be four electric forces, two attractions between the charges of different, and two repulsions between those of equal signs. Mossotti supposes the attractions to be somewhat greater than the repulsions, the difference between the two being precisely what we call gravitation. It is easily seen that such a difference might exist in cases where an action of a specific electric nature is not exerted.

Now, if the form of this theory is to be brought into harmony with the present state of electrical science, we must regard the four forces of Mossotti as the effect of certain states in the aether which are called forth by the positive and negative ions.

A positive ion, as well as a negative one, is the center of a dielectric displacement, and, in treating of electrical phenomena, these two displacements are considered as being of the same nature, so that, if in opposite directions and of equal magnitude, they wholly destroy each other.

If gravitation is to be included in the theory, this view must be modified. Indeed, if the actions exerted by positive and negative ions depended on vector-quantities of the same kind, in such a way that all phenomena in the neighborhood of a pair of ions with opposite charges were determined by the resulting vector, then electric actions could only be absent, if this resulting vector were 0, but, if such were the case, no other actions could exist; a gravitation, i.e. a force in the absence of an electric field, would be impossible.

I shall therefore suppose that the two disturbances in the aether, produced by positive and negative ions, are of a somewhat different nature, so that, even if they are represented in a diagram by equal and opposite vectors, the state of the aether is not the natural one. This corresponds in a sense to Mossotti's idea that positive and negative charges differ from each other to a larger extent, than may be expressed by the signs + and -.

After having attributed to each of the two states an independent and separate, existence, we may assume that, though both able to act on positive and negative ions, the one has more power over the positive particles and the other over the negative ones. This difference I will lead us to the same result that Mossotti attained by means [567] of the supposed inequality of the attractive and the repulsive forces.

§6. I shall suppose that each of the two disturbances of the aether is propagated with the velocity of light, and, taken by itself, obeys the ordinary laws of the electromagnetic field. These laws are expressed in the simplest form if, besides the dielectric displacement $\mathfrak{d}$, we consider the magnetic force $\mathfrak{H}$, both together determining, as we shall now say, *one* state of the aether or one field. In accordance with this, I shall introduce two pairs of vectors, the one $\mathfrak{d}$, $\mathfrak{H}$ belonging to the field that is produced by the positive ions, whereas the other pair $\mathfrak{d}'$, $\mathfrak{H}'$ serve to indicate the state of the aether which is called into existence by the negative ions. I shall write down two sets of equations, one for $\mathfrak{d}$, $\mathfrak{H}$, the other for $\mathfrak{d}'$, $\mathfrak{H}'$, and having the form which I have used in former papers[1] for the equations of the electromagnetic field, and which is founded on the assumption that the ions are perfectly permeable to the aether and that they can be displaced without dragging the aether along with them.

I shall immediately take this general case of moving particles.

Let us further suppose the charges to be distributed with finite volume-density, and let the units in which these are expressed be chosen in such a way that, in a body which exerts no electrical actions, the total amount of the positive charges has the same numerical value as that of the negative charges.

Let ρ be the density of the positive, and ρ' that of the negative charges, the first number being positive and the second negative.

Let υ (or υ') be the velocity of an ion.

Then the equations for the state $(\mathfrak{d}, \mathfrak{H})$ are[2]

$$\left.\begin{aligned}
&\mathrm{Div}\,\mathfrak{d} = \rho \\
&\mathrm{Div}\,\mathfrak{H} = 0 \\
&\mathrm{Curl}\,\mathfrak{H} = 4\pi\rho\upsilon + 4\pi\dot{\mathfrak{d}} \\
&4\pi V^2 \mathrm{Curl}\,\mathfrak{d} = -\dot{\mathfrak{H}}
\end{aligned}\right\} \qquad \text{(I)}$$

[568] I and those for the state $(\mathrm{d}', \mathrm{H}')$

$$\left.\begin{aligned}
&\mathrm{div}\,\mathfrak{d}' = \rho' \\
&\mathrm{div}\,\mathfrak{H}' = 0 \\
&\mathrm{Curl}\,\mathfrak{H}' = 4\pi\rho'\upsilon' + 4\pi\dot{\mathfrak{d}}' \\
&4\pi V^2 \mathrm{Curl}\,\mathfrak{d}' = -\dot{\mathfrak{H}}'
\end{aligned}\right\} \qquad \text{(II)}$$

---

1   Lorentz, La théorie électromagnétique de Maxwell at son application aux corps mouvants, *Arch. Néerl. XXV*, p. 363; Versuch einer Theorie der electrischen und optischen Erscheinungen in bewegten Körpern.

2   $\mathrm{Div}\,\mathfrak{d} = \dfrac{\partial \mathfrak{d}_x}{\partial x} + \dfrac{\partial \mathfrak{d}_y}{\partial y} + \dfrac{\partial \mathfrak{d}_z}{\partial z}$. $\mathrm{Curl}\,\mathfrak{d}$ is a vector, whose components are: $\dfrac{\partial \mathfrak{d}_z}{\partial y} - \dfrac{\partial \mathfrak{d}_y}{\partial z}$, etc.

In the ordinary theory of electromagnetism, the force acting on a particle, moving with velocity v, is

$$4\pi V^2 \mathfrak{d} + [v.\mathfrak{H}],$$

per unit charge.[3]

In the modified theory: we shall suppose that a positively electrified particle with charge e experiences a force

$$k_1 = \alpha\{4\pi V^2 \mathfrak{d} + [v.\mathfrak{H}]\}e \tag{10}$$

on account of the field $(\mathfrak{d}, \mathfrak{H})$, and a force

$$k_2 = \beta\{4\pi V^2 \mathfrak{d}' + [v.\mathfrak{H}']\}e \tag{11}$$

on account of the field $(\mathfrak{d}',\mathfrak{H}')$, the positive coefficients α and β having slightly different values.

For the forces, exerted on a negatively charged particle I shall write,

$$k_3 = \beta\{4\pi V^2 \mathfrak{d} + [v'.\mathfrak{H}]\}e' \tag{12}$$

and

$$k_4 = \alpha\{4\pi V^2 \mathfrak{d}' + [v'.\mathfrak{H}']\}e', \tag{13}$$

expressing by these formulae that e is acted on by $(\mathfrak{d}, \mathfrak{H})$ in the same way as e' by $(\mathfrak{d}',\mathfrak{H}')$, and vice versa.

§7. Let us next consider the actions exerted by a *pair* of oppositely charged ions, placed close to each other, and remaining so during their motion, For convenience of mathematical treatment, we may even reason as if the two charges penetrated each other, so that, if they are equal, $\rho' = -\rho$. |

On the other hand $v' = v$, hence, by (I) and (11),   [569]

$$\mathfrak{d}' = -\mathfrak{d} \qquad \text{and} \qquad \mathfrak{H}' = -\mathfrak{H}.$$

Now let us put in the field, produced by the pair of ions, a similar pair with charges e and $e' = -e$, and moving with the common velocity v. Then, by (10)–(13),

$$k_2 = -\frac{\beta}{\alpha}k_1, \qquad k_3 = -\frac{\beta}{\alpha}k_1, \qquad k_4 = k_1.$$

The total force on the positive particle will be

---

3   $[v . \mathfrak{H}]$ is the vector-product of v and $\mathfrak{H}$.

$$k_1 + k_2 = k_1\left(1 - \frac{\beta}{\alpha}\right)$$

and that on the negative ion

$$k_3 + k_4 = k_1\left(1 - \frac{\beta}{\alpha}\right).$$

These forces being equal and having the same direction, there is no force tending to *separate* the two ions, as would be the case in an *electric field*. Nevertheless, the pair is acted on by a resultant force

$$2k_1\left(1 - \frac{\beta}{\alpha}\right).$$

If now β be somewhat larger than α the factor $2(1 - \beta/\alpha)$ will have a certain negative value $-\varepsilon$, and our result may be expressed as follows:

If we wish to determine the action between two ponderable bodies, we may first consider the forces existing between the positive ions in the one and the positive ions in the other. We then have to reverse the direction of these forces, and to multiply them by the factor ε. Of course, we are led in this way to Newton's law of gravitation.

The assumption that all ponderable matter is composed of positive and negative ions is no essential part of the above theory. We might have confined ourselves to the supposition that the state of the aether which is the cause of gravitation is propagated in a similar way as that which exists in the electromagnetic field. |

[570]   Instead of introducing two pairs of vectors $(\mathfrak{d}, \mathfrak{H})$ and $(\mathfrak{d}', \mathfrak{H}')$, both of which come into play in the electromagnetic actions, as well as in the phenomenon of gravitation, we might have assumed one pair for the electromagnetic field and one for universal attraction.

For these latter vectors, say $\mathfrak{d}$, $\mathfrak{H}$, we should then have established the equations (I), ρ being the density of ponderable matter, and for the force acting on unit mass, we should have put

$$-\eta\{4\pi V^2 \mathfrak{d} + [\mathfrak{v}.\mathfrak{H}]\},$$

where η is a certain positive coefficient,

§8. Every theory of gravitation has to deal with the problem of the influence, exerted on this force by the motion of the heavenly bodies. The solution is easily deduced from our equations; it takes the same form as the corresponding solution for the electromagnetic actions between charged particles.[4]

I shall only treat the case of a body A, revolving around a central body M, this latter having a given constant velocity p. Let r be the line MA, taken in the direc-

---

4   See the second of the obeys cited papers.

tion from M towards A, x,y,z the relative coordinates of A with respect to M, to the velocity of A's motion relatively to M, ϑ the angle between w to and p, finally p_r, the component of p in the direction of r.

Then, besides the attraction

$$\frac{k}{r^2}, \tag{14}$$

which would exist if the bodies were both at rest, A will be subject to the following actions.

1st. A force

$$k \cdot \frac{p^2}{2V^2} \cdot \frac{1}{r^2} \tag{15}$$

in the direction of r.

2nd. A force whose components are

$$-\frac{k}{2V^2}\frac{\partial}{\partial x}\left(\frac{p_r^2}{r}\right), \; -\frac{k}{2V^2}\frac{\partial}{\partial y}\left(\frac{p_r^2}{r}\right), \; -\frac{k}{2V^2}\frac{\partial}{\partial z}\left(\frac{p_r^2}{r}\right). \tag{16}$$

| 3rd. A force   [571]

$$-\frac{k}{V^2}p \cdot \frac{1}{r^2}\frac{dr}{dt} \tag{17}$$

parallel to the velocity p.

4th. A force

$$\frac{k}{V^2}\frac{1}{r^2}pw\cos\vartheta \tag{18}$$

in the direction of r.

Of these, (15) and (16) depend only on the common velocity p, (17) and (18) on the contrary, on p and w to conjointly.

It is further to be remarked that the additional forces (15)-(18) are all of the second order with respect to the small quantities $\frac{p}{V}$ and $\frac{w}{V}$.

In so far, the law expressed by the above formulae presents a certain analogy with the laws proposed by Weber, Riemann and Clausius for the electromagnetic actions, and applied by some astronomers to the motions of the planets. Like the formulae of Clausius, our equations contain the absolute velocities, i. e. the velocities, relatively to the aether.

There is no doubt but that, in the present state of science, if we wish to try for gravitation a similar law as for electromagnetic forces, the law contained in (15)-(18) is to be preferred to the three other just mentioned laws.

§9. The forces (15)-(18) will give rise to small inequalities in the elements of a planetary orbit; in computing these, we have to take for p the velocity of the Sun's motion through space. I have calculated the *secular* variations, using the formulae communicated by Tisserand in his *Mécanique céleste*.

Let a be the mean distance to the Sun,

   e the eccentricity,

   φ the inclination to the ecliptic,

   θ the longitude of the ascending node,

   $\tilde{\omega}$ the longitude of perihelion,

κ' the mean anomaly at time, $t = 0$, in this sense that, if n I be the mean motion, as determined by a, the mean anomaly at time t is given by

$$\kappa' + \int_0^t n \, dt.$$

Further, let λ, μ and ν, be the direction-cosines of the velocity p with respect to: 1st. the radius vector of the perihelion, 2nd. a direction which is got by giving to that radius vector a rotation of $90°$, in the direction of the planet's revolution, 3rd. the normal to the plane of the orbit, drawn towards the side whence the planet is seen to revolve in the same direction as the hands of a watch.

Put $\omega = \tilde{\omega} - \theta$, $p/V = \delta$ and $na/V = \delta'$ (na is the velocity in a circular orbit of radius a).

Then I find for the variations *during one revolution*

$$\Delta a = 0$$

$$\Delta e = 2\pi\sqrt{(1-e^2)}\left\{\lambda\mu\delta^2\frac{(2-e^2)-2\sqrt{(1-e^2)}}{e^3} - \lambda\delta\delta'\frac{1-\sqrt{(1-e^2)}}{e^2}\right\}$$

$$\Delta\varphi = \frac{2\pi}{\sqrt{(1-e^2)}}\nu\left\{[-\lambda\delta^2\cos\omega + \delta(e\delta' - \mu\delta)\sin\omega]\frac{1-\sqrt{(1-e^2)}}{e^2} + \mu\delta^2\sin\omega\right\} \tag{19}$$

$$\Delta\theta = -\frac{2\pi}{\sqrt{(1-e^2)}\sin\varphi}\nu\left\{[\lambda\delta^2\sin\omega + \delta(e\delta' - \mu\delta)\cos\omega]\frac{1-\sqrt{(1-e^2)}}{e^2} + \mu\delta^2\cos\omega\right\}$$

$$\Delta\tilde{\omega} = \pi(\mu^2 - \lambda^2)\delta^2 \frac{(2 - e^2) - 2\sqrt{(1 - e^2)}}{e^4} + 2\pi\mu\delta\delta'\frac{\sqrt{(1 - e^2)} - 1}{e^3}$$

$$-\frac{2\pi tg\frac{1}{2}\varphi}{\sqrt{(1 - e^2)}}v\left\{[\lambda\delta^2\sin\omega + \delta(e\delta' - \mu\delta)\cos\omega]\frac{1 - \sqrt{(1 - e^2)}}{e^2} + \mu\delta^2\cos\omega\right\}$$

$$\Delta\kappa' = \pi(\lambda^2 - \mu^2)\delta^2\frac{(2 + e^2)\sqrt{(1 - e^2)} - 2}{e^4} - 2\pi\delta^2 - 2\pi\mu^2\delta^2$$

$$-2\pi\mu\delta\delta'\frac{(1 - e^2) - \sqrt{(1 - e^2)}}{e^3}$$

§10. I have worked out the case of the planet Mercury, taking 276° and +34° for the right ascension and declination of the apex of the Sun's motion. I have got the following results: |

$$\Delta a = 0 \qquad\qquad [573]$$

$$\Delta e = 0.018\delta^2 + 1.38\delta\delta'$$

$$\Delta\varphi = 0.95\delta^2 + 0.28\delta\delta'$$

$$\Delta\theta = 7.60\delta^2 - 4.26\delta\delta'$$

$$\Delta\tilde{\omega} = -0.09\delta^2 + 1.95\delta\delta'$$

$$\Delta\kappa' = -6.82\delta^2 - 1.93\delta\delta'$$

Now, $\delta' = 1.6 \times 10^{-4}$ and, if we put $\delta = 5.3 \times 10^{-5}$, we get

$$\Delta e = 117 \times 10^{-10}$$

$$\Delta\varphi = 51 \times 10^{-10}$$

$$\Delta\theta = -137 \times 10^{-10}$$

$$\Delta\tilde{\omega} = 162 \times 10^{-10}$$

$$\Delta\kappa' = -355 \times 10^{-10}.$$

The changes that take place in a century are found from these numbers, if we multiply them by 415, and, if the variations of φ, θ, $\tilde{\omega}$ and κ' are to be expressed in seconds, we have to introduce the factor 2.06×10^5. The result is, that the changes in φ, θ, $\tilde{\omega}$ and κ' amount to a few seconds, and that in e to 0.000005.

Hence we conclude that our modification of Newton's law cannot account for the observed inequality in the longitude of the perihelion—as Weber's law can to some extent do—but that, if we do not pretend to explain this inequality by an alteration of the law of attraction, there is nothing against the proposed formulae. Of course it will be necessary to apply them to other heavenly bodies, though it seems scarcely proba-

ble that there will be found any case in which the additional terms have an appreciable influence.

The special form of these terms may perhaps be modified. Yet, what has been said
is sufficient to show that gravitation may be attributed to actions which are propagated with no greater velocity than that of light.

As is well known, Laplace has been the first to discuss this question of the velocity of propagation of universal attraction, and later astronomers have often treated the
same problem. Let a body B be attracted by a body A, moving with the. velocity p.
Then, if the action is propagated with a finite velocity V, the influence which reaches
B at time t, will have been emitted by A at an anterior moment, say $t - \tau$. Let A_1
be the position of the acting body at this moment, A_2 that at time t. It is an easy mat-
[574] ter to calculate the distance between those positions. Now, if the action at time $| t$ is
calculated, as if A had continued to occupy the position A_1, one is led to an influence on the astronomical motions of the order p/V if V were equal to the velocity
of light, this influence would be much greater than observations permit us to suppose.
If, on the contrary, the terms with p/V are to have admissible values, V ought to be
many millions of times as great as the velocity of light.

From the considerations in this paper, it appears that this conclusion can be
avoided. Changes of state in the aether, satisfying equations of the form (I), are propagated with the velocity V; yet, no quantities of the first order p/V or w/V (§8),
but only terms containing p^2/V^2 and pw/V^2 appear in the results. This is brought
about by the peculiar way—determined by the equations—in which moving matter
changes the state of the aether; in the above mentioned case the condition of the
aether will *not* be what it would have been, if the acting body were at rest in the position A_1.

BENEDICT AND IMMANUEL FRIEDLAENDER

ABSOLUTE OR RELATIVE MOTION?

Originally published in German as a pamphlet "Absolute oder relative Bewegung?"
by Verlag von Leonhard Simion, Berlin 1896. 1: Immanuel Friedlaender; Berlin,
Spring 1896. 2: Dr. Benedict Friedlaender; Berlin, January 1896.

1. The Question of the Reality of Absolute Motion and a Means for its *Experimental* Resolution.
2. On the Problem of Motion and the Invertibility of Centrifugal Phenomena on the *Basis of Relative Inertia*.

PREFACE

The question treated in the following is old; in addition to the cited writings there   [3]
exists a rather extensive literature on it, about which we reserve comment to a later
time. Our work was conceived without knowledge of that literature. It once more illu-
minates the uncertainty of the basis of our mechanics and of all exact science in general—wherein it must necessarily deal with long-known material—and points the way
toward an attempt to solve the question, without any claim of already presenting the
solution. If experiments now in progress lead to a result, they may silence the frequently raised objection that the limitations on our experience do not permit a decision. But if the experiments do not succeed, we may perhaps lead others to successful
work in this direction by pointing out that the invertibility [*Umkehrbarkeit*] of the centrifugal force, considered already by Newton, urgently needs an experimental treat-
ment as well as a theoretical investigation by reduction to a law of relative inertia. |

1. THE QUESTION OF THE REALITY OF ABSOLUTE MOTION   [5]

A body whose position in space changes is said to be in motion. This is one of two
possible but fundamentally different definitions of motion. It presupposes that the
process of change in position of a *single* body in space, quite independent of actual or
possible relations to other bodies, has *content*, that is, that it differs from the state of
rest by *recognizable effects*. When we apply this concept to any mechanical problems
we have to refer observed motions to a coordinate system fixed in space in order to
represent them properly; to do this we must know the true motion of our Earth—from

Jürgen Renn (ed.). *The Genesis of General Relativity*, Vol. 3
Gravitation in the Twilight of Classical Physics: Between Mechanics, Field Theory, and Astronomy.
© 2007 Springer.

which all measurements originate—so that we can transform from its co-moving coordinates, which we have fixed in the Earth, to a system of coordinates at absolute rest in the universe.

This definition with its consequences forms the basis of our present-day mechanics, and the well-founded high regard for the achievements of this science often prevents critics from daring to examine the justification of this definition. Nevertheless many have already stated the suspicion that this definition of motion is wrong, [6] because only the relative motions of two or several bodies have *reality*, and I because the motion of one body, apart from its relations to other bodies, does *not* differ from a state of rest. This notion corresponds to our laws of thought and powers of imagination, but, as we shall see, not to the world of phenomena as understood according to the principles of mechanics.

Two groups of phenomena are relevant:
1. The appearance of the centrifugal force,
2. The stability of the free axes and of the plane of a Foucault pendulum.

The question of whether an absolute motion possesses reality, or whether there can be only relative motion, can be answered in different ways; to be precise there are *three* views that we want to consider here.[1] We will totally neglect those who deny the reality of absolute motion at the beginning of textbooks, give only the definition of relative motion, and then as they go along introduce a coordinate system fixed in space without allowing it to be noticed (without noticing it themselves?), particularly for the phenomena of rotation. Others, Kirchhoff for example, introduce *ab initio* coordinates at absolute rest, and in that way at least avoid self-contradiction. Only rarely is the question posed and discussed before it is answered.

Newton was the first to encounter the difficulty; he opines that it should be solved by experiment, and on the basis of an admittedly totally inadequate experiment[2] he [7] arrives at I the view that absolute motion is real (see Mach, *Geschichte der Mechanik*, p. 317).

Kant grasped the question in its full significance with perfect clarity. He arrives at a solution by construing a difference between purely mathematical, "phoronomical" motion and physical, "phenomenological" motion; about the first he states the following (*Metaphysische Anfangsgründe der Naturwissenschaft*, 1786, Phoronomie, Grundsatz 1):[1]

> Every motion, as object of a possible experience, can be viewed arbitrarily as motion of the body in a space at rest, or else as rest of the body, and, instead, as motion of the space in the opposite direction with the same speed.

By way of contrast, propositions 1 and 2 of the Phenomenology state:

---

1   In the following only the three main points of view will be briefly sketched, as seems necessary for understanding the experiment, without entering into the theoretical treatment of the question by numerous authors.
2   Cf. below, [p. 20 in the original].

1. The rectilinear motion of a matter with respect to an empirical space, as distinct from the opposite motion of the space, is a merely *possible* predicate. The same when thought in no relation at all to a matter external to it, that is *as absolute motion*, is *impossible*. [...]

2. The circular motion of a matter, as distinct from the opposite motion of the space, is an *actual* predicate of this matter; by contrast, the opposite motion of a relative space, assumed instead of the motion of the body, is no actual motion of the latter, but, if taken to be such, is mere semblance.

What Kant states here can be found in Budde's general mechanics[2] in a different form, which more nearly corresponds to the modern point of view and will therefore be more easily understood. He does admit (p. 6) that the determination of position is relative in nature, but after making the transition from *phoronomics to kinetics* he observes—presupposing the principle of inertia—that we are forced by the facts to assume an absolute coordinate system,

a I fundamental system within which the center of the Sun lies, with an angular velocity $-w$ relative to the Earth. The principle of inertia holds in this system F, and in every other system that is at rest or in uniform translation relative to it; but this principle is no longer valid in a different system that rotates with respect to F, as shown by our experience on Earth.   [8]

We can properly state this claim of Kant and Budde as follows: in the purely geometric phenomenon of motion, or in mathematical space, there are neither fixed directions nor a fixed system of coordinates, all motion here is only relative; but in the dynamical phenomena of motion, or in physical space, there is such a coordinate system, which however can be translated parallel to itself without resulting in any real change. In the following I will call this a *fixed system of directions*, and for purposes of visualization I wish to compare physical space with the interior of a crystal imagined to be infinite in extent, in which to a certain extent each rotation, but no translational motion, would have a physical meaning. Indeed, Budde concludes that space is filled with a *medium*, and that the principle of inertia is a property of matter relative to this medium.[3]

Finally, the third possible conception is taken by Mach (p. 322); namely, that only relative motion is real, that our present-day mechanics is incomplete, and that *"the mechanical principles can probably be put in such a form that centrifugal forces result also for relative rotation."*

In short, the three conceptions are:[4] I

1. There is absolute motion, translation as well as rotation; physical space, as opposed to purely geometrical space, possesses a *fixed coordinate system*.   [9]

---

3   Earlier authors already expressed similar views.

4   The first point of view is taken by Newton, Euler, Laplace, Lagrange, Poinsot, Poisson, Narr, the second, however partially in a quite different formulation, by Maxwell, Thomson-Tait, Streintz, Lange and others; the third especially by Mach. Some of these authors vacillate in their view or try, like Lange, to unite views 2 and 3.

2. There is no absolute translation, but there is absolute rotation. Real space, as opposed to that corresponding to our conceptual ability [*Denkvermögen*], has an *absolute system of directions*.

3. There is nothing but relative motion; *physical* space does *not differ* from *mathematical* space; but our present-day mechanics explains the phenomena of rotation incorrectly, or at least incompletely.

In the following we wish to treat the special case of centrifugal force first, in order to test the validity of the three views

Imagine a rigid system, consisting of a weightless rigid rod of length $2r$, which has two equal spheres, each of mass m, attached to its ends.

If we allow this system to rotate about an axis perpendicular to the rod at its midpoint—motions being referred to the Earth or to the fixed stars, because in the following we want to assume that the motion of the Earth relative to the fixed stars can be neglected as rather minor compared to that of our system—then there occurs a tension in the rod that equals mv^2/r, according to the well-known formula, where v is the velocity of the mass m. Now we treat this system, which we take to be located on the Earth, based on the three possible assumptions concerning motion. If we first proceed on the assumption that there is only relative motion, and that | our coordinate system can be fixed completely arbitrarily, then whenever we take the rod itself as a coordinate, for example, or choose any coordinate system rigidly connected to our system of bodies, there is no accounting for the occurrence of the centrifugal force, that is, the measurable stress in the rod. For the rotational motion would then be represented as a rotation of the Earth and the whole firmament of fixed stars about our rod with the two spheres.

[10]

However, the rotation of these external bodies cannot explain the existence of the stress in the rod according to any law of mechanics known to date.

Secondly, let us assume that space is constituted somewhat like an infinitely large crystal, that there is no fixed system of axes in it, but that it possesses a system of directions. In this case only the angular velocity and the direction of the rotational axis would be uniquely determined, and we would therefore be entitled to assume that our system is rotating about a fixed axis, perpendicular to the rod and at a distance $a \leq r$ from its midpoint. We then obtain on one side of the axis the stress

$$F_1 = \frac{mv^2(r-a)^2}{(r-a)r^2},$$

and on the other side,

$$F_2 = \frac{mv^2(r+a)^2}{(r+a)r^2},$$

where v denotes the *same* speed as in our earlier equation. Since the rotational axis just assumed—which is only imaginary—is actually free, the rod will take on a uniform stress along its entire extent, namely

$$F = \frac{F_1 m + F_2 m}{2m},$$

but this stress equals

$$\frac{F_1 + F_2}{2} = \frac{mv^2}{2r^2}(r - a + r + a) = \frac{mv^2}{r}.^5$$

 I What we have shown in this special example is valid in general according to [11] Budde's treatment: A system of directions in space, which suffices only to define the angular velocity of a system without yielding information on its translational motion, is enough to determine uniquely the centrifugal forces.

In our case just mentioned the motion would consist not only in rotation of one sphere about the other, which is simultaneously rotating about its axis with the same angular velocity and in the same sense, but at the same time the Earth together with the system of the world considered as a whole would move in a circle of radius r about our axis of rotation. This rotation would take place with the same angular velocity as that of our system and in the same sense, but such that the Earth and system of the world would always be only parallely displaced, and every line placed through the Earth and the firmament of fixed stars would maintain a constant direction.

Let us finally go on to the third point of view, the one which admittedly appeals least to us. Let us assume that there is absolute motion—even translation—that is, a space with a fixed coordinate system. Then an absolute rotation can be determined by using systems of masses such as our rod with the two spheres. Thereby we are enabled to refer our motion to a coordinate system of which we can claim that relative to space—relative to an absolute coordinate system—it does not rotate, but it may execute an arbitrary translational motion, for whose determination we have no basis. The whole edifice of our mechanics is based on this view, and by taking this view no contradiction with experience has yet turned up, as far as I know. But until one is in a position to exhibit an absolute translation, the known facts I agree equally [12] well with a space in which only a system of directions possesses reality. Experiments in this direction have been performed, but we do not want to go into that here.

So the phenomena of centrifugal force teach us—on the basis of the usual mechanical views, namely the formula mv^2/r —that there is an absolute space, or at least a system of directions, and that the motion of a rigid system in this space in certain cases *really* differs from rest in a perceptible way, that is, by the existence of the measurable centrifugal force. But this assumption means that there is a real and in certain circumstances mechanically effective space that significantly differs from the space of mathematics. The only attempt known to us to make this idea moderately

---

5   The tension in a thread at whose two ends the mass m_1 is being pulled under the influence of an acceleration g_1 by a force F_1, and m_2 with the acceleration g_2 and the force F_2, is given by

$$\frac{m_1 m_2 g_1 + m_1 m_2 g_2}{m_1 + m_2} = \frac{F_1 m_2 + F_2 m_1}{m_1 + m_2}.$$

understandable is Budde's above-mentioned attempt[6] to regard space as occupied, and to regard inertia as a consequence of relative motion with respect to the occupying medium.

The second group of phenomena mentioned above, namely the fact that free axes and the plane of the Foucault pendulum (at the pole) remain fixed with respect to the sky of fixed stars, can be treated more briefly here, since they behave in exactly the same way. Suppose we take the view that there is only relative motion. Then we cannot explain the phenomena mentioned above if we treat, say, the Earth as fixed. Why the Sun, Moon and stars should drag along the free axes or the plane of the pendulum upon their daily revolution about the Earth could not be explained by mechanics. If [13] we take the second view—that space I has a system of directions—then the phenomena mentioned above prove that the Earth possesses its own absolute rotation. They are explained and calculable under this assumption.

Of course, the same follows on the basis of a space in which not only absolute rotations, but also absolute translational motions are real.

We briefly summarize again what we have stated so far: our way of thinking is in accordance with a space in which one position as such does not differ from another position, nor one direction from another direction. That a body or a rigid system moves in this space, or that it rotates about an interior or exterior point or stays at rest, would then be only different ways of stating the same set of facts, depending on whether it pleases us to fix our system of coordinates by an external body, with respect to which the corresponding relative motion takes place, or by the system in question itself.

But from this point of view we cannot—as shown above—explain the two groups of phenomena of centrifugal force and preservation of axes on the basis of an arbitrary coordinate system.

We can then ascribe to space a system of directions, or assume that space is fixed in a sense and that we can recognize absolute rotation by the appearance of centrifugal phenomena—on the Earth, for example, by a decrease in g toward the equator, etc.; whereas, at least for the time being, we lack evidence for the recognition of any absolute translational motion of the Earth.

To repeat, our intuition opposes this result; the proper reason for this opposition is [14] that we see ourselves forced to admit a factual difference between mathematical I space and actual space, or differently put, between the space that corresponds to our conceptual ability and the space of phenomena perceived by our senses.

We can try to explain this result the way Budde did. Similar ideas are found already in Kant, who also tried to explain that the content of *vis viva* [*lebendige Kraft*] in a moving mass equals $mv^2/2$ by filling space with a medium.[7] But if we are not content with this, we must contest the result by disputing the premises; that is,

---

6   This idea was already formulated by Kant in the 1747 treatise regarding the true assessment of *vis viva* [*lebendige Kraft*] (which is in other respects full of obscurities), and it was later expressed by many others, in different forms.[3]

we must question the basis of the usual mechanical explanation of the rotational phenomena under discussion (namely, the law of inertia and the formula $mv^2/2$ for the *vis viva*).

Without knowing that Mach had already done this, I have for many years had doubts about the completeness of these foundations of mechanics, and in particular I have become convinced that the phenomenon of centrifugal force, properly mechanically understood, should also be explicable solely in terms of the relative motions of the system concerned without taking refuge in absolute motion. But I was well aware that the mere statement of these doubts does not amount to much, and that one must find either a new formulation of the expression for *vis viva* of a moving mass and thereby an improved version of the law of inertia, | or one must prove the inadequacy [15] of the prevailing view experimentally. The phenomena of the centrifugal force in particular seemed to me suited for such an experimental resolution of the question: if the centrifugal force that appears in a flywheel can be explained solely from its relative motion, then it must be possible to derive it also under the assumption that the flywheel is at rest, but that the Earth is turning with the same angular speed about the flywheel axis in the opposite sense. Now, just as the centrifugal force appeared in the resting wheel as a consequence of the rotation of the massive Earth together with the universe, so there should appear, I reasoned, in correspondingly smaller measure an action of centrifugal force in resting bodies in the vicinity of massive moving flywheels. If this phenomenon was verifiable, this would be the incentive for a reformulation of mechanics, and at the same time further insight would have been gained into the nature of gravity, since these phenomena must be due to the actions at a distance[8] of masses, and here in particular to the dependence of these actions at a distance on relative rotations.

In light of the smallness of the masses available for our experiment, I had little hope of an experimental solution, until I thought of a promising experimental arrangement in the fall of 1894. This arrangement consists of putting the most sensitive of all physical instruments, a torsion balance, in the extension of the axis of a heavy mass that rotates as rapidly as possible, namely a large flywheel, for example in a rolling-mill. If the beam of the balance, bearing two small spheres at its ends, is not parallel to the (vertical) plane of the flywheel when the latter is at rest but inclined by an angle of about 45° to it, then according to our | theory tensile forces that tend [16] to separate the spheres from the extension of the axis must appear, so as to make the balance parallel. However, a sensitive balance is a delicate instrument, and a rolling-

---

7   See footnote 6. Even if I think of space as filled, because otherwise action at a distance would remain inexplicable, I nevertheless strongly doubt that the phenomenon of inertia (respectively, of *vis viva*) derives from motion relative to the nearly massless aether, but I am of the opinion that motion relative to the aether as influenced by nearby gravitating masses, or equivalently motion relative to the gravitating masses themselves causes the inertia (respectively, the *vis viva*) of a moving mass through mediation by the aether.

8   Whether these actions are, as the writers believe, transmitted through a medium or not makes no difference to the matter.

mill is probably not the most comfortable and optimal location for precision measurements, and so due to the many sources of error my experiments, which I started
already in November of 1894 at the rolling-mill in Peine—with the kindest support
on the part of the management and engineers—have brought to light no incontestable
results that I would want to transmit to the public, even though a deviation in the
expected sense was established at the beginning and the end of the motion. But since
the balance that was used also reacted to other influences, in particular being deviated
by a burning candle at a distance of 4 meters in a room (the closer sphere being
"attracted"), and since at the rolling-mill at Peine certain furnace doors are opened
and closed in the same time intervals as the starting and stopping of the machines we
used, I do not yet consider my results unobjectionable. Experiments with a torsion
balance inside a double-walled box of copper with a layer of water about 14 mm
thick between the two walls, have shown that even this precaution does not suffice to
make the needle completely independent of external heat sources. Whether the heat
perturbations are to be explained simply by the circulation of air in the interior of the
box—which is always sealed airtight to the exterior—or whether effects as with a
radiometer are to be considered, I do not dare to decide for the time being. A new
instrument that is being prepared will, we hope, exclude all previous sources of error.
Although reliable results are not yet available, the constant occupation with the matter and the frequent discussions with my brother Dr. Benedict Friedlaender have led
[17] us to the conclusion that the matter is of sufficient importance | to publish our
thoughts already. My brother called my attention to Mach right at the start and in
joint work we have drawn many of the consequences that would result from our view.
The results of these considerations, as well as several opinions that I cannot fully
share, were put together by my brother in the second part of this treatise, and there he
has also attempted to state the law of inertia differently, so that one can derive from it
the relativity, hence also the invertibility, of centrifugal forces. But it seems to me that
the correct formulation of the law of inertia will be found only when the relative inertia, as an effect of masses on each other, and gravity, which is after all also an effect
of masses on each other, are reduced to a unified law.[9] The challenge to theoreticians
and calculators to attempt this will only be truly successful when the invertibility of
centrifugal force has been successfully demonstrated. |

[18]   ## 2. ON THE PROBLEM OF MOTION AND THE INVERTIBILITY OF
CENTRIFUGAL PHENOMENA

We are accustomed to regard mechanics as the most perfect of all natural sciences, and
efforts are made to reduce all other sciences to mechanics or, so to speak, to resolve

9   For this it would be very desirable to resolve the question of whether Weber's law applies to gravity,
as well as the question concerning gravity's speed of propagation. Regarding the latter question, one
might use an instrument that makes it possible to measure statically the diurnal variations of the
Earth's gravity as it depends upon the position of the celestial bodies.

them in mechanics. Mechanics is the most concrete and nevertheless also the mathematically (that is, for quantitative calculations) finest, clearest and most exactly developed science. We have no cause to examine to what extent this evaluation of mechanics is justified, and we recalled the above statements only in order to indicate that all considerations or matters of fact related to the foundations of mechanics may claim more than the usual interest and importance. Our subject at present is such a consideration, which truly concerns the foundations of mechanics and thereby those of our whole scientific worldview; however, this is no new subject, but surely a nearly forgotten and at any rate not always sufficiently respected problem, *the problem of motion*. This problem is probably connected with another one, which has been discussed far more frequently in former and more recent times, but which is apparently still a long way from an even moderately satisfactory solution, *the problem of gravitation.* |

Concerning gravity we have nothing more than the knowledge of superficially [19] perceptible facts, along with a purely *mathematical* theory, which is in no way physical. No mind thinking scientifically could ever have permanently and seriously believed in unmediated action at a distance; the apparent force at a distance can be nothing other than the result of the effects of forces that are transmitted in some way by the medium being situated between two gravitating bodies. But our presentations refer primarily only to the problem indicated first, that of motion, as we may briefly denote it, and not directly to the latter problem. For the science of motion with all its derivations and consequences contains an image, or rather a formulation that can *not* be imagined, which must be a hint of a present flaw to all who are convinced *that something that cannot be imagined also cannot be actual, i.e. acting*. Vis viva, for example, is defined to be proportional to the velocity squared ($mv^2/2$); and the velocity is defined as the measure of distance divided by that of time (l/t). Centrifugal phenomena are explained by the conflict between a constrained, curvilinear motion and inertia, which tends to maintain rectilinear motion. In both cases, and more generally, one thinks of motion—or let us say, rather, one defines motion—as *absolute* motion, as motion of a mass from one "absolute position" to another "absolute position"; for *thinking* about absolute motion is just what we can *not* do, and precisely this motivated the following considerations. The root of these considerations, as far as the author is concerned, lies in some difficulties and obscurities of none less than Newton, which | I came to know in the early 90's from quotes of Mach in his [20] *Geschichte der Mechanik.*

In themselves the arguments are as simple as they are unusual, which can easily lead to obscurity if one does not demonstrate and think through the matter step by step from the simplest case.

What we perceive of motion is always only relative motion, changes in position of masses *with respect to other masses*. Our hand moves against the rest of the body, considered as relatively motionless; we move on the deck of a ship, the ship moves on the surface of the Earth, it changes its distance from the continent, thought of as fixed. Our planet moves in the universe, namely with respect to a coordinate system that is considered fixed somewhere (in the Sun, for example), and so forth.

Now think (if you can) of a progressive motion of a *single* body in a universal space that is otherwise imagined to be totally empty; how can the motion be detected, i.e. distinguished from rest? By *nothing* we should think; indeed, the whole idea of such an absolute, progressive motion is meaningless. And still, in one case, namely when in motion, our absolutely moving body, thought of as isolated, is supposed to possess an amount of energy that is proportional to (half) the square of its (meaningless) velocity!

That this cannot be imagined is no new discovery; it is so apparent, and in connection with our astronomical knowledge the reasoning suggests itself so naturally, that many should have encountered it. (Cf., e.g., Wundt, *System der Philosophie* etc.)

More tenacious than the absurd idea of an absolute, so-called *translational* motion is that of the rotation of a sphere, for example, about an axis taken in its abso-
[21] lute sense. I To make the picture more concrete and impressive, let us immediately consider one of those well-known apparatuses used in schools for illustrating centrifugal effects. An approximately spherical framework of elastic brass blades can be rapidly rotated about its axis, so that all the blades are bent in such a way that this originally spherical frame suffers a polar flattening and an equatorial thickening. Here too, according to the usual interpretation, one thinks of the rotational motion as absolute and explains the whole phenomenon in the well-known way. Connected with this, or at least with an entirely equivalent example, is the difficulty that Newton already felt strongly, and which seduced that researcher into statements that seem to us more than merely risky. Newton had suspended a glass of water on two strings so that after twisting the strings about each other they put the glass into rapid rotation as they unwound. Since the friction between the glass and the water is rather small, and the inertia of the mass of water in the glass is quite considerable, at the beginning the glass turns nearly alone, whereas the water remains behind; only slowly does the mass of water take part in the rotation. One observes, as is well known, that no centrifugal effects whatever occur as long as the glass rotates alone (or almost alone); the surface of the water remains flat. But as the water takes part in the rotation more and more, it increasingly rises at the rim and falls at the center. Newton concludes from this that the centrifugal effects are a consequence of absolute—but not of relative—rotation; for in the beginning the glass turns compared to the objects "at rest" in the room, including also the mass of water in the glass; but afterwards, according to N., the water rotates "absolutely," and this absolute rotation brings centrifugal effects
[22] into play. It is not difficult to establish the I untenability or even the incorrectness of this Newtonian idea, as Mach did. In the beginning the glass turns with respect to the objects at rest in the room or with respect to the Earth, or more correctly the universe; whereas later the mass of water rotates with respect to the universe. In the first case there was relative rotation between *mass of water* (+ universe) and *glass*, in the second between *mass of water* (+ glass) and the *universe*; and only the latter, but not the former, produced the centrifugal effects. Mach justifiably points out that "absolute rotation" is a fiction, or more accurately, an unthinkable absurdity. The naive mind will immediately object that it is just not possible to hold the mass of water fixed and

now "let the universe rotate" about the same rotation axis; but the more acute mind will quickly agree that *both ideas are plainly identical, namely indistinguishable in any logical or practical way.* The true fact of the matter is just this, that the rotation of the glass with respect to the mass of water releases no centrifugal forces, but that the rotation of the universe with respect to the glass (or equivalently the rotation of the mass of water with respect to the universe) does do so. In the first case it is the very small mass of the glass's wall that rotates with respect to the mass of water, in the second case it is the universe; we should not be surprised that the vanishingly small mass of the glass does not call forth any noticeable centrifugal forces, but no one could know, as Mach aptly remarked, how the experiment would turn out if the thickness of the glass were significantly increased and its walls ultimately became several leagues thick. Of course, this experiment cannot be executed in this form. But there is still the question of whether an experimental arrangement is possible in practice that would amount to the same thing, and allow us to establish the *invertibility* or *relativity* of centrifugal effects. |

Now it was my brother Immanuel who devised and tried to test an experimental [23] arrangement of this type. Technical difficulties, some of them quite unexpected, have so far prevented the realization of a reliable result.

It is well known that the torsion balance is the most sensitive of all instruments. The large flywheels in rolling mills and other large factories are probably the largest rotating masses with which we can *experiment.* The centrifugal forces express themselves in a push causing recession from the axis of rotation. Thus, if we place a torsion balance at not too great a distance from a large flywheel, so that the point of suspension of the part of the torsion balance that can turn (the "needle") lies exactly or approximately on the extension of the flywheel's axis, then if the needle was not originally parallel to the plane of the flywheel it should tend to approach that position and show a corresponding deflection. Namely, the centrifugal force acts on each element of mass that is not on the rotation axis in a direction tending to move it away from the axis. It is immediately apparent that the greatest possible separation is reached when the needle becomes parallel. So far the difficulties opposing the experiment were that a sensitive torsion balance is also put into motion by perturbing influences —particularly effects of *heat*— as if, by the way, the warmer parts of the apparatus would have an *attractive* effect.

Now if we assume that the experiment were to work flawlessly, we would have thereby discovered a new mechanical-physical phenomenon, whose consequences would be extraordinarily far-reaching. Certainly the phenomenon would be explicable and predictable, as shall be shown immediately, if one were to recast motion and all concepts connected with it, including | *inertia* in particular, in such a way that *rel-* [24] *ative* motion would replace the present tacitly assumed concept of *absolute* motion. However, the predictability on the basis of this suggested recasting would be no objection to the claim that we would be dealing with a fundamentally new phenomenon; for precisely this recasting has not been carried out and tested by anyone. The law of inertia in the usual manner of representation can be transparently described by saying that every body opposes any change of its velocity (conceived as absolute) with a resistance proportional to its mass in the corresponding direction. Here, the remaining bodies of the universe are completely ignored; in fact, a point that must be especially emphasized is that the concept of mass is, except for its derivation from gravity (mg), derived precisely from the facts of inertia. Every change in velocity, i.e., every acceleration (in the simplest case, for example, the imparting of motion to a body previously at rest until it reaches a certain velocity) is held to be opposed by a *resistance*, the overcoming of which requires the quantity of energy that is afterwards, when the body is in motion, supposed to be contained in that motion as "kinetic energy." It is here to be noted once more that translational motion of a single body in space otherwise regarded as empty is an absurdity, namely, it does not differ from its opposite, rest. Thus, the creation of such a chimera should not require any energy; therefore, if in contrast the actual world does agree with our prerequisites of thought [*Denknotwendigkeiten*], it should surely make a difference with respect to which other bodies motion is to be created, in a word, what *relative* motion of previously nonmoving bodies is to be created. Accordingly, inertia is to be grasped rela-

[25] tively; one could formulate the law of relative inertia as follows: All | masses strive to maintain their *mutual* state of motion with respect to speed and direction; every change requires positive or negative energy expenditure, that is, work is either required—in the case of an increase in velocity—or is released—in the case of a decrease in velocity. The resistance to changes in velocity would then, as soon as we regard all motions as relative, be expressed not only in the one body that, as we are accustomed to say, we "set in motion" (that is, set in motion relative to the Earth) but also in all the others that we regard as being at rest in accordance with the usual conception. According to the usual conception, inertia occurs on a railway train that is to be brought from rest to a certain velocity, but not on the Earth; according to our view it occurs also on the Earth, even though this is not noticeable due to the extraordinarily much larger mass. But when we put very large masses into motion, to the extent permitted by our technology, and we can observe the behavior of very easily moved masses in the vicinity, it is possible that the relativity of inertia can be shown directly. Precisely that experiment of my brother, namely to find the needle in parallel orientation, which has been unsuccessful so far, would be, as is easily demonstrated, not only the proof of the invertibility of the centrifugal force, but also of the relativity of inertia. In our view, both are the same.

The application of the thought indicated here is very simple but unusual to a high degree. For if we consider the resistance to acceleration that some body exhibits, we do not have the slightest thought of other masses that are nearby! But if we do so and hold firmly to the guiding thought that the masses strive to maintain their *relative*

[26] velocity, it turns out that (for motion on a | straight line of body A relative to body B as the simplest case):

$$\left.\begin{array}{l}\text{accelerated approach and} \\ \text{decelerated withdrawal}\end{array}\right\} \text{must have a repulsive effect}$$

accelerated withdrawal and ⎱
decelerated approach ⎰ must have an attractive effect.

In the first two cases a recession, and conversely in the last two cases an approach of the second body B would satisfy the striving to maintain the relative velocity, that is, the inertia considered as relative. But since the second body B has to be treated as inert with respect to the Earth as well, the motion induced in it in this way by body A, having changing velocities, will not annul the relative velocity, but only reduce it; and no matter how large we may choose the mass of body A, it will always be extremely small compared to the Earth's mass, so the motion of body B with respect to the Earth can only be very small, and if it be detectable at all, then only by a sensitive apparatus. As further illustration one can say that due to its inertia with respect to A, body B strives to set itself in motion with respect to the Earth, but because of its inertia with respect to the Earth it strives to move relative to A rather than relative to the Earth.

Let us now apply these considerations to our flywheel and the torsion balance placed before it.

Let the circle $AFCBDF'A$ [in Fig. 1] represent the rim of the flywheel and P a readily movable body or mass point within the rim of the flywheel, as close as possible to its plane, namely a part of the mass of the arm of the torsion balance. For simplicity, let us assume that the point P actually lies within the plane of the flywheel, which of course cannot be realized with strict accuracy I for common wheels whose [27] spokes and rim lie in one plane.

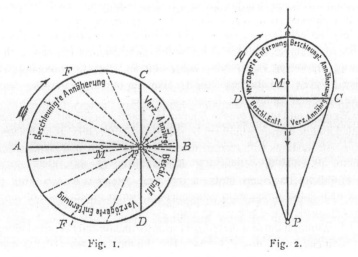

Fig. 1.   Fig. 2.

[Beschleunigte Annäherung–accelerated approach; Verz. Annäherg–decelerated approach; Beschl. Entf.–accelerated withdrawal; Verzögerte Entfernung–decelerated withdrawal.]

Let us now join point P to the center of the flywheel M, rotating in the direction of the arrow, and extend this line until it meets the rim at A on the left, at B on the right and also erect the perpendicular on AB at P, cutting the rim above at C, below at D. Then it is clear that every mass point of the rim on its way from A over C to B approaches the point P and then on the way from B over D to A recedes from it. However, the approach on the semicircle AB is accelerated up to C and then decelerated to B; similarly, the withdrawal on the semicircle BA is accelerated to D and decelerated from D to A. In view of the simplicity of the situation we can dispense with an analytic proof. But since in accordance with what we have said accelerated approach and decelerated withdrawal act in the same sense, namely both repulsively, while decelerated approach and accelerated withdrawal both act attractively, we see

[28] that we can divide the rim into two parts that differ I in their effect, namely, the part left of CD, which repels, and that part right of CD, which attracts the point P. But it is also easily understood that only the force components acting along the line AB are effective. Namely, opposite to every point F on the rim there is another, F', whose force components along the line AB reinforce each other, whereas the perpendicular components cancel, being equal and opposite. All forces that push from the left of CD and all those that pull from the right of CD act together along the line AB in the direction from A to B. Therefore, on the basis of the conception of the relativity of inertia an acceleration *away from the axis* is imparted to the point P, as our conception of the invertibility of centrifugal force requires. The relative rotation between the wall of Newton's bucket and the water contained in it would indeed generate appreciable centrifugal forces in the water if the wall were sufficiently massive to be no longer practically non-existent compared with the mass of the Earth.[10]

If the ideas sketched here are correct, many consequences will follow, some of which will admittedly seem very strange. The same amount of gunpowder, acting on the same cannon ball in the same cannon, would impart to the projectile a greater velocity on, for example, the Moon than on the Earth. Naturally, however, the greater velocity would represent, rather than a greater amount, the same amount of energy as the smaller velocity that the projectile receives on our more massive planet. This

[29] would reveal itself in the fact that despite the greater velocity, I the penetration capacity on the Moon would not be greater than on the Earth. For the $mv^2/2$ as measure of the so-called kinetic energy would not be the complete formula, as it fails to take into account the surroundings with regards to mass and distance, that is, the specification of the masses for which the velocity "v" holds.

One should not be overly hasty in rejecting our ideas as obviously incorrect because of their actual or apparent consequences. For example, Foucault's pendulum should be explicable according to our ideas as well. According to the conception of relativity of all motion the fact that the plane of the pendulum is carried along not by the Earth, assumed to be rotating, but rather by the universe when it is assumed to be

---

10   We originally said "universe" but now the "Earth." It is to be assumed that the Earth will probably play a much larger role than the more distant masses of the universe.

rotating; the plane notoriously follows the latter, and *not* the Earth. To resolve the difficulty one only has to assume that the forces that turn the plane have their origin in the one-sided action of the masses of Sun and Moon, whereas the uniformly distributed mass of the Earth has no effect.

The phenomena of the tides would also have to have a treatment different from the usual one; namely, in our figure we merely need to take our point P outside the circle and draw tangents from it to C and D [Fig. 2]; the circle then represents the Earth, the point P the Moon or the Sun, and PC and PD the axial section of a cone tangent to the Earth from the Moon, treated as a point. One will then see that the Earth will be divided by the approximately circular plane CD, which appears in the figure as a line, into two parts that, on account of the distance, are very nearly equal; of these, the half below CD, i.e., the part turned toward the Moon, will be attracted, while the part above CD, away from the Moon, must be repelled; the mobile water follows these attractive and repulsive forces and excites both the tidal waves that circle the Earth in the time between two culminations | of the Moon. Incidentally, this [30] explanation differs from the standard one essentially, rather than only in the point of view. Were it possible to connect the two objects rigidly, so that rotation but no approach or recession were possible, then a difference would result; according to the usual explanation the wave away from the Moon could *not* be realized; according to our explanation the occurrence of the tides would not be materially changed.

It is not our intention to dwell on the many other consequences, and likewise we forgo a more detailed interpretation of the analogies that suggest themselves. Still, let us mention the following parallels only in order to indicate the extent to which the problem of motion that we have raised and hypothetically solved here is related to that of the nature of gravity and at the same time comes rather close to the known effects of electric forces: a body that approaches a second one or moves away from it would be without influence on the latter as long as the velocity of approach (to be taken either with a positive or a negative sign) remains unchanged; any change of this velocity on the other hand would entail the above-demonstrated effect.

As is well known, the presence of a current in a conductor is not sufficient for the generation of induction effects, either the magnitude of the current or the distance must vary; in our case the *change* of distance, i.e. the motion, would not suffice for the generation of the attractive or repulsive effects, but rather the velocity itself has to change.

If we think of the effects in question as originating from some as yet unspecified waves, for example from longitudinal pressure waves (although most people would rather think first of transverse waves, due to the prevailing opinions!), then | the [31] breaking of the waves with equal speed would have no effect, whereas acceleration of the rhythm would induce a repulsive force as long as it lasts, and deceleration an attractive force. It should be noted that these last considerations are only hints and not completed developments that could be understood by attentive reading only. They are also hypothetical to a high degree, as are our main statements, which of course can be regarded as facts only if the experiments described above (or equivalent experiments) are successful. But let us finally emphasize that the success of the experiments would prove the presence of the type of action in question (even if not our interpretation), but its failure would not disprove this action. It remains questionable whether the effect is of an order of magnitude that would be reliably observable in the face of the experimental error sources. But if we deny the existence of the "inverse centrifugal force," for short, there would be consequences that would really be totally untenable. The incomprehensible would be deemed a fact, the logical absurdity of absolute motion would have to be regarded as having an effect, hence also as actual or real. To imagine and grasp this again in a concrete way, let the reader imagine being on a seat that is fixed on the axis of a rotational apparatus, freely rotating in otherwise empty space, in such a way that he must take part in the "rotation" of the apparatus. The "rotation" of the apparatus would then be accompanied by no change in position with respect to other heavenly bodies, it would not only be imperceptible *as such*, it would be totally unthinkable. It would be a logical monster. Nevertheless, according to the prevailing view that unthinkable "rotation," which cannot be differentiated from rest, is supposed to generate centrifugal forces so that the reader sitting on that seat can [32] observe the phantom "rotation" by the | equatorial bulge of his little speck of matter, and can even *measure* the motion, not to say the ghost of the motion, by the amount of the deformation.

The world as a whole, we dare say, is not made in a way that would be in conflict with our prerequisites of thought. And therefore the idea of the relativity of all motion and also the origin of the centrifugal effects in *relative* acceleration resistance may have a priori probability on its side, and not against it, in spite of all its unaccustomed and seemingly alarming consequences.

On the basis of our conception it is naturally also necessary to modify the interpretation of the astronomical facts. The Ptolomaic and the Copernican system are both equally "*correct*" as far as they both describe the actual motions of the celestial bodies truthfully; but this description takes a much *simpler* form if one puts the coordinate system at the larger so-called central body, rather than at the Earth. In accordance with the conception of the relativity of all motions, including therefore central motions, a revolution of the Earth can be completely replaced by an axial rotation of the Sun *insofar as only these two bodies come into consideration*. The circumstance that the Earth, despite the "attraction," does not plunge into the Sun, or the Moon into the Earth, is of course explained on the basis of the usual conception by the motion of revolution of the smaller celestial body; while, for example, the axial rotation of the Sun with respect to the universe is taken to be negligible, and plays no role at all. If our conception is correct, the so-called axial rotations are not irrelevant to the equilibrium of the world systems but must be equally taken into account like all other factors. Incidentally, the assumption of an attraction of the Earth by the Sun is not a felicitous interpretation of the factual situation insofar as the so-called *attractive* | [33] forces can only be adduced from the reduction of distance; naturally, this is not to say that the Sun would not attract the Earth if the relative motions of the two bodies were other than they actually are. However, as the facts stand, that true attraction does *not* obtain; in accordance with everything we know, it would indeed occur in the case of relative rest of the bodies and bring about the fall of the Earth into the Sun. The attraction is compensated by the existing relative motions, and this would correspond to the usual conception if it would take into account the relative motions instead of operating with the phantom of absolute rotation and inertia treated correspondingly as absolute.

It is also apparent that according to our conception the motions of the bodies of the solar system could be seen as pure *inertial motions*, whereas according to the usual view the inertial motion, or rather its permanent gravitationally modified tendency, would strive to produce a rectilinear-tangential motion.

The central point upon which our view differs from the conventional one can also be expressed precisely as follows. The prevailing view refers all locations, hence all motions and derived concepts such as accelerations and inertia in particular, to a coordinate system considered *absolutely fixed in space*; the absolutely fixed point in space would accordingly be not only an idea, but would have a most real meaning; it would be actual because it could act. It would be such although no criterion can be given for its being fixed. A sympathetic devotee of Kant objected in a private communication that my law of inertia is | not well defined, whereas the usual is definite and moreover [34] is merely an application of the law of causality to mechanics, with its (allegedly) *a priori* character. "No body changes its motion without cause." By no means do we acknowledge the *a priori* character of the law of causality; if so, the work of Galileo, to the extent that he discovered inertia, would then have been labor in vain and would have resulted only in trivialities, so to speak, which could have been obtained far more simply by deductively applying theorems that were certain *a priori*. To repeat, this we deem incorrect; yet our law of inertia may be given quite an analogous formulation, such as the statement that "no bodies change their *relative* motions without cause"; wherein the "old" law, you see, is only completed by the emphasized word "*relative.*" That Kantian objection is surprising, since it, in particular, further supports the objectively real meaning of spatial relations. This is also the basis for the interest that our treatment may perhaps claim, even in the case that it would for some reasons turn out to be untenable. Namely, in that case we would have shown that it is *possible* to proceed from the relativity of all motions, that one *can* explain inertia and centrifugal motions on the basis of the relativity hypothesis; but that upon following this chain of thought further one hits upon factual contradictions, which make the assumption of absolute motion necessary and therefore make manifest the real significance of a coordinate system taken to be absolutely fixed in space, and thereby with even greater probability make manifest the reality of the spatial relations.

From private objections I gather incidentally that if *incorrectly*, that is incompletely, applied, our hypothesis seems to include a violation of the principle of conservation of energy; when considered more exactly, however,| namely when our point [35] of view is completely implemented, this apparent contradiction disappears. It is true, as we already emphasized, that the formulas for the kinetic energy and everything depending on it are in need of an appropriate completion by respecting the other masses in the vicinity.

EDITORIAL NOTES

[1] Translation from: Henry Allison, Peter Heath (eds.): *Theoretical Philosophy after 1781. (The Cambridge Edition of the Works of Immanuel Kant.)* Cambridge University Press: Cambridge, 2002, 200, 261, 262.

[2] This reference is to Budde, E. *Allgemeine Mechanik der Punkte und starren Systeme: Ein Lehrbuch für Hochschulen.* First edition. Reimer: Berlin, 1890.

[3] This reference is to Kant, Immanuel: *Gedanken von der wahren Schätzung der lebendigen Kräfte und der Beurtheilung der Beweise derer sich Herr von Leibnitz und andere Mechaniker in dieser Streitsache bedienet haben, nebst einigen vorhergehenden Betrachtungen welche die Kraft der Körper überhaupt betreffen.* (1746). Published in: *Kants gesammelte Schriften*, vol. 1, Berlin: Königlich Preußische Akademie der Wissenschaften, 1910, 1–182.

AUGUST FÖPPL

ON ABSOLUTE AND RELATIVE MOTION

Originally published as "Über absolute und relative Bewegung" in Sitzungsberichte der Bayerischen Akademie der Wissenschaften, mathematisch-physikalische Klasse (1904) 34: 383–395 (submitted November 5, 1904). Excerpts already translated by Julian Barbour have been used. (Julian Barbour and Herbert Pfister (eds.)"Mach's Principle: From Newton's Bucket to Quantum Gravity.")

The most acute observations on the physical significance of the law of inertia and the related concept of absolute motion are due to Mach. According to him, in mechanics, just as in geometry, the assumption of an absolute space and, with it, an absolute motion in the strict sense is not permitted. Every motion is only comprehensible as a relative motion, and what one normally calls absolute motion is only motion relative to a reference system, a so-called inertial system, which is required by the law of inertia and has its orientation determined in accordance with some law by the masses of the universe.

Most authors are today in essential agreement with this point of view, as expressed most recently by Voss[1] and Poincaré[2] in particular. A different standpoint is adopted by Boltzmann,[3] who does not believe he can simply completely deny an absolute space and, with it, an absolute motion. Here, however, I shall proceed from Mach's view and attempt to add some further considerations to it. |

Mach summarizes his considerations in the following sentence:[4] "The natural standpoint for the natural scientist is still that of regarding the law of inertia provisionally as an adequate approximation, relating it in the spatial part to the heaven of fixed stars and in the time part to the rotation of the Earth, and to await a correction or refinement of our knowledge from extended experience." Now it seems to me not entirely impossible that just such an extended experience could now be at hand. In a recent publication of K. R. Koch[5] on the variation in time of the strength of gravity we read: "Accordingly, the assumption of a genuine variation of gravity, or, more pre- [384]

---

1   A. Voss, "Die Prinzipien der rationellen Mechanik" *Enzyklop. d. math. Wissensch.*, Band IV, 1, p. 39, 1901.
2   H. Poincaré, *Wissenschaft und Hypothese*. German translation by F. und L. Lindemann, Leipzig 1904.
3   L. Boltzmann, *Prinzipe der Mechanik*, II, p. 330, Leipzig 1904.
4   E. Mach, *Mechanik*, 4th ed. p. 252, Leipzig 1901.
5   K. R. Koch, Drude's *Annalen der Physik*, Band 15, p. 146, 1904.

Jürgen Renn (ed.). *The Genesis of General Relativity*, Vol. 3
Gravitation in the Twilight of Classical Physics: Between Mechanics, Field Theory, and Astronomy.
© 2007 Springer.

cisely, its difference between Stuttgart and Karlsruhe, seems to me appropriate." We shall naturally have to wait and see if this assertion stands up to further testing; at the least, we must now reckon with the real possibility that it is correct.

An explanation of such a phenomenon, if it is correct, would be very difficult on the basis of known causes. This circumstance encourages me to come forward now with a consideration that I have already developed earlier and long ago led me to the assumption that small periodic variations of gravity of measurable magnitude should be considered as a possibility.

Experience teaches us first that the inertial system required by the law of inertia can be taken to coincide with the heaven of the fixed stars to an accuracy adequate for practical purposes. It is also possible to choose a reference system differently, for example, fixed relative to the Earth, in order to describe the phenomena of motion. However, it is then necessary to apply to every material point the additional Coriolis [385] forces of relative motion if one is to predict the motions correctly. One can therefore | say that the inertial system is distinguished from any other reference system by the fact that in it one can dispense with the adoption of the additional forces. Rectilinear uniform translation of the chosen reference system can be left out of consideration here as unimportant.

However, it is obvious that the fixing of the inertial system relative to the heaven of the fixed stars cannot be regarded as fortuitous. Rather, one must ascribe it to the influence, expressed in some manner, of the masses out of which it is composed. We can therefore pose the question of the law in accordance with which the orientation of the inertial system is determined when the instantaneous form and relative motion of the complete system of masses, i.e., the values of the individual masses, their separations, and the differential quotients of these separations with respect to the time, are regarded as given.

The logical need for such a formulation of the problem if one wishes to avoid the assumption of an absolute space was also felt by Boltzmann when he referred in passing to the possibility[6] that the three principal axes of inertia of the complete universe could provide the required orientation. If this rather natural supposition could be maintained, the conceptual difficulties would indeed be overcome. However, I believe that this supposition is not admissible. Let us imagine, for example, a universe that is otherwise arranged like ours but with the only difference that there are no forces at all between the individual bodies in the universe. Then for the inertial system valid for this universe, all the bodies in it would move along straight lines. However, a calculation that is readily made shows us that under this assumption the principal axes of inertia of the complete system would in general execute rotations relative to the inertial system. It is therefore necessary to look for a different condition that can enable us to understand the fixing of the inertial system. |

[386]   If first we assume that all the bodies of the universe are at rest relative to each other except for a single mass point that I suppose is used to test the law of inertia,

---

6   Loc. cit., p. 333.

and which I will call the "test point," [*Aufpunkt*] then in accordance with the experiences we already have one could not doubt that the test point would, when no forces act on it, describe a straight path relative to a reference system rigidly fixed to the masses. In this case, the inertial system would be immediately fixed in space.

We can now imagine the case in which the bodies of the universe consist of two groups, one of which is "overwhelmingly" large compared to a smaller group and in which the masses within each group do not change their relative separations, whereas the smaller group, regarded in its totality, does carry out at the considered time a motion, say a rotation, relative to the larger group. If only one of the two groups were present, the inertial system would have to be fixed relative to it. Since the two work together, and one of the groups has been assumed to be much more "powerful" than the other, the inertial system will now be indeed very nearly at rest relative to the first group, but it will still execute a small motion relative to that group, which, of course, will be the consequence of the influence of the second, smaller group.

Given such a situation, what would be the most expedient way to proceed? I believe that one cannot be in doubt. One would fix the reference system exclusively using the first, overwhelming group and calculate as if this were the inertial system but take into account the influence of the second group by applying in this case to every test point the very weak additional forces of the relative motion that the chosen reference system executes relative to the true inertial system. If one makes such a decision, then these Coriolis forces no longer appear as mere computational quantities that arise from a coordinate transformation but as physically existing forces that are exerted by the masses of the smaller group | on every test point and arise because   [387] these masses have a motion relative to the chosen reference system.

To develop this idea further, one could start by investigating the case in which the second, smaller group that I just mentioned is represented by a single body. One then has the task of determining the magnitude and direction of the force, which will depend on the velocities of the single body and the test point relative to the reference system determined by the remaining bodies of the universe and on the separation between the single body and the test point. If we suppose that this problem has been solved for a single body, then, using the superposition law, we can also obtain the influence of a whole group of moving bodies.

The securely established observational results that are currently available are certainly not adequate to solve this fundamental problem; however, one does not therefore need to doubt that on the basis of further observations we could arrive at a solution.

After these preliminary considerations, I now turn to the case that corresponds to reality. Using the circumstance that the constellation of the fixed stars changes little in the course of several years or centuries, we can suppose that a reference system that more or less coincides with the inertial system is fixed relative to three suitably chosen stars. However, in order to take into account the small deviations that still remain, one must suppose that to each test point there are applied Coriolis forces, which, as we have just described, are to be interpreted as forces that depend on the velocities of the individual bodies in the universe and the velocity of the test point.

We are now in the position—and on this I put considerable value—to specify a condition meeting our requirement for causality that must be satisfied by the true inertial system required by the law of inertia. Namely, the true inertial system is the [388] reference system for which all the | velocity-dependent forces that arise from the individual bodies of the universe are in balance at the test point. Even if in practice it is clear that we have not gained very much through this statement, it does appear to me that we have thereby obtained a very suitable basis for forming a clear concept of what is known as absolute motion in mechanics. There is at the least a prospect opened up of a way of determining the inertial system once the law that establishes the velocity-dependent forces has been found. In other words, it will be possible to construct the absolute space that appears in the law of inertia without having to sacrifice the notion that ultimately all motions are merely relative.

Besides, in all these considerations my main aim is to make it at least plausible that if one is to find a satisfactory solution to the questions that relate to the law of inertia it will be necessary to assume the existence of forces between the bodies in the universe that depend on their velocities relative to the inertial system. If this is accepted, then there follows the task of looking for possible phenomena whose relation to the expected general law of nature could be such that the law governing the velocity-dependent forces could be inferred. These forces, which for brevity I shall in what follows simply call "velocity forces," have nothing to do with gravitational forces, which arise concurrently with them, and specifically they can—and probably will— follow a quite different law than the gravitational forces with regard to distance dependence.

At this point I should like to make a remark in order to divide this communication into two quite separate sections. I believe that I can defend with complete definiteness and confidence what I have said up to now. However, I regard what follows as merely [389] an attempt that could very | well fail; nevertheless, it is an attempt that at the least has a prospect of success and therefore must be brought forward at some time.

It seems to me that the most promising way of proving the existence of the postulated velocity forces and finding the law in accordance with which they act is to observe with the greatest possible accuracy phenomena associated with motions near the Earth that occur with great velocity. Just as the discovery of gravitation had as its starting point the observation of free fall, here too the first step to the solution of the puzzle could be obtained through observations of terrestrial motions and their correct interpretation. The immediate vicinity of the Earth's mass opens up some prospect of proving the existence of velocity forces more accurately than would be possible with the finest astronomical observations, which, as experience teaches, are certainly only very weak under normal circumstances.

This thought led me some time ago to make the gyroscope experiments that I reported to the Academy very nearly a year ago.[7] I expected then, as I explicitly said, to establish a behavior of the gyroscope that did not agree with the usual theory in the hope that the observed deviation could be attributed to the velocity forces I seek and that these would therefore be made accessible to experimental research. Now certain indications of a deviation were indeed discernible, but as a careful and conscientious experimentalist I could not put any weight on them and I was forced, as I did, to declare a negative result of the experiment as regards the direction that it was intended to follow in the first place. In the meanwhile, I have made some further experiments with the same apparatus, though admittedly few, since they are very laborious and time consuming. However, the result could do nothing but I strengthen [390] me in the view that the accuracy that can be achieved with this experimental arrangement is not sufficient to prove the existence of the velocity forces if they exist at all.

More promise of success probably lies in a further continuation of the free fall experiments, whose results to date can already be described as rather encouraging, after all. The ordinary theory, which does not take velocity forces into account, leads one to expect in the northern hemisphere, in addition to an easterly deviations of free fall motion from the vertical, a southerly deviation of such an extraordinarily small amount only that its experimental confirmation would be entirely out of the question. Nevertheless observers have time and again found southerly deviations of measurable magnitude, which are of a totally different order of magnitude (several hundred times larger and more) than those expected from theory. The newest observations in this area, due to the well-known American physicist E. H. Hall,[8] famous as an experimentalist and discoverer of "Hall's Phenomenon," have confirmed these experiences again. Indeed Hall considers further experiments on a larger scale (for greater heights of fall) necessary, and he holds out a prospect of such experiments. One may expect very valuable insights from them. Perhaps it will also serve to further the continued performance of such experiments if the hope for a positive result is strengthened by the theoretical considerations as I have offered here. For it takes indeed no small measure of courage to undertake painstaking and lengthy experiments, if the unanimous opinion of all theoreticians comes to this, that they cannot possibly lead to the expected result. This consideration has been for me the main motive to come forward with my views, although I I must admit that so far they are too deficient in an ade- [391] quate experimental basis to be likely to meet with much approval.

Now I come to the admittedly most doubtful conjecture that I formed in connection with the above, and which is connected with the observation of Koch mentioned in the beginning. One can understand immediately that I must also expect velocity forces that arise from the motion of the Earth with respect to the Sun. The Sun is a fixed star like others and it contributes its part to the determination of the inertial system; or, in other words, it exerts velocity forces, if we account for the motion relative to a system of reference that is established without regard to the Sun. Even if nothing is known about the dependence of these forces on distance we may nonetheless

7   *Sitzungsberichte* 1904, p. 5.
8   Edwin H. Hall, *Physical Review,* 17, p. 179 and p. 245, 1903; further *Proceedings of the American Acad.* 39, No. 15, p. 399, 1904.

regard it as probable that the influence of a closer body is larger than that of a much more distant one. Therefore nothing is more natural than the assumption of velocity forces of such a kind that could cause a small periodic change of gravity with a diurnal as well as an annual period.

One difficulty, a very serious and possibly insurmountable one, arises only when one assumes that these velocity forces could be of such a magnitude that they would be measurable on the surface of the Earth and that the observations of Koch could be evaluated in this sense. Then one necessarily encounters the astronomers' objections, who have noticed nothing of the occurrence of such forces in spite of the great accuracy with which they can predict the phenomena of motion in the solar system.

[392] This objection is so lucid that one would almost abandon the hope of being able to silence it. To be sure, as long as nothing is conclusive about the I laws of action of the velocity forces in other ways, one could retreat to the view that a remote possibility exists that this contradiction could be cleared up later. And in this hope one could first quietly wait and see what consequences derive from such observational results as those found by Koch, temporarily neglecting the contradiction. If for example the original observation of Koch were not only confirmed, but if also the daily period of gravity fluctuations were really found as are expected on the basis of the views proposed here, then one could regard this as a certain confirmation of the suggested theory in spite of all objections.

However, I understand that such a position would be untenable. If it is not possible even now to make it reasonably credible that the interpretation of Koch's observations that I regard as possible does not necessarily have to be in contradiction with astronomical experience, then no one will pay heed to my interpretation, and the danger could arise that the same fate awaits Koch's observations as so far has befallen the southerly deviation of falling bodies, that is, that one does not take it seriously and immediately tends to assign it to errors in observation, because it does not fit with the accepted theory.

Only with this intention and by no means in order to represent the several possibilities I am about to discuss as somehow particularly probable, I still mention the following.

Consider a planet that circles its central body in agreement with the first two laws of Kepler. Let the law of the velocity forces be of the form that the planet is subject to an attraction by its sun that is proportional to the velocity component orthogonal to the radius vector and inversely proportional to the first power of the distance. One immediately recognizes that under these circumstances one would not need any grav-

[393] itational force in addition to the velocity force in order to I explain the motion of the planet that is given by the observations. The astronomers of a solar system with only a single planet would have indeed no means to decide whether Newton's gravitational force or the velocity force adopted in the indicated manner were correct if they wished to restrict themselves to observation of the orbit alone. However, the difference would immediately be apparent when they took into account observations on their planet.

In accordance with Newton's gravitational law as well, there is, as is well known, a daily period of variation of the gravity force that gives rise to the contribution of the Sun to the motion of the tides but is too weak to be established by pendulum observations. However, if the astronomers of that solar system were to make the attempt to replace Newton's law of gravitation by the law of the velocity forces that we have mentioned, they would have to expect a much greater daily period, which, for the same relations between our Earth and the Sun, would be about 180 times greater than would be expected in the other case.

It should also be remarked here that the velocity law, which was chosen at random, is in fact only one of infinitely many that would all achieve the same, namely, the explanation of the motion of a single planet around its sun in agreement with Kepler's first two laws without having to invoke in addition Newton's gravitational force. All one needs to do is to allow the velocity component in the direction of the radius vector, which was hitherto assumed to be without influence, to participate as well in accordance with some arbitrary law and then arrange the law according to which the orthogonal velocity component acts on the force of attraction in such a way that the required motion results. There is also no need to make a restriction to the first power of the velocity; one could also consider the second or other powers. |

When a solar system has more than one planet, it is naturally much more difficult [394] to explain all the planetary orbits merely with the help of velocity forces, since it is now necessary to satisfy Kepler's third law as well. So far as I can see, one would then be forced to make quite artificial assumptions. Even if one could achieve success in a simpler manner than it now appears to me, it would still be questionable if one could also explain the disturbances of the planetary orbits, the motions of the moons, etc.

However, one should not forget the aim of this discussion. It is in no way my intention to replace Newton's law by a law of velocity forces. I only want to make it plausible that under certain circumstances the velocity forces by themselves could have effects very similar to those of the gravitational forces. If this is then granted, it immediately follows that in such an event it would be very difficult to separate out from the astronomical observations the part due, on the one hand, to gravitational forces and, on the other, to the velocity forces.

On the basis of this consideration, I believe it is best not to be deflected by the admittedly very weighty objections of the astronomers from seeking phenomena that could be related to velocity forces. If it does prove possible, following this entirely independent research approach, to derive a law of the velocity forces, it will still be possible to make, as the best test of the admissibility of the result, an accurate comparison with the astronomical observations, taking into account the error limits that are relevant.

Naturally, I would not recommend such a procedure if I did not have great confidence in the very existence of the velocity forces, even though I must leave it as an open question whether they have a magnitude such | that they are measurable in [395] motions accessible to our perception. If one will admit an absolute space, then, of course, every reason for the assumption of velocity forces disappears. However, in this point at least—that I do not recognize an absolute space—I am in agreement with the majority of natural scientists, and I therefore hope that I shall receive recognition among them, at least for the conclusions drawn in the first part of this communication.

AN ASTRONOMICAL ROAD TO
A NEW THEORY OF GRAVITATION

MATTHIAS SCHEMMEL

THE CONTINUITY BETWEEN CLASSICAL AND RELATIVISTIC COSMOLOGY IN THE WORK OF KARL SCHWARZSCHILD

1. KARL SCHWARZSCHILD: PIONEER OF RELATIVISTIC ASTRONOMY

Only a few weeks after Einstein had presented the successful calculation of Mercury's perihelion advance on the basis of his new theory of general relativity in late 1915, the German astronomer Karl Schwarzschild (1873–1916) published the first non-trivial exact solution of Einstein's field equations (Schwarzschild 1916a). The solution describes the spherically symmetric gravitational field in a vacuum and holds a central place in gravitation theory, comparable to that of the Coulomb potential in electrodynamics. It was not only an important point of departure for further theoretical research but also, up to recent times, the basis for all empirical tests of general relativity that proved not only the principle of equivalence but also the field equations themselves. Schwarzschild made a further substantial contribution to the theory when he found another exact solution describing the interior gravitational field of a sphere of fluid with uniform energy density (Schwarzschild 1916b). In this communication an important quantity makes its first appearance. It is the quantity that is later known as the *Schwarzschild radius*, which plays an important role in the theory of black holes many decades later.[1] But even long before the final theory of general relativity was established, Schwarzschild had already occupied himself with possible implications of its predecessors for astronomy; in 1913 he carried out observations of the solar spectrum in order to clarify if the gravitational redshift predicted by Einstein on the basis of the equivalence principle was detectable (Schwarzschild 1914).

In view of the fundamental role played by general relativity in astronomy, astrophysics, and cosmology today, it appears quite natural that an astronomer would engage in the study of this theory. Astronomical objects of all scales ranging from supermassive stars via galaxy nuclei and quasars to the universe as a whole are described on its basis. However, at the time when Schwarzschild made his contributions, the situation was quite different. None of the spectacular objects nowadays so successfully described by general relativity were in the focus of research, most of

---

1   For a thorough analysis of the early history of the interpretation of Schwarzschild's solutions and the Schwarzschild radius in particular, see (Eisenstaedt 1982; 1987; 1989). See also (Israel 1987), in particular sec. 7.7 on the *Schwarzschild 'Singularity'*.

Jürgen Renn (ed.). *The Genesis of General Relativity*, Vol. 3
Gravitation in the Twilight of Classical Physics: Between Mechanics, Field Theory, and Astronomy.
© 2007 Springer.

them not even known at all. Rather, the deviations from Newtonian theory that general relativity predicted were so small that in most cases they lay on the verge of detectability, even on astronomical scales. General relativity could thus easily be considered a physical theory—it was developed in the attempt to solve problems in physics such as the incompatibility of Newtonian gravitation theory and special relativity—with little implications on astronomy. And even as a physical theory it was still controversial, as is strikingly illustrated by the case of the physicist Max von Laue who as late as 1917 preferred Nordström's theory of gravitation to Einstein's.[2] Accordingly, at the time, astronomers showed little interest in general relativity. Einstein's plea to put the theory to an empirical test went unheard by most of them. In his attempts to provide empirical evidence for the theory, Erwin Freundlich, an outsider to the astronomical community, even met with hostility among Germany's most prominent astronomers.[3] Why was Schwarzschild an exception to this? What put him in the position to recognize so early the significance of general relativity?

The clue for answering these questions lies in the study of work Schwarzschild had done long before the rise of general relativity. In the course of the late 19th century, foundational questions surfaced in classical physics that had implicit consequences for astronomy: consequences that were often of a cosmological dimension. Mach's critique of Newton's absolute space, for example, immediately led to the question of an influence of distant stars on terrestrial physics. The deviation of the geometry of physical space from Euclidean geometry, to give another example, had become a possibility with the work of Gauss and Riemann and could be imagined to be measurable on cosmological scales. A further example is provided by the various attempts to modify Newton's law of gravitation. Such a modification would have consequences not only for planetary motion but also touches upon questions concerning the stability of the whole universe and the large-scale distribution of matter therein.[4] These foundational questions were, despite their astronomical implications, not on the agenda of contemporary astronomical research. Nevertheless, they were studied by a few individual scientists, among them Karl Schwarzschild.

In this paper it is argued that a continuity exists between Schwarzschild's prerelativistic work on foundational problems on the borderline of physics and astronomy and his occupation with general relativity. After a brief biographical introduction (sec. 2), Schwarzschild's prerelativistic considerations on the relativity of rotation (sec. 3) and on the non-Euclidean nature of physical space (sec. 4) are presented as they are documented in his publications as well as in his unpublished notes. On this background, Schwarzschild's reception of general relativity will then be shown to have been shaped to a large extent by his earlier experiences. In fact, what at first sight may appear to be a rather technical contribution to a physical theory—Schwarz-

---

2   See (Laue 1917).
3   For an account on Freundlich's work on empirical tests of general relativity and the astronomers' reaction to it, see (Hentschel 1997).
4   For a discussions of fundamental problems arising in Newtonian cosmology, see (Norton 1999).

schild's derivation of an exact solution of Einstein's field equations—turns out to have been motivated by Schwarzschild's concern for a consolidation of the connection between astronomy and the foundations of physics as established by Einstein's successful calculation of Mercury's perihelion motion (sec. 5). What is more, Schwarzschild was reexamining his prerelativistic cosmological considerations in the framework of the new theory of relativity as hitherto neglected manuscript evidence reveals for the case of the problem of rotation (sec. 6, a manuscript page from Schwarzschild's Nachlass is reproduced with annotations in the Appendix). Furthermore it turns out that, prepared by his earlier cosmological considerations, Schwarzschild was the first to consider a closed universe as a solution to Einstein's field equations (sec. 7). Summing up, Schwarzschild's road to general relativity may be called an astronomical one. Concluding this paper it will be argued that it was no coincidence that Schwarzschild of all astronomers took this road, but that this was the natural outcome of his interdisciplinary approach to the foundations of the exact sciences (sec. 8).

2. KARL SCHWARZSCHILD: ASTRONOMER, PHYSICIST AND ASTROPHYSICIST

Schwarzschild was born on October 9, 1873 in Frankfurt am Main, the eldest of seven children of a Jewish businessman.[5] He studied astronomy in Strasbourg and in Munich, where he obtained his doctoral degree in 1896 under Hugo von Seeliger (1849–1924), one of the most prominent German astronomers at the time. After having worked for three years at the Kuffner Observatory in Ottakring near Vienna, Schwarzschild obtained his post-doctoral degree (*Habilitation*) in Munich in 1899. On this occasion, Schwarzschild had to defend five theses, mostly concerned with foundational questions, that inspired him, as we will see, to much of the work relevant to our discussion. It is therefore interesting to question the extent to which Schwarzschild's teacher, von Seeliger, was involved in formulating these theses. While it may well be the case that Schwarzschild himself played some role in their creation, their exact wording makes it plausible that they were formulated by von Seeliger (see the discussion below). Thus this sheds some light on von Seeliger's ambivalent role in the early history of relativity. On one hand he was known to be very sceptical of relativity theory. For example, he severely criticized Erwin Freundlich's attempts to provide empirical evidence supporting general relativity. On the other hand he was interested in foundational questions of theoretical astronomy and apparently inspired Schwarzschild to much of the work discussed here.

In 1901, Schwarzschild was appointed professor of astronomy and director of the observatory of Göttingen University. He became closely associated with the circle of mathematicians and natural scientists around Felix Klein and furthered the integration of Göttingen astronomy with general scientific life.[6] Schwarzschild left Göttin-

5   For a short biographical account on Schwarzschild, see (Schwarzschild 1992, 1–25).

gen in 1909 and became director of the *Astrophysikalisches Institut* in Potsdam, but, for the short remainder of his life, he maintained the personal and scientific relationships established in Göttingen. Thus it was his Göttingen colleagues and acquaintances who, on several occasions, wrote him about the latest developments of Einstein's theory and pointed out the importance of its astronomical verification.[7] On May 11, 1916, Schwarzschild died an untimely death from a skin disease he contracted while serving at the Russian front.

Schwarzschild's scientific work is characterized by its rare breadth. The range of topics from physics and astronomy covered by his more than one hundred publications is hardly surpassed by any other single scientist of the twentieth century. Schwarzschild is further known to be one of the founders of astrophysics in Germany and was its most prominent exponent at the time. While disciplinary astrophysics itself was a rather specialized enterprise—using physical instruments for astronomical observation and applying physical theory to astronomical objects—Schwarzschild's interdisciplinary outlook on the foundations of science[8] enabled him to overcome the constraints imposed by specialization and deal with foundational problems on the borderline of physics and astronomy that were not in the focus of mainstream research.

3. SCHWARZSCHILD'S PRERELATIVISTIC CONSIDERATIONS ON THE RELATIVITY OF ROTATION

In 1897, while he was assistant at the Kuffner Observatory in Ottakring, Schwarzschild published a popular article entitled *Things at Rest in the Universe* (*Was in der Welt ruht,* Schwarzschild 1897). In this paper he discusses the relativity of motion and the problem of finding appropriate reference frames. In particular, he is concerned with the question of how fixed directions in space can be defined.

Schwarzschild's starting point is the observation that the motion of an object can only be perceived relative to other objects and that therefore any object may be considered at rest. The question of what thing is at rest in the universe should therefore be reformulated in a historical manner as "[w]hat things in the universe did one find useful to treat as being at rest, at different times [in history]?".[9] In the Copernican system, Schwarzschild explains, fixed directions in space were defined by reference

---

6   See (Blumenthal 1918).
7   See, in particular, Schwarzschild's correspondence with David Hilbert. On a postcard to Schwarz-schild from October 1915, for example, Hilbert wrote: "The astronomers, I think, should now leave everything aside and only strive to confirm or refute Einstein's law of gravitation." ("Die Astronomen, meine ich, müssten nun Alles liegen lassen u. nur danach trachten, das Einsteinsche Gravitationsgesetz zu bestätigen oder widerlegen!") Hilbert to Schwarzschild, 23 October, 1915, N Briefe 331, 6r. (This and all following translations are my own, a few of them are based on the companion volumes to the Einstein edition.) Examples of this kind are also found in Schwarzschild's correspondence with Arnold Sommerfeld. A selection of Sommerfeld's scientific correspondence has recently been published (Sommerfeld 2000–2004).
8   See sec. 8.

to the system of fixed stars. Towards the end of the 17th century it became clear however that the Copernican stipulation is not unambiguous: the stars perform motions relative to one another, the so-called proper motions. Schwarzschild therefore next considers the electromagnetic aether as a candidate for a material reference of rest but comes to the conclusion that the aether too cannot serve such a purpose since it is affected by ponderable matter moving through it. Schwarzschild concludes that there are no material objects in the universe that one could reasonably consider at rest and that one can only take resort to "certain conceptually defined points and directions that may serve as a substitute to a certain extent".[10]

In order to explain how fixed directions in space may be defined on the basis of the law of inertia, Schwarzschild refers to Foucault's pendulum. By accurate observations of the rotation of the pendulum's plane of oscillation, Schwarzschild explains, one could calculate the speed of rotation of the Earth, and would then have to describe as fixed the direction with respect to which the Earth rotates with the calculated speed.

In following this idea further, Schwarzschild establishes an interesting connection between inertia and gravitation in the following way. In regarding the planets orbiting around the Sun as gigantic, diagonally pushed pendulums, he conceives an astronomical realization of the physical model of the pendulum. In analogy to Foucault's pendulum, fixed directions in space are then given by the aphelia (or perihelia) of the orbits of the different planets. However, Schwarzschild explains, astronomical observations since the middle of the 19th century reveal that the directions singled out by the orbits of the different planets rotate with respect to each other at a very slow rate, so that it is "impossible to consider all as fixed."[11] Although Schwarzschild was aware of possible astronomical explanations, such as interplanetary friction, he considered it more probable that an explanation of these small anomalies has to go further, requiring a revision of the classical law of gravitation.

In this way, Schwarzschild established a relation between the two physical phenomena, inertia and gravitation, the integration of which was later to lie at the basis of Einstein's theory of general relativity. Moreover, the observational fact by which Schwarzschild links the two phenomena—the perihelion shift of the inner planets—was later to play a crucial role in the establishment of general relativity, for some years being the only empirical fact suggesting a superiority of general relativity over the Newtonian theory.

There are, of course, fundamental aspects of general relativity that have no analogue in Schwarzschild's prerelativistic considerations. Most notably, Schwarzschild did not consider a field theory of gravitation that unifies gravitation and inertia in one

---

9   "[w]as in der Welt hat man zu verschiedenen Zeiten als ruhend zu betrachten für gut befunden?"
     (Schwarzschild 1897, 514). All page numbers cited for this text refer to vol. 3 of the *Collected Works*
     edition (Schwarzschild 1992).

10   "[...] gewisse begrifflich definierte Punkte und Richtungen, die einigermaßen als Ersatz eintreten können [...]" (Schwarzschild 1897, 516).

11   "[...] unmöglich alle als fest betrachtet werden können." (Schwarzschild 1897, 520.)

single field. In fact, in this text, Schwarzschild does not even question the origin of inertia. Unlike Mach and Einstein, he does not search for a physical cause of inertia but rather assumes inertia to be given and, on its basis, defines fixed directions in space. Most probably he therefore thought of modifications of the Newtonian law of gravitation that do not affect inertial frames. For example, it was well known at the time that the change of the exponent in Newton's inverse square law yields perihelion motions.[12] Such a motion could have easily been subtracted from the observed motions in order to obtain the "true" inertial directions in space given by the planets' orbits. There are however notes found in Schwarzschild's manuscripts that show that he was concerned with the question of the origin of inertia and that, in this context, he considered the possibility of local inertial frames rotating with respect to one another. These notes, in which Schwarzschild was again using orbits of celestial bodies in order to determine inertial directions, shall now be discussed.

As explained in sec. 2, Schwarzschild had to defend five theses, probably formulated by his teacher von Seeliger, in order to obtain his post-doctoral degree in 1899. One of these theses read: "The existence of centrifugal forces is comprehensible only under the assumption of a medium pervading all of space."[13] In a notebook of 1899 (N 11:17),[14] we find Schwarzschild's tentative defense of this thesis. In a thought experiment reminiscent of Einstein's later ones, attempting to clarify the nature of rotation, Schwarzschild imagines two planets of identical constitution rotating with different angular velocity and having atmospheres that are so dense that the outer world cannot be observed.[15] An inhabitant of one of the planets travelling to the other would have no way of understanding how the difference in the "gravitational conditions" (*Schwereverhältnisse*) arises, since he would not notice the rotation. This shows clearly, Schwarzschild explains,

---

12 Thus, Schwarzschild's teacher Hugo von Seeliger wrote in a letter to Arnold Sommerfeld: "that the law of attraction $1/r^n$ ($n \neq 2$) [causes] perihelion shifts, that is known to any astronomer since time immemorial. [...] *Newton* already treated this case, or a quite similar one, in his 'Principia.'" ("[...] daß das Anziehungsgesetz $1/r^n$ ($n \neq 2$) Perihelbewegungen [hervorruft], das ist jedem Astronomen seit jeher bekannt. [...] Schon *Newton* hat diesen oder einen ganz ähnlichen Fall in den 'Prinzipien' behandelt.") Hugo von Seeliger to Arnold Sommerfeld, May 25, 1902, Arnold Sommerfeld Nachlass, Deutsches Museum, Munich, HS 1977-28/A, 321, 1-1.

13 "Die Existenz von Centrifugalkräften ist nur unter der Annahme eines den ganzen Raum erfüllenden Mittels zu begreifen." A document naming the five theses can be found in N 21 (for an explanation of this notation see the next footnote).

14 Here and in the following, references to Karl Schwarzschild's Nachlass in the Niedersächsische Staats- und Universitätsbibliothek Göttingen are indicated by an archival number following an 'N' (e.g. N 11:17).

15 Consider, for example, Einstein's thought experiment involving two identical fluid bodies rotating with respect to one another (Einstein 1916, 771–772). Schwarzschild's thought experiment is further reminiscent of a later one by Poincaré, who also considered a planet covered by clouds so that its inhabitants cannot see the sun or stars. They therefore, Poincaré argued, would have to wait longer than we did until a "Copernicus" arrived, who could explain the centrifugal and Coriolis forces by assuming that the planet rotates (Poincaré 1902, chap. 7).

that not only the internal relative circumstances but also the relations to the surrounding space have an influence on the processes in a system of bodies. Following Newton we could state that there is an absolute space and that the relation of motions to this absolute space has an influence on the forces appearing through this motion. Or, in other words: absolute space has an effect on the bodies. Now, we are used to thinking of anything having an effect as something real, namely something material, and from this it follows that, if we want to stick to the usual way of thinking, we have to imagine space, Newton's absolute space, filled with a substance.[16]

The hypothetical identification of space with a substance now puts Schwarzschild in a position to discuss the global validity of the locally distinguished directions:

> This substance does not have to be at absolute rest, but only in a state of motion that in some way distinguishes three fixed directions in space [...]. Then it is comprehensible that the centrifugal forces are based on a relation of the motion of the usual bodies to the motion of this substance.[17]

From the observation that the perihelia of double stars are at rest with respect to the directions that seem fixed inside the solar system, Schwarzschild concludes that the directions distinguished in their region of space have to be the same as in the solar system.

To sum up, while in the previous example, Schwarzschild had established a relation between inertia and gravitation, here he relates inertia to the structure of space, considering the possibility that the local inertial directions may vary on cosmological scales.

4. SCHWARZSCHILD'S PRERELATIVISTIC CONSIDERATIONS ON NON-EUCLIDEAN COSMOLOGY

A second example for Schwarzschild's prerelativistic treatment of foundational questions having cosmological implications is provided by his application of non-Euclidean geometry to physical space. Again, this work appears to have been inspired by one of the five theses Schwarzschild had to defend in order to attain his degree. This thesis reads: "The hypothesis that our space is curved should be rejected".[18] It is

---

16 "[...] daß auf die Vorgänge in einem Körpersystem nicht nur die inneren relativen Verhältnisse, sondern auch die Beziehungen zum Raum, der sie umgiebt, von Einfluß sind. Wir könnten mit Newton sagen, daß es einen absoluten Raum giebt und daß das Verhalten der Bewegungen zu diesem absoluten Raum auf die bei der Bewegung auftretenden Kräfte von Einfluß ist. Oder in anderen Worten: der absolute Raum hat eine Wirkung auf die Körper. Nun pflegen wir uns aber alles, was eine Wirkung hat, als etwas wirkliches, nämlich als etwas Materielles zu denken, und daraus folgt, daß wir, wenn wir überhaupt in der üblichen Denkweise bleiben wollen, uns den Raum, Newtons absoluten Raum durch einen Stoff erfüllt denken müssen." (N 11:17, 8v–9r.) There are no page numbers in this notebook. The page numbers given here refer to my pagination.

17 "Dieser Stoff muß nicht absolut ruhen, sondern nur eine Bewegungsform haben, welche auf irgend eine Weise drei besondere feste Richtungen auszeichnet [...]. Dann ist begreiflich, daß die Centrifugalkräfte auf einer Beziehung der Bewegung der gewöhnlichen Körper zur Bewegung dieses Stoffes beruhen." (N 11:17, 9r.)

18 "Die Hypothese einer Krümmung unseres Raumes ist zu verwerfen" (N 21).

plausible to assume that this thesis too was formulated by von Seeliger. In fact, the thesis seems to reflect von Seeliger's attitude toward the application of non-Euclidean geometry to physics and astronomy which was extremely sceptical as may be illustrated by the following passage from a talk by von Seeliger entitled *Remarks on the So-Called Absolute Motion, Space, and Time*:

> [...] the common and therefore very fatal misapprehension has emerged that one believed to be able to decide by measurement which geometry is the "true" one, or even, which space is the one in which we live. From the stand point taken here the latter formulation is by far the more dangerous one, since space in itself has no properties at all.[19]

In the above-mentioned notebook of Schwarzschild we find an entry in which the thesis is slightly reformulated as follows: "The assumption of a curvature of our space is without any advantage for the explanation of the structure of the system of fixed stars".[20] The note is accompanied by considerations and calculations in which Schwarzschild examines empirical consequences of a curvature of space, for example, on the parallaxes of stars. One year later, Schwarzschild published a more detailed account of these considerations, though now under a different perspective. While the notebook entries aimed at a rejection of the curvature of space—understandably so in view of their context, the defense of a thesis—the purpose of the published article is to estimate the degree of curvature that can be assumed without contradicting observation. The article bears the title *On the Permissible Scale of the Curvature of Space (Über das zulässige Krümmungsmaass des Raumes*, Schwarzschild 1900).[21]

In his article Schwarzschild mainly discusses two cases: hyperbolic space, having constant negative curvature, and spherical space, having constant positive curvature. He makes the assumption that light travels along geodesics.

As far as hyperbolic space is concerned, Schwarzschild is able to estimate a minimal radius of curvature with the help of the parallax. As is well known, the parallax of a star, for simplification assumed to be nearly perpendicular to the ecliptic, is defined as half the difference of the two angles under which the star is seen in an interval of half a year, $\pi = (\alpha - \beta)/2$ (see Fig. 1). (In Euclidean space this coincides with the angle under which the radius of the Earth's orbit is seen from that star.) In Euclidean space, therefore, a parallax of exactly zero implies that the star is infinitely far from the Earth, since parallel geodesics in Euclidean space intersect at infinity. In hyperbolic space, in contrast, neighboring geodesics diverge. Thus even stars infinitely remote from the Earth possess a certain parallax. This minimal paral-

19 "[Es] ist der verbreitete, aber gerade darum sehr verhängnisvolle Irrtum entstanden, daß man glaubte durch Messungen entscheiden zu können, welche Geometrie die "wahre" ist, oder gar, welcher Raum der ist, in dem wir leben. Von dem hier vertretenen Standpunkt aus ist die letztere Fassung die bei weitem gefährlichere, da der Raum an sich überhaupt keine Eigenschaften hat." (Seeliger 1913, 200–201.)

20 "Die Annahme einer Krümmung unseres Raumes ist ohne jeden Vorteil für die Erklärung des Baues des Fixsternsystems." (N 11:17, 4r.)

21 The article is based on a talk Schwarzschild held at the Heidelberg meeting of the *Astronomische Gesellschaft* in 1900.

lax decreases with an increase in the radius of curvature. Since for most stars no parallax can be observed, the minimal parallax is given by the accuracy of observation. This is given by Schwarzschild as 0.05″. From this he concludes that the curvature radius of hyperbolic space must be at least 4 million times the radius of the Earth's orbit.

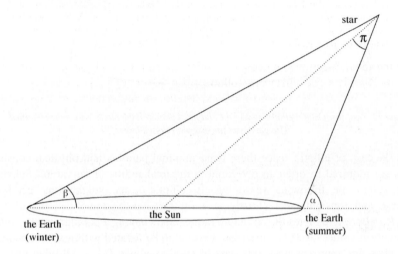

Figure 1: The annual parallax of a star nearly perpendicular to the ecliptic

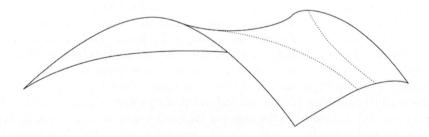

Figure 2: Two-dimensional hyperbolic space. Two geodesics are drawn as dotted lines

As concerns spherical space, Schwarzschild discusses the special case of an elliptic space. The latter can be obtained from usual spherical space by identifying antipodal points. As a consequence, two geodesics going around the world intersect at only one point. Schwarzschild's reason for preferring elliptic to spherical space is that, in the latter, light emitted at one point in space in different directions would converge on the antipodal point, a rather artificial-looking consequence which, according to Schwarzschild, one would not accept without being forced to.

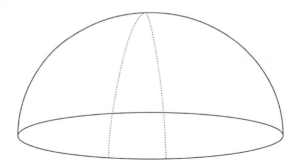

Figure 3: Two-dimensional elliptic space (the antipodal points on the circle c are to be identified).
Two geodesics are drawn as dotted lines

In the case of elliptic space there is no minimal parallax and physical consider-ations are required in order to determine a minimal radius of curvature. Schwarz-schild offers the following reasoning. In elliptic space neighboring geodesics converge and thus intersect already at a finite distance, namely at the distance $(\pi/2)R$, where R denotes the curvature radius. Stars having a parallax smaller than a certain given value, say $0.1''$, therefore have all to be located within a finite volume. Now, there are approximately 100 stars of parallax above $0.1''$. All other stars are thus to be found in this finite volume. If one assumes a uniform distribution of stars, one can determine a certain minimal radius of curvature. A weaker requirement, how-ever, is that the stars with parallax less than $0.1''$ occupy a volume large enough so that they do not influence each other in a way that could not have escaped observa-tion. Schwarzschild does, for instance, calculate that if the elliptic space had a curva-ture radius of about 30,000 times the radius of the Earth's orbit, stars at great distances from the Earth would be separated from one another by only about 40 times the radius of the Earth's orbit. The physical interactions between the stars resulting from this could hardly be concealed from observation. From these considerations Schwarzschild concludes that the minimal radius of curvature of elliptic space is of the order of 100 million times the radius of the Earth's orbit. Schwarzschild further argues that such a relatively small radius of curvature (roughly 1600 light years) is only a realistic possibility if one further assumes an absorption of the starlight of about 40 magnitudes in one circulation around the universe because it is only under this assumption that the appearance of a counter image of the Sun can be avoided.

In this article, as in a later one (Schwarzschild 1909), Schwarzschild expresses his preference for the elliptic space over the hyperbolic or even the Euclidean one, because its finiteness would make it possible in principle to investigate the macro-scopic world exhaustively. This idea would have a soothing effect on the mind.

While the differences between Schwarzschild's application of non-Euclidean geometry to physical space on one side and modern cosmology on the other are obvious (application to three-dimensional space rather than to four-dimensional space-time, no dynamics of geometry, consideration of scales that today are hardly considered cosmological), Schwarzschild's consideration also contains striking parallels to the modern treatment of the problem such as the idea that light proceeds along geodesics and, most notably, the possibility that the universe is spatially closed.[22]

5. THE PERIHELION BREAKTHROUGH

In sec. 3 we have seen how Schwarzschild put the perihelion anomalies of the inner planets into the context of the fundamental physical phenomena of inertia and gravitation. In view of this, Einstein's successful calculation of Mercury's perihelion motion on the basis of his new theory must have appeared to Schwarzschild as the realization of his earlier speculations on the relations between physics and astronomy. In this sec. it will be argued that it was indeed Einstein's perihelion result that instigated Schwarzschild's interest in general relativity. It turns out that even Schwarzschild's derivation of his first exact solution was motivated by his concern to consolidate Einstein's result.

As early as 1912 Schwarzschild had been confronted with the question of the observability of astronomical consequences of general relativity and its predecessors—in particular consequences of the principle of equivalence. Interestingly, in view of Schwarzschild's correspondence, it was not Einstein himself who first confronted Schwarzschild with the question of astronomical consequences of such a theory but rather one of his antagonists in the search for a new theory of gravitation: the theoretical physicist Max Abraham (1875–1922). Abraham, at that time holding a post as professor of rational mechanics at the University of Milan, was himself working on a new theory of gravitation on which he had already published.[23] Although Einstein and Abraham were in severe disagreement about the foundations the new theory of gravitation should build upon, some empirical consequences of Abraham's theory coincided with those Einstein had derived from his more general considerations. Thus, in his first publication on the matter, Abraham discusses the bending of light in a gravitational field that follows from Huygens' principle whenever the speed of light is assumed to be variable, and, in a footnote, points out that Einstein has drawn the astronomers' attention to the fact that the bending of star light in the gravitational field of the Sun may be observable (Abraham 1912, 2).

There are two remnants of Schwarzschild's correspondence with Abraham found in Schwarzschild's Nachlass. The first is a draft in Schwarzschild's hand of a letter most probably addressed to Abraham,[24] the second is a letter from Abraham to Schwarzschild, dated October 13, 1912 (N Briefe 5). From Schwarzschild's draft it

---

22 In an addendum to his article, Schwarzschild mentions a further possibility that later became a debated subject in relativity theory: the possible application of different topologies to physical space.

23 His first publication on that matter being (Abraham 1912).

becomes apparent that Abraham had previously raised the question whether there would be an effect recognizable through astronomical observation if the Sun's loss of inertial mass was proportional to the energy it radiates away, while the gravitational mass did not change in this proportion. In his letter from October 13, 1912, Abraham formulated another idea, arguing that the energy loss of the planets when cooling down in the process of the genesis of the solar system must have diminished the inertial and the gravitational mass in equal proportion, since otherwise Kepler's third law of planetary motion could not be valid. Finally, in their correspondence, the two discussed the possible shift of spectral lines in a gravitational field. Here, as elsewhere in his correspondence and writings prior to Einstein's perihelion result, Schwarzschild is very sceptical about the astronomical detectability of the predicted effects, although he appears to regard the nascent theory of relativity with openness. Thus, as concerns the redshift in the solar spectrum, Schwarzschild writes:

> The shift of the wavelengths on the Sun that Einstein demands, exists [...] due to a strange coincidence in exactly the right magnitude. There is, however, no doubt that it is to be blamed partly on pressure and partly on downwards motions in the solar atmosphere. To see more clearly in this respect, one has to study the lines at the different points of the solar disk. Until now, this has only been done in a really sufficient manner for the lines 3933 Å of calcium (St. John, Astrophysical J[ournal].] His results do in fact speak *against* the existence of the sought-after shift. Despite this, I do not want to claim that this is already an absolute veto against the theory. Before that, more lines would have to be equally well investigated.[25]

In 1913 Schwarzschild himself started an investigation of exactly the kind he had spoken of in his letter to Abraham, performing a series of observations of the band at 3883 Å in the solar spectrum. The continuation of these observations was foiled by the outbreak of war in 1914, but Schwarzschild reported on the results so far obtained in a communication that was presented to the Prussian Academy of Sciences by Einstein on November 5, 1914 (Schwarzschild 1914). In the introduction to this commu-

---

24 N Briefe 846. The draft is dated in another hand as "1912" and commented on as "possibly [to] W. Lorey." The contents however makes it most probable that it is the draft of a letter to Max Abraham. It may be dated September 29, 1912 (or a little earlier) on the basis of Abraham's letter to Schwarzschild from October 13, 1912, mentioning a letter from Schwarzschild from September 29: most probably the letter that was written on the basis of the draft.

25 "Die Verschiebung der Wellenlängen auf der Sonne, die Einstein fordert, besteht [...] durch einen merkwürdigen Zufall genau in der richtigen Größe. Es ist aber kein Zweifel, daß dieselbe zum Teil auf Druck, zum Teil auf absteigende Bewegungen in der Sonnenatmosphäre zu schieben sind. Um Klarheit darüber zu bekommen, muß man die Linien an den einzelnen Punkten der Sonnenscheibe studieren. Das ist in wirklich ausreichender Weise bisher nur für die Linien 3933 A. E. des Calciums geschehen (St. John, Astrophysical J. dessen Resultate sprechen durchaus *gegen* die Existenz der gesuchten Verschiebung. Trotzdem möchte ich nicht behaupten, daß hiermit schon ein absolutes Veto gegen die neue Theorie gegeben ist. Es müßten doch erst noch mehr Linien gleich gut untersucht werden." Draft of a letter from Schwarzschild to Abraham, probably September 29, 1912, N Briefe 846, 27v, 28r. St. John's publications in the *Astrophysical Journal* on the motion of calcium vapour in the solar atmosphere are (St. John 1910; 1910–11). For an account on St. John's later work on line shifts and its relation to general relativity, see (Hentschel 1993).

nication, Schwarzschild points out that the observation of the solar spectrum is of interest not only for the sake of solar physics, but "can, according to Mr. Einstein, inform us about the relativity of the world".[26] By referring to Einstein's article on the influence of gravitation on the propagation of light (Einstein 1911), Schwarzschild explicitly relates the gravitational redshift to the principle of equivalence. However, once more Schwarzschild does not conclude affirmatively, describing his preliminary results and other astronomers' observations he reports on as still being indecisive concerning the gravitational redshift.

In his correspondence with Max Planck in 1913, Schwarzschild even more clearly expresses his doubts concerning an astronomical verification of Einstein's theory. In a letter from January 31, 1913, Planck had asked Schwarzschild for an assessment of the feasibility and the expenses of the eclipse expedition that Erwin Freundlich was planning for the year 1914 and for which he was going to apply to the *Preussische Akademie der Wissenschaften* for funding (N Briefe 593, 2r, v). Freundlich intended to search for a deflection of starlight near the solar disk as predicted by Einstein. Schwarzschild commented on the observational side of the problem in the following way:

> In the problem itself I also have no particular confidence. The diminution of the frequency on the Sun and the shift to red of all spectral lines on the Sun that Einstein assumes can be regarded as refuted by the observations. The last word has not yet been spoken, but the shifts which for single lines are also to violet, can be too well interpreted as being due to pressure. Since this whole thing looks rather fishy, it won't be much different for the deflection of light rays by the Sun's gravitation.[27]

When Einstein succeeded in deriving the correct value for Mercury's perihelion shift from his theory,[28] Schwarzschild's appraisal of the new theory of relativity changed drastically. Einstein presented his calculation of Mercury's perihelion advance to the Prussian Academy of sciences on November 18, 1915. Schwarzschild was on leave from his military duties at the Russian front and attended the meeting.[29] Back in Russia, Schwarzschild wrote to Einstein:

> It is quite a wonderful thing that from such an abstract idea the Mercury anomaly emerges so stringently.[30]

---

26 "[...] kann nach Hrn. Einstein auch Auskunft über die Relativität der Welt geben." (Schwarzschild 1914, 1201.)

27 "Auch zum Probleme selbst habe ich kein besonderes Fiduz. Die Verminderung der Schwingungszahl auf der Sonne und die [...] Verschiebung aller Spektrallinien nach Rot auf der Sonne, die Einstein annimmt [...] kann als durch die Beobachtungen als widerlegt angesehn werden. Das letzte Wort ist noch nicht gesprochen, aber die [...] Verschiebungen, die [...] bei einzelnen Linien auch nach Violett gehen, lassen sich zu gut als Druckverschiebungen deuten. Da es hiermit ziemlich faul [?] steht, wird es mit der Ablenkung der Lichtstrahlen durch die Sonnengravitation auch nicht viel anders sein." Draft of a letter from Schwarzschild to Planck, after January 31, 1913, N Briefe 593, 6r. The passages omitted are crossed-out in Schwarzschild's manuscript.

28 On Einstein's derivation and its historical context, see (Earman and Janssen 1993).

29 See the minutes of the meeting on November 18, 1915, *Archiv der Berlin-Brandenburgischen Akademie der Wissenschaften* II–V, Vol. 91, 64–66.

In a letter of the same day to Arnold Sommerfeld, Schwarzschild even explicitly states that to him the perihelion result was much more convincing than the empirical consequences of Einstein's theory discussed earlier:

> Did you see Einstein's paper on the motion of Mercury's perihelion in which he obtains the observed value correctly from his last theory of gravitation? That is something much closer to the astronomers' heart than those minimal line shifts and ray bendings.[31]

It is a matter of course that the quantitatively appropriate explanation of an anomaly that had been detected by astronomers more than half a century earlier provided a stronger argument for the new theory than the hardly detectable effects that it also predicted. However, Einstein's successful calculation of the perihelion shift did not convince everybody to the same degree as Schwarzschild. More than one year after Einstein's calculation, Max von Laue still described the result as a "agreement of two single numbers"[32] which

> remarkable as it may be, does not seem to us to give sufficient reason to change the whole physical world picture in its foundations, as Einstein's theory does.[33]

In view of Schwarzschild's earlier contextualization of perihelion motions, it becomes understandable why, for him, Einstein's result signified much more than the "agreement of two single numbers." Furthermore, on the background of Schwarzschild's prerelativistic cosmological considerations that included the application of non-Euclidean geometry to physical space and the idea of mutually accelerated inertial systems, the changes brought about by Einstein's theory must have appeared less drastic to Schwarzschild than to most others, including von Laue.

Einstein's derivation of the perihelion advance which was based on an approximation had, however, one blemish: the uniqueness of the solution remained questionable. In order to consolidate Einstein's result, Schwarzschild tried to prove the uniqueness of the solution. In the above-mentioned letter to Sommerfeld, Schwarzschild reports:

> In Einstein's calculation the uniqueness of the solution remains doubtful. In the first approximation, which Einstein makes, the solution, when carried out completely, is even apparently ambiguous—one additionally gets the beginning of a divergent expansion. I have tried to derive an exact solution, and that was unexpectedly easy.[34]

---

30 "Es ist eine ganz wunderbare Sache, daß von so einer abstrakten Idee aus die Erklärung der Merkuranomalie so zwingend herauskommt." Schwarzschild to Einstein, December 22, 1915 (CPAE 8, Doc. 169).

31 "Haben Sie Einstein's Arbeit über die Bewegung des Merkurperihels gesehen, wo er den beobachteten Wert richtig aus seiner letzten Gravitationstheorie heraus bekommt? Das ist etwas, was den Astronomen viel tiefer zu Herzen geht, als die minimalen Linienverschiebungen und Strahlenkrümmungen." Schwarzschild to Sommerfeld, December 22, 1915, München, Deutsches Museum Archiv NL 89, 059, p. 1.

32 "Übereinstimmung zwischen zwei einzelnen Zahlen"

33 "[...] scheint uns, so bemerkenswert sie ist, doch kein hinreichender Grund, das gesamte physikalische Weltbild von Grund aus zu ändern, wie es die Einsteinsche Theorie tut." (Laue 1917, 269.)

And even in his publication that is today known for containing the first derivation of an exact non-trivial solution of Einstein's field equations, Schwarzschild emphasizes that, rather than the quest for an exact solution, it is the consolidation of Einstein's result which is of primary concern:

> It is always convenient to possess exact solutions of a simple form. More important is that the calculation yields, at the same time, the uniqueness of the solution about which Mr. Einstein's treatment remained doubtful and which arguably, in view of the way in which it emerges below, could hardly have been proven by such an approximative method.[35]

6. THE RELATIVITY OF ROTATION REVISITED

After his consolidation of the connection between general relativity and observational astronomy established by Einstein's perihelion calculation, Schwarzschild turned to other questions of theoretical astronomy for which general relativity appeared to provide the adequate framework. One of these questions concerned the relativity of rotation, a question we already encountered in Schwarzschild's prerelativistic work.

The major source documenting Schwarzschild's work on this question in the context of general relativity is a formerly unrecognized manuscript page in Schwarzschild's Nachlass in the University Library of Göttingen (N 2:2, 12r). A reproduction with explanations of the page is given in the Appendix. The page is full of calculations and contains hardly any text. It is found among a few similar pages, some of which contain notes on general relativity the purpose of which however is not obvious. The notes on the page under discussion are undated but obviously stem from the short period between Einstein's successful perihelion calculation in November 1915 and Schwarzschild's death in May 1916.

In these notes, Schwarzschild distinguishes an "inner" and an "outer" metric. The inner metric describes a Minkowski spacetime in a coordinate system rotating with constant angular velocity n. Using cylindrical coordinates, the outer metric can be written as

---

34 "Bei Einstein's Rechnung bleibt die Eindeutigkeit der Lösung noch zweifelhaft. In der ersten Annäherung, die Einstein macht, ist die Lösung sogar, wenn man sie vollständig macht, scheinbar mehrdeutig — man bekommt noch den Anfang einer divergenten Entwicklung herein. Ich habe versucht, eine strenge Lösung abzuleiten, und das ging unerwartet einfach." Schwarzschild to Sommerfeld, December 22, 1915, München, Deutsches Museum Archiv NL 89, 059, 1. In his letter to Einstein from December 22, 1915, Schwarzschild reports in even more detail about the motivation that led him to his exact solution, see (CPAE 8, Doc. 169).

35 "Es ist immer angenehm, über strenge Lösungen einfacher Form zu verfügen. Wichtiger ist, daß die Rechnung zugleich die eindeutige Bestimmtheit der Lösung ergibt, über die Hrn. Einsteins Behandlung noch Zweifel ließ, und die nach der Art, wie sie sich unten einstellt, wohl auch nur schwer durch ein solches Annäherungsverfahren erwiesen werden könnte." (Schwarzschild 1916a, 190.)

$$g_{\mu\nu_{outer}} = \begin{bmatrix} -f_1 & 0 & 0 & 0 \\ 0 & -f_2 & 0 & f \\ 0 & 0 & -1 & 0 \\ 0 & f & 0 & f_4 \end{bmatrix},$$

where x_1 is a radial coordinate, x_2 is an angle, x_3 is parallel to the symmetry axis of the cylinder, and x_4 is a time-like coordinate. f and f_i, $i = 1, 2, 4$, are functions of x_1 only. For the radial coordinate becoming infinitely large, Schwarzschild imposes the condition that the outer metric tends to the (non-rotating) Minkowski metric, rescaled in such a way that it satisfies the determinant condition, $|\det(g_{\mu\nu})| = 1$.[36] The spacetime Schwarzschild attempts to investigate thus consists of a cylindrical section of Minkowski space, the inner space, rotating with constant angular velocity n relative to an inertial frame at radius infinity and surrounded by an outer space, becoming Minkowskian for $x_1 \to \infty$.

Schwarzschild then obviously tries to find general expressions for the metric functions f and f_i, $i = 1, 2, 4$ (the g_{33}-component of the outer metric Schwarzschild had set to -1). In this he follows exactly the procedure he elaborated in his publication on the field of a point mass (Schwarzschild 1916a). First, he constructs the Lagrangian of a point particle

$$F = \frac{1}{2}\sum g_{\mu\nu}\frac{dx_\mu}{ds}\frac{dx_\nu}{ds}.$$

Next, he calculates $\partial F/\partial x_i$ and $\partial F/\partial \dot{x}_i$ for $i = 1, 2, 3, 4$ and puts the resulting terms into the Euler-Lagrange equation

$$\frac{\partial F}{\partial x_i} - \frac{d}{ds}\left(\frac{\partial F}{\partial \dot{x}_i}\right) = 0.$$

He then manipulates the resulting equations—further using the determinant condition $f_1(f_2 f_4 + f^2) = 1$—in such a way that he may read off the field strengths $\Gamma^\lambda_{\mu\nu}$ by a comparison of the coefficients with those appearing in the equations of motion of a point particle. He then puts the expressions for the field strengths into the field equations and manipulates them in order to determine the functions f and f_i by integra-

---

36   As in (Schwarzschild 1916a), Schwarzschild here employs the condition that the determinant of the metric tensor be unity. Einstein had introduced this condition in an addendum to his publication *Concerning the Theory of General Relativity (Zur allgemeinen Relativitätstheorie (Nachtrag)*, Einstein 1915a), and continued to use it in his paper on the perihelion motion of Mercury (Einstein 1915b). The field equations Einstein presented and used in these papers read $G_{\mu\nu} = -\kappa T_{\mu\nu}$, where $G_{\mu\nu}$ denotes the Ricci tensor and $T_{\mu\nu}$ denotes the energy-momentum tensor. These are not the field equations of the final theory, which contain a trace term on either the left or the right-hand side. However, since in vacuum the older field equations coincide with those of the final theory, Einstein and Schwarzschild's solutions still hold there.

tion. However, while in the case of the spherically symmetric vacuum field this procedure led to a simple result, in this case the differential equations become rather involved and no solution is obtained. The calculations on the page discussed end with a second order differential equation coupling f and f_1 (see the equation at the bottom right-hand side of Schwarzschild's manuscript page reproduced as Fig. 6 in the Appendix).

What is the purpose of Schwarzschild's notes? Clearly these notes are concerned with the problem of the relativity of rotation in general relativity. As was explained in sec. 3, Schwarzschild had, in his earlier work, considered the possibility of inertial reference frames rotating with respect to one another. The spacetime Schwarzschild considers here is a realization of such an arrangement in the framework of general relativity, one inertial system being given at radial infinity, the other within the rotating cylindrical section of spacetime. The difficulty in making any further-reaching statements about the physical situation Schwarzschild is trying to describe here arises from the fact that, in his notes, Schwarzschild does not specify the matter distribution he assumes.[37] It is, however, plausible to assume that the spacetime under consideration should function as a model for the spatially two-dimensional situation of a rotating disk of Minkowski space. Then Schwarzschild's calculations can be interpreted as the exploration of a simple model for the spacetime inside and outside the rotating system of fixed stars.[38] This would mean that Schwarzschild assumed the spacetime within the system to be approximately Minkowskian, an assumption that is consistent with his earlier observation that the perihelia of remote double stars do not rotate relative to the directions defined by the planetary orbits in the solar system and that therefore inertial frames inside the Galaxy do not rotate with respect to one another. At the same time, Schwarzschild must have assumed that in its rotational motion the matter content of the system of fixed stars—be it concentrated in a ring or distributed over an ellipsoidal volume—drags along the interior Minkowski space.[39]

---

37 A spacetime similar to the one Schwarzschild describes here is generated by an infinitely long, rotating, cylindrical shell of matter. The interior field of such a matter distribution is indeed Minkowskian; see (Davies et al. 1971). The exterior vacuum metric of such a matter distribution, including the dragging of inertial frames close to the shell, is discussed in (Frehland 1972). Non-local effects of such a matter distribution, corresponding to the Aharonov-Bohm effect in electrodynamics, are discussed in (Stachel 1983). The problem of the relativity of rotation was addressed in 1918 by Thirring who considered a rotating spherical mass shell rather than a cylindrical one (Thirring 1918, see also Lense and Thirring 1918).

38 In an earlier publication (Schwarzschild 1909, 41–42), Schwarzschild had described the solar system whose dynamics is dominated by a central mass as being of "monarchic constitution", and the Galaxy where every star is acted upon by all other stars as being of "republican constitution", and had speculated that the whole Universe might be built up from a hierarchical sequence of structures of these two basic types. While Schwarzschild's first exact solution to Einstein's field equations provides the basis for describing the monarchic constitution in the framework of general relativity, the notes considered here can be understood as an attempt to complement this with the description of the republican constitution within that framework.

39 The spacetime of an axially symmetric distribution of particles revolving with constant angular velocity was later derived by (Stockum 1937).

In fact, Schwarzschild could have hoped that, within general relativity, a problem concerning the rotation of the system of fixed stars could be resolved. On one hand, namely, Schwarzschild contended that, by analogy to other celestial motions, it must be assumed that the system of fixed stars as a whole rotates.[40] Yet, on the other hand, such a rotation had hardly been observed, as Schwarzschild explained in an earlier text:

> [...] it turns out that the average of those few thousand stars, whose proper motions are known, displays no evidence of rotation with respect to [the] directions [defined by the planetary orbits] [...].[41]

General relativity now provided a possible explanation of this phenomenon, if it was assumed that, together with the stars themselves, the global inertial system within the Galaxy was rotating. In searching for the functions f and f_i describing the outer metric, Schwarzschild would then have attempted to clarify in what sense one may speak in general relativity of a rotation of the system of fixed stars as a whole.

That Schwarzschild indeed considered the question of the rotation of the Galaxy in the context of general relativity is made evident by a letter from Einstein dated January 9, 1916.[42] In a preceding letter by Schwarzschild which is lost, Schwarzschild must have raised several questions, which Einstein answers one by one. Einstein's second point reads as follows:

> The statement that "the system of fixed stars" is free of rotation may retain a relative meaning, which is to be fixed by a comparison.
>
> The surface of the Earth is irregular, as long as I regard very small sections of it. However, it approaches the flat elementary shape when I regard larger sections of it, whose dimensions are still small in comparison to the length of the meridian. This elementary shape becomes a curved surface when I regard even larger sections of the Earth's surface.
>
> For the gravitational field things are similar. On a small scale the individual masses produce gravitational fields that, even with the most simplifying choice of the reference system, reflect the character of the quite irregular matter distribution on the small scale. If I consider larger regions, as astronomy presents them to us, the Galilean reference system provides me with the analogue to the flat elementary shape of the Earth's surface in the previous comparison. But if I consider even larger regions, there probably will be no continuation of the Galilean system to simplify the description of the universe to the same degree as on a small scale, that is, throughout which a mass point sufficiently remote from other masses moves uniformly in a straight line.[43]

Schwarzschild's response is consistent with the calculations as interpreted above:

> As concerns the inertial system, we are in agreement. You say that beyond the Milky Way system conditions may arise under which the Galilean system is no longer the simplest. I only hold that within the Milky Way system such conditions do not arise.[44]

---

40   See, for example, (Schwarzschild 1897, 519).
41   "[...] zeigt sich, dass der Durchschnitt aus jenen paar Tausend Sternen, deren Eigenbewegung man kennt, [...] keine Rotation gegen diese Richtungen aufweist." (Schwarzschild 1897, 520.)
42   Einstein to Schwarzschild, January 9, 1916, N 193, 3-5, see also (CPAE 8, Doc. 181).

In view of this exchange between Schwarzschild and Einstein, it appears obvious that Schwarzschild's calculations are related to the question he must have posed to Einstein: Does it make sense to speak of a rotation of the system of fixed stars? The calculations then document the attempt to explore the "even larger regions" outside the system of fixed stars, in which "there probably will be no continuation of the Galilean system."

7. A CLOSED UNIVERSE AS A SOLUTION OF EINSTEIN'S FIELD EQUATIONS

In his calculations, Schwarzschild had assumed the Universe to be asymptotically Minkowskian. In his correspondence with Einstein on the question of global frames of inertia, Schwarzschild mentions a further possibility. In direct continuation of the passage quoted above, he explicates:

> As concerns very large spaces, your theory has a quite similar position as Riemann's geometry, and you are certainly not unaware that one obtains an elliptic geometry from your theory, if one puts the entire universe under uniform pressure (energy tensor $-p, -p, -p, 0$). [45]

Thus, Schwarzschild was the first to entertain the possibility of a closed universe with an elliptic geometry as a solution to Einstein's field equations. Schwarzschild's remark that Einstein's theory had a similar position as Riemann's geometry thereby alludes to his prerelativistic application of elliptic geometry to the universe on the background of Riemannian geometry discussed in sec. 4.

Contrary to Schwarzschild's assumption, Einstein was, at the time, most probably unaware of such cosmological implications of his theory.[46] It was only through a debate with the Dutch astronomer Willem de Sitter (1872–1934) beginning in fall

---

43 "Die Aussage, dass "das Fixsternsystem" rotationsfrei sei, behält wohl einen relativen Sinn, der durch ein Gleichnis festgelegt sei.
Die Oberfläche der Erde ist, solange ich ganz kleine Teile derselben ins Auge fasse, unregelmässig. Sie nähert sich aber der ebenen Grundgestalt, wenn ich grössere Teile ins Auge fasse, deren Abmessungen aber immer noch klein sind gegen die Länge des Meridians. Diese Grundgestalt wird zu einer gekrümmten Fläche, wenn ich noch grössere Teile der Erdoberfläche ins Auge fasse.
So ähnlich ist es auch mit dem Gravitationsfeld. Im Kleinen liefern die einzelnen Massen Gravitationsfelder, welche auch bei möglichst vereinfachender Wahl des Bezugssystems den Charakter der ziemlich regellosen Verteilung der Materie im Kleinen widerspiegeln. Betrachte ich grössere Gebiete, wie sie uns die Astronomie bietet, so bietet mir das Galileische Bezugssystem das Analoge zu der ebenen Grundgestalt der Erdoberfläche beim vorigen Vergleich. Betrachte ich aber noch grössere Gebiete, so wird es wohl keine Fortsetzung des Galileischen Systems geben, welche in solchem Masse wie im Kleinen die Beschreibung der Welt einfach gestaltet d.h. in welchem überall der von anderen Massen hinlänglich entfernte Massenpunkt sich gradlinig gleichförmig bewegt." Einstein to Schwarzschild, January 9, 1916 (CPAE 8, Doc. 181).

44 "Was das Inertialsystem angeht, so sind wir einig. Sie sagen, daß jenseits des Milchstraßensystems sich Verhältnisse einstellen können, in denen das Galilei'sche System nicht mehr das einfachste ist. Ich behaupte nur, daß sich innerhalb des Milchstraßensystems solche Verhältnisse nicht einstellen." Schwarzschild to Einstein, February 6, 1916, N 193, 7–8, see also (CPAE 8, Doc. 188).

1916 that Einstein was led to consider a closed universe which he hesitantly proposed in 1917 (Einstein 1917).[47] The distinction between spherical and elliptic space had thereby remained obscure to him. De Sitter pointed the distinction out to Einstein, referring to Schwarzschild's 1900 paper on the curvature of space and the argument for preferring elliptic to spherical space given therein.[48]

In Einstein's debate with de Sitter, the question of the global geometry of the universe emerged from a discussion of the relativity of inertia. Strikingly, in Schwarzschild's correspondence with Einstein, the question of the global geometry of the Universe is brought up in exactly the same context.[49] It is therefore tempting to see here the commencement of an Einstein–Schwarzschild debate foreshadowing the later Einstein–de Sitter debate. Einstein appears, however, to have not yet been prepared to consider the cosmological implications of his theory at that time. And by the time he was slowly pushed into that direction in his exchange with the astronomer de Sitter, Schwarzschild had already died. Nevertheless, in view of Schwarzschild's deliberations discussed here, it seems safe to say that, had Schwarzschild lived longer, he could have made a substantial contribution to the cosmological debates emerging later.

8. SCHWARZSCHILD'S INTERDISCIPLINARY APPROACH TO THE FOUNDATIONS OF SCIENCE

Let us come back to the question raised at the beginning: Why did Schwarzschild recognize the significance of general relativity at such an early stage? Here it has been attempted to show that, already in his early astronomical work, Schwarzschild did not act as a specialist but attempted to meet the challenges resulting from the implica-

---

45 "Was die ganz großen Räume angeht, hat Ihre Theorie eine ganz ähnliche Stellung, wie Riemann's Geometrie, und es ist Ihnen gewiß nicht unbekannt, daß man die elliptische Geometrie aus Ihrer Theorie herausbekommt, wenn man die ganze Welt unter einem gleichförmigen Druck stehen läßt (Energietensor $-p, -p, -p, 0$)." Schwarzschild to Einstein, February 6, 1916, N 193, 7–8, see also (CPAE 8, Doc. 188). This energy tensor actually does not yield a spherical static universe. It does, however, yield the universe Schwarzschild is speaking of when the trace term in the field equations is neglected, i.e. in the context of the older field equations $G_{\mu\nu} = -\kappa T_{\mu\nu}$, where $G_{\mu\nu}$ denotes the Ricci tensor and $T_{\mu\nu}$ denotes the energy-momentum tensor (see footnote 36). It may be the case that Schwarzschild originally conceived of the tensor on the basis of these field equations and later did not modify it as the new field equations would have demanded. Schwarzschild continues his letter by explaining the solution inside a sphere of fluid with uniform energy density (energy tensor $-p, -p, -p, \rho_0$). Here, as well as in the corresponding publication (Schwarzschild 1916b, 431–432), Schwarzschild points out that inside the sphere spherical geometry applies.

46 In November 1916, Einstein still calls the question of the boundary conditions of the metric field "purely a matter of taste which will never attain a scientific meaning." ("eine reine Geschmacksfrage, die nie eine naturwissenschaftliche Bedeutung erlangen wird.") Einstein to de Sitter, November 4, 1916 (CPAE 8, Doc. 273).

47 On the Einstein–de Sitter debate, see (CPAE 8, 351–357) and the references given therein, in particular (Kerszberg 1989; 1989a).

48 De Sitter to Einstein, June 20, 1917 (CPAE 8, Doc. 355).

tions of foundational questions in physics on astronomy. Thus it comes as no surprise that Schwarzschild was also among the first to recognize that Einstein had—without being aware of it—provided the astronomers with the adequate framework for treating their questions. As a result, a clear continuity can be perceived in Schwarzschild's work on cosmology, prerelativistic and relativistic. In this context it is interesting to question the extent to which parallel cases are provided by the work of other pioneers of relativistic astronomy such as Willem de Sitter and Arthur Eddington (1882–1944). The study of this question, however, does not lie in the scope of this contribution. Here, in conclusion, it shall only be pointed out that it was no coincidence that Schwarzschild took an astronomical road to general relativity, but that this may rather be seen as the natural outcome of his interdisciplinary approach to the foundations of the exact sciences.

Indeed, not only is interdisciplinarity the hallmark of Schwarzschild's scientific work, but he also was quite aware of the general significance of interdisciplinarity for the progress of science. On many occasions Schwarzschild explained how he saw scientific progress emerging from the interplay of the different branches of science, for instance when, on the occasion of his inaugural lecture at the Prussian Academy of Sciences, he stated that "the greatest yet unsolved problem of celestial mechanics, the so-called many-body problem, most closely touches a problem of physics that concerns the foundations of its newest developments".[50] As a further example, consider the following passage from the same speech in which Schwarzschild describes the establishment of special relativity:

> [...] an important source for the electron and relativity theory lay in an astronomical problem. The astronomical aberration results from the finite propagation speed of light through the aether in combination with the Earth's motion in space. H.A. Lorentz occupied himself many times with the theory of aberration and searched for a satisfying picture of the aether's behavior when large masses, like the Earth, move through it, until he

---

49   In Einstein's letter to Schwarzschild from January 9, there is, in fact, a passage in which he expresses exactly the kind of strong Machian claims concerning his theory which later sparked off his debate with de Sitter. In direct continuation of the passage quoted above, Einstein explains: "According to my theory, inertia is an interaction between masses, in the end, not an effect in which, besides the mass under consideration, 'space' itself would be involved. The essence of my theory is precisely that no independent properties are attributed to space itself.
Jokingly one may put it this way. If I let all things in the world disappear, according to Newton the Galilean inertial space remains, according to my perception, however, *nothing* remains."
("Die Trägheit ist eben nach meiner Theorie im letzten Grunde eine Wechselwirkung der Massen, nicht eine Wirkung bei welcher ausser der ins Auge gefassten Masse der 'Raum' als solcher beteiligt ist. Das Wesentliche meiner Theorie ist gerade, dass dem Raum als solchem keine selbständigen Eigenschaften gegeben werden.
Man kann es scherzhaft so ausdrücken. Wenn ich alle Dinge aus der Welt verschwinden lasse, so bleibt nach Newton der Galileische Trägheitsraum, nach meiner Auffassung aber *nichts* übrig.") Einstein to Schwarzschild, January 9, 1916 (CPAE 8, Doc. 181).
50   "[...] berührt sich das höchste noch ungelöste Problem der Himmelsmechanik, das sogenannte Vielkörperproblem, aufs engste mit einem Problem der Physik, das an die Fundamente ihrer neuesten Entwicklung greift." (Schwarzschild 1913, 597.)

finally cut the knot by consistently implementing Fresnel's assumption that the aether is absolutely rigid and cannot be brought to flow by any force acting on it. In this way the path was cleared for the electron theory. Furthermore, the completely rigid aether stepped out of the circle of the objects that can be influenced and thus can be more closely perceived, so much so that relativity theory became possible, in which the concept of the aether only appears as a spacetime concept deepened by new experience.

Electron theory and relativity theory in turn have already posed various problems to astronomy as a consequence of the modifications of celestial mechanics they necessitate.[51]

Clearly Schwarzschild was equipped to take up these challenges posed to astronomy by its neighboring disciplines. He knew that scientific progress is not a matter of the advancement of isolated disciplines, as both the previous and the following quotations make clear:

Mathematics, physics, chemistry, astronomy march in line. Whichever lags behind is pulled forward. Whichever hastens ahead pulls the others forward. The closest solidarity exists between astronomy and the whole circle of exact sciences.[52]

ACKNOWLEDGEMENTS

This paper was previously published in the journal *Science in Context* (Schemmel 2005). I am indebted to Jürgen Renn who inspired and supported my work from its inception up to its present state. For helpful discussions and comments I would also like to thank Dieter Brill, Giuseppe Castagnetti, Peter Damerow, Jürgen Ehlers, Hubert Goenner, Michel Janssen, John Norton, and John Stachel.

---

51 "[...] eine wichtige Quelle für die Elektronen- und Relativitätstheorie in einem astronomischen Problem lag. Die astronomische Aberration ist eine Folge der endlichen Ausbreitungsgeschwindigkeit des Lichtes im Äther verbunden mit der Bewegung der Erde im Weltraum. H.A. Lorentz hat sich vielfach mit dem Problem der Aberration beschäftigt und nach einer befriedigenden Anschauung über das Verhalten des Äthers, wenn große Massen, wie die Erde, sich durch ihn hindurchbewegen, gesucht, bis er schließlich den Knoten zerhieb durch völlig konsequente Durchführung der alten Fresnelschen Annahme, daß der Äther absolut starr und durch keine auf ihn wirkende Kraft zum Fließen zu bringen sei. Dadurch war die Bahn frei geworden für die Elektronentheorie. Der völlig starre Äther trat ferner so sehr aus dem Kreis der beeinflußbaren und damit näher erkennbaren Objekte heraus, daß auch die Relativitätstheorie möglich wurde, bei welcher der Begriff des Äthers nur als ein durch neue Erfahrungen vertiefter Raum-Zeitbegriff erscheint.
Elektronentheorie und Relativitätstheorie haben auch Rückwärts der Astronomie schon wieder mancherlei Probleme gestellt infolge der Modifikationen der Himmelsmechanik, die sie notwendig machen." (Schwarzschild 1913, 598.)
52 "Mathematik, Physik, Chemie, Astronomie marschieren in einer Front. Wer zurückbleibt, wird nachgezogen. Wer vorauseilt, zieht die anderen nach. Es besteht die engste Solidarität der Astronomie mit dem ganzen Kreis der exakten Naturwissenschaften." (Schwarzschild 1913, 599.)

APPENDIX:
ANNOTATED REPRODUCTION OF SCHWARZSCHILD'S RELATIVISTIC NOTES ON THE PROBLEM OF ROTATION

In this appendix, the manuscript page that documents Schwarzschild's relativistic calculations on the problem of rotation referred to in sec. 6 is reproduced with annotations. It is preserved as a part of Schwarzschild's Nachlass in the Niedersächsische Staats- und Universitätsbibliothek Göttingen as page 12r of folder 2:2. The page is found among a few similar pages, some of which contain further notes on general relativity. For technical reasons this reproduction is divided into three parts shown in figs. 4, 5, and 6, respectively. I am grateful to the Niedersächsische Staats- und Universitätsbibliothek Göttingen for their permission to reproduce this page.

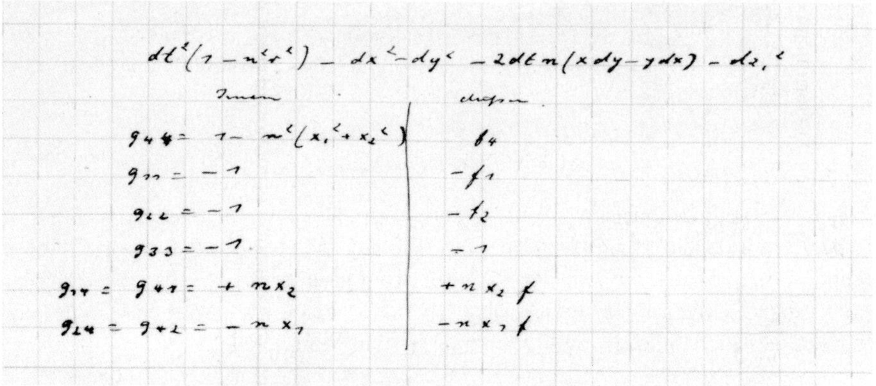

Figure 4: Schwarzschild N 2:2, 12r, upper part

Schwarzschild begins his considerations by writing down the metric for the "inner" and the "outer" spacetime in Cartesian coordinates,

$$g_{\mu\nu}{}_{\text{inner}} = \begin{bmatrix} -1 & 0 & 0 & nx_2 \\ 0 & -1 & 0 & -nx_1 \\ 0 & 0 & -1 & 0 \\ nx_2 & -nx_1 & 0 & 1-n^2(x_1^2+x_2^2) \end{bmatrix}; \quad g_{\mu\nu}{}_{\text{outer}} = \begin{bmatrix} -f_1 & 0 & 0 & nx_2f \\ 0 & -f_2 & 0 & -nx_1f \\ 0 & 0 & -1 & 0 \\ nx_2f & -nx_1f & 0 & f_4 \end{bmatrix}.$$

In the following calculation (figs. 5 and 6), he then shifts to cylindrical coordinates and absorbs the factors of n in the function f. The outer metric then reads

$$g_{\mu\nu}{}_{\text{outer}} = \begin{bmatrix} -f_1 & 0 & 0 & 0 \\ 0 & -f_2 & 0 & f \\ 0 & 0 & -1 & 0 \\ 0 & f & 0 & f_4 \end{bmatrix}.$$

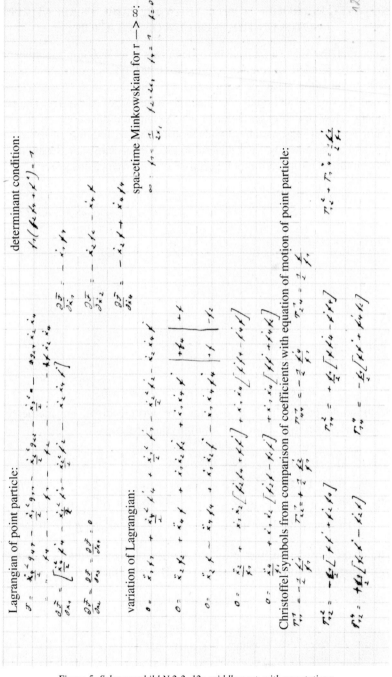

Figure 5: Schwarzschild N 2:2, 12r, middle part, with annotations

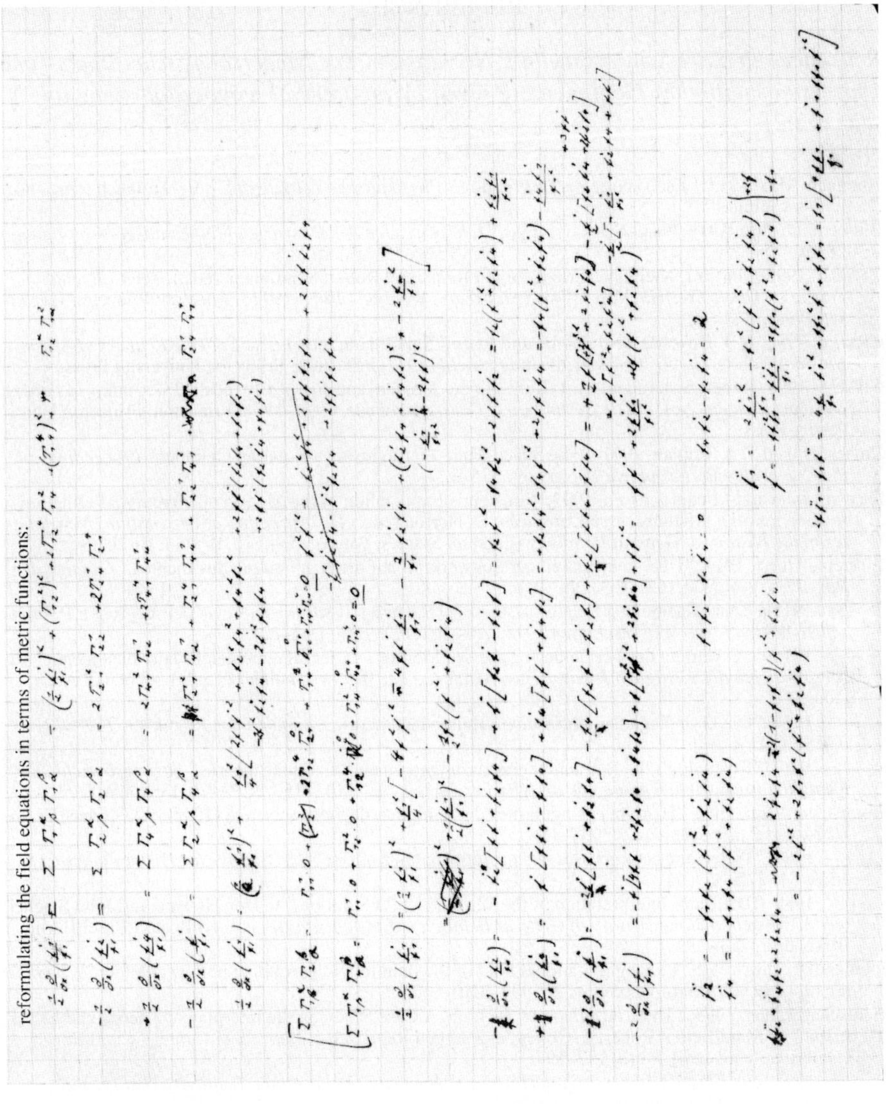

Figure 6: Schwarzschild N 2:2, 12r, lower part, with an annotation

REFERENCES

References to Karl Schwarzschild's Nachlass in the Niedersächsische Staats- und Universitätsbibliothek Göttingen are given by an archival number following an 'N' (e.g. N 2:2).

Abraham, Max. 1912. "Zur Theorie der Gravitation." *Physikalische Zeitschrift* 13: 1–4.(English translation in this volume.)
Blumenthal, Otto. 1918. "Karl Schwarzschild." *Jahresberichte der Deutschen Mathematiker-Vereinigung* 26: 56–75.
CPAE 3. 1993. Martin J. Klein, A. J. Kox, Jürgen Renn, and Robert Schulmann (eds.), *The Collected Papers of Albert Einstein*. Vol. 3. *The Swiss Years: Writings, 1909–1911*. Princeton: Princeton University Press.
CPAE 6. 1996. A. J. Kox, Martin J, Klein, and Robert Schulmann (eds.), *The Collected Papers of Albert Einstein*. Vol. 6. *The Berlin Years: Writings, 1914–1917*. Princeton: Princeton University Press.
CPAE 8. 1998. Robert Schulmann, A. J. Kox, Michel Janssen, and József Illy (eds.), *The Collected Papers of Albert Einstein*. Vol. 8. *The Berlin Years: Correspondence, 1914–1918*. Princeton: Princeton University Press.
Davies, H. and T. A. Caplan. 1971. "The Space-Time Metric inside a Rotating Cylinder." *Proceedings of the Cambridge Philosophical Society* 69: 325–327.
Earman, John and Michel Janssen. 1993. "Einstein's Explanation of the Motion of Mercury's Perihelion." In John Earman, Michel Janssen, and John D. Norton (eds.) *The Attraction of Gravitation: New Studies in the History of General Relativity (Einstein Studies*, vol. 5), 129–172. Boston: Birkhäuser.
Einstein, Albert. 1911. "Über den Einfluß der Schwerkraft auf die Ausbreitung des Lichtes." *Annalen der Physik* 35: 898–908, (CPAE 3, Doc. 23).
———. 1915a. "Zur allgemeinen Relativitätstheorie (Nachtrag)." *Sitzungsberichte der Königlich Preussischen Akademie der Wissenschaften* 1915: 799–801, (CPAE 6, Doc. 22).
———. 1915b. "Erklärung der Perihelbewegung des Merkur aus der allgemeinen Relativitätstheorie." *Sitzungsberichte der Königlich Preussischen Akademie der Wissenschaften* 1915: 831–839, (CPAE 6, Doc. 24).
———. 1916. "Die Grundlage der allgemeinen Relativitätstheorie." *Annalen der Physik* 49: 769–822, (CPAE 6, Doc. 30).
———. 1917. "Kosmologische Betrachtungen zur allgemeinen Relativitätstheorie." *Sitzungsberichte der Königlich Preussischen Akademie der Wissenschaften* 1917: 142–152, (CPAE 6, Doc. 43).
Eisenstaedt, Jean. 1982. "Histoire et singularités de la solution de Schwarzschild (1915-1923)." *Archive for History of Exact Sciences* 27: 157–198.
———. 1987. "Trajectoires et impasses de la solution de Schwarzschild." *Archive for History of Exact Sciences* 37: 275–357.
———. 1989. "The Early Interpretation of the Schwarzschild Solution." In Don Howard and John Stachel (eds.), *Einstein and the History of General Relativity*. (*Einstein Studies* vol. 1.) Boston/Basel/Berlin: Birkhäuser, 213–233.
Frehland, Eckart. 1972. "Exact gravitational field of the infinitely long rotating hollow cylinder." *Communications in mathematical physics* 26: 307–320.
Hentschel, Klaus. 1993. "The Conversion of St. John. A Case Study on the Interplay of Theory and Experiment." In Mara Beller, Robert S. Cohen, and Jürgen Renn (eds.), *Einstein in Context*. Cambridge: Cambridge University Press, 137–194.
———. 1997. *The Einstein Tower. An Intertexture of Dynamic Construction, Relativity Theory, and Astronomy*. Stanford: Stanford University Press.
Israel, Werner. 1987. "Dark Stars: the evolution of an idea." In Stephen Hawking and Werner Israel (eds.), *Three Hundred Years of Gravitation*. Cambridge: Cambridge University Press, 199–276.
Kerszberg, Pierre. 1989. "The Einstein–de Sitter Controversy of 1916–1917 and the Rise of Relativistic Cosmology." In Don Howard and John Stachel (eds.). *Einstein and the History of General Relativity*. (*Einstein Studies* vol. 1.) Boston/Basel/Berlin: Birkhäuser, 325–366.
———. 1989a. *The Invented Universe. The Einstein–De Sitter Controversy (1916–17) and the Rise of Relativistic Cosmology*. Oxford: Clarendon Press.
Laue, Max von. 1917. "Die Nordströmsche Gravitationstheorie." *Jahrbuch der Radioaktivität und Elektronik* 14: 263–313.
Lense, Joseph, and Hans Thirring. 1918. "Über den Einfluß der Eigenrotation der Zentralkörper auf die Bewegung der Planeten und Monde nach der Einsteinschen Gravitationstheorie." *Physikalische Zeitschrift* 19: 156–163.

Norton, John. 1999. "The Cosmological Woes of Newtonian Gravitation Theory." In Hubert Goenner, Jürgen Renn, Jim Ritter, and Tilman Sauer (eds.), *The Expanding Worlds of General Relativity*. (*Einstein Studies* vol. 7.) Boston/Basel/Berlin: Birkhäuser, 271–323.

Poincaré, Henri. 1902. *La Science et l'hypothèse*. Paris: E. Flammarion.

Schemmel, Matthias. 2005. "An Astronomical Road to General Relativity: The Continuity between Classical and Relativistic Cosmology in the Work of Karl Schwarzschild." *Science in Context* 18:451–478.

Schwarzschild, Karl. 1897. "Was in der Welt ruht." *Die Zeit*. Vol. 11, No. 142, 19 June 1897, Vienna, 181–183. (English translation in this volume.) [The page numbers given in the text refer to (Schwarzschild 1992) wherein this article is reprinted.]

———. 1900. "Über das zulässige Krümmungsmaass des Raumes." *Vierteljahrsschrift der Astronomischen Gesellschaft* 35: 337–347.

———. 1909. *Über das System der Fixsterne: Aus populären Vorträgen. Naturwissenschaftliche Vorträge und Schriften No.1*. Leipzig: Teubner, 39–43.

———. 1913. "Antrittsrede des Hrn. Schwarzschild." *Sitzungsberichte der königlich preussischen Akademie der Wissenschaften* 1913: 596-600.

———. 1914. "Über die Verschiebung der Bande bei 3883 Å im Sonnenspektrum." *Sitzungsberichte der Königlich Preussischen Akademie der Wissenschaften* 1914: 1201–1213.

———. 1916a. "Über das Gravitationsfeld eines Massenpunktes nach der Einsteinschen Theorie." *Sitzungsberichte der Königlich Preussischen Akademie der Wissenschaften* 1916: 189–196.

———. 1916b. "Über das Gravitationsfeld einer Kugel aus inkompressibler Flüssigkeit nach der Einsteinschen Theorie." *Sitzungsberichte der Königlich Preussischen Akademie der Wissenschaften* 1916: 424–434.

———. 1992. *Gesammelte Werke*, Vol. 3, Hans H. Voigt (ed.). Berlin: Springer.

Seeliger, Hugo von. 1913. "Bemerkungen über die sogenannte absolute Bewegung, Raum und Zeit." *Vierteljahrsschrift der Astronomischen Gesellschaft* 48: 195–201.

Sommerfeld, Arnold. 2000–2004. *Wissenschaftlicher Briefwechsel* (2 vols.), Michael Eckert and Karl Märker (eds.). Berlin: Verlag für Geschichte der Naturwissenschaften und der Technik.

St. John, Charles E. 1910. "The General Circulation of the Mean- and High-Level Calcium Vapour in the Solar Atmosphere." *Astrophysical Journal* 32: 36–82.

———. 1910–11. "Motion and Condition of Calcium Vapour over Sun-Spots and Other Special Regions." *Astrophysical Journal* 34: 57–78, 131–153.

Stachel, John. 1983. "The gravitational fields of some rotating and nonrotating cylindrical shells of matter." *Journal of Mathematical Physics* 25: 338–341.

Stockum, W. J. van. 1937. "The Gravitational Field of a Distribution of Particles Rotating about an Axis of Symmetry." *Proceedings of the Royal Society of Edinburgh* 57: 135–154.

Thirring, Hans. 1918. "Über die Wirkung rotierender ferner Massen in der Einsteinschen Gravitationstheorie." *Physikalische Zeitschrift* 19: 33–39.

KARL SCHWARZSCHILD

THINGS AT REST IN THE UNIVERSE

Originally published as "Was in der Welt ruht" in Die Zeit, Vienna, Vol. 11, No. 142, 19 June 1897, 181–183 (1897). Reprinted in Collected Works, vol. 3, H. H. Voigt, ed., Springer-Verlag, Berlin 1992, p. 514–521. Page numbers refer to reprinted version.

Is there anything in the universe that is at rest, around which or within which the rest of the universe is constructed, or is there no hold in the unending chain of motions in which everything seems to be caught up? It is worth considering the extent to which these questions are justified and how they can be answered.

On a clear evening a few weeks ago, many people claimed to have observed how the Moon rushed across the sky with most unusual speed. On that evening a light veil of mist seems to have been blown past the Moon by the wind, more subtly creating the same illusion one believes one observes every time broken clouds move quickly past it: one assumes that the clouds stand still and that the Moon rushes through them against the wind.

This phenomenon and other much more common ones, such as the apparent rotation of long furrows when one travels through a flat landscape by train, or the enchanted room at recent fairs which leads one to believe that one is upside down and walking on the ceiling, simply provide evidence to support the immediately comprehensible statement that all perceptible motion is relative, that one can say that an object moves relative to another one but never in absolute terms that an object moves or is at rest.

With this one can already see that neither of the original two questions has been properly formulated. Namely, if only the motion of different objects relative to each other can be perceived, then without actually contradicting experience one may ascribe a completely arbitrary motion to a particular object in the universe or, as a special case, declare it to be at rest. Thus, everything and nothing is at rest in the universe. Logically, there is nothing to stop someone from stipulating, for example, that the wing tips of a buzzing insect are still. In that case one would simply have to ascribe a buzzing motion in the opposite direction to one's own body, the Earth, and all celestial objects, in order to recover the directly observable relative motion of the insect's wings with respect to the rest of the universe.

Why, however, does one smile at he who would seriously wish to claim that the Moon really did rush across the sky, or that the fields rotated when one went through

Jürgen Renn (ed.). *The Genesis of General Relativity*, Vol. 3
Gravitation in the Twilight of Classical Physics: Between Mechanics, Field Theory, and Astronomy.
© 2007 Springer.

them on the train? The reasons are of a purely practical and utilitarian nature. A clumsy assumption regarding things at rest in the universe would result in an unspeakable confusion of ideas. And for this reason, the above questions are to be replaced by the following: What things in the universe did one find it useful to treat as being at rest, at different times? Each time it becomes necessary to redefine what can be treated as being at rest, an important stage is passed in the process of development of human ideas concerning the universe. It has often been forgotten, or not even rec-

[515]   ognized, that here we are | dealing with arbitrary assumptions, and at times the opinions have become doctrines which have even found their own martyrs.

When one talks of motion or rest in everyday life, one leaves out the object with respect to which motion or rest is ascribed, but one means: relative to the ground or to objects fixed with respect to the ground. It is useful to tacitly assume this particular completion of the concept of motion because most of what appears before our eyes during the course of the day is at rest relative to the ground, or at least we can name a special cause when something moves relative to the ground. However, since Copernicus science has begun to recognize other completions as useful, to ascribe motion to other objects, and to treat other things as "points at rest" in the universe.

Indeed, Copernicus suggested that all motion is relative to the center of the Sun and showed that all the complicated curves and ribbons traced by the planets in the sky found an explanation if one allows the planets to describe circles around the Sun, which is taken to be fixed.

But more precisely, Copernicus makes a further assumption, in that he stipulates that the fixed stars, located at a huge distance from our planetary system, should be considered to be at rest. In fact, the single assumption that the center of the Sun is fixed would not suffice as a definition of all motion. One could then recognize only changes in distance relative to this point but not rotations about it, just as one does not know whether a completely smooth ball, whose surface is the same all over, is stationary or rotating about its center.

Accordingly, in the Copernican system, one describes the motion of a body by giving its distance from the Sun, and direction in terms of a fixed star as seen from the Sun, at various times. The assumption that the fixed stars are at rest thereby forces one to attribute a daily rotation from west to east to the Earth, because relative to the Earth's surface the fixed stars apparently revolve from east to west within a 24 hour period. Consequently one must imagine that together with our surroundings we rush through space with the speed of a cannon ball, and that the direction of this motion continually changes. It cannot be said that at first such a way of thinking seems to be particularly in accordance with the principle of utility, which we must recognize as the single decisive principle for these questions. From this viewpoint, the reluctance of the Aristotelians at the time certainly does not seem to be as naive and ridiculous as it is generally made out to be. All the more worthy of admiration is a man such as Galileo, who, through his mechanical principles, made the adventurous elements of the new view as well as the presumed contradictions with everyday experience disappear, so that only its overwhelming advantages in the depiction of celestial motions remained. |

It was not until the end of the seventeenth century—when the use of the telescope [516] and progress in the production of finely divided circles, used to read off the position of stars, had led to an unexpected improvement in observation skills—that the Copernican stipulation of motion was recognized as not being wholly exact and unambiguous. At that time it was noticed that the fixed stars do not actually deserve their name: they all shift in position relative to each other, and in the very distant past Cygnus and Orion once formed quite different constellations. However, these shifts in position progress extremely slowly. If Stephan's Tower would list so far in the course of one year that its top shifted by one centimeter, then this shift, as observed from Kahlenberg, would be conspicuous compared to the shift in position of the majority of the fixed stars visible to the naked eye.[1] But as soon as the shifts are at all perceptible, Copernicus' use of the fixed star system as a basis for motion is no longer possible.

From those same observations concerning the change in position of the fixed stars relative to each other it has also emerged that the other Copernican point of reference, the center of the Sun, possesses no particular right to the privilege of being considered at rest. The "proper motions" of the fixed stars, such being the technical term, are indeed distributed irregularly so that neighboring stars in the sky move away in all possible directions, but on average out of several hundred stars one can see that when one looks into the sky in a certain direction, on the whole the constellations seem to spread, to get bigger, and in the opposite direction, to get smaller. From this it follows that the Sun and the former region of the sky come closer to each other. Although it would not be impossible to do, there would be no point in ascribing this motion wholly to the fixed stars and continuing to treat the Sun as being at rest, for we know that the Sun is a relatively subordinate member of the great family of stars.

Just realize into what insecurity the universe has fallen as a result, how imagination finds no place to drop anchor and no single rock in the world has a special right to be thought of as fixed and at rest.

Unfortunately we have still not reached the epoch of astronomy that will intervene offering a new fundamental definition. This will have come about when a pattern has been found in the seemingly so irregular proper motions. No effort is being spared in order to promote its appearance. Twenty observatories have joined together in order to produce a catalogue which details the position of 150,000 fixed stars relative to each other at the present time, and a major part of this huge task has already been completed. Just as many institutes have divided up the sky amongst themselves in order to establish the position of two million stars with even greater precision through photographic images. When this work is repeated I in a few decades time, [517] such a huge number of proper motions will be known to us that a law, if it exists, must reveal itself, and that a new Copernicus, who admittedly would have no prejudice to overcome, can show from where and how these motions appear ordered. Admittedly, it could also turn out that no general order exists within the army of fixed stars. In this case, all these spheres in space are to be compared with gas molecules, which fly around completely irregularly, so irregularly that the irregularity itself becomes a principle, according to which the effect of the gas mass as a whole can be derived through considerations of probabilities and averages.

But, for the time being, to what does the astronomer refer the heavenly motions?

Among the pieces of ponderable matter there is none, as we have seen, which can be reasonably distinguished as stationary. Thus among material things one's only chance is to look around for imponderables. As is well known, the universe is filled with an all-pervasive substance, which is weightless, neither solid nor fluid, neither visible nor invisible, and to which there is no physical description that really fits: this is the aether, the only imponderable of modern physics. Through the aether do the gentlest light waves shimmer, but the aether also mediates the mighty effects of electrical machinery. When the power of a distant waterfall is transferred to a central power station, then in a certain sense aether is the long rod impacted by the water over there and pushing the wheel of the machine here. The electrical cable is of only secondary importance; it simply maintains the energy flowing in the desired direction, rather than dispersing. Now, it would be very tempting from a philosophical standpoint to take the ever-present aether to be the basic stationary substance, and occasionally this has actually been done. But eventually, both optical and electrical phenomena have pressed upon us the conviction that the aether certainly cannot remain at rest where ponderable masses move through it, nor can it remain at rest in empty space, that rather it is traversed by internal currents. According to an investigation which Helmholtz carried out in the penultimate year of his life, the aether smoothly transfers energy from one particle to another without itself moving, as long as the energy is delivered to it evenly. Every build-up or acceleration in the energy supply, however, sets the aether into motion; and that will happen often enough.

Now that this hope has also come to naught, there remains no material object in the universe that one would have reason to consider at rest. There remain only certain conceptually defined points and directions that can serve as a substitute to a certain [518] extent. For bringing these to its attention, astronomy owes thanks to a | science to which it is intimately related, namely mechanics. Thus we arrive at the path along which research is currently moving.

One of the basic laws of mechanics, the law of inertia, goes as follows: Each body moves in a straight line at an unchanging speed, as long as no forces act on it. This statement, however, contains more than a law based on experience; it contains at the same time a certain definition of what should be considered as being at rest in space. Since we can only identify relative motions, we can in theory ascribe any motion we choose to an individual body, even when it is unaffected by other forces. The law of inertia prescribes: Take any body (or to be more exact, so that rotations can also be recognized, any three bodies remaining at unchanging distances from each other) on which no forces act, and ascribe to these bodies a straight path and a uniform speed. The same will then apply for every other body on which no forces act. The system of inertia has so innumerably many important applications near and far that it is certainly fit and proper to make those arbitrary assumptions which lend it its simplest and only natural form.

So let us look at which ideas concerning celestial motions astronomy has to develop on the basis of this stipulation implicit in the law of inertia.

In a system comprising many mutually attracting bodies, naturally no single body describes a straight line in general, but when such a system is isolated from the effects of any forces from external, foreign bodies, still its center of gravity will describe a straight line, as shown by a simple conclusion from the law of inertia. If one treats all the bodies of the universe as constituents of one system, there would certainly be no external influence because nothing else exists beyond this system, and so its center of gravity must proceed along a straight line. Recall further an experience from mechanics, that in a train even at the highest speed, every activity can be carried out just as it can when all is at rest, provided that the speed does not alter and the train does not go around curves. This too is a special case of a general conclusion drawn from the law of inertia, which can be stated as follows: The internal processes within an isolated system of bodies are exactly the same whether its center of gravity is stationary or whether it is moving in a straight line at uniform speed. As far as the internal processes in the star system are concerned, it is therefore equally irrelevant whether its center of gravity is stationary or moving steadily. Since, furthermore, the only things we can learn about are internal processes in the star system, it is sensible to stipulate that the center of gravity of all masses in the universe and in the entire star system is the ideal point of rest.

It will, of course, look bad for the practical realization of this stipulation if, as is plausible, an infinite number of bodies of finite mass exist in the universe, all of which we then I cannot hope ever to know. However, the center of gravity of the larg-   [519] est possible system of masses is to be taken as an approximation to the definition of this ideal point of rest, especially when this system of masses is separate and remote from other such systems and as such experiences nearly no forces due to them. It is likely that the millions of stars that can be seen through a medium-sized telescope form a special system, which has been given the name "Milky Way System." The shimmering band of the Milky Way is in reality a huge ring comprised of a countless number of stars, which forms the largest mass in this system and characterizes its form. What remains of the brighter stars appear to be scattered inside this ring or to form small external appendages. Maybe other similar specimens existing in unimaginably distant regions of space are co-ordinated with our own Milky Way, but perhaps it is the only one of its kind and beyond it exist only chaotic nebular masses.

The center of gravity of the Milky Way System should then for the time being be seen as a point at rest. One will know more about its position when the great undertakings mentioned above are completed and have been sifted. However, this falls into the next phase of astronomy; for now, one makes do with a rather poor substitute. To date one knows only the proper motions of a few thousand stars, and for now one assumes that the center of gravity of these stars is at rest. As calculation shows, a speed of approximately twelve kilometres per second must then be ascribed to the Sun, a speed of similar magnitude to that with which the planets move in their orbits around the Sun.

Now there is still the matter of fixing certain directions with respect to which the rotations of each body can be reckoned. As far as the ring of the Milky Way is concerned, by analogy with all known celestial motions we must assume that it rotates in some way; therefore, from the outset it cannot be used for this purpose. However, here again new consequences of the law of inertia provide help.

Foucault's pendulum experiment is well known. If one hangs a heavy ball from a thread and causes this pendulum to oscillate from east to west, then in the course of a few hours one notices a rotation of the direction of oscillation, gradually moving in the north-south direction and, if only the pendulum swings long enough, it eventually completes a full revolution in over a day. This extremely remarkable process is usually considered to be the best proof of the Earth's rotation. For in theory, under the assumptions of the law of inertia, the 24-hour rotation of the Earth, and the Earth's gravitational pull, one finds just the amount of rotation of the oscillation direction shown by observations; whereas in the case of a stationary Earth, no rotation should take place at all. Strictly speaking, however, the Foucault pendulum experiment [520] proves only I that a rotation of the Earth must be assumed whenever one wants to use the stipulations implicit in the law of inertia concerning that, to which motion is to be referred. Conversely, by accurate observations of the rotation of the pendulum, one could calculate the speed of rotation that has to be ascribed to the Earth according to the law of inertia, and one would then have to describe as fixed the direction with respect to which the Earth rotates with the calculated speed.

Because terrestrial pendulums are subject to too many disturbances due to air resistance and friction, one uses with a similar aim the larger pendulum experiments with which nature presents us. One can cause a pendulum to describe an elongated elliptical curve by pushing it at an angle, as in a well-known game of bowling. The planets are pendulum bobs of a similar kind. It is well known that the planets describe flattened curves, ellipses, around the Sun. The point in its path where a planet is furthest away from the Sun is called its aphelion. From the law of inertia and Newton's law of gravitation, it follows that for an isolated planet which revolves around the Sun, the direction of the aphelion is fixed in space. In reality, however, the planets are not isolated but exist in greater number, yet Newton's law allows the calculation of the small deviations from the elliptical path and the small rotations which the direction to the aphelion suffers as a result of the disturbing influences from other planets. Thus, after subtracting these disturbances from the observed aphelion directions, a direction fixed in space is obtained.

If one carries this out with the level of precision which was possible 50 years ago, then everything seems to fit together beautifully. The directions that, according to the theory, should turn out to be fixed for the different planets also appear to be fixed with respect to each other, and additionally it turns out that the average of those few thousand stars, whose proper motions we know, displays no evidence of rotation with respect to these directions within these limits of precision. According to this, the entire ring of the Milky Way can also rotate only extremely slowly. Today, however, things look different. With the level of precision which theory and observation have now achieved, one finds that the directions which one derives from the aphelions of the various planets and which one would expect to be fixed, actually perform minimal yet clearly recognizable shifts relative to each other, and thus it is impossible to treat them all as fixed. Our measure of time depends upon which directions one considers to be fixed. For we define a day to be the time it takes the Earth to revolve once around its axis and we must, of course, have a fixed direction within space in order to be able to judge when the Earth has completed a rotation. Two clocks, which agree at the beginning of a century, one tied to the direction of Jupiter's aphelion, the other to Mercury's aphelion, I would differ by three seconds at the end of it. One does not really know how to explain this difference. It is possible that friction plays a role in the case of these heavenly pendulums as well, for they do not swing through an absolutely empty space. However, it is more likely that Newton's law does not describe the attractive forces of the Sun and the planets with absolute accuracy. However, there are still insufficient clues to know how a correction of Newton's law should actually read, and so one must renounce the wish to determine fixed directions with greater precision from the planetary motions using mechanical theorems. At present one prefers to hold on to the other result which, from these considerations, proved itself to be reasonably correct, and to treat those few thousand stars as being without rotation on average. That is the provisional stipulation that, in a roundabout way, one has to choose also as regards the definition of rotations in the universe. [521]

Finally, a certain arbitrariness, with which each definition of fixed points and directions based on mechanical theorems is afflicted, is still to be pointed out. One can only base a conclusion on the law of inertia when all forces that act in a given case are known, or when it is known that no forces are present. Now there could always be forces acting in our surroundings, which spin us, together with all the neighboring stars, around arbitrarily in the universe, without however exerting any influence on the relative position of all these bodies. Such forces would, of course, completely elude our experience, and therefore the principle of utility, which alone guides us, commands us not to allow ourselves to become disconcerted by the thought of such a possibility when using theorems based on the law of inertia.

One senses a certain feeling of unease when one stops and thinks about all that is provisional, intermediate, undecided in present-day science concerning a point which is so important for a clear idea of the universe, as is the establishment of what is at rest in the universe. Yet, that is a characteristic of our times. The proud era of natural science, when it believed to have found absolute laws and to be able to give philosophy a real basis, is over. Similar instances of relativity as are found in the case of motion, exist everywhere, and with each broadening of experience, uncertainties turn up in previously accepted definitions. A profound scepticism has become fashionable, for one asks oneself even about the foundations of exact natural science, which are given the honorable name of "laws of nature": what can they be other than the most practical summary possible of what is most important for man within a limited field of experience?

EDITORIAL NOTE

[1] Mount Kahlenberg lies to the north of central Vienna, about 8 km from Stephan's Tower.

A NEW LAW OF GRAVITATION ENFORCED
BY SPECIAL RELATIVITY

SCOTT WALTER

BREAKING IN THE 4-VECTORS:
THE FOUR-DIMENSIONAL MOVEMENT IN
GRAVITATION, 1905–1910

INTRODUCTION

In July, 1905, Henri Poincaré (1854–1912) proposed two laws of gravitational attraction compatible with the principle of relativity and all astronomical observations explained by Newton's law. Two years later, in the fall of 1907, Albert Einstein (1879–1955) began to investigate the consequences of the principle of equivalence for the behavior of light rays in a gravitational field. The following year, Hermann Minkowski (1864–1909), Einstein's former mathematics instructor, borrowed Poincaré's notion of a four-dimensional vector space for his new matrix calculus, in which he expressed a novel theory of the electrodynamics of moving media, a space-time mechanics, and two laws of gravitational attraction. Following another two-year hiatus, Arnold Sommerfeld (1868–1951) characterized the relationship between the laws proposed by Poincaré and Minkowski, calling for this purpose both on space-time diagrams and a new 4-vector formalism.

Of these four efforts to capture gravitation in a relativistic framework, Einstein's has attracted the lion's share of attention, and understandably so in hindsight, but at the expense of a full understanding of what is arguably the most significant innovation in contemporary mathematical physics: the four-dimensional approach to laws of physics. In virtue of the common appeal made by Poincaré, Minkowski, and Sommerfeld to four-dimensional vectors in their studies of gravitational attraction, their respective contributions track the evolving form of four-dimensional physics in the early days of relativity theory.[1] The objective of this paper is to describe in terms of theorists' intentions and peer readings the emergence of a four-dimensional language for physics, as applied to the geometric and symbolic expression of gravitational action.

The subject of gravitational action at the turn of the twentieth century is well-suited for an investigation of this sort. This is not to say that the reform of Newton's

---

[1]   In limiting the scope of this paper to the methods applied by their authors to the problem of gravitation, four contributions to four-dimensional physics are neglected: that of Richard Hargreaves, based on integral invariants (Hargreaves 1908), two 4-vector systems due to Max Abraham (Abraham 1910) and Gilbert Newton Lewis (Lewis 1910a), and Vladimir Varičak's hyperbolic-function based approach (Varičak 1910).

Jürgen Renn (ed.). *The Genesis of General Relativity,* Vol. 3
Gravitation in the Twilight of Classical Physics: Between Mechanics, Field Theory, and Astronomy.
© 2007 Springer.

law was a burning issue for theorists. While several theories of gravitation claimed corroboration on a par with that of classical Newtonian theory, contemporary theoretical interest in gravitation as a research topic—including the Lorentz-invariant variety—was sharply curtailed by the absence of fresh empirical challenges to the inverse-square law. Rather, in virtue of the stability of the empirical knowledge base, and two centuries of research in celestial mechanics, the physics of gravitation was a well-worked, stable terrain, familiar to physicists, mathematicians and astronomers alike.[2]

The leading theory of gravitation in 1905 was the one discovered by Isaac Newton over two centuries earlier, based on instantaneous action at a distance. When Poincaré sought to bring gravitational attraction within the purview of the principle of relativity, he saw it had to propagate with a velocity no greater than that of light in empty space, such that a reformulation of Newton's law as a retarded action afforded a simple solution.

Newton's law was the principal model for Poincaré, but it was not the only one. With the success of Maxwell's theory in explaining electromagnetic phenomena (including the behavior of light) during the latter third of the nineteenth century, theories of contiguous action gained greater favor with physicists. In 1892, the Dutch theorist H. A. Lorentz produced a theory of mobile charged particles interacting in an immobile aether, that was an habile synthesis of Maxwell's field theory and Wilhelm Weber's particle theory of electrodynamics. After the discovery of the electron in 1897, and Lorentz's elegant explanation of the Zeeman effect, certain charged microscopic particles were understood to be electrons, and electrons the building-blocks of matter.[3]

In this new theoretical context of aether and electrons, Lorentz derived the force on an electron moving in microscopic versions of Maxwell's electric and magnetic fields. To determine the electromagnetic field of an electron in motion, Alfred Liénard and Emil Wiechert derived a formula for a potential propagating with finite velocity. In virtue of these two laws, both of which fell out of a Lagrangian from Karl Schwarzschild, the theory of electrons provided a means of calculating the force on a charged particle in motion due to the fields of a second charged particle in motion.[4]

An electron-based analogy to gravitational attraction of neutral mass points was then close at hand. Lorentz's electron theory was held in high esteem by early twenti-

2   For an overview of research on gravitation from 1850 to 1915, see (Roseveare 1982). On early 20th-century investigations of gravitational absorption, see de Andrade Martins (de Andrade Martins 1999). While only Lorentz-covariant theories are considered in this paper, the relative acceptance of the principle of relativity among theorists is understood as one parameter among several influencing the development of four-dimensional physics.

3   See (Buchwald 1985, 242; Darrigol 2000, 325; Buchwald and Warwick 2001).

4   Lorentz took the force per unit charge on a volume element of charged matter moving with velocity v in the electric and magnetic fields $\mathfrak{d}$ and $\mathfrak{h}$ to be $\mathfrak{f} = \mathfrak{d} + \frac{1}{c}[v \cdot \mathfrak{h}]$, where the brackets indicate a vector product (Lorentz 1904c, 2:156–7). For a comparison of electrodynamic Lagrangians from Maxwell to Schwarzschild, see (Darrigol 2000, app. 9).

eth-century theorists, including both Poincaré and Minkowski, who naturally catered to the most promising research program of the moment. They each proposed two force laws: one based on retarded action at a distance, the other appealing directly to contiguous action propagated in a medium. All four particle laws were taken up in turn by Sommerfeld.[5]

Several other writers have discussed Poincaré's and Minkowski's work on gravitation. Of the first four substantial synoptic reviews of the two theories, none employed the notation of the original works, although this fact itself reflects the rapid evolution of formal approaches in physics. Early comparisons were carried out with either Sommerfeld's 4-vector formalism (Sommerfeld 1910b; Kretschmann 1914), a relative coordinate notation (de Sitter 1911), or a mix of ordinary vector algebra and tensor calculus (Kottler 1922). No further comparison studies were published after 1922, excepting one summary by North (North 1965, 49–50), although since the 1960s, the work of Poincaré and Minkowski has continued to incite historical interest.[6] Sommerfeld's contribution, while it inflected theoretical practice in general, and contemporary reception of Lorentz-covariant gravitation theory in particular, has been neglected by historians.

The present study has three sections, beginning with Poincaré's contribution, moving on in the second section to Minkowski's initial response to Poincaré's theory, and a review of his formalism and laws of gravitation. A third section is taken up by Sommerfeld's interpretation of the laws proposed by Poincaré and Minkowski. The period of study is thus bracketed on one end by the discovery of special relativity in 1905, and on the other end by Sommerfeld's paper. While the latter work did not spell the end of either 4-vector formalisms or Lorentz-covariant theories of gravitation, it was the first four-dimensional vector algebra, and represents a point of closure for a study of the emergence of a conceptual framework for four-dimensional physics.

1. HENRI POINCARÉ'S LORENTZ-INVARIANT LAWS OF GRAVITATION

Poincaré's memoir on the dynamics of the electron (Poincaré 1906), like Einstein's relativity paper of 1905, contains the fundamental insight of the physical significance of the group of Lorentz transformations, not only for electrodynamics, but for all natural phenomena. The law of gravitation, to no lesser extent than the laws of electrodynamics, fell presumably within the purview of Einstein's theory, but this is not a point that Einstein, then working full time as a patent examiner in Bern, chose to elaborate upon immediately. Poincaré, on the other hand, as Professor of Mathematical Astron-

---

5   On the Maxwellian approach to gravitation, see (North 1965, chap. 3; Roseveare 1982, 129–31; Norton 1992, 32). The distinction drawn here between retarded action at a distance and field representations reflects that of Lorentz (Lorentz 1904b), for whom this was largely a matter of convenience. On nineteenth-century conceptions of the electromagnetic field, see (Cantor and Hodge 1981).

6   On Poincaré's theory see (Cunningham 1914, 173; Whitrow and Morduch 1965, 20; Harvey 1965, 452; Cuvaj 1970, app. 5; Schwartz 1972; Zahar 1989, 192;Torretti 1996, 132). On Minkowski's theory see (Weinstein 1914, 61; Pyenson 1985, 88; Corry 1997, 287).

omy and Celestial Mechanics at the Sorbonne, could hardly finesse the question of gravitation. In particular, his address to the scientific congress at the St. Louis World's Fair, on 24 September, 1904, had pinpointed Laplace's calculation of the propagation velocity of gravitation as a potential spoiler for the principle of relativity.[7]

There may have been another reason for Poincaré to investigate a relativistic theory of gravitation. In the course of his study of Lorentz's contractile electron, Poincaré noted that the required relations between electromagnetic energy and momentum were not satisfied in general. Raised earlier by Max Abraham, the problem was considered by Lorentz to be a fundamental one for his electron theory.[8]

Solving the stability problem of Lorentz's contractile electron was a trivial matter for Poincaré, as it meant transposing to electron theory a special solution to a general problem he had treated earlier at some length: to find the equilibrium form of a rotating fluid mass.[9] He postulated a non-electromagnetic, Lorentz-invariant "supplementary" potential that exerts a binding (negative) pressure inside the electron, and reduces the total energy of the electron in an amount proportional to the volume decrease resulting from Lorentz contraction. When combined with the electromagnetic field Lagrangian, this binding potential yields a total Lagrangian invariant with respect to the Lorentz group, as Poincaré required.

In accordance with the electromagnetic world-picture and the results of Kaufmann's experiments, Poincaré supposed the inertia of matter to be exclusively of electromagnetic origin, and he set out, as he wrote in §6 of his paper,

> to determine the total energy due to electron motion, the corresponding action, and the quantity of electromagnetic momentum, in order to calculate the electromagnetic masses of the electron.

---

7   Laplace estimated the propagation velocity of gravitation to be 10^6 times that of light, and Poincaré noted that such a signal velocity would allow inertial observers to detect their motion with respect to the aether (Poincaré 1904, 312).

8   See (Poincaré 1906, 153–154; Miller 1973, 230–233). Following Abraham's account (Abraham 1905, 205), the problem may be presented in outline as follows (using modified notation and units). Consider a deformable massless sphere of radius a and uniformly distributed surface charge, and assume that this is a good model of the electron. The longitudinal mass $m_{||}$ of this sphere may be defined as the quotient of external force and acceleration, $m_{||} = d|G|/d|v|$, where G is the electromagnetic momentum resulting from the electron's self-fields, and v is electron velocity. Defining the electromagnetic momentum to be $G = \int E \times B \, dV$, where E and B denote the electric and magnetic self-fields, and V is for volume, we let $c = 1$, and find the longitudinal mass for small velocities to be $m_{||} = \frac{e^2}{6\pi a}(1 - v^2)^{-3/2}$. Longitudinal electron mass may also be defined in terms of the electromagnetic energy W of the electron's self-fields, assuming quasistationary motion: $m_{||} = \frac{1}{|v|}\frac{dW}{d|v|}$, where $W = \frac{e^2}{6\pi a}(1 - v^2)^{-1/2} + \frac{e^2}{24\pi a}(1 - v^2)^{1/2}$. This leads, however, to an expression for longitudinal mass different from the previous one: $m_{||} = \frac{e^2}{6\pi a}\left[(1 - v^2)^{-3/2} + \frac{1}{4}(1 - v^2)^{-1/2}\right]$. From the difference in these two expressions for longitudinal mass, Abraham concluded that the Lorentz electron required the postulation of a non-electromagnetic force and was thereby not compatible with a purely electromagnetic foundation of physics.

9   See (Poincaré 1885, 1902a, 1902b). In the limit of null angular velocity, gravitational attraction can be replaced by electrostatic repulsion, with a sign reversal in the pressure gradient.

Non-electromagnetic mass does not figure in this analysis, and consequently, one would not expect the non-electromagnetic binding potential to contribute to the tensorial electromagnetic mass of the electron, although Poincaré did not state this in so many words. Instead, immediately after obtaining an expression for the binding potential, he derived the small-velocity, "experimental" mass from the electromagnetic field Lagrangian alone, neglecting a contribution from the binding potential. The mass of the slowly-moving Lorentz electron was then equal to the electrostatic mass, just as one would want for an electromagnetic foundation of mechanics. This fortuitous result, which revised Lorentz's electron mass value downward by a quarter, was obtained independently by Einstein, using a method that did not constrain electron structure (Einstein 1905, 917).[10] Although the question of electron mass was far from resolved, Poincaré had shown that the stability problem represented no fundamental obstacle to the pursuit of a new mechanics based on the concept of a contractile electron.

With this obstacle out of the way, Poincaré proceeded as if the laws of mechanics were applicable to the experimental mass of the electron.[11] Noting that the negative pressure deriving from his binding potential is proportional to the fourth power of mass, and furthermore, that Newtonian attraction is itself proportional to mass, Poincaré conjectured that

> there is some relation between the cause giving rise to gravitation and that giving rise to the supplementary potential.

On the basis of a formal relation between experimental mass and the binding potential, in other words, Poincaré predicted the unification of his negative internal electron pressure with the gravitational force, in a future theory encompassing all three forces.[12]

On this hopeful note, Poincaré began his memoir's ninth and final section, entitled "Hypotheses concerning gravitation." Lorentz's theory, Poincaré explained, promised to account for the observed relativity of motion:

> In this way Lorentz's theory would fully explain the impossibility of detecting absolute motion, if all forces were of electromagnetic origin.[13]

---

10   Poincaré also neglected the mass contribution of the binding potential in his 1906–1907 Sorbonne lectures, according to student notes (Poincaré 1953, 233). For reviews of Poincaré's derivation of the binding potential, see (Cuvaj 1970, app. 11) and (Miller 1973). On post-Minkowskian interpretations of the binding potential (also known as Poincaré pressure), see (Cuvaj 1970, 203; Miller 1981, 382, n. 29; Yaghjian 1992).

11   In this paper Poincaré made no distinction between inertial and gravitational mass.

12   As Cuvaj points out (Cuvaj 1968, 1112), Poincaré may have found inspiration for this conjecture in Paul Langevin's remark that gravitation stabilized the electron against Coulomb repulsion. Unlike Langevin, Poincaré anticipated a unified theory of gravitation and electrons, in the spirit of theories pursued later by Gustav Mie, Gunnar Nordström, David Hilbert, Hans Reissner, Hermann Weyl and Einstein; for an overview see (Vizgin 1994).

13   "Ainsi la théorie de Lorentz expliquerait complètement l'impossibilité de mettre en évidence le mouvement absolu, si toutes les forces étaient d'origine électromagnétique" (Poincaré 1906, 166).

The hypothesis of an electromagnetic origin of gravitational force had been advanced by Lorentz at the turn of the century. On the assumption that the force between "ions" (later "electrons") of unlike sign was of greater magnitude at a given separation than that between ions of like sign (following Mossotti's conjecture), Lorentz represented gravitational attraction as a field-theoretical phenomenon analogous to electromagnetism, reducing to the Newtonian law for bodies at rest with respect to the aether. Lorentz's theory tacitly assumed negative energy density for the "gravitational" field, and a gravitational aether of huge intrinsic positive energy density, two well-known sticking-points for Maxwell. Another difficulty stemmed from the dependence of gravitational force on absolute velocities.[14]

Neither Lorentz's gravitation theory nor Maxwell's sticking-points were mentioned by Poincaré in the ninth section of his memoir. Instead, he recalled a well-known empirical fact: two bodies that generate identical electromagnetic fields need not exert the same attraction on electrically neutral masses. Although Lorentz's theory clearly accounts for this fact, Poincaré concluded that the gravitational field was distinct from the electromagnetic field. What this tells us is that Poincaré's attention was not focused on Lorentz's theory of gravitation.[15]

To Poincaré's way of thinking, it was the impossibility of an electromagnetic reduction of gravitation that had driven Lorentz to suppose that all forces transform like electromagnetic ones:

> The gravitational field is therefore distinct from the electromagnetic field. Lorentz was obliged thereby to extend his hypothesis with the assumption that *forces of any origin whatsoever, and gravitation in particular, are affected by a translation* (or, if one prefers, by the Lorentz transformation) *in the same manner as electromagnetic forces*. (Poincaré 1906, 166.)[16]

---

14 See (Lorentz 1900; Havas 1979, 83; Torretti 1996, 131). On Lorentz's precursors see (Whittaker 1951–1953, 2:149; Zenneck 1903). Lorentz's theory of gravitation failed to convince Oliver Heaviside, who had carefully weighed the analogy from electromagnetism to gravitation (Heaviside 1893). In a letter to Lorentz, Heaviside called into question the theory's electromagnetic nature, by characterizing Lorentz's gravitational force as "action at a distance of a double kind" (18 July, 1901, Lorentz Papers, Rijksarchief in Noord-Holland te Haarlem). Aware of these difficulties, Lorentz eventually discarded his theory, citing its incompatibility with the principle of relativity (Lorentz 1914, 32).

15 In his 1906–1907 Sorbonne lectures (Poincaré 1953), Poincaré discussed a different theory (based on an idea due to Le Sage) that Lorentz had proposed in the same paper, without mentioning the Mossotti-style theory. His first discussion of the latter theory was in 1908, when he considered it to be an authentic relativistic theory, and one in which the force of gravitation was of electromagnetic origin (Poincaré 1908, 399).

16 Poincaré's account of Lorentz's reasoning should be taken with a grain of salt, as Lorentz made no mention of his theory of gravitation in the 1904 publication referred to by Poincaré, "Electromagnetic phenomena in a system moving with any velocity less than that of light." While the electron theory developed in the latter paper did not address the question of the origin of the gravitational force, it admitted the possibility of a reduction to electromagnetism (such as that of his own theory) by means of the additional hypothesis referred to in the quotation: all forces of interaction transformed in the same way as electric forces in an electrostatic system (Lorentz 1904a, §8). The contraction hypothesis formerly invoked to account for the null result of the Michelson-Morley experiments, Lorentz added, was subsumed by the new hypothesis.

It was the cogency of the latter hypothesis that Poincaré set out to examine in detail, with respect to gravitational attraction. The situation was analogous to the one Poincaré had encountered in the case of electron energy and momentum mentioned above, where he had considered constraining internal forces of the electron to be Lorentz-invariant. Such a constraint solved the problem immediately, but Poincaré recognized that it was inadmissible nonetheless, because it violated Maxwell's theory (p. 136). A similar violation in the realm of mechanics could not be ruled out in the case of gravitation, such that a careful analysis of the admissibility of the formal requirement of Lorentz-invariance was called for.

Poincaré set out to determine a general expression for the law of gravitation in accordance with the principle of relativity. A relativistic law of gravitation, he reasoned, must obey two constraints distinguishing it from the Newtonian law. First of all, the new force law could no longer depend solely on the masses of the two gravitating bodies and the distance between them. The force had to depend on their velocities, as well. Furthermore, gravitational action could no longer be considered instantaneous, but had to propagate with some finite velocity, so that the force acting on the passive mass would depend on the position and velocity of the active mass at some earlier instant in time. A gravitational propagation velocity greater than the speed of light, Poincaré observed, would be "difficult to understand," because attraction would then be a function of a position in space not yet occupied by the active mass (p. 167).

These were not the only conditions Poincaré wanted to satisfy. The new law of gravitation had also (1) to behave in the same way as electromagnetic forces under a Lorentz transformation, (2) to reduce to Newton's law in the case of relative rest of the two bodies, and (3) to come as close as possible to Newton's law in the case of small velocities. Posed in this way, Poincaré noted, the problem remains indeterminate, save in the case of null relative velocity, where the propagation velocity of gravitation does not enter into consideration. Poincaré reasoned that if two bodies have a common rectilinear velocity, then the force on the passive mass is orthogonal to an ellipsoid, at the center of which lies the active mass.

Undeterred by the indeterminacy of the question in general, Poincaré set about identifying quantities invariant with respect to the Lorentz group, from which he wanted to construct a law of gravitation satisfying the constraints just mentioned. To assist in the identification and interpretation of these invariants, Poincaré referred to a space of four dimensions. "Let us regard," he wrote,

$$x, \quad y, \quad z, \quad t\sqrt{-1}$$
$$\delta x, \quad \delta y, \quad \delta z, \quad \delta t\sqrt{-1}$$
$$\delta_1 x, \quad \delta_1 y, \quad \delta_1 z, \quad \delta_1 t\sqrt{-1},$$

as the coordinates of 3 points P, P', P'', in space of 4 dimensions. We see that the Lorentz transformation is merely a rotation in this space about the origin, regarded as fixed. Consequently, we will have no distinct invariants apart from the 6 distances between the 3 points P, P', P'', considered separately and with the origin, or, if one prefers, apart from the 2 expressions:

$$x^2 + y^2 + z^2 - t^2, \quad x\delta x + y\delta y + z\delta z - t\delta t,$$

or the 4 expressions of like form deduced by arbitrary permutation of the 3 points P, P', P''.[17]

Here Poincaré formed three quadruplets representing the differential displacement of two point masses, with respect to a certain four-dimensional vector space, later called a pseudo-Euclidean space.[18] By introducing such a 4-space, Poincaré simplified the task of identifying quantities invariant with respect to the Lorentz transformations, the line interval of the new space being formally identical to that of a Euclidean 4-space. He treated his three points P, P', and P'' as 4-vectors, the scalar products of which are invariant, just as in Euclidean space. In fact, Poincaré did not employ vector terminology or notation in his study of gravitation, but provided formal definitions of certain objects later called 4-vectors.

Poincaré's habit, and that of the overwhelming majority of his French colleagues in mathematical physics well into the 1920s, was to express ordinary vector quantities in Cartesian coordinate notation, and to forgo notational shortcuts when differentiating, writing these operations out in full.[19] Although he did not exclude symbols such as Δ or $\Box$ from his scientific papers and lectures, he employed them parsimoniously.[20] In line with this practice, Poincaré did little to promote vector methods from his chair at the Sorbonne. In twenty volumes of lectures on mathematical physics and celestial mechanics, there is not a single propaedeutic on quaternions or vector algebra.[21] Poincaré deplored the "long calculations rendered obscure by notational

---

17 "Regardons $x, y, z, t\sqrt{-1}, \delta x, \delta y, \delta z, \delta t\sqrt{-1}, \delta_1 x, \delta_1 y, \delta_1 z, \delta_1 t\sqrt{-1}$, comme les coordonnées de 3 points P, P', P'' dans l'espace à 4 dimensions. Nous voyons que la transformation de Lorentz n'est qu'une rotation de cet espace autour de l'origine, regardée comme fixe. Nous n'aurons donc pas d'autres invariants distincts que les six distances des trois points P, P', P'' entre eux et à l'origine, ou, si l'on aime mieux, que les 2 expressions: $x^2 + y^2 + z^2 - t^2, x\delta x + y\delta y + z\delta z - t\delta t$, ou les 4 expressions de même forme qu'on en déduit en permutant d'une manière quelconque les 3 points P, P', P'' " (Poincaré 1906, 168–9).

18 Poincaré's three points P, P', P'' may be interpreted in modern terminology as follows. Let the spacetime coordinates of the passive mass point be $A = (x_0, y_0, z_0, t_0)$, with ordinary velocity $\xi = (\delta x/\delta t, \delta y/\delta t, \delta z/\delta t)$, such that at time $t_0 + \delta t$ it occupies the spacetime point $A' = (x_0 + \delta x, y_0 + \delta y, z_0 + \delta z, t_0 + \delta t)$. Likewise for the active mass point, $B = (x_0 + x, y_0 + y, z_0 + z, t_0 + t)$, with ordinary velocity $\xi_1 = (\delta_1 x/\delta_1 t, \delta_1 y/\delta_1 t, \delta_1 z/\delta_1 t)$, such that at time $t_0 + t + \delta_1 t$, it occupies the spacetime point $B' = (x_0 + x + \delta_1 x, y_0 + y + \delta_1 y, z_0 + z + \delta_1 z, t_0 + t + \delta_1 t)$. Poincaré's three quadruplets may now be expressed as position 4-vectors: $P = B - A, P' = B' - B, P'' = A' - A$.

19 While the first German textbook on electromagnetism to employ vector notation systematically dates from 1894 (Föppl 1894), the first comparable textbook in French was published two decades later by Jean-Baptiste Pomey (1861–1943), instructor of theoretical electricity at the *École supérieure des Postes et Télégraphes* in Paris (Pomey 1914–1931, vol. 1).

20 The Laplacian was expressed generally as $\nabla^2 = \partial^2/\partial x^2 + \partial^2/\partial y^2 + \partial^2/\partial z^2$, but by Poincaré as Δ. The d'Alembertian, $\Box \equiv \partial^2/\partial x^2 + \partial^2/\partial y^2 + \partial^2/\partial z^2 - \partial^2/\partial t^2$, became in Poincaré's notation: $\Box \equiv \Delta - d^2/dt^2$. Poincaré employed $\Box$ in his lectures on electricity and optics (Poincaré 1901, 456), and was the first to employ it in a relativistic context.

complexity" in W. Voigt's molecular theory of light, and seems to have been of the opinion that in general, new notation only burdened the reader.[22]

The point of forming quadruplets was to obtain a set of Lorentz-invariants corresponding to the ten variables entering into the right-hand side of the new force law, representing the squared distance in space and time of the two bodies and their velocities $(\xi, \eta, \zeta, \xi_1, \eta_1, \zeta_1)$. How did Poincaré obtain his invariants? According to the method cited above, six invariants were to be found from the distances between P, P', P'', and the origin, or from the scalar products of P, P', and P''. These six intermediate invariants were then to be combined to obtain homogeneous invariants depending on the duration of propagation of gravitational action and the velocities of the two point masses. Poincaré skipped over the intermediate step and produced the following four invariants, in terms of squared distance, distance and velocity (twice), and the velocity product:

$$\sum x^2 - t^2, \quad \frac{t - \sum x\xi}{\sqrt{1 - \sum \xi^2}}, \quad \frac{t - \sum x\xi_1}{\sqrt{1 - \sum \xi_1^2}}, \quad \frac{1 - \sum \xi\xi_1}{\sqrt{(1 - \sum \xi^2)(\sqrt{1 - \sum \xi_1^2})}}. \tag{1}$$

The Lorentz-invariance and geometric significance of these quantities are readily verified.[23] These four invariants (1), the latter three of which were labeled A, B, and C, formed the core of Poincaré's constructive approach to the law of gravitation. (For convenience, I refer to Poincaré's four invariants [1] as his "kinematic" invariants.)

Inspection of the signs of these invariants reveals an inconsistency, the reason for which is apparent once the intermediate calculations have been performed. Instead of constructing his four invariants out of scalar products, Poincaré introduced an inversion for A, B, and C.[24] This sign inconsistency had no consequence on his search

---

21 Poincaré's manuscript lecture notes for celestial mechanics, however, show that he saw fit to introduce the quaternionic method to his students (undated notebook on quaternions and celestial mechanics, 32 pp., private collection, Paris; hpcd 76, 78, 93, Henri Poincaré Archives, Nancy).

22 Manuscript report of the PhD. thesis submitted by Henri Bouasse, 13 December, 1892, AJ[16]5535, *Archives Nationales*, Paris. From Poincaré's conservative habits regarding formalism, he appears as an unlikely candidate at best for the development of a four-dimensional calculus circa 1905; cf. H. M. Schwartz's counterfactual conjecture: if Poincaré had adopted the ordinary vector calculus by the time he wrote his *Rendiconti* paper, "he would have in all likelihood introduced explicitly ... the convenient four-dimensional vector calculus" (Schwartz 1972, 1287, n. 7).

23 The invariants (1) may be expressed in ordinary vector notation, letting $\Sigma x = x, \Sigma\xi = v, \Sigma\xi_1 = v_1$, and for convenience, $k = 1/\sqrt{1 - \Sigma\xi^2}, k_1 = 1/\sqrt{1 - \Sigma\xi_1^2}$, such that the four quantities (1) read as follows: $x^2 - t^2, k(t - xv), k_1(t - xv_1), kk_1(1 - vv_1)$.

24 Poincaré's four kinematic invariants (1) are functions of the following six intermediate invariants: $a = x^2 + y^2 + z^2 - t^2$, $b = x\delta x + y\delta y + z\delta z - t\delta t$, $c = x\delta_1 x + y\delta_1 y + z\delta_1 z - t\delta_1 t$, $d = \delta x\delta_1 x + \delta y\delta_1 y + \delta z\delta_1 z - \delta t\delta_1 t$, $e = \delta x^2 + \delta y^2 + \delta z^2 - \delta t^2$, $f = \delta_1 x^2 + \delta_1 y^2 + \delta_1 z^2 - \delta_1 t^2$. In terms of the latter six invariants, the four kinematic invariants (1) may be expressed as follows: $\Sigma x^2 - t^2 = a$, $A = -b/\sqrt{-e}$, $B = -c/\sqrt{-f}$, and $C = -d/(\sqrt{-e}\sqrt{-f})$. For a slightly different reconstruction of Poincaré's kinematic invariants, see (Zahar 1989, 193).

for a relativistic law of gravitation, although it affected his final result, and perplexed at least one of his readers, as I will show in section 3.

What Poincaré needed next for his force law was a Lorentz-invariant expression for the force itself. Up to this point, he had neither a velocity 4-vector nor a force 4-vector definition on hand. Presumably, the search for Lorentz-invariant expressions of force led him to define these 4-vectors. Earlier in his memoir (p. 135), Poincaré had determined the Lorentz transformations of force density, but now he was interested in the Lorentz transformations of force at a point. The transformations of force density:

$$X' = k(X + \varepsilon T), \qquad Y' = Y, \qquad Z' = Z, \qquad T' = k(T + \varepsilon X), \qquad (2)$$

where k is the Lorentz factor, $k = 1/\sqrt{1 - \varepsilon^2}$, and ε designates frame velocity, led Poincaré to define a fourth component of force density, T, as the product of the force density vector with velocity, $T = \sum X\xi$.[25] He gave the same definition for the temporal component of force at a point: $T_1 = \sum X_1\xi$.[26] Next, dividing force density by force at a point, Poincaré obtained the charge density ρ. Ostensibly from the transformation for charge density, Poincaré singled out the Lorentz-invariant factor:[27]

$$\frac{\rho}{\rho'} = \frac{1}{k(1 + \xi\varepsilon)} = \frac{\delta t}{\delta t'}. \qquad (3)$$

The components of a 4-velocity vector followed from the foregoing definitions of position and force density:

> The Lorentz transformation ... will act in the same way on $\xi, \eta, \zeta, 1$ as on $\delta x, \delta y, \delta z, \delta t$, with the difference that these expressions will be multiplied moreover by the *same* factor $\delta t/\delta t' = 1/k(1 + \xi\varepsilon)$.[28]

Concerning the latter definition, Poincaré observed a formal analogy between the force and force density 4-vectors, on one hand, and the position and velocity 4-vectors, on the other hand: these pairs of vectors transform in the same way, except that one member is multiplied by $1/k(1 + \xi\varepsilon)$. While this analogy may seem mathematically transparent, it merits notice, as it appears to have eluded Poincaré at first.

With these four kinematic 4-vectors in hand, Poincaré defined a fifth quadruplet Q with components of force density $(X, Y, Z, T\sqrt{-1})$. Just as in the previous case, the scalar products of his four quadruplets P, P', P'', and Q were to deliver four new Lorentz-invariants in terms of the force acting on the passive mass (X_1, Y_1, Z_1):[29]

---

25 This definition was remarked by (Pauli 1921, 637).

26 The same subscript denotes the *force* acting on the *passive* mass, ΣX_1, and the *velocity* of the *active* mass, ξ_1.

27 The ratio ρ/ρ' is equal to the Lorentz factor, since in Poincaré's configuration, $\varepsilon = -\xi$. Some writers hastily attribute a 4-current vector to Poincaré, the form $\rho(\xi, \eta, \zeta, i)$ being implied by Poincaré's 4-vector definitions of force density and velocity.

28 "La transformation de Lorentz ... agira sur $\xi, \eta, \zeta, 1$ de la même manière que sur $\delta x, \delta y, \delta z, \delta t$, avec cette différence que ces expressions seront en outre multipliées par le *même* facteur $\delta t/\delta t' = 1/k(1 + \xi\varepsilon)$ " (Poincaré 1906, 169).

$$\frac{\sum X_1^2 - T_1^2}{1 - \sum \xi^2}, \quad \frac{\sum X_1 x - T_1 t}{\sqrt{1 - \sum \xi^2}}, \quad \frac{\sum X_1 \xi_1 - T_1}{\sqrt{1 - \sum \xi^2}\sqrt{1 - \sum \xi_1^2}}, \quad \frac{\sum X_1 \xi - T_1}{\sqrt{1 - \sum \xi^2}}. \tag{4}$$

The fourth invariant in (4) was always null by definition of T_1, leaving only three invariants, denoted M, N, and P. (In order to distinguish these invariants from the kinematic invariants, I will refer to [4] as Poincaré's "force" invariants.)

Comparing the signs of the kinematic invariants (1) with those of the force invariants (4), we see that Poincaré obtained consistent signs only for the latter invariants. He must not have computed his force invariants in the same way as his kinematic invariants, for reasons that remain obscure. It is not entirely unlikely that in the course of his analysis of the transformations of velocity and force, Poincaré realized that he could compute the force invariants directly from the scalar products of four 4-vectors. Two facts, however, argue against this reading. In the first place, Poincaré did not mention that his force invariants were the scalar products of position, velocity and force 4-vectors. Secondly, he did not alter the signs of his kinematic invariants to make them correspond to scalar products of position and velocity 4-vectors.[30] The fact that Poincaré's kinematic invariants differ from products of 4-position and 4-velocity vectors leads us to believe that when forming these invariants he was *not* thinking in terms of 4-vectors.[31]

From this point on, Poincaré worked exclusively with arithmetic combinations of three force invariants (M, N, P) and four kinematic invariants $(\sum x^2 - t^2, A, B, C)$ in order to come up with a relativistic law of gravitation. He had no further use, in particular, for the four quadruplets he had identified in the process of constructing

29 The invariants (4) may be expressed in ordinary vector notation, recalling the definitions of note 23, and letting $\Sigma X_1 = f_1$, and $T_1 = f_1 v$: $k^2 f_1^2(1-v^2), k f_1(x-vt), k k_1 f_1(v_1-v), k^2 f_1(v-v)$. The fourth invariant is obviously null in this form.

30 Poincaré's force invariants (4) are functions of the following six intermediate invariants:
$m = k(X_1 \delta x + Y_1 \delta y + Z_1 \delta z - T_1 \delta t)$, $n = k(X_1 \delta_1 x + Y_1 \delta_1 y + Z_1 \delta_1 z - T_1 \delta_1 t)$, $o = k(X_1 x + Y_1 y + Z_1 z - T_1 t)$, $p = k^2(X_1^2 + Y_1^2 + Z_1^2 - T_1^2)$, $q = \delta x^2 + \delta y^2 + \delta z^2 - \delta t^2$, and $s = \delta_1 x^2 + \delta_1 y^2 + \delta_1 z^2 - \delta_1 t^2$. Let the four force invariants (4) be denoted by M, N, P, and S, then $M = p, N = o, P = n/\sqrt{-s}$, and $S = m/\sqrt{-q}$.
The same force invariants (4) are easily calculated using 4-vectors. Recalling the definitions in notes 23 and 29, let $\mathfrak{R} = (x, it)$, $U = k(v, i)$, $U_1 = k_1(v_1, i)$, and $F_1 = k(f_1, if_1 v)$, where $\sqrt{-1} = i$. Then the force invariants (4) may be expressed as scalar products of 4-vectors: $M = F_1 F_1$, $N = F_1 \mathfrak{R}$, $P = F_1 U_1$, and $S = F_1 U$.

31 The kinematic invariants (1) obtained by Poincaré differ from those obtained from the products of 4-position and 4-velocity, contrary to Zahar's account (Zahar 1989, 194). Recalling the 4-vectors $\mathfrak{R}, U, U_1$, from n. 30, we form the products: $\mathfrak{R}\mathfrak{R}, \mathfrak{R}U, \mathfrak{R}U_1$, and UU_1. In Poincaré's notation, the latter four products are expressed as follows:
$$\Sigma x^2 - t^2, -\frac{t - \Sigma x \xi}{\sqrt{1 - \Sigma \xi^2}}, -\frac{t - \Sigma x \xi_1}{\sqrt{1 - \Sigma \xi_1^2}}, -\frac{t - \Sigma \xi \xi_1}{\sqrt{(1 - \Sigma \xi^2)(1 - \Sigma \xi_1^2)}}.$$
These invariants differ from those of Poincaré (1) only by the sign of A, B, and C, as noted by (Sommerfeld 1910b, 686).

these same invariants (corresponding to modern 4-position, 4-velocity, 4-force-density and 4-force vectors), although in the end he expressed his laws of gravitation in terms of 4-force components.

To find a law applicable to the general case of two bodies in relative motion, Poincaré introduced constraints and approximations designed to reduce the complexity of his seven invariants and recover the form of the Newtonian law in the limit of slow motion ($\xi_1 \ll 1$). Poincaré naturally looked first to the velocity of propagation of gravitation. He briefly considered an emission theory, where the velocity of gravitation depends on the velocity of the source. Although the emission hypothesis was compatible with his invariants, Poincaré rejected this option because it violated his initial injunction barring a hyperlight velocity of gravitational propagation.[32] That left him with a propagation velocity of gravitation less than or equal to that of light, and to simplify his invariants Poincaré set it equal to that of light in empty space, such that $t = -\sqrt{\sum x^2} = -r$. This stipulation reduced the total number of invariants from seven to six.

With the propagation velocity of gravitation decided, Poincaré proceeded to construct a force law for point masses. He tried two approaches, the first of which is the most general. The basic idea of both approaches is to neglect terms in the square of velocity occurring in the invariants, and to compare the resulting approximations with their Newtonian counterparts. In the Newtonian scheme, the coordinates of the active mass point differ from those in the relativistic scheme (cf. note 18); Poincaré took the former to be $(x_0 + x_1, y_0 + y_1, z_0 + z_1)$ at the instant of time t_0, where the subscript 0 corresponds to the position of the passive mass point, and the coordinates with subscript 1 are found by assuming uniform motion of the source:

$$x = x_1 - \xi_1 r, \quad y = y_1 - \eta_1 r, \quad z = z_1 - \zeta_1 r, \quad r = r_1 - \sum x \xi_1. \tag{5}$$

In the first approach, Poincaré made use of both the kinematic and force invariants. Substituting the values (2) into the kinematic invariants $A, B,$ and C from (1) and the force invariants $M, N,$ and P from (33), neglecting terms in the square of velocity, Poincaré obtained their sought-after Newtonian counterparts. Replacing the force vector occurring in the transformed force invariants by Newton's law ($\sum X_1 = -1/r_1^2$), and rearranging, Poincaré obtained three quantities in terms of distance and velocity.[33] He then re-expressed these quantities in terms of two of his original kinematic invariants, A and B, and equated the three resulting kinematic invariants to their corresponding original force invariants (4). He now had the solution in hand; three expressions relate his force invariants (containing the force vector $\sum X_1$) to two of his kinematic invariants:

---

32   An emission theory was proposed a few years later by Walter Ritz; see (Ritz 1908).

33   Using (5), Poincaré found the transformed force invariants $1/r_1^4, -1/r_1 - \Sigma x_1(\xi - \xi_1)/r_1^2$, and $\Sigma x_1(\xi - \xi_1)/r_1^3$.

$$M = \frac{1}{B^4}, \quad N = \frac{+A}{B^2}, \quad P = \frac{A-B}{B^3}. \tag{6}$$

He noted that complementary terms could be entertained for the three relations (6), provided that they were certain functions of his kinematic invariants $A, B,$ and C. Then without warning, he cut short his demonstration, remarking that the gravitational force components would take on imaginary values:

> The solution (6) appears at first to be the simplest, nonetheless, it may not be adopted. In fact, since M, N, P are functions of X_1, Y_1, Z_1, and $T_1 = \sum X_1 \xi$, the values of X_1, Y_1, Z_1 can be drawn from these three equations (6), but in certain cases these values would become imaginary.[34]

The quoted remark seems to suggest that for selected values of the particle velocities, the force turns out to be imaginary. However, the real difficulty springs from the equation $M = 1/B^4$, which allows for a repulsive force. The general approach failed to deliver.[35]

The fact that Poincaré published the preceding derivation may be understood in one of two ways. On the one hand, there is a psychological explanation: Poincaré's habit, much deplored by his peers, was to present his findings more or less in the order in which he found them. The case at hand may be no different from the others. On the other hand, Poincaré may have felt it worthwhile to show that the general approach breaks down. From the latter point of view, Poincaré's result is a positive one.

For his second attack on the law of gravitation, Poincaré adopted a less general approach. He knew where his first approach had become unsuitable, and consequently, leaving aside his three force invariants, he fell back on the form of his basic force 4-vector, which he now wrote in terms of his kinematic invariants, re-expressed in terms of $r = -t$, $k_0 = 1/\sqrt{1-\xi^2}$, and $k_1 = 1/\sqrt{1-\xi_1^2}$.[36] He assumed the gravitational force on the passive mass (moving with velocity ξ, η, ζ) to be a function of the distance separating the two mass points, the velocity of the passive mass point, and the velocity of the source, with the form:

---

34 "Au premier abord, la solution (6) paraît la plus simple, elle ne peut néanmoins être adoptée; en effet, comme M, N, P sont des fonctions de X_1, Y_1, Z_1, et de $T_1 = \Sigma X_1 \xi$, on peut tirer de ces trois équations (6) les valeurs de X_1, Y_1, Z_1; mais dans certains cas ces valeurs deviendraient imaginaires" (Poincaré 1906, 172).

35 Replacing A and B in (6) by their definitions results in the three equations: $M = k^2 f_1^2(1-v^2) = 1/k^4(r+xv_1)^4$, $N = f_1(x+vr) = -(r+xv)/[k_1^2(r+xv_1)^2]$, $P = kk_1 f_1(v_1-v) = [k(r+xv)- -k_1(r+xv_1)]/k_1^3(r+xv_1)^3]$. Equations N and P imply an attractive force for all values of v and v_1, while M leads to the ambiguously-signed solution: $f_1 = \pm(1/[k^2(r+xv_1)^2])$. Presumably, the superfluous plus sign in (6) is an indication of Poincaré's preoccupation with obtaining a force of correct sign.

36 $A = -k_0(r+\Sigma x\xi)$, $B = -k_1(r+\Sigma x\xi_1)$, and $C = k_0 k_1(1-\Sigma x\xi\xi_1)$.

$$X_1 = x\frac{\alpha}{k_0} + \xi\beta + \xi_1\frac{k_1}{k_0}\gamma, \quad Z_1 = z\frac{\alpha}{k_0} + \zeta\beta + \zeta_1\frac{k_1}{k_0}\gamma,$$

$$Y_1 = y\frac{\alpha}{k_0} + \eta\beta + \eta_1\frac{k_1}{k_0}\gamma, \quad T_1 = -r\frac{\alpha}{k_0} + \beta + \frac{k_1}{k_0}\gamma,$$

(7)

where α, β, and γ denote functions of the kinematic invariants.[37] By definition, the component T_1 is the scalar product of the ordinary force and the velocity of the passive mass point, $T_1 = \sum X_1\xi$, such that the three functions α, β, γ satisfy the equation:

$$-A\alpha - \beta - C\gamma = 0. \tag{8}$$

Poincaré further assumed $\beta = 0$, thereby eliminating a term depending on the velocity of the passive mass, and fixing the value of γ in terms of α. Applying the same slow-motion approximation and translation (5) as in his initial approach, Poincaré found $X_1 = \alpha x_1$, and by comparison with Newton's law, α reduces to $-1/r_1^3$. In terms of the kinematic invariants (1), this relation was expressed as $\alpha = 1/B^3$, and the law of gravitation (7) took on the form:[38]

$$X_1 = \frac{x}{k_0 B^3} - \xi_1\frac{k_1}{k_0}\frac{A}{B^3 C}, \quad Z_1 = \frac{z}{k_0 B^3} - \zeta_1\frac{k_1}{k_0}\frac{A}{B^3 C},$$

$$Y_1 = \frac{y}{k_0 B^3} - \eta_1\frac{k_1}{k_0}\frac{A}{B^3 C}, \quad T_1 = \frac{r}{k_0 B^3} - \frac{k_1}{k_0}\frac{A}{B^3 C}.$$

(9)

Inspection of Poincaré's gravitational force (9) reveals two components: one parallel to the position 4-vector between the passive mass and the retarded source, and one parallel to the source 4-velocity. The law was not unique, Poincaré noted, and it neglected possible terms in the velocity of the passive mass.

Poincaré underlined the open-ended nature of his solution by proposing a second gravitational force law. Rearranging (9) and replacing the factor $1/B^3$ by C/B^3, such that the force depended linearly on the velocity of the passive mass, Poincaré arrived at a second law of gravitation:[39]

---

37 Using modern 4-vector notation, and denoting Poincaré's gravitational force 4-vector $F_1 = k_0(X_1, Y_1, Z_1, iT_1)$, equation (7) may be expressed: $F_1 = \alpha\Re + \beta U + \gamma U_1$, where $\Re$ denotes a light-like 4-vector between the mass points, α, β, γ stand for undetermined functions of the three kinematic invariants A, B, and C, while $U = k_0(v, i)$, $U_1 = k_1(v_1, i)$ designate the 4-velocities of the passive and active mass points, respectively.

38 In ordinary vector form, recalling the definitions in notes 23 and 29, the spatial part of Poincaré's law is expressed as follows: $f_1 = -[(x + rv_1) + v \times (v_1 \times x)]/[kk_1^2(r + xv_1)^3(1 - vv_1)]$. Cf. (Zahar 1989, 199).

39 This law may be reformulated using the vectors defined in notes 23 and 29, and neglecting (with Poincaré) the component T_1: $f_1 = -([(x + rv_1) + v \times (v_1 \times x)]/[k_1^2(r + xv_1)^3])$. Cf. (Zahar 1989, 199). Comparable expressions were developed by Lorentz and Kottler (Lorentz 1910, 1239; Kottler 1922, 169).

$$X_1 = \frac{\lambda}{B^3} + \frac{\eta\nu' - \zeta\mu'}{B^3},$$

$$Y_1 = \frac{\mu}{B^3} + \frac{\zeta\lambda' - \xi\nu'}{B^3}, \qquad (10)$$

$$Z_1 = \frac{\nu}{B^3} + \frac{\xi\mu' - \eta\lambda'}{B^3},$$

where

$$k_1(x + r\xi_1) = \lambda, \qquad k_1(y + r\eta_1) = \mu, \qquad k_1(z + r\zeta_1) = \nu,$$

$$k_1(\eta_1 z - \zeta_1 y) = \lambda', \qquad k_1(\zeta_1 x - \xi_1 z) = \mu', \qquad k_1(\xi_1 y - x\eta_1) = \nu'.$$

Poincaré neglected to write down the expression for T_1, probably because of its complicated form. (For the sake of simplicity, I refer to [9] and [10], including the latter's neglected fourth component, as Poincaré's first and second law.) The unprimed triplet $B^{-3}(\lambda, \mu, \nu)$ supports what Poincaré termed a "vague analogy" with the mechanical force on a charged particle due to an electric field, while the primed triplet $B^{-3}(\lambda', \mu', \nu')$ supports an analogy to the mechanical force on a charged particle due to a magnetic field. He identified the fields as follows:

> Now λ, μ, ν, or $\frac{\lambda}{B^3}, \frac{\mu}{B^3}, \frac{\nu}{B^3}$, is an electric field of sorts, while λ', μ', ν', or rather
>
> $\frac{\lambda'}{B^3}, \frac{\mu'}{B^3}, \frac{\nu'}{B^3}$, is a magnetic field of sorts.[40]

While Poincaré wrote freely of a "gravity wave" [*onde gravifique*], he abstained from speculating on the nature of the field referred to here. As one of the first theorists (with FitzGerald and Lorentz) to have employed retarded potentials in Maxwellian electrodynamics, Poincaré must have considered the possibility of introducing a corresponding gravitational 4-potential (Whittaker 1951–1953, 1: 394, n. 3).[41] But as matters stood when Poincaré submitted this paper for publication in July, 1905, he was not in a position to elaborate the physics of fields in four-dimensional terms, since he possessed neither a 4-potential nor a 6-vector.

Poincaré had realized the objective of formulating a Lorentz-invariant force of gravitation. As we have seen, he surpassed this objective by identifying not one but

---

40 "Alors λ, μ, ν, ou $\lambda/B^3, \mu/B^3, \nu/B^3$, est une espèce de champ électrique, tandis que λ', μ', ν', ou plutôt $\lambda'/B^3, \mu'/B^3, \nu'/B^3$, est une espèce de champ magnétique" (Poincaré 1906, 175).

41 A 4-potential corresponding to Poincaré's second law (10) was given by Kottler (Kottler 1922, 169). Additional assumptions are required in order to identify a "gravito-magnetic" field with a term arising from the Lorentz transformation of force: $v \times (v_1 \times x)$, or the second term of the 3-vector version of (10) (neglecting the global factor; see note 39). In particular, it must be assumed that when the sources of the "gravito-electric" field $B^{-3}(\lambda, \mu, \nu)$ are at rest, the force on a mass point m is $f = mB^{-3}(\lambda, \mu, \nu)$, independent of the velocity of m. For a detailed discussion, see (Jackson 1975, 578).

two such force laws. Designed to reduce to Newton's law in the first order of approximation in ξ_1 (or particle velocity divided by the speed of light), Poincaré's laws could diverge from Newton's only in second-order terms. The argument satisfied Poincaré, who did not report any precise numerical results, explaining that this would require further investigation. Instead, he noted that the disagreement would be ten thousand times smaller than a first-order difference stemming from the assumption of a propagation velocity of gravitation equal to that of light, "*ceteris non mutatis*" (p. 175). His result contradicted Laplace, who had predicted an observable first-order effect arising from just such an assumption. At the very least, Poincaré had demonstrated that Laplace's argument was not compelling in the context of the new dynamics.[42]

On several occasions over the next seven years, Poincaré returned to the question of gravitation and relativity, without ever comparing the predictions of his laws with observation. During his 1906–1907 Sorbonne lectures, for example, when he developed a general formula for perihelion advance, Poincaré used a Lagrangian approach, rather than one or the other of his laws (Poincaré 1953, 238). Student notes of this course indicate that he stopped short of a numerical evaluation for the various electron models (perhaps leaving this as an exercise). However, Poincaré later provided the relevant numbers in a general review of electron theory. Lorentz's theory called for an extra 7" centennial advance by Mercury's perihelion, a figure slightly greater than the one for Abraham's non-relativistic electron theory.[43] According to the best available data, Mercury's anomalous perihelial advance was 42", prompting Poincaré to remark that another explanation would have to be found in order to account for the remaining seconds of arc. Astronomical observations, Poincaré concluded soberly, provided no arguments in favor of the new electron dynamics (Poincaré 1908, 400).[44]

Poincaré capsulized the situation of his new theory in a fable in which Lorentz plays the role of Ptolemy, and Poincaré that of an unknown astronomer appearing sometime between Ptolemy and Copernicus. The unknown astronomer notices that all the planets traverse either an epicycle or a deferent in the same lapse of time, a regularity later captured in Kepler's second law. The analogy to electron dynamics turns on a regularity discovered by Poincaré in his study of gravitation:

> If we were to admit the postulate of relativity, we would find the same number in the law of gravitation and the laws of electromagnetism, which would be the velocity of light; and we would find it again in all the other forces of any origin whatsoever.[45]

---

42 Poincaré reviewed Laplace's argument in his 1906–1907 lectures (Poincaré 1953, 194). For a contemporary overview of the question of the propagation velocity of gravitation see (Tisserand 1889–1896, 1: 511).

43 Fritz Wacker, a student of Richard Gans in Tübingen, published similar results in (Wacker 1906).

44 Poincaré explained to his students that Mercury's anomalous advance could plausibly be attributed to an intra-Mercurial matter belt (Poincaré 1953, 265), an idea advanced forcefully by Hugo von Seeliger in 1906 (Roseveare 1982, 78). In a lecture delivered in September, 1909, Poincaré revised his estimate of the relativistic perihelial advance downward slightly to 6" (Poincaré 1909).

45 "[S]i nous admettions le postulat de relativité, nous trouverions dans la loi de gravitation et dans les lois électromagnétiques un nombre commun qui serait la vitesse de la lumière; et nous le retrouverions encore dans toutes les autres forces d'origine quelconque" (Poincaré 1906, 131).

This common propagation velocity of gravitational action, of electromagnetic fields, and of any other force, could be understood in one of two ways:

> Either everything in the universe would be of electromagnetic origin, or this aspect—shared, as it were, by all physical phenomena—would be a mere epiphenomenon, something due to our methods of measurement.[46]

If the electromagnetic worldview were valid, all particle interactions would be governed by Maxwell's equations, featuring a constant propagation velocity. Otherwise, the common propagation velocity of forces had to be a result of a measurement convention. In relativity theory, as Poincaré went on to point out, the measurement convention to adopt was one defining lengths as equal if and only if spanned by a light signal in the same lapse of time, as this convention was compatible with the Lorentz contraction. There was a choice to be made between the electromagnetic worldview (as realized in the electron models of Abraham and Bucherer-Langevin) and the postulate of relativity (as upheld by the Lorentz-Poincaré electron theory). Although Poincaré favored the latter theory, he felt that its destiny was to be superseded, just as Ptolemaic astronomy was superseded by Copernican heliocentrism.

The failure of his Lorentz-invariant law of gravitation to explain the anomalous advance of Mercury's perihelion probably fed Poincaré's dissatisfaction with the Lorentz-Poincaré theory in general, but what he found particularly troubling at the time was something else altogether: the discovery of magneto-cathode rays. There is no place in the Lorentz-Poincaré electron theory for rays that are both neutral (as Paul Villard reported in June, 1904) and deflected by electric and magnetic fields, which is probably why Poincaré felt the "entire theory" to be "endangered" by magneto-cathode rays.[47]

Uncertainty over the empirical adequacy of the Lorentz-Poincaré electron theory may explain why the *Rendiconti* memoir was Poincaré's last in the field of electron physics. But is it enough to explain his disinterest in the development of a four-dimensional formalism? One year after the publication of his article on electron dynamics, Poincaré commented:

> A translation of our physics into the language of four-dimensional geometry does in fact appear to be possible; the pursuit of this translation would entail great pain for limited profit, and I will just cite Hertz's mechanics, where we see something analogous. Meanwhile, it seems that the translation would remain less simple than the text and would always have the feel of a translation, and that three-dimensional language seems the best suited to the description of our world, even if one admits that this description may be carried out in another idiom.[48]

---

46 "Ou bien il n'y aurait rien au monde qui ne fût d'origine électromagnétique. Ou bien cette partie qui serait pour ainsi dire commune à tous les phénomènes physiques ne serait qu'une apparence, quelque chose qui tiendrait à nos méthodes de mesure" (Poincaré 1906, 131–132).

47 See (Poincaré 1906, 132; Stein 1987, 397, n. 29). On the history of magneto-cathode rays, see (Carazza and Kragh 1990).

Poincaré clearly saw in his own work the outline of a four-dimensional formalism for physics, yet he saw no future in its development, and this, entirely apart from the question of the empirical adequacy of the Lorentz-Poincaré theory.

Why did Poincaré discount the value of a language tailor-made for relativity? Three sources of disinterest in such a prospect spring to mind, the first of which stems from his conventionalist philosophy of science. Poincaré recognized an important role for notation in the exact sciences, as he famously remarked with respect to Edmond Laguerre's work on quadratic forms and Abelian functions that

> in the mathematical sciences, having the right notation is philosophically as important as having the right classification in the life sciences.[49]

More than likely, Poincaré was aware of the philosophical implications of a four-dimensional notation for physics, although he had yet to make his views public. But given his strong belief in the immanence of Euclidean geometry's fitness for physics, he must have considered the chances for success of such a language to be vanishingly small.[50]

A second source for Poincaré's disinterest in four-dimensional formalism is his practice of physics. As mentioned above, Poincaré dispensed with vectorial systems (and most notational shortcuts); he even avoided writing i for $\sqrt{-1}$. When considered in conjunction with his conventionalist belief in the suitability of Euclidean geometry for physics, this conservative habit with respect to notation makes Poincaré appear all the less likely to embrace a four-dimensional language for physics.

The third possible source of discontent is Poincaré's vexing experience with invariants of pseudo-Euclidean 4-space. As shown above (p. 205), Poincaré's first approach to the construction of a law of gravitation ended unsatisfactorily, and the failure of Poincaré's intuition in this instance may well have colored his view of the prospects for a four-dimensional physics.

An immediate consequence of Poincaré's refusal to work out the form of four-dimensional physics was that others could readily pick up where he left off. Roberto Marcolongo (1862–1945), Professor of Mathematical Physics in Messina, and a leading proponent of vectorial analysis, quickly discerned in Poincaré's paper a potential

---

48 "Il semble bien en effet qu'il serait possible de traduire notre physique dans le langage de la géométrie à quatre dimensions; tenter cette traduction ce serait se donner beaucoup de mal pour peu de profit, et je me bornerai à citer la mécanique de Hertz où l'on voit quelque chose d'analogue. Cependant, il semble que la traduction serait toujours moins simple que le texte, et qu'elle aurait toujours l'air d'une traduction, que la langue des trois dimensions semble la mieux appropriée à la description de notre monde, encore que cette description puisse se faire à la rigueur dans un autre idiome" (Poincaré 1907, 15). See also (Walter 1999b, 98), and for a different translation, (Galison 1979, 95). On Hertz's mechanics, see (Lützen 1999).

49 "[D]ans les Sciences mathématiques, une bonne notation a la même importance philosophique qu'une bonne classification dans les Sciences naturelles" (Poincaré 1898–1905, 1:x).

50 Poincaré's analysis of the concepts of space and time in relativity theory appeared in 1912 (Poincaré 1912). On the cool reception among mathematicians of Poincaré's views on physical geometry, see (Walter 1997).

for formal development. Marcolongo referred, like Poincaré, to a four-dimensional space with one imaginary axis, but defined the fourth coordinate as the product of time t and the negative square root of -1 (i.e., $-t\sqrt{-1}$ instead of $t\sqrt{-1}$). After forming a 4-vector potential out of the ordinary vector and scalar potentials, and defining a 4-current vector, he expressed the Lorentz-covariance of the equations of electrodynamics in matrix form. No other applications were forthcoming from Marcolongo, and a failure to produce further 4-vector quantities and functions limited the scope of his contribution, which went unnoticed outside of Italy (Marcolongo 1906).[51] Nothing further on Poincaré's method appeared in print until April, 1908, when Hermann Minkowski's paper on the four-dimensional formalism and its application to the problem of gravitation appeared in the *Göttinger Nachrichten*.

2. HERMANN MINKOWSKI'S SPACETIME LAWS OF GRAVITATION

The young Hermann Minkowski, second son of an immigrant family of Russian Jews, attended the Altstädtische Gymnasium in Königsberg (later Kaliningrad). Shortly after graduation, Minkowski submitted an essay for the Paris Academy's 1882 Grand Prize in Mathematical Sciences. His entry on quadratic forms shared top honors with a submission by the seasoned British mathematician Henry J. S. Smith, his senior by thirty-eight years.[52] The young mathematician went on to study with Heinrich Weber in Königsberg, and with Karl Weierstrass and Leopold Kronecker in Berlin. In the years following the prize competition, Minkowski became acquainted with Poincaré's writings on algebraic number theory and quadratic forms, and in particular, with a paper in Crelle's *Journal* containing some of the results from Minkowski's prize paper, still in press. To his friend David Hilbert he confided the "angst and alarm" brought on by Poincaré's entry into his field of predilection; with his "swift and versatile" energy, Poincaré was bound to bring the whole field to closure, or so it seemed to him at the time.[53] From the earliest, formative years of his scientific career, Minkowski found in Poincaré—his senior by a decade—a daunting intellectual rival.

While Minkowski had discovered in Poincaré a rival, he was soon to find that the Frenchman could also be a teacher, from whom he could learn new analytical skills and methods. Named Privatdozent in Bonn in 1887, Minkowski contributed to the abstract journal *Jahrbuch der Fortschritte der Mathematik*, and in 1892, took on the considerable task of summarizing the results of the paper for which Poincaré was

51 This paper later gave rise to a priority claim for a slightly different substitution: $u = it$ (Marcolongo to Arnold Sommerfeld, 5 May, 1913, Archives for History of Quantum Physics 32). On Marcolongo's paper see also (Maltese 2000, 135).

52 See (Rüdenberg 1973; Serre 1993; Strobl 1985).

53 Minkowski to Hilbert, 14 February, 1885, (Minkowski 1973, 30). Minkowski's fears turned out to be for naught, as Poincaré pursued a different line of research (Zassenhaus 1975, 446). On Minkowski's early work on the geometry of numbers see (Schwermer 1991); on later developments, see (Krätzel 1989).

awarded the King Oscar II Prize (Minkowski 1890–1893). The mathematics Poincaré created in his prize paper (the study of homoclinic points in particular) was highly innovative, and at the same time, difficult to follow. Among those whom we know had trouble understanding certain points of Poincaré's prize memoir were Charles Hermite, Gustav Mittag-Leffler, and Karl Weierstrass, who happened to constitute the prize committee.[54] Minkowski, however, welcomed the review as a learning opportunity, as he wrote to his friend and former teacher, Adolf Hurwitz:

> Poincaré's prize paper is also among the works I have to report on for the *Fortschritte*. I am quite fond of it. It is a fine opportunity for me to get acquainted with problems I have not worried about too much up to now, since I will naturally set a positive goal of making my case well.[55]

In the 1890s, building on his investigations of the algebraic theory of quadratic forms, Minkowski developed the geometric analog to this theory: geometrical number theory. A high point of his efforts in this new field, and one which contributed strongly to the establishment of his reputation in mathematical circles, was the publication of *Geometrie der Zahlen* (Minkowski 1896). The same year, Minkowski accepted a chair at Zurich Polytechnic, whereby he rejoined Hurwitz. Minkowski's lectures on mathematics and mathematical physics attracted a small following of talented and ambitious students, including the future physicists Walter Ritz and Albert Einstein, and the budding mathematicians Marcel Grossmann and Louis Kollros.[56]

Minkowski's lectures on mechanics in Zurich throw an interesting light on his view of symbolic methods in physics at the close of the nineteenth century. The theory of quaternions, he noted in 1897, was used nowhere outside of England, due to its "relatively abstract character and inherent difficulty."[57] Two of its fundamental concepts, scalars and vectors, had nevertheless gained broad approval among physicists, Minkowski wrote, and had found "frequent application especially in the theory of electricity."[58] Applications of quaternions to problems of physics were advanced in Germany with the publication of Felix Klein and Arnold Sommerfeld's *Theorie des Kreisels*, a work referred to in Minkowski's lecture notes of 1898–1899 (Klein and Sommerfeld 1879–1910).[59] Minkowski admired Klein and Sommerfeld's text,

---

54 See (Gray 1992) and the reception study by Barrow-Green (Barrow-Green 1997, chap. 6).

55 Minkowski to Hurwitz, 5 January, 1892, Cod. Ms. Math. Arch. 78: 188, Handschriftenabteilung, Niedersächsische Staats- und Universitätsbibliothek (NSUB). On Minkowski's report see also (Barrow-Green 1997, 143).

56 Minkowski papers, Arc. 4° 1712, Jewish National and University Library (JNUL); Minkowski to Hilbert, 11 March, 1901, (Minkowski 1973, 139).

57 Vorlesungen über analytische Mechanik, Wintersemester 1897/98, p. 29, Minkowski papers, Arc. 4° 1712, JNUL.

58 Loc. cit. note 57. The concepts of scalar and vector mentioned by Minkowski were those introduced by W. R. Hamilton (1805–1865), the founder of quaternion theory. Even in Britain, vectors were judged superior to quaternions for use in physics, giving rise to spirited exchanges in the pages of *Nature* during the 1890s, as noted by (Bork 1966) and (Crowe 1967, chap. 6). On the introduction of vector analysis as a standard tool of the physicist during this period, see (Jungnickel and McCormmach 1986, 2:342), and for a general history, see (Crowe 1967).

expressing "great interest" in the latter to Sommerfeld, along with his approval of the fundamental significance accorded to the concept of momentum. However, their text did not make the required reading list for Minkowski's course in mechanics.[60]

In 1899, at the request of Sommerfeld, who a year earlier had agreed to edit the physics volumes of Felix Klein's ambitious *Encyclopedia of the Mathematical Sciences including Applications* (hereafter *Encyklopädie*), Minkowski agreed to cover a topic in molecular physics he knew little about, but one perfectly suited to his skills as an analyst: capillarity.[61] The article that appeared seven years later represented his second contribution to physics, after a short note on theoretical hydrodynamics published in 1888, but which, ten years later, Minkowski claimed no one had read—save the abstracter (Minkowski 1888, 1907).[62]

When Minkowski accepted Göttingen's newly-created third chair of pure mathematics in the fall of 1902, the pace of his research changed brusquely. The University of Göttingen at the turn of the last century was a magnet for talented young mathematicians and physicists.[63] Minkowski soon was immersed in the activities of Göttingen's Royal Society of Science, its mathematical society, and research seminars. Several faculty members, including Max Abraham, Gustav Herglotz, Eduard Riecke, Karl Schwarzschild, and Emil Wiechert, actively pursued theoretical or experimental investigations motivated by the theory of electrons, and it was not long before Minkowski, too, took up the theory. During the summer semester of 1905 he co-led a seminar with Hilbert on electron theory, featuring reports by Wiechert and Herglotz, and by Max Laue, who had just finished a doctoral thesis under Max Planck's supervision.[64]

Along with seminars on advanced topics in physics and analytical mechanics, Göttingen featured a lively mathematical society, with weekly meetings devoted to presentations of work-in-progress and reports on scientific activity outside of Göttingen. The electron theory was a frequent topic of discussion in this venue. For instance, the problem of gravitational attraction was first addressed by Schwarzschild in December, 1904, in a report on Alexander Wilken's recent paper on the compatibility of Lorentz's electron theory with astronomical observations.[65]

---

59 Vorlesungen über Mechanik, Wintersemester 1898/99, 47, 59, Minkowski papers, Arc. 4° 1712, JNUL. Minkowski referred to Klein and Sommerfeld's text in relation to the concept of force and its anthropomorphic origins, the kinetic theory of gas, and the theory of elasticity.

60 Minkowski to Sommerfeld, 30 October, 1898, MSS 1013A, Special Collections, National Museum of American History. An extensive reading list of mechanics texts is found in Minkowski's course notes for the 1903–1904 winter semester, Mechanik I, 9, Minkowski papers, Arc. 4° 1712, JNUL.

61 Minkowski to Sommerfeld, 30 October, 1898, loc. cit. note 60; Minkowski to Sommerfeld, 18 November, 1899, Nachlass Sommerfeld, Arch HS1977-28/A, 233, Deutsches Museum München; research notebook, 12 December, 1899, Arc. 4° 1712, Minkowski papers, JNUL.

62 Minkowski to Sommerfeld, 30 October, 1898, loc. cit. note 60.

63 On Göttingen's rise to preeminence in these fields, see (Manegold 1970; Pyenson 1985, chap. 7; Rowe 1989, 1992).

64 Nachlass Hilbert 570/9, Handschriftenabteilung, NSUB; (Pyenson 1985, chap. 5).

65 *Jahresbericht der deutschen Mathematiker-Vereinigung* 14, 61.

A focal point of sorts for the mathematical society, Poincaré's scientific output fascinated Göttingen scientists in general, and Minkowski in particular, as mentioned above.[66] Minkowski reported to the mathematical society on Poincaré's publications on topology, automorphic functions, and capillarity, devoting three talks in 1905–1906 to Poincaré's 1888–1889 Sorbonne lectures on this subject (Poincaré 1895). Others reporting on Poincaré's work were Conrad Müller on Poincaré's St. Louis lecture on the current state and future of mathematical physics (31 January, 1905), Hugo Broggi on probability (27 October, 1905), Ernst Zermelo on a boundary-value problem (12 December, 1905), Erhard Schmidt on the theory of differential equations (19 December, 1905), Max Abraham on the Sorbonne lectures (6 February, 1906) and Paul Koebe on the uniformization theorem (19 November, 1907). One gathers from this list that the Göttingen mathematical society paid attention to Poincaré's contributions to celestial mechanics, mathematical physics, and pure mathematics, all subjects intersecting with the ongoing research of its members. It also appears that no other member of the mathematical society was quite as assiduous in this respect as Minkowski.[67]

When Einstein's relativity paper appeared in late September, 1905, it drew the attention of the Bonn experimentalist Walter Kaufmann, a former Göttingen Privatdozent and friend of Max Abraham, but neither Abraham nor any of his colleagues rushed to report on the new ideas to the mathematical society.[68] Poincaré's long memoir on the dynamics of the electron, published in January, 1906, fared better, although nearly two years went by before Minkowski found an occasion to comment on Poincaré's gravitation theory, and to present his own related work-in-progress. Minkowski's typescript has been conserved, and is the source referred to here.[69]

On the occasion of the 5 November, 1907, meeting of the mathematical society, Minkowski began his review of Poincaré's work by observing that gravitation remained an "important question" in relativity theory, since it was not yet known "how the law of gravitation is arranged for in the realm of the principle of relativity."[70] The basic problem of gravitation and relativity, in other words, had not been solved by Poincaré. Eliding mention of Poincaré's two laws, Minkowski recognized

---

66 Although Poincaré spoke on celestial mechanics in Göttingen in 1895 (Rowe 1992, 475) and was invited back in 1902, he did not return until 1909, a few months after Minkowski's sudden death. See Hilbert to Poincaré, 6 November, 1908 (Dugac 1986, 209); Klein to Poincaré, 14 Jan., 1902 (Dugac 1989, 124–125). Sponsored by the Wolfskehl Fund, Poincaré's 1909 lecture series took place during "Poincaré week", in the month of April. His lectures were published the following year (Poincaré 1910) in a collection launched in 1907, based on an idea of Minkowski's (Klein 1907, IV).

67 *Jahresbericht der deutschen Mathematiker-Vereinigung* 14: 128, 586; 15: 154–155; 17: 5.

68 On Kaufmann's cathode-ray deflection experiments, see (Miller 1981, 226) and (Hon 1995). Readings of Kaufmann's articles are discussed at length by Richard Staley (Staley 1998, 270).

69 Undated typescript of a lecture on a new form of the equations of electrodynamics, *Math. Archiv* 60: 3, Handschriftenabteilung, NSUB. This typescript differs significantly from the posthumously-published version (Minkowski 1915).

70 "Es entsteht die grosse Frage, wie sich denn das Gravitationsgesetz in das Reich des Relativitätsprinzipes einordnen lässt" (p. 15).

in his work only one positive result: by considering gravitational attraction as a "pure mathematical problem," he said, Poincaré had found gravitation to propagate with the speed of light, thereby overturning the standard Laplacian argument to the contrary.[71]

Minkowski expressed dissatisfaction with Poincaré's approach, allowing that Poincaré's was "only one of many" possible laws, a fact stemming from its construction out of Lorentz-invariants. Consequently, Poincaré's investigation "had by no means a definitive character."[72] A critical remark of this sort often introduces an alternative theory, but in this instance none was forthcoming, and as I will show in what follows, there is ample reason to doubt that Minkowski was actually in a position to improve on Poincaré's investigation. Nonetheless, at the end of his talk Minkowski set forth the possibility of elaborating his report.

Minkowski's lecture was not devoted entirely to Poincaré's investigation of Lorentz-invariant gravitation. The purpose of his lecture, according to the published abstract, was to present a new form of the equations of electrodynamics leading to a mathematical redescription of physical laws in four areas: electricity, matter, mechanics, and gravitation.[73] These laws were to be expressed in terms of the differential equations used by Lorentz as the foundation of his successful theory of electrons (Lorentz 1904a), but in a form taking greater advantage of the invariance of the quadratic form $x^2 + y^2 + z^2 - c^2 t^2$. Physical laws, Minkowski stated, were to be expressed with respect to a four-dimensional manifold, with coordinates $x_1, x_2,$ x_3, x_4, where units were chosen such that $c = 1$, the ordinary Cartesian coordinates $x, y,$ and z, went over into the first three, and the fourth was defined to be an imaginary time coordinate, $x_4 = it$. Implicitly, then, Minkowski took as his starting point the four-dimensional vector space described in the last section of Poincaré's memoir on the dynamics of the electron.

Minkowski acknowledged, albeit obliquely, a certain continuity between Poincaré's memoir and his own program to reform the laws of physics in four-dimensional terms. By formulating the electromagnetic field equations in four-dimensional notation, Minkowski said he was revealing a symmetry not realized by his predecessors, not even by Poincaré himself (Walter 1999b, 98). While Poincaré had not sought to modify the standard form of Maxwell's equations, Minkowski felt it was time for a change. The advantage of expressing Maxwell's equations in the new notation, Minkowski informed his Göttingen colleagues, was that they were then "easier to grasp" (p. 11).

His reformulation naturally began in the electromagnetic domain, with an expression for the potentials. He formed a 4-vector potential denoted (ψ) by taking the

---

71 Actually, Poincaré postulated the light-like propagation velocity of gravitation, as mentioned above, (p. 204).

72 "Poincaré weist ein solches Gesetz auf, indem er auf die Betrachtung von Invarianten der Lorentz-schen Gruppe eingeht, doch ist das Gesetz nur eines unter vielen möglichen, und die betreffenden Untersuchungen tragen in keiner Weise einen definitiven Charakter" (p. 16). See also (Pyenson 1973, 233).

73 *Jahresbericht der deutschen Mathematiker-Vereinigung* 17 (1908), *Mitt. u. Nachr.*, 4–5.

ordinary vector potential over for the first three components, and setting the fourth
component equal to the product of i and the scalar potential. The same method was
applied to obtain a four-component quantity for current density: for the first three
components, Minkowski took over the convection current density vector, ρw, or
charge density times velocity, and defined the fourth component to be the product of
i and the charge density. Rewriting the potential and current density vectors in this
way, Minkowski imposed what is now known as the Lorentz condition,
$\text{Div}(\psi) = 0$, where Div is an extension of ordinary divergence. This led him to the
following expression, summarizing two of the four Maxwell equations:

$$\Box \psi_j = -\rho_j \quad (j = 1, 2, 3, 4), \tag{11}$$

where $\Box$ is the d'Alembertian, employed earlier by Poincaré (cf. note 20).

Of the formal innovations presented by Minkowski to the mathematical society,
the most remarkable was what he called a *Traktor*, a six-component entity used to
represent the electromagnetic field.[74] He defined the six components via the 4-vector
potential, using a two-index notation: $\psi_{jk} = \partial \psi_k / \partial \psi_j - \partial \psi_j / \partial \psi_k$, noting the antisymmetry relation $\rho_{kj} = -\rho_{jk}$, and zeros along the diagonal $\psi_{jj} = 0$. In this way,
the Traktor components $\psi_{14}, \psi_{24}, \psi_{34}, \psi_{23}, \psi_{31}, \psi_{12}$ match up with the field quantities $-i\mathfrak{E}_x, -i\mathfrak{E}_y, -i\mathfrak{E}_z, \mathfrak{h}_x, \mathfrak{h}_y, \mathfrak{h}_z$.[75]

The Traktor first found application when Minkowski turned to his second topic:
the four-dimensional view of matter. Ignoring the electron theories of matter of
Lorentz and Joseph Larmor, Minkowski focused uniquely on the macroscopic electrodynamics of moving media.[76] For this subject he introduced a *Polarisationstrak-
tor*, (p), along with a 4-current-density, (σ), defined by the current density vector i
and the charge density ρ: $(\sigma) = i_x, i_y, i_z, i\rho$ (p. 9). Recalling (11), Minkowski wrote
Maxwell's source equations in covariant form:

$$\frac{\partial p_{1j}}{\partial x_1} + \frac{\partial p_{2j}}{\partial x_2} + \frac{\partial p_{3j}}{\partial x_3} + \frac{\partial p_{4j}}{\partial x_4} = \sigma_j + \Box \psi_j. \tag{12}$$

Minkowski's relativistic extension of Maxwell's theory was all the simpler in that it
elided the covariant expression of the constitutive equations, which involves 4-veloc-
ity.[77] While none of his formulas invoked 4-velocity, Minkowski acknowledged that
his theory required a "velocity vector of matter $(w) = w_1, w_2, w_3, w_4$" (p. 10).

---

74 The same term was employed by Cayley to denote a line which meets any given lines, in a paper of
1869 (Cayley 1869).

75 When written out in full, one obtains, for example, $\psi_{23} = \partial \psi_3 / \partial x_2 - \partial \psi_2 / \partial x_2 = \mathfrak{h}_x$. Minkowski
later renamed the Traktor a *Raum-Zeit-Vektor II. Art* (Minkowski 1908, §5) but it is better known as
either a 6-vector, an antisymmetric 6-tensor, or an antisymmetric, second-rank tensor. As the suite of
synonyms suggests, this object found frequent service in covariant formulations of electrodynamics.

76 For a comparison of the Lorentz and Larmor theories, see (Darrigol 1994).

77 On the four-dimensional transcription of Ohm's law see (Arzeliès and Henry 1959, 65–67).

In order to express the "visible velocity of matter in any location," Minkowski needed a new vector as a function of the coordinates x, y, z, t (p. 7). Had he understood Poincaré's 4-velocity definition (see above, p. 202), he undoubtedly would have employed it at this point. Instead, following the same method of generalization from three to four components successfully applied in the case of 4-vector potential, 4-current density, and 4-force density, Minkowski took over the components of the velocity vector $\mathfrak{w}$ for the spatial elements of the quadruplet designated w_1, w_2, w_3, w_4:

$$\mathfrak{w}_x, \quad \mathfrak{w}_y, \quad \mathfrak{w}_z, \quad i\sqrt{1 - \mathfrak{w}^2}. \tag{13}$$

There are two curious aspects to Minkowski's definition. First of all, its squared magnitude does not vanish when ordinary velocity vanishes; even a particle at rest with respect to a reference frame is described in that frame by a 4-velocity vector of nonzero length. This is also true of Poincaré's 4-velocity definition, and is a feature of relativistic kinematics. Secondly, the components of Minkowski's quadruplet do not transform like the coordinates x_1, x_2, x_3, x_4, and consequently lack what he knew to be an essential property of a 4-vector.[78]

The most likely source for Minkowski's blunder is Poincaré's paper. We recall that Poincaré's derivation of his kinematic invariants ignored 4-vectors (see above, p. 202), and what is more, his paper features a misleading misprint, according to which the spatial part of a 4-velocity vector is given to be the ordinary velocity vector.[79] Other sources of error can easily be imagined, of course.[80] It is strange that Minkowski did not check the transformation properties of his 4-velocity definition, but given its provenance, he probably had no reason to doubt its soundness.

Minkowski's mistake strongly suggests that at the time of his lecture, he did not yet conceive of particle motion in terms of a world line parameter. Such an approach to particle motion would undoubtedly have spared Minkowski the error, since it renders trivial the definition of 4-velocity.[81] As matters stood in November, 1907, however, Minkowski could proceed no further with his project of reformulation.[82] The development of four-dimensional mechanics was hobbled by Minkowski's spare

---

78 Minkowski mentions this very property on p. 6.

79 The passage in question may be translated as follows: "Next we consider $X, Y, Z, T\sqrt{-1}$, as the coordinates of a fourth point Q; the invariants will then be functions of the mutual distances of the five points O, P, P', P'', Q, and among these functions we must retain only those that are 0th degree homogeneous with respect, on one hand, to $X, Y, Z, T, \delta x, \delta y, \delta z, \delta t$ (variables that can be further replaced by $X_1, Y_1, Z_1, T_1, \xi, \eta, \zeta, 1$), and on the other hand, with respect to $\delta_1 x, \delta_1 y, \delta_1 z, 1$ (variables that can be further replaced by $\xi_1, \eta_1, \zeta_1, 1$)" (Poincaré 1906, 170). The misprint is in the next-to-last set of variables, where instead of 1 we should have $\delta_1 t$.

80 One other obvious source for Minkowski's error is Lorentz's transformation of charge density: $\rho' = \rho / \beta l^3$, where $1/\beta = \sqrt{1 - v^2/c^2}$, and l is a constant later set to unity (Lorentz 1904a, 813), although this formula was carefully corrected by Poincaré.

81 Let the differential parameter $d\tau$ of a world line be expressed in Minkowskian coordinates by $d\tau^2 = dx_1^2 + dx_2^2 + dx_3^2 + dx_4^2$. The 4-velocity vector U_μ is naturally defined to be the first derivative with respect to this parameter, $U_\mu = dx_\mu/d\tau$ ($\mu = 1, 2, 3, 4$).

stock of 4-vectors even more than that of electrodynamics. Although Minkowski defined a force-density 4-vector, the fourth component of which he correctly identified as the energy equation, he did not go on to define 4-force at a point.[83] Once again, the definition of a force 4-vector at a point would have been trivial, had Minkowski possessed a correct 4-velocity definition. No more than a review of Planck's recent investigation (Planck 1907), Minkowski's discussion of mechanics involved no 4-vectors at all. Likewise for the subsequent section on gravitation, which reviewed Poincaré's theory, as shown above (p. 214). Without a valid 4-vector for velocity, Minkowski's electrodynamics of moving media was severely hobbled; without a point force 4-vector, his four-dimensional mechanics and theory of gravitation could go nowhere.

The difficulty encountered by Minkowski in formulating a four-dimensional approach to physics is surprising in light of the account he gave later of the back ground to his discovery of spacetime (Minkowski 1909). Minkowski presented his spacetime view of relativity theory as a simple application of group methods to the differential equations of classical mechanics. These equations were known to be invariant with respect to uniform translations, just as the squared sum of differentials $dx^2 + dy^2 + dz^2$ was known to be invariant with respect to rotations and translations of Cartesian axes in Euclidean 3-space, and yet no one, he said, had thought of compounding the two corresponding transformation groups. When this is done properly (by introducing a positive parameter c), one ends up with a group Minkowski designated G_c, with respect to which the laws of physics are covariant. (The group G_c is now known as the Poincaré group.) Presumably, the four-dimensional approach appeared simple to Minkowski in hindsight, several months after his struggle with 4-velocity.

In summary, while Minkowski formulated the idea of a four-dimensional language for physics based on the form-invariance of the Maxwell equations under the transformations of the Lorentz group, his development of this program beyond electrodynamics was hindered by a misunderstanding of the four-dimensional counterpart of an ordinary velocity vector. This was to be only a temporary obstacle.

On 21 December, 1907, Minkowski presented to the Royal Society of Science in Göttingen a memoir entitled "The Basic Equations for Electromagnetic Processes in Moving Bodies," which I will refer to for brevity as the *Grundgleichungen*.[84] Minkowski's memoir revisits in detail most of the topics introduced in his 5 November lecture to the mathematical society, but employs none of the jargon of spaces,

---

82   The incongruity noted by Pyenson (Pyenson 1985, 84) between Minkowski's announcement of a four-dimensional physics on one hand, and on the other hand, a trifle of 4-vector definitions and expressions, is to be understood as a indication of Minkowski's gradual ascent of the learning curve of four-dimensional physics.

83   Minkowski defined the spatial components of the empty space force density 4-vector $\mathfrak{X}_j$ in terms of the ordinary force density components $\mathfrak{X}, \mathfrak{Y}, \mathfrak{Z}$, and their product with velocity: $\mathfrak{A} = \mathfrak{X}\mathfrak{w}_x, \mathfrak{Y}\mathfrak{w}_y, \mathfrak{Z}\mathfrak{w}_z$, such that $\mathfrak{X}_j = \mathfrak{X}, \mathfrak{Y}, \mathfrak{Z}, i\mathfrak{A}$. He also expressed the force density 4-vector as the product of 4-current-density and the Traktor: $\mathfrak{X}_j = \rho_1 \psi_{j1} + \rho_2 \psi_{j2} + \rho_3 \psi_{j3} + \rho_4 \psi_{j4}$.

geometries, and manifolds. What it emphasizes instead—in agreement with its title—is the achievement of the first theory of electrodynamics of moving bodies in full conformance to the principle of relativity. Also underlined is a second result described as "very surprising": the laws of mechanics follow from the postulate of relativity and the law of energy conservation alone. On the four-dimensional world and the new form of the equations of electrodynamics, both topics headlined in his November lecture, Minkowski remained coy. Curiously, the introduction mentions nothing about a new formalism, even though all but one of fourteen sections introduce and employ new notation or calculation rules (not counting the appendix).

The added emphasis on the laws of mechanics in Minkowski's introduction, on the other hand, reflects Minkowski's recent discovery of correct definitions of 4-velocity and 4-force, along with geometric interpretations of these entities. It was in the *Grundgleichungen* that Minkowski first employed the term "spacetime" [*Raumzeit*].[85] For example, he introduced 4-current density as the exemplar of a "spacetime vector of the first kind" (§5), and used it to derive a velocity 4-vector. Identifying $\rho_1, \rho_2, \rho_3, \rho_4$ with $\rho w_x, \rho w_y, \rho w_z, i\rho$, just as he had done in his lecture of 5 November, Minkowski wrote the transformation to a primed system moving with uniform velocity $q < 1$:

$$\rho' = \rho\left(\frac{-q w_z + 1}{\sqrt{1 - q^2}}\right), \quad \rho' w'_{z'} = \rho\left(\frac{w_z - q}{\sqrt{1 - q^2}}\right), \quad \rho' w'_{x'} = \rho w_x, \rho' w'_{y'} = \rho w_y. \quad (14)$$

Observing that this transformation did not alter the expression $\rho^2(1 - w^2)$, Minkowski announced an "important remark" concerning the relation of the primed to the unprimed velocity vector (§4). Dividing the 4-current density by the positive square root of the latter invariant, he obtained a valid definition of 4-velocity,

$$\frac{w_x}{\sqrt{1 - w^2}}, \quad \frac{w_y}{\sqrt{1 - w^2}}, \quad \frac{w_z}{\sqrt{1 - w^2}}, \quad \frac{i}{\sqrt{1 - w^2}}, \quad (15)$$

the squared magnitude of which is equal to -1. Minkowski seemed satisfied with this definition, naming it the spacetime velocity vector [*Raum-Zeit-Vektor Geschwindigkeit*].

---

84 Minkowski's manuscript was delivered to the printer on 21 February, 1908, corrected, and published on 5 April, 1908 (*Journal für die "Nachrichten" der Gesellschaft der Wissenschaften zu Göttingen, mathematische-naturwissenschaftliche Klasse 1894–1912, Scient.* 66, Nr. 1, 471, *Archiv der Akademie der Wissenschaften zu Göttingen*). I thank Tilman Sauer for pointing out this source to me.

85 While the published version of Minkowski's 5 November lecture refers on one occasion to a "*Raumzeitpunkt*" (Minkowski 1915, 934) the term occurs nowhere in the archival typescript. The source of this addition is unknown. A manuscript annotation of the first page of the typescript bears Sommerfeld's initials, and indicates that he compared parts of the typescript to the proofs, as Lewis Pyenson correctly points out (Pyenson 1985, 82). Pyenson errs, however, in attributing to Sommerfeld the authorship of the remaining annotations, which were all penned in Minkowski's characteristic cramped hand.

The significance of the spacetime velocity vector, Minkowski observed, lies in the relation it establishes between the coordinate differentials and matter in motion, according to the expression

$$\sqrt{-(dx_1^2 + dx_2^2 + dx_3^2 + dx_4^2)} = dt\sqrt{1 - \mathfrak{w}^2}. \tag{16}$$

The Lorentz-invariance of the right-hand side of (16), signaled earlier by both Poincaré and Planck, now described the relation of the sum of the squares of the coordinate differentials to the components of 4-velocity.

The latter relation plays no direct role in Minkowski's subsequent development of the electrodynamics of moving media, and in this it is unlike the 4-velocity definition. Rewriting the right-hand side of (16) as the ratio of the coordinate differential dx_4 to the temporal component of 4-velocity, w_4, Minkowski defined the spacetime integral of (16) as the "proper time" [*Eigenzeit*] pertaining to a particle of matter. The introduction of proper time streamlined Minkowski's 4-vector expressions, for instance, 4-velocity was now expressed in terms of the coordinate differentials, the imaginary unit, and the differential of proper time, $d\tau$:

$$\frac{dx}{d\tau}, \quad \frac{dy}{d\tau}, \quad \frac{dz}{d\tau}, \quad i\frac{dt}{d\tau}. \tag{17}$$

Along with the notational simplification realized by the introduction of proper time, Minkowski signaled a geometric interpretation of 4-velocity. Since proper time is the parameter of a spacetime line (or as he later called it, a world line), it follows that 4-velocity is equal to the slope of a world line at a given spacetime point, much like ordinary three-velocity is described by the slope of a displacement curve in classical kinematics. What Minkowski pointed out, in other words, is that 4-velocity is tangent to a world line at a given spacetime point (p. 108).

In order to develop his mechanics, Minkowski needed a workable definition of mass. He adapted Einstein's and Planck's notion of rest mass to the arena of spacetime by considering that a particle of matter sweeps out a hypertube in spacetime. Conservation of particle mass m was then expressed as invariance of the product of rest mass density with the volume slices of successive constant-time hypersurfaces over the length of the particle's world line, such that $dm/d\tau = 0$. Minkowski did not consider the case of variable rest mass density, which arises, for instance, in the case of heat exchange.

Minkowski's decision to adopt a constant rest mass density is linked to his view of the electrodynamics of moving media. Recall that he had introduced a six-vector in his 5 November lecture to represent the field. The product of the field and excitation six-vectors, he noted, leads to an interesting 4 by 4 matrix, combining the Maxwell stresses, Poynting vector, and electromagnetic energy density. He did not assign a name to this object, known later as the energy-momentum tensor, and often viewed as one of Minkowski's greatest achievements in electrodynamics.[86] Of special interest to Minkowski was the fact that the 4-divergence of this matrix, denoted lor S, is a 4-vector, K:[87]

$$K = \text{lor } S. \tag{18}$$

This 4-divergence (18) was used to define the "ponderomotive" force density, or generalized force per unit volume, neither mechanical nor non-mechanical in the pure sense of these terms. The 4-vector K is not normal, in general, to the velocity w of a given volume element, so to ensure that the ponderomotive force acts orthogonally to w, Minkowski added a component containing a velocity term:

$$K + (w\bar{K})w. \tag{19}$$

The parentheses in (19) indicate a scalar product, and $\bar{K}$ stands for the transpose of K. By defining the ponderomotive force density in this way, Minkowski effectively opted for an equation of motion in which 4-acceleration is normal to 4-velocity.[88] It appears that Minkowski let this view of force and acceleration guide his development of spacetime mechanics. In the latter domain, he formed a 4 by 4 matrix S in the force density and energy of an elastic media with the same transformation properties as the energy-momentum tensor S of (18), and used the 4-divergence of this tensor to express the equations of motion of a volume element of constant rest mass density v (p. 106):

$$v\frac{dw_h}{d\tau} = K_h + \kappa w_h \quad (h = 1, 2, 3, 4). \tag{20}$$

The factor κ was determined by the definition of 4-velocity to be equal to the scalar product $(K\bar{w})$, much like the definition of ponderomotive force (19). In sum, it may be supposed that the non-orthogonality with respect to a given volume element of the 4-divergence of Minkowski's asymmetric energy-momentum tensor for moving media led Minkowski to introduce a velocity term to his definition of ponderomotive force. This definition was then ported to spacetime mechanics, where for the sake of consistency, Minkowski held rest mass density constant in the equations of motion (20).

Minkowski's stipulation of constant rest mass density was eventually challenged by Max Abraham (Abraham 1909, 739) and others, for reasons that do not concern us here. Despite its obvious drawbacks, it greatly simplified the tasks of outlining the mechanics of spacetime and developing a theory of gravitation. For example, it per-

---

86 While Minkowski's tensor is traceless, it is also asymmetric, a fact which led to criticism and rejection by leading theorists of the day. His asymmetric tensor was later rehabilitated; for a technical discussion with reference to the original papers, see (Møller 1972, 219). In the absence of matter, his tensor assumes a symmetric form; in this form, it was hailed by theorists.

87 Minkowski defined the energy-momentum tensor S in two ways: as the product of six-vectors, $fF = S - L$, where L is the Lagrange density, and in component form via the equations for Maxwell stresses, the Poynting vector, and electromagnetic energy density (Minkowski 1908, 96).

88 Minkowski's alternative between a 4-force definition and the "natural" spacetime equations of motion was underlined by Pauli (Pauli 1921, 664).

mitted him to define the equations of motion of a particle in terms of the product of
rest mass and 4-acceleration, where the latter is the derivative of 4-velocity with
respect to proper time. Since 4-velocity is orthogonal to 4-acceleration, for constant
proper mass it is also orthogonal to a 4-vector Minkowski called a "driving force"
[*bewegende Kraft*, p. 108]. Minkowski wrote four equations defining this force:

$$m\frac{d}{d\tau}\frac{dx}{d\tau} = R_x, \quad m\frac{d}{d\tau}\frac{dy}{d\tau} = R_y, \quad m\frac{d}{d\tau}\frac{dz}{d\tau} = R_z, \quad m\frac{d}{d\tau}\frac{dt}{d\tau} = R_t. \tag{21}$$

The first three expressions differ from Planck's equations of motion, in that Planck
defined force as change in *momentum*, instead of mass times acceleration. It was only
a few months later that Minkowski explicitly defined four-momentum as the product
of 4-velocity with proper mass (Planck 1906, eq. 6; Minkowsi 1909, §4).[89] By dividing Minkowski's first three equations by a Lorentz factor, one obtains Planck's equations. Minkowski's fourth equation, R_t, formally dependent on the other three,
expresses the law of energy conservation.[90] From energy conservation and the relativity postulate alone, Minkowski concluded, one may derive the equations of
motion. This is the single "surprising" result of his investigation of relativistic
mechanics, referred to at the outset of his paper (see above, p. 218).

If Minkowski found few surprises in spacetime mechanics, many of his readers
were taken aback by his four-dimensional approach. For example, the first physicists
to comment on his work, Albert Einstein and Jakob Laub, rewrote Minkowski's
expressions in ordinary vector notation, sparing the reader the "sizable demands"
[*ziemlich große Anforderungen*] of Minkowski's mathematics (Einstein and Laub
1908, 532). They did not specify the nature of the demands, but the abstracter of their
paper pointed to the "special knowledge of the calculation methods" required for
assimilation of Minkowski's equations.[91] In other words, for Minkowski's readers, his
novel matrix calculus was the principal technical obstacle to overcome. Where
Poincaré pushed rejection of formalism to an extreme, Minkowski pulled in the other
direction, introducing a formalism foreign to the practice of physics. What motivated
this brash move is unclear, and his choice is all the more curious because he know-
ingly defied the German trend of vector notation in electrodynamics.[92] As mentioned
above, Minkowski was ill-disposed toward quaternions, although he admitted in print
that they could be brought into use for relativity instead of matrix calculus. He spoke
here from experience, as manuscript notes reveal that he used quaternions (in addition
to Cartesian-coordinate representation and ordinary vector analysis) to investigate the
electrodynamics of moving media.[93] In the end, however, he felt that for his purposes
quaternions were "too limited and cumbersome" [*zu eng und schwerfällig*, p. 79].

---

89 In the latter lecture, Minkowski proposed the modern definition of kinetic energy as the temporal
    component of 4-momentum times c^2, or $mc^2 dt/d\tau$. The "spatial" part of the driving force (21) was
    referred to by Lorentz (Lorentz 1910, 1237) as a "Minkowskian force" [*Minkowskische Kraft*], differ-
    ing from the Newtonian force by a Lorentz factor. Lorentz complemented the Minkowskian force
    with a "Minkowskian mass" [*Minkowskische Masse*].

As far as notation is concerned, Minkowski's treatment of differential operations broke cleanly with then-current practice. It also broke with the precedent of his 5 November lecture, where he had introduced, albeit parsimoniously, both □ and Div (see above, p. 215). For the *Grundgleichungen* he adopted a different approach, extending the ∇ to four dimensions, and labeling the resulting operator *lor*, already encountered above in (18). The name is short for Lorentz, and the effect is the operation: $|\partial/\partial x_1, \partial/\partial x_2, \partial/\partial x_3, \partial/\partial x_4|$. When applied to a 6-vector, *lor* results in a 4-vector, in what Minkowski described as an appropriate translation of the matrix product rule (p. 89); it also mimics the effect of the ordinary ∇. Transforming as a 4-vector, *lor* is liberally employed in the second part of the *Grundgleichungen*, to the exclusion of any and all particular 4-vector functions.[94] The use of *lor* made for a presentation of electrodynamics elegant in the extreme, at the expense of legibility for German physicists more used to thinking in terms of gradients, divergences, and curls (or rotations).

Minkowski's equations of electrodynamics departed radically in form with those of the old electrodynamics, shocking the thought patterns of physicists, and creating a phenomenon of rejection that took several years—and a new formalism—to overcome.[95] Why did Minkowski break with this tradition? Did he feel that the new physics of spacetime required a clean break with nineteenth-century practice? Perhaps,

---

90 Minkowski's argument may be summarized as follows. From the definition of a 4-vector, the following orthogonality relation holds for the driving force R:

$$R_x \frac{dx}{d\tau} + R_y \frac{dy}{d\tau} + R_z \frac{dz}{d\tau} = R_t \frac{dt}{d\tau}. \tag{22}$$

Integration of the rest-mass density over the hypersurface normal to the world line of the mass point results in the driving force components (21), but if the integration is to be performed instead over a constant-time hypersurface, proper time is replaced by coordinate time, such that the fourth equation reads: $md/dt(dt/d\tau) = R_t d\tau/dt$. From (22) we obtain an expression for R_t, which we multiply by $d\tau/dt$:

$$m \frac{d}{dt}\left(\frac{dt}{d\tau}\right) = \mathfrak{w}_x R_x \frac{d\tau}{dt} + \mathfrak{w}_y R_y \frac{d\tau}{dt} + \mathfrak{w}_z R_z \frac{d\tau}{dt}. \tag{23}$$

Minkowski reasoned that since the right-hand side of (23) describes the rate at which work is done on the particle, the left-hand side must be the rate of change of the particle's kinetic energy, such that (23) represents the law of energy conservation. He immediately related (23) to the kinetic energy of the particle:

$$m\left(\frac{dt}{d\tau} - 1\right) = m\left(\frac{1}{\sqrt{1-\mathfrak{w}^2}} - 1\right) = m\left(\frac{1}{2}|\mathfrak{w}|^2 + \frac{3}{8}|\mathfrak{w}|^4 + \cdots\right). \tag{24}$$

Minkowski did not justify the latter expression, but in virtue of his definition of proper time, $d\tau = dt\sqrt{1-\mathfrak{w}^2}$, the left-hand side of (23) may be rewritten as $m(d/dt)(1/\sqrt{1-\mathfrak{w}^2})$, such that upon integration the particle's kinetic energy is $m/\sqrt{1-\mathfrak{w}^2} + C$, where C is a constant. For agreement with the Newtonian expression of kinetic energy in case of small particle velocities ($\mathfrak{w} \ll 1$), we let $C = -m$, which accords both with (24) and the definition of kinetic energy given in a later lecture (cf. note 89).

91 *Jahrbuch über die Fortschritte für Mathematik* 39, 1908, 910.

but he must have recognized that the old methods would prove resistant to change. His own subsequent practice shows as much: after writing the *Grundgleichungen* Minkowski did not bother with *lor* during his private explorations of the formal side of electrodynamics, preferring the coordinate method.[96]

He also relied largely—but not exclusively—on a Cartesian-coordinate approach during his preliminary investigations of the subjects treated in the *Grundgleichungen*. His surviving research notes, made up almost entirely of symbolic calculations, shed an interesting light on Minkowski's path to both a theory of the electrodynamics of moving media, and a theory of gravitation, or more generally to his process of discovery. Notably, where the subjects of mechanics and gravitation are relegated to the appendix of the *Grundgleichungen*, these notes show that Minkowski pursued questions of electrodynamics and gravitation in parallel, switching from one topic to the other three times in the course of 163 carefully numbered pages. At least fifteen of these pages are specifically concerned with gravitation; the notes are undated, but those concerning gravitation are certainly posterior to the typescript of the 5 November lecture, because unlike the latter, they feature valid definitions of 4-velocity and 4-force.

Minkowski's attempt to capture gravitational action in terms of a 4-scalar potential is of particular interest. We recall that Minkowski had expressed Maxwell's equations in terms of a 4-vector potential (11) during his lecture of 5 November, and on this basis, it was natural for him to investigate the possibility of representing gravitational force on a point mass in a fashion analogous to that of the force on a point charge moving in an electromagnetic field. In his scratch notes, Minkowski defined a 4-scalar potential Φ, in terms of which he initially devised the law of motion:

---

92 This trend is described by Darrigol (Darrigol 1993, 270). The sharp contrast between the importance assigned to vector methods in France and Germany may be linked to the status accorded to applied mathematics in these two nations, as discussed by H. Gispert in her review of the French version of Klein's *Encyklopädie* (Gispert 2001).

93 At one point during his calculations Minkowski seemed convinced of the utility of this formalism, remarking that electrodynamics is "predestined for application of quaternionic calculations" (*Math. Archiv* 60: 6, 21, Handschriftenabteilung, NSUB).

94 A precedent for Minkowski's exclusive use of *lor* may be found in (Gibbs and Wilson 1901), where ∇ is similarly preferred to vector functions.

95 Cf. Max von Laue's remark that physicists understood little of Minkowski's work because of its unfamiliar mathematical expression (Laue 1951, 515) and Chuang Liu's account of the difficulty experienced by Max Abraham and Gunnar Nordström in applying Minkowski's formalism (Liu 1991, 66). While Minkowski's calculus is a straightforward extension of Cayley's formalism (for a summary, see (Cunningham 1914, chap. 8), the latter formalism was itself unfamiliar to physicists.

96 *Math. Archiv* 60: 5, Handschriftenabteilung, NSUB. This 82-page set of notes dates from 23 May to 6 October, 1908. A posthumously published paper on the electron-theoretical derivation of the basic equations of electrodynamics for moving media, while purported to be from Minkowski's Nachlass, was written entirely by Max Born, as he acknowledged (Minkowski and Born 1910, 527). In the latter publication *lor* makes only a brief appearance.

$$\frac{d}{d\tau}\frac{1}{\sqrt{1-v^2}} - \frac{\partial\Phi}{\partial t} = 0, \qquad \frac{d}{d\tau}\frac{-y'}{\sqrt{1-v^2}} - \frac{\partial\Phi}{\partial y} = 0,$$

$$\frac{d}{d\tau}\frac{-x'}{\sqrt{1-v^2}} - \frac{\partial\Phi}{\partial x} = 0, \qquad \frac{d}{d\tau}\frac{-z'}{\sqrt{1-v^2}} - \frac{\partial\Phi}{\partial z} = 0,$$

$$(25)$$

where constants are neglected, τ denotes proper time, and primes indicate differentiation with respect to coordinate time t (i.e. $x' = dx/dt$).[97] This generalization of the Newtonian potential to a 4-scalar potential appears to be one of the first paths explored by Minkowski in his study of gravitation, but his investigation is inconclusive. In particular, there is no indication in these notes of a recognition on Minkowski's part that a four-scalar potential conflicts with the postulates of invariant rest mass and light velocity.[98] Nor is there any evidence that he considered suspending either one of these postulates.

Likewise, in the *Grundgleichungen* there is no question of adopting either a variable mass density or a gravitational 4-potential. Once he had established the foundations of spacetime mechanics, Minkowski took up the case of gravitational attraction. The problem choice is significant, in that the same question had been treated at length by Poincaré (although not to Minkowski's satisfaction, as mentioned above, p. 215). Implicitly, Minkowski encouraged readers to compare methods and results. Explicitly, proceeding in what he described (in a footnote) as a "wholly different way" from Poincaré, Minkowski said he wanted to make "plausible" the inclusion of gravitation in the scheme of relativistic mechanics (p. 109). It will become clear in what follows that his project was more ambitious than the modest elaboration of a plausibility argument, as it was designed to validate his spacetime mechanics.

The point of departure for Minkowski's theory of gravitation was quite different from that of Poincaré, because the latter's results were integrated into the former's formalism. For example, where Poincaré initially assumed a finite propagation velocity of gravitation no greater than that of light, only to opt in the end for a velocity equal to that of light, Minkowski assumed implicitly from the outset that this velocity was equal to that of light. Similarly, Poincaré initially supposed the gravitational

---

97 *Math. Archiv* 60: 6, 10, Handschriftenabteilung, NSUB.

98 This "peculiar" consequence of Minkowski's spacetime mechanics was underlined by Maxwell's German translator, the Berlin physicist Max B. Weinstein (Weinstein 1914, 42). In Minkowski spacetime, 4-acceleration is orthogonal to 4-velocity: $U_\mu dU_\mu/d\tau = 0$, $\mu = 1, 2, 3, 4$, where τ is the proper time. We assume a 4-scalar potential Φ such that the gravitational 4-force $F_\mu = -m\partial\Phi/\partial x_\mu$. If we consider a point mass with 4-velocity U_μ subjected to a 4-force F_μ derived from this potential, we have $U_\mu F_\mu = -U_\mu m\partial\Phi/\partial x_\mu$. Writing 4-velocity as $dx_\mu/d\tau$, and substituting in the latter expression, we obtain

$$U_\mu F_\mu = -m\frac{dx_\mu}{d\tau}\frac{\partial\Phi}{\partial x_\mu} = -m\frac{d\Phi}{d\tau} = 0,$$

and consequently, $d\Phi/d\tau = 0$, which means that the law of motion describes the trajectory of the passive mass m only in the trivial case of constant Φ along its world line.

force to be Lorentz covariant, only to opt in the end for an analog of the Lorentz force, where Minkowski required implicitly from the outset that all forces transform like the Lorentz force.

Combining geometric and symbolic arguments, Minkowski's exposition of his theory of gravitation introduces a new geometric object, the three-dimensional "ray form" [*Strahlgebilde*] of a given spacetime point, known today as a light hypercone (or lightcone). For a fixed spacetime point $B^* = (x^*, y^*, z^*, t^*)$, the lightcone of B^* is defined by the sets of spacetime points $B = (x, y, z, t)$ satisfying the equation

$$(x - x*)^2 + (y - y*)^2 + (z - z*)^2 = (t - t*)^2, \quad t - t* \geqq 0. \tag{26}$$

For all the spacetime points B of this lightcone, B^* is what Minkowski called B's *lightpoint*. Any world line intersects the lightcone in one spacetime point only, Minkowski observed, such that for any spacetime point B on a world line there exists one and only one lightpoint B^*. Minkowski remarked in a later lecture that the lightcone divides four-dimensional space into three regions: time-like, space-like and light-like.[99]

Using this novel insight to the structure of four-dimensional space, in combination with the 4-vector notation set up in earlier in his memoir, Minkowski presented and applied his law of gravitational attraction in two highly condensed pages. Minkowski's geometric argument employs non-Euclidean relations that were unfamiliar to physicists, yet he provided no diagrams. Visually-intuitive arguments had fallen into disfavor with mathematicians by this time, with the rise of the axiomatic approach to geometry favored by David Hilbert (Rowe 1997), yet Minkowski never renounced the use of figures in geometry; he employed them in earlier works on number geometry, and went on to publish spacetime diagrams in the sequel to the *Grundgleichungen*.[100] For the purposes of my reconstruction, I refer to a spacetime diagram (Fig. 1) of the sort Minkowski employed in the sequel (reproduced in Fig. 3).[101]

---

99 Minkowski introduced the terms *zeitartig* and *raumartig* in (Minkowski 1909).

100 There is little agreement on where to situate Minkowski's work on relativity along a line running from the intuitive to the formal. Peter Galison (Galison 1979, 89) for example, underlines Minkowski's visual thinking (i.e., reasoning that appeals to figures or diagrams), while Leo Corry (Corry 1997, 275; 2004, chap. 4) considers Minkowski's work in the context of Hilbert's axiomatic program for physics.

101 Two spatial dimensions are suppressed in Fig. 1, and lightcones are represented by broken lines with slope equal to ± 1, the units being chosen so that the propagation velocity of light is unity ($c = 1$). In this model of Minkowski space, orthogonal coordinate axes appear oblique in general, for example, the spatial axes x^*, y^*, z^* are orthogonal to the tangent B^*C^* at spacetime point B^* of the central line of the filament F^* described by a particle of proper mass m^*.

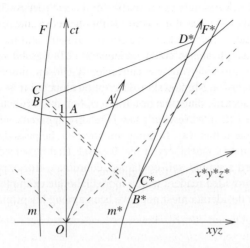

Figure 1: Minkowski's geometry of gravitation, with source in arbitrary motion.

On the assumption that the force of gravitation is a 4-vector normal to the 4-velocity of the passive mass m, Minkowski derived his law of attraction in the following way. The trajectories of two particles of mass m and m^* correspond to two spacetime filaments F and F^*, respectively. Minkowski's arguments refer to world lines he called central lines [*Hauptlinien*] of these filaments, which pass through points on the successive constant-time hypersurfaces delimited by the respective particle volumes. The central lines of the filaments F and F^* are shown in Fig. 1. An infinitesimal element of the central line of F is labeled BC, and the two lightpoints corresponding to the endpoints B and C are labeled B^* and C^* on the central line of F^*. From the origin of the rest frame O, a 4-vector parallel to B^*C^* intersects at A' the three-dimensional hypersurface defined by the equation $-x^2-y^2-z^2+t^2=1$. Finally, a space-like 4-vector BD^* extends from B to a point D^* on the world line tangent to the central line of F^* at B^*.

Referring to the latter configuration of seven spacetime points, two central lines, a lightcone and a calibration hypersurface, Minkowski expressed the spatial components of the driving force of gravitation exerted by m^* on m at B,

$$mm^*\left(\frac{OA'}{B^*D^*}\right)^3 BD^* . \qquad (27)$$

Minkowski's gravitational driving force is composed of the latter 4-vector (27) and a second 4-vector parallel to B^*C^* at B, such that the driving force is always orthogonal to the 4-velocity of the passive mass m at B. (For reasons of commodity, I will refer to this law of force as Minkowski's first law.)

The form of Minkowski's first law of gravitation is comparable to that of his ponderomotive force for moving media (19), in that the driving force has two components, only one of which depends on the motion of the test particle. In the gravitational case, however, Minkowski did not write out the 4-vector components in terms of matrix products. Instead, he relied on spacetime geometry and the definition of a 4-vector. The only way physicists could understand (27) was by reformulating it in terms of ordinary vectors referring to a conveniently chosen inertial frame, and even then, they could not rely on Minkowski's description alone, as it is incomplete.[102]

Even without spacetime diagrams or a transcription into ordinary vector notation, the formal analogy of (27) to Newton's law is readily apparent, and this is probably why Minkowski wrote it this way. In doing so, however, he passed up an opportunity to employ the new matrix machinery at his disposal. Had he seized this opportunity, he would have gained a simple, self-contained, coordinate-free expression of the law of gravitation, and provided readers with a more elaborate example of his calculus in action, but the latter desiderata must not have been among his primary objectives.[103]

---

102 The 4-vector OA' in (27) has unit magnitude by definition in all inertial frames, while $B*D*$ is a time-like 4-vector tangent to the central line of $F*$ at $B*$. Consequently, $B*D*$ may be taken to coincide with the temporal axis of a frame instantaneously at rest with $m*$ at $B*$, such that it has only one nonzero component: the difference in proper time between the points $B*$ and $D*$. It is assumed that the rest frame may be determined unambiguously for a particle in arbitrary motion, as asserted without proof by Minkowski in a later lecture (Minkowski 1909, §III); subsequently, Max Born (Born 1909, 26) remarked that any motion may be approximated by what he called hyperbolic motion, and noted that such motion is characterized by an acceleration of constant magnitude (as measured in an inertial frame). If we locate the origin of this frame at $B*$, and let $D* = (0, 0, 0, t)$, then $B*D* = (0, 0, 0, it)$, and $(B*D*)^3 = -it^3$. Likewise in this same frame, $A = A' = (0, 0, 0, 1)$, and $OA' = OA = (0, 0, 0, i)$. Minkowski understood the term $(OA'/B*D*)$ as the ratio [Verhältnis] of two parallel 4-vectors, an operation familiar from the calculus of quaternions, but one not defined for 4-vectors. While modern vector systems ignore vector division, in Hamilton's quaternionic calculus the quotient of vectors is unambiguously defined; see, for example, (Tait 1882–1884, chap. 2). Accordingly, the quotient in (27) is the ratio of lengths, $(OA'/B*D*) = 1/t$, and the cubed ratio is t^{-3}. The point B lies on the same constant-time hypersurface as $D*$, so we assign it the value $(x, y, z, t) = (r, t)$. This assignment determines the value of the 4-vector $BD* = (-x, -y, -z, 0) = (-r, 0)$. Since $B*$ is a lightpoint of B, we can apply (26) to obtain $x^2 + y^2 + z^2 = t^2 = r^2$, and consequently, $t^3 = r^3$. Substituting for t^3 results in $(OA'/B*D*)^3 = 1/t^3 = 1/r^3$. The 4-vector $B*D*$ is space-like, such that its projection on the constant-time hypersurface orthogonal to $B*D*$ at $D*$ is the ordinary vector $(-x, -y, -z) = -r$. In terms of ordinary vectors and scalars measured in the rest frame of $m*$, Minkowski's expression (27) is equivalent to Newton's law (neglecting the gravitational constant):

$$-mm*\frac{r}{r^3}. \tag{28}$$

Neither (27) nor (28) contains any velocity-dependent terms, while the time-like component of Minkowski's first law depends on the velocity of the passive mass. Newton's law (28) thus coincides with Minkowski's first law only in the case of relative rest.

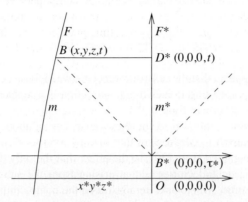

Figure 2: Minkowski's geometry of gravitation, with source in uniform motion.

Minkowski was not yet finished with his law of gravitation. Unlike Poincaré, after writing his law of gravitation, Minkowski went on to apply it to the particular case of uniform rectilinear motion of the source m^*. He considered the latter in a comoving frame, in which the temporal axis is chosen to coincide with the tangent to the central line of F^* at B^* (cf. the situation described in note 102). Referring to the reconstructed spacetime diagram in Fig. 2, the temporal axis is represented by a vertical line F^*, such that the origin is established in a frame comoving with m^*. To the retarded position of m^*, denoted B^*, Minkowski assigned the coordinates $(0, 0, 0, \tau^*)$, and to the position B of the passive mass m he assigned the coordinates (x, y, z, t). The geometry of this configuration fixes the location of D^* at $(0, 0, 0, t)$, from which the 4-vectors $BD^* = (-x, -y, -z, 0)$ and $B^*D^* = (0, 0, 0, i(t - \tau^*))$ are determined. In this case, Minkowski pointed out, (26) reduces to:

$$x^2 + y^2 + z^2 = (t - \tau^*)^2. \qquad (29)$$

Substituting the above values of BD^* and B^*D^* into Minkowski's formula (27), the spatial components of the 4-acceleration of the passive mass m at B due to the active mass m^* at B^* turn out to be:[104]

$$\frac{d^2x}{d\tau^2} = -\frac{m^*x}{(t-\tau^*)^3}, \quad \frac{d^2y}{d\tau^2} = -\frac{m^*y}{(t-\tau^*)^3}, \quad \frac{d^2z}{d\tau^2} = -\frac{m^*z}{(t-\tau^*)^3}. \qquad (30)$$

---

103 Minkowski's driving force may be expressed in his notation as a function of scalar products of 4-velocities and 4-position:

$$-mm^*\frac{(w\bar{w}^*)\mathfrak{R} - (w\bar{\mathfrak{R}})w^*}{(\mathfrak{R}\bar{w}^*)^3(w\bar{w}^*)}.$$

Here I let w and w^* designate 4-velocity at the passive and active mass points, while $\mathfrak{R}$ is the associated 4-position, the parentheses denote a scalar product, and the bar indicates transposition.

From (30) and (29), the corresponding temporal component at B may be determined:[105]

$$\frac{d^2t}{d\tau^2} = -\frac{m^*}{(t-\tau^*)^2}\frac{d(t-\tau^*)}{dt}.$$

(33)

Inspecting (30), it appears that the only difference between these acceleration components and those corresponding to Newtonian attraction is a replacement in the latter of coordinate time t by proper time τ. [106]

The formal similarity of (30) to the Newtonian law of motion under a central force probably suggested to Minkowski that his law induces Keplerian trajectories. With the knowledge gained from (30), to the effect that the only difference between classical and relativistic trajectories is that arising from the substitution of proper time for coordinate time, Minkowski demonstrated the compatibility of his relativistic law of gravitation with observation using only Kepler's equation and the definition of 4-velocity.

Writing Kepler's equation in terms of proper time yields:

$$n\tau = E - e\sin E,$$

(34)

where $n\tau$ denotes the mean anomaly, e the eccentricity, and E the eccentric anomaly. Minkowski referred to (34) and to the norm of a 4-velocity vector:

$$\left(\frac{dx}{d\tau}\right)^2 + \left(\frac{dy}{d\tau}\right)^2 + \left(\frac{dz}{d\tau}\right)^2 = \left(\frac{dt}{d\tau}\right)^2 - 1,$$

(35)

---

104 The intermediate calculations can be reconstructed as follows. Let the driving force be designated F_μ, $\mu = 1, 2, 3, 4$. Since $(OA'/B^*D^*)^3 = r^{-3}$, and $BD^* = (-x, -y, -z, 0)$, equations (21) and (27) yield: $F_1/m = d^2x/d\tau^2 = -m^*x/(t-\tau^*)^3$, $F_2/m = d^2y/d\tau^2 = -m^*y/(t-\tau^*)^3$, $F_3/m = d^2z/d\tau^2 = -m^*z/(t-\tau^*)^3$.

105 Minkowski omitted the intermediate calculations, which may be reconstructed in modern notation as follows. Let the 4-velocity of the passive mass point be designated $U_\mu = (dx/d\tau, dy/d\tau, dz/d\tau, idt/d\tau)$, while the first three components of its 4-acceleration, designated A_μ, at B due to the source m^* are given by (30). From the orthogonality of 4-velocity and 4-acceleration we have:

$$U_\mu A_\mu = -\frac{dx}{d\tau}\frac{m^*x}{(t-\tau^*)^3} - \frac{dy}{d\tau}\frac{m^*y}{(t-\tau^*)^3} - \frac{dz}{d\tau}\frac{m^*z}{(t-\tau^*)^3} - \frac{idt}{d\tau}\frac{id^2t}{d\tau^2} = 0.$$

Rearranging (31) results in an expression for the temporal component of 4-acceleration:

$$\frac{d^2t}{d\tau^2} = -\frac{m^*}{(t-\tau^*)^3}\left(\frac{xdx}{dt} + \frac{ydy}{dt} + \frac{zdz}{dt}\right).$$

Differentiating (29) with respect to dt results in $xdx/dt + ydy/dt + zdz/dt$ $=(t-\tau^*)d(t-\tau^*)/dt$, the right-hand side of which we substitute in (32) to obtain (33).

106 A young Polish physicist in Göttingen, Felix Joachim de Wisniewski later studied this case, but with equations differing from (30) by a Lorentz factor (de Wisniewski 1913a, 388). In a postscript to the second installment of his paper (Wisniewski 1913b, 676) he employed Minkowski's matrix notation, becoming, with Max Born, one of the rare physicists to adopt this notation.

in order to determine the difference between the mean anomaly in coordinate time nt and the mean anomaly in proper time $n\tau$. From (35), Minkowski deduced:[107]

$$\left(\frac{dt}{d\tau}\right)^2 - 1 = \frac{m^*}{ac^2}\frac{1 + e\cos E}{1 - e\cos E}. \tag{37}$$

Solving (37) for the coordinate time dt, expanding to terms in c^{-2}, and multiplying by n led Minkowski to the expression:

$$ndt = nd\tau\left(1 + \frac{1}{2}\frac{m^*}{ac^2}\frac{1 + e\cos E}{1 - e\cos E}\right). \tag{38}$$

Recalling (34), Minkowski managed to express the difference between the mean anomaly in coordinate time and proper time:[108]

$$nt + \text{const.} = \left(1 + \frac{1}{2}\frac{m^*}{ac^2}\right)n\tau + \frac{m^*}{ac^2}e\sin E. \tag{39}$$

Evaluating the relativistic factor $(m^*)/ac^2$ for solar mass and the Earth's semi-major axis to be 10^{-8}, Minkowski found the deviation from Newtonian orbits to be negligible in the solar system. On this basis, he concluded that

> a decision *against* such a law and the proposed modified mechanics in favor of the Newtonian law of attraction with Newtonian mechanics would not be deducible from astronomical observations.[109]

According to the quoted remark, there was more at stake here for Minkowski than just the empirical adequacy of his law of gravitational attraction, as his claim is for parity between Newton's law and classical mechanics, on one hand, and the *system*

---

107 The intermediate calculations were omitted by Minkowski, but figure among his research notes (*Math. Archiv* 60: 6, 126–127, Handschriftenabteilung, NSUB). Following the method outlined by Otto Dziobek (Dziobek 1888, 12), Minkowski began with the energy integral of Keplerian motion:

$$\left(\frac{dt}{dW}\right)^2 - 1 = \frac{2}{\ell^2}\left(\frac{M}{R} - C\right), \tag{36}$$

where ℓ denotes the velocity of light, M is the sum of the masses times the gravitational constant, $M = k^2(m + m^*)$, R is the radius, and C is a constant. The left-hand side of (36) is the same as the right-hand side of (35) for $W = \tau$. In order to express dt/dW (which is to say $dt/d\tau$) in terms of E, Minkowski considered a conic section in polar coordinates, with focus at the origin: $R = a(1 - e^2)/(1 + e\cos\varphi) = a(1 - \cos E)$, where a denotes the semi-major axis, and φ is the true anomaly. By eliminating φ in favor of E and e, and differentiating (34), Minkowski obtained an expression equivalent to (37).

108 I insert the eccentricity e in the second term on the right-hand side, correcting an obvious omission in Minkowski's paper (Minkowski 1908, 111, eq. 31).

109 "... eine Entscheidung *gegen* ein solches Gesetz und die vorgeschlagene modifizierte Mechanik zu Gunsten des Newtonschen Attraktionsgesetzes mit der Newtonschen Mechanik aus den astronomischen Beobachtungen nicht abzuleiten sein" (Minkowski 1908, 111).

composed of the law of gravitation and spacetime mechanics on the other hand. This new system, Minkowski claimed, was verified by astronomical observations at least as well as the classical system formed by the Newtonian law of attraction and Newtonian mechanics.

Instead of comparing his law with one or the other of Poincaré's laws, Minkowski noted a difference in *method*, as mentioned above. In light of Minkowski's emphasis on the methodological difference with Poincaré, and the hybrid geometric-symbolic nature of Minkowski's exposition, it is clear that the point of reexamining the problem of relativity and gravitation in the *Grundgleichungen* was not simply to make plausible the inclusion of gravitation in a relativistic framework. Rather, since gravitational attraction was the only example Minkowski provided of his formalism in action, his line of argument served to *validate* his four-dimensional calculus, over and above the requirements of plausibility.

From the latter point of view, Minkowski had grounds for satisfaction, although one imagines that he would have preferred to find that his law diverged from Newton's law just enough to account for the observed anomalies. It stands to reason that if Minkowski had been fully satisfied with his first law, he would not have proposed a second law in his next paper—which turned out to be the last he would finish for publication. The latter article developed out of a well-known lecture entitled "Space and Time" (*Raum und Zeit*), delivered in Cologne on 21 September, 1908, to the mathematics section of the German Association of Scientists and Physicians in its annual meeting (Walter 1999a, 49).

In the final section of his Cologne lecture, Minkowski took up the Lorentz-Poincaré theory, and showed how to determine the field due to a point charge in arbitrary motion. On this occasion, just as in his earlier discussion of gravitation in the *Grundgleichungen*, Minkowski referred to a spacetime diagram, but this time he provided the diagram (Fig. 3). Identifying the 4-vector potential components for the source charge on this diagram, Minkowski remarked that the Liénard-Wiechert law was a consequence of just these geometric relations.[110]

---

110 Minkowski's explanation of the construction of his spacetime diagram (Fig. 3) may be paraphrased in modern terminology as follows. Suppressing the z-axis, we associate two world lines with two point charges e_1 and e. The world line of e_1 passes through the point at which we wish to determine the field, P_1. To find the retarded position of the source e, we draw the retrograde lightcone (with broken lines) from P_1, which intersects the world line of e at P, where there is a hyperbola of curvature ρ with three infinitely-near points lying on the world line of e; it has its center at M. The coordinate origin is established at P, by letting the t-axis coincide with the tangent to the world line. A line from P_1 intersects this axis orthogonally at point Q; it is space-like, and if its projection on a constant-time hypersurface has length r, the length of the 4-vector PQ is r/c. The 4-vector potential has magnitude e/r and points in the direction of PQ (i.e., parallel to the 4-velocity of e at P). The x-axis lies parallel to QP_1, such that N is the intersection of a line through M normal to the x-axis.

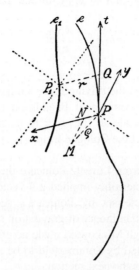

Figure 3: Minkowski's spacetime diagram of particle interaction (Minkowski 1909, 86).

Minkowski then described the driving force between two point charges. Adopting dot notation for differentiation with respect to proper time, he wrote the driving force exerted on an electron of charge e_1 at point P_1 by an electron of charge e:

$$-ee_1\left(\dot{t}_1 - \frac{\dot{x}_1}{c}\right)\mathfrak{K}, \tag{40}$$

where $\dot{t}_1$ and $\dot{x}_1$ are 4-velocity components of the test charge e_1 and $\mathfrak{K}$ is a certain 4-vector. This was the first such description of the electrodynamic driving force due to a 4-vector potential, the simplicity of which, Minkowski claimed, compared favorably with the earlier formulations of Schwarzschild and Lorentz.[111]

In the same celebratory tone, Minkowski finished his article with a discussion of gravitational attraction. The "reformed mechanics", he claimed, dissolved the disturbing disharmonies between Newtonian mechanics and electrodynamics. In order to provide an example of this dissolution, he asked how the Newtonian law of attraction would sit with his principle of relativity. Minkowski continued:

> I will assume that if two point masses m, m_1 describe world lines, a driving force vector is exerted by m on m_1, exactly like the one in the expression just given for the case of electrons, except that instead of $-ee_1$, we must now put in $+mm_1$.

---

111 Minkowski noted four conditions on $\mathfrak{K}$: it is normal to the 4-velocity of e_1 at P_1, $c\mathfrak{K}_t - \mathfrak{K}_x = 1/r^2$, $\mathfrak{K}_y = \ddot{y}/(c^2 r)$, and $\mathfrak{K}_z = 0$, where r is the space-like distance between the test charge e_1 at P_1 and the advanced position Q of the source e, and $\ddot{y}$ is the y-component of e's 4-acceleration at P. For a derivation of the 4-potential and 4-force corresponding to Minkowski's presentation, see (Pauli 1921, 644–645).

Applying the substitution suggested by Minkowski to (40), we obtain:

$$mm_1\left(\dot{t}_1 - \frac{\dot{x}_1}{c}\right)\Re, \tag{41}$$

where the coefficients m and m_1 refer to proper masses. Minkowski's new law of gravitation (41) fully expresses the driving force, unlike the formula (27) of his first law, which describes only one component. In addition, the 4-vectors are immediately identifiable from the notation alone. (In order to distinguish the law given in the *Grundgleichungen* from that of the Cologne lecture [41], I will call [41] Minkowski's second law.)

Since (40) was obtained from Lorentz-Poincaré theory via a 4-vector potential, the law of gravitation (41) ostensibly implied a 4-vector potential as well; in other words, following the example set by Poincaré's second law (10), Minkowski appealed in turn to a Maxwellian theory of gravitation similar to those of Heaviside, Lorentz, and Gans.[112] Although Minkowski made no effort to attach his law to these field theories, it was understood by Sommerfeld to be a formal consequence of just such a theory, as I will show in the next section.

What were the numerical consequences of this new law? Minkowski spared the reader the details, noting only that in the case of uniform motion of the source, the only divergence from a Keplerian orbit would stem from the replacement of coordinate time by proper time. He indicated that the numbers for this case had been worked out earlier, and his conclusion with respect to this new law was naturally the same: combined with the new mechanics, it was supported by astronomical observations to the same extent as the Newtonian law combined with classical mechanics.

Curiously enough, Minkowski offered no explanation of the need for a second law of attraction. Furthermore, by proposing two laws instead of one, Minkowski tacitly acknowledged defeat; despite his criticism of the Poincaré's approach (see above, p. 215), he could hardly claim to have solved unambiguously the problem of gravitation. It may also seem strange that Minkowski discarded the differences between his new law (41) and the one he had proposed earlier.[113]

Minkowski revealed neither the motivation behind a second law of gravitation, nor why he neglected the differences between his two laws, but there is a straightfor-

---

112 See above, p. 198, (Heaviside 1893), and (Gans 1905). Theories in which the gravitational field is determined by equations having the form of Maxwell's equations were later termed vector theories of gravitation by Max Abraham (Abraham 1914, 477). For a more recent version of such a theory, see (Coster and Shepanski 1969).

113 Minkowski's neglect of the differences between his two theories may explain why historians have failed to distinguish them. The principal difference between the two laws stems from the presence of acceleration effects in the second law. By 1905 it was known that accelerated electrons radiate energy, such that by formal analogy, a Maxwellian theory of gravitation should have featured accelerated point masses radiating "gravitational" energy. For a brief overview of research performed in the first two decades of the twentieth century on the energy radiated from accelerated electrons, see (Whittaker 1951–1953, 2:246).

ward way of explaining both of these mysteries. First, we recall the circumstances of his Cologne lecture, the final section of which Minkowski devoted to the theme of restoring unity to physics. What he wanted to stress on this occasion was that mechanics and electrodynamics harmonized in his four-dimensional scheme of things:

> In the mechanics reformed according to the world postulate, the disturbing disharmonies between Newtonian mechanics and modern electrodynamics fall out on their own.[114]

To support this view, Minkowski had to show that his reformed mechanics was a synthesis of classical mechanics and electrodynamics. A Maxwellian theory of gravitation fit the bill quite well, and consequently, Minkowski brought out his second law of gravitation (41). Clearly, this was not the time to point out the *differences* between his two laws. On the contrary, it was the perfect occasion to observe that a law of gravitation derived from a 4-vector potential formally identical to that of electrodynamics was observationally indistinguishable from Newton's law. Naturally, Minkowski seized this opportunity.

Sadly, Minkowski did not live long enough to develop his ideas on gravitation and electrodynamics; he died on 12 January, 1909, a few days after undergoing an operation for appendicitis. At the time, no objections to a field theory of gravitation analogous to Maxwell's electromagnetic theory were known, apart from Maxwell's own sticking-points (see above, p. 198). However, additional objections to this approach were raised by Max Abraham in 1912, after which the Maxwellian approach withered on the vine, as Gustav Mie and others pursued unified theories of electromagnetism and gravitation.[115]

Minkowski's first law of gravitation fared no better than his second law, but the four-dimensional language in which his two laws were couched had a bright future. The first one to use Minkowski's formal ideas to advantage was Sommerfeld, as we will see next.

3. ARNOLD SOMMERFELD'S HYPER-MINKOWSKIAN LAWS OF GRAVITATION

Neither Poincaré's nor Minkowski's work on gravitation and relativity drew comment until 25 October, 1910, when the second installment of Arnold Sommerfeld's vectorial version of Minkowski's calculus, entitled "Four-dimensional vector analysis" [*Vierdimensionale Vektoranalysis*], appeared in the *Annalen der Physik* (Sommerfeld 1910b). Sommerfeld's contribution differs from those of Poincaré and Minkowski in that it is openly concerned with the presentation of a new formalism, much as its title

---

114 "In der dem Weltpostulate gemäß reformierten Mechanik fallen die Disharmonien, die zwischen der Newtonschen Mechanik und der modernen Elektrodynamik gestört haben, von selbst aus" (Minkowski 1909, §5).

115 Abraham showed that a mass set into oscillation would be unstable due to the direction of energy flow (Norton 1992, 33). On the early history of unified field theories, see the reference in note 12.

indicates. In this section, I discuss Sommerfeld's interest in vectors, the salient aspects of his 4-vector formalism, and his portrayal of Poincaré's and Minkowski's laws of gravitation.

Sommerfeld displayed a lively interest in vectors, beginning with his editorship of the physics volume of Klein's six-volume *Encyklopädie* in the summer of 1898.[116] He imposed a certain style of vector notation on his contributing authors, including typeface, terminology, symbolic representation of operations, units and dimensions, and the choice of symbols for physical quantities. Articles 12 to 14 of the physics volume appeared in 1904, and were the first to implement the notation scheme backed by Sommerfeld, laid out the same year in the *Physikalische Zeitschrift*.[117] While Sommerfeld belonged to the Vector Commission formed at Felix Klein's behest in 1902, it was clear to him as early as 1901 that the article on Maxwell's theory (commissioned to Lorentz) would serve as a "general directive" for future work in electrodynamics.[118] His intuition turned out to be correct: the principal "vector" of influence was Lorentz's Article 13 (Lorentz 1904b), featuring sections on vector notation and algebra, which set a *de facto* standard for vector approaches to electrodynamics.

As mentioned above (p. 210), only one effort to extend Poincaré's four-dimensional approach beyond the domain of gravitation was published prior to Minkowski's *Grundgleichungen*. By 1910, the outlook for relativity theory had changed due to the authoritative support of Planck and Sommerfeld, the announcement of experimental results favoring Lorentz's electron theory, and the broad diffusion (in 1909) of Minkowski's Cologne lecture. Dozens of physicists and mathematicians began to take an interest in relativity, resulting in a leap in relativist publications.[119]

The principal promoter of Minkowskian relativity, Sommerfeld must have felt by 1910 that it was the right moment to introduce a four-dimensional formalism. He was not alone in feeling this way, for three other formal approaches based on Minkowski's work appeared in 1910. Two of these were 4-vector systems, similar in some respects to Sommerfeld's, and worked out by Max Abraham and the American physical chemist Gilbert Newton Lewis, respectively. A third, non-vectorial approach was proposed by the Zagreb mathematician Vladimir Varičak. Varičak's was a real, four-dimensional, coordinate-based approach relying on hyperbolic geometry. Sommerfeld probably viewed this system as a potential rival to his own approach; although he did not mention Varičak, he wrote that a non-Euclidean approach was

116 Sommerfeld's work on the *Encyklopädie* is discussed in an editorial note to his scientific correspondence (Sommerfeld 2001–2004, 1:40).
117 See (Reiff and Sommerfeld 1904; Lorentz 1904b, 1904c, Sommerfeld 1904). The scheme proposed by Sommerfeld differed from that published in articles 12 to 14 of the *Encyklopädie* only in that the operands of scalar and vector products were no longer separated by a dot.
118 Sommerfeld to Lorentz, 21 March, 1901, (Sommerfeld 2001–2004, 1:191). On Sommerfeld's participation on the Commission see (Reich 1996) and (Sommerfeld 2001–2004, 1:144).
119 For bibliometric data, and discussions of Sommerfeld's role in the rise of relativity theory, see (Walter 1999a, 68–73, 1999b, 96, 108).

possible but could not be recommended (Sommerfeld 1910a, 752, note 1). Of the three alternatives to Sommerfeld's system, the non-Euclidean style pursued by Varičak and others was the only one to obtain even a modest following. An investigation of the reasons for the contemporary neglect of these alternative four-dimensional approaches is beyond the purview of our study; for what concerns us directly, none of these methods was applied to the problem of gravitation.[120]

Sommerfeld's paper, like those of Abraham, Lewis, and Varičak, emphasized formalism, and in this it differed from the *Grundgleichungen*, as mentioned above. Like the latter work, it focused attention on the problem of gravitation. Following the example set by both Poincaré and Minkowski, Sommerfeld capped his two-part *Annalen* paper with an application to gravitational attraction, which consisted of a reformulation, comparison and commentary of their work in his own terms. Not only was Sommerfeld's comparison of Poincaré's and Minkowski's laws of gravitation the first of its kind, it also proved to be the definitive analysis for his generation.

Sommerfeld's four-dimensional vector algebra and analysis offered no new 4-vector or 6-vector definitions, but it introduced a suite of 4-vector functions, notation, and vocabulary. The most far-reaching modification with respect to Minkowski's calculus was the elimination of *lor* (cf. pp. 223–223) in favor of extended versions of ordinary vector functions. In Sommerfeld's notational scheme, the ordinary vector functions div, rot, and grad (used by Lorentz in his *Encyklopädie* article on Maxwell's theory) were replaced by 4-vector counterparts marked by a leading capital letter: Div, Rot, and Grad. These three functions were joined by a 4-vector divergence marked by German typeface, $\mathfrak{Div}$. Sommerfeld chose to retain $\square$ (cf. note 20), while noting the equivalence to his 4-vector functions: $\square$ = Div Grad. The principal advantage of the latter functions was that their meaning was familiar to physicists. In the same vein, Sommerfeld supplanted Minkowski's unwieldy terminology of "spacetime vectors of the first and second type" [*Raum-Zeit-Vektoren I^{ter} und II^{ter} Art*] with the more succinct "four-vector [*Vierervektor*] and "six-vector" [*Sechservektor*]. The result was a compact and transparent four-dimensional formalism differing as little as possible from the ordinary vector algebra employed in the physics volume of the *Encyklopädie*.[121]

To show how his formalism performed in action, Sommerfeld first took up the geometric interpretation and calculation of the electrodynamic 4-vector potential and 4-force. In the new notation, Sommerfeld wrote the electrodynamic 4-force $\mathfrak{K}$ between two point charges e and e_0 in terms of three components in the direction of the light-like 4-vector $\mathfrak{R}$, the source 4-velocity $\mathfrak{B}$, and the 4-acceleration $\dot{\mathfrak{B}}$:

---

120 See (Abraham 1910; Lewis 1910a, 1910b; Varičak 1910). On Varičak's contribution see (Walter 1999b).

121 Not all of Sommerfeld's notational choices were retained by later investigators; Laue, for instance, preferred a notational distinction between 4-vectors and 6-vectors. For a summary of notation used by Minkowski, Abraham, Lewis, and Laue, see (Reich 1994).

$$4\pi\mathfrak{K}_{\mathfrak{R}} = \frac{ee_0}{c(\mathfrak{R}\mathfrak{B})^2}\left(\frac{c^2-(\mathfrak{R}\dot{\mathfrak{B}})}{(\mathfrak{R}\mathfrak{B})}(\mathfrak{B}_0\mathfrak{B})+\mathfrak{B}_0\dot{\mathfrak{B}}\right)\mathfrak{R},$$

$$4\pi\mathfrak{K}_{\mathfrak{B}} = \frac{-ee_0}{c(\mathfrak{R}\mathfrak{B})^2}\frac{c^2-(\mathfrak{R}\dot{\mathfrak{B}})}{(\mathfrak{R}\mathfrak{B})}(\mathfrak{B}_0\mathfrak{R})\mathfrak{B}, \tag{42}$$

$$4\pi\mathfrak{K}_{\dot{\mathfrak{B}}} = \frac{-ee_0}{c(\mathfrak{R}\mathfrak{B})^2}\frac{c^2-(\mathfrak{R}\dot{\mathfrak{B}})}{(\mathfrak{R}\mathfrak{B})}(\mathfrak{B}_0\mathfrak{R})\dot{\mathfrak{B}},$$

where parentheses indicate scalar products. Sommerfeld was careful to note the equivalence between (42) and what he called Minkowski's "geometric rule" (40).

In the ninth and final section of his paper, Sommerfeld took up the law of electrostatics and the classical law of gravitation. The former was naturally considered to be a special case of (42), with two point charges relatively at rest. The same was true for the law of gravitation, as Sommerfeld noted that Minkowski had proposed a formal variant of (40) as a law of gravitational attraction (what I call Minkowski's second law, [41]). Sommerfeld's expression of the electrodynamic 4-force is unwieldy, but takes on a simpler form in case of uniform motion of the source ($\dot{\mathfrak{B}} = 0$). Neglecting the 4π factor, and substituting $-mm_0$ for $+ee_0$, Sommerfeld expressed the corresponding version of Minkowski's second law:

$$-mm_0c\frac{(\mathfrak{B}_0\mathfrak{B})\mathfrak{R}-(\mathfrak{B}_0\mathfrak{R})\mathfrak{B}}{(\mathfrak{R}\mathfrak{B})^3}. \tag{43}$$

The latter law is compact and self-contained, in that its interpretation depends only on the definitions and rules of the algebraic formalism. In this sense, (43) improves on the Minkowskian (41), even if it represents only a special case of the latter law.

Once Sommerfeld had expressed Minkowski's second law in his own terms, he turned to Poincaré's two laws. The transformation of Poincaré's first law was more laborious than the transformation of Minkowski's second law. First of all, Sommerfeld transcribed Poincaré's first law (9) into his 4-vector notation, while retaining the original designation of invariants. This step itself was not simple: in order to cast Poincaré's kinematic invariants as scalar products of 4-vectors, Sommerfeld had to adjust the leading sign of (9), to obtain:

$$\frac{k_0\mathfrak{K}}{mm'} = -\frac{1}{B^3C}\left(C\mathfrak{R}-\frac{1}{c}A\mathfrak{B}\right). \tag{44}$$

Sommerfeld noted the "correction" of what he called an "obvious sign error" in (9).[122] The difference is due to Poincaré's irregular derivation of the kinematic invariants (1), as mentioned above (p. 203), although from Sommerfeld's remark it is not clear that he saw it this way.

---

122 "Mit Umkehr des bei Poincaré offenbar versehentlichen Vorzeichens" (Sommerfeld 1910b, 686, n. 1).

The transformation of Poincaré's second law (10) was less straightforward. It appears that instead of deriving a 4-vector expression as in the previous case, Sommerfeld followed Poincaré's lead by eliminating the Lorentz-invariant factor C from the denominator on the right-hand side of the first law (44), which results in the equation:

$$\frac{k_0 \mathfrak{K}}{mm'} = -\frac{1}{B^3}\left(C\mathfrak{R} - \frac{1}{c}A\mathfrak{B}\right). \tag{45}$$

Sommerfeld expressed Poincaré's kinematic invariants A, B, and C as scalar products:

$$A = -\frac{1}{c}(\mathfrak{R}\mathfrak{B}_0), \quad B = -\frac{1}{c}(\mathfrak{R}\mathfrak{B}), \quad C = -\frac{1}{c^2}(\mathfrak{B}_0\mathfrak{B}). \tag{46}$$

He also replaced the mass term m' in (44) and (45) by the product of rest mass m_0 and the Lorentz factor k_0, i.e., $m' = m_0 k_0$. At this point, he could express Poincaré's two laws exclusively in terms of constants, scalars, and 4-vectors:

$$mm_0 c^3 \frac{(\mathfrak{B}_0\mathfrak{B})\mathfrak{R} - (\mathfrak{B}_0\mathfrak{R})\mathfrak{B}}{(\mathfrak{R}\mathfrak{B})^3(\mathfrak{B}_0 B)}, \tag{47}$$

$$-mm_0 c \frac{(\mathfrak{B}_0\mathfrak{B})\mathfrak{R} - (\mathfrak{B}_0\mathfrak{R})\mathfrak{B}}{(\mathfrak{R}\mathfrak{B})^3}. \tag{48}$$

In the latter form, Sommerfeld's (approximate) version of Minkowski's second law (43) matches exactly his (exact) version of Poincaré's second law (48). Sommerfeld pointed out this equivalence, and noted again that the difference between (47) and (48) amounted to a single factor, in the scalar product of 4-velocities: $C = -(\mathfrak{B}_0\mathfrak{B})/c^2$. (All six Lorentz-invariant laws of gravitation of Poincaré, Minkowski, and Sommerfeld are presented in Table 1.) Sommerfeld summed up his result by saying that when the acceleration of the active mass is neglected, Minkowski's special formulation of Newton's law (41) is subsumed by Poincaré's indeterminate formulation. In other words, the approximate form of Minkowski's second law was captured by Poincaré's remark that his first law (9) could be multiplied by an unlimited number of Lorentz-invariant quantities (within certain constraints).

The message of the basic equivalence of Poincaré's pair of laws to Minkowski's pair echoes the latter's argument in his Cologne lecture, to the effect that spacetime mechanics removed the disharmonies of classical mechanics and electrodynamics (see above, p. 234). This message was reinforced by Sommerfeld's graphical representation of the 4-vector components of these laws in a spacetime diagram, reproduced in Fig. 4. The 4-vector relations in (47) and (48) are shown in the figure; the world line of the active mass m appears on the left-hand side of the diagram, and the line OL (which coincides with $\mathfrak{R}$) lies on the retrograde lightcone from the origin O on the world line of the passive mass m_0. All three 4-vectors in (47) and (48), $\mathfrak{R}$, $\mathfrak{B}_0$, and $\mathfrak{B}_0$ are represented in the diagram, along with an angle ψ corresponding to the Lorentz-invariant $C = \cos\psi$ distinguishing (47) and (48).[123]

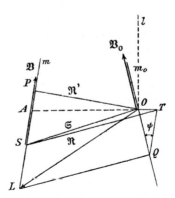

Figure 4: Sommerfeld's illustration of the two laws of gravitation (Sommerfeld 1910b, 687).

So far, Sommerfeld had dealt with three of the four laws of gravitation, leaving out only Minkowski's first law. Since Minkowski's presentation of his first law was a purely geometric affair, Sommerfeld had no choice but to reconstruct his argument with reference to a spacetime diagram describing the components of (27) in terms of the angle ψ and a fourth 4-vector, $\mathfrak{S}$. He showed the numerator in (47) and (48) to be equal to the product $(\mathfrak{B}_0\mathfrak{B})\mathfrak{S}$, and expressed the denominator of (48) in terms of the length R' of the 4-vector $\mathfrak{R}'$ in Fig. 4, to obtain the formula:

$$\mathfrak{K} = mm_0\cos\psi\frac{\mathfrak{S}}{R'^3},\tag{49}$$

which he showed to be equivalent to (47). Eliminating the factor $C = \cos\psi$ from the latter equation, Sommerfeld obtained an expression for (48) in terms of $\mathfrak{S}$:

$$\mathfrak{K} = \frac{mm_0\mathfrak{S}}{R'^3}.\tag{50}$$

The latter two driving force equations, (49) and (50), were thus rendered geometrically by Sommerfeld, facilitating the comprehension of their respective vector-symbolic expressions (47) and (48).

In general, the driving force of (49) is weaker, *ceteris paribus*, than that of (50) due to the cosine in the former, but Sommerfeld did not develop these results numerically, noting only that the four laws were equally valid from an empirical stand-

---

123 Sommerfeld explained Fig. 4 roughly as follows: two skew 4-velocities $\mathfrak{B}$ and $\mathfrak{B}_0$ determine a three-dimensional space, containing all the lines shown. Points $OLSAP$ are coplanar, while the triangles OQT and OTS, and the parallelogram $LQTS$ all generally lie in distinct planes. In particular, T lies outside the plane of $OLSAP$, and OT is orthogonal to $\mathfrak{B}_0$. The broken vertical line l represents the temporal axis of a frame with origin O; a space-like plane orthogonal to l at O intersects the world line of m at point A. The space-like 4-vector $\mathfrak{R}'$ is orthogonal to $\mathfrak{B}$, while $\mathfrak{S}$ is orthogonal to $\mathfrak{B}_0$; both $\mathfrak{R}'$ and $\mathfrak{S}$ intersect the origin, while $\mathfrak{B}$ and $\mathfrak{B}_0$ together form an angle ψ.

point.[124] He noted that Poincaré's analysis allowed for several other laws, but that in all cases, one sticking-point remained: there was no answer to the question of how to localize momentum in the gravitational field.

By rewriting Poincaré's and Minkowski's laws in his new 4-vector formalism, Sommerfeld effectively rationalized their contributions for physicists. The goal of his paper, announced at the outset, was to display the "remarkable simplification of electrodynamic concepts and calculations" resulting from "Minkowski's profound spacetime conception."[125] Actually, Sommerfeld's comparison of Poincaré's and Minkowski's laws of gravitation was designed to show *his* formalism in an attractive light. In realizing this comparison in his own formalism, Sommerfeld smoothed out the idiosyncrasies of Poincaré's method, inappropriately lending him a 4-vector approach. He felt that Poincaré had "already employed 4-vectors" (Sommerfeld 1910b, 685) although as shown in the first section, Poincaré's use of four-dimensional entities was tightly circumscribed by the objective of formulating Lorentz-invariants. In Thomas Kuhn's optical metaphor (Kuhn 1970, 112), Sommerfeld read Poincaré's theory through a Minkowskian lens; in other words, he read it as a spacetime theory. For Sommerfeld, no less than for Minkowski, the discussion of gravitation and relativity was modulated by the programmatic objective of promoting a four-dimensional formalism. Satisfying this objective without ignoring Poincaré's work, however, meant rationalizing Poincaré's contribution.[126]

Sommerfeld's reading of Minkowski's second law contrasts with its muted exposition in the original text (see above, pp. 234–234), in that he gave it pride of place with respect to the other three laws. This change in emphasis on Sommerfeld's part reflects his own research interests in electrodynamics, and his outlook on the future direction of physics.[127] But what originally motivated him to propose a 4-dimensional formalism? The inevitability of a 4-dimensional vector algebra as a standard tool of the physicist was probably a foregone conclusion for him by 1910, such that the promotion of the ordinary vector notation used in the *Encyklopädie* obliged him to propose essentially the same notation for 4-vectors. Sommerfeld referred modestly to his work as an "explanation of Minkowskian ideas" (Sommerfeld 1910a, 749) but as he explained to his friend Willy Wien, co-editor with Planck of the *Annalen der*

124 This view was confirmed independently by the Dutch astronomer W. de Sitter, who worked out the numbers for the one-body problem (de Sitter 1911). De Sitter found the second law to require a post-Newtonian centennial advance in Mercury's perihelion of 7″, while the first law required no additional advance. His figure for the second law agrees with the one given by Poincaré (see above, p. 208).
125 "In dieser und einigen anschließenden Studien möchte ich darstellen, wie merkwürdig sich die elektrodynamischen Begriffe und Rechnungen vereinfachen, wenn man sich dabei von der tiefsinnigen Raum-Zeit-Auffassung Minkowskis leiten läßt" (Sommerfeld 1910a, 749).
126 Faced with a similar situation in his Cologne lecture of September, 1908, Minkowski simply neglected to mention Poincaré's contribution; see (Walter 1999a, 56).
127 Sommerfeld later preferred Gustav Mie's field theory of gravitation. Such an approach was more promising than that of Poincaré and Minkowski, which grasped gravitation "to some extent as action at a distance" (Sommerfeld 1913, 73).

Physik, Minkowski's original 4-vector scheme had evolved. "The geometrical systematics" Sommerfeld announced, "is now hyper-Minkowskian."[128] In the same letter to Wien, Sommerfeld confessed that his paper had required substantial effort, and he expressed doubt that it would prove worthwhile. Sommerfeld displayed either pessimism or modesty here, but in fact his effort was richly rewarded, as his streamlined four-dimensional algebra and analysis quickly won both Einstein's praise and the confidence of his contemporaries.[129]

Sommerfeld's work was eagerly read by young theoretical physicists raised in the heady atmosphere of German vectorial electrodynamics. One of the early adepts of Sommerfeld's formalism was Philipp Frank (1884–1966), who was then a Privatdozent in Vienna. By way of introduction to his 1911 study of the Lorentz-covariance of Maxwell's equations, Frank described the new four-dimensional algebra as a combination of "Sommerfeld's intuitiveness with Minkowski's mathematical elegance" (Frank 1911, 600). He recognized, however, that of late, physicists had been overloaded with outlandish symbolic systems and terminology, and promised to stay within the boundaries of Sommerfeld's system, at least as far as this was possible.

Physicists were indeed inundated in 1910–1911 with a bewildering array of new symbolic systems, including an ordinary vector algebra (Burali-Forti and Marcolongo 1910), and a quaternionic calculus (Conway 1911), in addition to the hyperbolic-coordinate system and three 4-vector formalisms already mentioned. By 1911, 4-vector and 6-vector operations featured prominently in the pages of the *Annalen der Physik*. Out of the nine theoretical papers concerning relativity theory published in the *Annalen* that year, five made use of a four-dimensional approach to physics, either in terms of 4-vector operations, or by referring to spacetime coordinates. Four out of five authors of "four-dimensional" papers cited Minkowski's or Sommerfeld's work; the fifth referred to Max Laue's new relativity textbook (Laue 1911). This timely and well-written little book went far in standardizing the terminology and notation of four-dimensional algebra, such that by January of 1912, Max Abraham preferred the Sommerfeld-Laue notation to his own for the exposition of his theory of gravitation (Abraham 1910, 1912a, 1912b).

While young theorists were quick to pick up on the Sommerfeld-Laue calculus, textbook writers did not follow the trend. Of the four textbooks to appear on relativity in 1913–1914, only the second edition of Laue's book (Laue 1913) employed this formalism. Ebenezer Cunningham presented a 4-dimensional approach based on Minkowski's work, but explicitly rejected Sommerfeld's "quasi-geometrical language", which conflicted with his own purely algebraic presentation (Cunningham

---

128 "Die geometrische Systematik ist jetzt hyper-minkowskisch" (Sommerfeld to Wien, 11 July, 1910, Sommerfeld 2001–2004, 1:388).

129 Einstein to Sommerfeld, July, 1910, (CPAE 5, 243–247; Sommerfeld 2001–2004, 1:386–388). In light of Einstein and Laub's earlier dismissal of Minkowski's formalism (see above, pp. 222–223), Sommerfeld naturally supposed that Einstein would disapprove of his system, prompting the protest: "Wie können Sie denken, dass ich die Schönheit einer solchen Untersuchung nicht zu schätzen wüsste?"

1914, 99). A third textbook by Ludwik Silberstein (Silberstein 1914), a former student of Planck, gave preference to a quaternionic presentation, while the fourth, by Max B. Weinstein (Weinstein 1913), opted for Cartesian coordinates. Curiously enough, Weinstein dedicated his work to the memory of Minkowski. Apparently disturbed by this profession of fidelity, Max Born, who had briefly served as Minkowski's assistant, deplored the form of Weinstein's approach to relativity:

> [Minkowski] put perhaps just as much value on his presentation as on its content. For this reason, I do not believe that entrance to his conceptual world is facilitated when it is overwhelmed by an enormous surfeit of formulas.[130]

By this time, Born himself had dropped Minkowski's formalism in favor of the Sommerfeld-Laue approach, such that the target of his criticism was Weinstein's disregard for 4-dimensional methods in general, and not the neglect of Minkowski's matrix calculus.[131] What Born was pointing out here was that it had become highly impractical to study the theory of relativity without recourse to a 4-dimensional formalism. This may explain why Laue's was the only one of the four textbooks on relativity to be reedited, reaching a sixth edition in 1955.

In summary, the language developed by Sommerfeld for the expression of the laws of gravitation of Poincaré and Minkowski endured, while the laws themselves remained tentative at best. This much was clear as early as 1912, when Jun Ishiwara reported from Japan on the state of relativity theory. This theory, Ishiwara felt, had shed no light on the problem of gravitation, with a single exception: Minkowski and Sommerfeld's "formal mathematical treatment" (Ishiwara 1912, 588). The trend from Poincaré to Sommerfeld was one of increasing reliance on formal techniques catering to Lorentz-invariance; in the space of five years, the physical content of the laws of gravitation remained stable, while their formal garb evolved from Cartesian to hyper-Minkowskian.

4. CONCLUSION:
ON THE EMERGENCE OF THE FOUR-DIMENSIONAL VIEW

After a century-long process of accommodation to the use of tensor calculus and spacetime diagrams for analysis of physical interactions, the mathematical difficulties encountered by the pioneers of 4-dimensional physics are hard to come to terms with. Not only is the oft-encountered image of flat-spacetime physics as a trivial consequence of Einstein's special theory of relativity and Felix Klein's geometry consistent with such accommodation, it reflects Minkowski's own characterization of the back-

---

130 "[Minkowski] hat auf seine Darstellung vielleicht ebenso viel Wert gelegt, wie auf ihren Inhalt. Darum glaube ich nicht, daß der Zugang zu seiner Gedankenwelt erleichtert wird, wenn sie von einer ungeheuren [sic] Fülle von Formeln überschüttet wird" (Born 1914).
131 By the end of 1911 Born had already acknowledged that, despite its "formal simplicity and greater generality compared to the tradition of vectorial notation," Minkowski's calculus was "unable to hold its ground in mathematical physics" (Born 1912, 175).

ground of the four-dimensional approach (cf. p. 218). However, this description ought
not be taken at face value, being better understood as a rhetorical ploy designed to
induce mathematicians to enter the nascent field of relativistic physics (Walter 1999a).
When the principle of relativity was formulated in 1905, even for one as adept as
Henri Poincaré in the application of group methods, the path to a four-dimensional
language for physics appeared strewn with obstacles. Much as Poincaré had predicted
(above, p. 209), the construction of this language cost Minkowski and Sommerfeld
considerable pain and effort.

Clear-sighted as he proved to be in this regard, Poincaré did not foresee the emergence of forces that would accelerate the construction and acquisition of a four-
dimensional language. With hindsight, we can identify five factors favoring the use
and development of a four-dimensional language for physics between 1905 and 1910:
the elaboration of new concepts and definitions, the introduction of a graphic model
of spacetime, the experimental confirmation of relativity theory, the vector-symbolic
movement, and problem-solving performance.

In the beginning, the availability of workable four-dimensional concepts and definitions regulated the analytic reach of a four-dimensional approach to physics.
Poincaré's discovery of the 4-vectors of velocity and force in the course of his elaboration of Lorentz-invariant quantities, and Minkowski's initial misreading of Poincaré's
definitions underline how unintuitive these notions appeared to turn-of-the-century
mathematicians. The lack of a 4-velocity definition visibly hindered Minkowski's
elaboration of spacetime mechanics and theory of gravitation. It is remarkable that
even after Minkowski presented the notions of proper time, world line, rest-mass density, and the energy-momentum tensor, putting the spacetime electrodynamics and
mechanics on the same four-dimensional footing, his approach failed to convince
physicists. Nevertheless, all of these discoveries extended the reach of the four-dimensional approach, in the end making it a viable candidate for the theorist's toolbox.

Next, Minkowski's visually-intuitive spacetime diagram played a decisive role in
the emergence of the four-dimensional view. While the spacetime diagram reflects
some of the concepts mentioned above, its utility as a cognitive tool exceeded by far
that of the sum of its parts. In Minkowski's hands, the spacetime diagram was more
than a tool, it was a model used to present both of his laws of gravitation. Beyond
their practical function in problem-solving, spacetime diagrams favored the diffusion
in wider circles of both the theory of relativity and the four-dimensional view of this
theory, in particular among non-mathematicians, by providing a visually intuitive
means of grasping certain consequences of the theory of relativity, such as time dilation and Lorentz contraction. Minkowski's graphic model of spacetime thus
enhanced both formal and intuitive approaches to special relativity.

In the third place, the ultimate success of the four-dimensional view hinged on the
empirical adequacy of the theory of relativity. It is remarkable that the conceptual
groundwork, and much of the formal elaboration of the four-dimensional view was
accomplished during a time when the theory of relativity was less well corroborated
by experiment than its rivals. The reversal of this situation in favor of relativity theory

in late 1908 favored the reception of the existing four-dimensional methods, and provided new impetus both for their application and extension, and for the development of alternatives, such as that of Sommerfeld.

The fourth major factor influencing the elaboration of a four-dimensional view of physics was the vector-symbolic movement in physics and mathematics at the turn of the twentieth century (McCormmach 1976, xxxi). The participants in this movement, in which Sommerfeld was a leading figure, believed in the efficacy of vector-symbolic methods in physics and geometry, and sought to unify the plethora of notations employed by various writers. The movement's strength varied from country to country; it was largely ignored in France, for example, in favor of the coordinate-based notation favored by Poincaré and others. Poincaré's pronounced disinterest in the application and development of a four-dimensional calculus for physics was typical of contemporary French attitudes toward vector-symbolic methods. In Germany, on the other hand, electrodynamicists learned Maxwell's theory from the mid-1890s in terms of curl $\mathfrak{h}$ and div $\mathfrak{E}$. In Zürich and Göttingen during this period, Minkowski instructed students (including Einstein) in the ways of the vector calculus. Unlike Poincaré, Minkowski was convinced that a four-dimensional language for physics would be worth the effort spent on its elaboration, yet he ultimately abandoned the vector-symbolic model in favor of an elegant and sophisticated matrix calculus. This choice was deplored by physicists (including Einstein), and mooted by Sommerfeld's conservative extension of the standard vector formalism into an immediately successful 4-vector algebra and analysis. In sum, the vector-symbolic movement functioned alternatively as an accelerator of the elaboration of four-dimensional calculi (existing systems served as templates), and as a regulator (penalizing Minkowski's neglect of standard vector operations).

The fifth and final parameter affecting the emergence of the four-dimensional view of physics was problem-solving performance. From the standpoint of ease of calculation, any four-dimensional vector formalism at all compared well to a Cartesian-coordinate approach, as Weinstein's textbook demonstrated; the advantage of ordinary vector methods over Cartesian coordinates was less pronounced. As we have seen, Poincaré applied his approach to the problem of constructing a Lorentz-invariant law of gravitational attraction, and was followed in turn by Minkowski and Sommerfeld, both of whom also provided examples of problem-solving. In virtue of the clarity and order of Sommerfeld's detailed, coordinate-free comparison of the laws of gravitation of Poincaré and Minkowski, his 4-vector algebra appeared to be the superior four-dimensional approach, just when physicists and mathematicians were turning to relativity in greater numbers.

ACKNOWLEDGMENTS

This study was inaugurated with the encouragement and support of Jürgen Renn, during a stay at the Max Planck Institute for the History of Science (Berlin) in 1998. I am especially grateful to Urs Schoepflin and the Institute's library staff for their expert assistance. The themes explored here were presented at the Mathematisches Forschunginstitut Oberwolfach (January 2000), the University of Heidelberg (June 2000), the University of Paris 7–REHSEIS (April, 2001), the joint AMS/SMF meeting in Lyons (July, 2001), and the Eighth International Conference on the History of General Relativity (Amsterdam, June 2002). Several scholars shared their ideas with me, shaping the final form of the paper. In particular, I am indebted to Olivier Darrigol for insightful comments on an early draft. The paper has been improved thanks to a careful reading by Shaul Katzir, and discussions with John Norton and Philippe Lombard. The responsibility for any remaining infelicities is my own.

Table 1: Lorentz-Invariant Laws of Gravitation, 1906–1910

Poincaré (1906)[a]	Minkowski (1908)[b]	Sommerfeld (1910)[c]
$X_1 = \dfrac{x}{k_0 B^3} - \xi_1 \dfrac{k_1}{k_0} \dfrac{A}{B^3 C}$		
$Y_1 = \dfrac{y}{k_0 B^3} - \eta_1 \dfrac{k_1}{k_0} \dfrac{A}{B^3 C}$	$mm^* \left(\dfrac{OA'}{B^* D^*} \right)^3 BD^*$	$mm_0 c^3 \dfrac{(\mathfrak{B}_0 \mathfrak{B}) \mathfrak{R} - (\mathfrak{B}_0 \mathfrak{R}) \mathfrak{B}}{(\mathfrak{R} \mathfrak{B})^3 (\mathfrak{B}_0 \mathfrak{B})}$
$Z_1 = \dfrac{z}{k_0 B^3} - \zeta_1 \dfrac{k_1}{k_0} \dfrac{A}{B^3 C}$		
$T_1 = -\dfrac{r}{k_0 B^3} - \dfrac{k_1}{k_0} \dfrac{A}{B^3 C}$		
$X_1 = \dfrac{\lambda}{B^3} - \dfrac{\eta \nu' - \zeta \mu'}{B^3}$		
$Y_1 = \dfrac{\mu}{B^3} - \dfrac{\zeta \lambda' - \xi \nu'}{B^3}$	$mm_1 \left(\dot{t}_1 - \dfrac{\dot{x}_1}{c} \right) \mathfrak{R}$	$-mm_0 c \dfrac{(\mathfrak{B}_0 \mathfrak{B}) \mathfrak{R} - (\mathfrak{B}_0 \mathfrak{R}) \mathfrak{B}}{(\mathfrak{R} \mathfrak{B})^3}$
$Z_1 = \dfrac{\nu}{B^3} - \dfrac{\xi \mu' - \eta \lambda'}{B^3}$		

[a] Mass terms are neglected, such that the right-hand side of each equation is implicitly multiplied by the product of the two masses. When both sides of the four equations are multiplied by the factor k_0 they express components of a 4-vector, $k_0(X_1, Y_1, Z_1, iT_1)$. The constants k_0 and k_1 are defined as: $k_0 = 1/\sqrt{1 - \Sigma \xi^2}$ and $k_1 = 1/\sqrt{1 - \Sigma \xi_1^2}$. A, B, and C denote the last three Lorentz-invariants in (1): $A = \dfrac{t - \Sigma x \xi}{\sqrt{1 - \Sigma \xi^2}}$, $B = \dfrac{t - \Sigma x \xi_1}{\sqrt{1 - \Sigma \xi_1^2}}$, $C = \dfrac{1 - \Sigma \xi \xi_1}{\sqrt{(1 - \Sigma \xi^2)(1 - \Sigma \xi_1^2)}}$, where $\Sigma \xi$ and $\Sigma \xi_1$ designate the

ordinary velocities of the passive and active mass points, with components ξ, η, ζ, and ξ_1, η_1, ζ_1. The time t is set equal to the negative distance between the passive mass point and the retarded position of the active mass point, $t = -\sqrt{\Sigma x^2} = -r$. Poincaré's second law is shown in the bottom row; he neglected to write the fourth component T_1, determined from the first three by the orthogonality condition $T_1 = \Sigma X_1 \xi$. The new variables in the bottom row are:

$$\lambda = k_1(x + r\xi_1), \qquad \mu = k_1(y + r\eta_1), \qquad \nu = k_1(z + r\zeta_1),$$
$$\lambda' = k_1(\eta_1 z - \zeta_1 y), \qquad \mu' = k_1(\zeta_1 x - \xi_1 z), \qquad \nu' = k_1(\xi_1 y - x\eta_1).$$

[b] The formula in the top row describes the first three components of the driving force: the fourth component is obtained analytically. The constants m and m^* designate the passive and active proper mass, respectively, while the remaining letters stand for spacetime points, as reconstructed in Fig. 1 (p. 227). The formula in the bottom row represents the driving force of gravitation as described, but not formally expressed, in (Minkowski 1909). The constants m and m_1 designate the passive and active proper mass, $\dot{t}_1$ and $\dot{x}_1$ are 4-velocity components of the passive mass, c is the speed of light and $\mathfrak{K}$ is a 4-vector, for the definition of which see note 111.

[c] The constants m_0 and m designate the passive and active proper mass, respectively, c denotes the speed of light, $\mathfrak{B}_0$ and $\mathfrak{B}$ represent the corresponding 4-velocities, and $\mathfrak{R}$ stands for the light-like interval between the mass points.

REFERENCES

Abraham, Max. 1905. *Elektromagnetische Theorie der Strahlung*. (*Theorie der Elektrizität*, vol. 2.) Leipzig: Teubner.

———. 1909. "Zur elektromagnetischen Mechanik." *Physikalische Zeitschrift* 10: 737–741.

———. 1910. "Sull' elettrodinamica di Minkowski." *Rendiconti del Circolo Matematico di Palermo* 30: 33–46.

———. 1912a. "Das Elementargesetz der Gravitation." *Physikalische Zeitschrift* 13: 4–5.

———. 1912b. "Zur Theorie der Gravitation." *Physikalische Zeitschrift* 13: 1–4. (English translation in this volume.)

———. 1914. "Die neue Mechanik." *Scientia* (*Rivista di Scienza*) 15: 8–27.

Andrade Martins, Roberto de. 1999. "The search for gravitational absorption in the early 20th century." In (Goenner et al. 1999), 3–44.

Arzeliès, Henri, and J. Henry. 1959. *Milieux conducteurs ou polarisables en mouvement*. Paris: Gauthier-Villars.

Barrow-Green, June E. 1997. *Poincaré and the Three Body Problem*. (*History of Mathematics*, vol. 11.) Providence: AMS and LMS.

Bork, Alfred M. 1966. "'Vectors versus quaternions' — the letters in *Nature*." *American Journal of Physics* 34: 202–211.

Born, Max. 1909. "Die Theorie des starren Elektrons in der Kinematik des Relativitätsprinzips." *Annalen der Physik* 30: 1–56.

———. 1912. "Besprechung von Max Laue, Das Relativitätsprinzip." *Physikalische Zeitschrift* 13: 175–176.

———. 1914. "Besprechung von Max Weinstein, Die Physik der bewegten Materie und die Relativitätstheorie." *Physikalische Zeitschrift* 15: 676.

Buchwald, Jed Z. 1985. *From Maxwell to Microphysics*. Chicago: University of Chicago Press.

Buchwald, Jed Z., and Andrew Warwick. 2001. *Histories of the Electron: The Birth of Microphysics*. Dibner Institute Studies in the History of Science and Technology. Cambridge MA: MIT Press.

Burali-Forti, Cesare, and Roberto Marcolongo. 1910. *Éléments de calcul vectoriel*. Paris: Hermann.

Cantor, Geoffrey N., and Michael J. S. Hodge. 1981. *Conceptions of Ether: Studies in the History of Ether Theories 1740–1900*. Cambridge: Cambridge University Press.

Carazza, Bruno, and Helge Kragh. 1990. "Augusto Righi's magnetic rays: a failed research program in early 20th-century physics." *Historical Studies in the Physical and Biological Sciences* 21: 1–28.

Cayley, Arthur. 1869. "On the six coordinates of a line." *Transactions of the Cambridge Philosophical Society* 11: 290–323.

Conway, Arthur W. 1911. "On the application of quaternions to some recent developments of electrical theory." *Proceedings of the Royal Irish Academy* 29: 1–9.

Corry, Leo. 1997. "Hermann Minkowski and the postulate of relativity." *Archive for History of Exact Sciences* 51: 273–314.

———. 2004. *David Hilbert and the Axiomatization of Physics, 1898–1918: From "Grundlagen der Geometrie" to "Grundlagen der Physik".* Dordrecht: Kluwer.

Coster, H. G. L. and J. R. Shepanski. 1969. "Gravito-inertial fields and relativity." *Journal of Physics A* 2: 22–27.

CPAE 2. 1989. John Stachel, David C. Cassidy, Jürgen Renn, and Robert Schulmann (eds.), *The Collected Papers of Albert Einstein.* Vol. 2. *The Swiss Years: Writings, 1900–1909.* Princeton: Princeton University Press.

CPAE 5. 1993. Martin J. Klein, A. J. Kox, and Robert Schulmann (eds.), *The Collected Papers of Albert Einstein.* Vol. 5. *The Swiss Years: Correspondence, 1902–1914.* Princeton: Princeton University Press.

Cunningham, Ebenezer. 1914. *The Principle of Relativity.* Cambridge: Cambridge University Press.

Cuvaj, Camillo. 1968. "Henri Poincaré's mathematical contributions to relativity and the Poincaré stresses." *American Journal of Physics* 36: 1102–1113.

———. 1970. *A History of Relativity: The Role of Henri Poincaré and Paul Langevin.* Ph.D. dissertation, Yeshiva University.

Darrigol, Olivier. 1993. "The electrodynamic revolution in Germany as documented by early German expositions of 'Maxwell's theory'." *Archive for History of Exact Sciences* 45: 189–280.

———. 1994. "The electron theories of Larmor and Lorentz: a comparative study." *Historical Studies in the Physical and Biological Sciences* 24: 265–336.

———. 2000. *Electrodynamics from Ampère to Einstein.* Oxford: Oxford University Press.

Dugac, Pierre. 1986. "La correspondance d'Henri Poincaré avec des mathématiciens de A à H." *Cahiers du séminaire d'histoire des mathématiques* 7: 59–219.

———. 1989. "La correspondance d'Henri Poincaré avec des mathématiciens de J à Z." *Cahiers du séminaire d'histoire des mathématiques* 10: 83–229.

Dziobek, Otto F. 1888. *Die mathematischen Theorien der Planeten-Bewegungen.* Leipzig: Barth.

Einstein, Albert. 1905. "Zur Elektrodynamik bewegter Körper." *Annalen der Physik* 17: 891–921, (CPAE 2, Doc. 23).

———. 1907. "Relativitätsprinzip und die aus demselben gezogenen Folgerungen." *Jahrbuch der Radioaktivität und Elektronik* 4: 411–462, (CPAE 2, Doc. 47).

Einstein, Albert, and Jakob J. Laub. 1908. "Über die elektromagnetischen Grundgleichungen für bewegte Körper." *Annalen der Physik* 26: 532–540, (CPAE 2, Doc. 51).

Frank, Philipp. 1911. "Das Verhalten der elektromagnetischen Feldgleichungen gegenüber linearen Transformationen." *Annalen der Physik* 35: 599–607.

Föppl, August O. 1894. *Einführung in die Maxwell'sche Theorie der Elektricität.* Leipzig: Teubner.

Galison, Peter. 1979. "Minkowski's spacetime: from visual thinking to the absolute world." *Historical Studies in the Physical Sciences* 10: 85–121.

Gans, Richard. 1905. "Gravitation und Elektromagnetismus." *Jahresbericht der deutschen Mathematiker-Vereinigung* 14: 578–581.

Gibbs, Josiah W. and Edwin B. Wilson. 1901. *Vector Analysis.* New York: Charles Scribner's Sons.

Gispert, Hélène. 2001. "The German and French editions of the Klein-Molk Encyclopedia: contrasted images." In Umberto Bottazzini and Amy Dahan Dalmedico (eds.), *Changing Images in Mathematics: From the French Revolution to the New Millennium,* 93–112. *Studies in the History of Science, Technology and Medicine* 13. London: Routledge.

Goenner, Hubert, Jürgen Renn, Tilman Sauer, and Jim Ritter (eds.). 1999. *The Expanding Worlds of General Relativity.* (*Einstein Studies* vol. 7.) Boston/Basel/Berlin: Birkhäuser.

Gray, Jeremy. 1992. "Poincaré and the solar system." In Peter M. Harman and Alan E. Shapiro (eds.), *The Investigation of Difficult Things,* 503–524. Cambridge: Cambridge University Press.

Hargreaves, Richard. 1908. "Integral forms and their connexion with physical equations." *Transactions of the Cambridge Philosophical Society* 21: 107–122.

Harvey, A. L. 1965. "A brief review of Lorentz-covariant theories of gravitation." *American Journal of Physics* 33: 449–460.

Havas, Peter 1979. "Equations of motion and radiation reaction in the special and general theory of relativity." In Jürgen Ehlers (ed.), *Isolated Gravitating Systems in General Relativity,* 74–155. Proceedings of the International School of Physics "Enrico Fermi" 67. Amsterdam: North-Holland.

Heaviside, Oliver. 1893. "A gravitational and electromagnetic analogy." In *Electromagnetic Theory,* 3 vols., 1: 455–464. London: The Electrician.

Hon, Giora. 1995. "The case of Kaufmann's experiment and its varied reception." In Jed Z. Buchwald (ed.), *Scientific Practice: Theories and Stories of Doing Physics*, 170–223. Chicago: University of Chicago Press.

Ishiwara, Jun. 1912. "Bericht über die Relativitätstheorie." *Jahrbuch der Radioaktivität und Elektronik* 9: 560–648.

Jackson, J. David. 1975. *Classical Electrodynamics*. New York: Wiley, 2nd edition.

Jungnickel, Christa and Russell McCormmach. 1986. *Intellectual Mastery of Nature: Theoretical Physics from Ohm to Einstein*. Chicago: University of Chicago Press.

Klein, Felix. 1907. *Vorträge über den mathematischen Unterricht an den höheren Schulen*, Rudolf Schimmack (ed.). (*Mathematische Vorlesungen an der Universität Göttingen*, vol. 1.) Leipzig: Teubner.

Klein, Felix, and Arnold Sommerfeld. 1897–1910. *Über die Theorie des Kreisels*. 4 vols. Leipzig: Teubner.

Kottler, Felix. 1922. "Gravitation und Relativitätstheorie." In Karl Schwarzschild, Samuel Oppenheim, and Walther von Dyck (eds.), *Astronomie*, 2 vols, 2: 159–237. *Encyklopädie der mathematischen Wissenschaften mit Einschluss ihrer Anwendungen* 6. Leipzig: Teubner.

Kretschmann, Erich. 1914. *Eine Theorie der Schwerkraft im Rahmen der ursprünglichen Einsteinschen Relativitätstheorie*. Ph.D. dissertation, University of Berlin.

Krätzel, Ekkehard. 1989. "Kommentierender Anhang: zur Geometrie der Zahlen." In Ekkehard Krätzel and Bernulf Weissbach (eds.), *Ausgewählte Arbeiten zur Zahlentheorie und zur Geometrie*, 233–246. *Teubner-Archiv zur Mathematik* 12. Leipzig: Teubner.

Kuhn, Thomas S. 1970. *The Structure of Scientific Revolutions*. Chicago: University of Chicago Press, 2nd edition.

Laue, Max von. 1911. *Das Relativitätsprinzip*. (*Die Wissenschaft*, vol. 38.) Braunschweig: Vieweg.

———. 1913. *Das Relativitätsprinzip*. (*Die Wissenschaft*, vol. 38.) Braunschweig: Vieweg, 2nd edition.

———. 1951. "Sommerfelds Lebenswerk." *Naturwissenschaften* 38: 513–518.

Lewis, Gilbert N. 1910a. "On four-dimensional vector analysis, and its application in electrical theory." *Proceedings of the American Academy of Arts and Science* 46: 165–181.

———. 1910b. "Über vierdimensionale Vektoranalysis und deren Anwendung auf die Elektrizitätstheorie." *Jahrbuch der Radioaktivität und Elektronik* 7: 329–347.

Liu, Chuang. 1991. *Relativistic Thermodynamics: Its History and Foundations*. Ph.D. dissertation, University of Pittsburgh.

Lorentz, Hendrik A. 1900. "Considerations on Gravitation." *Proceedings of the Section of Sciences. Verslag Koninklijke Akademie van Wetenschapen* 2: 559–574. (Printed in this volume.)

———. 1904a. "Electromagnetic phenomena in a system moving with any velocity less than that of light." *Proceedings of the Section of Sciences. Verslag Koninklijke Akademie van Wetenschapen* 6: 809–831.

———. 1904b. "Maxwells elektromagnetische Theorie." In (Sommerfeld 1903–1926), 2: 63–144.

———. 1904c. "Weiterbildung der Maxwellschen Theorie; Elektronentheorie." In (Sommerfeld 1903–1926), 2: 145–280.

———. 1910. "Alte und neue Fragen der Physik." *Physikalische Zeitschrift* 11: 1234–1257. (English translation in this volume.)

———. 1914. "La gravitation." *Scientia (Rivista di Scienza)* 16: 28–59.

Lützen, Jesper. 1999. "Geometrising configurations: Heinrich Hertz and his mathematical precursors." In Jeremy Gray (ed.), *The Symbolic Universe: Geometry and Physics, 1890–1930*, 25–46. Oxford: Oxford University Press.

Maltese, Giulio. 2000. "The late entrance of relativity into Italian scientific community (1906–1930)." *Historical Studies in the Physical and Biological Sciences* 31: 125–173.

Manegold, Karl-Heinz. 1970. *Universität, Technische Hochschule und Industrie*. Berlin: Duncker & Humblot.

Marcolongo, Roberto. 1906. "Sugli integrali delle equazioni dell'elettrodinamica." *Rendiconti della Reale Accademia dei Lincei* 15: 344–349.

McCormmach, Russell. 1976. "Editor's foreword." *Historical Studies in the Physical Sciences* 7: xi–xxxv.

Miller, Arthur I. 1973. "A Study of Henri Poincaré's 'Sur la dynamique de l'électron'." *Archive for History of Exact Sciences* 10: 207–328.

———. 1981. *Albert Einstein's Special Theory of Relativity: Emergence (1905) and Early Interpretation*. Reading, MA: Addison-Wesley.

Minkowski, Hermann. 1888. "Über die Bewegung eines festen Körpers in einer Flüssigkeit." *Sitzungsberichte der königliche preußischen Akademie der Wissenschaften* 40: 1095–1110.

———. 1890–1893. "H. Poincaré, Sur le problème des trois corps et les équations de la dynamique." *Jahrbuch über die Fortschritte der Mathematik* 22: 907–914.

———. 1896. *Geometrie der Zahlen*. Leipzig: Teubner.

———. 1907. "Kapillarität." In (Sommerfeld 1903–1926), 1: 558–613.

————. 1908. "Die Grundgleichungen für die electromagnetischen Vorgänge in bewegten Körpern." *Nachrichten von der Königlichen Gesellschaft der Wissenschaften zu Göttingen*, 53–111. (English translation of the appendix "Mechanics and the Relativity Postulate" in this volume.)

————. 1909. "Raum und Zeit." *Jahresbericht der deutschen Mathematiker-Vereinigung* 18: 75–88.

————. 1915. "Das Relativitätsprinzip." *Jahresbericht der deutschen Mathematiker-Vereinigung* 24: 372–382.

————. 1973. *Briefe an David Hilbert*. Lily Rüdenberg and Hans Zassenhaus (eds.). Berlin: Springer-Verlag.

Minkowski, Hermann, and Max Born. 1910. "Eine Ableitung der Grundgleichungen für die elektromagnetischen Vorgänge in bewegten Körpern." *Mathematische Annalen* 68: 526–550.

Møller, Christian. 1972. *The Theory of Relativity*. Oxford: Oxford University Press, 2nd edition.

North, John D. 1965. *The Measure of the Universe: A History of Modern Cosmology*. Oxford: Oxford University Press.

Norton, John D. 1992. "Einstein, Nordström and the early demise of Lorentz-covariant theories of gravitation." *Archive for History of Exact Sciences* 45: 17–94.

Pauli, Wolfgang. 1921. "Relativitätstheorie." In (Sommerfeld 1903–1926), 2: 539–775.

Planck, Max. 1906. "Das Prinzip der Relativität und die Grundgleichungen der Mechanik." *Verhandlungen der Deutschen Physikalischen Gesellschaft* 8: 136–141.

————. 1907. "Zur Dynamik bewegter Systeme." *Sitzungsberichte der königliche preußischen Akad. der Wiss.*: 542–570.

Poincaré, Henri. 1885. "Sur l'équilibre d'une masse fluide animée d'un mouvement de rotation." *Acta mathematica* 7: 259–380.

————. 1895. *Capillarité*. J. Blondin (ed.). Paris: Georges Carré.

————. 1898–1905. "Préface." In Charles Hermite, Henri Poincaré, and Eugène Rouché (eds.), *Œuvres de Laguerre* 1: v–xv. Paris: Gauthier-Villars.

————. 1901. *Électricité et optique: la lumière et les théories électrodynamiques*. Jules Blondin and Eugène Néculcéa (eds.). Paris: Carré et Naud.

————. 1902a. *Figures d'équilibre d'une masse fluide*. Léon Dreyfus (ed.). Paris: C. Naud.

————. 1902b. "Sur la stabilité de l'équilibre des figures piriformes affectées par une masse fluide en rotation." *Philosophical Transactions of the Royal Society* A 198: 333–373.

————. 1904. "L'état actuel et l'avenir de la physique mathématique." *Bulletin des sciences mathématiques* 28: 302–324.

————. 1906. "Sur la dynamique de l'électron." *Rendiconti del Circolo Matematico di Palermo* 21: 129–176. (English translation of excerpt in this volume.)

————. 1907. "La relativité de l'espace". *Année psychologique* 13: 1–17.

————. 1908. "La dynamique de l'électron." *Revue générale des sciences pures et appliquées* 19: 386–402.

————. 1909. "La mécanique nouvelle." *Revue scientifique* 12: 170–177.

————. 1910. *Sechs Vorträge über ausgewählte Gegenstände aus der reinen Mathematik und mathematischen Physik*. (*Mathematische Vorlesungen an der Universität Göttingen*, vol. 4.) Leipzig/Berlin: Teubner.

————. 1912. "L'espace et le temps." *Scientia (Rivista di Scienza)* 12: 159–170.

————. 1953. "Les limites de la loi de Newton." *Bulletin astronomique* 17: 121–269.

Pomey, Jean-Baptiste. 1914–1931. *Cours d'électricité théorique*, 3 vols. Bibliothèque des Annales des Postes, Télégraphes et Téléphones. Paris: Gauthier-Villars.

Pyenson, Lewis. 1973. *The Goettingen Reception of Einstein's General Theory of Relativity*. Ph.D. dissertation, Johns Hopkins University.

————. 1985. *The Young Einstein: The Advent of Relativity*. Bristol: Hilger.

Reich, Karin. 1994. *Die Entwicklung des Tensorkalküls: vom absoluten Differentialkalkül zur Relativitätstheorie*. (*Science Networks Historical Studies*, vol. 11.) Basel/Boston: Birkhäuser.

————. 1996. "The emergence of vector calculus in physics: the early decades." In Gert Schubring (ed.), *Hermann Günther Graßmann (1809–1877): Visionary Mathematician, Scientist and Neohumanist Scholar*, 197–210. Dordrecht: Kluwer.

Reiff, Richard. and Arnold Sommerfeld. 1904. "Standpunkt der Fernwirkung: Die Elementargesetze." In (Sommerfeld 1903–1926), 2: 3–62.

Ritz, Walter. 1908. "Recherches critiques sur l'électrodynamique générale." *Annales de chimie et de physique* 13: 145–275.

Roseveare, N. T. 1982. *Mercury's Perihelion: From Le Verrier to Einstein*. Oxford: Oxford University Press.

Rowe, David E. 1989. "Klein, Hilbert, and the Göttingen Mathematical Tradition." In Katherina M. Olesko (ed.), *Science in Germany*, 186–213. Osiris 5. Philadelphia: History of Science Society.

————. 1992. *Felix Klein, David Hilbert, and the Göttingen Mathematical Tradition*. Ph.D. dissertation, City University of New York.

————. 1997. "In Search of Steiner's ghosts: imaginary elements in 19th-century geometry." In Dominique Flament (ed.), *Le nombre, un hydre à n visages: entre nombres complexes et vecteurs*, 193–208. Paris: Éditions de la Maison des Sciences de l'Homme.

Rüdenberg, Lily. 1973. "Einleitung: Erinnerungen an H. Minkowski." In (Minkowski 1973), 9–16.

Schwartz, Hermann M. 1972. "Poincaré's Rendiconti paper on relativity, III." *American Journal of Physics* 40: 1282–1287.

Schwermer, Joachim. 1991. "Räumliche Anschauung und Minima positiv definiter quadratischer Formen." *Jahresbericht der deutschen Mathematiker-Vereinigung* 93: 49–105.

Serre, Jean-Pierre. 1993. "Smith, Minkowski et l'Académie des Sciences." *Gazette des mathématiciens* 56: 3–9.

Silberstein, Ludwik. 1914. *The Theory of Relativity*. London: Macmillan.

de Sitter, Willem. 1911. "On the bearing of the principle of relativity on gravitational astronomy." *Monthly Notices of the Royal Astronomical Society* 71: 388–415.

Sommerfeld, Arnold. (ed). 1903–1926. *Physik*, 3 vols. (*Encyklopädie der mathematischen Wissenschaften mit Einschluss ihrer Anwendungen*, vol. 5.) Leipzig: Teubner.

Sommerfeld, Arnold. 1904. "Bezeichnung und Benennung der elektromagnetischen Grössen in der Enzyklopädie der mathematischen Wissenschaften V." *Physikalische Zeitschrift* 5: 467–470.

————. 1910a. "Zur Relativitätstheorie, I: Vierdimensionale Vektoralgebra." *Annalen der Physik* 32: 749–776.

————. 1910b. "Zur Relativitätstheorie, II: Vierdimensionale Vektoranalysis." *Annalen der Physik* 33: 649–689.

————. 1913. "Anmerkungen zu Minkowski, Raum und Zeit." In Otto Blumenthal (ed.), *Das Relativitätsprinzip; Eine Sammlung von Abhandlungen*, 69–73. (*Fortschritte der mathematischen Wissenschaften in Monographien*, vol. 2.) Leipzig: Teubner.

————. 2001–2004. *Wissenschaftlicher Briefwechsel*. Michael Eckert and Karl Märker (eds.). Diepholz: GNT-Verlag.

Staley, Richard. 1998. "On the histories of relativity: propagation and elaboration of relativity in participant histories in Germany, 1905–1911." *Isis* 89: 263–299.

Stein, Howard. 1987. "After the Baltimore Lectures: some philosophical reflections on the subsequent development of physics." In Robert Kargon and Peter Achinstein (eds.), *Kelvin's Baltimore Lectures and Modern Theoretical Physics: Historical and Philosophical Perspectives*, 375–398. Cambridge, MA: MIT Press.

Strobl, Walter. 1985. "Aus den wissenschaftlichen Anfängen Hermann Minkowskis." *Historia Mathematica* 12: 142–156.

Tait, Peter G. 1882–1884. *Traité élémentaire des quaternions*. Paris: Gauthier-Villars.

Tisserand, Francois-Félix. 1889–1896. *Traité de mécanique céleste*. Paris: Gauthier-Villars.

Torretti, Roberto. 1996. *Relativity and Geometry*. New York: Dover Publications, 2nd edition.

Varičak, Vladimir. 1910. "Anwendung der Lobatschefskijschen Geometrie in der Relativtheorie." *Physikalische Zeitschrift* 11: 93–96.

Vizgin, Vladimir P. 1994. *Unified Field Theories in the First Third of the 20th Century*. (*Science Networks Historical Studies*, vol. 3.) Basel: Birkhäuser.

Wacker, Fritz. 1906. "Über Gravitation und Elektromagnetismus." *Physikalische Zeitschrift* 7: 300–302.

Walter, Scott. 1997. "La vérité en géométrie: sur le rejet mathématique de la doctrine conventionnaliste." *Philosophia Scientiæ* 2: 103–135.

————. 1999a. "Minkowski, Mathematicians, and the Mathematical Theory of Relativity." In (Goenner et al. 1999), 45–86.

————. 1999b. "The Non-Euclidean Style of Minkowskian Relativity." In J. Gray (ed.), *The Symbolic Universe: Geometry and Physics, 1890–1930*, 91–127. Oxford: Oxford University Press.

Weinstein, Max B. 1913. *Die Physik der bewegten Materie und die Relativitätstheorie*. Leipzig: Barth.

————. 1914. *Kräfte und Spannungen: Das Gravitations- und Strahlenfeld*. (*Sammlung Vieweg*, vol. 8.) Braunschweig: Vieweg.

Whitrow, Gerald J., and George E. Morduch. 1965. "Relativistic theories of gravitation: a comparative analysis with particular reference to astronomical tests." *Vistas in Astronomy* 6: 1–67.

Whittaker, Edmund T. 1951–1953. *A History of the Theories of Aether and Electricity*. London: T. Nelson.

de Wisniewski, Felix J. 1913a. "Zur Minkowskischen Mechanik I." *Annalen der Physik* 40: 387–390.

————. 1913b. "Zur Minkowskischen Mechanik II." *Annalen der Physik* 40: 668–676.

Yaghjian, Arthur D. 1992. *Relativistic Dynamics of a Charged Sphere: Updating the Lorentz-Abraham Model*. (*Lecture Notes in Physics: Monographs*, vol. 11.) Berlin: Springer.

Zahar, Elie. 1989. *Einstein's Revolution: A Study in Heuristic*. La Salle, Ill.: Open Court.

Zassenhaus, Hans J. 1975. "On the Minkowski-Hilbert dialogue on mathematization." *Canadian Mathematical Bulletin* 18: 443–461.
Zenneck, Jonathan. 1903. "Gravitation." In (Sommerfeld 1903–1926), 1: 25–67. (Printed in this volume.)

HENRI POINCARÉ

ON THE DYNAMICS OF THE ELECTRON
(EXCERPTS)

Originally published as "Sur la dynamique de l'électron" in Rendiconti del Circolo Matematico di Palermo 21 (1906), pp. 129–175. Author's date: Paris, July 1905. The Introduction, §1, and §9 are translated here.[1]

INTRODUCTION

It seems at first that the aberration of light and related optical and electrical phenomena will provide us with a means of determining the absolute motion of the Earth, or rather its motion with respect to the aether, as opposed to its motion with respect to other celestial bodies. Fresnel pursued this idea, but soon recognized that the Earth's motion does not alter the laws of refraction and reflection. Analogous experiments, like that of the water-filled telescope, and all those considering terms no higher than first order relative to the aberration, yielded only negative results; the explanation was soon discovered. But Michelson, who conceived an experiment sensitive to terms depending on the square of the aberration, failed in turn.

It appears that this impossibility to detect the absolute motion of the Earth by experiment may be a general law of nature; we are naturally inclined to admit this law, which we will call the *Postulate of Relativity* and admit without restriction. Whether or not this postulate, which up to now agrees with experiment, may later be corroborated or disproved by experiments of greater precision, it is interesting in any case to ascertain its consequences.

An explanation was proposed by Lorentz and FitzGerald, who introduced the hypothesis of a contraction of all bodies in the direction of the Earth's motion and proportional to the square of the aberration. This contraction, which we will call the *Lorentzian contraction*, would explain Michelson's experiment and all others performed up to now. The hypothesis would become insufficient, however, if we were to admit the postulate of relativity in full generality.

Lorentz then sought to extend his hypothesis and to modify it in order to obtain perfect agreement with this postulate. This is what he succeeded in doing in his article entitled *Electromagnetic phenomena in a system moving with any velocity smaller than that of light (Proceedings of the Amsterdam Academy,* 27 May, 1904).

The importance of the question persuaded me to take it up in turn; the results I | obtained agree with those of Mr. Lorentz on all the significant points. I was led [130]

Jürgen Renn (ed.). *The Genesis of General Relativity,* Vol. 3
Gravitation in the Twilight of Classical Physics: Between Mechanics, Field Theory, and Astronomy.
© 2007 Springer.

merely to modify and extend them only in a few details; further on we will see the points of divergence, which are of secondary importance.

Lorentz's idea may be summed up like this: if we are able to impress a translation upon an entire system without modifying any observable phenomena, it is because the equations of an electromagnetic medium are unaltered by certain transformations, which we will call *Lorentz transformations*. Two systems, one of which is at rest, the other in translation, become thereby exact images of each other.

Langevin[*]) sought to modify Lorentz's idea; for both authors, the moving electron takes the form of a flattened ellipsoid. For Lorentz, two axes of the ellipsoid remain constant, while for Langevin, ellipsoid volume remains constant. The two scientists also showed that these two hypotheses are corroborated by Kaufmann's experiments to the same extent as the original hypothesis of Abraham (rigid-sphere electron).

The advantage of Langevin's theory is that it requires only electromagnetic forces, and bonds; it is, however, incompatible with the postulate of relativity. This is what Lorentz showed, and this is what I found in turn using a different method, which calls on principles of group theory.

We must return therefore to Lorentz's theory, but if we want to do this and avoid intolerable contradictions, we must posit the existence of a special force that explains both the contraction, and the constancy of two of the axes. I sought to determine this force, and found that *it may be assimilated to a constant external pressure on the deformable and compressible electron, whose work is proportional to the electron's change in volume.*

If the inertia of matter is exclusively of electromagnetic origin, as generally admitted in the wake of Kaufmann's experiment, and all forces are of electromagnetic origin (apart from this constant pressure that I just mentioned), the postulate of relativity may be established with perfect rigor. This is what I show by a very simple calculation based on the principle of least action.

But that is not all. In the article cited above, Lorentz judged it necessary to extend his hypothesis in such a way that the postulate remains valid in case there are forces of non-electromagnetic origin. According to Lorentz, all forces are affected by the Lorentz transformation (and consequently by a translation) in the same way as electromagnetic forces.

It was important to examine this hypothesis closely, and in particular to ascertain the modifications we would have to apply to the laws of gravitation.

[131]   We find first of all that it requires us to assume that gravitational propagation | is not instantaneous, but occurs with the speed of light. One might think that this is reason enough to reject the hypothesis, since Laplace demonstrated that this cannot be the case. In reality, however, the effect of this propagation is compensated in large

---

*   Langevin was anticipated by Mr. Bucherer of Bonn, who earlier advanced the same idea. (See: Bucherer, *Mathematische Einführung in die Elektronentheorie*, August, 1904. Teubner, Leipzig). [Poincaré's footnote.]

part by a different cause, in such a way that no contradiction arises between the proposed law and astronomical observations.

Is it possible to find a law satisfying Lorentz's condition, and reducing to Newton's law whenever the speeds of celestial bodies are small enough to allow us to neglect their squares (as well as the product of acceleration and distance) with respect to the square of the speed of light?

To this question we must respond in the affirmative, as we will see later.

Modified in this way, is the law compatible with astronomical observations?

It seems so on first sight, but the question will be settled only after an extended discussion.

Suppose, then, that this discussion is settled in favor of the new hypothesis, what should we conclude? If propagation of attraction occurs with the speed of light, it could not be a fortuitous accident. Rather, it must be because it is a function of the aether, and then we would have to try to penetrate the nature of this function, and to relate it to other fluid functions.

We cannot be content with a simple juxtaposition of formulas that agree with each other by good fortune alone; these formulas must, in a manner of speaking, interpenetrate. The mind will be satisfied only when it believes it has perceived the reason for this agreement, and the belief is strong enough to entertain the illusion that it could have been predicted.

But the question may be viewed from a different perspective, better shown via an analogy. Let us imagine a pre-Copernican astronomer who reflects on Ptolemy's system; he will notice that for all the planets, one of two circles—epicycle or deferent—is traversed in the same time. This fact cannot be due to chance, and consequently between all the planets there is a mysterious link we can only guess at.

Copernicus, however, destroys this apparent link by a simple change in the coordinate axes that were considered fixed. Each planet now describes a single circle, and orbital periods become independent (until Kepler reestablishes the link that was believed to have been destroyed).

It is possible that something analogous is taking place here. If we were to admit the postulate of relativity, we would find the same number in the law of gravitation and the laws of electromagnetism—the speed of light—and we would find it again in all other forces of any origin whatsoever. This state of affairs may be explained in one of two ways: either everything in the universe would be of electromagnetic origin, or this aspect—shared, as it were, by all physical phenomena—would be a mere epiphenomenon, something due to our methods of | measurement. How do we go about   [132] measuring? The first response will be: we transport objects considered to be invariable solids, one on top of the other. But that is no longer true in the current theory if we admit the Lorentzian contraction. In this theory, two lengths are equal, by definition, if they are traversed by light in equal times.

Perhaps if we were to abandon this definition Lorentz's theory would be as fully overthrown as was Ptolemy's system by Copernicus's intervention. Should that hap-

pen some day, it would not prove that Lorentz's efforts were in vain, because regardless of what one may think, Ptolemy was useful to Copernicus.

I, too, have not hesitated to publish these few partial results, even if at this very
moment the discovery of magneto-cathode rays seems to threaten the entire theory.

1. LORENTZ TRANSFORMATION

Lorentz adopted a certain system of units in order to do away with 4π factors in formulas. I will do the same, and in addition, select units of length and time in such a
way that the speed of light equals 1. Under these conditions, and denoting electric
displacement f, g, h, magnetic intensity α, β, γ, vector potential F, G, H, scalar
potential ψ, charge density ρ, electron velocity ξ, η, ζ, and current u, v, w, the fundamental formulas become:

$$\left.\begin{aligned}
u &= \frac{df}{dt} + \rho\xi = \frac{d\gamma}{dy} - \frac{d\beta}{dz}, \qquad \alpha = \frac{dH}{dy} - \frac{dG}{dz}, \qquad f = -\frac{dF}{dt} - \frac{d\psi}{dx}, \\
\frac{d\alpha}{dt} &= \frac{dg}{dz} - \frac{dh}{dy}, \qquad \frac{d\rho}{dt} + \sum\frac{d\rho\xi}{dx} = 0, \qquad \sum\frac{df}{dx} = \rho, \qquad \frac{d\psi}{dt} + \sum\frac{dF}{dx} = 0, \\
\Box &= \Delta - \frac{d^2}{dt^2} = \sum\frac{d^2}{dx^2} - \frac{d^2}{dt^2}, \qquad \Box\psi = -\rho, \qquad \Box F = -\rho\xi.
\end{aligned}\right\} \quad (1)$$

An elementary particle of matter of volume $dx\,dy\,dz$ is acted upon by a mechanical force, the components of which are derived from the formula:

$$X = \rho f + \rho(\eta\gamma - \zeta\beta). \tag{2}$$

These equations admit a remarkable transformation discovered by Lorentz, which
owes its interest to the fact that it explains why no experiment can inform us of the
absolute motion of the universe. Let us put:

$$x' = kl(x + \varepsilon t), \quad t' = kl(t + \varepsilon x), \quad y' = ly, \quad z' = lz, \tag{3}$$

where l and ε are two arbitrary constants, such that

$$k = \frac{1}{\sqrt{1 - \varepsilon^2}}.$$

[133]   Now if we put:

$$\Box' = \sum\frac{d^2}{dx'^2} - \frac{d^2}{dt'^2},$$

we will have:

$$\Box' = \Box l^{-2}.$$

Let a sphere be carried along with the electron in uniform translation, and let the
equation of this mobile sphere be:

$$(x - \xi t)^2 + (y - \eta t)^2 + (z - \zeta t)^2 = r^2,$$

and the volume of the sphere be $\frac{4}{3}\pi r^3$. [2]

The transformation will change the sphere into an ellipsoid, the equation of which is easy to find. We thus deduce easily from (3):

$$x = \frac{k}{l}(x' - \varepsilon t'), \quad t = \frac{k}{l}(t' - \varepsilon x'), \quad y = \frac{y'}{l}, \quad z = \frac{z'}{l}. \tag{3'}$$

The equation of the ellipsoid then becomes:

$$k^2(x' - \varepsilon t' - \xi t' + \varepsilon \xi x')^2 + (y' - \eta k t' + \eta k \varepsilon x')^2 + (z' - \zeta k t' + \zeta k \varepsilon x')^2 = l^2 r^2.$$

This ellipsoid is in uniform motion; for $t' = 0$, it reduces to

$$k^2 x'^2 (1 + \xi \varepsilon)^2 + (y' + \eta k \varepsilon x')^2 + (z' + \zeta k \varepsilon x')^2 = l^2 r^2.$$

and has a volume:

$$\frac{4}{3}\pi r^3 \frac{l^3}{k(1 + \xi \varepsilon)}.$$

If we want electron charge to be unaltered by the transformation, and if we designate the new charge density ρ' we will find:

$$\rho' = \frac{k}{l^3}(\rho + \varepsilon \rho \xi). \tag{4}$$

What will be the new velocity components ξ', η' and ζ'? We should have:

$$\xi' = \frac{dx'}{dt'} = \frac{d(x + \varepsilon t)}{d(t + \varepsilon x)} = \frac{\xi + \varepsilon}{1 + \varepsilon \xi},$$

$$\eta' = \frac{dy'}{dt'} = \frac{dy}{kd(t + \varepsilon x)} = \frac{\eta}{k(1 + \varepsilon \xi)}, \quad \zeta' = \frac{\zeta}{k(1 + \varepsilon \xi)},$$

whence:

$$\rho'\xi' = \frac{k}{l^3}(\rho\xi + \varepsilon\rho), \quad \rho'\eta' = \frac{1}{l^3}\rho\eta, \quad \rho'\zeta' = \frac{1}{l^3}\rho\zeta. \tag{4'}$$

Here is where I must point out for the first time a difference with Lorentz. In my notation, Lorentz put (loc. cit., page 813, formulas 7 and 8):

$$\rho' = \frac{1}{kl^3}\rho, \quad \xi' = k^2(\xi + \varepsilon), \quad \eta' = k\eta, \quad \zeta' = k\zeta. \text{ I}$$

In this way we recover the formulas: [134]

$$\rho'\xi' = \frac{k}{l^3}(\rho\xi + \varepsilon\rho), \quad \rho'\eta' = \frac{1}{l^3}\rho\eta, \quad \rho'\zeta' = \frac{1}{l^3}\rho\zeta;$$

although the value of ρ' differs.

It is important to notice that the formulas (4) and (4') satisfy the condition of continuity

$$\frac{d\rho'}{dt'} + \sum \frac{d\rho'\xi'}{dx'} = 0.$$

To see this, let λ be an undetermined coefficient and D the Jacobian of

$$t + \lambda\rho, \quad x + \lambda\rho\xi, \quad y + \lambda\rho\eta, \quad z + \lambda\rho\zeta \tag{5}$$

with respect to t, x, y, z. It follows that:

$$D = D_0 + D_1\lambda + D_2\lambda^2 + D_3\lambda^3 + D_4\lambda^4,$$

with $D_0 = 1, D_1 = \dfrac{d\rho}{dt} + \sum \dfrac{d\rho\xi}{dx} = 0.$

Let $\lambda' = l^4\lambda,$ [3] then the 4 functions

$$t' + \lambda'\rho', \quad x' + \lambda'\rho'\xi', \quad y' + \lambda'\rho'\eta', \quad z' + \lambda'\rho'\zeta' \tag{5'}$$

are related to the functions (5) by the same linear relationships as the old variables to the new ones. Therefore, if we denote D' the Jacobian of the functions (5') with respect to the new variables, it follows that:

$$D' = D, \quad D' = D'_0 + D'_1 \lambda' + \dots + D'_4 \lambda'^4,$$

and thereby:[4]

$$D'_0 = D_0 = 1, \quad D'_1 = l^{-4}D_1 = 0 = \frac{d\rho'}{dt'} + \sum \frac{d\rho'\xi'}{dx'}. \quad \text{Q. E. D.}$$

Under Lorentz's hypothesis, this condition would not be met since ρ' has a different value.

We will define the new vector and scalar potentials in such a way as to satisfy the conditions

$$\Box'\psi' = -\rho', \qquad \Box'F' = -\rho'\xi'. \tag{6}$$

From this we deduce:

$$\psi' = \frac{k}{l}(\psi + \varepsilon F), \qquad F' = \frac{k}{l}(F + \varepsilon\psi), \qquad G' = \frac{1}{l}G, \qquad H' = \frac{1}{l}H. \tag{7}$$

These formulas differ noticeably from those of Lorentz, although the divergence stems ultimately from the definitions employed.

New electric and magnetic fields are now chosen in order to satisfy the equations:

$$f' = -\frac{dF'}{dt'} - \frac{d\psi'}{dx'}, \qquad \alpha' = \frac{dH'}{dy'} - \frac{dG'}{dz'}. \tag{8}$$

[135]   I It is easy to see that:

$$\frac{d}{dt'} = \frac{k}{l}\left(\frac{d}{dt} - \varepsilon\frac{d}{dx}\right), \qquad \frac{d}{dx'} = \frac{k}{l}\left(\frac{d}{dx} - \varepsilon\frac{d}{dt}\right), \qquad \frac{d}{dy'} = \frac{1}{l}\frac{d}{dy}, \qquad \frac{d}{dz'} = \frac{1}{l}\frac{d}{dz},$$

and we deduce thereby:

$$f' = \frac{1}{l^2}f, \qquad g' = \frac{k}{l^2}(g + \varepsilon\gamma), \qquad h' = \frac{k}{l^2}(h - \varepsilon\beta), \\
\alpha' = \frac{1}{l^2}\alpha, \qquad \beta' = \frac{k}{l^2}(\beta - \varepsilon h), \qquad \gamma' = \frac{k}{l^2}(\gamma + \varepsilon g). \qquad (9)$$

These formulas are identical to those of Lorentz.

Our transformation does not alter (1). In fact, the condition of continuity, as well as (6) and (8) were already featured in (1) (neglecting the primes).

Combining (6) with the condition of continuity, we obtain:

$$\frac{d\psi'}{dt'} + \sum\frac{dF'}{dx'} = 0. \qquad (10)$$

It remains for us to establish:

$$\frac{df'}{dt'} + \rho'\xi' = \frac{d\gamma'}{dy'} - \frac{d\beta'}{dz'}, \qquad \frac{d\alpha'}{dt'} = \frac{dg'}{dz'} - \frac{dh'}{dy'}, \qquad \sum\frac{df'}{dx'} = \rho',$$

and it is easy to see that these are necessary consequences of (6), (8) and (10).

We must now compare forces before and after the transformation.

Let X, Y, Z be the force prior to the transformation, and X', Y', Z' the force after the transformation, both forces being per unit volume. In order for X' to satisfy the same equations as before the transformation, we must have:

$$X' = \rho'f' + \rho'(\eta'\gamma' - \zeta'\beta'),$$
$$Y' = \rho'g' + \rho'(\zeta'\alpha' - \xi'\gamma'),$$
$$Z' = \rho'h' + \rho'(\xi'\beta' - \eta'\alpha'),$$

or, replacing all quantities by their values (4), (4') and (9), and in light of (2):

$$X' = \frac{k}{l^5}\left(X + \varepsilon\sum X\xi\right), \\
Y' = \frac{1}{l^5}Y, \qquad\qquad (11) \\
Z' = \frac{1}{l^5}Z.$$

Instead of representing the components of force per unit volume by X_1, Y_1, Z_1, we now let these terms represent the force per unit electron charge, and we let X'_1, Y'_1, Z'_1 represent the latter force after transformation. It follows that:

$$X_1 = f + \eta\gamma - \zeta\beta, \quad X'_1 = f' + \eta'\gamma' - \zeta'\beta', \quad X = \rho X_1, \quad X' = \rho'X'_1$$

[136]   I and we obtain the equations:

$$
\left.
\begin{aligned}
X'_1 &= \frac{k}{l^5}\frac{\rho}{\rho'}(X_1 + \varepsilon\sum X_1\xi), \\[2mm]
Y'_1 &= \frac{1}{l^5}\frac{\rho}{\rho'}Y_1, \\[2mm]
Z'_1 &= \frac{1}{l^5}\frac{\rho}{\rho'}Z_1.
\end{aligned}
\right\}
\tag{11'}
$$

Lorentz found (page 813, equation (10) with different notation):

$$
\left.
\begin{aligned}
X_1 &= l^2 X'_1 - l^2\varepsilon(\eta'g' + \zeta'h'), \\[2mm]
Y_1 &= \frac{l^2}{k}Y'_1 + \frac{l^2\varepsilon}{k}\xi'g', \\[2mm]
Z_1 &= \frac{l^2}{k}Z'_1 + \frac{l^2\varepsilon}{k}\xi'h'.
\end{aligned}
\right\}
\tag{11''}
$$

Before going any further, it is important to locate the source of this significant divergence. It obviously springs from the fact that the formulas for ξ', η' and ζ' are not the same, while the formulas for the electric and magnetic fields are the same.

If electron inertia is exclusively of electromagnetic origin, and if electrons are subject only to forces of electromagnetic origin, then the conditions of equilibrium require that:

$$X = Y = Z = 0$$

inside the electrons.

According to (11), these relationships are equivalent to

$$X' = Y' = Z' = 0.$$

The electron's equilibrium conditions are therefore unaltered by the transformation.

Unfortunately, such a simple hypothesis is inadmissible. In fact, if we assume $\xi = \eta = \zeta = 0$, the condition $X = Y = Z = 0$ leads necessarily to $f = g = h = 0$, and consequently, to $\sum\dfrac{df}{dx} = 0$, i.e., $\rho = 0$. Similar results obtain for the most general case. We must then admit that in addition to electromagnetic forces there are either non-electromagnetic forces or bonds. Therefore, we need to identify the conditions that these forces or these bonds must satisfy for electron equilibrium to be undisturbed by the transformation. This will be the object of an upcoming section.

[...]

9. HYPOTHESES CONCERNING GRAVITATION.   [166]

In this way Lorentz's theory would fully explain the impossibility of detecting absolute motion, if all forces were of electromagnetic origin.

But there exist other forces to which an electromagnetic origin cannot be attributed, such as gravitation, for example. It may in fact happen, that two systems of bodies produce equivalent electromagnetic fields, i.e., exert the same action on electrified bodies and on currents, and at the same time, these two systems do not exert the same gravitational action on Newtonian masses. The gravitational field is therefore distinct from the electromagnetic field. Lorentz was obliged thereby to extend his hypothesis with the assumption that *forces of any origin whatsoever, and gravitation in particular, are affected by a translation* (or, if one prefers, by the Lorentz transformation) *in the same manner as electromagnetic forces.*

It is now appropriate to enter into the details of this hypothesis, and to examine it more closely. If we want the Newtonian force to be affected by the Lorentz transformation in this fashion, we can no longer suppose that it depends only on the relative position of the attracting and attracted bodies at the instant considered. The force should also depend on the velocities of the two bodies. And that is not all: it will be natural to suppose that the force acting on the attracted body at the instant t depends on the position and velocity of this body at this same instant t, but it will also depend on the position and velocity of the *attracting* body, not at the instant t, but at *an earlier instant*, as if gravitation had taken a certain time to propagate.

Let us now consider the position of the attracted body at the instant t_0 and let x_0, y_0, z_0 be its coordinates, and ξ, η, ζ its velocity components at this instant; let us consider also the attracting body at the corresponding instant $t_0 + t$ and let its coordinates be $x_0 + x, y_0 + y, z_0 + z$, and its velocity components be ξ_1, η_1, ζ_1 at this instant.

First we should have a relationship

$$\varphi(t, x, y, z, \xi, \eta, \zeta, \xi_1, \eta_1, \zeta_1) = 0 \tag{1}$$

in order to define the time t. This relationship will define the law of propagation of gravitational action (I do not constrain myself by any means to a propagation velocity equal in all directions).

Now let X_1, Y_1, Z_1 be the three components of the action exerted on the attracted body at the instant t_0; [5] we want to express X_1, Y_1, Z_1 as functions of

$$t, x, y, z, \xi, \eta, \zeta, \xi_1, \eta_1, \zeta_1. \tag{2}$$

What conditions must be satisfied? |

1° The condition (1) should not be altered by transformations of the Lorentz   [167] group.

2° The components X_1, Y_1, Z_1 should be affected by transformations of the Lorentz group in the same manner as the electromagnetic forces designated by the same letters, i.e., in accordance with (11') of section 1.

3° When the two bodies are at rest, the ordinary law of attraction will be recovered.

It is important to note that in the latter case, the relationship (1) vanishes, because if the two bodies are at rest the time t plays no role.

Posed in this fashion the problem is obviously indeterminate. We will therefore seek to satisfy to the utmost other, complementary conditions.

4° Since astronomical observations do not seem to show a sensible deviation from Newton's law, we will choose the solution that differs the least with this law for small velocities of the two bodies.

5° We will make an effort to arrange matters in such a way that t is always negative. Although we can imagine that the effect of gravitation requires a certain time in order to propagate, it would be difficult to understand how this effect could depend on the position *not yet attained* by the attracting body.

There is one case where the indeterminacy of the problem vanishes; it is the one where the two bodies are in mutual *relative* rest, i.e., where:

$$\xi = \xi_1, \qquad \eta = \eta_1, \qquad \zeta = \zeta_1;$$

this is then the case we will examine first, by supposing that these velocities are constant, such that the two bodies are engaged in a common uniform rectilinear translation.

We may suppose that the x–axis is parallel to this translation, such that $\eta = \zeta = 0$, and we will let $\varepsilon = -\xi$.

If we apply the Lorentz transformation under these conditions, after the transformation the two bodies will be at rest, and it follows that:

$$\xi' = \eta' = \zeta' = 0.$$

The components X'_1, Y'_1, Z'_1 should then agree with Newton's law and we will have, apart from a constant factor:

$$X'_1 = -\frac{x'}{r'^3}, \qquad Y'_1 = -\frac{y'}{r'^3}, \qquad Z'_1 = -\frac{z'}{r'^3}, \left.\begin{array}{c} \\ \\ \end{array}\right\}$$

$$r'^2 = x'^2 + y'^2 + z'^2. \tag{3}$$

But according to section 1 we have:

$$x' = k(x + \varepsilon t), \qquad y' = y, \qquad z' = z, \qquad t' = k(t + \varepsilon x),$$

$$\frac{\rho'}{\rho} = k(1 + \xi\varepsilon) = k(1 - \varepsilon^2) = \frac{1}{k}, \qquad \sum X_1\xi = -X_1\varepsilon, \text{ I}$$

$$X'_1 = k\frac{\rho}{\rho'}\left(X_1 + \varepsilon \sum X_1\xi\right) = k^2 X_1(1 - \varepsilon^2) = X_1,$$

$$Y'_1 = \frac{\rho}{\rho'}Y_1 = kY_1, \qquad Z'_1 = kZ_1.$$

We have in addition:

$$x + \varepsilon t = x - \xi t, \qquad r'^2 = k^2(x - \xi t)^2 + y^2 + z^2$$

and

$$X_1 = \frac{-k(x - \xi t)}{r'^3}, \qquad Y_1 = \frac{-y}{kr'^3}, \qquad Z_1 = \frac{-z}{r'^3}; \tag{4}$$

which may be written:

$$X_1 = \frac{dV}{dx}, \qquad Y_1 = \frac{dV}{dy}, \qquad Z_1 = \frac{dV}{dz}; \qquad V = \frac{1}{kr'}. \tag{4'}$$

It seems at first that the indeterminacy remains, since we made no hypotheses concerning the value of t, i.e., the transmission speed; and that besides, x is a function of t. It is easy to see, however, that the terms appearing in our formulas, $x - \xi t, y, z$, do not depend on t.

We see that if the two bodies translate together, the force acting on the attracted body is perpendicular to an ellipsoid, at the center of which lies the attracting body.

To advance further, we need to look for the *invariants of the Lorentz group*.

We know that the substitutions of this group (assuming $l = 1$) are linear substitutions that leave unaltered the quadratic form

$$x^2 + y^2 + z^2 - t^2.$$

Let us also put:

$$\xi = \frac{\delta x}{\delta t}, \qquad \eta = \frac{\delta y}{\delta t}, \qquad \zeta = \frac{\delta z}{\delta t};$$

$$\xi_1 = \frac{\delta_1 x}{\delta_1 t}, \qquad \eta_1 = \frac{\delta_1 y}{\delta_1 t}, \qquad \zeta_1 = \frac{\delta_1 z}{\delta_1 t};$$

we see that the Lorentz transformation will make $\delta x, \delta y, \delta z, \delta t$, and $\delta_1 x, \delta_1 y, \delta_1 z, \delta_1 t$ undergo the same linear substitutions as x, y, z, t.

Let us regard

$$x, \qquad y, \qquad z, \qquad t\sqrt{-1},$$

$$\delta x, \qquad \delta y, \qquad \delta z, \qquad \delta t \sqrt{-1},$$

$$\delta_1 x, \qquad \delta_1 y, \qquad \delta_1 z, \qquad \delta_1 t \sqrt{-1},$$

as the coordinates of 3 points P, P', P'' in space of 4 dimensions. We see that the Lorentz transformation is merely a rotation in this space about the origin, assumed fixed. Consequently, we will have no distinct invariants apart from the 6 distances between the 3 points P, P', P'', considered separately and with the origin, or, if one prefers, I apart from the two expressions   [169]

$$x^2 + y^2 + z^2 - t^2, \qquad x\delta x + y\delta y + z\delta z - t\delta t,$$

or the 4 expressions of like form deduced from an arbitrary permutation of the 3 points P, P', P''.

But what we seek are invariants that are functions of the 10 variables (2). Therefore, among the combinations of our 6 invariants we must find those depending only on these 10 variables, i.e., those that are 0th degree homogeneous with respect both to $\delta x, \delta y, \delta z, \delta t$, and to $\delta_1 x, \delta_1 y, \delta_1 z, \delta_1 t$. We will then be left with 4 distinct invariants:

$$\sum x^2 - t^2, \quad \frac{t - \sum x\xi}{\sqrt{1 - \sum \xi^2}}, \quad \frac{t - \sum x\xi_1}{\sqrt{1 - \sum \xi_1^2}}, \quad \frac{1 - \sum \xi\xi_1}{\sqrt{(1 - \sum \xi^2)(1 - \sum \xi_1^2)}}. \quad (5)$$

Next let us see how the force components are transformed; we recall the equations (11) of section 1, that refer not to the force X_1, Y_1, Z_1, considered at present, but to the force per unit volume: X, Y, Z.

We designate moreover

$$T = \sum X\xi;$$

we will see that (11) can be written ($l = 1$):

$$X' = k(X + \varepsilon T), \qquad T' = k(T + \varepsilon X), \\ Y' = Y, \qquad Z' = Z; \qquad \qquad (6)$$

in such a way that X, Y, Z, T undergo the same transformation as x, y, z, t. Consequently, the group invariants will be

$$\sum X^2 - T^2, \quad \sum Xx - Tt, \quad \sum X\delta x - T\delta t, \quad \sum X\delta_1 x - T\delta_1 t.$$

However, it is not X, Y, Z, that we need, but X_1, Y_1, Z_1, with

$$T_1 = \sum X_1 \xi.$$

We see that

$$\frac{X_1}{X} = \frac{Y_1}{Y} = \frac{Z_1}{Z} = \frac{T_1}{T} = \frac{1}{\rho}.$$

Therefore, the Lorentz transformation will act in the same manner on X_1, Y_1, Z_1, T_1 as on X, Y, Z, T, except that these expressions will be multiplied moreover by

$$\frac{\rho}{\rho'} = \frac{1}{k(1 + \xi\varepsilon)} = \frac{\delta t}{\delta t'}.$$

Likewise, the Lorentz transformation will act in the same way on $\xi, \eta, \zeta, 1$ as on $\delta x, \delta y, \delta z, \delta t$, except that these expressions will be multiplied moreover by the *same* factor:

$$\frac{\delta t}{\delta t'} = \frac{1}{k(1 + \xi\varepsilon)}.$$

Next we consider $X, Y, Z, T\sqrt{-1}$ as the coordinates of a fourth point I Q; the [170] invariants will then be functions of the mutual distances of the five points

$$O, \quad P, \quad P', \quad P'', \quad Q$$

and among these functions we must retain only those that are 0th degree homogeneous with respect, on one hand, to

$$X, \quad Y, \quad Z, \quad T, \quad \delta x, \quad \delta y, \quad \delta z, \quad \delta t$$

(variables that can be replaced further by $X_1, Y_1, Z_1, T_1, \xi, \eta, \zeta, 1$), and on the other hand, with respect to[6]

$$\delta_1 x, \quad \delta_1 y, \quad \delta_1 z, \quad \delta_1 t$$

(variables that can be replaced further by $\xi_1, \eta_1, \zeta_1, 1$).

In this way we find, beyond the four invariants (5), four distinct new invariants:

$$\frac{\sum X_1^2 - T_1^2}{1 - \sum \xi^2}, \quad \frac{\sum X_1 x - T_1 t}{\sqrt{1 - \sum \xi^2}}, \quad \frac{\sum X_1 \xi_1 - T_1}{\sqrt{1 - \sum \xi^2}\sqrt{1 - \sum \xi_1^2}}, \quad \frac{\sum X_1 \xi - T_1}{1 - \sum \xi^2}. \tag{7}$$

The latter invariant is always null according to the definition of T_1.

These terms being settled, what conditions must be satisfied?

1° The first term of (1), defining the velocity of propagation, has to be a function of the 4 invariants (5).

A wealth of hypotheses can obviously be entertained, of which we will examine only two:

A) We can have

$$\sum x^2 - t^2 = r^2 - t^2 = 0,$$

from whence $t = \pm r$, and, since t has to be negative, $t = -r$. This means that the velocity of propagation is equal to that of light. It seems at first that this hypothesis ought to be rejected outright. Laplace showed in effect that the propagation is either instantaneous or much faster than that of light. However, Laplace examined the hypothesis of finite propagation velocity *ceteris non mutatis*; here, on the contrary, this hypothesis is conjoined with many others, and it may be that between them a more or less perfect compensation takes place. The application of the Lorentz transformation has already provided us with numerous examples of this.

B) We can have

$$\frac{t - \sum x\xi_1}{\sqrt{1 - \sum \xi_1^2}} = 0, \qquad t = \sum x\xi_1.$$

The propagation velocity is therefore much faster than that of light, but in certain cases t could be positive, which, as we mentioned, seems hardly admissible.[7] *We will therefore stick with hypothesis* (A).

2° The four invariants (7) ought to be functions of the invariants (5).

[171]   3° When the two bodies are at absolute rest, X_1, Y_1, Z_1 ought to have the I values given by Newton's law, and when they are at relative rest, the values given by (4).

For the case of absolute rest, the first two invariants (7) ought to reduce to

$$\sum X_1^2, \qquad \sum X_1 x,$$

or, by Newton's law, to

$$\frac{1}{r^4}, \qquad -\frac{1}{r};$$

in addition, according to hypothesis (A), the 2nd and 3rd invariants in (5) become:

$$\frac{-r-\sum x\xi}{\sqrt{1-\sum \xi^2}}, \qquad \frac{-r-\sum x\xi_1}{\sqrt{1-\sum \xi_1^2}},$$

that is, for absolute rest,

$$-r, \qquad -r.$$

We may therefore admit, *for example*, that the first two invariants in (7) reduce to[8]

$$\frac{(1-\sum \xi_1^2)^2}{(r+\sum x\xi_1)^4}, \qquad -\frac{\sqrt{1-\sum \xi_1^2}}{r+\sum x\xi_1};$$

although other combinations are possible.

A choice must be made among these combinations, and furthermore, we need a 3rd equation in order to define X_1, Y_1, Z_1. In making such a choice, we should try to come as close as possible to Newton's law. Let us see what happens when we neglect the squares of the velocities ξ, η, etc. (still letting $t = -r$). The 4 invariants (5) then become:

$$0, \qquad -r-\sum x\xi, \qquad -r-\sum x\xi_1, \qquad 1$$

and the 4 invariants (7) become:

$$\sum X_1^2, \qquad \sum X_1(x+\xi r), \qquad \sum X_1(\xi_1-\xi), \qquad 0.$$

Before we can make a comparison with Newton's law, another transformation is required. In the case under consideration, $x_0 + x, y_0 + y, z_0 + z$, represent the coordinates of the attracting body at the instant $t_0 + t$, and $r = \sqrt{\sum x^2}$. With Newton's law we have to consider the coordinates of the attracting body $x_0 + x_1, y_0 + y_1, z_0 + z_1$ at the instant t_0, and the distance $r_1 = \sqrt{\sum x_1^2}$.

We may neglect the square of the time t required for propagation, and proceed, consequently, as if the motion were uniform; we then have:

$$x = x_1 + \xi_1 t, \quad y = y_1 + \eta_1 t, \quad z = z_1 + \zeta_1 t, \quad r(r - r_1) = \sum x \xi_1 t; \mathbf{\mid}$$

or, since $t = -r$,   [172]

$$x = x_1 - \xi_1 r, \quad y = y_1 - \eta_1 r, \quad z = z_1 - \zeta_1 r, \quad r = r_1 - \sum x \xi_1;$$

such that our 4 invariants (5) become:

$$0, \quad -r_1 + \sum x(\xi_1 - \xi), \quad -r_1, \quad 1$$

and our 4 invariants (7) become:

$$\sum X_1^2, \quad \sum X_1 [x_1 + (\xi - \xi_1) r_1], \quad \sum X_1(\xi_1 - \xi), \quad 0.$$

In the second of these expressions I wrote r_1 instead of r, because r is multiplied by $\xi - \xi_1$ and because I neglect the square of ξ.

For these 4 invariants (7), Newton's law would yield

$$\frac{1}{r_1^4}, \quad -\frac{1}{r_1} - \frac{\sum x_1(\xi - \xi_1)}{r_1^2}, \quad \frac{\sum x_1(\xi - \xi_1)}{r_1^3}, \quad 0.$$

Therefore, if we designate the 2nd and 3rd of the invariants (5) as A and B, and the first 3 invariants of (7) as M, N, P we will satisfy Newton's law to first-order terms in the square of velocity by setting:

$$M = \frac{1}{B^4}, \qquad N = \frac{+A}{B^2}, \qquad P = \frac{A - B}{B^3}. \tag{8}$$

This solution is not unique. Let C be the 4th invariant in (5); $C - 1$ is of the order of the square of ξ, and it is the same with $(A - B)^2$.

The solution (8) appears at first to be the simplest, nevertheless, it may not be adopted. In fact, since M, N, P are functions of X_1, Y_1, Z_1, and $T_1 = \sum X_1 \xi$, the values of X_1, Y_1, Z_1 can be drawn from these three equations (8), but in certain cases these values would become imaginary.

To avoid this difficulty we will proceed in a different manner. Let us put:

$$k_0 = \frac{1}{\sqrt{1 - \sum \xi^2}}, \qquad k_1 = \frac{1}{\sqrt{1 - \sum \xi_1^2}},$$

which is justified by analogy with the notation

$$k = \frac{1}{\sqrt{1 - \varepsilon^2}},$$

featured in the Lorentz substitution.

In this case, and in light of the condition $-r = t$, the invariants (5) become:

$$0, \quad A = -k_0\left(r + \sum x\xi\right), \quad B = -k_1\left(r + \sum x\xi_1\right), \quad C = k_0 k_1\left(1 - \sum \xi\xi_1\right). \ |$$

[173] Moreover, we notice that the following systems of quantities:

$$
\begin{array}{cccc}
x, & y, & z, & -r = t \\
k_0 X_1, & k_0 Y_1, & k_0 Z_1, & k_0 T_1 \\
k_0\xi, & k_0\eta, & k_0\zeta, & k_0 \\
k_1\xi_1, & k_1\eta_1, & k_1\zeta_1, & k_1
\end{array}
$$

undergo the *same* linear substitutions when the transformations of the Lorentz group are applied to them. We are led thereby to put:

$$
\left.
\begin{aligned}
X_1 &= x\frac{\alpha}{k_0} + \xi\beta + \xi_1\frac{k_1}{k_0}\gamma, \\
Y_1 &= y\frac{\alpha}{k_0} + \eta\beta + \eta_1\frac{k_1}{k_0}\gamma, \\
Z_1 &= z\frac{\alpha}{k_0} + \zeta\beta + \zeta_1\frac{k_1}{k_0}\gamma, \\
T_1 &= -r\frac{\alpha}{k_0} + \beta + \frac{k_1}{k_0}\gamma.
\end{aligned}
\right\}
\tag{9}
$$

It is clear that if α, β, γ are invariants, X_1, Y_1, Z_1, T_1 will satisfy the fundamental condition, i.e., the Lorentz transformations will make them undergo an appropriate linear substitution.

However, for equations (9) to be compatible we must have

$$\sum X_1\xi - T_1 = 0,$$

which becomes, replacing X_1, Y_1, Z_1, T_1 with their values in (9) and multiplying by k_0^2,

$$-A\alpha - \beta - C\gamma = 0. \tag{10}$$

What we would like is that the values of X_1, Y_1, Z_1 remain in line with Newton's law when we neglect (as above) the squares of velocities ξ, etc. with respect to the square of the velocity of light, and the products of acceleration and distance.

We could select

$$\beta = 0, \qquad \gamma = -\frac{A\alpha}{C}.$$

To the adopted order of approximation, we obtain

$$k_0 = k_1 = 1, \qquad C = 1, \qquad A = -r_1 + \sum x(\xi_1 - \xi), \qquad B = -r_1,$$

$$x = x_1 + \xi_1 t = x_1 - \xi_1 r.$$

The 1st equation in (9) then becomes

$$X_1 = \alpha(x - A\xi_1).$$

But if the square of ξ is neglected, $A\xi_1$ can be replaced by $-r_1\xi_1$ or I by $-r\xi_1$, [174] which yields:

$$X_1 = \alpha(x + \xi_1 r) = \alpha x_1.$$

Newton's law would yield

$$X_1 = -\frac{x_1}{r_1^3}.$$

Consequently, we must select a value for the invariant α, which reduces to $-\frac{1}{r_1^3}$ in the adopted order of approximation, that is, $\frac{1}{B^3}$. Equations (9) will become:

$$\left. \begin{aligned} X_1 &= \frac{x}{k_0 B^3} - \xi_1 \frac{k_1}{k_0} \frac{A}{B^3 C}, \\[2mm] Y_1 &= \frac{y}{k_0 B^3} - \eta_1 \frac{k_1}{k_0} \frac{A}{B^3 C}, \\[2mm] Z_1 &= \frac{z}{k_0 B^3} - \zeta_1 \frac{k_1}{k_0} \frac{A}{B^3 C}, \\[2mm] T_1 &= -\frac{r}{k_0 B^3} - \frac{k_1}{k_0} \frac{A}{B^3 C}. \end{aligned} \right\} \qquad (11)$$

We notice first that the corrected attraction is composed of two components: one parallel to the vector joining the positions of the two bodies, the other parallel to the velocity of the attracting body.

Remember that when we speak of the position or velocity of the attracting body, this refers to its position or velocity at the instant the gravitational wave takes off; for the attracted body, on the contrary, this refers to the position or velocity at the instant the gravitational wave arrives, assuming that this wave propagates with the velocity of light.

I believe it would be premature to seek to push the discussion of these formulas further; I will therefore confine myself to a few remarks.

1° The solutions (11) are not unique; we may, in fact, replace the global factor $\frac{1}{B^3}$, by

$$\frac{1}{B^3} + (C - 1)f_1(A, B, C) + (A - B)^2 f_2(A, B, C),$$

where f_1 and f_2 are arbitrary functions of A, B, C. Alternatively, we may forgo setting β to zero, but add any complementary terms to α, β, γ that satisfy condition (10) and are of second order with respect to the ξ for α and of first order for β and γ.

2° The first equation in (11) may be written:

$$X_1 = \frac{k_1}{B^3 C}[x(1 - \sum \xi\xi_1) + \xi_1(r + \sum x\xi)] \qquad (11')$$

and the quantity in brackets itself may be written:

$$(x + r\xi_1) + \eta(\xi_1 y - x\eta_1) + \zeta(\xi_1 z - x\zeta_1), \qquad (12)$$

[175] I such that the total force may be separated into three components corresponding to the three parentheses of expression (12); the first component is vaguely analogous to the mechanical force due to the electric field, the two others to the mechanical force due to the magnetic field; to extend the analogy I may, in light of the first remark, replace $1/B^3$ in (11) by $C/(B^3,)$ in such a way that X_1, Y_1, Z_1 are linear functions of the attracted body's velocity ξ, η, ζ, since C has vanished from the denominator of (11').

Next we put:

$$\left.\begin{array}{lll} k_1(x + r\xi_1) = \lambda, & k_1(y + r\eta_1) = \mu, & k_1(z + r\zeta_1) = \nu, \\ k_1(\eta_1 z - \zeta_1 y) = \lambda', & k_1(\zeta_1 x - \xi_1 z) = \mu', & k_1(\xi_1 y - x\eta_1) = \nu'; \end{array}\right\} \qquad (13)$$

and since C has vanished from the denominator of (11'), it will follow that:

$$\left.\begin{array}{l} X_1 = \dfrac{\lambda}{B^3} + \dfrac{\eta\nu' - \zeta\mu'}{B^3}, \\[2mm] Y_1 = \dfrac{\mu}{B^3} + \dfrac{\zeta\lambda' - \xi\nu'}{B^3}, \\[2mm] Z_1 = \dfrac{\nu}{B^3} + \dfrac{\xi\mu' - \eta\lambda'}{B^3}; \end{array}\right\} \qquad (14)$$

and we will have moreover:

$$B^2 = \sum \lambda^2 - \sum \lambda'^2. \qquad (15)$$

Now λ, μ, ν, or $\dfrac{\lambda}{B^3}, \dfrac{\mu}{B^3}, \dfrac{\nu}{B^3}$, is an electric field of sorts, while λ', μ', ν', or rather $\dfrac{\lambda'}{B^3}, \dfrac{\mu'}{B^3}, \dfrac{\nu'}{B^3}$, is a magnetic field of sorts.

3° The postulate of relativity would compel us to adopt solution (11), or solution (14), or any solution at all among those derived on the basis of the first remark. However, the first question to ask is whether or not these solutions are compatible with astronomical observations. The deviation from Newton's law is of the order of ξ^2, i.e., 10000 times smaller than if it were of the order of ξ, i.e., if the propagation were to take place with the velocity of light, *ceteris non mutatis*; consequently, it is legitimate to hope that it will not be too large. To settle this question, however, would require an extended discussion.

EDITORIAL NOTES

[1] Translated by Scott Walter from *Rendiconti del Circolo Matematico di Palermo* 21, 1906, 129–176. The original notation is faithfully reproduced, including the use of "*d*" for both ordinary and partial differentiation. The translator's endnote calls are bracketed. For alternative translations of Poincaré's memoir, see C. W. Kilmister (*Special Theory of Relativity*, Oxford: Pergamon, 1970, 145–185), and by H. M. Schwartz (*American Journal of Physics* 39:1287–1294; 40:862–872, 1282–1287).

[2] The original reads: "$\frac{4}{3}\pi r^2$".

[3] The original reads: "$\lambda' = l^2\rho'$".

[4] The original reads: "$D'_1 = l^{-2}D_1$".

[5] The original reads: "à l'instant *t*".

[6] The original reads: "$\delta_1 x, \quad \delta_1 y, \quad \delta_1 z, \quad 1.$"

[7] The original reads: "*t* pourrait être négatif."

[8] The original has (4) instead of (7).

HERMANN MINKOWSKI

MECHANICS AND THE RELATIVITY POSTULATE

Originally published as "Mechanik und Relativitätspostulat," appended to "Die Grundgleichungen für elektromagnetischen Vorgänge in bewegten Körpern" in Nachrichten der Königlichen Gesellschaft der Wissenschaften zu Göttingen, Mathematisch-Physikalische Klasse, 1908, pp. 53–111.

It would be highly unsatisfactory if the new conception of the notion of time, which is [98] characterized by the freedom of Lorentz transformations, could be accepted as valid for only a subfield of physics.

Now, many authors state that classical mechanics is opposed to the relativity postulate, which has been taken here to be the foundation of electrodynamics.

In order to assess this we focus on a special Lorentz transformation as represented by eqs. (10), (11), (12),[1] with a non-zero vector v of arbitrary direction, and magnitude q that is <1. But for the moment we will pretend that the ratio of the unit of length and the unit of time has not yet been established, and accordingly we will write in these equations $ct, ct', \frac{q}{c}$ instead of t, t', q, where c is a certain positive constant, and we must have $q < c$. The equations referred to then become

$$\mathfrak{r}_{\bar{v}}' = \mathfrak{r}_{\bar{v}}, \qquad \mathfrak{r}_{v}' = \frac{c(\mathfrak{r}_{v} - qt)}{\sqrt{c^2 - q^2}}, \qquad t' = \frac{-q\mathfrak{r}_{v} + c^2 t}{c\sqrt{c^2 - q^2}};$$

we recall that $\mathfrak{r}$ means the spatial vector x, y, z, and $\mathfrak{r}'$ means the spatial vector x', y', z'.

In these equations we fix v and pass to the limit $c = \infty$, with the result

$$\mathfrak{r}_{\bar{v}}' = \mathfrak{r}_{\bar{v}}, \qquad \mathfrak{r}_{v}' = \mathfrak{r}_{v} - qt, \qquad t' = t.$$

I These new equations would mean a transition from the system of spatial coordinates [99] x, y, z to another spatial coordinate system x', y', z' with parallel axes, whose origin moves with respect to the first in a straight line with constant velocity, while the time parameter would be totally unaffected.

On the basis of this remark one may state:

Classical mechanics postulates covariance of the laws of physics for the group of homogeneous linear transformations of the expression

Jürgen Renn (ed.). *The Genesis of General Relativity*, Vol. 3
Gravitation in the Twilight of Classical Physics: Between Mechanics, Field Theory, and Astronomy.
© 2007 Springer.

$$-x^2 - y^2 - z^2 + c^2t^2 \qquad\qquad (1)$$

into itself, with the specification $c = \infty$.

Now it would be downright confusing to find in one subfield of physics a covariance of the laws for transformations of expression (1) into itself with a certain finite c, but with $c = \infty$ in another subfield. That Newtonian mechanics could only claim this covariance for $c = \infty$, and could not devise it with c equal to the speed of light, needs no explanation. But is it not legitimate today to try and regard that traditional covariance for $c = \infty$ as only an approximation, gained from preliminary experience, to a more exact covariance of the laws of nature for a certain finite c?

I want to explain in detail that by *reforming mechanics, replacing Newton's relativity postulate with* $c = \infty$ *by one with a finite* c, the axiomatic construction of mechanics even seems to attain considerable perfection.

The ratio of the unit of time to the unit of length shall be normalized so that the relevant relativity postulate has $c = 1$.

When I now want to transfer geometrical pictures to the manifold of the four variables x, y, z, t, it may be convenient for ease of understanding what follows to exclude totally y, z from consideration at first, and to interpret x and t as arbitrary oblique-angled rectilinear coordinates in a plane.

A spacetime origin $O(x, y, z, t = 0, 0, 0, 0)$ remains fixed under the Lorentz transformations. The object

$$-x^2 - y^2 - z^2 + t^2 = 1, \qquad t > 0, \qquad\qquad (2)$$

[100]   I a *hyperbolic shell*, includes the spacetime point $A(x, y, z, t = 0, 0, 0, 1)$ and all spacetime points A' that appear as $(x', y', z', t' = 0, 0, 0, 1)$ in the new components x', y', z', t' after some Lorentz transformation.

The direction of a radius vector OA' from O to a point A' of (2) and the directions of the tangents to (2) at A' shall be called *normal* to each other.

Let us follow a definite point in the matter on its orbit for all times t. I call the totality of the spacetime points x, y, z, t, that correspond to this point at different times t a *spacetime line* [world line].[2]

The problem of determining the motion of matter is to be understood in this way: *For every spacetime point the direction of the spacetime line running through it is to be determined.*

To *transform* a spacetime point $P(x, y, z, t)$ to *rest* means to introduce by a Lorentz-transformation a system of reference x', y', z', t' such that the t'-axis OA' acquires that direction which is exhibited by the spacetime line running through P. The space $t' = $ const., which is to be laid down through P, shall be called the *normal* space to the spacetime line at P. To the increment dt of time t starting at P there corresponds the increment[1]

---

1   We resume the earlier conventions (see sections 3 and 4) for the notation with indices and the symbols
     $\mathfrak{w}, w$. [3]

$$d\tau = \sqrt{dt^2 - dx^2 - dy^2 - dz^2} = dt\sqrt{1 - \mathfrak{w}^2} = \frac{dx_4}{w_4} \qquad (3)$$

of the parameter t' to be introduced in this construction. The value of the integral

$$\int d\tau = \int \sqrt{-(dx_1^2 + dx_2^2 + dx_3^2 + dx_4^2)},$$

calculated over the spacetime line from any fixed initial point P^0 to an endpoint P, taken to be variable, shall be called the *proper time* of the corresponding point of the matter at the spacetime point P. (This is a generalization of the concept of *local time* introduced by Lorentz for uniform motion.)

If we take a spatially extended body R^0 at a particular time t^0, then the region of all the spacetime lines passing through the spacetime points R^0, t^0 shall be called a *spacetime thread* [pencil].

If we have an analytic expression $\Theta(x, y, z, t)$, such that $\Theta(x, y, z, t) = 0$ is met at one point by every spacetime line of the thread, where |

$$-\left(\frac{\partial\Theta}{\partial x}\right)^2 - \left(\frac{\partial\Theta}{\partial y}\right)^2 - \left(\frac{\partial\Theta}{\partial z}\right)^2 + \left(\frac{\partial\Theta}{\partial t}\right)^2 > 0, \qquad \frac{\partial\Theta}{\partial t} > 0 \qquad [101]$$

then we will call the set Q of the meeting points involved a *cross section* of the tread. At every point $P(x, y, z, t)$ of such a cross section we can introduce a system of reference x', y', z', t' by means of a Lorentz-transformation such that thereafter we have

$$\frac{\partial\Theta}{\partial x'} = 0, \qquad \frac{\partial\Theta}{\partial y'} = 0, \qquad \frac{\partial\Theta}{\partial z'} = 0, \qquad \frac{\partial\Theta}{\partial t'} > 0$$

The direction of the corresponding unique t'-axis shall be called the *upper normal* [future-pointing normal] of the cross section Q at the point P, and the value $dJ = \iiint dx' dy' dz'$ for a neighborhood of P within the cross section shall be called an *element of volume content* [*Inhaltselement*] of the cross section. In this sense R^0, t^0 can be called the thread's cross section $t = t^0$ normal to the t-axis, and the volume of the body R^0 can be called the *volume content* [*Inhalt*] of this cross section.

By allowing the space R^0 to converge to a point we arrive at the concept of an *infinitely thin* spacetime thread. In such a thread we will always think of *one* spacetime line as a somehow distinguished *central line*. Also, we understand by the *proper time of the thread* the proper time measured on this central line; by the *normal section* of the thread its intersections with the spaces that are normal to the central line at its points.

Now we formulate the *principle of conservation of mass*.

To every region R at a time t there belongs a positive quantity, the *mass within R at time t*. If R converges to a point x, y, z, t, then the quotient of this mass and the volume of R shall approach a limit $\mu(x, y, z, t)$, the *mass density* at the spacetime point x, y, z, t.

The principle of mass conservation states: *For an infinitely thin spacetime thread the product μdJ of the mass density μ at one point x, y, z, t of the thread (i.e., of the central line of the thread) and the volume content dJ of the normal section to the t-axis passing through the point is always constant along the entire thread.*

Now the volume content dJ_n of the thread's normal section passing through x, y, z, t is calculated as |

[102]

$$dJ_n = \frac{1}{\sqrt{1-\mathfrak{w}^2}}dJ = -i w_4 dJ = \frac{dt}{d\tau}dJ, \qquad (4)$$

and therefore let us define

$$v = \frac{\mu}{-i w_4} = \mu\sqrt{1-\mathfrak{w}^2} = \mu\frac{d\tau}{dt} \qquad (5)$$

as the *rest-mass density* at the point x, y, z, t. Then the principle of mass conservation can also be formulated as follows:

For an infinitely thin spacetime thread the product of the rest-mass density and the volume content of the normal section at a point of the thread is always constant along the entire thread.

In an arbitrary spacetime thread, fix an initial cross section Q^0 and then a second cross section Q^1, which has those and only those points in common with Q^0 that lie on the thread's boundary, and let the spacetime lines within the thread assume larger values t on Q^1 than on Q^0. The finite region bounded by Q^0 and Q^1 together shall be called a *spacetime sickle* [lens], with Q^0 the *lower* and Q^1 the *upper* boundary of the sickle.

Think of the thread decomposed into many very thin spacetime threads; then to every entrance of a thin thread in the lower boundary of the sickle there corresponds an exit through the upper boundary, where the product $v dJ_n$ formed in the sense of (4) and (5) takes the same value each time. Therefore the difference of the two integrals $\int v dJ_n$ vanishes, when the first extends over the upper, the second over the lower boundary of the sickle. According to a well-known theorem of integral calculus this difference equals the integral[4]

$$\iiiint \mathrm{lorv}\overline{w}\,dx\,dy\,dz\,dt,$$

taken over the whole region of the sickle, where we have (cf. (67) in section 12)[5]

$$\mathrm{lorv}\overline{w} = \frac{\partial v w_1}{\partial x_1} + \frac{\partial v w_2}{\partial x_2} + \frac{\partial v w_3}{\partial x_3} + \frac{\partial v w_4}{\partial x_4}.$$

If the sickle is contracted to one spacetime point x, y, z, t, the consequence of this is the differential equation

$$\text{lor}\, v\overline{w} = 0,\tag{6}$$

that is, the *continuity condition* |

$$\frac{\partial \mu w_x}{\partial x} + \frac{\partial \mu w_y}{\partial y} + \frac{\partial \mu w_z}{\partial z} + \frac{\partial \mu}{\partial t} = 0.\qquad\qquad [103]$$

Further we form the integral

$$\mathbf{N} = \iiiint v\, dx\, dy\, dz\, dt.\tag{7}$$

taken over the whole region of a spacetime sickle. We divide the sickle into thin spacetime threads, and further divide each thread into small elements $d\tau$ of its proper time, which are however still large compared to the linear dimension of the normal section; we put the mass of such a thread $v\, dJ_n = dm$ and write τ^0 and τ^1 for the proper time of the thread on the lower resp. upper boundary of the sickle; then the integral (7) can also be interpreted as

$$\iint v\, dJ_n\, d\tau = \int (\tau^1 - \tau^0)\, dm$$

over all the threads in the sickle.

Now I consider the spacetime lines within a spacetime sickle as curves made of a substance, consisting of substance-points, and I imagine them subjected to a continuous change in position within the sickle of the following type: The whole curves shall be displaced arbitrarily but with *fixed end points on the lower and upper boundary* of the sickle, and each substance-point on them shall be guided so that it is always displaced *normal to its curve*. The whole process shall be capable of analytic representation by means of a parameter ϑ and the value $\vartheta = 0$ shall correspond to the curves that actually run as spacetime lines inside the sickle. Such a process shall be called a *virtual displacement within the sickle*.

Let the point x, y, z, t in the sickle for $\vartheta = 0$ arrive at $x + \delta x, y + \delta y, z + \delta z, t + \delta t$ for the parameter value ϑ; then the latter quantities are functions of x, y, z, t, ϑ. Let us again consider an infinitely thin spacetime thread at the location x, y, z, t with a normal section of volume content dJ_n, and let $dJ_n + \delta dJ_n$ be the volume content of the normal section at the corresponding location of the varied thread; then we will take the *principle of conservation of mass* into account by assuming at this varied location a rest mass density $v + \delta v$ according to

$$(v + \delta v)(dJ_n + \delta dJ_n) = v\, dJ_n = dm,\tag{8}$$

| where v is understood to be the actual rest mass density at x, y, z, t. According to [104] this convention the integral (7), extended over the region of the sickle, then varies upon the virtual displacement as a definite function $\mathbf{N} + \delta\mathbf{N}$ of ϑ, and we will call this function $\mathbf{N} + \delta\mathbf{N}$ the *mass action* associated with the virtual displacement.

Using index notation we will have:

$$d(x_h + \delta x_h) = dx_h + \sum_k \frac{\partial \delta x_h}{\partial x_k} dx_k + \frac{\partial \delta x_h}{\partial \vartheta} d\vartheta \qquad \binom{k = 1, 2, 3, 4}{h = 1, 2, 3, 4}. \tag{9}$$

Now it soon becomes evident by reason of the remarks already made above that the value of $N + \delta N$ for parameter value ϑ will be:

$$N + \delta N = \iiiint v \frac{d(\tau + \delta \tau)}{d\tau} dx\,dy\,dz\,dt, \tag{10}$$

taken over the sickle, where $d(\tau + \delta \tau)$ means the quantity derived from

$$\sqrt{-(dx_1 + d\delta x_1)^2 - (dx_2 + d\delta x_2)^2 - (dx_3 + d\delta x_3)^2 - (dx_4 + d\delta x_4)^2}$$

by means of (9) and

$$dx_1 = w_1 d\tau, \quad dx_2 = w_2 d\tau, \quad dx_3 = w_3 d\tau, \quad dx_4 = w_4 d\tau, \quad d\vartheta = 0$$

thus we have

$$\frac{d(\tau + \delta \tau)}{d\tau} = \sqrt{-\sum_h \left(w_h + \sum_k \frac{\partial \delta x_h}{\partial x_k} w_k \right)^2} \qquad \binom{k = 1, 2, 3, 4}{h = 1, 2, 3, 4}. \tag{11}$$

Now we want to subject the value of the differential quotient

$$\left(\frac{d(N + \delta N)}{d\vartheta} \right)_{(\vartheta = 0)} \tag{12}$$

to a transformation. Since every δx_h as function of the arguments $x_1, x_2, x_3, x_4, \vartheta$ vanishes in general for $\vartheta = 0$, we also have in general $\frac{\partial \delta x_h}{\partial x_k} = 0$ for $\vartheta = 0$. If we now put

$$\left(\frac{\partial \delta x_h}{\partial \vartheta} \right)_{\vartheta = 0} = \xi_h \qquad (h = 1, 2, 3, 4), \tag{13}$$

then it follows by reason of (10) and (11) for the expression (12):

$$-\iiiint v \sum_h w_h \left(\frac{\partial \xi_h}{\partial x_1} w_1 + \frac{\partial \xi_h}{\partial x_2} w_2 + \frac{\partial \xi_h}{\partial x_3} w_3 + \frac{\partial \xi_h}{\partial x_4} w_4 \right) dx\,dy\,dz\,dt.$$

[105] For the systems x_1, x_2, x_3, x_4 on the boundary of the sickle, $\delta x_1, \delta x_2, \delta x_3, \delta x_4$ are to vanish for all values of ϑ, and I therefore also $\xi_1, \xi_2, \xi_3, \xi_4$ are everywhere zero. Accordingly the last integral changes by partial integration into

$$\iiiint \sum_h \xi_h \left(\frac{\partial v w_h w_1}{\partial x_1} + \frac{\partial v w_h w_2}{\partial x_2} + \frac{\partial v w_h w_3}{\partial x_3} + \frac{\partial v w_h w_4}{\partial x_4} \right) dx\, dy\, dz\, dt.$$

Herein the expression in parentheses is

$$= w_h \sum_k \frac{\partial v w_k}{\partial x_k} + v \sum_k w_k \frac{\partial w_h}{\partial x_k}.$$

Here the first sum vanishes due to the continuity condition (6), the second one can be represented as

$$\frac{\partial w_h}{\partial x_1}\frac{dx_1}{d\tau} + \frac{\partial w_h}{\partial x_2}\frac{dx_2}{d\tau} + \frac{\partial w_h}{\partial x_3}\frac{dx_3}{d\tau} + \frac{\partial w_h}{\partial x_4}\frac{dx_4}{d\tau} = \frac{dw_h}{d\tau} = \frac{d}{d\tau}\left(\frac{dx_h}{d\tau}\right)$$

where $d/d\tau$ indicates differential quotients in the direction of the spacetime line of one location. Hence this finally results in *the expression for the differential quotient* (12)

$$\iiiint v \left(\frac{dw_1}{d\tau}\xi_1 + \frac{dw_2}{d\tau}\xi_2 + \frac{dw_3}{d\tau}\xi_3 + \frac{dw_4}{d\tau}\xi_4 \right) dx\, dy\, dz\, dt. \qquad (14)$$

For a virtual displacement in the sickle we had demanded in addition that the points, considered as points of a substance, should proceed normal to the curves that are formed from them; this means for $\vartheta = 0$ that the ξ_h have to satisfy the *condition*

$$w_1\xi_1 + w_2\xi_2 + w_3\xi_3 + w_4\xi_4 = 0. \qquad (15)$$

If we recall the Maxwell stresses in the electrodynamics of bodies at rest and consider on the other hand our results in the sections 12 and 13, then a certain *adaptation of Hamilton's principle* for continuous extended elastic media to the *relativity postulate* suggests itself.

Let there be specified at every spacetime point (as in section 13) a spacetime matrix of the second kind

$$S = \begin{vmatrix} S_{11}, S_{12}, S_{13}, S_{14}, \\ S_{21}, S_{22}, S_{23}, S_{24}, \\ S_{31}, S_{32}, S_{33}, S_{34}, \\ S_{41}, S_{42}, S_{43}, S_{44}, \end{vmatrix} = \begin{vmatrix} X_x, & Y_x, & Z_x, & -iT_x \\ X_y, & Y_y, & Z_y, & -iT_y \\ X_z, & Y_z, & Z_z, & -iT_z \\ -iX_t, & -iY_t, & -iZ_t, & T_t \end{vmatrix} \qquad (16)$$

where $X_x, Y_x, \ldots Z_z, T_x, \ldots X_t, \ldots T_t$ are real quantities. |
For a virtual displacement in a spacetime sickle with the notation as above, let the [106] value of the integral

$$W + \delta W = \iiint \left(\sum_{h,k} S_{hk} \frac{\partial(x_k + \delta x_k)}{\partial x_h} \right) dx\, dy\, dz\, dt, \tag{17}$$

over the region of the sickle be called the *stress action* associated with the virtual displacement.

The sum that occurs here is, written out in more detail and with real quantities

$$X_x + Y_y + Z_z + T_t$$
$$+ X_x \frac{\partial \delta x}{\partial x} + X_y \frac{\partial \delta x}{\partial y} + \dots + Z_z \frac{\partial \delta z}{\partial z}$$
$$- X_t \frac{\partial \delta x}{\partial t} - \dots + T_x \frac{\partial \delta t}{\partial x} + \dots + T_t \frac{\partial \delta t}{\partial t}.$$

Now we will postulate the following *principle of least action* [*Minimalprinzip*] *for mechanics:*

Whenever any spacetime sickle is delimited, then for each virtual displacement in the sickle the sum of the mass action and of the stress action shall always be an extremum for the actually occurring behavior of the spacetime lines in the sickle.

The point of this statement is that for each virtual displacement we shall have (using the notation explained above)

$$\left(\frac{d(\delta N + \delta W)}{d\vartheta} \right)_{\vartheta = 0} = 0. \tag{18}$$

By the methods of variational calculus the following four differential equations follow immediately by transformation (14) from this principle of least action, taking into account the condition (15)

$$\nu \frac{dw_h}{d\tau} = K_h + \kappa w_h \qquad (h = 1, 2, 3, 4), \tag{19}$$

where

$$K_h = \frac{\partial S_{1h}}{\partial x_1} + \frac{\partial S_{2h}}{\partial x_2} + \frac{\partial S_{3h}}{\partial x_3} + \frac{\partial S_{4h}}{\partial x_4} \tag{20}$$

[107] are the components of the spacetime vector of the first kind $K = \text{lor} S$ and κ is a factor to be determined from $w\bar{w} = -1$. By multiplying (19) by w_h and then I summing over $h = 1, 2, 3, 4$ one finds $\kappa = K\bar{w}$ and $K + (K\bar{w})w$ clearly becomes a spacetime vector of the first kind *normal* to w. Writing the components of this vector as

$$X, Y, Z, iT,$$

we arrive at the following *laws for the motion of matter:*

$$v\frac{d}{d\tau}\frac{dx}{d\tau} = X,$$

$$v\frac{d}{d\tau}\frac{dy}{d\tau} = Y,$$

$$v\frac{d}{d\tau}\frac{dz}{d\tau} = Z,$$
(21)

$$v\frac{d}{d\tau}\frac{dt}{d\tau} = T.$$

Here we have

$$\left(\frac{dx}{d\tau}\right)^2 + \left(\frac{dy}{d\tau}\right)^2 + \left(\frac{dz}{d\tau}\right)^2 = \left(\frac{dt}{d\tau}\right)^2 - 1$$

and

$$X\frac{dx}{d\tau} + Y\frac{dy}{d\tau} + Z\frac{dZ}{d\tau} = T\frac{dt}{d\tau},$$

and by reason of these relations the fourth of the equations (21) could be viewed as a consequence of the first three of them.

From (21) we further derive the laws for the motion of a *point mass*, that is for the course of an infinitely thin spacetime thread.

Let x, y, z, t denote a point of the center line, defined arbitrarily within the thread. We form the equations (21) for the points of the thread's *normal section* through x, y, z, t and integrate their product by the volume element [*Inhaltselement*] of the section over the entire region of the normal section. Let the integrals of the right-hand sides be R_x, R_y, R_z, R_t and m be the constant mass of the thread, yielding

$$m\frac{d}{d\tau}\frac{dx}{d\tau} = R_x,$$

$$m\frac{d}{d\tau}\frac{dy}{d\tau} = R_y,$$

$$m\frac{d}{d\tau}\frac{dz}{d\tau} = R_z,$$
(22)

$$m\frac{d}{d\tau}\frac{dt}{d\tau} = R_t.$$

| Here R, with the components R_x, R_y, R_z, iR_t, is again a spacetime vector of the first   [108]
kind, which is *normal* to spacetime vector of the first kind w, the velocity of the mass point, with components

$$\frac{dx}{d\tau}, \qquad \frac{dy}{d\tau}, \qquad \frac{dz}{d\tau}, \qquad i\frac{dt}{d\tau}.$$

We will call this vector R the *accelerating force* [*bewegende Kraft*] of the mass point.

If, on the other hand, one integrates the equations not over the thread's normal section but correspondingly over a cross section *normal to the t-axis* and passing through x, y, z, t then one obtains (see (4)) the equations (22) multiplied by $d\tau/dt$; in particular the last equations is

$$m\frac{d}{dt}\left(\frac{dt}{d\tau}\right) = \mathfrak{w}_x R_x \frac{d\tau}{dt} + \mathfrak{w}_y R_y \frac{d\tau}{dt} + \mathfrak{w}_z R_z \frac{d\tau}{dt}.$$

The right-hand side will have to be interpreted as the *work done* on the point mass per unit time. The equation itself will then be viewed as the *energy theorem* for the motion of the point mass, and the expression

$$m\left(\frac{dt}{d\tau} - 1\right) = m\left(\frac{1}{\sqrt{1 - \mathbf{w}^2}} - 1\right) = m\left(\frac{1}{2}|\mathfrak{w}|^2 + \frac{3}{8}|\mathfrak{w}|^4 + \cdots\right)$$

will be identified with the *kinetic energy* of the point mass.

Since one always has $dt > d\tau$, one could characterize the quotient $(dt - d\tau)/(d\tau)$ as the advancement of time with respect to proper time of the point mass, and then put it as follows: the kinetic energy of the point mass is the product of its mass and the advancement of time with respect to its proper time.

The *quadruplet* of equations (22) again shows the full symmetry in x, y, z, it as demanded by the relativity postulate, *where the fourth equation*, analogous to what we have encountered in electrodynamics, *can be said, as it were, to be more highly evident physically*. On the basis of this symmetry the triplet of the first three equations is to be constructed according to the pattern of the fourth equation, and in view of this circumstance the claim is justified: *If the relativity postulate is put at the head [109] of mechanics, then the complete | laws of motion follow solely from the energy theorem.*

I would not like omit making it plausible that a contradiction to the assumptions of the relativity postulate is not to be expected from the phenomena of *gravitation.*[2]

For $B^*(x^*, y^*, z^*, t^*)$ a fixed spacetime point, the region of all spacetime points $B(x, y, z, t)$ satisfying

$$(x - x^*)^2 + (x - y^*)^2 + (z - z^*)^2 = (t - t^*)^2, \qquad t - t^* \geqq 0 \tag{23}$$

shall be called the *ray object* [future lightcone] of the spacetime point B^*.

This object intersects any spacetime line only in a single spacetime point B, as follows on the one hand from the *convexity* of the object, and on the other hand from the circumstance that all possible directions of the spacetime line are directed from B^* only toward the concave side of the object. B^* shall then be called a *light point* of B.

---

2   H. Poincaré *Rend. Circ. Matem. Palermo*, Vol. XXI (1906), p. 129 [in this volume] has attempted to make the Newtonian law of attraction compatible with the relativity postulatem, along quite different lines than what I present here.

If the point $B(x, y, z, t)$ in the condition (23) is considered as fixed, and the point $B^*(x^*, y^*, z^*, t^*)$ as variable, then the same relation represents the region of all spacetime points B^* that are light points of B, and one shows in an analogous way that on any spacetime line there always occurs just a single point B^* that is a light point of B.

Let a mass point F of mass m experience an accelerating force, in the presence of another mass point F^* of mass m^* according to the following law. Consider the spacetime threads of F and F^* with center lines contained in them. Let BC be an infinitesimal element of the center line of F, further let B^* and C^* be the light points of B and C, respectively, on the center line of F^*, and let OA' be the radius vector parallel to B^*C^* of the fundamental hyperbolic shell (2), and finally let D^* be the intersection point of the line B^*C^* with its normal space that passes through B. *The accelerating force of the point mass F at the spacetime point B shall now be that spacetime vector of the first kind, normal to BC, which is formed additively from the vector*

$$mm^*\left(\frac{OA'}{B^*D^*}\right)^3 BD^* \tag{24}$$

in the direction BD^* and a suitable vector ∣ in the direction B^*C^*. Here OA'/B^*D^* [110] is understood to be the ratio of the two parallel vectors concerned.

It is clear that this definition is to be characterized as covariant with respect to the Lorentz group.

Now we ask how the spacetime thread of F behaves according to these considerations if the point mass F^* executes uniform translational motion, so that the center line of the thread of F^* is a straight line. We shift the spacetime origin O to it and can then introduce this straight line as the t-axis by means of a Lorentz transformation. Now let x, y, z, t mean the point B and τ^* the proper time of the point B^*, with origin at O. Our definition then leads to the equations

$$\frac{d^2x}{d\tau^2} = -\frac{m^*x}{(t-\tau^*)^3}, \qquad \frac{d^2y}{d\tau^2} = -\frac{m^*y}{(t-\tau^*)^3}, \qquad \frac{d^2z}{d\tau^2} = -\frac{m^*z}{(t-\tau^*)^3} \tag{25}$$

and

$$\frac{d^2t}{d\tau^2} = -\frac{m^*}{(t-\tau^*)^2}\frac{d(t-\tau^*)}{dt}, \tag{26}$$

where we have

$$x^2 + y^2 + z^2 = (t-\tau^*)^2 \tag{27}$$

and

$$\left(\frac{dx}{d\tau}\right)^2 + \left(\frac{dy}{d\tau}\right)^2 + \left(\frac{dz}{d\tau}\right)^2 = \left(\frac{dt}{d\tau}\right)^2 - 1. \tag{28}$$

In view of (27) the three equations (25) read the same as the equations for the motion of a mass point under attraction by a fixed center according to Newton's law, except that the time t is replaced by the proper time τ of the mass point. The fourth equation (26) then gives the connection between proper time and time for the point mass.

Now let the orbit of the space point x, y, z for different τ be an ellipse with semi-major axis a and eccentricity e, and let E be its eccentric anomaly, T the increase in proper time for a full execution of an orbit, and finally $nT = 2\pi$, so that for a suitable origin of τ Kepler's equation

$$n\tau = E - e \sin E \tag{29}$$

holds. If we also change the unit of time and denote the speed of light by c, then (28) becomes

$$\left(\frac{dt}{d\tau}\right)^2 - 1 = \left(\frac{m^*}{ac^2}\frac{1 + e\cos E}{1 - e\cos E}\right). \tag{30}$$

[111]   I By neglecting c^{-4} compared to 1 it follows that

$$ndt = nd\tau\left(1 + \frac{1}{2}\frac{m^*}{ac^2}\frac{1 + e\cos E}{1 - e\cos E}\right),$$

which by using (29) results in[6]

$$nt + \text{const.} = \left(1 + \frac{1}{2}\frac{m^*}{ac^2}\right)n\tau + \frac{m^*}{ac^2}e\sin E. \tag{31}$$

Here the factor $\dfrac{m^*}{ac^2}$ is the square of the ratio of a certain mean speed of F in its orbit to the speed of light. If we substitute for m^* the mass of the Sun and for a the semi-major axis of the Earth's orbit, then this factor amounts to 10^{-8}.

A law of attraction for masses according to the formulation exhibited above in connection with the relativity postulate would also imply *propagation of gravitation with the speed of light*. In view of the smallness of the periodic term in (31) a decision *against* such a law and the suggested modified mechanics in favor of Newton's law of attraction with Newtonian mechanics should not be derivable from astronomical observations.

EDITORIAL NOTES

[1] The equations (10) to (12) in the actual body of the text (not reproduced here) represent the Lorentz transformations and read

$$\mathfrak{r}'_\mathfrak{v} = \frac{\mathfrak{r}_v - qt}{\sqrt{1 - q^2}} \text{ for the direction of } \mathfrak{v}, \tag{10}$$

$$\mathfrak{r}'_\mathfrak{v} = \mathfrak{r}_{\bar{\mathfrak{v}}} \text{ for any direction } \bar{\mathfrak{v}} \text{ perpendicular to } \mathfrak{v}, \tag{11}$$

$$\text{and further } t' = \frac{-q\mathfrak{r}_\mathfrak{v} + t}{\sqrt{1 - q^2}}. \tag{12}$$

[2] For some of Minkowski's technical terms, the English equivalent according to present-day physics is given in square brackets.

[3] The references are to sections in the actual body of the text. $\mathfrak{w}$ denotes the three-dimensional velocity vector of matter at the given spacetime point, w denotes the corresponding four-vector:

$$\left(\frac{\mathfrak{w}}{\sqrt{1 - \mathfrak{w}^2}}, \frac{i}{\sqrt{1 - \mathfrak{w}^2}} \right).$$

[4] Minkowski defines lor as the 1×4 matrix

$$\left(\frac{\partial}{\partial x_1}, \frac{\partial}{\partial x_2}, \frac{\partial}{\partial x_3}, \frac{\partial}{\partial x_4} \right).$$

[5] Equation (67) in the body of the text reads

$$\text{lor}\,\bar{s} = \frac{\partial s_1}{\partial x_1} + \frac{\partial s_2}{\partial x_2} + \frac{\partial s_3}{\partial x_3} + \frac{\partial s_4}{\partial x_4}. \tag{67}$$

[6] The eccentricity e in the second term on the right-hand side of eq. (32) has been inserted, correcting an obvious omission.

HENDRIK A. LORENTZ

OLD AND NEW QUESTIONS IN PHYSICS
(EXCERPT)

Originally published as "Alte und neue Fragen der Physik" in Physikalische Zeitschrift, 11, 1910, pp. 1234–1257. Based on the Wolfskehl lectures given in Göttingen, 24–29 October 1910. The second and third lectures are translated here (pp. 1236–1244). Lorentz made use of Born's notes from the lecture and submitted the paper on 2 November 1910.

SECOND LECTURE

To discuss Einstein's *principle of relativity* here at Göttingen, where Minkowski was active, seems to me a particularly welcome task.

The significance of this principle can be illuminated from several different angles. We will not speak here of the mathematical aspect of the question, which was given such a splendid treatment by Minkowski, and which was further developed by Abraham, Sommerfeld and others. Rather, after some epistemological remarks about the concepts of space and time, the physical phenomena that may contribute to an experimental test of the principle shall be discussed.

The principle of relativity claims the following: If a physical phenomenon is described by certain equations in the system of reference x, y, z, t, then a phenomenon will also exist that can be described by the same equations in another system of reference x', y', z', t'. Here the two systems of reference are connected by relations containing the speed of light c and expressing the motion of one system with a uniform velocity relative to the other.

If observer A is located in the first, and B in the second system of reference, and each is supplied with measuring rods and clocks at rest in his system, then A will measure the values of x, y, z, t, and B the values of x', y', z', t', where it should be noted that A and B can also use one and the same measuring rod and the same clock. We have to assume that when the first observer somehow hands his rod and clock over to the second observer, they automatically assume the proper length and the proper rate so that B arrives at the values x', y', z', t' from his measurements. Either one will then find the same value for the speed of light, and will quite generally be able to make the same observations.

Jürgen Renn (ed.). *The Genesis of General Relativity,* Vol. 3
Gravitation in the Twilight of Classical Physics: Between Mechanics, Field Theory, and Astronomy.
© 2007 Springer.

Assume there is an aether; then among all systems x, y, z, t a single one would be distinguished by the state of rest of its coordinate axes as well as its clock in the aether. If one associates with this the idea (also held tenaciously by the speaker) that space and time are totally different from each other and that there is a "true time" (simultaneity would then exist independent of location, corresponding to the circumstance that we are able to imagine infinitely large velocities) then it is easily seen that this true time should be shown precisely by clocks that are at rest in the aether. Now, if the principle of relativity were generally valid in nature, then we would of course not be in a position to determine whether the system of reference being used at the moment is the distinguished one. Thus one arrives at the same results as those found when one denies the existence of the aether and of the true time, and regards all systems of reference as equivalent, following Einstein and Minkowski. It is surely up to each individual which of the two schools of thought he wishes to identify with.

[1237]   In order to discuss the physical aspect of the question, we first have to establish the transformation formulas, limiting ourselves to a special form I already used in the year 1887 by W. Voigt in his treatment of the Doppler principle; namely,

$$x' = x, \qquad y' = y, \qquad z' = az - bct, \qquad t' = at - \frac{b}{c}z;$$

where the constants $a > 0$, b satisfy the relation

$$a^2 - b^2 = 1,$$

which entails the identity $x'^2 + y'^2 + z'^2 - c^2 t'^2 = x^2 + y^2 + z^2 - c^2 t^2$. The origin of the system x', y', z' moves with respect to the system x, y, z in the z-direction with speed $(b/a)c$, which is always less than c. In general we have to assume that every velocity is less than c.

All state variables of any phenomenon, measured in one or the other system are connected by certain transformation formulas. For example, for the speed of a point these are

$$v_x' = \frac{v_x}{\omega}, \qquad v_y' = \frac{v_y}{\omega}, \qquad v_z' = \frac{av_z - bc}{\omega},$$

where

$$\omega = a - \frac{bv_z}{c}.$$

Further we consider a system of points whose velocity is a continuous function of the coordinates. Let dS be a volume element surrounding the point $P(x, y, z)$ at time t; to this value of t there corresponds according to the transformation equations a point P in time t' in the other system of reference, and every point lying in dS at time t has certain [coordinates] x', y', z' for this fixed value of t'. The points x', y', z' fill a volume element dS', which is related to dS as follows[1]

$$dS' = \frac{dS}{\omega}.$$

Let us imagine some agent (matter, electricity etc.) connected with the points, and let us assume that the observer B has occasion to associate with each point the same amount of the agent as the observer A, then the spatial densities must obviously be in the inverse ratio as the volume elements, that is

$$\rho' = \omega\rho.$$

All these relations are reciprocal, that is, the primed and unprimed letters may be interchanged if at the same time b is replaced by $-b$.

The basic equations of the electromagnetic field retain their form under the transformation if the following quantities are introduced:

$$\mathfrak{d}_x' = a\mathfrak{d}_x - b\mathfrak{h}_y, \qquad \mathfrak{d}_y' = a\mathfrak{d}_y + b\mathfrak{h}_x, \qquad \mathfrak{d}_z' = \mathfrak{d}_z,$$
$$\mathfrak{h}_x' = a\mathfrak{h}_x + b\mathfrak{d}_y, \qquad \mathfrak{h}_y' = a\mathfrak{h}_y - b\mathfrak{d}_x, \qquad \mathfrak{h}_z' = \mathfrak{h}_z.$$

Thus in the system x', y', z', t' the following equations hold between these quantities, the transformed space density ρ' and the transformed velocity $\mathfrak{v}'$:

$$\text{div } \mathfrak{d}' = \rho',$$
$$\text{div } \mathfrak{h}' = 0,$$
$$\text{curl } \mathfrak{h}' = \frac{1}{c}(\dot{\mathfrak{d}}' + \rho'\mathfrak{v}'),$$
$$\text{curl } \mathfrak{d}' = -\frac{1}{c}\dot{\mathfrak{h}}'.$$

With this the field equations of the electron theory satisfy the principle of relativity; but there is still the matter of harmonizing the equations of motion of the electrons themselves with this principle.

We will consider somewhat more generally the motion of an arbitrary material point. Here it is useful to introduce the concept of "proper time," Minkowski's beautiful invention. According to this there belongs to each point a time of its own, as it were, which is independent of the system of reference chosen; its differential is defined by the equation

$$d\tau = \sqrt{1 - \frac{v^2}{c^2}}dt.$$

The expressions formed with the aid of the proper time τ,

$$\frac{d}{d\tau}\frac{dx}{d\tau}, \qquad \frac{d}{d\tau}\frac{dy}{d\tau}, \qquad \frac{d}{d\tau}\frac{dz}{d\tau},$$

which are linear homogeneous functions of the components of the ordinary acceleration, will be called the components of the "Minkowskian acceleration." We describe
the motion of a point by the equations:

$$m\frac{d}{d\tau}\frac{dx}{d\tau} = \Re_x, \text{ etc.,}$$

where m is a constant, which we call the "Minkowskian mass." We designate the
vector $\Re$ as the "Minkowskian force."

   It is then easy to derive the transformation formulas for this acceleration and this
force; we leave m unchanged. Thus we have

$$\Re_x{}' = \Re_x, \qquad \Re_y{}' = \Re_y, \qquad \Re_z{}' = a\Re_z - \frac{b}{c}(\mathfrak{v} \cdot \Re).$$

The essential point is the following. The principle of relativity demands that if for an
actual phenomenon the Minkowskian forces depend in a certain way on the coordinates, velocities, etc. in one system of reference, then the transformed Minkowskian
forces I depend in the same way on the transformed coordinates, velocities etc. in the
other system of reference. This is a special property that must be shared by all forces
in nature if the principle of relativity is to be valid. Presupposing this we can calculate
the forces acting on moving bodies if we know them for the case of rest. For example,
if an electron of charge e is in motion, we consider a system of reference in which it
is momentarily at rest. Then the electron in this system is under the influence of the
Minkowskian force

$$\Re = c\mathfrak{d};$$

from this it follows by application of the transformation equations for $\Re$ and $\mathfrak{d}$ that
the Minkowskian force acting on the electron that moves with velocity $\mathfrak{v}$ in an arbitrary coordinate system amounts to

$$\Re = c\frac{\mathfrak{d} + \dfrac{1}{c}[\mathfrak{v} \cdot \mathfrak{h}]}{\sqrt{1 - \dfrac{\mathfrak{v}^2}{c^2}}}.$$

This formula does not agree with the usual ansatz of the electron theory, because of
the presence of the denominator. The difference is due to the fact that usually one
does not operate with our Minkowskian force, but with the "Newtonian force" $\mathfrak{F}$,
and we see that these two forces are related as follows for an electron:

$$\mathfrak{F} = \Re\sqrt{1 - \frac{\mathfrak{v}^2}{c^2}}.$$

It is to be assumed that this relation is valid for arbitrary material points.

Thus the phenomena of motion can be treated in two different ways, using either the Minkowskian or the Newtonian force. In the latter case the equations of motion take the form

$$\mathfrak{F} = m_1 \mathfrak{j}_1 + m_2 \mathfrak{j}_2,$$

and here $\mathfrak{j}_1$ means the ordinary acceleration in the direction of motion, $\mathfrak{j}_2$ the ordinary normal acceleration, and the factors

$$m_1 = \frac{m}{\sqrt{\left(1 - \frac{v^2}{c^2}\right)^3}}, \qquad m_2 = \frac{m}{\sqrt{1 - \frac{v^2}{c^2}}},$$

are called the "longitudinal" and the "transverse mass."

Just like the Minkowskian forces, the Newtonian forces that occur in nature must also fulfill certain conditions in order to satisfy the relativity principle. This is the case if, for example, a normal pressure of a constant magnitude p per unit area acts on a surface regardless of the state of motion; then in the transformed system a normal pressure of the same magnitude acts on the corresponding moving surface element. Since we have already recognized the invariance of the field equations, the question of whether the motions in an electron system satisfy the relativity principle amounts merely to an experimental test of the formulas for the longitudinal and transverse masses m_1, m_2; although the experiments of Bucherer and Hupka seem to confirm these formulas, one has not yet arrived at a definitive decision.

Concerning the mass of the electron, one should remember that this is electromagnetic in nature; so it will depend on the distribution of charge within the electron. Therefore the formulas for the mass can be correct only if the charge distribution, and hence also the shape of the electron, vary in a definite way with the velocity. One must assume that an electron, which is a sphere when at rest, becomes an ellipsoid that is flattened in the direction of motion as a result of translation; the amount of flattening is

$$\sqrt{1 - \frac{v^2}{c^2}}.$$

If we assume that the shape and size of the electron are regulated by internal forces, then to agree with the relativity principle these forces must have properties such that this flattening occurs by itself when in motion. Regarding this Poincaré has made the following hypothesis. The electron is a charged, expandible skin, and the electrical repulsion of the different points of the electron is balanced by an inner normal tension of unchangeable magnitude. Indeed, according to the above such normal tensions satisfy the relativity principle.

In the same way all molecular forces acting within ponderable matter, as well as the quasi-elastic and resistive forces acting on the electron, have to satisfy certain conditions in order to be in accord with the relativity principle. Then every moving body will be unchanged for a co-moving observer, but for an observer at rest it will experience a change in dimensions, which is a consequence of the change in molecular forces demanded by these conditions. This also leads automatically to the contraction of bodies, which was already I devised earlier to explain the negative outcome of Michelson's interferometer experiment and of all similar experiments that were to determine an influence of the Earth's motion on optical phenomena.

[1239]

Concerning rigid bodies, as investigated by Born, Herglotz, Noether, and Levi-Civita, the difficulties occurring in the consideration of rotation can surely be relieved by ascribing their rigidity to the action of particularly intense molecular forces.

Finally let us turn to *gravitation*. The relativity principle demands a modification of Newton's law, foremost a propagation of the effect with the speed of light. The possibility of a finite speed of propagation of gravity was already discussed by Laplace, who imagined as the cause of gravity a fluid streaming toward the Sun, which pushes the planets toward the Sun. He found that the speed c of this fluid must be assumed to be at least 100 million times larger than that of light, so that the calculations remain in agreement with the astronomical observations. The necessity of such a large value of c is due to the occurrence of v/c to the first power in his final formulas, where v is the speed of a planet. But if the propagation speed c of gravity is to have the value of the speed of light, as demanded by the relativity principle, then a contradiction with the observations can only be avoided if only quantities of second (and higher) order in v/c occur in the expression for the modified law of gravitation.

Restricting oneself to quantities of second order, one can, on the basis of a suggestive electron-theoretic analogy, easily give a condition that determines the modified law in a unique way. Namely, if one considers the force acting on an electron that moves with a velocity $\mathfrak{v}$,

$$e\left(\mathfrak{d} + \frac{1}{c}[\mathfrak{v} \cdot \mathfrak{h}]\right),$$

then the vectors $\mathfrak{d}$ and $\mathfrak{h}$ depend, in addition, on the velocities $\mathfrak{v}'$ of the electrons that produce the field; in the vector product $[\mathfrak{v} \cdot h]$, products of the form $\mathfrak{v}\mathfrak{v}'$ do occur, but not the square $\mathfrak{v}^2$ of the speed of the electron under consideration. Accordingly let us assume that in the expression for the attraction acting on the point 1 due to point 2 there is no term in the square of the velocity $\mathfrak{v}_1^2$ of point 1. Then all velocities whatsoever must drop out in a system of reference in which point 2 is at rest $(\mathfrak{v}_2 = 0)$; therefore the law will reduce to the usual Newtonian one in this system. Now making the transition by transforming to an arbitrary coordinate system, one finds that the force acting on point 1 is composed of two parts, the first, an attraction in the direction of the line connecting them of magnitude

$$R + \frac{1}{c^2}\left\{\frac{1}{2}\mathfrak{v}_2^2 R + \frac{1}{2}\mathfrak{v}_{2r}^2\left(r\frac{dR}{dr} - R\right) - (\mathfrak{v}_1 \cdot \mathfrak{v}_2)R\right\},$$

the second, a force in the direction of $\mathfrak{v}_2$ of magnitude

$$\frac{1}{c^2} v_{1r} \, R v_2 \, ;$$

here r means the distance between two simultaneous positions of the two points, v_r the component of v along the connecting line drawn from 1 to 2, and R that function of r which represents the law of attraction in the case of rest ($R = k/r^2$ for Newtonian attraction, $R = kr$ for quasielastic forces). Note that "force" is always understood to be the "Newtonian force," not the "Minkowskian" one. Minkowski, by the way, has given a somewhat different expression for the law of gravity. The latter as well as the one described above can be found in Poincaré.

THIRD LECTURE

At the end of the previous lecture a modified law of gravitation was given, which is in agreement with the relativity principle. Concerning this one should note that the principle of equality of action and reaction is not satisfied.

Now the perturbations that can arise due to those additional second order terms will be discussed. Besides many short-period perturbations, which have no significance, there is a secular motion of the planets' perihelia. Prof. de Sitter computes $6.69''$ per century for Mercury's perturbations. Since Laplace, it has been known that Mercury has an anomalous perihelion motion of $44''$ per century; although this anomaly has the right sign, it is much too large to be explained by those additional terms. Instead, Seeliger attributes it to a perturbation by the carrier of the zodiacal light, | whose mass one can suitably determine in a plausible way. So, from this one [1240] can arrive at no decision, as long as the accuracy of astronomical measurements is not significantly increased. To be absolutely accurate one would also have to take into account the difference between the Earth's "proper time" and the time of the solar system.

A different method to test the validity of the modified law of gravitation can be based on a procedure suggested by Maxwell to decide whether the solar system moves through the aether. If this were the case, then the *eclipses of Jupiter's moons* should be advanced or delayed depending on Jupiter's position with respect to the Earth.

For if the Jupiter-Earth distance is a and the component of the solar system's velocity in the aether in the direction of the line connecting Jupiter to Earth is v, then the time required to cover the distance a in the case of rest, a/c, would be changed to $a/(c+v)$; thus the motion brings about an advance or delay, which amounts to av/c^2 up to terms of second order, and which takes on different values according to the value of the velocity component v, which of course depends on the position of the two planets. Now it is clear that such a dependence of the phenomena on the motion through the aether contradicts the relativity principle.

In order to clear up this contradiction let us simplify the situation schematically. We suppose that the Sun S has a mass that is infinitely large compared to that of the planet. Let the velocity of the solar system coincide with the z- axis, which we lay through the Sun. The intersection points of the planet's orbit with the z- axis are denoted as the upper resp. lower transit, A resp. B. (Fig. 1)

We place the observer at the Sun. At each transit of the planet through the z- axis a signal will propagate towards the Sun. The period of revolution shall be T. When the Sun is at rest the time between an upper and lower transit will be $(1/2)T$ for the assumed circular motion; the same is true for the time between the arrivals of the two light signals. By contrast, if the Sun is in motion in the z- direction, the signal from the upper transit must suffer an advance av/c^2, that from the lower transit a delay of the same amount; if the uniform orbital motion (assumed as self-evident by Maxwell) is preserved without perturbation, the time interval between the arrivals of the light signals of two successive passes would appear alternately increased and decreased by $2av/c^2$. Preservation of the uniform circular motion during a translation through the aether, as is assumed above, is, however, impossible according to the relativity principle. For if we describe the process in a coordinate system that does not take part in the motion, the modified law of gravitation will have to be applied, and this results in a non-uniform planetary motion, due to which the difference in time intervals between the arrivals of the light signals exactly cancels.

Therefore the determination of whether an advance or delay of the eclipses in fact occurs can be used to decide in favor or against the relativity principle. However, the numerical situation is again rather unfavorable. Thus Mr. Burton, who has access to 330 photometric observations of eclipses of Jupiter's first moon made at the Harvard observatory, estimates the probable error of the final result for v as 50 km/sec; on the other hand, one has observed speeds of stars of 70 km/sec, and the speed of the solar system with respect to the fixed stars is estimated at 20 km/sec. The relativity principle is therefore hardly supported by Burton's calculations; at best they could invalidate it, namely if, for example, the final result were a value exceeding 100 km/sec.

Let us leave it undecided whether or not the new mechanics will receive confirmation by astronomical observations. But we will not fail to familiarize ourselves with some of its basic formulas.

[1241]   If one defines work as the scalar product of "Newtonian force" and I displacement, then the equations of motion yield the *energy principle* in its usual form, so that the work done per unit of time equals the increase in energy ε :

$$\mathfrak{F}_x \frac{dx}{dt} + \mathfrak{F}_y \frac{dy}{dt} + \mathfrak{F}_z \frac{dz}{dt} = \frac{d\varepsilon}{dt}.$$

Here *energy* is expressed by

$$\varepsilon = mc^2 \left(\frac{1}{\sqrt{1 - \dfrac{v^2}{c^2}}} - 1 \right);$$

this agrees up to second order terms with the value of the kinetic energy in customary mechanics:

$$\varepsilon = \frac{1}{2}mv^2.$$

Furthermore, from the equations of motion one can derive *Hamilton's principle*

$$\int_{t_1}^{t_2} (\delta L + \delta A)\,dt = 0;$$

here δA is the work of the "Newtonian force" upon a virtual displacement and L is the *Lagrangian*, which takes the form

$$L = -mc^2 \left(\sqrt{1 - \frac{v^2}{c^2}} - 1 \right).$$

From Hamilton's principle one can conversely obtain the equations of motion. The quantities

$$\frac{\partial L}{\partial \dot{x}}, \qquad \frac{\partial L}{\partial \dot{y}}, \qquad \frac{\partial L}{\partial \dot{z}}$$

are to be identified as the *components of the momentum*.

All these formulas can be verified by the electromagnetic equations of motion for an electron. One then has to take the following value for the "Minkowskian mass" m,

$$m = \frac{e^2}{6\pi Rc^2},$$

and to add to the electric and magnetic energy the energy of those internal stresses which determine the shape of the electron, as we saw above. Thus from the general principle of least action for arbitrary electromagnetic systems, discussed in the first lecture, one can obtain Hamilton's principle for a point mass as given above by specialization to an electron, but the work of those internal stresses must be taken into account.

We now go over to a discussion of the *equations of the electromagnetic field for ponderable bodies*. These have been written down purely phenomenologically by Minkowski, then M. Born and Ph. Frank showed that they can also be derived from the ideas of electron theory; by the latter procedure Lorentz himself also found these equations, in a slightly different technical form.

To obtain relations between observable quantities one must smear out the details of the phenomena due to individual electrons by averaging over a large number of them. One is lead to the following equations (which are identical to those of the usual Maxwell theory):

$$\operatorname{div}\mathfrak{D} = \rho_l,$$

$$\operatorname{div}\mathfrak{B} = 0,$$

$$\operatorname{curl}\mathfrak{H} = \frac{1}{c}(\mathfrak{C} + \dot{\mathfrak{D}}),$$

$$\operatorname{curl}\mathfrak{C} = -\frac{1}{c}\dot{\mathfrak{B}}.$$

Here $\mathfrak{D}$ is the dielectric displacement, $\mathfrak{B}$ the magnetic induction, $\mathfrak{H}$ the magnetic force, $\mathfrak{C}$ the electric force, $\mathfrak{C}$ the electric current, and ρ_l the density of the observable electric charges. Denoting mean values by an overbar we have, for example,

$$\mathfrak{C} = \bar{\mathfrak{d}}, \qquad \mathfrak{B} = \bar{\mathfrak{h}},$$

where $\mathfrak{d}$, $\mathfrak{h}$ have their former meaning; further we have

$$\mathfrak{D} = \mathfrak{C} + \mathfrak{P},$$

$$\mathfrak{H} = \mathfrak{B} - \mathfrak{M} - \frac{1}{c}[\mathfrak{P} \cdot \mathfrak{w}],$$

where $\mathfrak{P}$ is the electric moment and $\mathfrak{M}$ the magnetization per unit volume, and $\mathfrak{w}$ denotes the velocity of matter. When deriving these formulas one divides the electrons into three types. The first type, the polarization electrons, generate the electric moment $\mathfrak{P}$ by their displacement; the second type, the magnetization electrons, generate the magnetic state $\mathfrak{M}$ by their orbital motion; the third type, the conduction electrons, move freely within the matter and generate the observable charge density ρ_l and the current $\mathfrak{C}$. The latter is additionally to be divided into two parts; for if $\mathfrak{u}$ is the relative velocity of the electrons with respect to the matter, then the total velocity of the electrons is $v = \mathfrak{w} + \mathfrak{u}$, hence the current carried by them is

$$\mathfrak{C} = \overline{\rho v} = \overline{\rho \mathfrak{w}} + \overline{\rho \mathfrak{u}};$$

$\bar{\rho}$ is the observable charge ρ_l, $\bar{\rho}\mathfrak{w}$ is the convection current, and $\overline{\rho \mathfrak{u}}$ the conduction current proper.

[1242]   There are transformation formulas for all these quantities, I and we give a few of them below:

$$\mathfrak{C}_x' = \mathfrak{C}_x, \qquad \mathfrak{C}_y' = \mathfrak{C}_y, \qquad \mathfrak{C}_z' = a\mathfrak{C}_z - bc\rho_l,$$

$$\rho_l' = a\rho_l - \frac{b}{c}\mathfrak{C}_z,$$

$$\mathfrak{P}_x' = a\mathfrak{P}_x - \frac{b}{c}(\mathfrak{w}_z\mathfrak{P}_x - \mathfrak{w}_x\mathfrak{P}_z) + b\mathfrak{M}_y,$$

$$\mathfrak{P}_y' = a\mathfrak{P}_y - \frac{b}{c}(\mathfrak{w}_z\mathfrak{P}_y - \mathfrak{w}_y\mathfrak{P}_z) + b\mathfrak{M}_x,$$

$$\mathfrak{P}_z' = \mathfrak{P}_z.$$

Further, the following auxiliary vectors are useful:

$$\mathfrak{H}_1 = \mathfrak{H} - \frac{1}{c}[\mathfrak{w}\cdot\mathfrak{D}], \qquad \mathfrak{B}_1 = \mathfrak{B} - \frac{1}{c}[\mathfrak{w}\cdot\mathfrak{C}],$$

$$\mathfrak{C}_1 = \mathfrak{C} + \frac{1}{c}[\mathfrak{w}\cdot\mathfrak{B}], \qquad \mathfrak{D}_1 = \mathfrak{D} + \frac{1}{c}[\mathfrak{w}\cdot\mathfrak{H}].$$

Now the field equations given above must still be completed by establishing the relations that exist between the vectors $\mathfrak{C}, \mathfrak{H}$ and $\mathfrak{D}, \mathfrak{B}$. These relations can be obtained in two ways.

The first, phenomenological method proceeds as follows: One considers an arbitrarily moving point of matter and introduces a system of reference in which it is at rest; if the element of volume surrounding the point is isotropic in the rest system, the equation appropriate for bodies at rest (between $\mathfrak{C}$ and $\mathfrak{D}$, for example)

$$\mathfrak{D} = \varepsilon\mathfrak{C}$$

holds; or equally well

$$\mathfrak{D}_1 = \varepsilon\mathfrak{C}_1,$$

because the auxiliary vectors $\mathfrak{D}_1, \mathfrak{C}_1$ are identical with $\mathfrak{D}, \mathfrak{C}$ when $\mathfrak{w} = 0$ But $\mathfrak{D}_1$ and $\mathfrak{C}_1$ transform in the same way, and this implies that also in the original system the equation

$$\mathfrak{D}_1 = \varepsilon\mathfrak{C}_1,$$

and correspondingly

$$\mathfrak{B}_1 = \mu\mathfrak{H}_1,$$

remains valid. Concerning the conduction current we remark only that it depends on $\mathfrak{C}_1$.

The second method has its roots in the mechanics of electrons. Just as the equation $\mathfrak{D} = \varepsilon\mathfrak{C}$ for bodies at rest turns out to be a consequence of the assumption of quasielastic forces, which restore the electrons to their rest positions, so one will obtain the equation $\mathfrak{D}_1 = \varepsilon\mathfrak{C}_1$ for moving bodies if one ascribes to the quasi-elastic forces the properties demanded by the relativity principle. The latter will be satisfied if one takes for these forces the expression of the generalized law of attraction, where R must be taken proportional to r.

The explanation of the resistance to conduction proceeds similarly. A satisfactory electron-theoretic explanation of the magnetic properties of matter is presently not at hand.

Finally the significance of the above equations shall be elucidated in three remarkable cases.

The *first remark* is connected with the equation

$$\rho_l' = a\rho_l - \frac{b}{c}\mathfrak{C}_z.$$

According to this, ρ_l' can vanish without having $\rho_l = 0$ if a current $\mathfrak{C}$ is present; that is, an observer A will declare a body to be charged that must be treated as uncharged by B moving relative to him. One can understand this by noting that every body contains an equal number of positive and negative electrons, which compensate in uncharged bodies. When the body moves with velocity $\mathfrak{w}$ in the presence of a conduction current, the two types of electrons will attain different total velocities, therefore the quantity $\omega = a - b(\mathfrak{v}_z/c)$ will also have different values for the two types. When an observer B moving with the body calculates the mean charge density $\bar{\rho}' = \overline{\omega\rho}$ for both types of electrons he can obtain zero for the sum, even when for an observer A in whose reference frame the body is moving the mean values $\bar{\rho}$ of the positive and negative electrons do not compensate.

This circumstance calls forth the memory of an old question. Around the year 1880 there was a great discussion among physicists about *Clausius' fundamental law* of electrodynamics. One attempted to derive a contradiction between this law and observations by concluding that according to the law a current-carrying conductor on the Earth would have to exert an influence on a co-moving charge e due to the motion of the Earth, which could have been observed. That the law actually does not demand this influence was noted by Budde; this is because the current due to the Earth's motion acts on itself and causes a "compensating charge" in the current-carrying conductor, which exactly cancels the first influence. The electron theory leads to similar conclusions, and Lorentz finds |

[1243]

$$\frac{1}{c^2}\mathfrak{w}_z\,\mathfrak{C}_z;$$

for the density of the compensating charge, if the velocity is in the direction of the z-axis; this must be assumed by an observer A who does not take part in the motion of the Earth, whereas it does not exist for a co-moving observer B. The value given above agrees exactly with the formula derived from the relativity principle; for if $\rho_l' = 0$, one finds from this formula

$$\rho_l = \frac{b}{ac}\mathfrak{C}_z,$$

and since, according to what was said in the second lecture on p. 288 [p. 1237 in the original], $\mathfrak{w}_z = bc/a$ is the speed of the two systems of reference with respect to each other, one indeed finds

$$\rho_l = \frac{1}{c^2}\mathfrak{w}_z \, \mathfrak{C}_z.$$

The *second remark* starts from the transformation equations for the electric moment $\mathfrak{P}$ p. 296 [p. 1241 in the original] in which the presence of the magnetization $\mathfrak{M}$ lets us recognize the impossibility of differentiating precisely between polarization- and magnetization electrons. Rather, in a magnetized body ($\mathfrak{M} \neq 0$) $\mathfrak{P} = 0$, can vanish when judged from one system of reference, whereas in another $\mathfrak{P}'$ differs from zero. This will now be applied to a special case, where we confine attention to quantities of first order. The body we consider (such as a steel magnet) shall contain only conduction electrons and electrons that produce an $\mathfrak{M}$ but no $\mathfrak{P}$ when the body is at rest; it shall have the shape of an infinitely extended plate, bounded by two planes a, b; the middle plane shall be the yz- plane. (Fig. 2) When it is at rest a constant magnetization $\mathfrak{M}_y$ shall be present, whereas $\mathfrak{P} = 0$. When the body is given a speed v in the z- direction an observer who does not take part in the motion will observe the electric polarization

$$\mathfrak{P}_x = -\frac{v}{c}\mathfrak{M}_y.$$

Now we imagine at either side of the body two conductors c, d, which form together with it two equal condensers, and these shall be shorted out by a wire (from c to d). When there is motion, charges will be created on c and d, which can be calculated as follows. Since a current is clearly impossible in the x- direction, we have $\mathfrak{C}_{lx} = 0$ or $\mathfrak{C}_x = (v/c)\mathfrak{B}_y$. Since the process is stationary we have $\dot{\mathfrak{B}} = 0$; then the existence of a potential φ follows from $\mathrm{curl}\mathfrak{C} = 0$. If Δ is the thickness of the slab one has

$$\varphi_a - \varphi_b = \frac{v}{c}\Delta\mathfrak{B}_y.$$

From the symmetry of the arrangement it clearly follows that

$$\varphi_d - \varphi_a = \varphi_b - \varphi_c \, ,$$

and because the plates c, d are shorted out, we must have

$$\varphi_d = \varphi_c \, ;$$

this implies

$$\varphi_d - \varphi_a = -\frac{v}{2c}\Delta\mathfrak{B}_y.$$

If γ is the capacity of one of the two condensers, the charge of plate d becomes

$$-\frac{v}{2c}\gamma\Delta\mathfrak{B}_y,$$

and c receives the equal and opposite amount.

Now we compare this process with the inverse case, that the magnet a, b is at rest and the plates c, d move with the opposite velocity. According to the relativity principle everything would have to be the same as in the first case. Indeed one finds at once from the usual law of induction exactly the amount of charge on plate d given above. But this charge on d must now induce an equal and opposite one on the plane a of the magnet at rest, and corresponding statements must hold for b and c. Since no current can flow ($\mathfrak{C} = 0$), there must be the same charges on the magnet, whether the magnet is moving and the plates are at rest or conversely. So we have to think how it happens that in the first case the opposite charge appears on the plane a of the moving magnet as on the plate d; this becomes possible only due to the polarization $\mathfrak{P}_x = -(v/c)\mathfrak{M}_y$ produced by the motion. For one has

[1244]

$$\mathfrak{D}_x = \mathfrak{C}_x + \mathfrak{P}_x = \frac{v}{c}\mathfrak{B}_y - \frac{v}{c}\mathfrak{M}_y;$$

since here $\mathfrak{P}$ is to be neglected to first order in the velocity, that is the term $[\mathfrak{P} \cdot \mathfrak{w}]$, we have

$$\mathfrak{B} - \mathfrak{M} = \mathfrak{H},$$

But $\mathfrak{H}$ is zero because we assume the plate to be infinitely extended. This implies

$$\mathfrak{D}_x = 0,$$

i.e., in the moving plate there is no dielectric displacement, so the charge on a corresponds to that on d, as the relativity principle demands.

The *last remark* concerns again the circumstance that according to the relativity principle the motion of the Earth cannot have any influence on electromagnetic processes. But Liénard has pointed out a phenomenon where such an influence is to be expected, an influence of first order in magnitude; Poincaré has also discussed this case in his book *Electricité et Optique*. It concerns the ponderomotive force on a conductor. To determine this force one may make the suggestive ansatz for the force acting on the conduction electrons per unit charge:

$$\mathfrak{C}_1 = \mathfrak{C} + \frac{1}{c}[\mathfrak{v} \cdot \mathfrak{B}];$$

then this results in the force caused by the Earth's motion on the conductor in the direction of the motion by an amount

$$\frac{1}{c^2}(\mathfrak{C}_l \cdot \mathfrak{C})\mathfrak{w}_z;$$

since $(\mathfrak{C}_l \cdot \mathfrak{C})$ is the heat generated by the conduction current $\mathfrak{C}$ this expression is easily calculated numerically (which admittedly results in a value inaccessible to observation).

If one now asks oneself, how this result that contradicts the relativity principle can come about, one finds that indeed one has not calculated the force acting on the matter of the conductor, but on the electrons moving inside the conductor. The latter force must first be transferred to the matter by forces, which are unknown to us in detail, and that happens without change of magnitude only if action equals reaction for the forces between matter and electrons. But for moving bodies action does not equal reaction in this case according to the relativity principle, and this circumstance exactly compensates Liénard's force.

In summary, one can say that there is little prospect for an experimental confirmation of the principle of relativity; except for a few astronomical observations, only measurements of the electron mass are worth considering. But one must not forget that the outcome of the negative experiments, such as Michelson's interference experiment and the experiments to find a double refraction caused by the Earth's motion, can only be explained by the relativity principle.

EDITORIAL NOTE

[1] In the original, the denominator is missing from the right-hand side.

THE PROBLEM OF GRAVITATION
AS A CHALLENGE FOR THE
MINKOWSKI FORMALISM

JÜRGEN RENN

THE SUMMIT ALMOST SCALED:
MAX ABRAHAM AS A PIONEER OF A
RELATIVISTIC THEORY OF GRAVITATION

1. THE FRAGILE LADDER OF THE MINKOWSKI FORMALISM

1.1 Abraham's Bold Step

Today Max Abraham is known mainly for his achievements in the field of electrodynamics and, in particular, for the successful series of textbooks associated with his
name.[1] He is, however, largely forgotten as a pioneer of a relativistic theory of gravitation. The papers he dedicated to the subject between 1911 and 1915 are mainly
remembered for the controversy with Einstein that they document.[2] In hindsight it is
clear that Abraham's approach to a relativistic theory of gravitation—an attempt to
formulate a field theory of gravitation in the framework of Minkowski's formalism—
would lead to a dead end. However, only by exploring the consequences of this
approach did it eventually become clear that it did not lead anywhere. And it was,
after all, the failure of Abraham's bold step which encouraged others to either pursue
his endeavor through more appropriate means, as was the case with Gunnar Nordström, or to take even bolder steps than Abraham and attempt even higher summits,
as was the case with Einstein.[3] In fact, it was largely due to Abraham's efforts that
Einstein became familiar with the limits of Minkowski's formalism and also learned
how to overcome them.

In the following, we will first review Abraham's attempt to take up the challenge
of using Minkowski's formalism as the framework for a relativistic theory of gravitation and then show how this bold step led to an ardent controversy with Einstein
which, for Abraham, eventually led to a rejection of relativity theory altogether. We
will then consider some of the insights and achievements that Abraham attained in
the course of his research, which have been largely forgotten because his approach
turned out to lead to a dead end. In short, we will portray Abraham as someone who

---

1   See (Abraham and Föppl 1904–1908) and the subsequent editions of this work.
2   See, in particular (Abraham 1912e, 1912f) and for a historical discussion (Cattani and De Maria
    1989), to which the following account is much indebted.
3   See the papers by Nordström in this volume.

Jürgen Renn (ed.). *The Genesis of General Relativity*, Vol. 3
Gravitation in the Twilight of Classical Physics: Between Mechanics, Field Theory, and Astronomy.
© 2007 Springer.

almost scaled the summit, who aimed high but failed to reach the goal he envisioned. As a consequence of his failure and of criticism by others, he eventually gave up mountain climbing altogether but, at the same time, encouraged others to attempt the summit he had failed to reach, not least because of the magnificent vistas about which, on the basis of his experience, he could report.

The starting point for Abraham's work on a relativistic theory of gravitation were some of the insights that Einstein had attained on the basis of his equivalence principle, in particular the idea of a variable speed of light.[4] Precisely because of this insight, Einstein did not consider Minkowski's formalism to be a useful tool for building up a relativistic theory of gravitation, since he took the constancy of the speed of light to be one of its fundamental principles. Moreover, Einstein was skeptical of such a path for other reasons as well.[5] Abraham, on the other hand, took the bold step of modifying Minkowski's framework by accommodating it to the assumption of a variable speed of light. In this way, he succeeded in overcoming the unacceptably restrictive conditions imposed by the constancy of the speed of light on a straightforward, special relativistic theory of gravitation. Abraham thus became the first to exploit the mathematical potential of Minkowski's four-dimensional formalism for a theory of gravitation.

In the introduction to a paper submitted in December 1911 and published in January 1912, Abraham first of all acknowledges his debt to Einstein's idea concerning the relation between a variable speed of light and the gravitational potential:

> In a recently published paper A. Einstein proposed the hypothesis that the speed of light (c) depends on the gravitational potential (Φ). In the following note, I develop a theory of the gravitational force which satisfies the principle of relativity and derive from it a relation between c and Φ, which in first approximation is equivalent to Einstein's.[6]

Abraham then adopts Minkowski's formalism for his purposes:

> Following Minkowski's presentation we regard
>
> $$x, y, z \qquad \text{and} \qquad u = il = ict \qquad (1)$$
>
> as the coordinates of a four-dimensional space.[7]

---

4 See "The First Two Acts" and "The Third Way to General Relativity" (in vol. 1 and vol. 3 of this series respectively).

5 See, for instance, Einstein's comments in his contemporary letters, e.g., to Wilhelm Wien, 11 March 1912 and 17 May 1912, (CPAE 5, Doc. 371 and 395) to which we shall refer later. See also the historical discussion in the "Editorial Note" in (CPAE 4, 122–128).

6 "In einer vor kurzem erschienenen Arbeit hat A. Einstein die Hypothese aufgestellt, daß die Geschwindigkeit des Lichtes (c) vom Gravitationspotential (Φ) abhänge. In der folgenden Note entwickle ich eine Theorie der Schwerkraft, welche dem Prinzip der Relativität genügt, und leite aus ihr eine Beziehung zwischen c und Φ ab, die in erster Annäherung mit der Einsteinschen gleichwertig ist." See (Abraham 1912h, 1). A complete English translation of this paper is given in this volume.

7 "In dem wir der Darstellung Minkowskis folgen, betrachten wir (x, y, z) und $(u = il = ict)$ als Koordinaten eines vierdimensionalen Raumes." (Abraham 1912h, 1)

He immediately proceeds to state a differential equation for the gravitational potential, which essentially corresponds to a four-dimensional generalization of the classical Poisson equation:

> Let the "rest density" ν, as well as the gravitational potential Φ, be scalars in this space, and let them be linked through the differential equation:

$$\frac{\partial^2 \Phi}{\partial x^2} + \frac{\partial^2 \Phi}{\partial y^2} + \frac{\partial^2 \Phi}{\partial z^2} + \frac{\partial^2 \Phi}{\partial u^2} = 4\pi\gamma\nu. \tag{2}$$

> (γ is the gravitational constant.)[8]

Somewhat later Abraham introduced further Minkowskian concepts and terminology:

> We write $\dot{x}$ $\dot{y}$ $\dot{z}$ $\dot{u}$ for the first derivatives of the coordinates of a material "world point" with respect to its "proper time" τ, i.e., for the components of the "velocity" four-vector $\mathfrak{V}$, and $\ddot{x}$ $\ddot{y}$ $\ddot{z}$ $\ddot{u}$ for the second derivatives, i.e., for the components of the "acceleration" four-vector $\mathfrak{B}$.[9]

He then observes that, in Minkowski's formalism, the first derivatives must satisfy a certain relation involving the speed of light:

> Between the first derivatives the following identity holds:

$$\dot{x}^2 + \dot{y}^2 + \dot{z}^2 + \dot{u}^2 = -c^2, \tag{3}$$

> or

$$\dot{l}^2 \left\{ \left(\frac{dx}{dl}\right)^2 + \left(\frac{dy}{dl}\right)^2 + \left(\frac{dz}{dl}\right)^2 - 1 \right\} = -c^2 \dots [10] \tag{4}$$

Finally Abraham derives a relation corresponding to the orthogonality relation between four-velocity and acceleration in ordinary Minkowski space but now under the assumption of a variable speed of light:

> Now, by differentiating eq. (4) [eq. (3)] with respect to proper time, Minkowski obtains the condition of "orthogonality" of the velocity and acceleration four-vectors. However, if c is considered to be variable, the place of that condition is taken by the following:

$$\dot{x}\ddot{x} + \dot{y}\ddot{y} + z'\ddot{z} + \dot{u}\ddot{u} = -c\frac{dc}{d\tau}, \tag{5}$$

> as the differentiation of eq. (4) [eq. (3)] shows.[11]

---

8 "Die "Ruhdichte" ν sei ein Skalar in diesem Raume, und ebenso das Schwerkraftpotential Φ; sie mögen miteinander verknüpft sein durch die Differentialgleichung: ...(1) [eq. (2)]. (γ ist die Gravitationskonstante.)" (Abraham 1912h, 1)

9 "Wir schreiben $\dot{x}$ $\dot{y}$ $\dot{z}$ $\dot{u}$ für die ersten Ableitungen der Koordinaten eines materiellen "Weltpunktes" nach seiner "Eigenzeit" τ, d. h. für die Komponenten des Vierervektors "Geschwindigkeit" $\mathfrak{V}$, und $\ddot{x}$ $\ddot{y}$ $\ddot{z}$ $\ddot{u}$ für die zweiten Ableitungen, d. h. für die Komponenten des Vierervektors "Beschleunigung" $\mathfrak{B}$." (Abraham 1912h, 1)

10 "Es besteht zwischen den ersten Ableitung die Identität: ...(4) [eq. (3)] oder ... [eq. (4)]." (Abraham 1912h, 1–2)

In this way, the introduction of the hypothesis of a variable speed of light makes it possible for Abraham to circumvent a problematic restriction imposed by Minkowski's formalism on a relativistic theory of gravitation.[12] In fact, in a theory with variable c, the four-vectors for velocity and for acceleration no longer have to be orthogonal to each other. It follows that the gravitational potential also no longer has to be constant along the world line of a particle, contrary to the conclusion reached in the usual Minkowski's formalism. Although Abraham's line of attack was clearly stimulated by Einstein's earlier use of a variable speed of light, it thus emerges as being so closely associated with a plausible modification of the four-dimensional formalism that this approach may also be conceived as exploring an independent possibility offered by the contemporary state of the gravitation problem.

In Einstein's papers of 1907 and 1911, the variable speed of light was linked to the gravitational potential via the concept of time in an accelerated frame of reference, i.e., via an essentially kinematic relation, independently of the use of the equivalence principle for transferring the results obtained in an accelerated system to a system with a gravitational field. For Abraham, on the other hand, an analogous relation between the speed of light and the gravitational potential follows if the second derivative terms in equation (5) are identified with the acceleration due to the gravitational force, i.e., from an essentially dynamic relation. Indeed, for the relation between four-dimensional gravitational force and four-dimensional gravitational potential Abraham assumed:

$$F_x = -\frac{\partial \Phi}{\partial x} \qquad F_y = -\frac{\partial \Phi}{\partial y}$$

$$F_z = -\frac{\partial \Phi}{\partial z} \qquad F_u = -\frac{\partial \Phi}{\partial u} \tag{6}$$

He further assumed that the relation between four-acceleration and four-force is analogous to Newton's second law:

$$\ddot{x} = F_x \qquad \ddot{y} = F_y$$

$$\ddot{z} = F_z \qquad \ddot{u} = F_u \tag{7}$$

With these two relations, equation (5) can now be written as a relation between the speed of light and the gravitational potential:

$$\dot{x}\frac{\partial \Phi}{\partial x} + \dot{y}\frac{\partial \Phi}{\partial y} + \dot{z}'\frac{\partial \Phi}{\partial z} + \dot{u}\frac{\partial \Phi}{\partial u} = c\frac{dc}{d\tau} \tag{8}$$

---

11 "Nun erhält Minkowski, indem er Gl. (4) [eq. (3)] nach der Eigenzeit differenziert, die Bedingung der "Orthogonalität" der Vierervektoren Geschwindigkeit und Beschleunigung. Wenn jedoch c als veränderlich angesehen wird, tritt an Stelle jener Bedingung die folgende: ... (5) [eq. 5] wie die Differentiation von Gl. (4) [eq. (3)] zeigt." (Abraham 1912h, 2)

12 See "Einstein, Nordström and the Early Demise of Scalar, Lorentz Covariant Theories of Gravitation," (in vol. 3 of this series).

which can be rewritten as:

$$\frac{d\Phi}{d\tau} = c\frac{dc}{d\tau}.$$ (9)

The last equation in turn can be integrated to yield:

$$\frac{c^2}{2} - \frac{c_0^2}{2} = \Phi - \Phi_0,$$ (10)

where c_0 and Φ_0 are the speed of light and the gravitational potential at the origin of coordinates. Abraham comments on this formula:

> The increase of half the square of the speed of light is equal to the increase of the gravitational potential.

> Instead of this relation, which is exactly valid according to our theory, one can, neglecting the square of the quotient of Φ and c^2, take Einstein's formula (loc. cit. p. 906):

$$c = c_0\left(1 + \frac{\Phi - \Phi_0}{c^2}\right).$$ (11)

> However, eq. (6) [eq. (10)] better serves to manifest the independence from the arbitrarily chosen origin of coordinates.[13]

Abraham had thus achieved all essential elements of Einstein's research in the years 1907 to 1911, albeit in an entirely different way which, in addition, offered a wealth of mathematical resources for the further elaboration of a full-fledged theory of gravitation.

1.2 Abraham's Mathematics versus Einstein's Physics

While the exploitation of Minkowski's formalism had opened up new mathematical possibilities to Abraham, the physical interpretation of his results had yet to be explored. Whereas Einstein had been aware from the beginning that he was transgressing the limits of his original theory of relativity, Abraham seemed to have been initially convinced that he had found the relativistic theory of gravitation that was called for after the establishment of the principle of relativity, as is clear from his introductory remark quoted above. But since the constancy of the speed of light was one of the foundational elements of special relativity, it was questionable with which right Abraham could make use of relations derived from Minkowski's reformulation of special relativity for a theory in which the speed of light depends on the gravita-

---

13 "*Der Zuwachs des halben Quadrats der Lichtgeschwindigkeit ist gleich dem Zuwachs des Schwerkraftpotentials.* Anstelle dieser, nach unserer Theorie exakt gültigen Beziehung kann man, bei Vernachlässigung des Quadrats des Quotienten aus Φ und c^2, die Einsteinsche Formel setzten (loc. cit. S. 906): ... [eq. 11]. Indessen läßt die Formel (6) besser die Unabhängigkeit von dem willkürlich wählbaren Koordinatenursprung hervortreten." [Reference is to (Einstein 1911).] (Abraham 1912h, 2)

tional potential. It seems that Einstein, in a personal communication now lost, immediately brought Abraham's attention to this conflict between the formalism of the
latter's theory and one of its fundamental assumptions, the variability of the speed of
light.[14]

Abraham's use of the straightforward special relativistic generalization of the
Poisson equation (2), as well as of other elements of Minkowski's formalism, without
really bothering about the gravity of energy, made Einstein skeptical about Abraham's claims. For Einstein, the gravity of energy and the variability of the speed of
light were closely connected:

> It is a great pity that the gravitation theory is leading to so little which is observable. But
> it nevertheless must be taken seriously because the theory of relativity requires such a
> further development with urgency since the gravitation vector cannot be integrated into
> the relativity theory with constant c if one requires the *gravitational* mass of energy.[15]

In Einstein's opinion, a special relativistic framework was hence not the appropriate
starting point for coping with the concept of a gravity of energy and for predicting
effects such as the gravitational deflection of light, while Abraham, in fact, claimed
that he could do so:

> But as far as the bending of light rays in the gravitational field is concerned, which can be
> derived from (6) [eq. (10)] with the help of Huygens' principle, it is identical with the
> bending of the trajectories of those light particles. This is one of the numerous incom-
> plete analogies between the modern theory of radiation and the emission theory of
> light.[16]

Einstein, on the other hand, criticized Abraham's theory for not really explaining the
bending of light, probably because it so closely resembled a special relativistic theory:

> I am having a controversy with Abraham because of his theory of gravitation. The latter
> does in reality not give an account of a bending of light rays.[17]

---

14   For evidence of this personal communication see the "Correction" ("Berichtigung") to (Abraham
     1912h), quoted in note 20.
15   "Es ist sehr schade, dass die Gravitationstheorie zu so wenig Beobachtbarem führt. Aber sie muss
     trotzdem ernst genommen werden, weil die Relativitätstheorie eine derartige Weiterentwicklung
     gebieterisch verlangt, indem der Gravitationsvektor in die Rel. Theorie mit konstantem c sich nicht
     einfügen lässt, wenn man die *schwere* Masse der Energie fordert." Einstein to Wilhelm Wien, 17 May
     1912, (CPAE 5, Doc. 395, 465).
16   "Was aber die Krümmung der Lichtstrahlen im Schwerkraftfelde anbelangt, die aus (6) [eq. (10)] mit
     Hilfe des Huygensschen Prinzips sich ableiten läßt, so ist sie identisch mit der Krümmung der Bahn-
     kurve jener Lichtteilchen. Es ist dies eine der Zahlreichen unvollständigen Analogien zwischen der
     modernen Strahlungstheorie und der Emissionstheorie des Lichtes." (Abraham 1912h, 2) He referred
     to Einstein's 1911 paper and also undertook a comparison with light deflection in an emission theory
     of light.
17   "Ich habe eine Kontroverse mit Abraham wegen dessen Theorie der Gravitation. Dieselbe gibt in
     Wahrheit von einer Krümmung der Lichtstrahlen nicht Rechenschaft." Einstein to Wilhelm Wien,
     27 January 1912, (CPAE 5, Doc. 343, 394).

In later comments Einstein acknowledged that light deflection actually follows from Abraham's theory and concentrated his criticism more generally on what he perceived as an incoherent use of the mathematical formalism of special relativity:

> The theory of Abraham, according to which light is also curved, just as it is in my case, is inconsequent from the point of view of the theory of invariants.[18]

What Einstein meant precisely becomes clearer from another contemporary comment:

> The matter is not, however, as simple as Abraham believes it to be. In particular the principle of the constant c and hence the equivalence of the four dimensions is lost.[19]

Einstein must also have addressed such criticism to Abraham directly. The latter, in any case, reacted to Einstein's arguments by pursuing his modification of Minkowski's formalism in greater depth. In a short note published on 15 February 1912 as a reply to Einstein's critique, Abraham revoked the lines with which he had earlier referred to Minkowski's formalism, instead introducing an infinitesimal line element with variable metric, thus effectively extending Minkowski's spacetime to a more general semi-Riemannian manifold:

> In lines 16, 17 of my note "On the Theory of Gravitation" an oversight has to be corrected which was brought to my attention by a friendly note from Mr. A. Einstein. One should read there: 'we consider dx, dy, dz and $du = idl = icdt$ as components of a displacement $\vec{ds}$ in four-dimensional space'.
>
> Hence
>
> $$ds^2 = dx^2 + dy^2 + dz^2 - c^2dt^2 \qquad (12)$$
>
> is the square of the four-dimensional line element where the speed of light c is determined by equation (6) [eq. (10)].[20]

In this way, Abraham had effectively introduced the mathematical representation of the gravitational potential that was to be at the core of later general relativity, the general four-dimensional line element involving a variable metric tensor. However, for the time being, Abraham's expression remained an isolated mathematical formula without context and physical meaning which, at this point, was indeed neither provided by Abraham's nor by Einstein's physical understanding of gravitation. Abraham's expression in particular was neither related to insights about coordinate

---

18 "Die Theorie von Abraham, nach welcher das Licht ebenfalls ebenso wie bei mir gekrümmt ist, ist vom invariantentheoretischen Standpunkt inkonsequent." Einstein to Erwin Freundlich, mid-August 1913, (CPAE 5, Doc. 468, 550).

19 "So einfach, wie Abraham meint, ist die Angelegenheit aber nicht. Insbesondere geht das Prinzip des konstanten c und damit die Gleichwertigkeit der 4 Dimensionen verloren." Einstein to Wilhelm Wien, 11 March 1912, (CPAE 5, Doc. 371, 430).

20 "Auf Z. 16, 17 meiner Note "Zur Theorie der Gravitation" ist ein Versehen zu berichtigen, auf welches ich durch eine freundliche Mitteilung des Herrn A. Einstein aufmerksam geworden bin. Man lese daselbst: 'betrachten wir dx, dy, dz und $du = idl = icdt$ als Komponenten einer Verschiebung $\vec{ds}$ im vierdimensionalen Raume'. Es ist also $ds^2 = dx^2 + dy^2 + dz^2 - c^2dt^2$ das Quadrat des vierdimensionalen Linienelementes, wobei die Lichtgeschwindigkeit c durch G. (6) bestimmt ist." ("Berichtigung," Abraham 1912h, 176)

systems nor to any ideas about a generalized inertial motion, as is the case for the line element in general relativity. It therefore comes as no surprise that Abraham's formulation did not constitute any noteworthy "discovery" and that it was not even taken very seriously by either Abraham or Einstein.[21]

Einstein did acknowledge the impressive formal advance owed to Abraham's bold "mathematical" approach, in particular when compared with his own sluggish progress; but he quickly realized the theory's impoverished physical meaning. In a letter written shortly after the publication of Abraham's first paper on 27 January 1912, he remarked:

> Abraham has supplemented my gravitation thing, making it into a closed theory, but he made considerable errors in reasoning so that the thing is probably incorrect. This is what happens when one operates formally, without thinking physically![22]

However, he must have been initially impressed by the elegance with which Abraham's formalism yielded essentially the same results as his own, mathematically more pedestrian efforts, and also more than what he himself had achieved. This is evident from a comment Einstein made about two months later, when he was already convinced that Abraham's theory was not tenable:

> Abraham's theory has been created out of thin air, i.e., out of nothing but considerations of mathematical beauty, and is completely untenable. I find it hard to understand how this intelligent man allowed himself to get carried away with such superficiality. *At first (for 14 days!) I too was completely "bluffed" by the beauty and simplicity of his formulas*. [my emphasis][23]

At the beginning of February, Einstein was apparently still willing to concede to Abraham the benefit of the doubt. He wrote:

> Abraham has further developed the new gravitation theory; we are corresponding about this since we are not completely of the same opinion.[24]

As early as mid-February, however, Einstein had formed his firm, negative judgement on Abraham's theory:

> Abraham's theory is completely untenable.[25]

---

21 For a critical view on the notion of discovery, see (Renn et al. 2001).

22 "Abraham hat meine Gravitationssache zu einer geschlossenen Theorie ergänzt, aber bedenkliche Denkfehler dabei gemacht, sodass die Sache wohl unrichtig ist. Das kommt davon, wenn man formal operiert, ohne dabei physikalisch zu denken!" Einstein to Heinrich Zangger, 27 January 1912, (CPAE 5, Doc. 344, 395).

23 "Abrahams Theorie ist aus dem hohlen Bauche, d. h. aus blossen mathematischen Schönheitserwägungen geschöpft und vollständig unhaltbar. Ich kann gar nicht begreifen, wie sich der intelligente Mann zu solcher Oberflächlichkeit hat hinreissen lassen können. Im ersten Augenblick (14 Tage lang!) war ich allerdings auch ganz "geblüfft" durch die Schönheit und Einfachheit seiner Formeln." Einstein to Michele Besso, 26 March 1912, (CPAE 5, Doc. 377, 436–437).

24 "Abraham hat die neue Gravitationstheorie weiter ausgeführt; wir korrespondieren darüber, weil wir nicht vollkommen gleicher Meinung sind." Einstein to Michele Besso, 4 February 1912, (CPAE 5, Doc. 354, 406).

Einstein's harsh judgement may have also been related to the progress he himself was meanwhile making on a theory of gravitation based on the equivalence principle. In fact, the first reference to this theory, later published in Einstein's papers on the static gravitational field, is found in a letter from mid-February:

> The second thing concerns the relationship: gravitational field—acceleration field—velocity of light. Simple and beautiful things emerge here quite automatically. The velocity of light c is variable. It determines the gravitational force. A stationary point with mass 1 is acted upon by the force
>
> $$-\frac{\partial c}{\partial x} - \frac{\partial c}{\partial y} - \frac{\partial c}{\partial z}.$$
>
> In empty space c satisfies Laplace's equation. The inertial mass of a body is m/c, that is, it decreases with the gravitational potential. The equations of motion for the material point agree essentially with those of the customary theory of relativity. Abraham's theory is unfounded in every respect if there really is an equivalence between the gravitational field and the "acceleration field."[26]

It thus seems that around the end of January, Einstein, impressed by "the beauty and simplicity" of Abraham's formulas, had taken up work on a gravitation theory of his own and that he had then, in mid February, after an exchange of letters with Abraham, come to the conclusion that the latter's theory must be untenable, not least because its essential achievements could also be obtained on a physically much more sound basis from the equivalence principle. The relation between Einstein's own work on a gravitation theory and his negative judgement of Abraham's theory is also confirmed by another contemporary comment:

> In the course of my research on gravitation I discovered that Abraham's theory (1st issue of *Phys. Zeitschr.*) is completely untenable.[27]

After having initially limited himself to private communications, Einstein then prepared himself for a public controversy with Abraham:

---

25 "Abrahams Theorie der Gravitation ist ganz unhaltbar." Einstein to Paul Ehrenfest, 12 February 1912, (CPAE 5, Doc. 357, 408).

26 "Die zweite Sache betrifft die Beziehung Gravitationsfeld—Beschleunigungsfeld—Lichtgeschwindigkeit. Es kommen da einfache und schöne Dinge ganz zwangläufig heraus. Die Lichtgeschwindigkeit c ist variabel. Sie bestimmt die Gravitationskraft. Auf einen ruhenden Punkt von der Masse 1 wirkt die Kraft

$$-\frac{\partial c}{\partial x} - \frac{\partial c}{\partial y} - \frac{\partial c}{\partial z}.$$

c erfüllt im leeren Raume die Laplace'sche Gleichung. Die träge Masse eines Körpers ist m/c, sinkt also mit dem Schwerepotential. Die Bewegungsgleichungen des materiellen Punktes stimmen mit denen der gewöhnlichen Relativitätstheorie im Wesentlichen überein. Die Theorie Abrahams ist in allen Teilen unzutreffend, wenn die Aequivalenz zwischen Schwerefeld und "Beschleunigungsfeld" wirklich besteht." See Einstein to Hendrik A. Lorentz, 18 February 1912, (CPAE 5, Doc. 360, 413).

27 "Bei meiner Untersuchung über Gravitation entdeckte ich, dass Abrahams Theorie (1. Heft der phys. Zeitschr.) ganz unhaltbar ist." Einstein to Heinrich Zangger, before 29 February 1912, (CPAE 5, Doc. 366, 421).

> Abraham's theory is completely wrong. I will probably get into a heavy ink fight with
> him.[28]

As no letters by Abraham on this issue have survived, only the controversy, published later, allows one to infer how he defended himself against Einstein's criticism in his letters. This issue is of importance since Einstein's published criticism of February 1912 addresses a point that is not explicitly made in Abraham's early papers, the admissibility of Lorentz transformations in the infinitesimally small. But Abraham's later papers do suggest that he viewed the correction published in February 1912 as showing that the Lorentz transformation can be at least locally upheld. In the introductory section of a paper published in September 1912 he wrote for example:

> I have just availed myself of the language of the theory of relativity. But it will become
> clear that this theory cannot be brought into agreement with the views on the force of
> gravity presented here, in particular because the axiom of the constancy of the speed of
> light is relinquished. In my earlier papers on gravitation I have attempted to preserve at
> least in the infinitesimally small the invariance with respect to the Lorentz transforma-
> tions.[29]

Similarly Abraham wrote in June 1912:

> I had given to the expressions of the gravitation tensor as well as to the equations of
> motion of the material point in the gravitational field a form which in the infinitesimally
> small is invariant with respect to Lorentz transformations.[30]

Indeed, in the series of four papers (plus one correction) Abraham published between January and March 1912, starting from his basic paper "On the Theory of Gravitation," via "The Elementary Law of Gravitation," the "Correction," and "The Free Fall," up to "The Conservation of Energy and Matter in the Gravitational Field"[31] Abraham made free use of infinitesimal Lorentz transformations, and even of their integration,[32] without ever explicitly justifying that this procedure is legitimate in the case of a variable speed of light.

In February 1912, Einstein published his criticism of Abraham in the context of his own first paper on a field theory of gravitation. To the editor of the *Annalen der Physik* he wrote:

---

28   "Abrahams Theorie ist ganz falsch. Ich werde wohl ein schweres Tintenduell mit ihm bekommen."
     Einstein to Ludwig Hopf, after 20 February 1912, (CPAE 5, Doc. 364, 418).
29   " Ich habe mich soeben der Sprache der Relativität bedient. Doch wird sich zeigen, daß diese Theorie
     mit den hier vorgetragenen Ansichten über die Schwerkraft nicht zu vereinbaren ist, schon darum
     nicht weil das Axiom von der Konstanz der Lichtgeschwindigkeit aufgegeben wird. Ich habe in mei-
     nen früheren Arbeiten über die Gravitation versucht, wenigstens in unendlich kleinen die Invarianz
     gegenüber den Lorentz-Transformationen zu bewahren." (Abraham1912b, 793–794)
30   "Ich hatte den Ausdrücken des Gravitationstensor, sowie den Bewegungsgleichungen des materiellen
     Punktes im Schwerefelde eine Form gegeben, die im unendlich kleinen gegenüber Lorentztransfor-
     mationen invariant ist." (Abraham 1912f, 1057)
31   See (Abraham 1912h, 1912a, 1912c, 1912d) respectively.
32   See (Abraham 1912c, 310).

I hereby send you a paper for the *Annalen*. Many a drop of sweat is hanging on it, but I now have complete confidence in the matter. Abraham's theory is completely unacceptable. How could anybody have the luck to guess effortlessly equations which are correct! Now I am looking for the dynamics of gravitation. But this will not happen so quickly![33]

In his paper, the aim of criticizing Abraham's theory is immediately announced in the introduction:

Since then, Abraham has constructed a theory of gravitation which contains the consequences drawn in my first paper as special cases. But we will see in the following that Abraham's system of equations cannot be brought into agreement with the equivalence hypothesis, and that his conception of time and space cannot be maintained even from the purely mathematically formal point of view.[34]

Einstein's argument against Abraham's theory is given in §4 of his paper "General Remarks on Space and Time":

Which is now the relation of the above theory to the old theory of relativity (i.e. to the theory of the universal c)? According to Abraham's opinion the transformation equations of Lorentz are still valid in the infinitesimally small, that is, there shall be an $x - t$ transformation so that:

$$dx' = \frac{dx - v\,dt}{\sqrt{1 - \dfrac{v^2}{c^2}}},$$

$$dt' = \frac{-\dfrac{v}{c^2}dx + dt}{\sqrt{1 - \dfrac{v^2}{c^2}}}$$

(13)

are valid. dx' and dt' must be complete differentials. Therefore the following equations must be valid:

$$\frac{\partial}{\partial t}\left\{\frac{1}{\sqrt{1 - \dfrac{v^2}{c^2}}}\right\} = \frac{\partial}{\partial x}\left\{\frac{-v}{\sqrt{1 - \dfrac{v^2}{c^2}}}\right\},$$

$$\frac{\partial}{\partial t}\left\{\frac{-\dfrac{v}{c^2}}{\sqrt{1 - \dfrac{v^2}{c^2}}}\right\} = \frac{\partial}{\partial x}\left\{\frac{1}{\sqrt{1 - \dfrac{v^2}{c^2}}}\right\}.$$

(14)

---

33 "Ich sende Ihnen hier eine Arbeit für die Annalen. Hängt mancher Schweisstropfen daran, aber ich habe jetzt alles Vertrauen zu der Sache. Abrahams Theorie der Gravitation ist ganz unannehmbar. Wie könnte einer auch das Glück haben Gleichungen mühelos zu erraten, die richtig sind! Nun suche ich nach der Dynamik der Gravitation. Es wird aber nicht schnell damit gehen!" Einstein to Wilhelm Wien, 24 February 1912, (CPAE 5, Doc. 365, 420).

34 "Seitdem hat Abraham eine Theorie der Gravitation aufgestellt, welche die in meiner ersten Arbeit gezogenen Folgerungen als Spezialfälle enthält. Wir werden aber im folgenden sehen, daß sich das Gleichungssystem Abrahams mit der Äquivalenzhypothese nicht in Einklang bringen läßt, und daß dessen Auffassung von Zeit und Raum sich schon vom rein mathematisch formalen Standpunkte aus nicht aufrecht erhalten läßt." (Einstein 1912, 355)

In the unprimed system the gravitational field shall now be static. Then c is an arbitrarily given function of x, but independent of t. If the primed system shall be a 'uniformly' moved one, then for fixed x, v must be in any case independent of t. The left-hand sides of the equations, and hence also the right-hand sides must therefore vanish. But the latter is impossible, since for c given in terms of arbitrary functions of x, both right-hand sides cannot be made to vanish by appropriately choosing v in dependence on x. Therefore it is shown that even for infinitesimally small regions of space and time one cannot adhere to the Lorentz transformation as soon as one gives up the universal constancy of c. [35]

In modern terminology, Einstein's argument amounts to showing that it is generally impossible to transform a coordinate system in which the metric does not have the Minkowskian form into another, well-defined one by means of a Lorentz transformation. In a letter somewhat later Einstein summarized his negative attitude toward Abraham's treatment of space and time as follows:

I have now finished my studies on the statics of gravitation and have great confidence in the results. But the generalization appears to be very difficult. My results are not in agreement with those of Abraham. The latter has worked here, contrary to his usual style, rather superficially. Already his treatment of space and time is untenable. [36]

In spite of Einstein's criticism, Abraham nevertheless continued to use Minkowski's framework for elaborating consequences of his theory of gravitation, some of which will be discussed below. Einstein, in turn, became gradually convinced that it was worthwhile after all to take a closer look at the utility of a modification of this formalism for his version of a gravitational field theory as well. Driven by Abraham's bold and occasionally stubborn persistence, Einstein in May 1912 thus finally recognized that a generalized line element, as suggested by Abraham's note of three months earlier, indeed represents the key to a generally relativistic gravitation theory. [37]

---

35 "In was für einem Verhältnis steht nun die vorstehende Theorie zu der alten Relativitätstheorie (d. H. zu der Theorie des universellen c)? Nach Abrahams Meinung sollen die Transformationsgleichungen von Lorentz nach wie vor im unendlich Kleinen gelten, d. h. es soll eine $x - t$ –Transformation geben, so daß [eq. (13)] gelten. dx' und dt' müssen vollständige Differentiale sein. Es sollen also die Gleichungen gelten [eq. (14)]. Es sei nun im ungestrichenen System das Gravitationsfeld ein statisches. Dann ist c eine beliebig gegebene Funktion von x, von t aber unabhängig. Soll das gestrichene System ein "gleichförmig" bewegtes sein, so muß v bei festgehaltenem x jedenfalls von t unabhängig sein. Es müssen daher die linken Seiten der Gleichungen, somit auch die rechten Seiten verschwinden. Letzteres ist aber unmöglich, da bei beliebig in Funktionen von x gegebenem c nicht beide rechten Seiten zum Verschwinden gebracht werden können, indem man v in Funktion von x passend wählt. Damit ist also erwiesen, daß man auch für unendlich kleine Raum-Zeitgebiete nicht an der Lorentztransformation festhalten kann, sobald man die universelle Konstanz von c aufgibt." (Einstein 1912, 368)

36 "Die Untersuchungen über die Statik der Gravitation habe ich nun fertig und setze grosses Vertrauen in die Resultate. Aber die Verallgemeinerung scheint sehr schwierig zu sein. Meine Ergebnisse sind mit denen von Abraham nicht im Einklang. Dieser hat gegen seine sonstige Gewohnheit hier recht oberflächlich gearbeitet. Schon seine Behandlung von Raum und Zeit ist unhaltbar." Einstein to Ludwig Hopf, 12 June 1912, (CPAE 5, Doc. 408, 483).

37 See "Classical Physics in Disarray ..." and "The First Two Acts" (both in vol. 1 of this series).

1.3 Abraham's Rejection of Relativity

As time went by, Abraham's reaction to the controversy with Einstein became more and more acrid, until eventually he rejected the theory of relativity altogether. This is in stark contrast to Einstein's ever more resolved search for a generalization of the relativity principle as familiar from classical physics. In fact, Einstein had concluded his criticism of Abraham with the following comment:

> To me the spacetime problem seems to lie as follows. If one limits oneself to a region of constant gravitational potential, the natural laws take on an outstandingly simple and invariant form if one refers them to a spacetime system of that manifold which are connected to each other by the Lorentz transformations with constant c. If one does not restrict oneself to regions of constant c, then the manifold of equivalent systems, just as the manifold of the transformations that leave the natural laws unchanged, will become a larger one, but the laws will, on the other hand, become more complicated.[38]

In his published response to Einstein's critique, Abraham, on the other hand, wrote (the first sentence is quoted above):

> I had given to the expressions of the gravitation tensor as well as to the equations of motion of the material point in the gravitational field a form which in the infinitesimally small is invariant with respect to Lorentz transformations. In the restriction to the infinitesimally small it is already implicit that this invariance shall not be maintained in the finite. In fact, if the gravitational field influences the speed of light, then it is clear from the outset that there is an essential difference between a reference system $\Sigma(xyzt)$ in which the gravitational field is a static one and a reference frame $\Sigma(x'y'z't')$ which is uniformly moved with respect to the former, in which the gravitational field, and hence also the speed of light, is changing with time. There can be no talk about any kind of relativity, i.e., about a correspondence between the two systems, which would express itself in equations between their spacetime parameters $xyzt$ and $x'y'z't'$. Indeed the differential equations between dx', dt' and dx, dt, which contain the Lorentz-transformation in the infinitesimally small, are, as Mr. Einstein observes, not integrable.[39]

Abraham thus drew from the same mathematical fact a consequence that is diametrically opposed to that drawn by Einstein. Instead of calling for an extension of the relativity principle, as Einstein did, Abraham called for a complete rejection of this principle. He therefore continued:

> But in this inherently correct observation I cannot find any justification for the claim that "my conception of time and space cannot be maintained even from the purely mathematically formal point of view." However, any relativistic spacetime conception which would find its expression in relations between spacetime parameters of Σ and Σ' becomes untenable. Such a relativistic spacetime conception is, on the other hand,

---

38 "Mir scheint das Raum-Zeitproblem wie folgt zu liegen. Beschränkt man sich auf ein Gebiet von konstantem Gravitationspotential, so werden die Naturgesetze von ausgezeichnet einfacher und invarianter Form, wenn man sie auf ein Raum-Zeitsystem derjenigen Mannigfaltigkeit bezieht, welche durch die Lorentztransformationen mit konstantem c miteinander verknüpft sind. Beschränkt man sich nicht auf Gebiete von konstantem c, so wird die Mannigfaltigkeit der äquivalenten Systeme, sowie die Mannigfaltigkeit der die Naturgesetze ungeändert lassenden Transformationen eine größere werden, aber es werden dafür die Gesetze komplizierter werden." (Einstein 1912, 368–369)

entirely far fetched to me. As I have already mentioned elsewhere [Abraham 1912g], to
me, the interpretation in the sense of an absolute theory seems rather the appropriate one.
If among all reference frames that one is privileged in which the gravitational field is
static, or quasi static, then it is permissible to refer to a motion related to this system as
'absolute'.[40]

He proceeds by explaining the relation of his conception to traditional physical ideas
about preferred systems of reference, such as that identified by Neumann's "body
α," and concludes with the remark:

Who wishes to do so, may interpret this conception as an argument for the 'existence of
the aether'.[41]

Abraham became ever more skeptical about Einstein's attempts to extend the
principle of relativity, as his later publications show. Although he actively contributed
to developing tools and concepts of a relativistic theory of gravitation, such as expressions for stresses and energy in a gravitational field, he never dealt in detail with the
revisions of the concept of time implied by his own theory, let alone those of the concepts of space and time associated with Einstein's rival theories.[42] In a paper submitted in July 1912, Abraham pointed to Einstein's difficulties with implementing the
equivalence hypothesis in his own theory of gravitation, but merely restricted himself
to the following brief remark on the subject:

Here I would therefore prefer to develop the new gravitation theory without entering the
spacetime problem.[43]

---

39   "Ich hatte den Ausdrücken des Gravitationstensors, sowie den Bewegungsgleichungen des materiellen
     Punktes im Schwerefelde eine Form gegeben, die im unendlich kleinen gegenüber Lorentztransfor-
     mationen invariant ist. In der Beschränkung auf das unendlich kleine liegt schon implicite enthalten,
     daß im endlichen diese Invarianz nicht bestehen soll. In der Tat, wenn das Gravitationsfeld die Licht-
     geschwindigkeit beeinflußt, so ist es von vornherein klar, daß ein wesentlicher Unterschied zwischen
     einem Bezugssystem $\Sigma(xyzt)$ besteht, in welchem das Schwerkraftfeld ein statisches ist, und einem
     gegen dieses gleichförmig bewegten Bezugssystem $\Sigma(x'y'z't')$, in welchem das Schwerkraftfeld,
     und mithin auch die Lichtgeschwindigkeit, sich zeitlich verändert. Es kann von irgend einer Art von
     Relativität, d. h. von einer Korrespondenz der beiden Systeme, die sich in Gleichungen zwischen
     ihren Raum-Zeit-Parametern $xyzt$ uns $x'y'z't'$ ausdrücken würde, keine Rede sein. In der Tat sind,
     wie Hr. Einstein bemerkt, die Differenzialgleichungen zwischen dx', dt' und dx, dt, welche die
     Lorentztransformation im unendlich kleinen enthalten, nicht integrabel." (Abraham 1912f, 1057)
40   "In dieser an sich zutreffenden Bemerkung kann ich freilich keine Rechtfertigung für die Behauptung
     finden, daß "meine Auffassung von Zeit und Raum sich schon vom rein mathematisch formalen
     Standpunkt aus nicht aufrecht erhalten läßt". Unhaltbar wird allerdings jede relativistische Raum-
     Zeit-Auffassung, die in Beziehungen zwischen den Raum-Zeit-Parametern von Σ und Σ' ihren Aus-
     druck finden würde. Eine solche relativistische Raum-Zeit-Auffassung liegt mir indessen ganz fern.
     Mir scheint vielmehr, wie ich bereits an anderem Orte erwähnt habe [Abraham 1912g], die Deutung
     im Sinne einer Absoluttheorie die passende zu sein. Wenn unter allen Bezugssystemen dasjenige aus-
     gezeichnet ist, in welchem das Schwerefeld statisch, oder quasi-statisch ist, so ist es erlaubt, eine auf
     dieses System bezogene Bewegung "absolut" zu nennen." (Abraham 1912f, 1057)
41   "Wer will, mag diese Vorstellung als Argument für die 'Existenz des Äthers' deuten. (Abraham 1912f,
     1058)

Abraham did not just ignore the consequences of the new gravitation theories for the concepts of space and time, he actually rejected them, with no lesser consequence than that with which Einstein pursued them. In short, the same challenging task of creating a gravitation theory compatible with relativity theory found complementary responses in Abraham and Einstein. While both not only acknowledged the difficulties of making them compatible, but also returned to a revision of special relativity, Einstein did so in order to complete the relativity revolution, and Abraham to undo it. Commenting on Einstein's later attempts at a relativistic theory of gravitation, Abraham wrote in a popular review of 1914:

> Hence, at the cliff of gravitation every theory of relativity fails, the special one of 1905, as well as the general one of 1913. The relativistic ideas are obviously not sufficiently advanced to serve as a framework for a complete worldview.
>
> But historical merit does remain for the theory of relativity with regard to its critique of the concepts of space and time. It has taught us that these concepts depend on the ideas we form concerning the behavior of the measurement rods and clocks that we use for the measurement of lengths and intervals of time, and which are subject to change with them [the measurement instruments]. This will secure an honorable funeral for the theory of relativity.[44]

Although Abraham did not accept the new concepts of space and time introduced by Einstein, he remained an acute critic of the latter's ongoing search for a relativistic

---

42 In March 1911 Abraham submitted a paper (Abraham 1912c) in which he used the distinction between proper time and coordinate time in order to draw far-reaching cosmological consequences from his theory. In this paper Abraham explained his definition of time coordinates:
"τ denotes the 'proper time' of the moved point, which is related to the time t measured in the reference system as follows:

$$\frac{du}{d\tau} = ic\frac{dt}{d\tau} = \frac{ic}{\kappa}$$

$$\kappa = \sqrt{1 - \frac{v^2}{c^2}}$$

(v magnitude of velocity of the material point).
Clearly, Abraham's explanation of the relation between proper time and coordinate time, although formally analogous to Minkowski's formalism, can neither be based on this formalism, as the speed of light is assumed to be variable, nor be brought into harmony with Einstein's physical motivation for these two notions of time. Contrary to Einstein it appears from the remainder of this paper that Abraham considered only the coordinate time t to have any physical meaning.

43 "Ich möchte es daher hier vorziehen, die neue Gravitationstheorie zu entwickeln, ohne auf das Raum-Zeit-Problem einzugehen." (Abraham 1912b, 794)

44 "An der Klippe der Schwerkraft scheitert also jede Relativitätstheorie, sowohl die spezielle von 1905, wie die allgemeine von 1913. Die relativistischen Ideen sind offenbar nicht weit genug, um einem vollständigen Weltbilde als Rahmen zu dienen. Doch bleibt der Relativitätstheorie ein historisches Verdienst um die Kritik der Begriffe von Raum und Zeit. Sie hat uns gehehrt, dass diese Begriffe von den Vorstellungen abhängen, die wir uns von dem Verhalten der zur Messung von Längen und Zeitintervallen dienenden Massstäbe und Uhren bilden, und die mit ihnen dem Wandel unterworfen sind. Dies sichert der Relativitätstheorie ein ehrenvolles Begräbnis." (Abraham 1914, 26)

theory of gravitation, which, apart from Abraham's reaction, found only little reso-
nance in the contemporary physics community. In a letter to his friend Michele
Besso, Einstein wrote:

> Physicists have such a passive attitude toward my work on gravitation. Abraham is still
> the one who shows the most comprehension. It is true that he complaints violently in
> 'Scientia' against anything to do with relativity, but with understanding.[45]

In spite, and occasionally perhaps because of his intellectual distance, Abraham
noticed problematic features of Einstein's attempts, which eventually became important issues in the development of general relativity, such as the question of the sense
in which Einstein's mathematical requirement of a generalized covariance actually
also realizes a generalization of the relativity principle as he claimed. In a review
paper of 1915, Abraham expressed his doubts concerning Einstein's claim that his
Entwurf theory of 1913, which is covariant with regard to general linear transformations, actually represents a generalization of the relativity principle of the special theory of 1905:

> The significance of this transformation group lies in the fact that it contains the Lorentz
> transformations; in the earlier theory of relativity the covariance with respect to this
> group gave expression to the equivalence of systems of reference in translatory motion
> with respect to each other. Is this presently also the case in the "generalized theory of rel-
> ativity"? Does the covariance of the field equations with respect to linear orthogonal
> transformations imply that in a finite system of mutually gravitating bodies the course of
> the relative motions is not altered by a uniform translation of the entire system? That this
> is so has so far not been proven.[46]

Abraham thus pinpoints a distinction that is today considered as being crucial for
understanding the covariance properties of a gravitational field equation, the distinction between general covariance as a property of the mathematical formulation of
such a field equation, and the symmetry group under which the theory remains invariant, as is the case for the Lorentz group of special relativity. He thus anticipates a
conceptual clarification that is usually attributed to Kretschmann.[47] But in spite of
this significant insight, Abraham's further explanation of his skeptical attitude shows
signs of a still immature understanding of the kinematics within a generic four-

45 "Zur Gravitationsarbeit verhält sich die physikalische Menschheit ziemlich passiv. Das meiste Verständnis hat wohl Abraham dafür. Er schimpft zwar in der "Scienza" kräftig über alle Relativität, aber
   mit Verstand." Einstein to Michele Besso after 1 January 1914, (CPAE 4, Doc. 499). See (Cattani and
   De Maria 1989, 171).
46 "Die Bedeutung dieser Transformationsgruppe beruht darauf, daß sie die Lorentzschen Transformationen enthält; die Kovarianz ihnen gegenüber brachte in der früheren Relativitätstheorie die Gleichberechtigung translatorisch gegeneinander bewegter Bezugssysteme zum Ausdruck. Ist dies nun auch
   in der "verallgemeinerten Relativitätstheorie" der Fall? Bedingt es die Kovarianz der Feldgleichungen
   gegenüber linearen orthogonalen Transformationen, daß in einem endlichen Systeme gegeneinander
   gravitierender Körper der Ablauf der relativen Bewegungen durch eine gleichförmige Translation des
   ganzen Systems nicht geändert wird? Daß dem so sei, ist bisher nicht bewiesen worden." (Abraham
   1915, 515)
47 See (Kretschmann 1917) and, for historical discussion, (Norton 1992, 1993; Rynasiewicz 1999).

dimensional spacetime manifold, in particular, concerning his lack of understanding of the relation between different coordinate representations of one and the same physical magnitude:

> Such a proof may already be impossible to conduct for the reason that the concept of "uniform motion" of a finite system is completely up in the air in the new theory of relativity. Since the "natural" space and time measurement is influenced by the values of the space and time potential, observers at different locations in the gravitational field will ascribe different velocities to the same material point. Only for an infinitesimally small region of four-dimensional space—i.e. for one in which the potentials $g_{\mu\nu}$ can be considered a constant—is "velocity" defined at all.[48]

Abraham then returns to the problem of interpreting the covariance property of the field equations in terms of a relativity principle:

> Presumably, only within such an infinitesimal region may the covariance of the gravitational equations with respect to linear orthogonal transformations be interpreted in the sense of an equivalence of systems of reference moving with respect to each other. But if relativity of motion no longer exists for finite systems of gravitating masses in Einstein's theory, with what right does he then assign such great importance to the formal connection to the earlier theory of relativity?[49]

Abraham's criticism was evidently colored by his growing hostility toward any theory of relativity, and motivated, without doubt, by his own failure to successfully establish a relativistic theory of gravitation. But because it raised such crucial issues as the physical significance of the unfamiliar mathematical objects introduced in the course of Einstein's search for such a theory, it nevertheless represents an important intellectual context of the emergence of general relativity. Abraham's sometimes ardent polemics reveal not only the challenging difficulties with which Einstein confronted his contemporaries, but also contributed, at the time, to anchoring his highflying mathematical artifices in the shared knowledge of contemporary physics, a process made particularly ungainly by the hesitant and cool reaction to Einstein's labors of other contemporary scientists.

---

48 "Ein solcher Beweis dürfte auch schon aus dem Grunde sich nicht führen lassen, weil der Begriff der "gleichförmigen Bewegung" eines endlichen Systems in der neuen Relativitätstheorie völlig in der Luft schwebt. Denn da die "natürliche" Raum- und Zeitmessung durch die lokalen Potentialwerte beeinflußt wird, so werden Beobachter an verschiedenen Orten im Schwerefelde demselben materiellen Punkte verschiedene Geschwindigkeiten zuschreiben. Nur für ein unendlich kleines Gebiet des vierdimensionalen Raumes—d. h. für ein solches, in welchem die Potentiale $g_{\mu\nu}$ als konstant gelten könne—ist die "Geschwindigkeit überhaupt definiert." (Abraham 1915, 515)

49 "Vermutlich dürfte nur innerhalb eines solchen infinitesimalen Gebietes die Kovarianz der Gravitationsgleichungen gegenüber orthogonalen linearen Transformationen im Sinne einer Gleichberechtigung gegeneinander bewegter Bezugssysteme zu deuten sein. Wenn aber in Einsteins Theorie für endliche Systeme gravitierender Massen keine Bewegungsrelativität mehr besteht, mit welchem Rechte legt er dann dem formalen Anschluß an die frühere Relativitätstheorie eine so große Wichtigkeit bei?" (Abraham 1915, 515–516)

2. VISTAS FROM JUST BELOW THE SUMMIT

2.1 A New Theory of Gravitation Meets the Shared Knowledge

Abraham's insights into the nature and effects of the gravitational field conceived in a relativistic context are today even less well known than his theoretical approach. It may seem that they have been rightly forgotten because of their apparent lack of impact on the further development of the theory. It was, however, precisely Abraham's untiring efforts to elaborate his theoretical approach and to draw physical consequences from it that turned his theory into a valuable touchstone and standard of comparison for other contemporary attempts at a relativistic theory of gravitation. Some of the insights that Abraham achieved in the course of his research, such as the possibility and essential properties of gravitational waves remain to this day a standard for a relativistic theory of gravitation.

In the first paper on his theory, Abraham did not go into much detail about the physical consequences of his new theory of gravitation. But he did, of course, have to confront the shared physical knowledge of his time. This ranged from the knowledge embodied in Newton's theory of gravitation—including the deviations from it—via energy and momentum conservation to some of the unusual insights attained by Einstein on the basis of the equivalence principle, such as the deflection of light, which possibly had to be incorporated in this new theory of gravitation as well.

In his paper, Abraham first discusses the equations of motion in a gravitational field which he formulates, as we have seen above, in analogy to Newtonian dynamics. From equations (6) and (7) he obtains:

$$ m\frac{d\dot{x}}{d\tau} = -m\frac{\partial\Phi}{\partial x}, \qquad m\frac{d\dot{y}}{d\tau} = -m\frac{\partial\Phi}{\partial y}, \qquad m\frac{d\dot{z}}{d\tau} = -m\frac{\partial\Phi}{\partial z}. \tag{15} $$

These three equations he interpreted as representing momentum conservation. He then added:

In contrast, the last of the equations (3) [i.e. eq. (7)]:

$$ m\frac{dl}{d\tau} = -mi\mathfrak{F}_u = mi\frac{\partial\Phi}{\partial u} = m\frac{\partial\Phi}{\partial l} \tag{16} $$

expresses the law of conservation of *vis viva* in Minkowskian mechanics.[50]

Abraham had thus dealt with one of the inescapable requirements imposed on a new theory by the accumulated knowledge of classical and special relativistic physics. He also went into some detail about the energy balance of physical processes in a gravitational field.

Concerning the observable consequences of his theory, he did not even mention the gravitational redshift predicted by Einstein, probably because at that point he

---

50 See (Abraham 1912h, 2).

could not or did not want to follow Einstein's introduction of two notions of time in a gravitational field.[51] But he did take up Einstein's prediction of a gravitational light deflection, which he compared to that predicted by the emission theory of light. Abraham had, after all, developed his theory of gravitation in reaction to Einstein's 1911 paper and, in particular, to the introduction of a variable speed of light which this paper emphasized. Commenting on equations (10) and (11) he wrote (as partly quoted above):

> It is instructive to compare the relation obtained with the emission theory of light. Let us imagine that the particles emitted from the light source move according to the laws of Galilean mechanics and that they are subject to the gravitational force. They would then experience an increase of *vis viva* (kinetic energy) which is equal to the *decrease* of potential energy. The increase of *vis viva* (kinetic energy) of the light particles calculated according to (6) [i.e. eq. (10)] is, by contrast, equal to the *increase* of their potential energy, i.e., of the same amount but of opposite sign as that following from the emission theory. But as far as the bending of light rays in the gravitational field is concerned, which can be derived from (6) [i.e. eq. (10)] with the aid of Huygens' principle,[1)] it is identical with the bending of the trajectory of those light particles. This is one of the numerous incomplete analogies between the modern theory of radiation and the emission theory of light.[52]

To this passage, after "with the aid of Huygens' principle," Abraham appended the following footnote in which he refers to Einstein:

> A. Einstein has shown that a light ray passing the surface of the Sun is deflected towards the Sun and has drawn the attention of the astronomers to this consequence of the theory which can perhaps serve as its verification.[53]

Apart from this footnote Abraham did not discuss specific observational consequences of his theory. This, however, does not imply that he was not interested in them. On the contrary, at about the same time as Einstein, he independently and actively pursued the astronomical consequences of a relativistic theory of gravitation as is evident from a letter he wrote to Schwarzschild.[54]

---

51 Gravitational redshift is mentioned for the first time with reference to Einstein, in (Abraham 1913, 197).

52 "Es ist lehrreich, die erhaltene Beziehung der Emissionstheorie des Lichts gegenüberzustellen. Wir wollen uns vorstellen, daß die von der Lichtquelle emittierten Teilchen sich gemäß den Gesetzen der Galileischen Mechanik bewegen, und daß sie der Schwerkraft unterworfen seien. Dann würden sie einen Zuwachs an lebendiger Kraft erfahren, welcher gleich der Abnahme der potentiellen Energie ist. Dagegen ist der nach (6) [eq. 10] berechnete Zuwachs der Lichtteilehen an lebendiger Kraft gleich der Zunahme ihrer potentiellen Energie, d. h. von gleichem Betrage, aber von entgegengesetztem Vorzeichen, wie der aus der Emissionstheorie folgende. Was aber die Krümmung der Lichtstrahlen im Schwerkraftfelde anbelangt, die aus (6) [eq. 10] mit Hilfe des Huygensschen Prinzips sich ableiten läßt, so ist sie identisch mit der Krümmung der Bahnkurve jener Lichtteilchen. Es ist dies eine der zahlreichen unvollständigen Analogien zwischen der modernen Strahlungstheorie und der Emissionstheorie des Lichtes." (Abraham 1912h, 2)

53 "A. Einstein hat gezeigt, daß ein die Sonnenoberfläche passierender Lichtstrahl nach der Sonne hin abgelenkt wird, und hat die Aufmerksamkeit der Astronomen auf diese Konsequenz der Theorie gelenkt, die vielleicht zu ihrer Prüfung dienen kann." (Abraham 1912h, 2, fn.1)

In a publication following immediately, "The Elementary Law of Gravitation,"[55] which was also submitted in December 1911, Abraham returned to the question of the relation of his theory to experience. The principal aim of this paper is the derivation from the theory published in the preceding paper of an expression of the gravitational force between two "world points" P and P_0, "an elementary law of gravitation," as Abraham called it. For this purpose he integrated the field equation of his theory, equation (2), using the Cauchy method of residues and following a procedure by Herglotz.[56] Such an elementary law allowed for a comparison not only with Newton's law of gravitation but also with earlier adaptations of action-at-a-distance laws to a relativistic framework. Furthermore, it allowed Abraham, at least in principle, to address the astronomical consequences of his theory and to explore deviations from classical predictions.

He concluded his publication with the following summary:

> According to this elementary law the moving force exerted by P on P_0 is represented as a sum of two four-vectors, of which one is parallel to the radius $\mathfrak{R}$, drawn from the world point P_0 to P, the other to the velocity vector $\mathfrak{B}$ of P. This corresponds to the approaches of Poincaré and Minkowski. But our elementary law is simpler, insofar as it does not involve the velocity of the attracted point, and more general, because it also takes into account the acceleration of the attracting point. Its comparison with astronomical observation could serve for the examination of the theory of gravitation developed in the previous note.[57]

Abraham, however, did not specify which astronomical phenomena might be compared to the consequences of his elementary law. But modifications of Newton's elementary law had earlier been referred to the perihelion anomaly of Mercury[58] so that this may well have been one of the astronomical consequences Abraham had in mind. In other words, the deviation of planetary motion from the implications of Newtonian mechanics was also part of the generally shared knowledge that a new gravitation theory had to confront. On the basis of estimating the order of magnitude of deviations from Newton's law, Abraham himself later remarked:

---

54 Abraham to Schwarzschild, see "The Continuity between Classical and Relativistic Cosmology in the Work of Karl Schwarzschild," p. 165, (in this volume).
55 See (Abraham 1912a).
56 See (Herglotz 1904), and also (Sommerfeld 1910, 665; 1911, 51) where according to Abraham the analogous electrical problem is treated.
57 "Diesem Elementargesetz zufolge stellt sich die von P auf P_0 ausgeübte bewegende Kraft als Summe zweier Vierervektoren dar, von denen der eine dem vom Weltpunkte P_0 nach P gezogenen Fahrstrahl $\mathfrak{B}$, der andere dem Geschwindigkeitsvektor $\mathfrak{B}$ von P parallel ist. Dieses entspricht den Ansätzen von Poincaré und von Minkowski. Doch unser Elementargesetz einfacher, insofern als die Geschwindigkeit des angezogenen Punkts nicht eingeht, und allgemeiner, weil es auch die Beschleunigung des anziehenden Punkts berücksichtigt. Seine Vergleichung mit der astronomischen Beobachtung könnte zur Prüfung der in der vorigen Note entwickelten Theorie der Schwerkraft deinen." (Abraham 1912a, 5)
58 See (Zenneck 1903), compare also (Poincaré 1906 and Minkowski 1909).

The correction of Newton's law introduced in $(16)^{59}$ is hence likely to be too small for
noticeably influencing the planetary motion. (Abraham 1912b, 797)

Nevertheless, in the same year 1912, G. Pavanini calculated the perihelion shift of
Mercury according to Abraham's theory, finding a value of $14''$, 52, that is, approximately one third of the observed value.[60] Abraham's theory thus made a more accurate prediction than the much more elaborate theory of gravitation Einstein and
Grossmann published in 1913.[61]

2.2 Audacious Outlooks

The empirical consequences of Abraham's theory of gravitation were by no means
limited to those already envisaged by Einstein, or to those immediately obvious from
the body of knowledge covered by the Newtonian theory. Abraham was a master of
classical electrodynamics, and, for this reason, was not only used to elaborating in
depth and detail the consequences of a complex theory, but also had a specific model
that suggested where to look for analogies and differences between the two field theories—electromagnetic and gravitational. In the following, we will briefly look at
two of the outstanding achievements that resulted from Abraham's efforts to draw far-
reaching physical consequences from his theory. Both concern subjects of great interest to present research in general relativity, spacetime singularities and gravitational
waves. Until now, the name Max Abraham has played no role in the history of these
subjects, although his contributions are unlikely to have gone entirely unnoticed by
those who continued the search for a relativistic theory of gravitation at the time he
gave up.

In a paper submitted in March 1912, "The Free Fall" (Abraham 1912c), Abraham
considered the motion of free fall in a homogeneous gravitational field, i.e., in exactly
the same kind of gravitational field that was used in the formulation of Einstein's
equivalence principle, and drew some far-reaching cosmological consequences from
his calculations. These consequences essentially depend on equations (10), relating
the speed of light and the gravitational potential, and on the relation between coordinate time and proper time. From the law of fall, which Abraham derived from his
equations of motion in a gravitational field, he first concluded that there is a point in
(coordinate) time, corresponding to a certain distance of fall, at which the speed of
light becomes zero. This point corresponds to a singularity in the relation between
proper time and coordinate time. From the condition that the speed of light must
always remain larger than zero, he then concluded, with the help of (10), that:[62]

---

59   A law of the form $Force = \dfrac{A}{r^2} - \dfrac{B}{r^3}$.

60   See (Pavanini 1912). See also the brief discussion of these results in (Abraham 1915, 488). For the
     corresponding calculation in Nordström's theory, see (Behacker 1913, 989).
61   See (Einstein and Grossmann 1913). See also "What Did Einstein Know ..." (in vol. 2 of this series).
62   See (Abraham 1912c, 311).

$$\Phi_0 - \Phi < \frac{c_0^2}{2}. \tag{17}$$

As a first step, he interpreted this relation as implying that the existence of a homogeneous gravitational field of infinite extension is excluded and then turned to further, as he called them, "cosmogonic" consequences of this relation.
He considered a star of mass m with the potential:

$$\Phi = -\gamma \cdot \frac{m}{r} \tag{18}$$

and compared the potential difference between an infinite distance and the surface of the star at distance $r = a$:

$$\Phi_0 - \Phi = \gamma \cdot \frac{m}{a}. \tag{19}$$

From equation (17) he then inferred that there is a maximum for the quotient of mass and radius:

Therefore

$$\gamma \cdot \frac{m}{a} < \frac{c_0^2}{2} \left(= \frac{9}{2} \cdot 10^{20} \right) \tag{20}$$

must hold. For the Sun one has in *cgs* units

$$\left(\gamma \cdot \frac{m}{a} \right)_s = 2 \cdot 10^{15}. \tag{21}$$

Thus, the *quotient of mass and radius for an arbitrary star must satisfy the inequality*

$$\frac{m}{a} < 2,25 \cdot 10^5 \cdot \left(\frac{m}{a} \right)_s. \tag{22}$$

For stars whose mean density is equal to that of the Sun, the quotients m/a are proportional to the squares of the radii. Hence, the following must hold

$$a < 500(a)_s, \tag{23}$$

and thus

$$m < 10^8(m)_s. \tag{24}$$

That is, *the mass of a star whose mean density is equal to that of the Sun cannot become larger than one hundred million times the mass of the Sun.*

Since this limit is rather high, no difficulties arise from this for our theory.[63]

Abraham, in March 1912, was hence the first to hit upon a singularity in a field theory of gravitation and to calculate what was later called the "Schwarzschild radius." Although his understanding of the relation between proper time and coordinate time does not correspond to the modern one, his consideration of their relation for a freely falling object in the field of a point mass corresponds to a procedure still applied in modern general relativity. From his introductory reference to the "cosmogonic" implications of his theory one may further infer that he did not consider this singularity to be fictitious, contrary to the first explorers of the Schwarzschild singularity within general relativity, Eddington and Lemaitre.[64] Abraham did not, however, directly relate his result to light deflection, although the vanishing of the speed of light at the singularity clearly implies that light cannot escape from it. But Abraham did not make this consequence explicit and rather limited himself to interpreting the singularity in terms of limits on the size of stars.

In a lecture presented in October 1912 and published the following year (Abraham 1913) Abraham was also the first to discuss the possibility of gravitational waves in a relativistic field theory of gravitation. There he wrote:

> According to our theory, light and gravitation have the same speed of propagation; but whereas light waves are transverse, gravitational waves are longitudinal. Incidentally, the problem of the oscillating particle can be treated in a similar manner as that of the oscillating electron; the strength of the emitted gravitational waves depends on the product of gravitational mass and the acceleration of the particle. Is it possible to detect these gravitational waves?

> This hope is futile. Indeed, to impart an acceleration to one particle, another particle is necessary which, according to the law of action and reaction, is driven in the opposite side. But now, the strength of the emitted gravitational waves depends on the sum of the products of the gravitational mass and the acceleration of the two particles, while, according to the reaction principle, the sum of the products of inertial mass and acceleration is equal to zero.[65] Therefore, although the existence of gravitational waves is compatible with the assumed field mechanism, through the equality of gravitational and inertial mass the possibility of its production is practically excluded. It follows from this that the planetary system does not lose its mechanical energy through radiation, whereas an analogous system consisting of negative electrons circling around a positive nucleus gradually radiates its energy away. The life of the planetary system is thus not threatened by such a danger.[66]

Abraham's argument amounts to showing that, because of momentum conservation, there can be no dipole moment in gravitational waves; it is an argument still used in

---

63 Es muß also sein: [(eq. 20)]. Für die Sonne hat man, in C.G.S.Einheiten [(eq. 21)]. Daher muß *der Quotient aus Masse und Radius für einen beliebigen Stern der Ungleichung genügen* [(eq. 22)]. Für Sterne, deren mittlere Dichte derjenigen der Sonne gleich ist, stehen die Quotienten m/a im Verhältnis der Quadrate der Radien; es muß dann sein [(eq. 23)] und somit [(eq. 24)]. D h. *die Masse eines Sternes, dessen mittlere Dichte derjenigen der Sonne gleich ist, kann nicht größer werden als das Hundertmillionenfache der Masse der Sonne.* Da diese Grenze recht hoch ist, so entsteht hieraus keine Schwierigkeit für unsere Theorie." (Abraham 1912c, 311)

64 See (Eisenstaedt 1989).

65 *Abraham's footnote*: Here the momentum of the gravitational field has, however, been neglected; but this [momentum] practically comes as little into consideration as the energy of the field. ("Hierbei ist allerdings der Impuls des Schwerefeldes unberücksichtigt geblieben; aber dieser kommt ebensowenig wie die Energie des Feldes praktisch in Betracht.")

modern textbooks on general relativity.[67] His allusion to an electrical analogue is clearly a reference to a Bohr-like atomic model, which classically would, of course, be unstable.

Abraham nevertheless took the idea of gravitational waves so seriously that he published another detailed study comparing electromagnetic and gravitational waves, including quantitative comparisons.[68] In a later review he took up this earlier study and discussed the possibility that gravitational waves are emitted during the emission of α particles by radioactive atoms:

> One could now surmise that during the emission of α particles by radioactive atoms, in which very large accelerations occur, the hypothetical gravitational waves are excited in noticeable strength. But at the same time there are electrical waves which are excited and, as the quantitative discussion shows, the force which is exerted by the gravitational waves emitted by an α particle upon another α particle amounts to maximally 10^{-36} of the electrical force.[69]

If, or rather when, gravitational waves are one day directly verified, Abraham's papers will certainly not constitute a relevant theoretical reference point. From the point of view of a history of knowledge, however, they do represent a reference point for gauging the possibilities open to the development of a theory of gravitation around 1912.[70] While Abraham's efforts are hardly suited to detracting from Einstein's triumph of late 1915, they do make evident the extent to which this solitary tri-

---

66   "Nach unserer Theorie haben Licht und Schwere die gleiche Fortpflanzungsgeschwindigkeit; aber während die Lichtwellen transversal sind, sind die Schwerewellen longitudinal. Übrigens kann das Problem des schwingenden Massenteilchens in ähnlicher Weise behandelt werden wie dasjenige des schwingenden Elektrons; die Stärke der ausgesandten Gravitationswellen hängt von dem Produkt aus schwerer Masse und Beschleunigung des Teilchens ab. Ist es möglich, diese Schwerewellen zu entdecken?
Diese Hoffnung ist vergeblich. In der Tat, um einem Teilchen eine Beschleunigung zu erteilen, bedarf es eines anderen Teilchens, welches, vermöge des Gesetzes von Wirkung und Gegenwirkung, nach der entgegengesetzten Seite getrieben wird. Nun hängt aber die Stärke der ausgesandten Schwerewelle von der Summe der Produkte aus schwerer Masse und Beschleunigung der beiden Teilchen ab, während nach dem Gegenwirkungsprinzip die Summe der Produkte aus träger Masse und Beschleunigung gleich null ist.[1)] [see previous note] Es ist also zwar die Existenz der Gravitationswellen mit dem angenommenen Feldmechanismus verträglich, aber durch die Identität von schwerer und träger Masse wird die Möglichkeit ihrer Erzeugung praktisch ausgeschlossen. Hieraus geht hervor, daß das Planetensystem nicht seine mechanische Energie durch Strahlung verliert, während ein analoges System, bestehend aus negativen Elektronen, die um einen positiven Kern kreisen, allmählich seine Energie ausstrahlt. Das Leben des Plantensystems ist also nicht durch eine solche Gefahr bedroht." (Abraham 1913, 208–209)
67   See, e.g., (Wald 1984, 83).
68   See (Abraham 1912g).
69   "Man könnte nun vermuten, daß während der Emission von α -Strahlen durch radioaktive Atome, bei der sehr große Beschleunigungen auftreten, die hypothetischen Schwerkraftwellen in merklicher Stärke erregt werden. Indessen werden dabei gleichzeitig elektrische Wellen erregt und, wie die quantitative Diskussion zeigt, beträgt die Kraft, welche die von einem α -Teilchen entsandten Schwerewellen auf ein zweites α -Teilchen ausüben, höchstens 10^{-36} der elektrischen Kraft." (Abraham 1915, 487)

umph merely realized a potential inherent in the shared knowledge of the time. After all, even if considered in hindsight, Abraham's thoughts on maximal sizes of stars and gravitational waves represent the grandiose vistas offered by an outlook from considerable heights, even if it is also clear in hindsight that Abraham was not the one who reached the highest summit.

REFERENCES

Abraham, Max. 1912a. "Das Elementargesetz der Gravitation." *Physikalische Zeitschrift* (13) 4–5.
———. 1912b. "Das Gravitationsfeld." *Physikalische Zeitschrift* 13: 793–797.
———. 1912c. "Der freie Fall." *Physikalische Zeitschrift* 13: 310–311. (English translation in this volume.)
———. 1912d. "Die Erhaltung der Energie und der Materie im Schwerkraftfelde." *Physikalische Zeitschrift* (13) 311–314.
———. 1912e. "Nochmals Relativität und Gravitation. Bemerkungen zu A. Einsteins Erwiderung." *Annalen der Physik* (38) 444–448.
———. 1912f. "Relativität und Gravitation. Erwiderung auf eine Bemerkung des Herrn A. Einstein." *Annalen der Physik* (38) 1056–1058.
———. 1912g. "Sulle onde luminose e gravitazionali." *Nuovo Cimento* (6), 3: 211.
———. 1912h. "Zur Theorie der Gravitation." *Physikalische Zeitschrift* 13: 1–4, "Berichtigung," p. 176. (English translation in this volume.)
———. 1913. "Eine neue Gravitationstheorie." *Archiv der Mathematik und Physik* (20): 193–209. (English translation in this volume.)
———. 1914. "Die neue Mechanik." *Scientia* 15: 8–27.
———. 1915. "Neuere Gravitationstheorien." *Jahrbuch der Radioaktivität und Elektronik* 11: 470–520. (English translation in this volume.)
Abraham, Max and August Föppl. 1904–1908. *Theorie der Elektrizität* (2 vols). Leipzig: Teubner.
Behacker, M. 1913. "Der freie Fall und die Planetenbewegung in. Nordströms Gravitationstheorie." *Physikalische Zeitschrift*: 14: 989–992.
Cattani, Carlo, and Michelangelo De Maria. 1989. "Max Abraham and the Reception of Relativity in Italy: His 1912 and 1914 Controversies with Einstein." In D. Howard and J. Stachel (eds.), *Einstein and the History of General Relativity*. (*Einstein Studies* vol. 1.) Boston: Birkhäuser, 160–174.
CPAE 3. 1993. Martin J. Klein, A. J. Kox, Jürgen Renn, and Robert Schulmann (eds.), *The Collected Papers of Albert Einstein*. Vol. 3. *The Swiss Years: Writings, 1909–1911*. Princeton: Princeton University Press.
CPAE 4. 1995. Martin J. Klein, A. J. Kox, Jürgen Renn, and Robert Schulmann (eds.), *The Collected Papers of Albert Einstein*. Vol. 4. *The Swiss Years: Writings, 1912–1914*. Princeton: Princeton University Press.
CPAE 5. 1993. Martin J. Klein, A. J. Kox, and Robert Schulmann (eds.), *The Collected Papers of Albert Einstein*. Vol. 5. *The Swiss Years: Correspondence, 1902–1914*. Princeton: Princeton University Press.
Einstein, Albert. 1911. "Über den Einfluß der Schwerkraft auf die Ausbreitung des Lichtes." *Annalen der Physik* 35: 898–908, (CPAE 3, Doc. 23).
———. 1912. "Lichtgeschwindigkeit und Statik des Gravitationsfeldes." *Annalen der Physik* 38: 355–369, (CPAE 4, Doc. 3).
Einstein, Albert, and Marcel Grossmann. 1913. "Entwurf einer verallgemeinerten Relativitätstheorie und einer Theorie der Gravitation." *Zeitschrift für Mathematik und Physik* (62) 3: 225–261.
Eisenstaedt, Jean. 1989. "The Early Interpretation of the Schwarzschild Solution." In D. Howard and J. Stachel (eds.), *Einstein and the History of General Relativity*, (*Einstein Studies* vol. 1). Boston: Birkhäuser, 213–233.
Herglotz, G. 1904. "Zur Elektronentheorie." *Nachrichten von der Königlichen Gesellschaft der Wissenschaften zu Göttingen. Mathematisch-physikalische Klasse* 357–382.
Kennefick, Daniel. 1999. "Controversies in the History of the Radiation Reaction Problem in General Relativity." In H. Goenner, J. Renn, J. Ritter and T. Sauer (eds.), *The Expanding Worlds of General Relativity*. (*Einstein Studies* vol. 7.) Boston: Birkhäuser.

---

70   For further discussion, see (Kennefick 1999, 2006).

————. 2006. *Traveling at the Speed of Thought: Einstein and the Quest for Gravitational Waves.* Princeton: Princeton University Press.

Kretschmann, Erich. 1917. "Über den physikalischen Sinn der Relativitätspostulate, A. Einsteins neue und seine ursprüngliche Relativitätstheorie." *Annalen der Physik* (53) 16: 575–614.

Minkowski, Hermann. 1908. "Die Grundgleichungen für elektromagnetischen Vorgänge in bewegten Körpern." *Nachrichten der Königlichen Gesellschaft der Wissenschaften zu Göttingen, Mathematisch-Physikalische Klasse*, 53–111. (English translation of the appendix "Mechanics and the Relativity Postulate" in this volume.)

Norton, John. 1992. "The Physical Content of General Covariance." In J. Eisenstaedt and A. J. Kox (eds.), *Studies in the History of General Relativity*, (*Einstein Studies* vol. 3). Boston: Birkhäuser, 281–315.

————. 1993. "General Covariance and the Foundations of General Relativity: Eight Decades of Dispute." *Reports on Progress in Physics* (56) 791–858.

Pavanini, G. 1912. "Prime consequenze d'una recente teoria della gravitazione." *Rendic. d. R. Acc. dei Lincei* XXI ^{2}, 648; XXII^1, 369, 1913.

Poincaré, Henri. 1906. "Sur la dynamique de l'électron." *Rendiconto del Circolo Matematico di Palermo* 21: 129–175. (English translation in this volume.)

Renn, Jürgen, Peter Damerow and Simone Rieger. 2001. "Hunting the White Elephant: When and How did Galileo Discover the Law of Fall? (with an Appendix by Domenico Giulini)." In J. Renn (ed.), *Galileo in Context*. Cambridge: Cambridge University Press, 29–149.

Rynasiewicz, Robert. 1999. "Kretschmann's Analysis of Covariance and Relativity Principles." In H. Goenner, J. Renn, J. Ritter and T. Sauer (eds.), *The Expanding Worlds of General Relativity*, (*Einstein Studies* vol. 7). Boston: Birkhäuser, 431–462.

Sommerfeld, Arnold. (ed). 1903–1926. *Physik*, 3 vols. (*Encyklopädie der mathematischen Wissenschaften mit Einschluss ihrer Anwendungen*, vol. 5.) Leipzig: Teubner.

Sommerfeld, Arnold. 1910. "Zur Relativitätstheorie II. Vierdimensionale Vektoranalysis." *Annalen der Physik* (33) 649–689.

————. 1911 *Sitzungsberichte d. Bayer. Akad. d. Wissensch.* p. 51.

Wald, Robert M. 1984. *General Relativity.* Chicago: Chicago University Press.

Zenneck, Jonathan. 1903. "Gravitation." In (Sommerfeld 1903–1926), 1: 25–67. (Printed in this volume.)

MAX ABRAHAM

ON THE THEORY OF GRAVITATION

Originally published in Rendiconti della R. Accademia dei Lincei. German translation by the author published as "Zur Theorie der Gravitation" in Physikalische Zeitschrift 13, 1912, pp. 1–4; correction, p. 176. Received December 14, 1911. Author's date: Milan, 1911. English translation taken from the German.

In a recently published paper A. Einstein[1] proposed the hypothesis that the speed of light (c) depends on the gravitational potential (Φ). In the following note, I develop a theory of the gravitational force which satisfies the principle of relativity and derive from it a relation between c and Φ, which in first approximation is equivalent to Einstein's. This theory attributes values to the densities of the energy and the energy flux of the gravitational field which differ from those hitherto assumed.

Following Minkowski's[2] presentation, we consider

$$x, y, z \qquad \text{and} \qquad u = il = ict$$

as the coordinates of a four-dimensional space. Let the "rest density" ν, as well as the gravitational potential Φ, be scalars in this space, and let them be linked through the differential equation:

$$\frac{\partial^2 \Phi}{\partial x^2} + \frac{\partial^2 \Phi}{\partial y^2} + \frac{\partial^2 \Phi}{\partial z^2} + \frac{\partial^2 \Phi}{\partial u^2} = 4\pi\gamma\nu. \tag{1}$$

(γ is the gravitational constant.)

The "accelerating force" [*bewegende Kraft*] acting on the unit mass in the gravitational field is a four-vector

$$\mathfrak{F} = -\text{Grad}\,\Phi \tag{2}$$

with components

$$\mathfrak{F}_x = -\frac{\partial \Phi}{\partial x}, \quad \mathfrak{F}_y = -\frac{\partial \Phi}{\partial y}, \quad \mathfrak{F}_z = -\frac{\partial \Phi}{\partial z}, \quad \mathfrak{F}_u = -\frac{\partial \Phi}{\partial u}. \tag{2a}$$

---

1   A. Einstein, *Ann. d. Phys.* 35, p. 898, 1911.
2   H. Minkowski, *Göttinger Nachr.* 1908, p. 53.

Jürgen Renn (ed.). *The Genesis of General Relativity*, Vol. 3
Gravitation in the Twilight of Classical Physics: Between Mechanics, Field Theory, and Astronomy.
© 2007 Springer.

According to eqs. (1) and (2), the gravitational force propagates with the speed of light as required by the principle of relativity; whereas, however, light waves are transverse, the *gravitational waves are longitudinal*.

We write $\dot{x}, \dot{y}, \dot{z}, \dot{u}$ for the first derivatives of the coordinates of a material "world point" with respect to its "proper time" τ, i.e., for the components of the "velocity" four-vector $\mathfrak{V}$ and $\ddot{x}, \ddot{y}, \ddot{z}, \ddot{u}$ for the second derivatives, i.e., for the components of the "acceleration" four-vector $\mathfrak{B}$. Then the *equations of motion*[3] are:

$$\ddot{x} = \mathfrak{F}_x, \quad \ddot{y} = \mathfrak{F}_y, \quad \ddot{z} = \mathfrak{F}_z, \quad \ddot{u} = \mathfrak{F}_u. \tag{3}$$

Between the first derivatives exist the identity: |

[2]
$$\dot{x}^2 + \dot{y}^2 + \dot{z}^2 + \dot{u}^2 = -c^2, \tag{4}$$

or

$$\dot{l}^2 \left\{ \left(\frac{dx}{dl}\right)^2 + \left(\frac{dy}{dl}\right)^2 + \left(\frac{dz}{dl}\right)^2 - 1 \right\} = -c^2.$$

Therefore, if one sets

$$\beta^2 = \left(\frac{dx}{dl}\right)^2 + \left(\frac{dy}{dl}\right)^2 + \left(\frac{dz}{dl}\right)^2, \quad \kappa = \sqrt{1 - \beta^2}, \tag{4a}$$

it follows that

$$\dot{l} = \frac{c}{\sqrt{1 - \beta^2}} = c\kappa^{-1}. \tag{4b}$$

Now, by differentiating eq. (4) with respect to proper time, Minkowski obtains the condition of the "orthogonality" of the velocity and acceleration four-vectors. However, if c is considered to be variable, the place of that condition is taken by the following:

$$\dot{x}\ddot{x} + \dot{y}\ddot{y} + \dot{z}\ddot{z} + \dot{u}\ddot{u} = -c\frac{dc}{d\tau}, \tag{5}$$

as the differentiation of eq. (4) shows. By introducing here, instead of the acceleration, the accelerating force according to (3) and (2), we obtain

$$\dot{x}\frac{\partial\Phi}{\partial x} + \dot{y}\frac{\partial\Phi}{\partial y} + \dot{z}\frac{\partial\Phi}{\partial z} + \dot{u}\frac{\partial\Phi}{\partial u} = c\frac{dc}{d\tau}$$

---

3   As I have shown, when energy of a non-mechanical nature is supplied to matter, the Minkowskian equations of motion require modification. See this journal 10, 737, 1909; 11, 527, 1910. But, now, we are dealing with purely mechanical effects so that this modification need not be considered.

or

$$\frac{d\Phi}{d\tau} = c\frac{dc}{d\tau}.$$

The integration yields

$$\frac{c^2}{2} - \frac{c_0^2}{2} = \Phi - \Phi_0, \tag{6}$$

where c_0 and Φ_0 represent the speed of light and the gravitational potential at the origin of coordinates. Equation (6) implies: *The increase of half the square of the speed of light is equal to the increase of the gravitational potential.*

Instead of this relation, which is exactly valid according to our theory, one can, neglecting the square of the quotient of Φ and c^2, take Einstein's formula (loc. cit. p. 906):

$$c = c_0\left(1 + \frac{\Phi - \Phi_0}{c^2}\right).$$

However, eq. (6) better serves to manifest the independence from the arbitrarily chosen origin of coordinates.

It is instructive to compare the relation obtained with the emission theory of light. Let us imagine that the particles emitted from the light source move according to the laws of Galilean mechanics and that they are subject to the gravitational force. They would then experience an increase of *vis viva* [*lebendiger Kraft*] equal to the *decrease* of potential energy. The increase of *vis viva* of the light particles calculated according to (6) is, by contrast, equal to the *increase* of their potential energy, i.e., of the same amount but of opposite sign as that following from the emission theory. But as far as the bending of light rays in the gravitational field is concerned, which can be derived from (6) with the aid of Huygens' principle,[4] it is identical with the bending of the trajectories of those light particles. This is one of the numerous incomplete analogies between the modern theory of radiation and the emission theory of light.

We consider the motion of a material point of mass m in a gravitational field. The first three equations of motion yield:

$$m\frac{d\dot{x}}{d\tau} = -m\frac{\partial\Phi}{\partial x}, \qquad m\frac{d\dot{y}}{d\tau} = -m\frac{\partial\Phi}{\partial y}, \qquad m\frac{d\dot{z}}{d\tau} = -m\frac{\partial\Phi}{\partial z}; \tag{7}$$

they contain the law of conservation of momentum [*Impulssatz*]. In contrast, the last of the eqs. (3):

---

4   A. Einstein has shown that a light ray passing the surface of the Sun is deflected towards the Sun and has drawn the attention of the astronomers to this consequence of the theory which can perhaps serve as its verification.

$$m\frac{dl}{d\tau} = -mi\widetilde{\mathfrak{F}}_u = mi\frac{\partial\Phi}{\partial u} = m\frac{\partial\Phi}{\partial l} \tag{8}$$

expresses the law of conservation of *vis viva* in Minkowskian mechanics. In particular, if the gravitational field depends on the position but not on the time, then one obtains, multiplying (8) by c and taking (4b) into consideration:

$$mc\frac{d}{d\tau}(c\kappa^{-1}) = 0. \tag{9}$$

Now, by considering c as constant, Minkowski interprets $m(c^2\kappa^{-1} - 1)$ as the kinetic energy of the material point. In contrast, in the theory developed here, which takes c to be variable, this would not be permissible. It also seems impossible in this case to give a general expression for the energy of the material point whose decrease would be precisely equal to the energy extracted from the gravitational field.

However, we can convince ourselves that, at least for small velocities, the theorem of conservation of energy, which is confirmed by experience in this realm, follows from (9). From (9) one obtains

$$mc\kappa^{-1} = \text{const.} \tag{9a}$$

[3] I By neglecting squares and products of β^2 and $\dfrac{\Phi-\Phi_0}{c^2}$, according to (4a) we write:

$$\kappa^{-1} = (1-\beta^2)^{-1/2} = 1+\frac{1}{2}\beta^2, \qquad \beta^2 = \frac{v^2}{c_0^2},$$

and, according to (6):

$$c = \{c_0^2 + 2(\Phi-\Phi_0)\}^{1/2} = c_0 + \frac{\Phi-\Phi_0}{c_0}.$$

Then, we obtain

$$mc\kappa^{-1} = mc_0\left(1+\frac{1}{2}\beta^2\right) + m\frac{\Phi-\Phi_0}{c_0}.$$

Therefore, if we multiply (9a) by c_0, it follows that

$$\frac{1}{2}mv^2 + m\Phi = \text{const.}, \tag{9b}$$

i.e., the law of the conservation of energy in the usual form. *Thus, in the limiting case of small velocities, the new mechanics coincides with the old.* As a consequence of the relation (6) between the velocity of light c and the gravitational potential Φ, it also now emerges clearly that not only the *"kinetic" energy* $1/2\,mv^2$, but also the *potential energy* $m\Phi$ *are associated with the material point.*

We now consider two material points of masses m and m_0 moving with small velocities in a stationary gravitational field. Each of the two points possesses a potential energy, of which the part depending upon the mutual distance r is:

$$-\gamma \frac{m_0 m}{r} = -E.$$

Therefore, the total of this variable part of the potential energy for the two points amounts to $-2E$. Hence, if, under their mutual attraction, the two points approach one another, then the *increase in their total vis viva is equal to half of the decrease of their total potential energy.*

Where now does the other half reside? Obviously in the gravitational field. Indeed, as we shall see, in this case, our theory ascribes to the field energy the value E, which is equal and opposite to the one previously assumed. In this way, the difficulty emphasized by Maxwell[5]—that the energy density in a gravitational field, when set to zero for vanishing forces, would become negative elsewhere—disappears. *The expression (13) for the energy density in a gravitational field at which we will arrive is strictly positive.* But, besides the field energy E, the total energy of the system also contains the energy of the matter whose potential part in the stationary field is $-2E$.

The credit for having extended the concept of energy flux to the gravitational field goes to V. Volterra.[6] However, his expression for the energy flux is based on Maxwell's assumption concerning the energy distribution in the field. Accordingly, the theory developed here will thus lead to different results also with respect to energy flux.

The fictitious stresses, the energy flux as well as the densities of energy and momentum for a field depend on a *"four-dimensional tensor"*.[7] We write for the *ten components of the gravitational tensor*:

5   [J.] Clerk Maxwell, *Scientific papers* 1, 570.
6   V. Volterra, *Nuovo Cimento* 337, 1899, 1.
7   Regarding these "world tensors" or "ten-tensors" see M. Abraham, *Rendiconti del circolo matematico di Palermo* 1910, 1; A. Sommerfeld, *Ann. d. Phys.* 32, 749, 1910; M. Laue, "Das Relativitätsprinzip", Braunschweig 1911, p. 73.

$$X_x = \frac{1}{4\pi\gamma}\left\{-\left(\frac{\partial\Phi}{\partial x}\right)^2 + \Psi\right\}$$

$$Y_y = \frac{1}{4\pi\gamma}\left\{-\left(\frac{\partial\Phi}{\partial y}\right)^2 + \Psi\right\}$$

$$Z_z = \frac{1}{4\pi\gamma}\left\{-\left(\frac{\partial\Phi}{\partial z}\right)^2 + \Psi\right\}$$

$$X_y = Y_x = -\frac{1}{4\pi\gamma}\frac{\partial\Phi}{\partial x}\frac{\partial\Phi}{\partial y}$$

$$Y_z = Z_y = -\frac{1}{4\pi\gamma}\frac{\partial\Phi}{\partial y}\frac{\partial\Phi}{\partial z}$$

$$Z_x = X_z = -\frac{1}{4\pi\gamma}\frac{\partial\Phi}{\partial z}\frac{\partial\Phi}{\partial x}$$

(10)

$$X_u = U_x = -\frac{1}{4\pi\gamma}\frac{\partial\Phi}{\partial u}\frac{\partial\Phi}{\partial x}$$

$$Y_u = U_y = -\frac{1}{4\pi\gamma}\frac{\partial\Phi}{\partial u}\frac{\partial\Phi}{\partial y}$$

$$Z_u = U_z = -\frac{1}{4\pi\gamma}\frac{\partial\Phi}{\partial u}\frac{\partial\Phi}{\partial z}$$

(10a)

$$U_u = \frac{1}{4\pi\gamma}\left\{-\left(\frac{\partial\Phi}{\partial u}\right)^2 + \Psi\right\}.$$

(10b)

Here, Φ and

$$\Psi = \frac{1}{2}\left\{\left(\frac{\partial\Phi}{\partial x}\right)^2 + \left(\frac{\partial\Phi}{\partial y}\right)^2 + \left(\frac{\partial\Phi}{\partial z}\right)^2 + \left(\frac{\partial\Phi}{\partial u}\right)^2\right\}$$

(10c)

are four-dimensional scalars. Thus, it becomes immediately clear that the components (10), (10a), (10b) of the ten-tensor indeed transform like the squares and products of the coordinates x, y, z, u. From them, [4] the components of the *accelerating force* per unit volume are derived as follows:

$$
\left.
\begin{aligned}
v\mathfrak{F}_x &= \frac{\partial X_x}{\partial x} + \frac{\partial X_y}{\partial y} + \frac{\partial X_z}{\partial z} + \frac{\partial X_u}{\partial u} \\[4pt]
v\mathfrak{F}_y &= \frac{\partial Y_x}{\partial x} + \frac{\partial Y_y}{\partial y} + \frac{\partial Y_z}{\partial z} + \frac{\partial Y_u}{\partial u} \\[4pt]
v\mathfrak{F}_z &= \frac{\partial Z_x}{\partial x} + \frac{\partial Z_y}{\partial y} + \frac{\partial Z_z}{\partial z} + \frac{\partial Z_u}{\partial u} \\[4pt]
v\mathfrak{F}_u &= \frac{\partial U_x}{\partial x} + \frac{\partial U_y}{\partial y} + \frac{\partial U_z}{\partial z} + \frac{\partial U_u}{\partial u}
\end{aligned}
\right\}
\tag{11}
$$

By substituting the above expressions for the components of the ten-tensor and by taking into consideration the differential equation (1), one arrives at the value

$$
v\mathfrak{F} = -v\,\mathrm{Grad}\,\Phi.
\tag{12}
$$

for the accelerating force per unit volume, in agreement with (2).

We discuss the formulae for the components of the gravitational tensor. The first six components (10) determine the *"fictitious stresses"* in the gravitational field. In the *stationary field* there exist the *normal stresses*:

$$
\left.
\begin{aligned}
X_x &= \frac{1}{8\pi\gamma}\left\{-\left(\frac{\partial\Phi}{\partial x}\right)^2 + \left(\frac{\partial\Phi}{\partial y}\right)^2 + \left(\frac{\partial\Phi}{\partial z}\right)^2\right\}, \\[6pt]
Y_y &= \frac{1}{8\pi\gamma}\left\{\left(\frac{\partial\Phi}{\partial x}\right)^2 - \left(\frac{\partial\Phi}{\partial y}\right)^2 + \left(\frac{\partial\Phi}{\partial z}\right)^2\right\}, \\[6pt]
Z_z &= \frac{1}{8\pi\gamma}\left\{\left(\frac{\partial\Phi}{\partial x}\right)^2 + \left(\frac{\partial\Phi}{\partial y}\right)^2 - \left(\frac{\partial\Phi}{\partial z}\right)^2\right\},
\end{aligned}
\right\}
\tag{12a}
$$

and the *shear stresses*

$$
\left.
\begin{aligned}
X_y = Y_x &= -\frac{1}{4\pi\gamma}\frac{\partial\Phi}{\partial x}\frac{\partial\Phi}{\partial y}, \\[6pt]
Y_z = Z_y &= -\frac{1}{4\pi\gamma}\frac{\partial\Phi}{\partial y}\frac{\partial\Phi}{\partial z}, \\[6pt]
Z_x = X_z &= -\frac{1}{4\pi\gamma}\frac{\partial\Phi}{\partial z}\frac{\partial\Phi}{\partial x}.
\end{aligned}
\right\}
\tag{12b}
$$

They correspond to a *pressure along the lines of force and to a pull perpendicular to the lines of force, and their total value is proportional to the field strength and, for a stationary field, is equal to the energy density* (ε) *(see eq. (13)). For a temporally varying field there is still a hydrostatic pressure to be added:*

$$p = \frac{1}{8\pi\gamma}\left(\frac{\partial\Phi}{\partial l}\right)^2, \tag{12c}$$

which has to be subtracted from the normal stress (12a).

The *energy density* in the *gravitational field* is:

$$\varepsilon = U_u = \frac{1}{8\pi\gamma}\left\{\left(\frac{\partial\Phi}{\partial x}\right)^2 + \left(\frac{\partial\Phi}{\partial y}\right)^2 + \left(\frac{\partial\Phi}{\partial z}\right)^2 + \left(\frac{\partial\Phi}{\partial l}\right)^2\right\}. \tag{13}$$

This turns out to be *strictly positive* and, in particular, to be *proportional to the square of the gravitational force* $\mathfrak{F}$ in the *stationary field*.

The *energy flux* has the components:[8]

$$\mathfrak{S}_x = icU_x, \qquad \mathfrak{S}_y = icU_y, \qquad \mathfrak{S}_z = icU_z,$$

with which, from (10a), follows:

$$\mathfrak{S} = -\frac{1}{4\pi\gamma}\frac{\partial\Phi}{\partial t}\operatorname{grad}\Phi = \frac{1}{4\pi\gamma}\frac{\partial\Phi}{\partial t}\cdot\mathfrak{F}, \tag{14}$$

i.e. the energy flux has the direction of the gravitational force; its magnitude is proportional to the product of the magnitude of gravitational force and the temporal increase of the potential.

If one multiplies the last of the eqs. (11) by $-ic$, then it becomes:[8]

$$-icv\mathfrak{F}_u = -\operatorname{div}\mathfrak{S} - \frac{\partial\varepsilon}{\partial t}. \tag{14a}$$

Therefore

$$-icv\mathfrak{F}_u = icv\frac{\partial\Phi}{\partial u} = v\frac{\partial\Phi}{\partial l} \tag{14b}$$

yields the energy, which, per unit volume and time, is extracted from the field and is supplied to the matter.

Since, furthermore, as can be seen from (11),[8]

$$\frac{i}{c}X_u, \qquad \frac{i}{c}Y_u, \qquad \frac{i}{c}Z_u,$$

---

[8]   Here, the speed of light c is considered as constant, and therefore the influence of the potential on the speed of light discussed above is neglected.

are the components of the momentum density $(\mathfrak{g})$ of the gravitational field, it follows from (10a) that:

$$\mathfrak{g} = \frac{1}{c^2}\mathfrak{S} = \frac{1}{4\pi\gamma c^2}\frac{\partial\Phi}{\partial t}\cdot\mathfrak{F}. \tag{15}$$

The symmetry of the ten-tensor implies this relation between the energy flux and the momentum density.[9]

CORRECTION

In lines 8 and 9 [lines 16 and 17 in the original] of my note "On the Theory of Gravi-   [176]
tation" an oversight has to be corrected which was brought to my attention by a friendly note from Mr. A. Einstein. Hence one should read there "we consider dx, dy, dz and $du = idl = icdt$ as components of a displacement ds in four-dimensional space."

Hence:

$$ds^2 = dx^2 + dy^2 + dz^2 - c\ dt^2$$

is the square of the four-dimensional line element, where the speed of light c is determined by eq. (6).

---

9   M. Planck, this journal, 9, 828, 1908, has put forth the claim that also a mechanical energy flux always
      implies a corresponding momentum.

MAX ABRAHAM

THE FREE FALL

Originally published in Rendiconti del R. Instituto Lombardo di scienze e lettere.
German translation by the author published as "Der freie Fall" in Physikalische
Zeitschrift 13, 1912, pp. 310–311. Received March 11, 1912. Author's date: Milan,
March 1912. English translation taken from the German.

In a recently published communication[1] I have developed a new theory of gravitation.
In this theory, the speed of light c is linked to the gravitational potential Φ through
the relation:

$$\frac{1}{2}c^2 - \frac{1}{2}c_0^{\,2} = \Phi - \Phi_0. \tag{1}$$

The equations of motion of a material point in a gravitational field are:

$$\frac{d^2x}{d\tau^2} = -\frac{\partial\Phi}{\partial x}, \qquad \frac{d^2y}{d\tau^2} = -\frac{\partial\Phi}{\partial y}, \qquad \frac{d^2z}{d\tau^2} = -\frac{\partial\Phi}{\partial z}, \tag{2}$$

$$\frac{d^2u}{d\tau^2} = -\frac{\partial\Phi}{\partial u}; \tag{2a}$$

τ denotes the "*proper time*" of the moving point and is related to the time t measured in the reference frame as follows:

$$\frac{du}{d\tau} = ic\frac{dt}{d\tau} = \frac{ic}{\kappa}, \tag{3}$$

$$\kappa = \sqrt{1 - \frac{v^2}{c^2}} \tag{4}$$

(v is the magnitude of the velocity of the material point).

In the static field, the last (2a) of the equations of motion

---

1   This journal, 13, 1, 1912.

Jürgen Renn (ed.). *The Genesis of General Relativity*, Vol. 3
Gravitation in the Twilight of Classical Physics: Between Mechanics, Field Theory, and Astronomy.
© 2007 Springer.

$$\frac{d}{d\tau}\left(\frac{ic}{\kappa}\right) = \frac{i}{c}\frac{\partial\Phi}{\partial t}$$

assumes the form

$$\frac{c}{\kappa} = \text{const};\tag{5}$$

it corresponds to the law of conservation of energy.

In the following, the *free fall in empty space* shall be treated on the basis of this theory. The gravitational field is assumed to be homogeneous and parallel to the x- axis. Therefore

$$-\frac{\partial\Phi}{\partial x} = g, \qquad -\frac{\partial\Phi}{\partial y} = -\frac{\partial\Phi}{\partial z} = 0.$$

Then, eqs. (2) yield:

$$\frac{d^2x}{d\tau^2} = g, \qquad \frac{d^2y}{d\tau^2} = 0, \qquad \frac{d^2z}{d\tau^2} = 0.\tag{6}$$

These equations of motion differ from those of Galilean mechanics merely by the proper time τ of the falling point taking the place of t. From these it follows immediately that *the trajectory of the point is a parabola also in the new theory.*

We want to restrict ourselves to the consideration of motion parallel to the x- axis, i.e. along the field, and, in particular, with the following initial conditions:

$$\tau = 0: \qquad x = 0, \qquad \frac{dx}{d\tau} = 0.\tag{7}$$

The gravitational potential may be referred to the plane $(x = 0)$:

$$\Phi_0 = 0, \qquad \Phi = -gx;$$

then, according to (1), one has to set:

$$c^2 = c_0^2 - 2gx;\tag{8}$$

where c_0 denotes the speed of light for $x = 0$. Since we assume the initial speed of the falling point to be equal to zero, initially, we have, according to (4), $\kappa = 1$. Therefore eq. (5) reads:

$$\frac{c}{\kappa} = c_0.\tag{9}$$

From this and from (3) follows the relation

$$\frac{dt}{d\tau} = \frac{1}{\kappa} = \frac{c_0}{c},\tag{10}$$

which connects the time t with the proper time τ; as a consequence of (8) it can be written as:

$$\frac{dt}{d\tau} = \frac{1}{\sqrt{1 - \dfrac{2gx}{c_0^2}}}.$$ (11)

By integrating the first of the equations of motion (6), subject to the initial conditions (7), we obtain

$$\frac{dx}{d\tau} = g\tau,$$ (12)

$$x = \frac{1}{2}g\tau^2,$$ (13)

i.e. the free fall velocity is proportional to the proper time, the free fall distance to half its square.

The task is now to replace the proper time τ by introducing the time t measured in the coordinate system. In order to determine t as a function of τ, we use eq. (11); taking into account (13) yields

$$\frac{dt}{d\tau} = \frac{1}{\sqrt{1 - \left(\dfrac{g\tau}{c_0}\right)^2}},$$ (14)

and thus

$$t = \int_0^\tau \frac{d\tau}{\sqrt{1 - \left(\dfrac{g\tau}{c_0}\right)^2}} = \frac{c_0}{g}\,\text{arc }\sin\left(\frac{g\tau}{c_0}\right).$$ (15)

By taking the inverse of this functional relation we express τ through t: |

$$\tau = \frac{c_0}{g}\sin\left(\frac{gt}{c_0}\right).$$ (16)   [311]

Therefore it follows that

$$\frac{d\tau}{dt} = \cos\left(\frac{gt}{c_0}\right).$$ (16a)

Thus, the *free fall velocity* becomes:

$$v = \frac{dx}{dt} = \frac{dx\,d\tau}{d\tau\,dt} = g\tau\frac{d\tau}{dt} = c_0\sin\left(\frac{gt}{c_0}\right)\cos\left(\frac{gt}{c_0}\right).$$ (17)

On the other hand, since according to (10):

$$c = c_0 \frac{d\tau}{dt} = c_0 \cos\left(\frac{gt}{c_0}\right),$$ (17a)

we then have:

$$\frac{v}{c} = \sin\left(\frac{gt}{c_0}\right)$$ (18)

for the quotient of the velocity of fall and of the speed of light at the point in question. Finally, the *distance of fall* follows from (13) and (16):

$$x = \frac{c_0^2}{2g} \sin^2\left(\frac{gt}{c_0}\right).$$ (19)

It is understood that all of these relations are valid only for

$$gt < c_0 \cdot \frac{\pi}{2}$$ (20)

Indeed, for $gt = c_0 \cdot \frac{\pi}{2}$, one would have

$$gx = \frac{c_0^2}{2},$$

and therefore

$$\Phi - \Phi_0 = -gx = -\frac{c_0^2}{2}.$$

However, for this value of x, the speed of light c would, according to (1), become zero which is inadmissible.

It is obvious that, by imposing the condition

$$\Phi_0 - \Phi < \frac{c_0^2}{2}$$ (21)

on the gravitational potential, the fundamental relation (1) excludes the existence of a homogeneous gravitational field of infinite extent

From the condition (21) one can derive similar consequences of the new theory of gravitation which are of interest for cosmogony.

The potential of a star of mass m is

$$\Phi = -\gamma \frac{m}{r}$$ (22)

(γ gravitational constant). Thus, the potential difference between infinite distance ($r = \infty$), and the surface of the star ($r = a$) is:

$$\Phi_0 - \Phi = \gamma \frac{m}{a}.$$

Therefore,

$$\gamma \frac{m}{a} < \frac{c_0^2}{2} \left(= \frac{9}{2} \cdot 10^{20} \right) \tag{23}$$

must hold. For the Sun one has in *cgs* units

$$\left(\gamma \cdot \frac{m}{a} \right)_s = 2 \cdot 10^{15}.$$

Thus, the *quotient of mass and radius for an arbitrary star must satisfy the inequality*

$$\frac{m}{a} < 2,25 \cdot 10^5 \cdot \left(\frac{m}{a} \right)_s. \tag{24}$$

For stars whose mean density is equal to that of the Sun, the quotients m/a are proportional to the squares of the radii. Hence, the following must hold

$$a < 500(a)_s, \tag{24a}$$

and thus

$$m < 10^8 (m)_s. \tag{24b}$$

That is, *the mass of a star whose mean density is equal to that of the Sun cannot become larger than one hundred million times the mass of the Sun.*

Since this limit is rather high, no difficulties arise from this for our theory.

MAX ABRAHAM

A NEW THEORY OF GRAVITATION

Lecture presented on October 19, 1912 to the Societa italiana per il progresso delle scienze. German translation by the author published as "Eine neue Gravitationstheorie" in Archiv der Mathematik und Physik. Third series 20, 1913, pp. 193–209. English translation taken from the German.

Modern physics does not allow forces that propagate with an infinite speed. It does not hold that Newton's law is the true fundamental law of gravitation; rather, it endeavors to obtain this action-at-a-distance law from differential equations attributing a finite speed of propagation to the gravitational force.

A model of such a theory of local action is provided to us by Maxwell's theory of the electromagnetic field. Its fundamental laws are differential equations connecting the electric vector to the magnetic vector. The electromagnetic energy is, according to this theory, distributed throughout the field. When the field changes in time, an energy flow determined by the Poynting vector obtains. If, for example, an electron starts to oscillate, it sends out electromagnetic waves. With the waves the electromagnetic energy flows from the neighborhood of the electron to the initially undisturbed regions of space; this radiation of energy results in the damping of the oscillations of the electron.

The analogy between Coulomb's and Newton's law suggests a similar interpretation of the gravitational force. It was Maxwell himself who developed a theory of gravitation, modelled on electrostatics, and emphasized the difficulties of such a theory.[1] The different signs of the forces—attraction of masses compared to repulsion of charges of equal sign—entails that the energy density of the gravitational field becomes negative as soon as one assumes that when the field vanishes so does the energy. One would need to drop this last assumption and imagine | that if there were no [194] gravitational field, a certain amount of energy resides in the aether, which decreases upon excitation of the field. But even so, one still does not avoid all objections.

Let us consider for example a material particle which is initially at rest and then is set into oscillation. In theories of the electromagnetic type it emits waves similar to light waves, i.e. transverse waves propagating with the speed of light. With the waves, the gravitational field enters into previously undisturbed regions of space. Hence, the

---

1   J. Cl. Maxwell, *Scientific papers* I, p. 570.

Jürgen Renn (ed.). *The Genesis of General Relativity*, Vol. 3
Gravitation in the Twilight of Classical Physics: Between Mechanics, Field Theory, and Astronomy.
© 2007 Springer.

energy density of the aether decreases in these regions; i.e. the energy flows towards the oscillating particle whose energy increases at the expense of the aether energy. This influx of energy results in an increase of the oscillation of the particle; its equilibrium is therefore unstable.

Similar difficulties arise in this way for all gravitational theories of Maxwellian type; among these, especially the theory of H. A. Lorentz[2] and R. Gans[3] has found followers among physicists. Starting from the hypothesis that matter consists of positive and negative electrons and that the attraction between electrons of opposite charge is slightly greater than the repulsion between those of like charge, it arrives at field equations of Maxwellian character connecting the gravitational vector, which corresponds to the electric vector, to a second one analogous to the magnetic vector. In this theory, the energy flux of the gravitational field is expressed by a vector corresponding to the Poynting vector but with opposite sign.[4] Here, the above mentioned difficulty appears; indeed, R. Gans found that the radiation reaction force has the opposite sign as in the dynamics of the electron. Thus, as a result of the emitted radiation, the acceleration of a neutral particle would increase rather than decrease as for electrons. Hence, the equilibrium of the particle would not be stable.

Therefore, we have to dispense with the close analogy between gravitation and electromagnetism without thereby relinquishing the essential notions of Maxwell's theory, namely: *The fundamental laws must be differential equations describing the excitation and propagation of the gravitational field; associated with this field is a positive energy density as well as an energy flux.* |

[195]   The problem of gravitation is even more urgent, as modern physics has discovered interesting relations between mass (m) and energy (E). According to the theory of relativity one should have

$$E = mc^2 \quad (c \text{ is the speed of light in the vacuum}). \tag{1}$$

Since we still lack a satisfactory theory of gravitation, one has been able to derive this relation only for the inertial mass. Is it valid also for the gravitational mass?

We displace a body in a gravitational field. Its potential energy, and consequently the first term in equation (1), changes with the gravitational potential. It follows that one of the factors on the right-hand side, or even both, must depend on the gravitational potential. We want to consider the hypothesis that c, *the speed of light, depends on the gravitational potential.* This hypothesis was first enunciated by A. Einstein.[5] Taking it as a starting point, I undertook to develop a theory of the gravitational field[6] which I then, in constructive competition with Mr. A. Einstein,[7] gave a

2   H. A. Lorentz, *Verlag. Akad. v. Wetensch. te Amsterdam* 8, 1900, p. 603.
3   R. Gans. *Physik. Zeitschrift* 1905, p. 803.
4   R. Gans, *H. Weber-Festschrift* 1912, p. 75.
5   A. Einstein, *Ann. d. Physik* 35, (1911), p. 898.
6   M. Abraham, *Physik. Zeitschrift* 1912, p. 1 and p. 311.
7   A. Einstein, *Ann. d. Physik* 38 (1912), p. 355 and p. 433.

more satisfactory form.[8] In the following, I want to present the essential features of the new theory of gravitation.

Postulate I. The surfaces c = constant *coincide with the equipotential surfaces of the gravitational field.* Or, in other words: *The negative gradient of* c *gives the direction of the gravitational force.*

If the speed of light varies in the gravitational field, then, according to Huygens' principle, a light ray in this field will be refracted as in an inhomogeneous medium. This consequence was derived by A. Einstein; he showed that a light ray passing the surface of the Sun must be deflected, in fact, just as if it were attracted by the Sun. However, this deflection, only observable at a total eclipse of the Sun, lies at the limit of observability.

For bodies at rest in the gravitational field we apply the usual geometry; hence, we assume that the unit of length, the meter, is independent of c and can thus serve to measure length in arbitrary regions of the gravitational field. We now want to consider two regions in which c may have different values, c_1 and c_2. We bring an antenna having a length of one meter from the | first region into the second. Since the length of the antenna remains constant, obviously the periods of its electromagnetic normal modes change in inverse proportion to the speed of light c:

$$\tau_1 : \tau_2 = c_2 : c_1. \tag{2}$$

[196]

If one could construct a clock whose rate were independent of c, then one could determine the change in period by transporting the clock, with the antenna, from one region into the other. However, we rule out the possibility of such a clock by putting forward the following postulate:

Postulate II. An observer belonging to a material system is unable to perceive that he, together with the system, is brought into a region in which c *has a different value.*

From this postulate,[9] it follows that the duration of an arbitrary process changes in the same proportion to c as does the period of the normal modes of the antenna; because otherwise the observer would be able to determine the latter change. Hence, from the second postulate follows the general theorem: The duration of an arbitrary process in a system changes in inverse proportion to c if the system's location in a gravitational field changes. Or more briefly:
"The times are of degree c^{-1} *."*

Within this theory the unit of time has only a local significance, whereas a universal validity will be attributed to the unit of length, at least for the state of rest.

---

8   M. Abraham, *Physik. Zeitschrift* 1912, p. 793.

9   It is understood that this postulate refers only to the value of c itself, and not to its derivative; the value of c has no influence on the events in the system which present themselves to an observer belonging to that system. The gradient of c, however, i.e. the gravitational force, of course influences these events.

According to postulate II, there exists a certain relativity. Let us imagine that the Earth reaches locations in space in which the gravitational potential, and therefore also c, has a different value. According to the latter postulate it would not be possible for us to determine this fact through any terrestrial measurement. All measurements, and hence also all the constants of physics, would remain unchanged; also the measurement of the speed of light, e.g. according to the method of Fizeau, would produce the same result as before since with respect to a terrestrial clock its changes are compensated for by the changes of the angular velocity of the gear wheel. Obviously, all

[197] velocities change generally in proportion to the light I velocities, i.e. they are of degree c, because the lengths are of degree c^0 and the times of degree c^{-1}.

It is, however, in no way ruled out that an observer not belonging to the system discovers this particular influence of the gravitational potential on the periods. For example, a terrestrial observer measuring the Fraunhofer lines of the Sun and comparing them to the corresponding lines of terrestrial sources, should find that their frequencies behave as

$$\nu_2 : \nu_1 = \tau_1 : \tau_2 = c_2 : c_1.$$

From this would follow a relative shift of the Sun's lines:

$$\frac{\nu_2 - \nu_1}{\nu_1} = \frac{c_2 - c_1}{c_1},$$

namely towards the red end of the spectrum, because the gravitational potential on the Sun, and hence also c, has a smaller value than on Earth. For this relative displacement of the Sun's lines Einstein found the value $2 \cdot 10^{-6}$, which according to Doppler's principle would correspond to a speed of 0.6 kilometers per second. Astrophysicists are now well able to measure shifts of this order, and indeed they have found such shifts, and precisely in the sense required by our theory. However, they interpret the shift partially as the Doppler effect of descending flows of absorbing gases, and partially as pressure effects. But perhaps the totality of the phenomena on the surface of the Sun can be better interpreted if one takes the predicted gravitational shifts of the lines into account.

We now turn from kinematics to dynamics. The second postulate implies that *all mechanical quantities of the same class are of the same degree in* c. For example, a system of forces which maintain themselves in equilibrium in a region, where c has the value c_1, must still do so when the system is brought into a region of the field in which c is equal to c_2. Therefore, all forces must change in the same proportion with c. All forms of energy must be of the same degree in c as well, because if two forms of energy, e.g. the kinetic and the potential, were of different degree, the conversion of energy from one form into the other would give rise to a periodic process whose frequency is not of the correct degree, i.e. that of c. Similar considerations apply for other dynamical quantities. Since all these quantities are constituted from mass,

[198] length and time, it is sufficient to know the *degree of the mass* to determine all I their degrees since we already know the degree of the lengths (c^0) and of the times (c^{-1}).

I raised the question earlier whether, as for the inertial mass, the gravitational mass is proportional to the energy. What would take place during the transformation of radioactive elements if their continual loss of energy as heat would cause a decrease in their inertial mass, but not in their weight? Obviously, uranium and its daughter element radium suspended from strings of equal lengths would result in pendulums with unequal periods of oscillations. Even the equilibrium position of such pendulums would be different since the attraction of the Earth acts on the gravitational, whereas the centrifugal force acts on the inertial mass. This contradicts experience. Either one must give up any relation between mass and energy, or assume that the weight too, like the inertia, is proportional to energy. The first alternative would mean the collapse of the new mechanics. We prefer the second and hence propose the following postulate.

Postulate III. The forces which are acting on two bodies at the same location of the gravitational field are in proportion to their energy.

Here, one has to take into consideration not only the potential and kinetic energy of the molar and molecular motion, but also the chemical and electromagnetic energy. For example, the electrons in a metal carry electromagnetic energy; hence they are subject to gravitation.[10] Similarly, the thermal radiation in the interior of a cavity will also acquire weight. If one interprets the third postulate in this manner, then *the laws of the conservation of energy and of the conservation of weight merge into one.*

Proceeding with the mathematical presentation, we start with the expression for the *Lagrangian function*, which for the dynamics of the electron is:

$$L = -mc^2 f\left(\frac{v}{c}\right);$$ (3)

v denotes the velocity, and m the rest mass of the electron.[11] From | the *Lagrangian function* we derive in the well known manner the values for the momentum and energy as follows: [199]

$$G = \frac{\partial L}{\partial v},$$ (3a)

$$E = v\frac{\partial L}{\partial v} - L,$$ (3b)

while the *equations of motion* are:

---

10 This is also supported by comments of J. Koenigsberger (*Verh. d. D. physik. Ges. XIV* (1912), p. 185).

11 In order for m to have this meaning, we must have $f''(0) = -f(0) = -1$ (this holds for example in the theory of relativity, where $f(v/c) = \sqrt{1-(v/c)^2}$); or else, an insignificant numerical factor becomes associated with m in equation (3).

352   MAX ABRAHAM

$$\frac{d}{dt}\left(\frac{\partial L}{\partial \dot{x}}\right) - \frac{\partial L}{\partial x} = 0 \text{ etc.} \tag{4}$$

The new mechanics has, in the discussion of these equations, so far restricted itself to the case of constant c; in this case the Lagrangian function depends only on the velocity but not on the location. Hence, only the first terms containing the time derivatives of the momentum components enter into the equations of motion:

$$\frac{d}{dt}\left(\frac{\partial L}{\partial v} \cdot \frac{\dot{x}}{v}\right) = \frac{d}{dt}\left(G \cdot \frac{\dot{x}}{v}\right) = \frac{d\mathfrak{G}_x}{dt} \text{ etc.} \tag{4a}$$

In our theory of gravitation however, through the speed of light c, the Lagrangian function also depends on the coordinates; hence one must retain the second terms in Lagrange's equations:

$$\frac{\partial L}{\partial x} = \frac{\partial L}{\partial c}\frac{\partial c}{\partial x} \text{ etc.} \tag{4b}$$

They represent the components of a force proportional to the gradient of c. According to postulate I, *this force* is now precisely *the gravitational force*. In vector form, the equations of motion (4) are:

$$\frac{d\mathfrak{G}}{dt} = \frac{\partial L}{\partial c}\operatorname{grad} c. \tag{5}$$

One recognizes that the Lagrangian equations are nothing else but the analytic expression of our first postulate. (They hold exactly for the free motion of material points in the gravitational field, but can also be applied to systems whose extensions are so small that they can be considered equivalent to a material point.) Especially now, while the world of mathematics prepares for the centennial celebration of the great founder of analytical mechanics, we do not want to omit drawing attention to the significance of his work for the new mechanics.

[200]   Now we want to introduce the third postulate, which for a given location in the field, sets the gravitational force proportional to the energy of the I moving point; in order to fulfill it we must write:

$$\frac{\partial L}{\partial c} = -\chi \cdot E, \tag{6}$$

where χ depends only on c, but not on v. Then the second term of (5), i.e. the gravitational force, becomes:

$$\mathfrak{R} = -\chi(c) \cdot E \cdot \operatorname{grad} c. \tag{6a}$$

Setting

$$\log\varphi(c) = \int^c dc\chi(c), \tag{6b}$$

it follows from (6) that

$$E = -\frac{1}{\chi(c)}\frac{\partial L}{\partial c} = -\varphi\frac{\partial L}{\partial \varphi}. \tag{6c}$$

This and (3b) yields

$$L = v\frac{\partial L}{\partial v} + \varphi\frac{\partial L}{\partial \varphi}. \tag{7}$$

On the basis of a well known theorem of Euler we infer: the Lagrangian function is a homogeneous linear function of v and φ. We wish to write it in the form:

$$L = -M\varphi \cdot f\left(\frac{v}{\varphi}\right). \tag{7a}$$

M denotes a constant (mass constant) belonging to the material mass point in question that is independent of φ and hence of c. The comparison with the expression (3) yields:

$$\varphi = c, \quad M = mc. \tag{7b}$$

From this follows the value for the rest mass m:

$$m = \frac{M}{c}. \tag{8}$$

Therefore, the rest mass is of degree c^{-1}.

Since we now know the degree of the lengths (c^0), times (c^{-1}) and masses (c^{-1}), we are in a position to derive the degree of each class of dynamical quantities. For example, *energies are of degree* c, *likewise forces are of degree* c, *actions* (having dimension of energy times time) *of degree* c^0.

From (6b) and (7b) it follows that

$$\chi(c) = \frac{1}{c},$$

so that the expression (6a) for the *gravitational force* becomes:

$$\Re = -\frac{E}{c}\operatorname{grad}c. \tag{9}$$

I As one sees, the first postulate is satisfied *because the force is proportional to the* [201] *negative gradient of* c, as well as the third, because it is *proportional to the energy*.

A potential energy is associated with a material point resting in a gravitational field; one obtains it from (3b) and (3) by setting v equal to zero:

$$E = -L = +mc^2;$$

therefore, according to (8):

$$E = M \cdot c. \tag{9a}$$

From this expression for the *energy* of a *resting point* follows, according to (9), the value for the *gravitational force* for the case of rest:

$$\Re = -M \operatorname{grad} c. \tag{9b}$$

The work done by this force is, as it must be, equal to the decrease of the potential energy (9a).

We are now in a position to determine for a given gravitational field, i.e. for a given field of the scalar c, the force acting on a material point. But, we still have not solved the problem initially posed, to find the gravitational field which corresponds to a given distribution of matter. Now, the interrelation between action and reaction suggests we assume that like the attracted, so also the *attracting mass is proportional to the energy*, and hence that we consider the energy as the source of the gravitational field. However, the question arises: Besides the energy of the matter, is the energy of the gravitational field itself to be taken into account? If we knew the energy of the field, then, through the application of the principle of virtual work, without further ado, at least the statics of the gravitational field can be derived. Consequently, we prefer to start from reasonable assumptions [*Ansätze*] for the field energy and the energy flux, and to obtain from these the relations connecting the gravitational vector with the density of matter and energy, respectively.

The simplest assumption would be that the energy density of the static field is proportional to the square of the gradient of c, on which, according to (9b), the field strength depends. However, as we have seen, the energy and hence also the density is of degree c, while the square of the gradient of c is obviously of degree c^2. This consideration leads to the introduction of the auxiliary variable

$$u = \sqrt{c} \tag{10}$$

[202] and to assign to the *energy density of the gravitational field* the value:

$$\varepsilon = \frac{1}{2\alpha}\left\{ \left(\frac{\partial u}{\partial x}\right)^2 + \left(\frac{\partial u}{\partial y}\right)^2 + \left(\frac{\partial u}{\partial z}\right)^2 + \left(\frac{1}{c}\frac{\partial u}{\partial t}\right)^2 \right\} \tag{11}$$

(α denotes a universal constant of degree c^0).

In our theory, this expression is valid also for the dynamical field; it is supplemented by the ansatz for the *energy flux in the gravitational field*

$$\mathfrak{S} = -\frac{1}{\alpha}\frac{\partial u}{\partial t} \cdot \operatorname{grad} u. \tag{12}$$

Since according to (11) the field energy is always positive, and vanishes only when the field vanishes, one sees that the energy always flows with the wave by which the disturbance is propagated. Let us imagine for example that such a disturbance, for which $u < u_0$, enters a region in which initially $u = u_0$. Then, the gradient is pointing from u towards the undisturbed region, whereas, during the passing of the wave, the time derivative of u has a negative sign. Hence, the expression (12) for the energy flux has the correct sign. It retains it when $u > u_0$ for the disturbance, because then the gradient as well as the time derivative change sign. *In the theory presented here, the energy flux always corresponds to an emission of energy from the disturbed region*; thus the above mentioned objection is not raised against it. The radiation reaction always implies a decrease of the acceleration of the material particle; its *equilibrium* is hence *not unstable*.

We now want to consider a field containing matter at rest. As we are dealing with a continuous distribution of matter over a volume V, we set

$$M = \int \mu dV, \qquad (13)$$

and call

$$\mu = \lim_{V=0} \left(\frac{M}{V} \right)$$

the "*specific density*" of the matter. Since M does not depend on c, then for an incompressible fluid, whose particles do not change in volume, μ too is independent of the location in the gravitational field.

The rest energy of matter with respect to the unit volume is according to (9a),

$$\eta = \lim_{V=0} \left(\frac{Mc}{V} \right) = \mu c$$

or, according to (10)

$$\eta = \mu u^2. \qquad (14)$$

| Since μ is constant for the case of rest, from this follows: [203]

$$\frac{\partial \eta}{\partial t} = 2\mu u \frac{\partial u}{\partial t} \qquad (14a)$$

as the temporal increase of the rest energy per unit volume, caused by a temporal variation of the gravitational field.

Now we apply the energy equation which demands that the convergence of the energy flux is equal to the sum of the increases of the energy densities of the field (ε) and of the matter (η):

$$-\mathrm{div}\, \mathfrak{S} = \frac{\partial \varepsilon}{\partial t} + \frac{\partial \eta}{\partial t}. \qquad (15)$$

From the expressions (11) and (12) we derive

$$\frac{\partial \varepsilon}{\partial t} = \frac{1}{\alpha}\left\{\frac{\partial u}{\partial x}\frac{\partial^2 u}{\partial x \partial t} + \frac{\partial u}{\partial y}\frac{\partial^2 u}{\partial y \partial t} + \frac{\partial u}{\partial z}\frac{\partial^2 u}{\partial z \partial t} + \frac{1}{c}\frac{\partial u}{\partial t}\frac{\partial}{\partial t}\left(\frac{1}{c}\frac{\partial u}{\partial t}\right)\right\}, \tag{15a}$$

$$-\mathrm{div}\,\mathfrak{S} = \frac{1}{\alpha}\left\{\frac{\partial^2 u}{\partial x \partial t}\frac{\partial u}{\partial x} + \frac{\partial^2 u}{\partial y \partial t}\frac{\partial u}{\partial y} + \frac{\partial^2 u}{\partial z \partial t}\frac{\partial u}{\partial z} + \frac{\partial u}{\partial t}\Delta u\right\}, \tag{15b}$$

with

$$\Delta u = \mathrm{div}\,\mathrm{grad}\,u = \frac{\partial^2 u}{\partial x^2} + \frac{\partial^2 u}{\partial y^2} + \frac{\partial^2 u}{\partial z^2}. \tag{15c}$$

If, in addition, we set

$$\Box u = \Delta u - \frac{1}{c}\frac{\partial}{\partial t}\left(\frac{1}{c}\frac{\partial u}{\partial t}\right), \tag{16}$$

then from (15), taking into account (15a, b) and (14a), it follows that

$$\Box u = 2\alpha\mu u. \tag{17}$$

Here, the second term relates to matter at rest. According to the previous discussion, for the case of motion one has to introduce its energy density, and replace (17) by the more general formulation:

$$\Box u = 2\alpha\frac{\eta}{u}. \tag{17a}$$

This is the fundamental equation relating the gravitational field to the energy of the matter. It can function as the analytic expression of the fourth postulate of our theory, which relates the attracting mass of a body to its energy. For a static field, according to (16), the fundamental equation takes the form:

$$\Delta u \equiv \mathrm{div}\,\mathrm{grad}\,u = 2\alpha\frac{\eta}{u} = 2\alpha\mu u. \tag{17b}$$

Therefore, in this case the *divergence of the gradient of $u = \sqrt{c}$ is proportional to the energy density of matter.*

Now, what about the energy of the gravitational field itself? At least for the static field (17b), it is easy to put the basic equation I into a form in which the energy density of the field appears on the right-hand side. One has

[204]

$$\frac{\partial^2 c}{\partial x^2} = \frac{\partial^2 u^2}{\partial x^2} = 2u\frac{\partial^2 u}{\partial x^2} + 2\left(\frac{\partial u}{\partial x}\right)^2 \text{ etc.},$$

thus,

$$\Delta c = 2u\Delta u + 2(\mathrm{grad}\,u)^2;$$

according to (11), the energy density of the static field is now:

$$2\alpha\varepsilon = (\mathrm{grad}\,u)^2.$$

Therefore, we can write (17b) as

$$\Delta c \equiv \mathrm{div}\ \mathrm{grad}\,c = 4\alpha(\eta + \varepsilon). \qquad (18)$$

Hence, in the static field, the divergence of the gravitational vector, i.e. of the gradi-ent of c, *is proportional to the density of the total energy.* Integrating (18) over a vol-ume V bounded by the surface f, we obtain

$$\int df\frac{\partial c}{\partial n} = 4\alpha\int dV(\eta + \varepsilon) = 4\alpha E, \qquad (19)$$

i.e. in the static gravitational field, *the flux of the gravitational vector through a closed surface is proportional to the enclosed energy.*

To highlight the significance of these statements, we consider a special case: *a sphere at rest* (e.g. the Sun), which is composed of homogeneous concentric layers. The basic equation (17b) yields:

$$\Delta u = 2\alpha\mu u \text{ for } r < a, \qquad (20)$$

i.e. for the interior of the sphere,

$$\Delta u = 0 \text{ for } r > a, \qquad (20a)$$

i.e. for the exterior of the sphere.

This last equation is Laplace's differential equation; its symmetric integral

$$u = u_0\left(1 - \frac{\vartheta}{r}\right) \qquad (r > a) \qquad (21)$$

determines the gravitational field outside the attracting sphere (where u_0 denotes the value of $u = \sqrt{c}$ for $r^{-1} = 0$, ϑ a different constant).
From (21) follows

$$\frac{du}{dr} = \frac{u_0 \cdot \vartheta}{r^2}. \qquad (21a)$$

Thus, the radial gradient of c becomes:

$$\frac{dc}{dr} = 2u\frac{du}{dr} = \frac{2c_0\vartheta}{r^2}\left(1 - \frac{\vartheta}{r}\right). \qquad (22)$$

| Now, according to (9b) the gravitational force is proportional to this gradient. *It fol-*   [205]
lows from this that Newton's law is not strictly valid in our theory. The center of the Sun attracts the planet (considered as a material point) with a force which contains besides the term with r^{-2} *also a term with* r^{-3}.

According to the theorem (eq. (19)) just developed, the flux of the gradient of c passing through a sphere of radius r is proportional to the total energy enclosed by the sphere:

$$4\pi r^2 \frac{dc}{dr} = 4\alpha E. \tag{23}$$

A sphere of infinite radius encloses the total energy (E_t) of the sphere and of its gravitational field; hence, from (22) and (23) it follows

$$E_t = \frac{2\pi c_0}{\alpha} \cdot \vartheta. \tag{23a}$$

In contrast, the sphere of radius $r = a$ contains only the internal energy of the sphere; therefore,

$$E_i = \frac{2\pi c_0}{\alpha} \cdot \vartheta \left(1 - \frac{\vartheta}{a}\right). \tag{23b}$$

Hence, the energy of the external field has the value

$$E_a = E_t - E_i = \frac{\vartheta}{a} E_t. \tag{23c}$$

We write

$$\psi = \frac{\vartheta}{a} = \frac{E_a}{E_t}. \tag{24}$$

This quantity ψ, i.e. the ratio of the external to the total energy, enters into expression (22) for the value of the gravitational force outside of the sphere:

$$\frac{dc}{dr} = \frac{2c_0 a \psi}{r^2} \left(1 - \psi \frac{a}{r}\right) = \frac{\alpha E_t}{\pi r^2} \left(1 - \frac{E_a}{E_t} \cdot \frac{a}{r}\right). \tag{24a}$$

We can therefore say: *The deviation from Newton's law is caused by the energy of the external gravitational field.*

To determine the quotient ψ, one must integrate equation (20), which is valid for the interior of the sphere. We want to compare this differential equation with Poisson's; in Poisson's equation, the right-hand side depends only on the density of the attracting masses, whereas on the right of (20), the potential (u) itself enters as a factor. This implies that in our theory the contributions of the individual mass elements do not superimpose; rather, along with the value of u at a specific location, the contribution of the I masses at the same location decreases as well. Since now the neighboring masses cause a decrease of the potential, it is apparent that large accumulations of matter produce here a smaller attraction than they would according to the usual theory. However, as we will see, the difference is, even for the Sun, still not noticeable.

[206]

Taking into account the symmetry of the field, the differential equation (20) yields:

$$\frac{1}{r^2}\frac{d}{dr}\left(r^2\frac{du}{dr}\right) = 2\alpha\mu u,$$

or

$$\frac{d^2(ru)}{dr^2} = 2\alpha\mu \cdot ur \qquad (\text{for } r \leq a). \tag{25}$$

To integrate the equation, one must know the distribution of the specific density μ along a radius. We want to restrict ourselves to the specific case of a sphere consisting of an incompressible fluid (μ = constant). By setting

$$\kappa^2 = 2\alpha\mu, \tag{25a}$$

we find u expressed through the hyperbolic sine

$$u = \frac{A}{r} \cdot \sinh\Phi(\kappa r). \tag{26}$$

The integral of (25) remains finite for $r = 0$. The constant A is determined by the condition that for $r = a$, the values of u and of

$$\frac{du}{dr} = A\left\{\frac{\kappa}{r}\cosh(\kappa r) - \frac{1}{r^2}\sinh(\kappa r)\right\} \tag{26a}$$

agree with those valid outside the sphere, which are given by (21) and (21a) respectively:

$$\left.\begin{array}{l}\dfrac{A}{a}\sinh(\kappa a) = u_0\left(1 - \dfrac{\vartheta}{a}\right), \\[12pt] \dfrac{A}{a}\{\kappa a\cosh(\kappa a) - \sinh(\kappa a)\} = \dfrac{u_0\vartheta}{a}.\end{array}\right\} \tag{26b}$$

In addition, these two equations still determine the value of the quotient ψ (cf. 24); only this is of interest to us here. We find

$$\frac{\psi}{1-\psi} = \frac{\kappa a\cosh(\kappa a) - \sinh(\kappa a)}{\sinh(\kappa a)} = \frac{\kappa a - \tanh(\kappa a)}{\tanh(\kappa a)} = \xi$$

and therefore,

$$\psi = \frac{\xi}{1+\xi} = 1 - \frac{\tanh(\kappa a)}{\kappa a}. \tag{27}$$

The gravitational force (24a) which will be exerted by the incompressible sphere on an external point depends on this quantity, which determines the ratio of the external to the total energy of the sphere. |

[207]   We first want to evaluate ψ for the realistic case:

I. κa *small*: One has

$$\frac{\tanh(\kappa a)}{\kappa a} = \frac{1 + \dfrac{(\kappa a)^2}{3!} + \cdots}{1 + \dfrac{(\kappa a)^2}{2!} + \cdots} = 1 - \frac{(\kappa a)^2}{3},$$

and thus

$$\psi = \frac{\kappa^2 a^2}{3} = \frac{2\alpha\mu a^2}{3} = \frac{\alpha M}{2\pi a} \qquad \left(M = \mu \cdot \frac{4\pi a^3}{3} \right). \tag{27a}$$

Since in this case ψ is very small, then so also is the proportion of the external field energy to the total energy of the sphere; ignoring it, we find from (24a) and (27a) the force acting on a resting material point P':

$$M'\frac{dc}{dr} = \frac{\alpha c_0}{\pi}\frac{MM'}{r^2}.$$

Going over to the usual units, we set

$$M = cm, \quad M' = c'm'$$

and neglect the difference between c, c' and c_0. The force then becomes

$$\frac{\alpha}{\pi}c_0^3\frac{mm'}{r^2} = \gamma\frac{mm'}{r^2},$$

where γ is the usual gravitational constant. Therefore, one has

$$\alpha = \frac{\pi\gamma}{c_0^3}$$

and hence, according to (27a):

$$\psi = \frac{\gamma m}{2ac_0^2}, \tag{27b}$$

which for the Sun gives the value:

$$\psi = 10^{-6}.$$

This is thus, indeed, still the limiting case of a small ψ, or rather, κa. For the purposes of astronomy, the deviation from Newton's law due to the external field energy is to be neglected, and the Sun is to be replaced by a mass point.[12] |

II. Although it is apparent from what has been said that the *converse limiting case*   [208]
does not correspond to reality, it shall be briefly considered:

$$\kappa a \text{ large } \tanh(\kappa a) = 1.$$

Here, according to (27)

$$\psi = \frac{E_a}{E_t} = 1 - \frac{1}{\kappa a} \tag{27c}$$

becomes only slightly less than one, i.e. nearly the entire energy resides in the external gravitational field. In the interior resides only the small fraction

$$1 - \psi = \frac{E_i}{E_t} = \frac{1}{\kappa a}.$$

For the gravitational force, from (24a) it follows that

$$\frac{dc}{dr} = \frac{2c_0 a}{r^2}\left(1 - \psi \frac{a}{r}\right).$$

Thus, at large distances the gravitational force is not proportional to the volume of the sphere, but to its radius. Here, the above mentioned screening effect of large accumulations of masses becomes apparent (i.e., of a very large value of the radius a or of the density μ).

We want to return to the basic equation (17a):

$$\frac{\partial^2 u}{\partial x^2} + \frac{\partial^2 u}{\partial y^2} + \frac{\partial^2 u}{\partial z^2} - \frac{1}{c}\frac{\partial}{\partial t}\left(\frac{1}{c}\frac{\partial u}{\partial t}\right) = 2\alpha \frac{\eta}{u}.$$

From it one derives the disturbance in the gravitational field caused by the motion of matter. Outside the matter, it indicates a propagation of the disturbance with the speed of light (c). However, a rigorous treatment of the problem of propagation is made more difficult because the disturbance of the field itself influences the value of u, and thereby the value of the speed of propagation, c. The same difficulty appears as well with the propagation of sound, whose speed depends on pressure, and thereby varies upon the passing of the sound wave. But in all practical cases, the variation of u is so minute that one can consider the speed of propagation of gravitation to be constant.

---

12 With reference to the principle of action and reaction, one also obtains from this a clue as to the range of the domain of validity of the above considerations based on Lagrange's equations, wherein the object was replaced by a material point, and where, upon the calculation of the mass, only the self energy, and not the energy of the gravitational field produced by the body, was taken into consideration. As one can now see, this procedure is still permissible for objects of the order of the fixed stars.

According to our theory, light and gravitation have the same speed of propagation; but whereas light waves are transverse, gravitational waves are longitudinal. Incidentally, the problem of the oscillating particle can be treated in a similar manner as that of the oscillating electron; the strength of the emitted gravitational waves depends on the product of the gravitational mass and the acceleration of the particle. Is it possible to detect these gravitational waves? |

[209]   This hope is futile. Indeed, to impart an acceleration to one particle, another particle is necessary which, according to the law of action and reaction, is driven in the opposite direction. But now, the strength of the emitted gravitational wave depends on the sum of the products of the gravitational mass and the acceleration of the two particles, while, according to the reaction principle, the sum of the products of inertial mass and acceleration is equal to zero.[13] Therefore, although the existence of gravitational waves is compatible with the assumed field mechanism, through the equality of gravitational and inertial mass the possibility of its production is practically excluded. It follows from this that the planetary system does not lose its mechanical energy through radiation, whereas an analogous system consisting of negative electrons circling around a positive nucleus gradually radiates its energy away. The life of the planetary system is thus not threatened by such a danger.

Our theory of gravitation based on the assumption of a variable c contradicts from the outset the second axiom of the theory of relativity. However, in a vacuum, the invariance with respect to the Lorentz transformations is preserved as is shown by the form of the fundamental equation (17a) which applies there:

$$u \square u = 0.$$

Hence, outside of matter, the Lorentz group still applies in the infinitesimally small. *It is the matter which breaks the invariance under the Lorentz group*, because in the equation

$$u \square u = 2\alpha\eta$$

the first term is an invariant of the group, whereas the second term, proportional to the energy density, is not. It is precisely the very plausible hypothesis that the attracting mass is proportional to the energy that forces us to abandon the Lorentz group in the infinitesimally small as well.

Thus, Einstein's relativity theory of 1905 turns to dust. Will there, like the Phoenix rising out of the ashes, emerge a new, more general principle of relativity? Or, will one return to absolute space, and beckon back the much scorned aether, so that it can support, in addition to the electromagnetic, also the gravitational field?

---

13   Here however, the momentum of the gravitational field has not been taken into consideration; but this is of just as little practical importance as the energy of the field.

MAX ABRAHAM

RECENT THEORIES OF GRAVITATION

Originally published as "Neuere Gravitationstheorien" in Jahrbuch der Radioaktivität und Elektronik, 11, 1915, pp. 470–520. Received 15 December 1914. Author's Date: Milan, December 1914.

CONTENTS

I. Introduction. A. Vector theory of gravitation. B. Law of conservation of momentum and law of conservation of energy; World tensors. C. Inertia and gravity.[1]

II. Scalar theories. A. Energy density and energy flux in a gravitational field. B. Abraham's first theory. C. Abraham's second theory. D. Theories of Nordström and Mie. E. Nordström's second theory. F. Kretschmann's theory.

III. Tensor theories. A. Einstein's theory of the static gravitational field. B. The generalized theory of relativity of A. Einstein and M. Grossmann.

Notation. References.

I. INTRODUCTION   [473]

A. The Vector Theory of Gravitation

Since I. Newton postulated his action-at-a-distance law of attraction of masses, theoretical physics has endeavored to reduce this law to one of local interactions. In this attempt the older theories of gravitation[1] were based on concepts derived from the theory of elasticity, hydrodynamics and the kinetic theory of gases. Electromagnetic theories of the gravitational field appeared only at the end of the nineteenth century, encouraged by Maxwell's local field theory of electrodynamics. The best received theory was probably one developed by H. A. Lorentz[2] which adapted a hypothesis

---

1   P. Drude, *Wied. Ann.* 62, 1, 1897 and J. Zenneck, *Enzyklopädie d. math. Wissensch.* V, 1, article 2 [in this volume], give overviews of the state of the theory of gravitation at the end of the last century.

2   H. A. Lorentz, *Verslag. Akad. v. Wetenschapen te Amsterdam*, 8, 603, 1900.

Jürgen Renn (ed.). *The Genesis of General Relativity*, Vol. 3
Gravitation in the Twilight of Classical Physics: Between Mechanics, Field Theory, and Astronomy.
© 2007 Springer.

already pursued by Aepinus, Mosotti and Zöllner to the framework of electron theory. According to this hypothesis, the attraction of unlike electric charges should be somewhat larger than the repulsion of like ones. For electrically neutral matter, whose atoms are supposed to consist of positive and negative electrons, Newton's law for the attraction of masses at rest follows from Coulomb's law of electrostatic forces. The force $(\mathfrak{C}^g)$ acting on a unit of mass of the stationary matter arises thus as the resultant of two electric forces of opposite direction and nearly equal magnitude, act-

[474]   ing on the | positive and negative electrons. However, if the matter moves, then a magnetic force is associated with each of these two electric forces, related to them by the electromagnetic field equations. The two magnetic forces act on the convection current of the positive and the negative electrons, respectively. The resultant $(\mathfrak{H}^g)$ of the two magnetic vectors, likewise opposite in direction and of nearly equal magnitude, determines a mechanical force acting on the moving matter. All in all, it follows that if v denotes the velocity vector and c the speed of light, then the expression for the gravitational force [*Schwerkraft*] per unit mass is

$$\mathfrak{F}^g = \mathfrak{C}^g + \frac{1}{c}[\mathfrak{v}\mathfrak{H}^g]. \tag{1}$$

This expression corresponds precisely to the one which represents the force $(\mathfrak{F}^e)$ per unit charge in the electron theory. The results of the Lorentzian theory of gravitation thus can be summarized as follows: with the electromagnetic vector pair $(\mathfrak{C}, \mathfrak{H})$ is associated a second pair, $(\mathfrak{C}^g, \mathfrak{H}^g)$, characterizing the gravitational field, which determines the gravitational force in accordance with (1). This pair is linked to the density and velocity of matter by differential equations that agree with the field equations of the electron theory, except for the signs of the charge and mass density respectively.

We will call a theory that represents the gravitational field by means of two electromagnetically interrelated vectors simply a "*vector theory of gravitation*." As is apparent from eq. (1), in such a theory the force acting on a body at a given point in the gravitational field depends not only on its mass, but also on its velocity; a moving material point acts on another point just like as one moving charge on another. However, since such an influence of the state of motion on the gravitational force has been discovered neither in physics nor astronomy, the vector $\mathfrak{H}^g$ plays a merely hypothetical role. In order to explain the absence of the force arising from $\mathfrak{H}^g$, and represented by the second term in (1), the vector theory appeals to the smallness of the velocity of the bodies in comparison to the speed of light c. Indeed, for celestial bodies the quotient v : c is of the order of 10^{-4}. In addition, the quotient of the magnitudes of the vectors $\mathfrak{H}^g$ and $\mathfrak{C}^g$ is itself of the same order, so that even according to the vector

[475]   theory the planetary system | satisfies the Newtonian law up to terms of order 10^{-8}; deviations of this order are of course permissible astronomically as well.

According to the vector theory, the precise law of interaction of two moving mass points should agree up to sign with the fundamental law of electrodynamics. Therefore, a counterpart to the "forces of induction" of electrodynamics ought to exist in mechanics. In the same way as the acceleration of an electric charge produces forces which act on a neighboring charge in a sense to oppose the acceleration, so should the acceleration of a body generate induced gravitational forces which, because of the different signs, act to accelerate neighboring bodies. Accordingly, it seems possible that a system of masses set in motion by a small force further accelerates by itself through internal forces. Thus, the equilibrium of a gravitating system of masses would not be stable.

This instability is related to a difficulty which arises in the vector theory of gravitation, already noted by Maxwell.[3] The differential equations of the static gravitational field are here:

$$\operatorname{div} \mathfrak{E}^g = -\mu, \tag{2}$$

$$\mathfrak{E}^g = -\operatorname{grad}\varphi, \tag{3}$$

where μ denotes the mass density and φ the gravitational potential in suitably chosen units.

Now, the expression for the potential energy of a system of gravitating masses in the action-at-a-distance theory is:

$$E = \int dV \frac{1}{2}\mu\varphi. \tag{4}$$

If, as in electrostatics, one transforms this equation, through integrating by parts, into the following form:

$$E = -\int dV \frac{1}{2}(\mathfrak{E}^g)^2, \tag{4a}$$

then one sees that the interpretation in the sense of the local field theory, leads to a distribution of energy in the field with a density

$$\eta^g = -\frac{1}{2}(\mathfrak{E}^g)^2. \tag{4b}$$

Hence, in contrast to electrostatics, and due to I the opposite sign in eq. (2), *in the* [476] *vector theory the energy density of the gravitational field turns out to be negative.* The same applies for a changing gravitational field whose energy density, according to the vector theory, should be

$$\eta^g = -\frac{1}{2}\{(\mathfrak{E}^g)^2 + (\mathfrak{H}^g)^2\}. \tag{4c}$$

Accordingly, a region of space, if it does contain a gravitational field, ought to contain less energy than if it were without a field. And when the gravitational field—

---

3   J. Cl. Maxwell, *Scientific Papers,* Vol. I, p. 570.

spreading, say, in the nature of a wave—enters into a previously field-free region, energy ought to flow in a direction opposite to that of the propagation of the waves. This paradoxical conclusion is peculiar to the vector theory of the gravitational field.

The so-called "*theory of relativity*" arose from the Lorentzian electrodynamics of moving bodies. This theory demands, that in empty space all forces propagate with the same speed as light. Therefore, according to the theory of relativity, the gravitational force as well must propagate with the speed of light. H. Poincaré[2] raised the question as to how this requirement can be brought into agreement with the view of Laplace, that the speed of propagation of gravitation, if at all finite, must be far greater than that of light (c). He remarked, that this view applies only, if quantities of first order (with respect to the quotient $v : c$) enter into the fundamental law of attraction of two masses. But if one formulates the fundamental law so that it fits into a relativistic scheme, yet deviates only in terms of second order and higher from Newton's, then Laplace's reservations lose their significance. Poincaré gives such formulations of the fundamental law of gravitation. These approaches contain also the force term of H. Minkowski,[3] which, by the way, can be obtained by transferring the fundamental law of the theory of electrons to gravitation.[4] The interaction laws for the attraction of masses developed by these pioneers of the theory of relativity are accordingly quite readily compatible with the vector theory of gravitation sketched above. On the other hand, that vector theory gave also an account of those processes in the field through whose mediation one mass transfers energy and momentum to the other; these fundamental relativistic laws, however, lack the derivation from the field equations. Thus the difficulties associated with the vector theory were not resolved but only concealed. What remained unresolved was the problem of developing field I equations of the gravitational field that yield a propagation of the gravitational force with the speed of light and also ascribe to the field a positive energy, transferred via an energy flux, and a momentum transferred by means of fictitious stresses.

[477]

B. Conservation Laws of Momentum and Energy: World Tensors

The above mentioned work by H. Poincaré[2] already contains the beginnings of the four-dimensional vector calculus, which was then further developed by H. Minkowski[3] In our notation, we will mainly follow the presentation of A. Sommerfeld.[4] The four-dimensional formulation of the theory of relativity, as is well known, interprets the group of Lorentz transformations as a rotation group of a four-dimensional space, whose coordinates are the Cartesian coordinates x, y, z of ordinary space and $u = ict$; these transformations leave the expression

$$x^2 + y^2 + z^2 + u^2$$

invariant. Poincaré had already dealt with four-vectors whose components transform like the coordinates x, y, z, u. H. Minkowski introduced the concept of the six-vector, and characterized the electromagnetic field $\{\mathfrak{E}, \mathfrak{H}\}$ in a vacuum through such a vector. The six-vector $\{\mathfrak{E}, \mathfrak{H}\}$ is derived from the four-potential $\{\mathfrak{A}, \Phi\}$, whose 4 components are given by the electromagnetic vector- and scalar potential. The vector theory of the gravitational force sketched above, which, of course, is formally identical with the electromagnetic field theory, accordingly also derives the six-vector of the gravitational field $\{\mathfrak{E}^g, \mathfrak{H}^g\}$ from a four-potential. Also in this four-dimensional sense—namely insofar as the gravitational potentials form a four-vector—one will accordingly have to denote that theory of gravitation as a "vector theory."

Besides the two kinds of "world vectors," the *"world tensors"* are important for the theory of relativity.[4] The ten components of such a symmetric tensor T transform like the squares and products of the four coordinates x, y, z, u. From it one derives a four-vector which we want to call "the divergence of the ten-tensor T," and whose components are to be formed according to the following scheme: |

$$\operatorname{div} T = \left\{ \begin{array}{l} \dfrac{\partial T_{xx}}{\partial x} + \dfrac{\partial T_{xy}}{\partial y} + \dfrac{\partial T_{xz}}{\partial z} + \dfrac{\partial T_{xu}}{\partial u}, \\[2mm] \dfrac{\partial T_{yx}}{\partial x} + \dfrac{\partial T_{yy}}{\partial y} + \dfrac{\partial T_{yz}}{\partial z} + \dfrac{\partial T_{yu}}{\partial u}, \\[2mm] \dfrac{\partial T_{zx}}{\partial x} + \dfrac{\partial T_{zy}}{\partial y} + \dfrac{\partial T_{zz}}{\partial z} + \dfrac{\partial T_{zu}}{\partial u}, \\[2mm] \dfrac{\partial T_{ux}}{\partial x} + \dfrac{\partial T_{uy}}{\partial y} + \dfrac{\partial T_{uz}}{\partial z} + \dfrac{\partial T_{uu}}{\partial u}. \end{array} \right\} \tag{5}$$

[478]

In the Maxwell-Lorentz electrodynamics, the electromagnetic force acting on a unit volume is determined by the four-vector

$$\rho \mathfrak{F}^e = \operatorname{div} T^e, \tag{6}$$

where T_e designates the *"electromagnetic ten-tensor."* Its components

$$T^e_{xx}, T^e_{yy}, T^e_{zz}$$

represent the fictitious normal stresses, and

$$T^e_{xy} = T^e_{yx}, \quad T^e_{yz} = T^e_{zy}, \quad T^e_{zx} = T^e_{xz} \tag{6a}$$

represent the shear stresses; the symmetry relations (6a) imply the vanishing of the torques of these fictitious surface forces [*Flächenkräfte*]. The three remaining symmetry conditions of the ten-tensor T^e

$$T^e_{xu} = T^e_{ux}, \quad T^e_{yu} = T^e_{uy}, \quad T^e_{zu} = T^e_{uz} \tag{6b}$$

---

4   M. Abraham, *Rendiconti del circolo matematico di Palermo*, XXVIII^2, 17, 1909. M. Laue, *Das Relativitätsprinzip*, p. 4 of the 2nd ed. (1913).

have the following significance:

$$T_{xu}^e = -ic g_x^e, \quad T_{yu}^e = -ic g_y^e, \quad T_{zu}^e = -ic g_z^e \tag{6c}$$

determine the components of the momentum density (g^e) of the electromagnetic field,

$$T_{ux}^e = -\frac{i}{c}\mathfrak{S}_x^e, \quad T_{uy}^e = -\frac{i}{c}\mathfrak{S}_y^e, \quad T_{uz}^e = -\frac{i}{c}\mathfrak{S}_z^e \tag{6d}$$

determine those of the electromagnetic energy flux ($\mathfrak{S}^e$); the eqs. (6b) contain therefore the "*theorem of the momentum of electromagnetic energy flux*":

$$g^e = \frac{\mathfrak{S}^e}{c^2}. \tag{6e}$$

Since according to (6c)

$$\frac{\partial T_{xu}^e}{\partial u} = \frac{\partial T_{xu}^e}{ic\partial t} = -\frac{\partial g_x^e}{\partial t} \qquad \text{etc.},$$

the first three of the equations (6), to be formed according to the scheme (5), express the *law of conservation of momentum* for the electromagnetic field, as they derive from the stresses and the momentum density of the field the momentum given by the field to the unit of volume of matter. I The last of the eqs. (6), however, contains the law of *conservation of energy*. Indeed, if one equates the last tensor component to the electromagnetic energy density:

[479]

$$T_{uu}^e = \eta^e = \frac{1}{2}(\mathfrak{E}^2 + \mathfrak{H}^2), \tag{6f}$$

and furthermore

$$\rho \mathfrak{F}_u^e = \frac{i\,dA^e}{c\,dt}, \tag{6g}$$

where dA^e is the work done on a unit volume, then the last of eqs. (6) reads, taking into consideration (6d):

$$\frac{dA^e}{dt} = -\operatorname{div}\mathfrak{S}^e - \frac{\partial \eta^e}{\partial t}. \tag{6h}$$

It therefore expresses the law of conservation of energy for the electromagnetic field, because it relates the energy transferred from the field to a unit volume of matter to the energy flux and the energy density of the field.

One would now wish to retain the validity of the conservation laws of momentum and energy, and of the theorem of the momentum of energy flux for the gravitational field by deriving the gravitational force per unit volume of matter as a four-vector

$$\mu \mathfrak{F}^g = \operatorname{div} T^g \tag{7}$$

from a symmetric "*gravitational tensor T^g*." The vector theory of gravitation satisfies this demand by expressing the tensor T^g through the vectors $\mathfrak{E}^g$, $\mathfrak{H}^g$, in the same way in which $-T^e$ is determined through the electromagnetic vectors $\mathfrak{E}$, $\mathfrak{H}$. We have shown above that precisely the difference in sign causes difficulties for the vector theory. In the following, we will become acquainted with other possibilities for representing the gravitational tensor T^g.

C. Inertia and Gravity[1]

If the theorem of the momentum of energy flux is valid, then a momentum, and thus an inertial mass, is associated with a convectively moving quantity of energy. A well-known example is the "electromagnetic mass" of the electron which, in the dynamics of the electron, is derived from the momentum of the energy flux flowing in the electromagnetic field of the moving electron. If one considers the theorem of the momentum of energy flux as valid for arbitrary kinds of energy flows, then it follows that the momentum of a uniformly moving body isolated from external effects is |

$$\mathfrak{G} = \frac{\mathfrak{v}}{c^2} E. \tag{8}$$ [480]

To this corresponds an inertial rest-mass of value

$$m_0 = \frac{E_0}{c^2}. \tag{8a}$$

Here, E_0 denotes the "rest-energy" of the body, i.e. its energy with respect to a coordinate system Σ_0 in which the body is at rest; m_0 measures the inertia of the body accelerated from rest. Equation (8a) expresses "*the theorem of the inertia of energy*."

If all forces in nature can be fitted into the scheme of a symmetric world tensor, then the theorem of the momentum of energy flux and that which is derived from it, the theorem of the inertia of energy, gain general validity. The inertial mass of a body is then proportional to its energy content; it can be increased by the gain of energy, or decreased through the loss of energy. However, the denominator c^2 in (8a) entails that the changes in energy that occur in the usual chemical reactions are too minute to cause measurable changes in mass. However, the radioactive transformations, in view of their enormous heat production, should be accompanied by a noticeable decrease in mass. Such changes in mass—still quite small at any rate—could, however, be determined only with a scale. The scale, however, does not measure inertia but weight [*Schwere*]. Hence, the question arises: Is there also gravitational mass associated with energy? Is there, as a counterpart to the theorem of the inertia of energy, a theorem of the weight of energy? This question leads us back to the problem of gravitation.

Experience teaches us that all bodies fall with equal acceleration in a vacuum and that the period of a pendulum is independent of its chemical composition. Therefore, *the gravitational mass is proportional to the inertial mass*. The most precise test of the law of proportionality is due to B. Eötvös;[5] with the aid of a torsion balance, this researcher investigated whether the gravitational force acts on all bodies on the surface of the Earth in the same direction. Since the gravitational force is the resultant of the Earth's attraction of masses [*Massenanziehung*] and of the centrifugal force, this resultant, i.e. the vertical, would have a different direction for different bodies if strict proportionality between the two masses did not hold. This is because | the attraction of masses is determined by the gravitational mass and the centrifugal force by the inertial mass. Although Eötvös recently refined his measurements so that deviations of the order 10^{-8} would not have escaped him, he nevertheless could not find such. With the corresponding accuracy, the law of proportionality of gravitational and inertial mass has been shown to be valid.

For the time being, the investigations of Eötvös do not cover radioactive bodies. If at all, one should most likely expect a departure in the behavior of gravitational and inertial mass in radioactive transmutations, namely in the case that the former remains constant while the latter decreases because of the emission of energy. Uranium oxide for example transmutes by radioactive decay into lead oxide, while emitting 8 α-particles per atom, during which the fraction $2.3 \cdot 10^{-4}$ of the energy is emitted.[6] According to the theorem of the equivalence of inertia and energy, the inertial mass should decrease by the same fraction; if the gravitational mass did not remain constant, or if it changed in a different ratio, then the proportionality between gravitational and inertial mass could not be maintained under radioactive transformations. However, L. Southerns,[7] using pendulum observations to which he ascribed an accuracy of $5 \cdot 10^{-6}$, discovered no difference with respect to the mass ratio between uranium oxide and lead oxide. Thus, here too, the law of proportionality is valid.

Classical mechanics introduces the proportionality of gravitational mass and inertia as an empirical law without deeper justification. The theorem of the inertia of energy suggests that, in the new mechanics, the equivalence of the two masses be explained by the gravitational mass also being proportional to the energy. Then furthermore, the law of conservation of (gravitational) mass, which takes an isolated position in the traditional physics and chemistry, would merge with the law of conservation of energy into a single one. However, if not the energy itself, but another quantity were the determining factor for the gravitational mass of a body, then still, in all practical cases, this quantity should be proportional to the inertial mass with the above stated accuracy. It is the merit of the newer theories of gravitation, on which we want to report in the following, to have put the discussion of the relation between

[481]

5　B. Eötvös, *Mathem. u. naturwiss. Ber. aus Ungarn*, 8, 65, 1890.
6　See also R. Swinne, *Physik. Zeitschr.*, 14, 145, 1913.
7　L. Southerns, *Proc. Roy. Soc. London*, 84, 325, 1910.

inertia, gravitation and energy on a rational foundation and, in so doing, to have prepared the way for the exploration of these relations. I

II. SCALAR THEORIES

[482]

A. Energy Density and Energy Flux in the Gravitational Field

In the vector theory of gravitation, the expression (4b) gave the energy density of the static gravitational field. If one assumes that a similar expression is also valid for time varying fields, when only the static force $\mathfrak{E}^g$ is replaced by the dynamic gravitational force $\mathfrak{F}^g$, then

$$\eta^g = -\frac{1}{2}(\mathfrak{F}^g)^2. \tag{9}$$

In a remarkable paper on the energy flux in a gravitational field, V. Volterra[1][2] takes this expression as the basis for the energy density. He divides the energy flux into two parts:

$$\mathfrak{S} = \mathfrak{S}^g + \mathfrak{S}^m, \tag{10}$$

of which the first

$$\mathfrak{S}^g = -\frac{\varphi \partial \mathfrak{F}^g}{\partial t} \tag{10a}$$

is caused by the variation of the gravitational field with respect to time, whereas the second

$$\mathfrak{S}^m = \varphi \mu \upsilon \tag{10b}$$

represents the energy transport by moving matter. The mass density μ is related via

$$\mathrm{div}\,\mathfrak{F}^g = -\mu \tag{11}$$

with the gravitational force per unit mass, which, by virtue of

$$\mathfrak{F}^g = -\mathrm{grad}\,\varphi \tag{11a}$$

is derived from the scalar gravitational potential φ. Insofar as the existence of a scalar potential is also assumed in a dynamic gravitational field, Volterra's theory is to be counted among the "scalar theories of gravitation" (in a three-dimensional sense).

We want to convince ourselves that Volterra's approaches satisfy the energy equation. According to (9) one has

$$-\frac{\partial \eta^g}{\partial t} = \left(\mathfrak{F}^g \cdot \frac{\partial \mathfrak{F}^g}{\partial t}\right),$$

whereas from (10), (10a), and (10b) it follows that

$$-\text{div}\mathfrak{S} = \text{div}\varphi\frac{\partial\mathfrak{F}^g}{\partial t} - \text{div}\varphi\mu\upsilon$$

$$= \varphi\text{div}\frac{\partial\mathfrak{F}^g}{\partial t} + \left(\frac{\partial\mathfrak{F}^g}{\partial t}\cdot\text{grad}\varphi\right) - \varphi\,\text{div}\mu\upsilon - (\mu\upsilon\cdot\text{grad}\varphi),$$

and taking into consideration (11), (11a) and the equation of continuity of matter |

$$\frac{\partial\mu}{\partial t} = -\text{div}\mu\upsilon,$$

$$-\text{div}\mathfrak{S} = -\left(\mathfrak{F}^g\cdot\frac{\partial\mathfrak{F}^g}{\partial t}\right) + (\mu\upsilon\cdot\mathfrak{F}^g).$$

Therefore, one obtains

$$(\mu\mathfrak{F}^g\cdot\upsilon) = -\text{div}\mathfrak{S} - \frac{\partial\eta^g}{\partial t}. \tag{11b}$$

Here, on the left hand side, is the work done by the gravitational force per unit volume and time; on the right-hand side, is the energy gain caused by the influx of energy and by the temporal decrease of the energy density. The law of conservation of energy is indeed satisfied.

However, it is not possible, by means of constructing a corresponding world tensor T^g, to reconcile the approaches of Volterra with the first three of the eqs. (7), which formulate the laws of conservation of momentum. It is also remarkable that in (10b) an energy transport by moving matter is introduced, without a corresponding energy density of the matter. If one takes account of such an energy density

$$\eta^m = \varphi\mu \tag{12}$$

then one obtains the correct value for the total energy of a system of masses at rest

$$E = \int dV\frac{1}{2}\mu\varphi = E^m + E^g,$$

if one sets

$$\eta^g = \frac{1}{2}(\mathfrak{F}^g)^2 \tag{12a}$$

for the *energy density of the gravitational field*.

Indeed, according to (11) and (11a), the identity

$$\int dV\frac{1}{2}\mu\varphi = -\int dV\frac{1}{2}(\mathfrak{F}^g)^2$$

holds, and it follows that

$$\int dV \left\{ \mu\varphi + \frac{1}{2}(\mathfrak{F}^g)^2 \right\} = \int dV \frac{1}{2}\mu\varphi = E.$$

The value (12a) for the *energy density is positive*. Hence, the difficulty appearing in the vector theory of gravitation is eliminated if one succeeds in pairing that energy density with an energy flow.

We equate[6] the *energy flux in a gravitational field to the gravitational force per unit mass multiplied by the time derivative of the gravitational potential:*

$$\mathfrak{S}^g = \frac{\partial\varphi}{\partial t}\mathfrak{F}^g, \tag{13}$$

I whereas, for the energy flux of matter, we retain the expression (10b). Then the *total* [484] *energy flux* becomes

$$\mathfrak{S} = \mathfrak{S}^m + \mathfrak{S}^g = \varphi\mu v + \frac{\partial\varphi}{\partial t}\mathfrak{F}^g, \tag{14}$$

whereas the *total energy density,* is according to (12, 12a)

$$\eta = \eta^m + \eta^g = \varphi\mu + \frac{1}{2}(\mathfrak{F}^g)^2. \tag{14a}$$

One can easily convince oneself that the energy equation is satisfied. One has

$$\frac{\partial\eta}{\partial t} = \varphi\frac{\partial\mu}{\partial t} + \mu\frac{\partial\varphi}{\partial t} + \left(\mathfrak{F}^g \cdot \frac{\partial\mathfrak{F}^g}{\partial t}\right),$$

$$\mathrm{div}\,\mathfrak{S} = \varphi\,\mathrm{div}\,\mu v + (\mu v \cdot \mathrm{grad}\varphi) + \frac{\partial\varphi}{\partial t}\,\mathrm{div}\,\mathfrak{F}^g + \left(\mathfrak{F}^g \cdot \mathrm{grad}\frac{\partial\varphi}{\partial t}\right),$$

and from this, taking into consideration (11, 11a) and the equation of continuity, the law of conservation of energy follows

$$(\mu v \cdot \mathfrak{F}^g) = -\mathrm{div}\,\mathfrak{S} - \frac{\partial\eta}{\partial t}. \tag{14b}$$

As we will see immediately, the expressions (12a) and (13) for the energy density and the energy flux in a gravitational field, fit readily into the scheme of a symmetric gravitational tensor.[8]

B. Abraham's First Theory

So far we have been talking about "scalar theories" of gravitation merely in the context of a three-dimensional vector calculus, and taking them to be theories which derive the gravitational force from a scalar potential φ also for a dynamic field. In the four-dimensional context, *a theory of gravitation is to be designated as a scalar theory, in which the gravitational potential is a scalar, i.e. an invariant with respect to rotation of the four-dimensional space of the* (x, y, z, u). The gravitational force with respect to a unit volume, considered as a four-vector, shall have the components |

$$\mu\mathfrak{F}_x^g = -\nu\frac{\partial\varphi}{\partial x}, \qquad \mu\mathfrak{F}_y^g = -\nu\frac{\partial\varphi}{\partial y}, \\ \mu\mathfrak{F}_z^g = -\nu\frac{\partial\varphi}{\partial z}, \qquad \mu\mathfrak{F}_u^g = -\nu\frac{\partial\varphi}{\partial u}; \Bigg\} \tag{15}$$

ν is an, for now, undetermined scalar factor. As the comparison with (6g) teaches,

$$\frac{c}{i}\mu\mathfrak{F}_u^g = ic\nu\frac{\partial\varphi}{\partial u} = \nu\frac{\partial\varphi}{\partial t} \tag{15a}$$

represents the energy that is transferred per unit space and time from the gravitational field to the matter. In the symbolism of the four-dimensional vector analysis[4], the eqs. (15) are written as

$$\mu\mathfrak{F}^g = -\nu\,\mathrm{grad}\varphi. \tag{15b}$$

Now, corresponding to the scheme (7), the gravitational force should be derived from a symmetric gravitational tensor T^g, such that the expressions (12a, 13) for the energy density and the energy flux correspond to the appropriate tensor components. M. Abraham[6] defines the ten-tensor T^g in the following manner:

---

8   Here, I have presented the train of thought which I pursued in the development of my first theory of gravitation[6] in such detail, because G. Jaumann (*Physik. Zeitschr.*, 15, 159, 1914) formulated an inappropriate hypothesis concerning the psychological origin of this theory. Jaumann's theory (*Wien. Ber.*, 121, 95, 1912), of which I became aware only after that first publication, is so far removed from the conceptual realm of the investigations summarized in this report, that it appears to me that this is not the place for its discussion.

$$T^g_{xx} = -\frac{1}{2}\left(\frac{\partial\varphi}{\partial x}\right)^2 + \frac{1}{2}\left(\frac{\partial\varphi}{\partial y}\right)^2 + \frac{1}{2}\left(\frac{\partial\varphi}{\partial z}\right)^2 + \frac{1}{2}\left(\frac{\partial\varphi}{\partial u}\right)^2,$$

$$T^g_{yy} = \frac{1}{2}\left(\frac{\partial\varphi}{\partial x}\right)^2 - \frac{1}{2}\left(\frac{\partial\varphi}{\partial y}\right)^2 + \frac{1}{2}\left(\frac{\partial\varphi}{\partial z}\right)^2 + \frac{1}{2}\left(\frac{\partial\varphi}{\partial u}\right)^2,$$

$$T^g_{zz} = \frac{1}{2}\left(\frac{\partial\varphi}{\partial x}\right)^2 + \frac{1}{2}\left(\frac{\partial\varphi}{\partial y}\right)^2 - \frac{1}{2}\left(\frac{\partial\varphi}{\partial z}\right)^2 + \frac{1}{2}\left(\frac{\partial\varphi}{\partial u}\right)^2,$$

$$T^g_{uu} = \frac{1}{2}\left(\frac{\partial\varphi}{\partial x}\right)^2 + \frac{1}{2}\left(\frac{\partial\varphi}{\partial y}\right)^2 + \frac{1}{2}\left(\frac{\partial\varphi}{\partial z}\right)^2 - \frac{1}{2}\left(\frac{\partial\varphi}{\partial u}\right)^2,$$

$$T^g_{xy} = T^g_{yx} = -\frac{\partial\varphi}{\partial x}\frac{\partial\varphi}{\partial y},$$

$$T^g_{yz} = T^g_{zy} = -\frac{\partial\varphi}{\partial y}\frac{\partial\varphi}{\partial z},$$

$$T^g_{zx} = T^g_{xz} = -\frac{\partial\varphi}{\partial z}\frac{\partial\varphi}{\partial x},$$ (16)

$$T^g_{xu} = T^g_{ux} = -\frac{\partial\varphi}{\partial x}\frac{\partial\varphi}{\partial u},$$

$$T^g_{yu} = T^g_{uy} = -\frac{\partial\varphi}{\partial y}\frac{\partial\varphi}{\partial u},$$

$$T^g_{zu} = T^g_{uz} = -\frac{\partial\varphi}{\partial z}\frac{\partial\varphi}{\partial u}.$$

As the comparison with (6c, d, e, f) shows, the appropriate expressions for the *momentum density, energy flux* and *energy density* are accordingly |

$$c^2 \mathfrak{g}^g = \mathfrak{S}^g = -\frac{\partial\varphi}{\partial t}\,\mathrm{grad}\,\varphi,$$ (16a) [486]

$$\eta^g = T^g_{uu} = \frac{1}{2}\left\{\left(\frac{\partial\varphi}{\partial x}\right)^2 + \left(\frac{\partial\varphi}{\partial y}\right)^2 + \left(\frac{\partial\varphi}{\partial z}\right)^2 + \frac{1}{c^2}\left(\frac{\partial\varphi}{\partial t}\right)^2\right\}.$$ (16b)

Equation (16a) agrees with (13) and, in addition, contains the theorem of the momentum of energy flux. Equation (16b) becomes (12a) for static fields; but also for the dynamic field the energy density calculated according to (16b), is always positive. The fictitious stresses contained in T^g agree for the static field, apart from signs, with the Maxwellian electrostatic stresses. Let us write the system (16), which derives a ten-tensor T^g from the scalar φ, symbolically as

$$T^g = \mathrm{Ten}\,\varphi.$$ (16c)

We now apply the scheme (5), and obtain for the divergence of the gravitational tensor:

$$\operatorname{div} T^g = \operatorname{div} \operatorname{ten} \varphi = -\Box \varphi \cdot \operatorname{grad} \varphi, \tag{17}$$

using the abbreviation

$$\Box \varphi = \operatorname{div} \operatorname{grad} \varphi = \frac{\partial^2 \varphi}{\partial x^2} + \frac{\partial^2 \varphi}{\partial y^2} + \frac{\partial^2 \varphi}{\partial z^2} + \frac{\partial^2 \varphi}{\partial u^2}. \tag{17a}$$

The laws of conservation of momentum and energy summarized in (7) thus yield

$$\mu \mathfrak{F}^g = -\Box \varphi \cdot \operatorname{grad} \varphi. \tag{17b}$$

In order to obtain agreement with (15b), the gravitational potential is required to satisfy the *field equation*

$$\Box \varphi = \nu, \tag{18}$$

which is to be taken as a generalization of Poisson's equation.

The theory of relativity demands that ν be a scalar in the four-dimensional sense, like the gravitational potential φ and the operator $\Box$; the "rest-mass density" is such a scalar. For this reason, in his first communication,[6] Abraham used this for ν. By integrating (18), he calculates the gravitational potential of a moving mass point and thus obtains the "*fundamental law of gravitation*."[7] This law is simpler than the fundamental law of electrodynamics, insofar as the *direction of the attracting force is independent of the state of motion of the attracted point*. This is a peculiarity of the scalar theories, which determine the gravitational force by means of a single vector $\mathfrak{F}^g$.

In the vacuum, ν is equal to zero and the gravitational potential therefore satisfies the differential equation |

$$\Box \varphi = \frac{\partial^2 \varphi}{\partial x^2} + \frac{\partial^2 \varphi}{\partial y^2} + \frac{\partial^2 \varphi}{\partial z^2} - \frac{1}{c} \frac{\partial}{\partial t} \left(\frac{1}{c} \frac{\partial \varphi}{\partial t} \right) = 0, \tag{18a}$$

the so-called "wave equation." This results in the possibility of gravitational waves, which propagate with the speed of light. However, in the scalar theories the *gravitational waves are longitudinal*, whereas in the vector theory they are considered as transverse, in analogy with the electromagnetic waves. According to both theories, the factor determining the amplitude of the gravitational wave emitted by an oscillating mass particle is the product of gravitational mass and acceleration. One could now surmise that during the emission of α-rays from radioactive atoms, where very large accelerations occur, the hypothetical gravitational waves will be excited in noticeable strength. However, electric waves are simultaneously excited and, as the quantitative discussion[10] shows, the force excited by gravitational waves from one emitted α-particle on a second α-particle is at most 10^{-36} of the electric force. That the gravitational waves play no role in the balance of nature has however still a deeper reason. If one mass particle imparts an acceleration to another through collision or long range forces, then the second particle also acts to accelerate the first in such a way that the vector sum of the products of inertial mass and acceleration is equal to zero. What determines the amplitude of the gravitational wave emitted from the system of the two particles is the vector sum of the products of gravitational mass and acceleration. Due to the proportionality of inertial and gravitational mass, this sum is also equal to zero. Thus, even theoretically, one cannot provide a means to excite gravitational waves of noticeable strength.

One aspect of Abraham's first theory has so far not been mentioned. Following an hypothesis already proposed by A. Einstein,[5] one considers the gravitational potential φ to be a function of the speed of light c, and, in particular, the following relation between the two quantities results from the assumed form of the equation of motion of material points:

$$\varphi - \varphi_0 = \frac{c^2}{2} - \frac{c_0^2}{2}. \tag{19}$$

Since, according to this, c is variable in a gravitational field, Einstein and Abraham thus give up the postulate of the constancy of the speed of light, which was a fundamental requirement of Einstein's theory of relativity I of 1905. The covariance of the physical laws with respect to the Lorentz transformations demanded by that theory exists then only in infinitesimally small spacetime regions in which c can be considered to be constant. Only with regard to the processes in an infinitesimal region are two uniformly moving systems of reference to be considered as equivalent. But if one regards finite systems of gravitating masses, then there exists no equivalence whatsoever between such systems of reference. Indeed, G. Pavanini[18] was able to show that, according to Abraham's theory, secular perturbations of the Keplerian motion occur in a two-body system, which depend on the "absolute" state of motion of the system. If the center of mass of the system is at rest, then only a secular motion of the perihelion occurs; for Mercury it would amount to 14″, 52 in a hundred years, that is only about one third of the actual motion of the perihelion. [488]

Our planetary system moves within the gravitational field of the other celestial bodies. The particular system of reference in which the external gravitational field can be taken as static, plays a distinguished role according to Abraham[10] [13] and is to coincide with the "absolute" system of reference, which, at least in the case of rotational motion, makes itself felt through centrifugal and Coriolis forces, and which the action-at-a-distance theory of C. Neumann anchors in a hypothetical "body α." This view, however, did not meet with the approval of Einstein. He attempted to salvage the principle of relativity of 1905[14], wherein, however, he had to restrict the equivalence between systems of reference in uniform translatory motion with respect to each other to processes in "isolated systems," and to the "limiting case of constant gravitational potential." As Abraham remarked,[15] to the contrary, it is not possible to shield a system against the gravitational force, and the limiting case of constant gravitational potential corresponds to the absence of gravitational force. If, on the other hand, those theoreticians who define the gravitational potential through c consider gravitation as an essential property of matter, then they must abandon "yesterday's theory of relativity."

On the occasion of that dispute, Einstein,[14] incidentally, revealed the prospect of a more general "tomorrow's principle of relativity," encompassing accelerated and rotational motion. To what extent the "generalized theory of relativity," published by him in association with M. Grossmann the following year (1913), fulfills that promise is to be discussed in detail later in (III B). |

[489]   *C. Abraham's Second Theory*

In that discussion there emerged at least this much agreement, that in the development of the theory of gravitation attention has to be paid to the relation between weight and energy. In a lecture[16] at the international congress of mathematics at Cambridge (August 1912), M. Abraham gave the highest priority to the *postulate of the weight of energy*. He proved that the gravitational force on a moving point mass can be strictly proportional to its energy only if its Lagrangian function is a linear homogeneous function of the velocity and of the gravitational potential. The Lagrangian of the earlier theory of relativity

$$\mathfrak{L} = -m_0 c^2 \sqrt{1 - \left(\frac{v}{c}\right)^2} \tag{20}$$

is to be adapted to this demand by interpreting the speed of light c itself as a potential and considering

$$m_0 c = M \text{ (mass constant)} \tag{20a}$$

as independent of v and c. Then the *Lagrangian of the mass-point* becomes

$$\mathfrak{L} = -M \cdot \sqrt{c^2 - v^2}, \tag{21}$$

and the *rest-energy*

$$E_0 = -\mathfrak{L}_0 = Mc. \tag{21a}$$

Hence, the rest-energy decreases with decreasing c, whereas the *rest-mass*

$$m_0 = \frac{E_0}{c^2} = \frac{M}{c}$$

increases with decreasing c.

Now one must demand that all mechanical quantities of the same class possess the same degree in c, thus all energies possess the degree c, all masses the degree c^{-1}. Later the author derives this requirement[17] from the following postulate:[3] "An observer belonging to a mechanical system must not perceive that he, together with the system, is brought into a region in which c has a different value" Since lengths should not depend on c, time intervals are of degree c^{-1}. The energy density as well as the remaining components of the ten-tensor T^g are of degree c; but since they are of the second degree in the derivative of φ with respect to the coordinates (cf. 16), so φ is to be replaced not by c, but by $\sqrt{c}$. Correspondingly, Abraham chooses the gravitational tensor T^g according to the scheme (16, 16c):

$$T^g = \text{ten}\sqrt{c}. \tag{22}$$

I Then from (5) one obtains, similar to (17),

[490]

$$\text{div}\, T^g = \text{div ten}\sqrt{c} = -\Box(\sqrt{c}) \cdot \text{grad}\sqrt{c}, \tag{22a}$$

where we have set

$$\Box(\sqrt{c}) = \frac{\partial^2 \sqrt{c}}{\partial x^2} + \frac{\partial^2 \sqrt{c}}{\partial y^2} + \frac{\partial^2 \sqrt{c}}{\partial z^2} - \frac{1}{c}\frac{\partial}{\partial t}\left(\frac{1}{c}\frac{\partial \sqrt{c}}{\partial t}\right). \tag{22b}$$

In order for the postulate of the weight of energy to be valid, the *field equation* is to be formulated as follows

$$\sqrt{c} \cdot \Box(\sqrt{c}) = 2\eta^m. \tag{23}$$

Then, according to (22a), the laws of conservation of momentum and energy, summarized in (7), yield

$$\mu\mathfrak{F}^g = -\frac{2\eta^m}{\sqrt{c}}\text{grad}\sqrt{c} = -\frac{\eta^m}{c}\text{grad}\, c. \tag{23a}$$

for the gravitational force per unit volume. Integration over a body of sufficiently small volume yields the *gravitational force*

$$\mathfrak{K}^g = -\frac{E^m}{c}\text{grad}\, c. \tag{24}$$

Therefore, the weight of a body is proportional to its energy; by the way, in (24), the gravitational field determined by the gradient of c has been assumed to be homogeneous over the extension of the body, and correspondingly the gravitational field produced by the body itself and its energy has been neglected.

In a lecture[17] presented at the congress of the "Società italiana per il progresso delle scienze," the author treats the role of the self-excited field and its energy for the case of rest. Since here η^m denotes the density of rest-energy of matter, it is useful to define μ as the mass constant with respect to the unit volume, and correspondingly to write (21a) as

$$\eta^m = \mu c, \tag{24a}$$

so that according to (23a)

$$\mathfrak{F}^g = -\mathrm{grad}\, c \tag{24b}$$

denotes the gravitational force per unit of mass as measured by M. For incompressible fluids, for example, μ is a constant also independent of c.

For the static gravitational field, the differential eq. (23) is written as

$$\Delta(\sqrt{c}) = \frac{2\eta^m}{\sqrt{c}} = 2\mu\sqrt{c}. \tag{25}$$

Since, according to (16b), the energy density of the static field is

$$\eta^g = \frac{1}{2}(\mathrm{grad}\,\sqrt{c})^2,$$

[491] and since the identity

$$\Delta c = 2\sqrt{c}\Delta\sqrt{c} + 2(\mathrm{grad}\,\sqrt{c})^2$$

applies, one can write (25) as

$$\Delta c = 4(\eta^m + \eta^g) = 4\eta. \tag{25a}$$

Thus, the divergence of the gradient of c is proportional to the density of the total energy, i.e., according to Gauss' theorem: *The flux of the gravitational force vector $\mathfrak{F}^g$ (cf. 24b) through a closed surface is equal to four times the enclosed energy.*

For a stationary homogeneous sphere of radius a we have, according to (25),

$$\Delta(\sqrt{c}) = 0 \qquad \text{for} \quad r > a, \tag{26}$$

$$\Delta(\sqrt{c}) = 2\mu\sqrt{c} \quad \text{for} \quad r < a. \tag{26a}$$

The integration is not difficult to perform.[17] According to (26), the gradient of $\sqrt{c}$ decreases outside the sphere as r^{-2}; the gradient of c, the determining factor for the gravitational force per unit mass according to (24b), is obtained as

$$-\mathfrak{F}_r^g = \frac{dc}{dr} = \frac{E}{\pi r^2}\left(1 - \psi\frac{a}{r}\right), \tag{26b}$$

where

$$\psi = E_a/E \tag{26c}$$

denotes the quotient of the energy E_a of the external gravitational field and the total energy E of the sphere and of its gravitational field. This result is in agreement with the above theorem concerning the flux of the vector $\mathfrak{F}^g$. The energy of the gravitational field outside the sphere is responsible for the *deviation from Newton's law* because it causes the difference in the force flux through two concentric spheres. Incidentally, eq. (26b) applies also to a mass distribution homogeneous within concentric layers.

The integration of (26a) for the interior of a homogeneous sphere leads to hyperbolic functions. For a body with the mass and the radius of the Sun the quantity ψ, given in (26c), turns out to be of the order of 10^{-6}. This provides an idea of the order of magnitude of the quotient of the external and of the total energy, as well as of the order of magnitude of the deviation from the Newtonian law determined by it. A celestial body may be considered as a material point if that quotient can be neglected under the given circumstances.

We return to the Lagrangian function (21) of the material point; it implies the values for *momentum* and *energy* |

$$\mathfrak{G} = \frac{\partial \mathfrak{L}}{\partial v} \cdot \frac{v}{v} = \frac{Mv}{\sqrt{c^2 - v^2}}, \qquad (27) \quad [492]$$

$$E = v\frac{\partial \mathfrak{L}}{\partial v} - \mathfrak{L} = \frac{Mc^2}{\sqrt{c^2 - v^2}}. \qquad (27a)$$

Lagrange's equations for the free motion of a point mass

$$\frac{d}{dt}\left(\frac{\partial \mathfrak{L}}{\partial x}\right) - \frac{\partial \mathfrak{L}}{\partial x} = 0 \qquad \text{etc.}$$

can be written

$$\frac{d\mathfrak{G}}{dt} = \operatorname{grad}\mathfrak{L} = \frac{\partial \mathfrak{L}}{\partial c}\operatorname{grad}c = \mathfrak{K}^g; \qquad (28)$$

They contain the inertial force as well as the gravitational force, which is

$$\mathfrak{K}^g = \frac{\partial \mathfrak{L}}{\partial c}\operatorname{grad}c = -\frac{Mc}{\sqrt{c^2 - v^2}}\operatorname{grad}c, \qquad (28a)$$

or, according to (27a),

$$\mathfrak{K}^g = -\frac{E}{c}\operatorname{grad}c. \qquad (28b)$$

For a given material point in a gravitational field, the gravitational force is, also in the case of motion, thus proportional to its energy; this corresponds to the primary postulate of the weight of energy.

The substitution of (27) and (28a) in (28) yields, upon the cancellation of the constant M, the *equation of motion of free material points in a gravitational field*

$$\frac{d}{dt}\left(\frac{v}{\sqrt{c^2 - v^2}}\right) = -\frac{c}{\sqrt{c^2 - v^2}}\operatorname{grad}c. \qquad (29)$$

The energy equation, which connects the temporal increase of the energy E of the material point with the local temporal differential quotient of c with respect to time, is, by the way, joined to the momentum eq. (28) as a fourth equation. In a static field, where c depends only on position, E remains constant; in this case, according to (27a), the *energy equation* is

$$\frac{c^2}{\sqrt{c^2 - v^2}} = \text{const.} \tag{29a}$$

If one divides the equation of motion (29) by this constant expression, which then can be moved under the derivative, one obtains

$$\frac{d}{dt}\left(\frac{v}{c^2}\right) = -\frac{1}{c}\operatorname{grad} c \tag{29b}$$

[493] | as the differential equation of motion of free material points in a static gravitational field. The equations of motion (29) and (29b), incidentally, have been given first by A. Einstein,[11] and the energy eq. (29a) previously by M. Abraham.[8]

From it [eq, (29b)] one can derive the *theory of the free fall*. If v_0 is the initial velocity, and c_0 the speed of light appropriate to the initial location, then

$$v^2 = c^2\left(1 - \frac{c^2}{c_0^2} + c^2\frac{v_0^2}{c_0^4}\right), \tag{30}$$

and, in particular, if the initial velocity v_0 is equal to zero

$$v = c\sqrt{1 - \frac{c^2}{c_0^2}}. \tag{30a}$$

While c decreases from the initial value c_0, the velocity of the fall v increases at first, reaches a maximum

$$v_m = \frac{1}{2}c_0 \quad \text{for} \quad c = \frac{c_0}{\sqrt{2}}, \tag{30b}$$

and finally tends towards the limit c as c continues to decrease.

In order to determine the distance of fall as a function of time, c must be given as a function of z. In his note about the free fall,[9] M. Abraham puts

$$c^2 = c_0^2 + 2gz, \tag{31}$$

corresponding to a homogeneous mass-free field in his first theory. Then (30a) is satisfied by

$$\left.\begin{array}{c} \dfrac{v}{c} = \sin\!\left(\dfrac{gt}{c_0}\right) \\[2em] \dfrac{c}{c_0} = \cos\!\left(\dfrac{gt}{c_0}\right) \end{array}\right\} \qquad 0 \le t < \dfrac{\pi c_0}{2\, g}. \tag{31a), (31b}$$

If one integrates the velocity of fall

$$-\frac{dz}{dt} = v = c_0 \cos\!\left(\frac{gt}{c_0}\right)\sin\!\left(\frac{gt}{c_0}\right) \tag{31c}$$

with respect to time, then the distance of fall becomes

$$-z = \frac{c_0^{\,2}}{2g}\sin^2\!\left(\frac{gt}{c_0}\right), \tag{31d}$$

which also satisfies (31). At the time $t = \pi c_0/2g$, i.e. for the distance of fall $c_0^2/2g$, c and v would both become zero. Of course, one cannot produce homogeneous fields of such extension. Practically, only times of fall have to be considered that are small l compared to the limiting time; then the above laws of fall become the Galilean ones.   [494]

B. Caldonazzo[(23)] has concerned himself with the trajectories of freely point masses in homogeneous gravitational fields. He compares them to light rays whose paths follow from Fermat's principle of minimum light travel time,

$$\delta t = \delta\!\int\frac{ds}{c} = 0, \tag{32}$$

if c is given as a function of position. The differential equation for a *light ray trajectory* is, with t denoting a tangential unit vector

$$\frac{d}{ds}\!\left(\frac{\mathbf{t}}{c}\right) = \operatorname{grad}\!\left(\frac{1}{c}\right). \tag{32a}$$

On the other hand, the equation of motion (29b) implies the equation for the *trajectories of a material points* in a static gravitational field

$$\frac{d}{ds}\!\left(\frac{v\mathbf{t}}{c^2}\right) = \frac{c}{v}\operatorname{grad}\!\left(\frac{1}{c}\right). \tag{32b}$$

An examination of the two differential equations (32a, b) shows that the second goes over into the first if one replaces v by c. From this remark follows an interesting relation between the trajectories of material points and those of light-points. Let us follow a material point P and a light-point L, which both emerge from O in the direction defined by the unit vector $\mathbf{t}_0$. The initial speed of L is the speed of light c_0

associated with the point O, whereas we let that of P be equal to v_0. The trajectory of L is then uniquely determined by (32a), the one of P by (32b), where for v one has to use the function of position given by the energy eq. (30). Different initial speeds v_0 of P correspond to different trajectories; this set of trajectories has as the limiting curve the one corresponding to the initial speed c_0. But for $v_0 = c_0$ eq. (30) implies $v = c$, whence (32b) goes over into (32a). *Thus, the set of trajectories of material points shot from a point O in a given direction with different initial speed v_0, and moving in a static gravitational field, contain as a limiting curve $(v_0 = c_0)$ the light path that emerges from O in the same direction.*

Caldonazzo treats in more detail the case of the homogeneous gravitational field, where the equipotential surfaces, $c =$ const., are horizontal | planes. If the initial velocity was horizontal, it follows from (29b) that

[495]

$$v_x = \pm v_0 \frac{c^2}{c_0^2} \tag{33}$$

for the horizontal velocity component, and, hence, on account of (30), the vertical velocity component of the material points becomes

$$v_z = -\sqrt{v^2 - v_x^2} = -c\sqrt{1 - \frac{c^2}{c_0^2}}\,. \tag{33a}$$

The velocity components of the light-point are, according to the theorem just proved, obtained from this by setting $v_0 = c_0$:

$$c_x = \pm \frac{c^2}{c_0}, \tag{33b}$$

$$c_z = -c\sqrt{1 - \frac{c^2}{c_0^2}}\,. \tag{33c}$$

It follows that

$$v_z = c_z, \qquad v_x = \frac{v_0}{c_0}\, c_x. \tag{34}$$

Therefore, all the material points shot horizontally with different initial speeds v_0 have the same velocity of fall as the light-point, whereas the horizontal velocities are in the same ratio as the initial velocities. Thus all these points fall by equal distances during equal times; the horizontal projections of paths at equal time are, however, related to that of the light path as the initial velocities v_0 and c_0. Thus the trajectories of material points are obtained from the light curve to which they are tangent at the vertex by a contraction in the horizontal direction. Based on this result, Caldonazzo constructs, in a suitably chosen coordinate scale, the trajectories for the following three cases:

$$c = \sqrt{z}.$$ (I)

This corresponds to a homogeneous mass-free field in Abraham's first theory, where $c^2 = z$ should satisfy the Laplace equation (cf. (18a)). The light curves are cycloids. The trajectories resulting from them by contraction in the horizontal direction are so called "Fermat cycloids."

$$c = z.$$ (II)

Here, c itself satisfies Laplace's equation. According to the Einstein's equivalence hypothesis (III A), this should be the case in mass-free fields. I The light curves are circles, hence the trajectories of material points are ellipses. [496]

$$c = z^2.$$ (III)

$\sqrt{c}$ satisfies Laplace's equation as demanded by the postulate of the weight of energy. The light-curves are elastic curves [*elastische Kurven*], the trajectories of material points are affine to them.

If one considers c as variable in the gravitational field, how are the electromagnetic field equations to be formulated? According to I. Ishiwara,[19] as follows:

$$
\left.
\begin{aligned}
\operatorname{curl}(\mathfrak{H}\sqrt{c}) &= \frac{\partial}{\partial t}\left(\frac{\mathfrak{E}}{\sqrt{c}}\right) + \rho\frac{\upsilon}{\sqrt{c}}, \\[4pt]
\operatorname{div}\left(\frac{\mathfrak{H}}{\sqrt{c}}\right) &= 0, \\[4pt]
-\operatorname{curl}(\mathfrak{E}\sqrt{c}) &= \frac{\partial}{\partial t}\left(\frac{\mathfrak{H}}{\sqrt{c}}\right), \\[4pt]
\operatorname{div}\left(\frac{\mathfrak{E}}{\sqrt{c}}\right) &= \frac{\rho}{\sqrt{c}}.
\end{aligned}
\right\}
$$ (35)

By retaining the usual expressions for the components of the electromagnetic tensor T^e, he obtains for its divergence, i.e., for the energy-momentum transfer of the electromagnetic field, the expression:

$$\rho\mathfrak{F}^e + \frac{\overset{e}{\eta}}{c}\operatorname{grad}c = \operatorname{div}T^e.$$ (35a)

The first term represents the electromagnetic four-force per unit volume; it corresponds to a transfer of momentum and energy from the electromagnetic field to the electrically charged matter. The second term, however, *shows a transfer of momentum and energy from the electromagnetic field to the gravitational field, which should take place everywhere where there exists electromagnetic energy.* This conception corresponds to the postulate of the weight of energy; and, according to it, the *electro-*

magnetic energy, just like the energy of matter, must *generate a gravitational field*, so that the field equation (23) is written as

$$\sqrt{c}\,\square\,(\sqrt{c}) = 2(\eta^m + \eta^e).\tag{36}$$

Then, (cf. (23a)) the momentum and energy transferred from the gravitational field per unit volume and time becomes

$$-\left(\frac{\eta^m + \eta^e}{c}\right)\operatorname{grad}c = \operatorname{div}T^g.\tag{36a}$$

[497]   I Here, the term containing η^m shows the transfer of momentum and energy to matter, the term containing η^e the transfer to the electromagnetic field; the latter is equal and opposite to the corresponding term appearing in (35a), justifying the above interpretation. Upon adding of the divergences of the tensors T^e and T^g it $[\eta^e]$ cancels. The *total four-force acting on the unit volume of matter* is[4]

$$\rho\mathfrak{F}^e - \frac{\eta^m}{c}\operatorname{grad}c = \operatorname{div}(T^e + T^g).\tag{36b}$$

It is composed of the electromagnetic force acting on the charge, and of the gravitational force acting on the energy of matter.

The field equations (35) thus agree with the postulate of the weight of energy, and so fit into the theory presented here. The coupling of the two fields is, by the way, not a simple one. For, according to (35) the gravitational potential $\sqrt{c}$ influences the electromagnetic field; on the other hand, according to (36), the gravitational potential depends on the distribution of energy within the electromagnetic field.

D. Theories of Nordström and Mie

In Abraham's first theory (II B), the postulate of the constancy of the speed of light is renounced, and thus the validity of the theory of relativity is restricted to infinitely small spacetime regions. Abraham's second theory gives up the validity of the principle of relativity even in this restricted sense, because the two sides of its field equation (23) exhibit different behavior under Lorentz transformation.

The theories of gravitation of G. Nordström and G. Mie, however, view the speed of light as constant; their equations are invariant with respect to Lorentz transformations.

Following Abraham's first theory, G. Nordström[20] derives the gravitational tensor T^g from a scalar gravitational potential φ according to the scheme (16). This [potential] satisfies the field equation (18), in which ν denotes the "rest-mass density" of matter. Then, as in (II B, eq. (17), (17b)), on the basis of the conservation laws of momentum and energy there results,

$$\mu\mathfrak{F}^g = -\nu\operatorname{grad}\varphi\tag{37}$$

as the *gravitational force per unit of volume*.

We integrate over the volume of a moving body. Since I its volume elements expe-   [498]
rience a Lorentz contraction as a result of the motion, we have

$$dV = dV_0 \sqrt{1 - \frac{v^2}{c^2}} \qquad (V_0 \text{ rest-volume}). \qquad (38)$$

If, furthermore, one takes the rest-mass of the body to be

$$m_0 = \int v dV_0, \qquad (38a)$$

then, in case the body is so small that the gravitational field is considered to be homogeneous over its extent, one obtains the expression

$$\Re^g = \int \mu \mathfrak{F}^g dV = -\frac{m_0}{c} \sqrt{c^2 - v^2} \cdot \text{grad} \varphi \qquad (39)$$

for the *resultant gravitational force*. It becomes readily apparent that the appearance
of the square root as a factor in the expression for the gravitational force is a necessary consequence following from the theory of relativity, because that theory cannot
avoid assigning to the scalar φ, by means of (18), a v invariant under Lorentz transformations. *In the theory of relativity the gravitational force of a moving body is
therefore not proportional to its energy, but to its Lagrangian function* (cf. (20)):

$$\mathfrak{L} = -m_0 c \sqrt{c^2 - v^2}. \qquad (40)$$

At low speeds this means that not the sum but the *difference* of the *potential* and
kinetic energy should be the determining factor for the weight of a body.

As c is constant, the gravitational potential φ can enter the Lagrangian function
(40) only through the multiplicative constant m_0. This must take place in such a way
that in the law of conservation of momentum formulated in the Lagrangian manner
(see (28)),

$$\frac{d\mathfrak{B}}{dt} = \text{grad} \mathfrak{L} = \frac{\partial \mathfrak{L}}{\partial \varphi} \text{grad} \varphi, \qquad (40a)$$

the right-hand side, i.e., the gravitational force

$$\Re^g = \frac{\partial \mathfrak{L}}{\partial \varphi} \text{grad} \varphi = -\frac{dm_0}{d\varphi} c \sqrt{c^2 - v^2} \cdot \text{grad} \varphi, \qquad (40b)$$

agrees with (39). Therefore, for the rest-mass, the differential equation

$$\frac{dm_0}{d\varphi} = \frac{m_0}{c^2} \qquad (40c)$$

must apply, whose integral is

$$m_0 = \frac{M}{c} e^{\frac{\varphi}{c^2}};$$ (41)

| M denotes a "mass-constant" which is independent of φ. Thus, the gravitational potential enters exponentially into the *Lagrangian function of the first theory of Nordström*,

$$\mathfrak{L} = -Me^{\frac{\varphi}{c^2}} \cdot \sqrt{c^2 - v^2},$$ (41a)

and enters similarly into the *expressions for momentum* and *energy*

$$\mathfrak{G} = \frac{\partial \mathfrak{L}}{\partial v} \frac{v}{v} = \frac{M}{\sqrt{c^2 - v^2}} e^{\frac{\varphi}{c^2}} \cdot v,$$ (41b)

$$E = v \frac{\partial \mathfrak{L}}{\partial v} - \mathfrak{L} = \frac{Mc^2}{\sqrt{c^2 - v^2}} \cdot e^{\frac{\varphi}{c^2}},$$ (41c)

and into the *expression for the gravitational force* (40b)

$$\mathfrak{K}^g = -\frac{M}{c^2} e^{\frac{\varphi}{c^2}} \sqrt{c^2 - v^2} \cdot \text{grad}\varphi.$$ (41d)

The equation of motion of material points, or equivalent bodies, in a gravitational field

$$\frac{d\mathfrak{G}}{dt} = \mathfrak{K}^g$$ (42)

takes, according to (41b, d), the form

$$\frac{d}{dt}\left(e^{\frac{\varphi}{c^2}} \cdot \frac{v}{\sqrt{c^2 - v^2}}\right) = -\frac{1}{c^2} \cdot e^{\frac{\varphi}{c^2}} \cdot \sqrt{c^2 - v^2} \cdot \text{grad}\varphi.$$ (42a)

In a static field we have the energy integral (cf. (41c))

$$\frac{e^{\frac{\varphi}{c^2}}}{\sqrt{c^2 - v^2}} = \text{const.},$$ (42b)

so that in Nordström's theory, the *equation of motion of material points in a static field* is:[22]

$$\frac{dv}{dt} = -\left(1 - \frac{v^2}{c^2}\right)\text{grad}\varphi.$$ (42c)

It is apparent that in a homogeneous gravitational field—in contrast with (II C)—the greater the velocity of horizontally projected bodies, the slower they fall. The light-curve is, however, here too to be considered as a limiting curve of the family of trajectories; because for $v = c$, the acceleration is zero, thus the motion is uniform and rectilinear, like that of a light-point under the presupposition of the constancy of the speed of light.

For *free fall* in a homogeneous mass-free field, the equation of motion

$$\frac{dv}{dt} = g - \frac{gv^2}{c^2} \tag{42d}$$

applies. | It agrees with the differential equation that one uses in classical mechanics [500] for the fall in air under the assumption of a law of friction proportional to the square [of the speed]; therefore its integration poses no difficulties and becomes, if one sets the initial velocity equal to zero,

$$v = c\tanh\left(\frac{gt}{c}\right). \tag{42e}$$

If the field extended sufficiently far, the speed would asymptotically approach the speed of light.

M. Behacker[25] has treated horizontal projection [of bodies] and planetary motion on the basis of Nordström's first theory. The deviations from the laws of classical mechanics are here too of second order in $v : c$.

The theory of G. Mie,[21] though developed independently, is closely related to Nordström's first theory. It forms part of an extensive investigation of the "foundations of a theory of matter," for whose presentation this is not the appropriate place. Within matter, Mie differentiates between two gravitational four-vectors, which, however, coincide in the case of an "ideal vacuum." Here, as with Abraham, the differential equations of longitudinal waves apply. The gravitational waves emitted by an oscillating mass-particle are treated in more detail by Mie. He further emphasizes that, from the point of view of the theory of relativity, one necessarily arrives at the notion that the weight of a body is not to be set proportional to its energy, but to its Lagrangian function. Accordingly, the kinetic energy of the thermal motion of the molecules makes a negative contribution to the weight. Since, on the other hand, in the theory of relativity the theorem of the inertia of energy is valid, the gravitational mass of a warm body is thus not precisely proportional to its inertial mass, but with increasing temperature the quotient of gravitational and inertial mass decreases. However, the proof of the decrease demanded by Mie's theory would require pendulum measurements with an accuracy of the order of 10^{-11}; this theoretical deviation from the proportionality law is not subjectable to experiment.

Greater difficulties arise for the Mie-Nordström theory from observations by Southerns, which demonstrate the constancy of the quotient of inertial and gravitational mass during radioactive transformations (I C). These observations can only be brought into agreement with that theory through rather artificial assumptions. One could for example imagine that the entire energy loss during the transformation

[501] takes place at the expense of only the electric potential energy, so that the kinetic (magnetic) energy has the same value before and after the transformation of the uranium atom. Or, one can assume with Kretschmann[36] that the total energy emitted from the uranium atom in the form of heat, β - and γ -particles does not come from those particles that later form the lead atom, but from the eight escaped helium atoms. Then, however, helium would have to show a correspondingly greater deviation from the proportionality law.

G. Mie also investigated the relation between the gravitational potential and electromagnetic processes. An electromagnetic field possesses a gravitational mass, which is proportional to its Lagrangian function, i.e. proportional to the difference of its electric and magnetic energy. A plane electromagnetic wave, in which the electric and the magnetic energy density is known to have the same value, accordingly has no weight in this theory (corresponding to the rectilinear propagation of light). An electrostatic field, however, possesses a positive weight and a magnetostatic field a negative weight. On the other hand, the gravitational potential enters exponentially into the electromagnetic quantities, as well as into the mechanical ones. Nevertheless, the value of the potential in a system escapes detection by observers belonging to that system. This is expressed by Mie's theorem of the "*relativity of the gravitational potential*": If two physical systems differ merely by the value of the gravitational potential, then this does not have the least effect on the size and the form of the electrons and of the other material elementary particles, on their charge, their laws of oscillation and motion, and on the velocity of light, indeed on all physical relations and processes.

E. Nordström's Second Theory

With the intention of satisfying the law of the proportionality of the gravitational and inertial mass, as far as it is possible within the framework of the theory of relativity, G. Nordström[26] later made a change in his theory. Before we turn to a discussion of this change, we must briefly return to the properties of the world tensors (I B), and mention several theorems of von Laue regarding "complete static systems."

We form the "*diagonal sum*" of a world tensor T

$$D = T_{xx} + T_{yy} + T_{zz} + T_{uu};$$ (43)

[502] since the components entering this sum transform like the square of the four coordinates, the diagonal sum is an invariant; it is frequently called the "Laue scalar."

The diagonal sum of the *electromagnetic world tensor* T^e is identically equal to zero:

$$D^e = T^e_{xx} + T^e_{yy} + T^e_{zz} + T^e_{uu} = 0.$$ (43a)

In contrast, our *gravitational tensor* T^g yields, according to (16), a diagonal sum different from zero:

$$D^g = \left(\frac{\partial \varphi}{\partial x}\right)^2 + \left(\frac{\partial \varphi}{\partial y}\right)^2 + \left(\frac{\partial \varphi}{\partial z}\right)^2 + \left(\frac{\partial \varphi}{\partial u}\right)^2. \tag{43b}$$

The matter is the carrier of a third world tensor T^m; its last component is the energy density of matter. For a body consisting of discrete and freely movable mass particles, the momentum density and the kinetic stresses determine the remaining components of the "*material tensor*" T^m. In a continuously connected body, elastic stresses enter in addition into the consideration, as they also cause an energy flux and thus a momentum density. It appears useful to include in T^m the "stress tensor." The diagonal sum of the material tensor

$$D^m = T^m_{xx} + T^m_{yy} + T^m_{zz} + T^m_{uu} \tag{43c}$$

then equals the sum of the three (kinetic and elastic) principal stresses plus the total energy density (η^m) of the matter.

Now, if

$$T = T^m + T^e + T^g \tag{44}$$

is the *total ten-tensor, resultant of the material, the electromagnetic and of the gravitational tensor*, then the *laws of conservation of momentum and energy* can be summarized as (cf. (5))

$$0 = \operatorname{div} T. \tag{44a}$$

Namely the momentum and energy extracted from the electromagnetic field and from the gravitational field are transformed into momentum and energy of matter.

By a "*complete static system*," one understands, according to M. Laue, an isolated physical system that is in equilibrium in an appropriately chosen system of reference Σ_0. Therefore, in Σ_0 the components

$$T^0_{xu}, \quad T^0_{yu}, \quad T^0_{zu}$$

of the resultant tensor, which determine the energy flux and the momentum density, [503] are all zero; one can derive from this[9] that in Σ_0 the volume integrals of the resulting normal stresses are equal to zero:

$$\int dV_0 T^0_{xx} = \int dV_0 T^0_{yy} = \int dV_0 T^0_{zz} = 0. \tag{45}$$

If one is only concerned about the temporal mean values, then a system in static equilibrium can also be taken as a completely static system. It suffices that the tempo-

---

9   M. Laue, *Das Relativitätsprinzip*, 2nd. ed., p. 209.

ral mean value of the resultant energy flux vanish everywhere. Thus, for example, a hot gas in static equilibrium, together with its container, forms a complete static system. For the measurable stresses, the eqs. (45) apply; the negative contributions arising from the kinetic pressure of the gas, and the positive, arising from the stresses in the walls of the container, cancel each other in the integrals extended over the entire system.

Because it is an invariant, the diagonal sum (43) of the resultant world tensor does not depend on the choice of the system of reference. Its value in a complete static system can thus be determined by referring it to Σ_0:

$$D = D^0 = T^0_{xx} + T^0_{yy} + T^0_{zz} + T^0_{uu}. \tag{45a}$$

In view of (45), the volume integral yields

$$\int dV_0 D = \int dV_0 T^0_{uu} = \int dV_0 \eta_0 = E_0, \tag{45b}$$

where E_0 denotes the *total rest-energy* of the system.

After these preparations, we turn to Nordström's second theory of gravitation. It too is based on the tensor-scheme of (16); however, the following takes the place of the field equation (18),

$$\varphi \Box \varphi = D^m. \tag{46}$$

Since the diagonal sum D^m of the material tensor, as well as φ and the operator $\Box$, are four-dimensional scalars, this approach corresponds to the relativistic scheme. The gravitational force per unit volume now follows from (17b):

$$\mu \mathfrak{F}^g = -\frac{D^m}{\varphi} \cdot \mathrm{grad}\varphi. \tag{46a}$$

The integration over the volume of a body moving with respect to Σ_0, taking into account the Lorentz contraction (38), yields the gravitational force acting on the body.

[504]

$$\mathfrak{K}^g = -M\sqrt{c^2 - v^2} \cdot \mathrm{grad}\varphi, \tag{46b}$$

where

$$Mc = \int \frac{dV_0 D^m}{\varphi} = \int \frac{dV_0 D^m_0}{\varphi_0} \tag{46c}$$

denotes the gravitational mass of the body for the case of rest ($v = 0$). Indeed, on the one hand, according to (46b), the gravitational force on the resting body is equal to the negative gradient of the potential φ multiplied by Mc. On the other hand, when rest reigns in Σ_0, it follows from (46) that

$$\Delta\varphi_0 = \text{div grad}\varphi_0 = \frac{D_0^m}{\varphi_0}.$$

Therefore, according to Gauss' law, the force flux through a surface f_0 enclosing the body, which determines the attracting mass, becomes:

$$\int df_0 \frac{\partial \varphi_0}{\partial n} = \int \frac{dV_0 D_0^m}{\varphi_0} = Mc. \tag{46d}$$

The gravitating mass Mc, and thus also M, is considered to be a *constant*, i.e. as independent of φ, in Nordström's second theory.

Now, what relation exists between the gravitational mass and the rest-energy for a complete static system? Equation (45b) yields:

$$E_0 = \int dV_0 D_0. \tag{47}$$

Taking into consideration (43a, b), one obtains for the diagonal sum of the resultant world tensor, since equilibrium reigns in Σ_0,

$$D_0 = D_0^m + (\text{grad}\varphi_0)^2. \tag{47a}$$

And since, furthermore,

$$\varphi_0 \Delta\varphi_0 + (\text{grad}\varphi_0)^2 = \text{div}(\varphi_0 \text{grad}\varphi_0)$$

is an identity, the field equation (46) can then be brought into the form

$$\text{div}(\varphi_0 \text{grad}\varphi_0) = D_0 \tag{47b}$$

and from (47) it follows, on the basis of Gauss' theorem, that

$$E_0 = \int dV_0 D_0 = \int df_0 \varphi_0 \frac{\partial \varphi_0}{\partial n}, \tag{47c}$$

where the integration is to be carried out over a surface wholly enclosing the complete static system and its gravitational field. However, on such a surface, far removed from the system, one has to set for φ_0, the potential φ_a of the external masses not belonging to the system, | which is constant there. Namely, if φ_a were not constant [505] on f_0, the external field would superpose on the system's own; the energy of its own field could then not be separated from the external field, and the system could thus not be considered isolated (i.e., enclosed by the surface f_0). Now, it follows from (47c) and (46d) that

$$E_0 = \varphi_a \int df_0 \frac{\partial \varphi_0}{\partial n} = \varphi_a Mc. \tag{48}$$

Accordingly, the total *rest-energy of the complete static system is equal to its gravitational mass multiplied by the external gravitational potential*. On the other hand, according to the theorem of the inertia of energy (8a),

$$m_0 = \frac{E_0}{c^2} = \varphi_a \frac{M}{c} \tag{48a}$$

is the *inertial rest-mass*.

In Nordström's second theory, the law of proportionality between inertial and gravitational mass applies for a complete static system, at least for the rest-masses. Of course, for moving bodies, the postulate of the weight of energy is not satisfied in Nordström's second theory any more than in his first, since, in the expression (46b) for the gravitational force, the root factor characteristic of the theory of relativity, and entering into the denominator of the expression for the energy, appears again in the numerator. However, admittedly one cannot determine the weight of a moving body with a scale; thus, as far as weighing is concerned, the theorem of the weight of energy therefore applies in Nordström's theory, and accordingly also the theorem that weight is conserved. However, the strict validity of these theorems, and thus also that of the law of proportionality of inertial and gravitational mass, fails when the masses are in motion. For example, in pendulum measurements, one would expect deviations of the order of $(v/c)^2$, which are however unmeasurable of course.

But what is now the situation in an isolated system in a state of statistical equilibrium, e.g., in a hot gas? The thermal motion of the molecules causes here, too, a reduction of weight corresponding to the negative kinetic stresses entering into D^m, and thus also into the expression for the gravitational mass (cf. (46d)). This, however, is compensated by the positive contributions which the stresses in the wall of the vessel contribute to D^m, and thus to the weight. Thus, for Nordström, the proportionality of weight and energy comes about merely through a compensation of the contributions of the individual parts of the complete static system. | It is in no way based on a fundamental property of matter or energy.

[506]

A complete system of sufficiently small dimensions can be considered as equivalent to a material point. The motion in a given gravitational field is then obtained from Lagrange's equations (40a)

$$\frac{d\mathfrak{G}}{dt} = \frac{d\mathfrak{L}}{\partial \varphi}\mathrm{grad}\varphi = \mathfrak{K}^g. \tag{49}$$

As a comparison with (46b) reveals, one has to set for the *Lagrangian function*

$$\mathfrak{L} = -M\varphi\sqrt{c^2 - v^2}. \tag{49a}$$

This Lagrangian of Nordström's second theory turns into the one (41a) of his first theory, if φ is replaced by the exponential function $e^{\frac{\varphi}{c^2}}$. However, through this substitution, the field equation (46) does not change into agreeing with (18) of the first theory; therefore, the two theories are not equivalent.

The momentum and energy can be derived from the Lagrangian function (49a) in the usual manner:

$$\mathfrak{B} = \frac{\partial \mathfrak{L} v}{\partial v\, v} = \frac{M\varphi}{\sqrt{c^2 - v^2}} \cdot v, \tag{49b}$$

$$E = v\frac{\partial \mathfrak{L}}{\partial v} - \mathfrak{L} = \frac{Mc^2\varphi}{\sqrt{c^2 - v^2}}. \tag{49c}$$

Through substitution of (49b) into (49) one obtains the *equations of motion of material points* and of the equivalent complete systems

$$\frac{d}{dt}\left(\frac{\varphi \mathrm{v}}{\sqrt{c^2 - v^2}}\right) = -\sqrt{c^2 - v^2} \cdot \operatorname{grad}\varphi. \tag{50}$$

In a static gravitational field, the energy (49c) is constant:

$$\frac{\varphi}{\sqrt{c^2 - v^2}} = \text{const.} \tag{50a}$$

Therefore, the equation of motion can be brought into the form[31]

$$\frac{dv}{dt} = -\left(\frac{c^2 - v^2}{\varphi}\right)\operatorname{grad}\varphi. \tag{50b}$$

For a homogeneous mass-free gravitational field, whose lines of force are parallel to the negative z-axis, one must set, corresponding to the field equation (46),

$$\varphi = \varphi_0 + gz. \tag{50c}$$

| The free fall from the rest position $z = 0$ in this field can be treated on the basis of [507] the energy equation (50a), which yields

$$\varphi = \frac{\varphi_0}{c} \cdot \sqrt{c^2 - v^2}. \tag{50d}$$

It is satisfied by

$$z = -\frac{\varphi_0}{g}\left(1 - \cos\left(\frac{gct}{\varphi_0}\right)\right); \tag{51}$$

because according to (50c)

$$\varphi = \varphi_0 \cos\left(\frac{gct}{\varphi_0}\right), \tag{51a}$$

and, on the other hand, the velocity of fall is

$$\upsilon = -\frac{dz}{dt} = c\sin\left(\frac{gct}{\varphi_0}\right), \tag{51b}$$

so that (50d) is indeed satisfied.

G. Nordström,[31] incidentally, also treats oblique projection [of bodies] and planetary motion; the area law proves to be valid. The deviations from the laws of classical mechanics are minute.

In Nordström's theory, the rest-energy is proportional to the external gravitational potential (cf. (48)). In Abraham's theory, a similar behavior was found; the rest-energy (cf. (21a)) was proportional to the speed of light, which there played the role of the gravitational potential. Thus, in either theory, the rest-energy of a body decreases as a result of the approach of external masses. The rest-mass, in contrast, exhibits a different behavior in the two theories; for Abraham (cf. (21b)) the rest-mass is inversely proportional to c, for Nordström (cf. (48a)) it is proportional to φ_a. Hence, for Abraham, the inertial mass of a body increases upon the approach of an external body, while, for Nordström, it decreases.

Incidentally, according to Nordström, the units of time (τ) and length (λ) should depend in the following way on the gravitational potential:

$$\tau\varphi = \text{const.}, \tag{52}$$

$$\lambda\varphi = \text{const.}, \tag{52a}$$

i.e., the rate of a portable clock and the length of a portable measuring rod should be inversely proportional to the gravitational potential. Thus, local spatial and temporal measurements do not allow the construction of the universal system of reference in which light should propagate rectilinearly. This conception of space and time can [508] hardly be made compatible with the theory of relativity of Minkowski. | However, it can be readily fitted into the Einstein-Grossmann "generalized theory of relativity," as has been shown by A. Einstein and A. D. Fokker.[34]

We must refrain here from entering into Nordström's five-dimensional interpretation of his theory.[33] [37]

F. Kretschmann's Theory

Among the newest relativistic theories of gravitation, the one by E. Kretschmann[36] deserves to be mentioned even if only briefly. This theory is to be counted among the scalar theories, as its assumes that the gravitational force is determined by the gradient of an "aether pressure" p, which, in a vacuum, satisfies the equation

$$\Box p = 0.$$

Matter is assumed to consist of elementary drops, which obey the equations of state of "ideal fluids of smallest compressibility and least expendability" developed by E. Lamla in his inaugural dissertation "Über die Hydrodynamik des Relativitätsprinzips." These drops are assumed to move irrotationally and each to carry a positive electric elementary charge. The divergence of the gradient of p in the interior of the drops, which determines the attracting mass, turns out to be proportional to the square of their acceleration. On the other hand, the attracted mass of a body is proportional to the sum over the volume of its elementary drops. In order to obtain the proportionality of the attracted, the attracting and of the inertial mass, the author assumes not only that the drops are all of the same kind, but that they also move with the same mean velocity and acceleration; then, of course, all three masses simply become proportional to the number of the particles. Following A. Korn and H. A. Lorentz, the author explains, that such an energy equilibrium of the elementary drops is produced by means of a universal radiation state, or state of oscillation, which is supposed to immediately smooth out disturbances in the energy equilibrium.

The contrast between this theory and the other theories of gravitation discussed here is strikingly clear. The latter introduce merely mechanical and energetic quantities, since the speed of light, too, is a quantity of relativistic mechanics. Kretschmann's theory, in contrast, is based on quite particular concepts, which are virtually without significance for the final result; for not the particular properties, but only the number of | particles is to be considered in the derivation of the proportionality of inertial and gravitational mass. [509]

III. TENSOR THEORIES

A. Einstein's Theory of the Static Gravitational Field

In I C we discussed the relation between gravity,[2] inertia, and energy, and the necessity of assigning to them a place in the worldview of modern physics. These relations also formed the starting point of the investigations of A. Einstein.

His first paper[5] is based on the so called *equivalence hypothesis*: "Equivalence exists between two systems of reference of which one is at rest in a homogeneous gravitational field and the other is uniformly accelerated in a field free of gravitation." Indeed, in classical mechanics, such an equality exists because the inertial force under uniform acceleration is equivalent to a constant gravitational force; the equivalence hypothesis results here in the identity of inertial and gravitational mass. But it is questionable whether the equivalence hypothesis can be maintained for other physical processes, namely for those that satisfy the principle of relativity of 1905; if this were the case, then from this and from the theorem of the inertia of energy, valid in that theory, the theorem of the weight of energy would follow immediately. Without a critical examination of whether his earlier principle of relativity is compatible with the new equivalence hypothesis, Einstein connects these two trains of thought and arrives thus at the following remarkable result:

The gravitational potential influences the speed of light; if the values of the speeds of light, c and c_0, correspond to the potentials, φ and φ_0, then

$$c = c_0\left(1 + \frac{\varphi - \varphi_0}{c^2}\right). \tag{53}$$

By application of Huygens' principle this implies a curvature of the light rays grazing the surface of the Sun; the total deviation of such a ray would amount to 0.83 seconds of arc, by which the angular distance of a star appears to be increased with respect to the center of the Sun. Einstein considers it possible to observe this devia-

[510]   tion on the occasion of an eclipse of the Sun. |

Furthermore, the gravitational potential changes the frequency of periodic processes as determined by the following formula

$$\frac{\nu_0 - \nu}{\nu_0} = \frac{\varphi_0 - \varphi}{c^2}. \tag{53a}$$

If φ refers to the surface of the Sun and φ_0 to the Earth, then the right-hand side is equal to $2 \cdot 10^{-6}$. Thus, the frequency ν of the oscillations of light should be somewhat smaller on the surface of the Sun than the frequency ν_0 of the corresponding spectral line of a terrestrial light source: i.e., compared to the terrestrial lines, the solar ones should be displaced to the right-hand side of the spectrum. Astrophysicists have indeed found displacements of the Fraunhofer lines in this sense, but have mostly ascribed them to pressure effects.

In that first study, Einstein had investigated the effect of the gravitational field on radiation phenomena without, however, making use of the connection found between the gravitational potential and the speed of light for the theory of the gravitational field itself. Only after the first publication by Abraham did Einstein also turn to the problem of gravitation.[11] There, again, he started from the equivalence hypothesis and derived the equations of motion ((29), (29b)) of material points in static gravitational fields. He also concluded from that hypothesis that the speed of light c in mass-free static fields must satisfy Laplace's equation. Shortly thereafter,[12] however, he convinced himself that in order to avoid contradictions with the law of conservation of momentum, one has to begin from Laplace's equation for $\sqrt{c}$ rather than for c, and that, hence, the equivalence hypothesis does not form a firm foundation for the theory of the gravitational field. This led to the necessity for a new foundation for the theory. M. Abraham (II C) made the theorem of the weight of energy the starting point of his second theory, and derived from it the field equations and the equations of motion. For static fields, this theory coincides with Einstein's. The essential difference appears only upon the extension to time varying fields. Abraham's theory considers such fields as still being determined solely by the four derivatives of the velocity of light, whereas Einstein's theory derives the dynamic field from a tensor potential.

B. The Generalized Theory of Relativity of A. Einstein and M. Grossmann

The basic idea of the tensor theory of the gravitational field can I be understood as   [511]
follows. The energy density, which in a static field is determined by the divergence of
the gradient of the gravitational potential, plays in the theory of relativity merely the
role of one component of the resulting world tensor T; it is joined by nine other tensor components which characterize the energy flux and the stresses. The tensor theory
assumes that, like the energy density (T_{44}), the remaining nine components $T_{\mu\nu}$
$(\mu, \nu = 1, 2, 3, 4)$ generate gravitational fields whose potentials $g_{\mu\nu}$ form a ten-tensor themselves. With the introduction of such a tensor potential it aims to give the differential equations of the gravitational field a form satisfying the scheme of the
principle of relativity. That it appears possible to fit tensor theories of gravitation into
the framework of Minkowski's conception of spacetime has been shown by
G. Mie.[28] However, the theory of gravitation sketched by A. Einstein and
M. Grossmann[24] supersedes the framework of the earlier theory of relativity by
closely relating these tensor potentials $g_{\mu\nu}$ to a generalized relativistic spacetime
doctrine.

In Minkowski's four-dimensional representation of the theory of relativity, the
square of the four-dimensional distance between two neighboring spacetime points
was given by the following quadratic differential form:

$$ds^2 = dx^2 + dy^2 + dz^2 - c^2dt^2, \tag{54}$$

in which c, the speed of light, was constant. In a static gravitational field, the very
expression (54) should, according to Einstein, express the "natural distance" ds of
two neighboring world points, but where c is now a function of x, y, z. Its application to the general case of a dynamic gravitational field would, however, imply a pref-
erential treatment of the time coordinate $(x_4 = t)$ over the spatial coordinates
$(x_1 = x, x_2 = y, x_3 = z)$, which would not be reconcilable with the relativistic
ideas about space and time. It appears natural to replace the special quadratic form
(54) by the most general homogeneous function of second degree in the coordinate
differentials:

$$ds^2 = \sum_{\mu\nu} g_{\mu\nu}dx_\mu dx_\nu \qquad (\mu, \nu = 1, 2, 3, 4). \tag{55}$$

This form was used by E. and G. for the natural distance between two neighboring
world points. The system of the ten coefficients $g_{\mu\nu}$ in the form (55) form a world
tensor, and it is said to be identical to the tensor potential of the dynamic gravitational
field, similar to the way c determines the gravitational potential in the static case. I   [512]

According to (55) the natural length of a portable, infinitely small measuring rod
is not determined solely by the coordinate differentials dx_1, dx_2, dx_3 of its end
points; rather, the six potentials

$$g_{11}, \qquad g_{22}, \qquad g_{33}, \qquad g_{12} = g_{21}, \qquad g_{23} = g_{32}, \qquad g_{31} = g_{13}$$

enter into it, and, in particular, in a way that depends on the direction of the measuring rod. That means that apparently rigid bodies are stretched and twisted in a gravitational field. The measure of time too is influenced by the gravitational potential g_{44} (this occurs already in the static field) in such a way that the natural distance of two neighboring points in time, measured with the aid of a portable clock, is different from the differential dx_4 of the time coordinate. Thus, the coordinates x_1, x_2, x_3, x_4 have no direct physical meaning. In order to determine from their differentials the natural distance between neighboring spacetime points, the values of the ten gravitational potentials $g_{\mu\nu}$ must be known. Regarding, by the way, the three potentials,

$$g_{14} = g_{41}, \qquad g_{24} = g_{42}, \qquad g_{34} = g_{43},$$

they arise, for example, if the system of reference rotates in a static gravitational field, and then determine the velocity of the point in question. The complete system of the $g_{\mu\nu}$ characterizes the state of deformation of four-dimensional space.

In the same way as the invariant differential form (54) is related to the group of rotations in four dimensions (Lorentz transformations), and hence to the vector calculus of Minkowski's theory of relativity, so the invariance of the more general differential form (55) is related to a more general group of transformations. The study of the behavior of the different geometric objects (vectors, tensors) with respect to these transformations forms the mathematical basis of the "generalized theory of relativity." The authors of the *Entwurf* found the required mathematical tools already fully developed in the "absolute differential calculus" of G. Ricci and T. Levi-Civita. Based on this, in the second part of the *Entwurf*, M. Grossmann outlines the fundamental concepts of the four-dimensional vector calculus, which corresponds to that general group of transformations. This is not the place to enter into the mathematical aspects of the *Entwurf*. It may, however, be noted that Grossmann designates the "four-vectors" of the usual theory of relativity as "first rank tensors," whereas by "second rank tensors" he understands the quantities which by us are called simply "tensors." |

[513]

In the first, physical, part of the *Entwurf*, written by A. Einstein, the equations of motion of material points are derived in a generally covariant form from Hamilton's principle

$$\delta\int ds = 0. \tag{56}$$

The gravitational force appears as dependent on the forty derivatives of the ten potentials $g_{\mu\nu}$; here, however, the so called "fictitious forces of relative motion" are counted among the gravitational forces.

Just like the equations of motion, the differential equations of the electromagnetic field are also brought to a generally covariant form. Yet, the attempt to develop generally covariant equations for the gravitational field fails. The *field equations* put down by Einstein read as follows:

$$\sum_{\alpha\beta\mu}\frac{\partial}{\partial x_\alpha}\left(\sqrt{-g}\gamma_{\alpha\beta}g_{\sigma\mu}\frac{\partial\gamma_{\mu\nu}}{\partial x_\beta}\right) = S^m_{\sigma\nu} + S^g_{\sigma\nu}. \tag{57}$$

Here, g is the determinant of the $g_{\mu\nu}$, $\gamma_{\mu\nu}$ is the cofactor (*adjungierte Unterdeterminante*) of $g_{\mu\nu}$ divided by g. On the right-hand side are quantities that are linear functions of the components of the material tensor T^m and of the gravitational tensor T^g:

$$S^m_{\sigma\nu} = \sqrt{-g}\cdot\sum_\mu g_{\sigma\mu}T^m_{\mu\nu}, \tag{57a}$$

$$S^g_{\sigma\nu} = \sqrt{-g}\cdot\sum_\mu g_{\sigma\mu}T^g_{\mu\nu}. \tag{57b}$$

If one places the quantities $S^g_{\sigma\nu}$:

$$S^g_{\sigma\nu} = \sqrt{-g}\cdot\left\{\sum_{\beta\tau\rho}\gamma_{\beta\nu}\frac{\partial g_{\tau\rho}}{\partial x_\sigma}\frac{\partial\gamma_{\tau\rho}}{\partial x_\beta} - \frac{1}{2}\sum_{\alpha\beta\tau\rho}\delta_{\sigma\nu}\gamma_{\alpha\beta}\frac{\partial g_{\tau\rho}}{\partial x_\alpha}\frac{\partial\gamma_{\tau\rho}}{\partial x_\beta}\right\}, \tag{57c}$$

(in which one has to set $\delta_{\sigma\nu} = 1$, for $\sigma = \nu$, $\delta_{\sigma\nu} = 0$ for $\sigma \neq \nu$) on the left hand side of (57), then there arise ten second order partial differential equations for the ten quantities $\gamma_{\mu\nu}$, respectively $g_{\mu\nu}$, in the right-hand side of which the components of the material tensor T^m enter as field generating quantities. If an electromagnetic field is present, then the components of the electromagnetic tensor T^e are of course to be introduced in the same manner.

With regard to the invariance properties of his gravitational field equations, Einstein still appears to have hoped, during the writing of the *Entwurf*, to obtain covariance, if not for the more general group of transformations associated with the form (55), but at least for a group encompassing acceleration transformations; thereby, his earlier | "equivalence hypothesis" was to be supported mathematically. In his Vienna lecture,[27] however, he states that it is probable, and later[29] that it is certain, that these field equations are only covariant with respect to linear transformations. Lately however, with the aid of M. Grossmann, he was able to prove[35] that only the laws of conservation of momentum and energy are responsible for restricting the diversity of "allowed" transformations. These conservation laws, which we expressed symbolically by

$$\mathrm{div}(T^m + T^g) = 0$$

become in the generalized theory of relativity, upon the introduction of the quantities (57a, b):

$$\sum_\nu\frac{\partial}{\partial x_\nu}(S^m_{\sigma\nu} + S^g_{\sigma\nu}) = 0. \tag{58}$$

Those four equations, together with the field equations (57), yield four third order partial differential equation

$$\sum_{\alpha\beta\mu\nu} \frac{\partial^2}{\partial x_\nu \partial x_\alpha}\left(\sqrt{-g}\gamma_{\alpha\beta}g_{\sigma\mu}\frac{\partial\gamma_{\mu\nu}}{\partial x_\beta}\right) = 0. \tag{58a}$$

Only such coordinate systems are allowed that are characterized by $g_{\mu\nu}$ satisfying this system of equations; only transformations that transform such "adapted" systems of reference into each other are "allowed." For all allowed transformations, the field equations (57) now turn out[35] to be covariant. A covariance in a further sense can hardly be demanded.

Although from a mathematical point of view the question appears to have been settled, the physicist still wishes to obtain an insight into the extent of the group of "allowed" transformations. Do these contain, besides the linear ones, still other real transformations? And do these have a physical significance, perhaps as transformations describing acceleration or rotation? Only in this case would one be justified in speaking of a "generalized theory of relativity" in the sense that the equivalence of different systems of reference, which the principle of relativity of 1905 postulates for systems in translatory motion with respect to each other, is now extended to such systems that are in accelerated or rotational motion with respect to one another. Such an extension of the relativity of motion does not appear to be achievable.

Also, the physical significance of covariance with respect to linear transformations is not sufficiently discussed in the *Entwurf*. | According to G. Mie,[28] it expresses the following theorem: "The observable laws of nature do not depend on the absolute values of the gravitational potentials $g_{\mu\nu}$ at the location of the observer." G. Mie showed already earlier[21] that this *"theorem of the relativity of the gravitational potential"* is valid in his own theory (II D). However, in Mie's theory, based on the spacetime doctrine of Minkowski, the space and time measurements as well as the dimensions and periods of oscillation of elementary particles remain unchanged, even in a gravitational field. Einstein, in contrast, and also Nordström (II E), achieve the independence of the physical processes from the values of the potentials only because, together with the units of length and time, the dimensions and the periods of oscillation of the atoms and electrons depend on the gravitational potentials $g_{\mu\nu}$.

A. Einstein places such an importance on the circumstance that his gravitational equations are invariant with respect to linear orthogonal substitutions that, in his Vienna lecture,[27] he believed he was allowed to leave theories unmentioned whose field equations do not have such a covariance. The significance of this transformation group rests on the fact that it contains the transformations; in the early theory of relativity the covariance with respect to this group gave expression to the equivalence of systems of reference in translatory motion with respect to each other. Is this presently also the case in the "generalized theory of relativity"? Does the covariance of the field equations with respect to linear orthogonal transformations imply that in a finite system of mutually gravitating bodies the course of the relative motions is not altered by a uniform translation of the entire system? That this is so has so far not been proven. Such a proof may already be impossible to construct for the reason that the concept of "uniform motion" of a finite system is completely up in the air in the new theory of relativity. Since the "natural" space and time measurement is influenced by the values of the local potential, observers at different locations in the gravitational field will ascribe different velocities to the same material point. Only for an infinitesimally small region of four-dimensional space, i.e. for one in which the potentials $g_{\mu\nu}$ can be considered a constant—is "velocity" defined at all. Presumably, only within such an infinitesimal region may the covariance of the gravitational equations with respect to linear orthogonal transformations be interpreted in the sense of an l equivalence of [516] systems of reference moving with respect to each other. But if relativity of motion no longer exists for finite systems of gravitating masses in Einstein's theory, with what right does he assign such great importance to the formal connection to the earlier theory of relativity?

As one can see, the new theory of relativity indeed generalizes the concept of space and time, but thereby restricts the relativity of motion to infinitesimally small spacetime regions. Therefore, its endeavors to achieve relativity with respect to rotational or accelerated motion also do not appear to be promising. Since the velocities of the system of reference enter into the $g_{\mu\nu}$, a spacetime region is only to be considered as infinitesimally small, if the velocity is spatially and temporally constant in it. Therefore it makes no sense to speak of a rotational or accelerated motion of an "infinitesimally small region," and, for instance, to claim that the "equivalence hypothesis is valid in the infinitesimally small."

To what extent is the theorem of the weight of energy valid in the Einstein-Grossmann theory of gravitation? The field equations (57) make it apparent that the quantities $S_{\sigma\nu}$, according to (57a, b) linear functions of the components $T_{\mu\nu}$ of the resultant world tensor, are the ones that generate the gravitational field. The gravitational forces, which a body exerts and experiences, are accordingly defined by the volume integral of these tensor components:

$$\int dV T_{\mu\nu} = m_{\mu\nu}. \tag{59}$$

Therefore, when dealing with a moving and stressed body, one generally has to differentiate between ten "masses" $m_{\mu\nu}$. There also exist ten "gravitational forces," which are derived from the ten potentials $g_{\mu\nu}$. If one transports bodies with differing stress states and different states of motion to the same world point of a variable gravitational field, then the resultant gravitational forces acting on them have in general different directions, and their magnitudes do not depend solely on the energy of the bodies, but also on all the ten $m_{\mu\nu}$.

This intricate mechanism fortunately simplifies since, in relation to the gravitational force derived from g_{44}, the other nine gravitational forces are very small as all the remaining nine masses in relation to the mass m_{44} determined by the energy. Especially for "static complete systems" (II E), all l masses (59) except for m_{44} are [517] equal to zero, and only the gravitational force acting on it remains effective. *Thus, the total energy alone is the factor determining the gravitational mass of complete static systems.* Similarly as in Nordström's second theory, this result is caused by the compensation of the forces that act on the different parts of the system, and as there it can be applied to closed systems in static equilibrium.

Let us consider for example a hot gas which is enclosed in a cylinder. Forces act on the individual molecules of the gas, whose magnitudes are neither proportional to the energy of the molecules, nor have directions agreeing exactly with the vertical. Nevertheless, the gravitational forces acting on the molecules and on the volume elements of the cylinder sum to a resultant, whose temporal mean value is proportional to the total energy and has a vertical direction. The relation between gravity[2] and energy (respectively, inertia) are accordingly represented in Einstein's tensor theory to the same degree as in Nordström's scalar theory. However, one will have to agree with G. Mie,[28] when he denies that these theories manifest an "physical unity of essence" of gravitational and inertial mass.

In his detailed critique of Einstein's theory, G. Mie[28] attempts to prove that the proportionality of gravitational mass and energy for complete static systems occurs there only as the result of certain specific assumptions, which, incidentally, are said to contradict one another. In his reply,[29] Einstein states the belief—without refuting Mie's objections one by one—that he can trace those contradictions to Mie's demand of covariance only with respect to Lorentz transformations, thereby introducing "preferred systems of reference." That "special assumptions of any kind are not used" in the establishment of his theory, as Einstein asserts there, does not appear credible. In his just-published summary presentation[38] of the "general theory of relativity" Einstein derives the differential equations expressing his theory of gravitation from a variational principle, on the basis of certain specializations whose physical significance is however not elucidated.

The *integration of the field equations* (57) is extraordinarily difficult. Only the method of successive approximations promises success. In this one will usually take [518] as a first approximation the solution that treats the field I as static. Here, Einstein's theory becomes identical with Abraham's theory; therefore the solution of the problem of the sphere, for example, given in (II C), remains valid here as well.

In his Vienna lecture,[27] A. Einstein takes the normal values of the $g_{\mu\nu}$ as the first approximation:

$$g_{11} = g_{22} = g_{33} = 1, \qquad g_{44} = -c^2, \qquad g_{\mu\nu} = 0 \text{ for } \mu \neq \nu;$$

he considers the deviations $g^*_{\mu\nu}$ from these normal values as quantities of first order, and arrives, by neglecting quantities of second order, at the following differential equations:

$$\Box g^*_{\mu\nu} = T^m_{\mu\nu}. \tag{60}$$

For incoherent motions of masses, the last (T^m_{44}) among the components of the material tensor T^m is the most important; it determines the potential $g^*_{44} = \Phi^g$. Then follow the components $T^m_{14}, T^m_{24}, T^m_{34}$, which are of first order in v/c; these determine the potentials $g^*_{14}, g^*_{24}, g^*_{34}$, which can be viewed as the components of a space vector $-(1/c)\mathfrak{A}^g$. The remaining components of T^m are of second order in v/c. If one neglects quantities of this order, then one only needs to consider those four potentials, and obtains for them the differential equations

$$\square \Phi^g = c^2\mu, \tag{60a}$$

$$\square \mathfrak{A}^g = c^2\mu \cdot \frac{v}{c}, \tag{60b}$$

where μ is the mass density.

Here the analogy to electrodynamics catches one's eye. Except for the sign, the field equations (60a, b) agree with those that must be satisfied in the theory of electrons by the "electromagnetic potentials", the scalar one (Φ) and the vectorial one (A). In this approximation, the Einstein-Grossmann tensor theory of the gravitational field leads to the same results as the vector theory sketched in (I A). Also concerning the expression for the *gravitational force per unit mass*, there exists a far reaching analogy. Einstein obtains:

$$\mathfrak{F}^g = -\frac{1}{2}\mathrm{grad}\,\Phi^g - \frac{1}{c}\frac{\partial \mathfrak{A}^g}{\partial t} + \frac{1}{c}[v,\,\mathrm{curl}\,\mathfrak{A}^g]. \tag{61}$$

This expression agrees with the one (1) from the vector theory:

$$\mathfrak{F}^g = \mathfrak{E}^g + \frac{1}{c}[v\mathfrak{H}^g], \tag{61a}$$

I if one writes: [519]

$$\mathfrak{E}^g = -\frac{1}{2}\mathrm{grad}\,\Phi^g - \frac{1}{c}\frac{\partial \mathfrak{A}^g}{\partial t}, \tag{61b}$$

$$\mathfrak{H}^g = \mathrm{curl}\,\mathfrak{A}^g. \tag{61c}$$

Apart from the factor $1/2$ in the gradient of Φ^g, the formulas (61b, c) are identical with the well known formulas which derive the two vectors of the electromagnetic field from the electromagnetic potentials.

The Einstein-Grossmann theory of gravitation is related to the vector theory to the extent that the approximation applied is satisfactory. As in the vector theory, induced gravitational forces are generated on neighboring bodies by the acceleration of a body, forces that act in the direction of the acceleration. Einstein attaches great importance to the existence of these induced gravitational forces in connection with the so called "*hypothesis of the relativity of inertia.*"

This hypothesis, already advocated by E. Mach, states that the inertia of a body is only a consequence of the relative acceleration with respect to the totality of the remaining bodies. If this hypothesis applies, then the inertial mass of a body will depend on its position with respect to the remaining bodies. This is now the case in the Einstein's theory because the rest-mass

$$m_0 = \frac{M}{c} \qquad (M \text{ mass constant})$$

is inversely proportional to the speed of light c, whose value in a static gravitational field is reduced upon the approach of external masses. The inertia of a body is thus increased by the accumulation of mass in its vicinity; through this circumstance, Einstein believes himself justified in considering inertia as a consequence of the presence of the remaining masses in the context of that relativity hypothesis.

This consideration looses cogency however, if one calculates by how much c is decreased, and hence m_0 increased, by the masses of celestial objects. The mass of the Sun, at its surface, engenders a relative change in c of the order of 10^{-6}, which then decreases approximately inversely proportional to the distance from the center of the Sun. But in order for the mass of a material point to be considered as being a consequence of the presence of the remaining matter, it mass should vanish upon the removal of that matter. The mass to be removed from the vicinity of that point for that [520]   purpose would now be a millionfold larger than the I mass of the Sun. Accordingly, the hypothesis of the relativity of inertia can only be justified from the point of view of Einstein's theory, if, in addition to the masses of the visible bodies, one assumes also hidden masses. But with this, that hypothesis looses all concrete physical meaning. In the end it does not matter if one anchors the system of reference in "hidden masses," in the "aether" or in a "body α."

Upon the mutual approach of two bodies A and B, their inertial masses grow. But according to the hypothesis of the relativity of inertia, this increase in mass should not occur if A and B are accelerated in unison. Indeed, according to the theorem of the inertia of energy, the inertial resistance of the system of the two bodies should even be smaller than the sum of the inertial masses of the two bodies while separated, because the energy of the system decreases upon the approach of the two bodies. Here, the gravitational forces, induced by the temporal variation of the vector potential $\mathfrak{A}^g$, now come into play, forces that the bodies A and B exert on each other as a result of their acceleration. Through them, the additional inertial forces are overcompensated, so that the inertial mass of the system $A + B$ becomes smaller than the sum of the inertial masses of the bodies A and B.

Einstein sees a significant advantage of this tensor theory over the scalar theories in that the hypothesis of the relativity of inertia fits into his theory, because the scalar theories lack the vector potential $\mathfrak{A}^g$, from which the induced forces are derived. However, we saw that the theory of Einstein and Grossmann also fails to provide a satisfactory quantitative basis for Mach's bold hypothesis, unless one appeals to hidden masses. In view of the enormous complication engendered by the ten-fold multiplication of the gravitational potentials and by the distortion of the four-dimensional world, one is likely, for reasons of the Machian "economy of thought," to prefer the scalar theories, as long as the supposition, that there exist ten gravitational forces instead of one, is not supported by experience.

APPENDIX

Notation: [472]

t time, c the speed of light.

$x, y, z, u = ict$, the coordinates of four-dimensional space.

$$\Box \equiv \frac{\partial^2}{\partial x^2} + \frac{\partial^2}{\partial y^2} + \frac{\partial^2}{\partial z^2} + \frac{\partial^2}{\partial u^2}.$$

$\mathfrak{v}$ velocity, v its magnitude.

$\mathfrak{E}, \mathfrak{H}$ electric and magnetic vector.

$\mathfrak{F}^e$ force per unit charge.

ρ charge density.

μ mass density, ν rest-mass density.

M mass constant, m_0 rest-mass.

V volume.

$\mathfrak{E}^g, \mathfrak{H}^g$ gravitational vectors in the vector theory.

$\mathfrak{F}^g$ gravitational force per unit mass.

$\mathfrak{K}^g$ gravitational force, φ gravitational potential.

$\mathfrak{L}$ Lagrangian function. |

$\mathfrak{G}$ momentum.   [473]

E energy, E_0 rest-energy.

E^m, E^e, E^g energy of matter, of the electromagnetic and of the gravitational field.

T^m, T^e, T^g world tensors of matter, of the electromagnetic and of the gravitational field.

D^m, D^e, D^g sum of their diagonal components.

T, D resultant of world tensors and of diagonal sums respectively.

η^m, η^e, η^g energy density of matter, of the electromagnetic and of the gravitational field.

g^m, g^e, g^g the momentum density of matter, of the electromagnetic and of the gravitational field.

$\mathfrak{S}^m, \mathfrak{S}^e, \mathfrak{S}^g$ the energy flux of matter, of the electromagnetic and of the gravitational field.

$g_{\mu\nu}$ Einstein's gravitational potentials.

[470]

REFERENCES

1. V. Volterra, "Sul flusso di energia meccanica." *Nuovo Cimento* (4), X, 337, 1899.

2. H. Poincaré, "Sur la dynamique de l'électron." *Rendiconti del circolo matematico di Palermo,* XXI, 129, 1906 I.

3. H. Minkowski, "Die Grundgleichungen für die elektromagnetischen Vorgänge in bewegten Körpern." *Göttinger Nachrichten,* 1908, p. 53.

4. A. Sommerfeld, "Zur Relativitätstheorie." *Ann. d. Phys.,* 32, 749; 33, 649, 1910. I

[471] 5. A. Einstein, "Über den Einfluß der Schwerkraft auf die Ausbreitung des Lichtes." *Ann. d. Phys.,* 35, 898, 1911.

6. M. Abraham, "Sulla teoria della gravitazione." *Rendiconti della R. Acc. dei Lincei,* XX², 678, 1911; XXI¹, 27, 1912. German: *Physik. Zeitschr.,* 13, 1, 1912.

7. M. Abraham, "Sulla legge elementare della gravitazione." *Rendic. d. R. Acc. dei Lincei,* XXI¹, 94, 1912. German: *Physik. Zeitschr.,* 13, 4, 1912.

8. M. Abraham, "Sulla conservazione dell' energia e della materia nel campo gravitazionale." *Rendic. d. R. Acc. d. Lincei,* XXI¹, 432, 1912. German: *Physik. Zeitschr.,* 13, 311, 1912.

9. M. Abraham, "Sulla caduta libera." *Rondic d. R. Instituto Lombardo* (2), XLV, 290, 1912. German: *Physik. Zeitschr.,* 13, 310, 1912.

10. M. Abraham, "Sulle onde luminose e gravitazionali." *Nuovo Cimento* (6), III, 211, 1912.

11. A. Einstein, "Lichtgeschwindigkeit und Statik des Gravitationsfeldes." *Ann. d. Phys.,* 38, 355, 1912.

12. A. Einstein, "Zur Theorie des statischen Gravitationsfeldes." *Ann. d. Phys.,* 38, 443, 1912.

13. M. Abraham, "Relativität und Gravitation." *Ann. d. Phys.,* 38, 1056, 1912.

14. A. Einstein, "Relativität und Gravitation." *Ann d. Phys.,* 38, 1059, 1912.

15. M. Abraham, "Nochmals Relativität und Gravitation." *Ann. d. Phys.,* 39, 444, 1912.

16. M. Abraham, "Das Gravitationsfeld." *Physik. Zeitsch.,* 13, 793, 1912.

17. M. Abraham, "Una nuova teoria della gravitazione." *Nuovo Cimento* (6), IV, 459, 1912. German: *Archiv der Mathematik u. Physik* (3), XX, 193, 1912.

18. G. Pavanini, "Prime conseguence d'una recente teoria della gravitazione." *Rendic. d. R. Acc. dei Lincei,* XXI², 648, 1912; XXII¹, 369, 1913.

19. I. Ishiwara, "Zur Theorie der Gravitation." *Physik. Zeitschr.,* 13, 1189, 1912.

20. G. Nordström, "Relativitätsprinzip und Gravitation." *Physik. Zeitschr.,* 13, 1126, 1912.

21. G. Mie, "Grundlagen einer Theorie der Materie." Chapter V. *Ann. d. Phys.*, 40, 25, 1913.

22. G. Nordström, "Träge und schwere Masse in der Relativitätsmechanik." *Ann. d. Phys.*, 40, 856, 1913.

23. B. Caldonazzo, "Traiettorie dei raggi luminosi e dei punti materiali nel campo gravitazionale." *Nuovo Cimento* (6), V, 267, 1913.

24. A. Einstein and M. Grossmann, *Entwurf einer verallgemeinerten Relativitätstheorie und einer Theorie der Gravitation.* Leipzig, B. G. Teubner 1913.

25. M. Behacker, "Der freie Fall und die Planetenbewegung in Nordströms Gravitationstheorie." *Physik. Zeitschr.*, 14, 989, 1913.

26. G. Nordström, "Zur Theorie der Gravitation vom Standpunkt des Relativitätsprinzips." *Ann. d. Phys.*, 42, 533, 1913.

27. A. Einstein, "Zum gegenwärtigen Stande des Gravitationsproblems." *Physik. Zeitschr.*, 14, 1249, 1913. |

28. G. Mie, "Bemerkungen zu der Einsteinschen Gravitationstheorie." *Physik. Zeitschr.*, 15, 115 and 169, 1914.   [472]

29. A. Einstein, "Prinzipielles zur verallgemeinerten Relativitätstheorie und Gravitationstheorie." *Physik. Zeitschr.*, 15, 176, 1914.

30. I. Ishiwara, "Grundlagen einer relativistischen elektromagnetischen Gravitationstheorie." *Physik. Zeitschr.*, 15, 294, 1914.

31. G. Nordström, "Die Fallgesetze und Planetenbewegungen in der Relativitätstheorie." *Ann. d. Phys.*, 43, 1101, 1914.

32. G. Nordström, "Über den Energiesatz in der Gravitationstheorie." *Physik. Zeitschr.*, 15, 375, 1914.

33. G. Nordström, "Über die Möglichkeit, das elektromagnetische Feld und das Gravitationsfeld zu vereinigen." *Physik. Zeitschr.*, 15, 504, 1914.

34. A. Einstein and A. D. Fokker, "Die Nordströmsche Gravitationstheorie vom Standpunkt des absoluten Differentialkalküls." *Ann. d. Phys.*, 44, 321, 1914.

35. A. Einstein and M. Grossmann, "Kovarianzeigenschaften der Feldgleichungen der auf die verallgemeinerte Relativitätstheorie gegründeten Gravitationstheorie." *Zeitschr. f. Mathem. u. Phys.*, 63, 215, 1914.

36. E. Kretschmann, *Eine Theorie der Schwerkraft im Rahmen der ursprünglichen Einsteinschen Relativitätstheorie.* Doctoral Thesis. Berlin 1914.

37. G. Nordström, "Zur Elektrizitäts- und Gravitationstheorie." *Finska V. S. Förhandlingar*, LVII. A. No. 4. 1914.

38. A. Einstein, "Die formale Grundlage der allgemeinen Relativitätstheorie." *Sitzungsber. d. Berl. Ak.*, Nov. 1914.

EDITORIAL NOTES

[1] Here "*Schwere*" has been translated as "gravity," but elsewhere it has been translated as "weight," since in these other occurrences Abraham apparently uses "*Schwere*" and "*Gewicht*" interchangeably.

[2] Raised numbers in parentheses refer to the numbers of the above chronological list of references.

[3] Note that the wording Abraham uses here differs slightly from that of the earlier work to which he refers.

[4] The superscript ^{g} in T^g in eq. (36b) is missing in the original.

A FIELD THEORY OF GRAVITATION
IN THE FRAMEWORK
OF SPECIAL RELATIVITY

JOHN D. NORTON

EINSTEIN, NORDSTRÖM, AND THE EARLY DEMISE OF SCALAR, LORENTZ COVARIANT THEORIES OF GRAVITATION

1. INTRODUCTION

The advent of the special theory of relativity in 1905 brought many problems for the physics community. One, it seemed, would not be a great source of trouble. It was the problem of reconciling Newtonian gravitation theory with the new theory of space and time. Indeed it seemed that Newtonian theory could be rendered compatible with special relativity by any number of small modifications, each of which would be unlikely to lead to any significant deviations from the empirically testable consequences of Newtonian theory.[1] Einstein's response to this problem is now legend. He decided almost immediately to abandon the search for a Lorentz covariant gravitation theory, for he had failed to construct such a theory that was compatible with the equality of inertial and gravitational mass. Positing what he later called the principle of equivalence, he decided that gravitation theory held the key to repairing what he perceived as the defect of the special theory of relativity—its relativity principle failed to apply to accelerated motion. He advanced a novel gravitation theory in which the gravitational potential was the now variable speed of light and in which special relativity held only as a limiting case.

It is almost impossible for modern readers to view this story with their vision unclouded by the knowledge that Einstein's fantastic 1907 speculations would lead to his greatest scientific success, the general theory of relativity. Yet, as we shall see, in

---

1   In the historical period under consideration, there was no single label for a gravitation theory compatible with special relativity. The Einstein of 1907 would have talked of the compatibility of gravitation and the *principle* of relativity, since he then tended to use the term "principle of relativity" where we would now use "theory of relativity". See (CPAE 2, 254). Minkowski (1908, 90) however, talked of reform "in accordance with the world postulate." Nordström (1912, 1126), like Einstein, spoke of "adapting ... the theory of gravitation to the principle of relativity" or (Nordström 1913, 872) of "treating gravitational phenomena from the standpoint of the theory of relativity," emphasizing in both cases that he planned to do so retaining the constancy of the speed of light in order to distinguish his work from Einstein's and Abraham's. For clarity I shall describe gravitation theories compatible with special relativity by the old-fashioned but still anachronistic label "Lorentz covariant." It describes exactly the goal of research, a gravitation theory whose equations are covariant under Lorentz transformation. For a simplified presentation of the material in this chapter, see also (Norton 1993).

Jürgen Renn (ed.). *The Genesis of General Relativity*, Vol. 3
Gravitation in the Twilight of Classical Physics: Between Mechanics, Field Theory, and Astronomy.
© 2007 Springer.

1907 Einstein had only the slenderest of grounds for judging all Lorentz covariant gravitation theories unacceptable. His 1907 judgement was clearly overly hasty. It was found quite soon that one could construct Lorentz covariant gravitation theories satisfying the equality of inertial and gravitational mass without great difficulty. Nonetheless we now do believe that Einstein was right in so far as a thorough pursuit of Lorentz covariant gravitation theories does lead us inexorably to abandon special relativity. In the picturesque wording of Misner et al. (1973, Ch.7) "gravity bursts out of special relativity."

These facts raise some interesting questions. As Einstein sped towards his general theory of relativity in the period 1907–1915 did he reassess his original, hasty 1907 judgement of the inadequacy of Lorentz covariant gravitation theories? In particular, what of the most naturally suggested Lorentz covariant gravitation theory, one in which the gravitational field was represented by a scalar field and the differential operators of the Newtonian theory were replaced by their Lorentz covariant counterparts? Where does this theory lead? Did the Einstein of the early 1910s have good reason to expect that developing this theory would lead outside special relativity?

This paper provides the answers to these questions. They arise in circumstances surrounding a gravitation theory, developed in 1912–1914, by the Finnish physicist Gunnar Nordström. It was one of a number of more conservative gravitation theories advanced during this period. Nordström advanced this most conservative scalar, Lorentz covariant gravitation theory and developed it so that it incorporated the equality of inertial and gravitation mass. It turned out that even in this most conservative approach, odd things happened to space and time. In particular, the lengths of rods and the rates of clocks turn out to be affected by the gravitational field, so that the spaces and times of the theory's background Minkowski spacetime ceased to be directly measurable. The *dénouement* of the story came in early 1914. It was shown that this conservative path led to the same sort of gravitation theory as did Einstein's more extravagant speculations on generalizing the principle of relativity. It lead to a theory, akin to general relativity, in which gravitation was incorporated into a dynamical spacetime background. If one abandoned the inaccessible background of Minkowski spacetime and simply assumed that the spacetime of the theory was the one revealed by idealized rod and clock measurements, then it turned out that the gravitation theory was actually the theory of a spacetime that was only conformally flat—gravitation had burst out of special relativity. Most strikingly the theory's gravitational field equation was an equation strongly reminiscent to modern readers of the field equations of general relativity:

$$R = kT$$

where R is the Riemann curvature scalar and T the trace of the stress-energy tensor. This equation was revealed before Einstein had advanced the generally covariant field equations of general relativity, at a time in which he believed that no such field equations could be physically acceptable.

What makes the story especially interesting are the two leading players other than Nordström. The first was Einstein himself. He was in continued contact with Nordström during the period in which the Nordström theory was developed. We shall see that the theory actually evolved through a continued exchange between them, with Einstein often supplying ideas decisive to the development of the theory. Thus the theory might more accurately be called the "Einstein-Nordström theory." Again it was Einstein in collaboration with Adriaan Fokker who revealed in early 1914 the connection between the theory and conformally flat spacetimes.

The second leading player other than Nordström was not a person but a branch of special relativity, the relativistic mechanics of stressed bodies. This study was under intensive development at this time and had proven to be a locus of remarkably non-classical results. For example it turned out that a moving body would acquire additional energy, inertia and momentum simply by being subjected to stresses, even if the stresses did not elastically deform the body. The latest results of these studies—most notably those of Laue—provided Einstein and Nordström with the means of incorporating the equality of inertial and gravitational mass into their theory. It was also the analysis of stressed bodies within the theory that led directly to the conclusion that even idealized rods and clocks could not measure the background Minkowski spacetime directly but must be affected by the gravitational field. For Einstein and Nordström concluded that a body would also acquire a gravitational mass if subjected to non-deforming stresses and that one had to assume that such a body would alter its size in moving through the gravitational field on pain of violating the law of conservation of energy.

Finally we shall see that the requirement of equality of inertial and gravitational mass is a persistent theme of Einstein's and Nordström's work. However the requirement proves somewhat elastic with both Einstein and Nordström drifting between conflicting versions of it. It will be convenient to prepare the reader by collecting and stating the relevant versions here. On the observational level, the equality could be taken as requiring:

- Uniqueness of free fall:[2] The trajectories of free fall of all bodies are independent of their internal constitution.

Einstein preferred a more restrictive version:

- Independence of vertical acceleration: The vertical acceleration of bodies in free fall is independent of their constitutions and horizontal velocities.

In attempting to devise theories compatible with these observational requirements, Einstein and Nordström considered requiring equality of gravitational mass with

- inertial rest mass
- the inertial mass of closed systems
- the inertial mass of complete static systems
- the inertial mass of a a complete stationary systems[3]

---

2   This name is drawn from (Misner et al. 1973, 1050).

3   The notions of complete static and complete stationary systems arise in the context of the mechanics of stressed bodies and are discussed in Sections 9 and 12 below.

More often than not these theoretical requirements failed to bring about the desired observational consequences. Unfortunately it is often unclear precisely which requirement is intended when the equality of inertial and gravitational mass was invoked.

2. THE PROBLEM OF GRAVITATION IMMEDIATELY AFTER 1905

In the years immediately following 1905 it was hard to see that there would be any special problem in modifying Newtonian gravitation theory in order to bring it into accord with the special theory of relativity. The problem was not whether it could be done, but how to choose the best of the many possibilities perceived, given the expectation that relativistic corrections to Newtonian theory might not have measurable consequences even in the very sensitive domain of planetary astronomy. Poincaré (1905, 1507–1508;1906, 166–75), for example, had addressed the problem in his celebrated papers on the dynamics of the electron. He limited himself to seeking an expression for the gravitational force of attraction between two masses that would be Lorentz covariant[4] and would yield the Newtonian limit for bodies at rest. Since this failed to specify a unique result he applied further constraints including the requirement[5] of minimal deviations from Newtonian theory for bodies with small velocities, in order to preserve the Newtonian successes in astronomy. The resulting law, Poincaré noted, was not unique and he indicated how variants consistent with its constraints could be derived by modifying the terms of the original law.

Minkowski (1908, 401–404; 1909, 443–4) also sought a relativistic generalization of the Newtonian expression for the gravitational force acting between two bodies. His analysis was simpler than Poincaré's since merely stating his law in terms of the geometric structures of his four dimensional spacetime was sufficient to guarantee automatic compatibility with special relativity. Where Poincaré (1905, 1508; 1906, 175) had merely noted his expectation that the deviations from Newtonian astronomical prediction introduced by relativistic corrections would be small, Minkowski (1908, 404) computed the deviations due to his law for planetary motions and concluded that they were so small that they allowed no decision to be made concerning the law.

Presumably neither Poincaré nor Minkowski were seeking a fundamental theory of gravitation, for they both considered action-at-a-distance laws at a time when field theories were dominant. Rather the point was to make *plausible*[6] the idea that some slight modification of Newtonian gravitational law was all that was necessary to bring it into accord with special relativity, even if precise determination of that modification was beyond the reach of the current state of observational astronomy.

---

4   More precisely he required that the law governing propagation of gravitational action be Lorentz covariant and that the gravitational forces transform in the same way as electromagnetic forces.
5   Also he required that gravitational action propagate forward in time from a given body.
6   The word is Minkowski's. He introduced his treatment of gravitation (Minkowski 1908, 401) with the remark "I would not like to fail to make it plausible that nothing in the phenomena of gravitation can be expected to contradict the assumption of the postulate of relativity."

3. EINSTEIN'S 1907 REJECTION OF LORENTZ COVARIANT GRAVITATION THEORIES

In 1907 Einstein's attention was focussed on the problem of gravitation and relativity theory when he agreed to write a review article on relativity theory for Johannes Stark's *Jahrbuch der Radioaktivität und Elektronik*. The relevant parts of the review article (Einstein 1907a, 414; Section V, 454–62) say nothing of the possibility of a Lorentz covariant gravitation theory. Rather Einstein speculates immediately on the possibility of extending the principle of relativity to accelerated motion. He suggests the relevance of gravitation to this possibility and posits what is later called the principle of equivalence as the first step towards the complete extension of the principle of relativity.

It is only through later reminiscences that we know something of the circumstances leading to these conclusions. The most informative are given over 25 years later in 1933 when Einstein gave a sketch of his pathway to general relativity.[7] In it he wrote (Einstein 1933, 286–87):

> I came a step nearer to the solution of the problem [of extending the principle of relativity] when I attempted to deal with law of gravity within the framework of the special theory of relativity. Like most writers at the time, I tried to frame a *field-law* for gravitation, since it was no longer possible, at least in any natural way, to introduce direct action at a distance owing to the abolition of the notion of absolute simultaneity.

> The simplest thing was, of course, to retain the Laplacian scalar potential of gravity, and to complete the equation of Poisson in an obvious way by a term differentiated with respect to time in such a way that the special theory of relativity was satisfied. The law of motion of the mass point in a gravitational field had also to be adapted to the special theory of relativity. The path was not so unmistakably marked out here, since the inert mass of a body might depend on the gravitational potential. In fact this was to be expected on account of the principle of the inertia of energy.

While Einstein's verbal description is brief, the type of gravitation theory he alludes to is not too hard to reconstruct. In Newtonian gravitation theory, with a scalar potential ϕ, mass density ρ and G the gravitation constant, the gravitational field equation—the "equation of Poisson"— is[8]

$$\nabla^2\phi = \left(\frac{\partial^2}{\partial x^2} + \frac{\partial^2}{\partial y^2} + \frac{\partial^2}{\partial z^2}\right)\phi = \frac{\partial^2\phi}{\partial x_i \partial x_i} = 4\pi G\rho. \tag{1}$$

The force $\mathbf{f}$, with components (f_1, f_2, f_3), on a point mass m is given by $-m\nabla\phi$ so that

$$f_i = m\frac{dv_i}{dt} = -m\left(\frac{\partial}{\partial x_i}\right)\phi \tag{2}$$

---

7   A similar account is given more briefly in (Einstein 1949, 58–63).
8    $(x, y, z) = (x_1, x_2, x_3)$ are the usual spatial Cartesian coordinates. The index i ranges over 1, 2, 3. Here and henceforth, summation over repeated indices is implied.

is the law of motion of a point mass m with velocity $v_i = dx_i/dt$ in the gravitational field ϕ.

The adaptation of (1) to special relativity is most straightforward. The added term, differentiated with respect to the time coordinate, converts the Laplacian operator ∇^2 into a Lorentz covariant d' Alembertian $\Box^2$. so that the field equation alluded to by Einstein would be

$$\Box^2\phi = \left(\frac{1}{c^2}\frac{\partial^2}{\partial t^2} - \nabla^2\right)\phi = -4\pi G\nu. \tag{3}$$

For consistency ϕ is assumed to be Lorentz invariant and the mass density ρ must be replaced with a Lorentz invariant, such as the rest mass density ν used here.

The modification to the law of motion of a point mass is less clear. The natural Lorentz covariant extension of (2) is most obvious if we adopt the four dimensional spacetime methods introduced by Minkowski (1908). Einstein could not have been using these methods in 1907. However I shall write the natural extension here since Einstein gives us little other guide to the form of the equation he considered, since the properties of this equation fit exactly with Einstein's further remarks and since this equation will lead us directly to Nordström's work. The extension of (2) is

$$F_\mu = m\frac{dU_\mu}{d\tau} = -m\frac{\partial\phi}{\partial x_\mu} \tag{4}$$

where F_μ is the four force on a point mass with rest mass m, $U_\mu = dx_\mu/d\tau$ is its four velocity, τ is the proper time and $\mu = 1, 2, 3, 4$.[9] Following the practice of Nordström's papers, the coordinates are $(x_1, x_2, x_3, x_4) = (x, y, z, u = ict)$, for c the speed of light.

Simple as this extension is, it turns out to be incompatible with the kinematics of a Minkowski spacetime. In a Minkowski spacetime, the constancy of c entails that the four velocity U_μ along a world line is orthogonal to the four acceleration $dU_\mu/d\tau$. For we have $c^2 d\tau^2 = -dx_\mu dx_\mu$, so that $c^2 = -U_\mu U_\mu$ and the orthogonality now follows from the constancy of c

$$-\frac{1}{2}\frac{dc^2}{d\tau} = U_\mu\frac{dU_\mu}{d\tau} = 0 \tag{5}$$

(4) and (5) together entail

$$F_\mu U_\mu = -m\frac{\partial\phi}{\partial x_\mu}\frac{dx_\mu}{d\tau} = -m\frac{d\phi}{d\tau} = 0$$

so that the law (4) can only obtain in a Minkowski spacetime in the extremely narrow case in which the field ϕ is constant along the world line of the particle, i.e.

---

9   Throughout this paper, Latin indices $i, k, \ldots$ range over $1, 2, 3$ and Greek indices $\mu, \nu, \ldots$ range over $1, 2, 3, 4$.

$$\frac{d\phi}{d\tau} = 0. \tag{6}$$

We shall see below that one escape from this problem published by Nordström involves allowing the rest mass m to be a function of the potential ϕ. Perhaps this is what Einstein referred to above when he noted of the law of motion that the "path was not so unmistakably marked out here, since the inert mass of a body might depend on the gravitational potential."

Whatever the precise form of the modifications Einstein made, he was clearly unhappy with the outcome. Continuing his recollections, he noted:

> These investigations, however, led to a result which raised my strong suspicions. According to classical mechanics, the vertical acceleration of a body in the vertical gravitational field is independent of the horizontal component of its velocity. Hence in such a gravitational field the vertical acceleration of a mechanical system or of its center of gravity works out independently of its internal kinetic energy. But in the theory I advanced, the acceleration of a falling body was not independent of its horizontal velocity or the internal energy of the system.

The result Einstein mentions here is readily recoverable from the law of motion (4) in a special case in which it is compatible with the identity (5). The result has more general applicability, however. The modifications introduced by Nordström to render (4) compatible with (5) vanish in this special case, as would, presumably, other natural modifications that Einstein may have entertained. So this special case is also a special case of these more generally applicable laws.

We consider a coordinate system in which:

(i) the field is time independent ($\partial\phi/\partial t = 0$) at some event and

(ii) the motion of a point mass m in free fall at that event is such that the "vertical" direction of the field, as given by the acceleration three vector dv_i/dt, is perpendicular to the three velocity v_i, so that

$$v_i \cdot \frac{dv_i}{dt} = 0 \tag{7}$$

and the point's motion is momentarily "horizontal."

Condition (7) greatly simplifies the analysis, since it entails that the t derivative of any function of $v^2 = v_i v_i$ vanishes, so that we have

$$\frac{d}{dt}\frac{dt}{d\tau} = \frac{d}{dt}\left(\frac{1}{\sqrt{1 - \frac{v^2}{c^2}}}\right) = 0. \tag{8}$$

Notice also that in this case (7) entails that $v_i \frac{\partial\phi}{\partial x_i} = 0$ so that

$$\frac{d\phi}{d\tau} = \frac{dt}{d\tau}\left(\frac{\partial\phi}{\partial t} + v_i\frac{\partial\phi}{\partial x_i}\right) = 0$$

and (6) and then also (5) are satisfied for this special case. Finally an expression for the acceleration of the point mass now follows directly from (4) and is[10]

$$\frac{dv_i}{dt} = -\left(1 - \frac{v^2}{c^2}\right)\frac{\partial\phi}{\partial x_i}. \tag{9}$$

According to (9), the greater the horizontal velocity v, the less the vertical acceleration, so that this acceleration is dependent on the horizontal velocity as Einstein claimed.

Einstein also claims in his remarks that the vertical acceleration would not be independent of the internal energy of the falling system. This result is suggested by equation (9), which tells us that the vertical acceleration of a point mass diminishes with its kinetic energy if the velocity generating that kinetic energy is horizontally directed. If we apply this result to the particles of a kinetic gas, we infer that in general each individual particle will fall slower the greater its velocity. Presumably this result applies to the whole system of a kinetic gas so that the gas falls slower the greater the kinetic energy of its particles, that is, the greater its internal energy. This example of a kinetic gas was precisely the one given by Einstein in an informal lecture on April 14, 1954 in Princeton according to lecture notes taken by J. A. Wheeler.[11]

Einstein continued his recollections by explaining that he felt these results so contradicted experience that he abandoned the search for a Lorentz covariant gravitation theory.

> This did not fit with the old experimental fact that all bodies have the same acceleration in a gravitational field. This law, which may also be formulated as the law of the equality of inertial and gravitational mass, was now brought home to me in all its significance. I was in the highest degree amazed at its existence and guessed that in it must lie the key to a deeper understanding of inertia and gravitation. I had no serious doubts about its strict validity even without knowing the results of the admirable experiments of Eötvös, which—if my memory is right—I only came to know later. I now abandoned as inadequate the attempt to treat the problem of gravitation, in the manner outlined above, within the framework of the special theory of relativity. It clearly failed to do justice to the most fundamental property of gravitation.

Einstein then recounted briefly the introduction of the principle of equivalence, upon which would be based his continued work on gravitation and relativity, and concluded

---

10 Since $\frac{dU_i}{d\tau} = \frac{dt}{d\tau}\frac{d}{dt}\left(v_i\frac{dt}{d\tau}\right)$, (9) follows directly from (4) using (8).

11 Wheeler's notes read "I had to write a paper about the content of special relativity. Then I came to the question how to handle gravity. The object falls with a different acceleration if it is moving than if it is not moving. ... Thus a gas falls with another acceleration if heated than if not heated. I felt this is not true ..." (Wheeler 1979, 188).

> Such reflections kept me busy from 1908 to 1911, and I attempted to draw special con-
> clusions from them, of which I do not propose to speak here. For the moment the one
> important thing was the discovery that a reasonable theory of gravitation could only be
> hoped for from an extension of the principle of relativity.

Our sources concerning Einstein's 1907 renunciation of Lorentz covariant gravitation theories are largely later recollections so we should be somewhat wary of them.
Nonetheless they all agree in the essential details:[12] Einstein began his attempts to
discover a Lorentz covariant theory of gravitation as a part of his work on his 1907
Jahrbuch review article. He found an inconsistency between these attempts and the
exact equality of inertial and gravitational mass, which he found sufficiently disturbing to lead him to abandon the search for such theories.

We shall see shortly that Einstein's 1907 evaluation and dismissal of the prospects
of a Lorentz covariant gravitation theory—as reconstructed above—was far too hasty.
Within a few years Einstein himself would play a role in showing that one could construct a Lorentz covariant gravitation theory that was fully compatible with the exact
equality of inertial and gravitational mass. We can understand why Einstein's 1907
analysis would be hurried, however, once we realize that the he could have devoted
very little time to contemplation of the prospects of a Lorentz covariant gravitation
theory. He accepted the commission of the *Jahrbuch*'s editor, Stark, to write the
review in a letter of September 25, 1907 (EA 22 333) and the lengthy and completed
article was submitted to the journal on December 4, 1907, a little over two months
later. This period must have been a very busy one for Einstein. As he explained to
Stark in the September 25 letter, he was not well read in the current literature perti-
nent to relativity theory, since the library was closed during his free time. He asked
Stark to send him relevant publications that he might not have seen.[13] During this
period, whatever time Einstein could have spent privately contemplating the prospects of a Lorentz covariant gravitation theory would have been multiply diluted.
There were the attractions of the principle of equivalence, whose advent so dazzled
him that he called it the "happiest thought of [his] life".[14] Its exploitation attracted all
the pages of the review article which concern gravitation and in which the prospects
of a Lorentz covariant gravitation theory are not even mentioned. Further diluting his
time would be the demands of the remaining sections of the review article. The section devoted to gravitation filled only nine of the article's fifty two pages. Finally, of

---

12 See also the 1920 recollections of Einstein on p. 23, "Grundgedanken und Methoden der Relativitätstheorie in ihrer Entwicklung dargestellt," unpublished manuscript, control number 2 070, Duplicate
Einstein Archive, Mudd Manuscript Library, Princeton, NJ. (Henceforth "EA 2 070".) Einstein
recalls:
"When, in the year 1907, I was working on a summary essay concerning the special theory of relativity for the *Jahrbuch für Radioaktivität und Elektronik* [sic], I had to try to modify Newton's theory of
gravitation in such a way that it would fit into the theory [of relativity]. Attempts in this direction
showed the possibility of carrying out this enterprise, but they did not satisfy me because they had to
be supported by hypotheses without physical basis." Translation from (Holton 1975, 369–71).
13 Einstein thanked him for sending papers in a letter of October 4, 1907 (EA, 22 320).
14 In Einstein's 1920 manuscript (EA 2 070, 23–25).

course, there were the obligations of his job at the patent office. It is no wonder that
he lamented to Stark in a letter of November 1, 1907, that he worked on the article in
his "unfortunately truly meagerly measured free time" (EA, 22 335).

4. EINSTEIN'S ARGUMENT OF JULY 1912

If the Einstein of 1907 had not probed deeply the prospects of Lorentz covariant
gravitation theories, we might well wonder if he returned to give the problem more
thorough treatment in the years following. We have good reason to believe that as late
as July 1912, Einstein had made no significant advance on his deliberations of
1907.[15] Our source is an acrimonious dispute raging at this time between Einstein
and Max Abraham. In language that rarely appeared in the unpolluted pages of *Annalen der Physik*, Abraham (1912c, 1056) accused Einstein's theory of relativity of having "exerted an hypnotic influence especially on the youngest mathematical
physicists which threatened to hamper the healthy development of theoretical physics." He rejoiced especially in what he saw as major retractions in Einstein's latest
papers on relativity and gravitation. Einstein (1911) involved a theory of gravitation
which gave up the constancy of the velocity of light and Einstein (1912a, 1912b) even
dispensed with the requirement of the invariance of the equations of motion under
Lorentz transformation. These concessions, concluded Abraham triumphantly, were
the "death blow" for relativity theory.

Einstein took this attack very seriously. His correspondence from this time, a simple gauge of the focus of his thoughts, was filled with remarks on Abraham. He
repeatedly condemned Abraham's (1912a, 1912b) new theory of gravitation, which
had adopted Einstein's idea of a variable speed of light as the gravitational potential.
"A stately beast that lacks three legs," he wrote scathingly of the theory to Ludwig
Hopf.[16] He anticipated the dispute with Abraham with some relish, writing to Hopf
earlier of the coming "difficult ink duel."[17] The public dispute ended fairly quickly,
however, with Einstein publishing a measured and detailed reply (Einstein 1912d)
and then refusing to reply to Abraham's rejoinder (Abraham 1912d). Instead Einstein
published a short note (Einstein 1912e) indicating that both parties had stated their
views and asking readers not to interpret Einstein's silence as agreement. Nonetheless Einstein continued to hold a high opinion of Abraham as a physicist, lamenting
in a letter to Hopf that Abraham's theory was "truly superficial, contrary to his [Abra-
ham's] usual practice."[18]

---

15 This is a little surprising. Einstein had neglected gravitation in 1908–1911, possibly because of his
    preoccupation with the problem of quanta. (See Pais (1983, 187–90.)) However he had returned to
    gravitation with vigor with his June 1911 submission of Einstein (1911) and by July 1912, the time of
    his dispute with Abraham, he had completed at least two more novel papers on the subject, (Einstein
    1912a, 1912b), and possibly a third, (Einstein 1912c).
16 Einstein to Ludwig Hopf, 16 August 1912, (EA 13 288).
17 Einstein to Ludwig Hopf, December 1911 (?), (EA 13 282).

Under these circumstances, Einstein had every incentive to make the best case for his new work on gravitation. In particular, we would expect Einstein to advance the best arguments available to him to justify his 1907 judgement of the untenability of Lorentz covariant gravitation theories, for it was this conclusion that necessitated the consideration of gravitation theories that went beyond special relativity. What he included in his response shows us that as late as July 4, 1912—the date of submission of his response (Einstein 1912d)—his grounds for this judgement had advanced very little beyond those he recalled having in 1907. He wrote (pp. 1062–63)

> One of the most important results of the theory of relativity is the realization that every energy E possesses an inertia (E/c^2) proportional to it. Since each inertial mass is at the same time a gravitational mass, as far as our experience goes, we cannot help but ascribe to each energy E a gravitational mass E/c^2. [19] From this it follows immediately that gravitation acts more strongly on a moving body than on the same body in case it is at rest.
>
> If the gravitational field is to be interpreted in the sense of our current theory of relativity, this can happen only in two ways. One can conceive of the gravitation vector either as a four-vector or a six-vector. For each of these two cases there are transformation formulae for the transition to a uniformly moving reference system. By means of these transformation formulae and the transformation formulae for ponderomotive forces one can find for both cases the forces acting on moving material points in a static gravitational field. However from this one arrives at results which conflict with the consequences mentioned of the law of the gravitational mass of energy. Therefore it seems that the gravitation vector cannot be incorporated without contradiction in the scheme of the current theory of relativity.

Einstein's argument is a fairly minor embellishment of the reflections summarized in Section 3 above. Einstein has replaced a single theory, embodied in equations such as (3) and (4), with two general classes of gravitation theory, the four-vector and six-vector theory. In both classes of gravitation theory, in the case of moving masses, Einstein claims that the gravitational field fails to act on them in proportion to their total energy, in effect violating the requirement of equality of inertial and gravitational mass.

---

18 Einstein to Ludwig Hopf, 12 June 1912, (EA 13 286). Einstein retained his high opinion of Abraham as a physicist. Late the following year, after his work had advanced into the first sketch of the general theory of relativity, Einstein conceded to his confidant Besso that "Abraham has the most understanding [of the new theory]." Einstein to Michele Besso, end of 1913, in (Speziali 1972, 50). For further mention of Abraham in correspondence from this period see Einstein to Heinrich Zangger, 27 January 1912, (EA 39 644); Einstein to Wilhelm Wien, 27 January 1912, (EA 23 548); Einstein to Heinrich Zangger, 29 February 1912, (EA 39 653); Einstein to Wilhelm Wien, 24 February 1912, (EA 23 550); Einstein to Heinrich Zangger, 20 May 1912, (EA 39 655); Einstein to Michele Besso, 26 March 1912, (EA 7 066); Einstein to Heinrich Zangger, summer 1912, (EA 39 657); Einstein to Arnold Sommerfeld, 29 October 1912, (EA 21 380). See also (Pais 1982, 231–32).

19 At this point, Einstein inserts the footnote:
    Hr. Langevin has orally called my attention to the fact that one comes to a contradiction with experience if one does not make this assumption. That is, in radioactive decay large quantities of energy are given off, so that the *inertial* mass of the matter must diminish. If the gravitational mass were not to diminish proportionally, then the gravitational acceleration of bodies made out of different elements would have to be demonstrably different in the same gravitational field.

Einstein does not give a full derivation of the result claimed. However we can reconstruct what he intended from the derivation sketch given. The two types of force fields correspond to the "spacetime vectors type I and II" introduced by Minkowski (1908, §5), which soon came to be known as four- and six-vector fields, respectively (Sommerfeld 1910, 750). They represented the two types of force fields then examined routinely in physics. The four-vector corresponds to the modern vector of a four-dimensional manifold. The gravitational four-force F_μ acting on a body with rest mass m in a four-vector theory is

$$F_\mu = mG_\mu. \tag{10a}$$

An example of such a theory is given by (4) above in which the gravitation four-vector G_μ is set equal to $-\partial\phi/\partial x_\mu$. The six-vector corresponds to our modern antisymmetric second rank tensor which has six independent components. The classic example of a six-vector is what Sommerfeld called "the six-vector ... of the electromagnetic field" (Sommerfeld 1910, 754). We would now identify it as the Maxwell field tensor. Presumably Einstein intended a six-vector gravitation theory to be modelled after electrodynamics, so that the gravitational four-force F_μ acting on a body with rest mass m and four-velocity U_ν in such a theory would be given by

$$F_\mu = mG_{\mu\nu}U_\nu. \tag{11a}$$

The gravitation six-vector, $G_{\mu\nu}$, satisfies the antisymmetry condition $G_{\mu\nu} = -G_{\nu\mu}$. This antisymmetry guarantees compatibility with the identity (5) since it forces $F_\mu U_\mu = 0$.

Einstein claims that one needs only the transformation formulae for four and six-vectors and for ponderomotive forces. to arrive at the results. However, since both (10a) and (11a) are Lorentz covariant, application of the transformation formulae to these equations simply returns equations of identical form—an uninformative outcome. We do recover results of the type Einstein claims, however, if we apply these transformation formulae to non-covariant specializations of (10a) and (11a).

We consider arbitrary four and six-vector gravitational fields G_μ and $G_{\mu\nu}$. In each there is a body of mass m in free fall. In each case, select and orient a coordinate system $S'(x', y', z', u' = ict')$ in such a way that each mass is instantaneously at rest and is accelerating only in $y' = x'_2$ direction. For these coordinate systems, the three spatial components of the four-force, F'_i, are equal to the three components of the three force, f'_i, acting on the masses. In particular the $x'_2 = y'$ component f'_2 of the three force is given in each case by

$$f'_2 = mG'_2, \tag{10b}$$

$$f'_2 = mG'_2{}_\nu U'_\nu = mG'_{24} ic \tag{11b}$$

since $U'_\mu = (0, 0, 0, ic)$. If we now transform from S' to a reference system $S(x, y, z, u = ict)$ moving at velocity v in the $x'_1 = x'$ direction S', then the relevant Lorentz transformation formulae are

$$f'_2 = \frac{f_2}{\sqrt{1 - \dfrac{v^2}{c^2}}}, \qquad G'_{24} = \frac{G_{24} - (iv/c)G_{21}}{\sqrt{1 - \dfrac{v^2}{c^2}}}.$$

Thus far we have not restricted the choice of four or six-vector fields G_μ and $G_{\mu\nu}$. The considerations that follow are simplified if we consider a special case of the six-vector field $G_{\mu\nu}$ in which $G_{21} = 0$. [20] Substituting with these transformation formulae for f'_2 and G'_{24} in this special in (10b) and (11b), we recover

$$f_2 = m\sqrt{1 - \frac{v^2}{c^2}}G_2 \qquad (10c)$$

$$f_2 = mG_{24}ic. \qquad (11c)$$

These two equations describe the component of gravitational three-force, f_2, in the "vertical" $x_2 = y$ direction on a mass m moving with velocity v in the "horizontal" $x_1 = x$ direction. In his 1912 argument, Einstein noted that the inertia of energy and the equality of inertial and gravitational mass leads us to expect that "gravitation acts more strongly on a moving body than on the same body in case it is at rest." We read directly from equations (10c) and (11c) that both four and six-vector theories fail to satisfy this condition. The gravitational force is independent of velocity in the six-vector case and actually decreases with velocity in the four-vector case. To meet Einstein's requirements, the gravitational force would need to increase with velocity, in direct proportion to the mass's energy $mc^2/\sqrt{1 - (v^2/c^2)}$.

We can also confirm that (10c) and (11c) lead to the result that the vertical acceleration of the masses is not independent of their horizontal velocities. To see this, note that, were the masses of (10c) and (11c) instantaneously at rest, the vertical forces exerted by the two fields would be respectively

$$f_2^{\text{rest}} = mG_2, \quad f_2^{\text{rest}} = mG_{24}ic.$$

In all cases, the three velocity and three-accelerations are perpendicular, so that condition (8) holds. Therefore we have

---

20   This restriction does not compromise the generality of Einstein's claim. If a Lorentz covariant theory proves inadequate in a special case, that is sufficient to demonstrate its general inadequacy. A natural instance of a six-vector field $G_{\mu\nu}$ in which $G_{21} = 0$ is easy to construct. Following the model of electromagnetism, we assume that $G_{\mu\nu}$ is generated by a vector potential A_μ, according to

$$G_{\mu\nu} = \frac{\partial A_\mu}{\partial x_\nu} - \frac{\partial A_\nu}{\partial x_\mu}.$$

We choose a "gravito-static" field in $S(x, y, z, u)$, that is, one that is analogous to the electrostatic field, by setting $A_\mu = (0, 0, 0, A_4)$. Since A_1 and A_2 are everywhere vanishing, $G_{21} = 0$. Finally note that Einstein does explicitly restrict his 1912 claim to static gravitational fields. Perhaps he also considered simplifying special examples of this type.

$$f_i = \frac{1}{\sqrt{1 - \dfrac{v^2}{c^2}}} m \frac{dv_i}{dt}, \quad f_i^{\text{rest}} = m \left(\frac{dv_i}{dt} \right)^{\text{rest}}.$$

Combining these results with (10c) and (11c) we recover expressions for the vertical acceleration dv_2/dt of the masses in terms of the acceleration $(dv_2/dt)^{\text{rest}}$ they would have had if they had no horizontal velocity

$$\frac{dv_2}{dt} = \left(1 - \frac{v^2}{c^2} \right) \left(\frac{dv_2}{dt} \right)^{\text{rest}} \tag{10d}$$

$$\frac{dv_2}{dt} = \sqrt{1 - \frac{v^2}{c^2}} \left(\frac{dv_2}{dt} \right)^{\text{rest}} \tag{11d}$$

We see that in both four and six-vector cases the vertical acceleration decreases with horizontal velocity, with equation (10d) generalizing the result in equation (9).

5. A GRAVITATION THEORY MODELLED AFTER MAXWELL'S ELECTROMAGNETISM?

Einstein's mention of a six-vector theory of gravitation in his 1912 response to Abraham raises the question of Einstein's attitude to a very obvious strategy of relativization of Newtonian gravitation theory. With hindsight one can view the transition from the theory of Coulomb electrostatic fields to full Maxwell electromagnetism as the first successful relativization of a field theory. Now Newtonian gravitation theory is formally identical to the theory of electrostatic fields excepting a change of sign needed to ensure that gravitational masses attract where like electric charges repel. This suggests that one can relativize Newtonian gravitation theory by augmenting it to a theory formally identical to Maxwell theory excepting this same change of sign.

While it is only with hindsight that one sees the transition from electrostatics to electromagnetism as a relativization, Einstein had certainly developed this hindsight by 1913. In his (Einstein 1913, 1250) he noted that Newtonian theory has sufficed so far for celestial mechanics because of the smallness of the speeds and accelerations of the heavenly bodies. Were these motions to be governed instead by electric forces of similar magnitude, one would need only Coulomb' s law to calculate these motions with great accuracy. Maxwell's theory would not be required. The problem of relativizing gravitation theory, Einstein continued, corresponded exactly to this problem: if we knew only experimentally of electrostatics but that electrical action could not propagate faster than light, would we be able to develop Maxwell electromagnetics? In the same paper Einstein proceeded to show (p. 1261) that his early 1913 version of general relativity reduced in suitable weak field approximation to a theory with a four-vector field potential that was formally analogous to electrodynamics. It was this approximation that yielded the weak field effects we now label as "Machian." The previous year, when seeking similar effects in his 1912 theory of static gravitational fields, Einstein demonstrated that he then expected a relativized gravitation theory to be formally analogous to electrodynamics at some level. For then he wrote a paper with the revealing title "Is there a gravitational effect that is analogous to electrodynamic induction?" (Einstein 1912c).

The celebrated defect of a theory of gravitation modelled after Maxwell electromagnetism was first pointed out by Maxwell himself (Maxwell 1864, 571). In such a theory, due to the change of signs, the energy density of the gravitational field is negative and becomes more negative as the field becomes stronger. In order not to introduce net negative energies into the theory, one must then suppose that space, in the absence of gravitational forces, must contain a positive energy density sufficiently great to offset the negative energy of any possible field strength. Maxwell professed himself baffled by the question of how a medium could possess such properties and renounced further work on the problem. As it turns out it was Einstein's foe, Abraham, shortly after his exchange with Einstein, who refined Maxwell's concern into a more telling objection. In a lecture of October 19, 1912, he reviewed his own gravitation theory based on Einstein's idea of using the speed of light as a gravitational potential. (Abraham 1912e) He first reflected (pp. 193–94), however, on a gravitation theory modelled after Maxwell electromagnetism. In such a theory, a mass, set into oscillation, would emit waves analogous to light waves. However, because of the change of sign, the energy flow would not be away from the mass but towards it, so that the energy of oscillation would increase. In other words such an oscillating mass would have no stable equilibrium. Similar difficulties were reported by him for gravitation theories of Maxwellian form due to H.A. Lorentz and R. Gans.

What was Einstein's attitude to such a theory of gravitation? He was clearly aware of the formal possibility of such a theory in 1912 and 1913. From his failure to exploit such a theory, we can only assume that he did not think it an adequate means of relativizing gravitation.[21] Unfortunately I know of no source from that period through which Einstein states a definite view on the matter beyond the brief remarks in his exchange with Abraham. We shall see that Einstein is about to renounce the conclusion of his reply to Abraham, that a Lorentz covariant theory cannot capture the equality of inertial and gravitational mass, at least for the case of Nordström's theory of gravitation. Did Einstein have other reservations about six-vector theories of gravitation? How seriously, for example, did he regard the negative field energy problem in such a theory?

The idea of an analogy between a relativized gravitation theory and electrodynamics seems to play no significant role in the methods Einstein used to generate relativized gravitation theories. The effect analogous to electrodynamic induction of

---

21   Notice these reservations must have amounted to more than the observation that such a theory fails to extend the relativity of motion to acceleration. In (Einstein 1913), immediately after his remarks on the similarity between the problems of relativizing gravitation and electrostatics, he considers Lorentz covariant gravitation theories. The only theory taken seriously in this category is a version of Nordström's theory of gravitation.

(Einstein 1912c), for example, was derived fully within Einstein's 1912 theory of static gravitational fields and the analogy to electrodynamics appeared only in the description of the final result. In general, the mention of an analogy to electrodynamics seems intended solely to aid Einstein's readers in understanding the enterprise and physical effects appearing in the relativized theories of gravitation by relating them to an example familiar to his readers. That we have any surviving, written remarks by Einstein directly on this matter we owe to J.W. Killian. Some thirty years later, in a letter of June 9, 1943 (EA 14 261) to Einstein, Killian proposed a gravitation theory modelled after Maxwell electromagnetism.[22] Einstein's reply of June 28, 1943, gives a fairly thorough statement of his attitude at that time to this theory.[23]

Because there was no question of experimental support for the theory, Einstein proposed to speak only to its formal properties. To begin, he noted, Maxwell's equations only form a complete theory for parts of space free of source charges, for the theory cannot determine the velocity field of the charge distribution without further assumption. After Lorentz, to form a complete theory, it was assumed that charges were carried by ponderable masses whose motions followed from Newton's laws. What Einstein called "real difficulties" arise only in explaining inertia. These difficulties result from the negative gravitational field energy density in the theory. Assuming, apparently, that the energy of a mass in the theory would reside in its gravitational field, Einstein pointed out that the kinetic energy of a moving mass point would be negative. This negativity would have to be overcome by a device entirely arbitrary from the perspective of the theory's equations, the introduction of a compensating positive energy density located within the masses. This difficulty is more serious for the gravitational version of the theory, for, in the electromagnetic theory, the positivity of electromagnetic field energy density allows one to locate all the energy of a charge in its electromagnetic field.

Calling the preceding difficulty "the fundamental problem of the wrong sign," Einstein closed his letter with brief treatment of two further and, by suggestion, lesser difficulties. The proposed theory could not account, Einstein continued, for the proportionality of inertial and gravitational mass. Here we finally see the concern that drove Einstein's work on Lorentz covariant gravitation theory in the decade following 1907. Yet Einstein does not use the transformation arguments of this early period to establish the failure of the proposed theory to yield this proportionality. Instead he continues to imagine that the energy and therefore inertia of a mass resides in its gravitational field. Some fixed quantity of gravitational mass could be configured in many different ways. Thus it follows that the one quantity of gravitational mass could be associated with many different gravitational fields and thus many different inertial masses, in contradiction with the proportionality sought.[24] Finally Einstein remarked

22 Einstein addresses his reply to "Mr. J.W. Killian, Dept. of Physics, Rockerfeller Hall, Ithaca N.Y."
23 EA 14 265 is an autograph draft of the letter in German. EA 14 264 is an unsigned typescript of the English translation. There are some significant differences of content between the two, indicating further editing of content presumably by Einstein between the draft and typescript.

that the proposed theory allows no interaction between electromagnetic and gravitational fields other than through charged, ponderable masses. Thus it could not explain the bending of starlight in a gravitational field.

6. NORDSTRÖM'S FIRST THEORY OF GRAVITATION

The dispute between Einstein and Abraham was observed with interest by a Finnish physicist, Gunnar Nordström.[25] In a paper submitted to *Physikalische Zeitschrift* in October 1912 (Nordström 1912), he explained that Einstein's hypothesis that the speed of light c depends on the gravitational potential led to considerable problems such as revealed in the Einstein-Abraham dispute. Nordström announced (p. 1126) that he believed he had found an alternative to Einstein's hypothesis which would

> ... leave c constant and still adapt the theory of gravitation to the relativity principle in such a way that gravitational and inertial masses are equal.

The theory of gravitation which Nordström developed was a slight modification of the theory embodied in equations (3) and (4) above. Selecting the commonly used coordinates $x, y, z, u = ict$, Nordström gave his version of the field equation (3):

$$\frac{\partial^2 \Phi}{\partial x^2} + \frac{\partial^2 \Phi}{\partial y^2} + \frac{\partial^2 \Phi}{\partial z^2} + \frac{\partial^2 \Phi}{\partial u^2} = 4\pi f \gamma \tag{12}$$

where Φ is the gravitational potential, γ the rest density of matter and f the gravitational constant. As he noted, this field equation was identical to the one advanced by Abraham (1912a, equation (1)) in the latter's gravitation theory.

Where Nordström differed from Abraham, however, was in the treatment of the force equation (4). This equation, as Nordström pointed out, is incompatible with the constancy of c. We saw above that force equation (4), in conjunction with the constancy of c in equation (5) entails the unphysical condition (6). Abraham had resolved the problem by invoking Einstein's hypothesis that c not be constant but vary with gravitational potential so that condition (5) no longer obtains. Thus Abraham's gravitation theory was no longer a special relativistic theory. Nordström, determined to preserve special relativity and the constancy of c, offered a choice of two modified versions of (4).

First, one could allow the rest mass m of a body in a gravitational field to vary with gravitational potential. Defining

---

24   In the German autograph draft (EA 14 265), Einstein imagines some fixed quantity of gravitational mass distributed between two bodies. The field strength they generate, and therefore their energy and inertial mass, would increase as the bodies were concentrated into smaller regions of space. (We may conjecture here that Einstein is ignoring the fact the field energy becomes more *negative* as the field strength increases.) The translated typescript (EA 14 264) simplifies the example by imagining that the gravitational mass is located in a single corpuscle, whose field and thence inertia varies with the radius of the corpuscle.

25   For a brief account of Nordström's life and his contribution to gravitation theory see (Isaksson 1985).

$$\mathfrak{F}_\mu = -\frac{\partial \Phi}{\partial x_\mu},$$

the four force on a body of mass m is

$$m\mathfrak{F}_\mu = -m\frac{\partial \Phi}{\partial x_\mu} = \frac{d}{d\tau}(m\mathfrak{a}_\mu) = m\frac{d\mathfrak{a}_\mu}{d\tau} + \mathfrak{a}_\mu\frac{dm}{d\tau} \tag{13}$$

where $\mathfrak{a}_\mu$ is the mass' four velocity and τ proper time.[26] The dependence of m on Φ introduces the additional, final term in $dm/d\tau$, which prevents the derivation of the disastrous condition (6). In its place, by contracting (13) with $\mathfrak{a}_\mu = \frac{dx_\mu}{dt}$ and noting that $\mathfrak{a}_\mu\mathfrak{a}_\mu = -c^2$, Nordström recovered the condition

$$m\frac{d\Phi}{d\tau} = c^2\frac{dm}{d\tau} \tag{14}$$

which yields an expression for the Φ dependence of m upon integration

$$m = m_0\exp\left(\frac{\Phi}{c^2}\right) \tag{15}$$

where m_0 is the value of m when $\Phi = 0$. Using (14) to substitute $\frac{dm}{d\tau}$ in (13), Nordström then recovered an equation of motion for a mass point independent of m

$$-\frac{\partial \Phi}{\partial x_\mu} = \frac{d\mathfrak{a}_\mu}{d\tau} + \frac{\mathfrak{a}_\mu d\Phi}{c^2 d\tau}. \tag{16}$$

---

26 *Note on notation*: The notation used in the sequence of papers discussed here varies. I shall follow the notation of the original papers as it changes, with one exception for brevity. Where the components of an equation such as (13) were written out explicitly as four equations

$$-m\frac{\partial \Phi}{\partial x} = \frac{d}{d\tau}(m\mathfrak{a}_x) = m\frac{d\mathfrak{a}_x}{d\tau} + \mathfrak{a}_x\frac{dm}{d\tau},$$

$$-m\frac{\partial \Phi}{\partial y} = \frac{d}{d\tau}(m\mathfrak{a}_y) = m\frac{d\mathfrak{a}_y}{d\tau} + \mathfrak{a}_y\frac{dm}{\tau},$$

etc.

I silently introduce the coordinates $x_\mu = (x_1, x_2, x_3, x_4) = (x, y, z, u) = ict$ and corresponding index notation as in equation (13) above.

Nordström's second alternative to force equation (4) preserved the independence

of m from the potential. The quantity $m\mathfrak{F}_\mu = -m\dfrac{\partial \Phi}{\partial x_\mu}$ could not be set equal to the

gravitational four-force on a mass m, $\dfrac{d}{d\tau}(m\mathfrak{a}_\mu)$, for that would be incompatible with

the orthogonality (5) of four velocity and four acceleration. However one can retain
compatibility with this orthogonality if one selects as the four force only that part of
$m\mathfrak{F}_\mu$ which is orthogonal to the four-velocity $\mathfrak{a}_\mu$. This yields the second alternative
for the force equation

$$\frac{d}{d\tau}(m\mathfrak{a}_\mu) = m\mathfrak{F}_\mu + m\frac{\mathfrak{a}_\mu}{c^2}(\mathfrak{F}_\nu\mathfrak{a}_\nu) = \begin{pmatrix} \text{part of } m\mathfrak{F}_\mu \\ \text{orthogonal} \\ \text{to } \mathfrak{a}_\mu \end{pmatrix}, \tag{17}$$

Nordström somewhat casually noted that he would use the first alternative, since it
corresponded to "the position of most researchers in the domain of relativity theory."
(p. 1126) Indeed Nordström proceeded to show that both force equations lead to
exactly the same equation of motion (16) for a point mass, planting the suggestion
that the choice between alternatives could be made arbitrarily.

Regular readers of *Physikalische Zeitschrift*, however, would know that Nordström's decision between the two alternatives could not have been made so casually
by him. For in late 1909 and early 1910, Nordström had engaged in a lively public
dispute with none other than Abraham on a problem in relativistic electrodynamics
that was in formal terms virtually the twin of the choice between the force laws (13)
and (17). (Nordström 1909, 1910; Abraham 1909, 1910.) The problem centered on
the correct expression for the four force density on a matter distribution in the case of
Joule heating. The usual formula for the four force density $\mathfrak{K}_\mu$ on a mass distribution
with rest mass density v is, in the notation of Abraham (1910),

$$\mathfrak{K}_\mu = v\frac{d}{d\tau}\left(\frac{dx_\mu}{d\tau}\right)$$

with τ proper time and coordinates $x_\mu = (x, y, z, u = ict)$. In the case of Joule heating, it turns out that this expression leads to a contradiction with the orthogonality
condition (5). The two escapes from this problem at issue in the dispute are formally
the same as the two alternative gravitational force laws. Nordström defended
Minkowski's approach, which took the four force density to be that part of $\mathfrak{K}_\mu$

orthogonal to the matter four velocity $\left(\dfrac{dx_\mu}{d\tau}\right)$ —the counterpart of force law (17).

Abraham concluded that the rest mass density v would increases in response to
the energy of Joule heat generated. He showed a consistent system could be achieved

if one now imported this variable v into the scope of the $d/d\tau$ operator in the expression for $\mathfrak{K}_\mu$ —an escape that is the counterpart of (13). Abraham's escape was judged the only tenable one when he was able to show that it yielded the then standard Lorentz transformation formula for heat whereas the Nordström-Minkowski formula did not.[27]

The connection between Nordström's 1912 gravitation theory and this earlier dispute surfaced only in extremely abbreviated form in Nordström (1912). In a brief sentence in the body of the paper, Nordström noted gingerly that (p. 1127)

> the latter way of thinking [alternative (17)] corresponds to Minkowski's original, that treated first [alternative (13)] to that held by Laue and Abraham.

That Nordström had any stake in the differing viewpoints is only revealed in a footnote to this sentence in which the reader is invited to consult "the discussion between Abraham and the author," followed by a citation to the four papers forming the dispute. Nordström then closes with the remark[28]

> I now take the position then taken up by Abraham.

Nordström now continued his treatment of gravitation by extending the discussion from isolated point masses to the case of continuous matter distributions—an area in which he had some interest and expertise (Nordström 1911). He derived a series of results in a straightforward manner. They included expressions for gravitational four force density on a continuous mass distributions and the corresponding equations of motion, expressions for the energy density and flux due to both gravitational field and matter distribution, the gravitational field stress-energy tensor and the laws of conservation of energy and momentum.

The last result Nordström derives concerns point masses. He notes that the field equation (12) admits the familiar retarded potential as a solution for a matter distribution with rest density γ

$$\Phi(x_0, y_0, z_0, t) = -f\int\frac{dx\,dy\,dz}{r}\gamma_{t-r/c} + \text{constant}$$

where $\gamma_{t-r/c}$ is γ evaluated at time $t-r/c$, the integration extends over all of three dimensional space and $r = \sqrt{(x-x_0)^2 + (y-y_0)^2 + (z-z_0)^2}$. It follows from the factor of $1/r$ in the integral that the potential Φ at a true point mass would be $-\infty$. Allowing for the dependence of mass on potential given in (15), it follows that the mass of such a point would have to be zero so that true point masses cannot exist.

---

27 Thus the authoritative judgement of Pauli's Teubner Encyklopädie article (Pauli 1921, 108) is that "Nordström's objections cannot be upheld." For a lengthy discussion of this debate and an indication that the issues were not so simple, see (Liu 1991).

28 His earlier work (Nordström 1911) had explicitly employed Abraham's "force concept," although Nordström had then noted very evasively that, in using it, he "wish[es] to assert no definite opinion on the correctness of one or other of the two concepts" (p. 854).

Nordström concluded with confidence, however, that he could see no contradictions arising from this result.

We may wonder at Nordström's lack of concern over this result. It would be thoroughly intelligible, however, if Nordström were to agree—as Nordström's later (Nordström 1913a, 856) suggests —with Laue's view on the relation between the theory of point masses and of continua. Laue had urged that the former ought to be derived from the latter (Laue 1911a, 525). Under this view, the properties of extended masses are derived from consideration of discrete volumes in a continuous matter distribution, not from the accumulated behavior of many point masses. So the impossibility of point masses in Nordström's theory would present no obstacle in his generation of the behavior of extended bodies.

7. EINSTEIN REPLIES

In advancing his theory, Nordström had claimed to do precisely what Einstein had claimed impossible: the construction of a Lorentz covariant theory of gravitation in which the equality of inertial and gravitational mass held. We need not guess whether Einstein communicated his displeasure to Nordström, for Einstein's missive was sufficiently swift for Nordström to acknowledge it in an addendum (p. 1129) to his paper which read

> *Addendum to proofs.* From a letter from Herr Prof. Dr. A. Einstein I learn that he had already earlier concerned himself with the possibility used above by me for treating gravitational phenomena in a simple way. He however came to the conviction that the consequences of such a theory cannot correspond with reality. In a simple example he shows that, according to this theory, a rotating system in a gravitational field will acquire a smaller acceleration than a non-rotating system.

Einstein's objection to Nordström is clearly an instance of his then standard objection to Lorentz covariant theories of gravitation: in such theories the acceleration of fall is not independent of a body's energy so that the equality of inertial and gravitational mass is violated. It is not hard to guess how Einstein would establish this result for a spinning body in Nordström's theory. It would seem to follow directly from the familiar equation (9) which holds in Nordström's theory and which says, loosely speaking, that a body falls slower if it has a greater horizontal velocity. Indeed, as we shall see below, this is precisely how Nordström shortly establishes the result in his next paper on gravitation theory.

Nordström continued and completed his addendum with a somewhat casual dismissal of Einstein's objection.

> I do not find this result dubious in itself, for the difference is too small to yield a contradiction with experience. Of course, the result under discussion shows that my theory is not compatible with Einstein's principle of equivalence, according to which an unaccelerated reference system in a homogeneous gravitational field is equivalent to an accelerated reference system in a gravitation free space.
>
> In this circumstance, however, I do not see a sufficient reason to reject the theory. For, even though Einstein's hypothesis is extraordinarily ingenious, on the other hand it still

provides great difficulties. Therefore other attempts at treating gravitation are also desirable and I want to provide a contribution to them with my communication.

Nordström's reply is thoroughly reasonable. The requirement of exact equality of inertial and gravitational mass was clearly an obsession of Einstein's thinking at this time and not shared by Einstein's contemporaries. The now celebrated Eötvös experiment had not yet been mentioned in the publications cited up to this point. We can also see from equation (9) that the failure of the equality of inertial and gravitational mass implied by Nordström's theory would reside in a second order effect in v/c. Nordström clearly believed it to be beyond recovery from then available experiments.

Finally we should note that Einstein and Nordström are using quite different versions of the requirement of the equality of inertial and gravitational mass. At this time, Einstein presumed that the total inertial mass would enter into the equality with the expectation that it would yield the independence of the vertical acceleration of a body in free fall from its horizontal velocity. We can only conjecture the precise sense Nordström had in mind, when he promised his theory would satisfy the equality, for he does not explain how the equality is expressed in his theory. My guess is that he took the rest mass to represent the body's inertial mass in the equality, for Nordström's equation (16) clearly shows that the motion of a massive particle in free fall is independent of its rest mass m. Under this reading Nordström's version of the equality entails the weaker observational requirement of the "uniqueness of free fall" defined above in Section 1.[29]

8. NORDSTRÖM'S FIRST THEORY ELABORATED

Nordström's first paper on his gravitation theory was followed fairly quickly by another (Nordström 1913a), submitted to *Annalen der Physik* in January 1913. This new paper largely ignored Einstein's objection although the paper bore the title "Inertial and gravitational mass in relativistic mechanics." The closing sections of this paper recapitulated the basic results of (Nordström 1912) with essentially notational differences only. In Section 6, Nordström's original field equation (12) was rewritten as

$$\frac{\partial^2 \Phi}{\partial x^2} + \frac{\partial^2 \Phi}{\partial y^2} + \frac{\partial^2 \Phi}{\partial z^2} + \frac{\partial^2 \Phi}{\partial u^2} = g\nu. \qquad (12')$$

As before, Φ was the gravitational potential. The rest density of matter was now represented by ν and Nordström explicitly named the new constant g the "gravitation factor." The force equation was presented as a density in terms of the gravitational force per unit volume of matter $\mathfrak{R}_\mu^g$

---

29   The equality of inertial and gravitational mass and the uniqueness of free fall are distinct from the principle of equivalence. Einstein's version of the principle has been routinely misrepresented since about 1920 in virtually all literatures. See (Norton 1985). It is stated correctly, however, in Nordström's addendum.

$$\mathfrak{K}_\mu^g = -g v \frac{\partial \Phi}{\partial x_\mu}. \tag{18}$$

The only difference between this expression and the analogous one offered in the previous paper was the presence of the gravitation factor g. Since this factor was a constant and thus did not materially alter the physical content of either of the basic equations, Nordström might well have anticipated his readers' puzzlement over its use. He hastened to explain that, while g was a constant here, nothing ruled out the assumption that g might vary with the inner constitution of matter. The paper continued to derive the Φ dependence of rest mass m of equation (15). The new version of the relation now contained the gravitation factor g and read

$$m = m_0 \exp\left(\frac{g\Phi}{c^2}\right).$$

Nordström no longer even mentioned the possibility of avoiding this dependence of m on Φ by positing the alternative force equation (17). The section continued with a brief treatment of the gravitational field stress-energy tensor and related quantities. It closed with a statement of the retarded potential solution of the field equation.

The final section 7 of the paper analyzed the motion of a point mass in free fall in an arbitrary static gravitational field. The analysis was qualified by repetition of his earlier observation that true point masses are impossible in his theory (Nordström 1912, 1129). In addition he noted that the particle's own field must be assumed to be vanishingly weak in relation to the external field. The bulk of the section is given over to a tedious but straightforward derivation of the analog of equation (9). Nordström considered a static field, that is one in which $\partial \Phi / \partial t = 0$, where t is the time coordinate of the coordinate system $(x, y, z, u = ict)$. He assumed the field homogeneous and acting only in the z-direction of the coordinate system. A point mass in free fall moves according to

$$\frac{dv_z}{dt} = -\left(1 - \frac{v^2}{c^2}\right) g \frac{\partial \phi}{\partial z}, \quad \frac{dv_x}{dt} = 0, \quad \frac{dv_y}{dt} = 0, \tag{19}$$

where v_x, v_y and v_z are the components of the mass' velocity v.[30] At this point in the paper, readers of (Nordström 1912) might well suspect that the entire purpose of developing equation (19) was to enable statement of the objection of Einstein reported in that last paper's addendum. For, after observing that this result (19) tells us that a body with horizontal velocity falls slower than one without, he concluded immediately that a rotating body must fall slower than a non-rotating body.

---

30   Notice that this result is more general than result (9), since it is not restricted to masses with vanishing vertical velocity, that is, masses whose motion satisfies the condition (7). Curiously Nordström's condition that the field be homogeneous, so that $\partial \Phi / \partial z$ = constant, is invoked nowhere in the derivation or discussion of the result.

Because this example will be reappraised shortly, it is worth inserting the steps that Nordström must have assumed to arrive at this conclusion. In the simplest case, the axis of rotation of the body is aligned vertically in the static field. Each small element of the spinning body has a horizontal motion, due to the rotation. If each such element were independent, then the vertical acceleration of each would be given by the equation (19), so that each element would fall slower because of the horizontal velocity imparted by the rotation. If this result holds for each element, it seems unproblematic to conclude that it obtains for the whole, so that the vertical acceleration of fall of the body is diminished by its rotation.

While Nordström urged that this effect is much too small to be accessible to observation, he was more sanguine about the analogous effect on the acceleration of fall of a body by the independent motions of its molecules. Its possibility could not be denied, he said. However, in the penultimate paragraph of the paper, he anticipated that such an effect could be incorporated into his theory by allowing the gravitation factor g to depend on the molecular motion of the body. He pointed out that the rest energy of a body would also be influenced by this molecular motion.

The results Nordström recapitulated in Section 6 and 7 were not the major novelties of the paper. In fact the paper was intended to address a quite precise problem. The field equation (12') contained a density term v. Nordström's problem was to identify what this term should be. The term—or, more precisely, gv —represented the gravitational field source density. According to Nordström's understanding of the equality of inertial and gravitational mass, v must also represent the inertial properties of the source matter. The selection of such a term was not straightforward. For, drawing upon his own work and that of Laue and others, he knew that stressed bodies would exhibit inertial properties that were not reducible to the inertial properties of any individual masses that may compose them. Thus Nordström recognized that his gravitation theory must be developed by means of the theory of relativistic continua, in which stresses were treated. This had clearly been his program from the start. In a footnote to the first paragraph of his first paper on gravitation, Nordström (1912), at the mention of the equality inertial and gravitational mass, he foreshadowed his next paper

> By the equality of inertial and gravitational mass, I do not understand, however, that every inertial phenomenon is caused by an inertial and gravitational mass. For elastically stressed bodies, according to Laue ..., one recovers a quantity of motion [momentum] that cannot at all be reduced back to a mass. I will return to this question in a future communication.

The special behavior of stressed bodies proved to be of decisive importance for the development of Nordström's theory. Therefore, in the following section, I review the understanding of this behavior at the time of Nordström's work on gravitation. I will then return to (Nordström 1913a).

9. LAUE AND THE BEHAVIOR OF STRESSED BODIES

By 1911 it was apparent that a range of problems in the theory of relativity had a common core—they all involved the behavior of stressed bodies—and that a general theory of stressed bodies should be able to handle all of these problems in a unified format. The development of this general theory was largely the work of Laue and came from a synthesis and generalization of the work of many of his predecessors, including Einstein, Lorentz, Minkowski and Planck. The fullest expression of this general theory came in Laue (1911a) and was also incorporated into Laue (1911b), the first text book published on the new theory of relativity.[31] Three problems treated in Laue's work give us a sense of the range of problems that Laue's work addressed.

9.1 Three Problems for Relativity Theory

In 1909, in a remarkably prescient paper, Lewis and Tolman (1909) set out to develop relativistic mechanics in a manner that was independent of electromagnetic theory using simple and vivid arguments. At this time relativity theory was almost invariably coupled with Lorentzian electrodynamics and its content was accessible essentially only to those with significant expertise in electrodynamics. Their exposition was marred, however, by an error in its closing pages (pp. 520–21). By this point, they had established the Lorentz transformation for forces transverse to the direction of motion. Specifically, if the force is f^0_{trans} in the rest frame, then the force f_{trans} measured in a frame moving at a fraction β of the speed of light is

$$f_{trans} = \sqrt{1 - \beta^2} f^0_{trans}. \tag{20}$$

To recover the transformation formula for forces parallel to the direction of motion, the "longitudinal" direction, they considered the rigid, right angled lever of Fig. 1.
The arms ab and bc are of equal length and pivot about point b. In its rest frame two equal forces f act at points a and c, the first in direction bc, the second in direction ba. The level will not turn since there is no net turning couple about its pivot b. They then imagined the whole system in motion in the direction bc. They conclude—presumably directly from the principle of relativity—that the system must remain in equilibrium. Therefore the net turning couple about b must continue to vanish for the moving system, so that

$$\left(\begin{matrix}\text{force} \\ \text{at } c\end{matrix}\right)\left(\begin{matrix}\text{length} \\ bc\end{matrix}\right) + \left(\begin{matrix}\text{force} \\ \text{at } a\end{matrix}\right)\left(\begin{matrix}\text{length} \\ ab\end{matrix}\right) = 0.$$

Now, according to (20), the transverse force at c is diminished by the factor $\sqrt{1 - \beta^2}$. The length of its lever arm bc is also contracted by the same factor,

---

31   Presumably the two works were prepared together. (Laue 1911a) was submitted on 30 April, 1911. The introduction to (Laue 1911b) is dated May, 1911.

whereas the arm ab, being transverse to the motion, is uncontracted. Lewis and Tolman now concluded that equilibrium can only be maintained if the longitudinal force f_{long} at a transforms according to

$$f_{long} = (1 - \beta^2) f^0_{long} . \tag{21}$$

This conclusion comes from an argument so simple that one would hardly suspect it. What they did not point out, however, was that its conclusion (21) contradicted the then standard expositions of relativity theory (e.g. Einstein 1907a, 448) according to which (20) is correct but (21) should be replaced by

$$f_{long} = f^0_{long}. \tag{22}$$

We now see the problem in its starkest form. If we apply the standard transformation formulae (20) and (22) to the case of Lewis and Tolman's bent lever we seem driven to a curious conclusion. We have a system at equilibrium in its rest frame which now forfeits that equilibrium in a moving frame through the appearance of a non-vanishing turning couple. Indeed we seem to have a violation of the principle of relativity, for the presence of this turning couple should yield an experimental indication of the motion of the system.

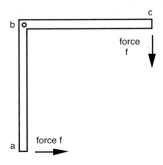

Figure 1: Lewis and Tolman's Bent Lever

The second problem is, at first glance, quite unrelated to the Lewis and Tolman bent lever. Under a classical analysis, one expects that a charged, parallel plate condenser can experience a net turning couple if it is set in motion through the aether. In a classic experiment, Trouton and Noble (1903) sought to detect the turning couple acting on a charged condenser due to its motion with the Earth. Their null result is celebrated. Just as in the case of the Lewis and Tolman bent lever, the problem is to see how relativity theory allows one to predict this null result, which otherwise would contradict the principle of relativity. In fact, as Laue (1911c, 517) and others soon pointed out, the two problems were closely connected. In its rest frame, the Trouton-Noble condenser was simply a rigid system of two parallel plates with an electric force acting on each plate in such a way that the entire system was in equilibrium. If that equilibrium system was set into motion, under either a classical or relativistic analysis, the electric forces would transform according to the Lorentz transformation (20) and (22). Unless the direction of motion imparted was parallel or exactly perpendicular to the plates, the net effect would be exactly the same as the Lewis and Tolman bent lever. A non-vanishing turning couple is predicted which deprives the system of equilibrium. The couple ought to be detectable in violation of the principle of relativity.[32]

The third problem concerns the theory of electrons. The decade preceding 1911 had seen considerable work on the problem of providing a model for the electron. Best known of these were the models of Lorentz and Abraham, which depicted electrons as electrically charged spheres with varying properties. The general problem was to show that the relativistic dynamics of an acceptable model of the electron would coincide with the relativistic dynamics of a point mass. There were a range of difficulties to be addressed here. In introducing his (Laue 1911a, 524–25), Laue recalled a brief exchange between Ehrenfest and Einstein. In a short note, Ehrenfest (1907) had drawn on work of Abraham that raised the possibility of troubling behavior by an electron of non-spherical or non-ellipsoidal shape when at rest. It was suggested that such an electron cannot persist in uniform translational motion unless forces are applied to it.[33] We might note that such a result would violate not only the principle of inertia in the dynamics of point masses but also the principle of relativity. Einstein's reply (1907b) was more a promise than resolution, although he ultimately proved correct. He pointed out that Ehrenfest's model of the electron was incomplete. One must also posit that the electron's charge was carried by a rigid frame, stressed to counteract the forces of self repulsion of the charge distribution. Ehrenfest's problem could not be solved until a theory of such frames was developed.

Finally another aspect of the problem of the relativistic dynamics of electrons was the notorious question of electromagnetic mass. If one computed the total momentum and energy of the electromagnetic field of an electron, the result universally accepted at this time was the one reported in (Laue 1911b, 98):

$$\begin{pmatrix} \text{Total field} \\ \text{momentum} \end{pmatrix} = \frac{4}{3}\frac{1}{c^2}\begin{pmatrix} \text{Total field} \\ \text{energy} \end{pmatrix}\begin{pmatrix} \text{electron} \\ \text{velocity} \end{pmatrix}. \tag{23}$$

The conflict with the relativistic dynamics of point masses arose if one now posited that all the energy and momentum of the electron resides in its electromagnetic field.

For one must then identify $\frac{1}{c^2}\begin{pmatrix} \text{Total field} \\ \text{energy} \end{pmatrix}$, the electromagnetic mass of the electron,

---

32   For further extensive discussion of the Trouton-Noble experiment and its aether theoretic treatment by Lorentz, see (Janssen 1995).

33   In a footnote, Ehrenfest pointed out the analogy to the turning couple induced on a charged condenser and reviewed the then current explanation of it absence in terms of molecular forces.

as the total inertial mass of the electron, so that equation (23) tells us that the momentum of an electron is $4/3$ the product of its mass and velocity. The canonical resolution of this difficulty, as stated for example in (Pauli 1921, 185–86), is that such a purely electromagnetic account of the dynamics of the electron is inadmissible. As Einstein (1907b) urged, there must be also stresses of a non-electromagnetic character within the electron.[34] The puzzle Laue addressed in 1911 was to find very general circumstances under which the dynamics of such an electron would agree with the relativistic dynamics of point masses.

9.2 The General-Stress Energy Tensor

The focus of Laue's treatment of stressed bodies in his (1911a) and (1911b) lay in a general stress-energy tensor.[35] While Minkowski (1908, §13) had introduced the four dimensional stress-energy tensor at the birth of four dimensional methods in relativity theory, his use of the tensor was restricted to the special case of the electromagnetic field. Laue's 1911 work concentrated on extending the use of this tensor to the most general domain. The properties of this tensor and its behavior under Lorentz transformation summarized a great deal of the then current knowledge of the behavior of stressed bodies. Laue (1911a) uses a coordinate system $(x, y, z, l = ict)$ so that the components of the stress energy tensor $T_{\mu\nu}$ have the following interpretations:

$$\begin{pmatrix} T_{xx} = p_{xx} & T_{xy} = p_{xy} & T_{xz} = p_{xz} & T_{xl} = -ic g_x \\ T_{yx} = p_{yx} & T_{yy} = p_{yy} & T_{yz} = p_{yz} & T_{yl} = -ic g_y \\ T_{zx} = p_{zx} & T_{zy} = p_{zy} & T_{zz} = p_{zz} & T_{zl} = -ic g_z \\ T_{lx} = \frac{i}{c}\mathfrak{S}_x & T_{ly} = \frac{i}{c}\mathfrak{S}_y & T_{lz} = \frac{i}{c}\mathfrak{S}_z & T_{ll} = -W \end{pmatrix}.$$

The three dimensional tensor $p_{ik}(i, k = 1, 2, 3)$ is the familiar stress tensor. The vector $\mathfrak{g} = (\mathfrak{g}_x, \mathfrak{g}_y, \mathfrak{g}_z)$ represents the momentum density. The vector $\mathfrak{S} = (\mathfrak{S}_x, \mathfrak{S}_y, \mathfrak{S}_z)$ is the energy flux. W is the energy density.

The most fundamental result of relativistic dynamics is Einstein's celebrated inertia of energy according to which every quantity of energy E is associated with an inertial mass (E/c^2). The symmetry of Laue's tensor entails a result closely con-

---

34   While these stresses are needed to preserve the mechanical equilibrium of the electron, Rohrlich (1960) showed that they were not needed to eliminate the extraneous factor of 4/3 in equation (23). He showed that the standard derivation of (23) was erroneous and that the correct derivation did not yield the troubling factor of 4/3.

35   The label "stress-energy tensor" is anachronistic. Laue had no special name for the tensor other than the generic "world tensor," which, according to the text book exposition of Laue (1911b, §13) described any structure which transformed as what we would now call a second rank, symmetric tensor. Notice that the term "tensor" was still restricted at this time to what we would now call second rank tensors and even then usually to symmetric, second rank tensors. See (Norton 1992a, Appendix).

nected with Einstein's inertia of energy and attributed to Planck by Laue (1911a, 530). We have $(T_{xl}, T_{yl}, T_{zl}) = (T_{lx}, T_{ly}, T_{lz})$ which immediately leads to

$$g = \frac{1}{c^2} \mathfrak{S} \tag{24}$$

This tells us that whenever there is an energy flux $\mathfrak{S}$ present in a body then there is an associated momentum density g.

As emphasized in (Laue 1911c), this result is already sufficient to resolve the first of the three problems described above in Section 9.1, the Lewis and Tolman bent lever. Notice first that Fig. 1 does not display all the forces present. There must be reaction forces present at the pivot point b to preserve equilibrium in the rest frame. See Fig. 2, which also includes the effect of the motion of the system at velocity v in the direction bc. When the lever moves in the direction bc, then work is done by the force f at point a which acts in the direction bc. The energy of this work is transmitted along the arm ab as an energy current of magnitude fv and is lost at the pivot point b as work done against the reaction force that acts in the direction cb. This energy current fv in the arm ab must be associated with a momentum fv/c^2, according to (24) when integrated over the volume of the arm ab, and this momentum will be directed from a towards b. As Laue (1911c) showed, a short calculation reveals that this momentum provides precisely the additional turning couple needed to return the moving system to equilibrium. Notice that the force at c and its associated reaction force at b are directed transverse to the motion so no work is done by them.[36]

The essential and entirely non-classical part of the analysis resides in the result that the there is an additional momentum present in the moving arm ab because it is under the influence of a shear stress due to the force f at a and the corresponding reaction force at b. As Laue (1911c, 517) and Pauli (1921, 128–29) point out, exactly this same relativistic effect explains the absence of net turning couple in the Trouton-Noble condenser. The condenser's dielectric must be stressed in reaction to the attractive forces between the oppositely charged plates. The additional momentum associated with these non-electromagnetic stresses provides the additional turning couple required to preserve the equilibrium of the moving condenser.

---

36 Assume that the level arms are of unit length at rest so that the arm bc contracts to length $\sqrt{1 - v^2/c^2}$ when the system moves at velocity v in direction bc. The turning couple about a point b' at rest and instantaneously coincident with the moving pivot point b, due to the applied forces alone is $f(1 - v^2/c^2) - f = -f(v^2/c^2)$ where a positive couple is in the clockwise direction. The relativistic momentum fv/c^2 generates angular momentum about the point b'. Since the distance of the arm ab from b' is growing at the rate of v, this angular momentum is increasing at a rate $v(fv/c^2) = f(v^2/c^2)$, which is exactly the turning couple needed to balance the couple due to the applied forces.

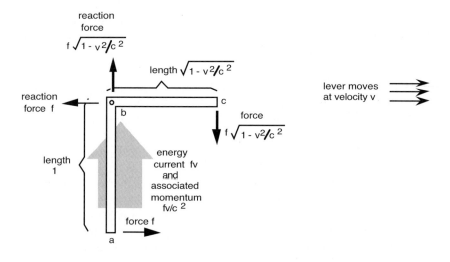

*Figure 2: Lewis and Tolman's Bent Lever Showing Reaction Forces
and Effects of Motion*

While this analysis satisfactorily resolves at least the case of the Lewis and Tolman lever, more complicated cases will require a clearer statement of the relationship between the stresses in a moving body and the momentum associated with it.
These results were derived directly by Laue (1911a, 531–32) from the Lorentz transformation of the components of the stress-energy tensor. In the rest frame of the
stressed matter distribution, the matter has energy density W^0 and a stress tensor
p_{ik}^0, for $i, k = 1, 2, 3$. The momentum density $\mathfrak{g}$ and energy flux $\mathfrak{S}$ vanish. Transforming to a frame of reference moving at velocity q in the x direction, from a
direct application of the Lorentz transformation formula for a tensor, Laue recovered
results that included

$$\mathfrak{g}_x = \frac{q}{c^2 - q^2}(p_{xx}^0 + W^0), \quad \mathfrak{g}_y = \frac{q}{c\sqrt{c^2 - q^2}}p_{xy}^0, \quad W = \frac{c^2 W^0 + q^2 p_{xx}^0}{c^2 - q^2}. \quad (25)$$

The first two equations show that, in a moving body, there is a momentum density
associated with both normal stresses (p_{xx}^0) and with shear stresses (p_{xy}^0). The Lewis
and Tolman lever is a case in which a momentum density is associated with shear
stresses in a moving body in accord with (25). The third equation shows that there is
an energy density associated with normal stresses in moving body.

This last result, in a form integrated over a whole stressed body, had already been
investigated and clearly stated by Einstein (1907c, §1), as a part of his continuing
analysis of the inertia of energy. He gave the result a very plausible, intuitive basis,
relating it directly to the relativity of simultaneity. He imagined a rigid body in uni-

form translational motion. At some single instant in the body's rest frame, an equilibrium state of stress appears in the body. Since it appears at a single instant, the new forces do not alter the state of motion of the body. However in the frame in which the body moves, because of the relativity of simultaneity, these new forces do not appear simultaneously over the entire body. Thus there is a brief period of disequilibrium of forces during which net work is done on the body. This new work is exactly the energy associated with the stresses p_{xx}^0 in the equation (25). Einstein (1907a, §2) continued with an example similar in structure to the Trouton-Noble condenser—a rigid body, in uniform motion, carrying an electric charge distribution. The forces between the charges carried stress the rigid body, so that there is an energy associated with these stresses. Einstein showed that this latter energy was essential. Otherwise the energy of the moving body would depend on the direction of its motion which would lead to a contradiction.[37]

Laue (1911, §2) continued his treatment of the transformation formulae (25) by restating them for an extended body. In particular, integration of (25) over such a body revealed relationships between the rest energy E^0 of the body and its energy E and momentum $\mathfrak{G}$ in the frame of reference in which the body moves at velocity q. Writing the three components of the body's ordinary velocity as (q_1, q_2, q_3), he recovered[38]

$$E = \frac{c}{\sqrt{c^2 - q^2}}\left\{E^0 + \frac{1}{c^2}q_i q_k \int p_{ik}^0 dV^0\right\}$$

$$\mathfrak{G}_i = \frac{q_i}{c\sqrt{c^2 - q^2}}\left\{E^0 + \frac{1}{q^2}q_j q_k \int p_{jk}^0 dV^0\right\} \tag{26}$$

$$+ \frac{1}{c^2}\left\{q_k \int p_{ik}^0 dV^0 - \frac{q_i}{q^2}q_j q_k \int p_{jk}^0 dV^0\right\}.$$

The expression for momentum had an immediate and important consequence. In general, whenever the body was stressed so that the stress tensor p_{ik}^0 does not vanish, the momentum $\mathfrak{G}_i$ of the body will not be in the same direction as its velocity q_i. This was exemplified in the Lewis and Tolman lever. Although it was set in motion in the direction bc, the presence of stresses in the arm ab led to a momentum in that arm

---

37 Specifically, in the body's rest frame, the body can rotate infinitely slowly without application of any forces. By the principle of relativity, this same motion will be possible if the body is in uniform translational motion as well. However in this latter case the kinetic energy of the body would alter according to its orientation as it rotates. Since no forces were applied, this would violate the "energy principle," the law of conservation of energy. Notice that the rotation is infinitely slow, so that it does not contribute to the body's kinetic energy.

38 V^0 is the rest volume of the body and $i, j, k = 1, 2, 3$. I have simplified Laue's opaque notation by introducing an index notation, where Laue used round and square brackets to represent various products. For example, where I would write $q_i q_k p_{ik}^0$, he would write "$(q[qp^0])$."

directed transverse to the motion. If this momentum is added vectorially to the momentum of the inertial mass of the lever, the resultant total momentum vector will not be parallel to the direction of motion.

Laue was now in a position to restate the analysis given for the Lewis and Tolman lever in a way that would apply to general systems. This was the principle of burden of (Laue 1911a, §3 and §4). To begin, Laue introduced a new three dimensional stress tensor. In a body at rest, the time rate of change of momentum density $\partial \mathfrak{g}_i / \partial t$ is given by the negative divergence of the tensor p_{ik}:

$$\frac{\partial \mathfrak{g}_i}{\partial t} = -\frac{\partial p_{ik}}{\partial x_k}.$$

However if one wishes to investigate the time rate of change of momentum density in a moving body, one must replace the partial time derivative $\frac{\partial}{\partial t}$ with a total time derivative coordinated to the motion, $\frac{d}{dt} = \frac{\partial}{\partial t} + q_x \frac{\partial}{\partial x} + q_y \frac{\partial}{\partial y} + q_z \frac{\partial}{\partial z}$. Laue was able to show that the relevant time rate of change of momentum was given as the negative divergence of a new tensor t_{ik}

$$\frac{d \mathfrak{g}_i}{dt} = -\frac{\partial t_{ik}}{\partial x_k}.$$

where this "tensor of elastic stresses" was defined by

$$t_{ik} = p_{ik} + \mathfrak{g}_i q_k.$$

Note in particular that t_{ik} will not in general be symmetric since the momentum density $\mathfrak{g}_i$ will not in general be parallel to the velocity q_i.

The lack of symmetry of t_{ik} is a cause of momentary concern, for it is exactly the symmetry of p_{ik} that enables recovery, in effect, of the law of conservation of angular momentum. More precisely, the symmetry of the stress tensor is needed for the standard derivation of the result that the time rate of change of angular momentum of a body is equal to the total turning couple impressed on its surface. Laue proceeds to show, however, that this asymmetry does not threaten recovery of this law and is, in fact essential for it.[39] He writes the time rate of change of total angular momentum $\mathfrak{L}_i$ of a moving body as[40]

---

39 Laue calls §4, which contains this discussion, "the area law." I presume this is a reference to Kepler's second law of planetary motion, which amount to a statement of the conservation of angular momentum for planetary motion.

40 dV is a volume element of the body. I make no apology at this point for shielding the reader from Laue's notation, which has become more than opaque. Laue now uses square brackets to represent vector products, where earlier they represented an inner product of vector and tensor. ε_{ikl} is the fully antisymmetric Levi-Civita tensor, so that $\varepsilon_{ikl} A_k B_l$ is the vector product of two vectors A_k and B_l.

$$\frac{d\mathfrak{L}_i}{dt} = \int \varepsilon_{ikl}\mathfrak{r}_k\frac{d\mathfrak{g}_l}{dt} + \varepsilon_{ikl}q_k\mathfrak{g}_l dV = \int -\varepsilon_{ikl}\mathfrak{r}_k\frac{\partial t_{lm}}{\partial x_m} + \varepsilon_{ikl}q_k\mathfrak{g}_l dV. \qquad (27)$$

Were the tensor t_{ik} symmetric, then the integration of the first term alone could be carried out using a version of Gauss' theorem. One could then arrive at the result that the total rate of change of momentum of the body is given by the turning couple applied to its surface. Thus if there is no applied couple, angular momentum would be conserved. However the tensor t_{ik} is not symmetric, and so an integral over the first term leaves a residual rate of change of angular momentum even when no turning couple is applied to the body. Fortunately there is a second term in the integrals of (27) that results from allowing for the use of the total time derivative. It is the vector product of the velocity q_i and momentum density $\mathfrak{g}_i$. This term would not be present in a classical analysis since these two vectors would then be parallel so that their vector product would vanish. In the relativistic context, this is not the case. This term corresponds exactly to the stress induced momentum in the arm ab of the Lewis and Tolman lever. This extra term exactly cancels the residual rate of change of angular momentum of the first term, restoring the desired result, the rate of change of angular momentum equals the externally applied turning couple.

9.3 Laue's "Complete Static Systems"

The last of the group of results developed in (Laue 1911a, §5) proved to be the most important for the longer term development of Nordström's theory of gravitation. Laue had shown clearly just how different the behavior of stressed and unstressed bodies in relativity theory could be. He now sought to delineate circumstances in which the presence of stresses within a body would not affect its overall dynamics. Such was the case of a "complete static system," which Laue defined as follows:

> We understand by this term such a system which is in static equilibrium in any justified
> reference system K^0, without sustaining an interaction with other bodies.

This definition is somewhat elusive and the corresponding definition in (Laue 1911b, 168–69) is similar but even briefer. In both cases, however, Laue immediately gave the same example of such a system, "an electrostatic field including all its charge carriers." This example and the definition leaves open the question of whether a body spinning at constant speed and not interacting with any other bodies is a complete static system. Such a body is in equilibrium and static in the sense that its properties are not changing with time, especially if the body spins around an axis of rotational symmetry. Tolman (1934, 81) gave a clearer definition:

> And in general we shall understand by a complete static system, an entire structure
> which can remain in a permanent state of rest with respect to a set of proper coordinates
> S^0 without the necessity for any forces from the outside.

He clearly understood this definition to rule out rotating bodies, for he noted a few lines later "the velocity of all parts of the system is zero in these coordinates $[S^0]$".

Tolman used this to justify the condition that the momentum density in the rest frame vanishes at every point

$$\mathfrak{g}^0 = 0. \tag{28}$$

Presumably Laue agreed for he also invoked this condition. From it, both Laue and Tolman derived the fundamental result characteristic of complete static systems:

$$\int p_{ik}^0 dV^0 = 0, \tag{29}$$

where the integral extends over the rest volume V^0 of the whole body. Laue allowed, in effect, that his conception of a complete static system could be relaxed without compromising the recovery of (29). For in a footnote (Laue 1911a, 540) to the example of an electrostatic field with its charge carriers, he noted that one could also consider the case of electrostatic-magnetostatic fields. Even though (28) failed to obtain for this case, the time derivative of $\mathfrak{g}^0$ did vanish which still allowed the derivation of (29).

This fundamental property (29) of complete static systems greatly simplified the expression (26) for the energy and momentum of a stressed body. Through (29) all the terms explicitly dependent on stresses vanish so that

$$E = \frac{c}{\sqrt{c^2 - q^2}} E^0,$$

$$\mathfrak{B}_i = \frac{q_i}{c\sqrt{c^2 - q^2}} E^0. \tag{30}$$

As Laue pointed out, these expressions coincide precisely with those of a point mass with rest mass $m^0 = (E^0/c^2)$. Moreover under quasi-stationary acceleration—that is acceleration in which "the inner state $(E^0, \mathbf{p}^0)$ is not noticeably changed"—a complete static system will behave exactly like a point mass.

Laue could now offer a full resolution of the remaining problems described above in Section 9.1. An electron together with its field is a complete static system, he noted, no matter how it may be formed. As a result it will behave like a point mass, as long as its acceleration is quasi-stationary. In particular it will sustain inertial motion without the need for impressed forces. While Laue did not explicitly mention the problem of relating the electron's total field momentum to its inertial mass, Laue's result (30) resolves whatever difficulty might arise for the overall behavior of an electron. For however the electron may be constructed, as long as it forms a complete static system, equation (30) shows that the extraneous factor of 4/3 in equation (23) cannot appear. Finally, the Trouton-Noble condenser is a complete static system. While neither the momentum of its electromagnetic field or of its stressed mechanical structure will lie in the direction of its motion, equation (30) shows that the combined momentum $\mathfrak{B}_i$ will lie parallel to the velocity q_i, so that there is no net turning couple acting on the condenser.[41]

10. THE DEFINITION OF INERTIAL MASS IN NORDSTRÖM'S FIRST THEORY

What had emerged clearly from Laue's work was that the inertial properties of bodies could not be explained solely in terms of their rest masses and velocities, if the bodies were stressed. For Laue's equation (26) showed that the momentum of a moving body would be changed merely by the imposition of a stress, even though that stress need not deform the body or perform net work on it. Nordström clearly had results such as these in mind when he laid out the project of his (Nordström 1913a, 856–57). Laue and Herglotz, he reported, had constructed the entire mechanics of extended bodies without exploiting the concept of inertial mass. That concept, he continued, was neither necessary nor sufficient to represent the inertial properties of stressed matter. This now seems to overstate the difficulty, for Laue's entire system depended upon Einstein's result of the inertia of energy. Nonetheless nowhere did Laue's mechanics of stressed bodies provide a single quantity that represented *the* inertial mass of a stressed body.

It was to this last omission that Nordström planned to direct his paper. It was important, he urged, to develop a notion of the inertial mass of matter for the development of a gravitation theory. Such a theory must be based on the "unity of essence"[42] of inertia and gravity. He promised to treat the relativistic mechanics of deformable bodies in such a way that it would reveal a concept of inertial mass suitable for use in a theory of gravitation.

Nordström's analysis was embedded in a lengthy treatment of the mechanics of deformable bodies whose details will not be recapitulated here. Its basic supposition, however, was that the stress energy tensor $T_{\mu\nu}$ of a body with an arbitrary state of motion and stress would be given as the sum of two symmetric tensors (p. 858)

$$T_{\mu\nu} = p_{\mu\nu} + \nu\mathfrak{B}_\mu\mathfrak{B}_\nu. \tag{31}$$

The second tensor, $\nu\mathfrak{B}_\mu\mathfrak{B}_\nu$, he called the "material tensor." It represented the contribution to the total stress tensor from a matter distribution with rest mass density ν and four velocity $\mathfrak{B}_\mu$. The first tensor, $p_{\mu\nu}$, he called the "elastic stress tensor." It represented the stresses in the matter distribution. In the rest frame of the matter distribution, Nordström wrote the elastic stress tensor as (p. 863)

---

41 However this result did not end Laue's analysis of the Trouton-Noble experiment. See (Laue 1912).

42 *Wesenseinheit.* The term is sufficiently strong and idiosyncratic for it to be noteworthy that, so far as I know, Einstein was the only other figure from this period who used even a related term in connection with inertia and gravitation. In a paper cited earlier in (Nordström 1912, 1126), Einstein (1912d, 1063) had talked of the "equality of essence" (*Wesensgleichheit*) of inertial and gravitational mass. Einstein used the term again twice in later discussion. See (Norton 1985, 233).

$$\begin{bmatrix} p_{xx}^0 & p_{xy}^0 & p_{xz}^0 & 0 \\ p_{yx}^0 & p_{yy}^0 & p_{yz}^0 & 0 \\ p_{zx}^0 & p_{zy}^0 & p_{zz}^0 & 0 \\ 0 & 0 & 0 & p_{uu}^0 \end{bmatrix}$$

The six zero-valued components in this matrix represent the momentum density and energy current due to the presence of stresses. They must vanish, Nordström pointed out, since stresses cannot be responsible for a momentum or energy current in the rest frame.[43] In particular, Nordström identified the component p_{uu}^0 as a Lorentz invariant.[44]

In the crucial Section 4, "Definition of Inertial Mass," Nordstrom turned his attention to the (uu) component of equation (31) in the rest frame. This equation gave an expression for the Lorentz invariant rest energy density Ψ in terms of the sum of two invariant quantities[45]

$$\Psi = -p_{uu}^0 + c^2 v. \tag{32}$$

This equation gave simplest expression to the quantity of fundamental interest to Nordström's whole paper, the density v, which would provide the source for the gravitational field equation. This density would be determined once Ψ and p_{uu}^0 were fixed. However, while the rest energy density Ψ was a "defined quantity," it was not so clear how p_{uu}^0 was to be determined. It represented an energy density associated with the stresses. Clearly if there were no stresses in the material, then this energy would have to be zero. But what if there were stresses?

To proceed Nordström considered a special case, a body in which there is an isotropic, normal pressure. In this case, Nordström continued, it is possible to fix the value of p_{uu}^0 in such a way that the density v can be determined. The elastic stress tensor could be generated out of a single scalar invariant, which I will write here as p, so that the elastic stress tensor in the rest frame is given by

---

43 In Section 5, Nordström augmented his analysis by considering the effect of heat conduction. This was represented by a third symmetric tensor, $w_{\mu\nu}$, whose *only* non-zero components in the rest frame were exactly these six components. Thus heat conduction was represented by an energy current and associated momentum density which did not arise from stresses and which had no associated energy density in the rest frame.

44 This followed easily from the fact that the tensor $p_{\mu\nu}$ twice contracted with the four velocity $\mathfrak{B}_\mu$ yields a Lorentz invariant, $p_{\mu\nu}\mathfrak{B}_\mu\mathfrak{B}_\nu$, which can be evaluated in the rest frame, where $\mathfrak{B}_\mu = (0, 0, 0, ic)$, and turns out to be $-c^2 p_{uu}^0$.

45 The presence of the negative sign follows from the use of a coordinate system in which the fourth coordinate is $x_4 = u = ict$. Therefore $T_{44} = T_{uu} = -$ (energy density).

$$\begin{bmatrix} p & 0 & 0 & 0 \\ 0 & p & 0 & 0 \\ 0 & 0 & p & 0 \\ 0 & 0 & 0 & p \end{bmatrix} \tag{33}$$

With this particular choice of stress tensor, Nordström pointed out, there is no momentum density associated with the stresses when the body is in motion. We can confirm this conclusion merely by inspecting the matrix (33). Since p is an invariant, the matrix will transform back into itself under Lorentz transformation. Therefore in all frames of reference, the six components (p_{14}, p_{24}, p_{34}) and (p_{41}, p_{42}, p_{43}) will remain zero. But these six components between them represent the momentum density and energy current due the stresses. Thus any momentum density present in the body will be due to the density v.[46]

At this point, the reader might expect Nordström to recommend that one set p_{uu}^0 in the general case in such a way that there are no momentum densities associated with stresses. Nordström informs us, however, that he could find no natural way of doing this. As a result, he urged that the "simplest and most expedient definition" lies in setting

$$c^2 v = \Psi \tag{34}$$

so that $p_{uu}^0 = 0$. To quell any concern that this choice had been made with undue haste, Nordström continued by asserting that the factual content of relativistic mechanics is unaffected by the choice of quantity that represents inertial mass. It is only when weight is assigned to inertial masses, such as in his gravitation theory, that the choice becomes important.

The reader who has followed the development of Nordström's argument up to this point cannot fail to be perplexed at the indirectness of what is the core of the entire paper! There are three problems. First, the choice of v as given in (34) seems unchallengeable as the correct expression for the rest density of inertial mass. It merely sets this density equal to $1/c^2$ times the rest energy density—exactly as one would expect from Einstein's celebrated result of the inertia of energy. Indeed, any other division of total rest energy Ψ between the two terms of (32) would force us to say that v does not represent the total inertial rest mass density, for there would be another part of the body's energy it does not embrace. Second, no argument is given for the claim that the choice of v has no effect on the factual content of relativistic mechanics.[47] Finally, even if this second point is correct, it hardly seems worth much attention since the choice of expression for v does significantly affect the factual content of gravitation theory.

The clue that explains the vagaries of Nordström's analysis lies in his citation of his own (Nordström 1911). There one finds an elaborate analysis of the relativistic

---

46   Unless, of course, heat conduction is present.

mechanics of the special case of a body with isotropic normal stresses—exactly the special case considered above. The analysis began by representing the stresses through a tensor of form (33). Nordström then showed the effect of arbitrarily reset-ting the value of p_{uu}^0. Reverting to the notation of (Nordström 1911), Nordström imagined that the p_{uu}^0 term of (33) is replaced by some arbitrary U_0. He showed that the effect of this substitution is simply to replace the rest mass density γ (the analog of v in the 1913 paper) in the equations of the theory with an augmented

$$\gamma' = \gamma + \frac{p + U_0}{c^2}$$

without otherwise altering the theory's relations. He was able to conclude that setting $U_0 = p$ "is not a specialization of the theory, but only a specialization of concepts." In the introduction to (Nordström 1911), he had announced his plan to extend this analysis to the more general case with tangential stresses in another paper. Presum-ably the discussion of Section 4 in (Nordström 1913a) was intended to inform his readers that he was now unable to make good on his earlier plan. Indeed the remarks that seemed puzzling are merely a synopsis of some of the major points of (Nord-ström 1911). That the choice of (34) does not affect the factual content of relativistic mechanics is merely an extension of the result developed in detail in (Nordström 1911). It had become something of a moot point, however, in the context of gravita-tion theory.

To sum up, Nordström's choice of source density v was given by equation (34) and it was this result that gave meaning to the quantity v in the final sections of the paper in which his gravitation theory was recapitulated. We can give this quantity more transparent form by writing it in a manifestly covariant manner[48]

$$v = -\frac{1}{c^2} T_{\mu\nu} \mathfrak{B}_\mu \mathfrak{B}_\nu.$$ (35)

Natural as this choice seemed to Nordström, it was Einstein who shortly proclaimed that another term derived from the stress energy tensor was the only viable candidate and that this unique candidate led to disastrous results.

---

47 On reflection, however, I think the result not surprising. Barring special routes such as might be pro-
     vided through gravitation theory, we have no independent access to the energy represented by the term
     p_{uu}^0. For example, in so far as this energy is able to generate inertial effects, such as through genera-
     tion of a momentum density, it is only through its contribution to the sum $T_{uu} = -p_{uu}^0 + c^2 v$. The
     momentum density follows from the Lorentz transformation of the tensor $T_{\mu\nu}$. How we envisage the
     energy divided between the two terms of this sum will be immaterial to the final density yielded.
48 While $-c^2 v$ is the trace of the material tensor $v \mathfrak{B}_\mu \mathfrak{B}_\nu$, this quantity $-c^2 v$ is not the trace of the full
     tensor $T_{\mu\nu}$ as given in (31). This latter trace would contain terms in p_{xx}^0, etc.

11. EINSTEIN OBJECTS AGAIN

By early 1913, Einstein's work on his own gravitation theory had taken a dramatic turn. With his return to Zurich in August 1912, he had begun a collaboration with his old friend Marcel Grossmann. It culminated in the first sketch of his general theory of relativity, (Einstein and Grossmann 1913), the so-called "*Entwurf*" paper. This work furnished his colleagues all the essential elements of the completed theory of 1915, excepting generally covariant gravitational field equations.[49] While we now know that this work would soon be Einstein' s most celebrated achievement, the Einstein of 1913 could not count on such a jubilant reception for his new theory. He had already survived a bitter dispute with Abraham over the variability of the speed of light in his earlier theory of gravitation. And Einstein sensed that the lack of general covariance of his gravitational field equations was a serious defect of the theory which would attract justifiable criticism.

There was one aspect of the theory which dogged it for many years, its very great complexity compared with other gravitation theories. In particular, in representing gravitation by a metric tensor, Einstein had, in effect, decided to replace the single scalar potential of gravitation theories such as Newton's and Nordström's, with ten gravitational potentials, the components of the metric tensor. This concern was addressed squarely by Einstein in Section 7 of his part of the *Entwurf* paper. It was entitled "Can the gravitational field be reduced to a scalar?" Einstein believed he could answer this question decisively in the negative, thereby, of course, ruling out not just Nordström's theory of gravitation, but any relativistic gravitation theory which represented the gravitational field by a scalar potential.

Einstein's analysis revealed that he agreed with Nordström's assessment of the importance of Laue's work for gravitation theory. However he felt that Laue's work, in conjunction with the requirement of the equality of inertial and gravitational mass, pointed unambiguously to a different quantity as the gravitational source density. That was the trace of the stress-energy tensor. He proposed that a scalar theory would be based on the equation of motion for a point mass

$$\delta \left\{ \int \Phi ds \right\} = 0 \qquad (36)$$

where Φ is the gravitational potential, *ds* is the spacetime line element of special relativity and δ represents a variation of the mass' world line. He continued, tacitly comparing the scalar theory with his new *Entwurf* theory:

> Here also material processes of arbitrary kind are characterized by a stress-energy tensor $T_{\mu\nu}$. However in this approach a scalar determines the interaction between the gravitational field and material processes. This scalar, as Herr Laue has made me aware, can only be

---

49   For an account of this episode, see (Norton 1984).

$$\sum_{\mu} T_{\mu\mu} = P$$

I want to call it "Laue's scalar." Then one can do justice to the law of the equivalence of inertial and gravitational mass here also up to a certain degree. That is, Herr Laue has pointed my attention to the fact that, for a closed system,

$$\int P dV = \int T_{44} d\tau.$$

From this one sees that the weight of a closed system is determined by its total energy according to this approach as well.

Recall that Einstein's version of the requirement of the equality of inertial and gravitational mass seeks to use the total energy of a system as a measure of it as a gravitational source. The selection of P, the trace of the stress energy tensor, does this for the special case of one of Laue's complete static systems. For such a system, the integral of the trace P over the spatial volume V of the system is equal to the negative value of the total energy of the system[50] $\int T_{44} dV$ since

$$\int P dV = \int T_{11} + T_{22} + T_{33} + T_{44} dV = \int T_{44} dV. \tag{37}$$

The three terms in T_{11}, T_{22} and T_{33} in the integral vanish because of the fundamental property (29) of complete static systems. Notice that Einstein can only say he does justice to the equality of inertial and gravitational mass "up to a certain degree," since this result is known to hold only for complete static systems and then only in their rest frames.

Einstein's wording indicates direct personal communication from Laue. concerning the stress-energy tensor and complete static systems. Such personal communication is entirely compatible with the fact that both Einstein and Laue were then in Zurich, with Einstein at the ETH and Laue at the University of Zurich. Below, in Section 15, I will argue that there is evidence that, prior to his move to Zurich, Einstein was unaware of the particular application of Laue's work discussed here by him.

Einstein continued his analysis by arguing that this choice of gravitational source density was disastrous. It leads to a violation of the law of conservation of energy. He wrote:

The weight of a system that is not closed would depend however on the orthogonal stresses T_{11} etc. to which the system is subjected. From this there arise consequences which seem to me unacceptable as will be shown in the example of cavity radiation.[51]

---

50   Presumably Einstein's "$d\tau$" is a misprint and should read dV. In a coordinate system in which $x_4 = ict$, $T_{44} = -$(energy density), so that $\int T_{44} dV$ is the negative value of the total energy.
51   [JDN] One might well think that only Einstein could seriously ask after the gravitational mass of such an oddity in gravitation theory as radiation enclosed in a massless, mirrored chamber. Yet Planck (1908, 4) had already asked exactly this question.

For radiation in a vacuum it is well known that the scalar P vanishes. If the radiation is enclosed in a massless, mirrored box, then its walls experience tensile stresses that cause the system, taken as a whole, to be accorded a gravitational mass $\int P d\tau$ which corresponds with the energy E of the radiation.

Now instead of the radiation being enclosed in an empty box, I imagine it bounded
1. by the mirrored walls of a fixed shaft S,
2. by two vertically moveable, mirrored walls W_1 and W_2, which are firmly fixed to one another by a rod. (See Fig. 3.)

In this case the gravitational mass $\int P d\tau$ of the moving system amounts to only a third part of the value which arises for a box moving as a whole. Therefore, in raising the radiation against a gravitational field, one would have to expend only a third part of the work as in the case considered before, in which the radiation is enclosed in a box. This seems unacceptable to me.

Einstein's objection bears a little expansion. He has devised two means of raising and lowering some fixed quantity of radiation in a gravitational field. Notice that in either case the radiation by itself has no gravitational mass, since the trace of the stress-energy tensor of pure electromagnetic radiation vanishes. What introduces such a mass is the fact that the radiation is held within an enclosure upon which it exerts a pressure, so that the enclosure is stressed. Even though the members of the enclosure are assumed massless, it turns out that a gravitational mass must still be ascribed to them simply because they are stressed. The beauty of Einstein's argument is that the gravitational masses ascribed in each of the two cases can be inferred essentially without calculation.

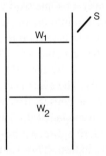

Figure 3: Rendering of Figure in Einstein's Text

In the first case, the radiation is moved in a mirrored box. The radiation and enclosing box form a complete static system. Therefore the gravitational mass of the box together with the radiation is given by the total energy of the radiation, $\int T_{44}^{em} dV$, in its rest frame, where I now write the stress-energy tensor of the electromagnetic radiation as $T_{\mu\nu}^{em}$. For ease of transition to the second case, it is convenient to imagine that each of the three pairs of opposing wall of the box is held in place *only* by a connecting rod, that the faces of the box are aligned with the x, y and z axes and that the disposition of the system is identical in these three directions. Each connecting rod will be stressed in reaction to the radiation pressure. I write $T^X_{\mu\nu}$ for the stress-energy tensor of the rod aligned in the x direction and the two stressed walls that this rod connects. $T^Y_{\mu\nu}$ and $T^Z_{\mu\nu}$ represent the other two corresponding systems. We can then infer directly from (37) that the gravitational mass of the entire system in its rest frame is proportional to

$$\int T^{\mathrm{em}}_{44} dV = \int T^{\mathrm{em}}_{\mu\mu} dV + \int T^X_{\mu\mu} dV + \int T^Y_{\mu\mu} dV + \int T^Z_{\mu\mu} dV = 3\int T^Z_{\mu\mu} dV.$$

The second equality follows from $T^{\mathrm{em}}_{\mu\mu} = 0$ and from the symmetry of the three axes, which entails

$$\int T^X_{\mu\mu} dV = \int T^Y_{\mu\mu} dV = \int T^Z_{\mu\mu} dV.$$

In Einstein's second case the radiation is trapped between sliding, mirrored baffles in a mirrored shaft aligned, let us say, in the z direction. The only component of the moveable system carrying a gravitation mass in this case will be the stressed rod and the stressed baffles it connects. Its gravitational mass in its rest frame will simply be proportional to the volume integral of the trace of its stress energy tensor. This integral is equal to $\int T^Z_{\mu\mu} dV$ and is one third of the corresponding integral for the first case, as Einstein claimed.

We now combine the two cases into a cycle. We lower the radiation inside the cube into the gravitational field, recovering some work, since the system has a gravitational mass. We then transfer the radiation into the baffle system and raise it. Only one third of the work released in the first step is needed to elevate the radiation because of the baffle system's reduced gravitational mass.[52] The mirrored cube and baffles are weightless once they are unstressed by the release of the radiation so they can be returned to their original positions. The cycle is complete with a net gain of energy.

That a theory should violate the conservation of energy is one of the most serious objections that Einstein could raise against it. Notice that he did not mention another possible objection that would derive directly from the vanishing of the trace of the stress-energy tensor of pure radiation. This vanishing entails that light cannot be deflected by a gravitational field. However, in 1913, prior to the experimental determination of this effect, Einstein could hardly have expected this last objection to have any force.

As devastating and spectacular as Einstein's objection was, he had at this time developed the unfortunate habit of advancing devastating arguments to prove conclusion he later wished to retract, see (Norton 1984, §5). This objection to all theories of

---

52 Since the estimates of gravitational mass are made in the system's rest frames, these motions would have to be carried out infinitely slowly.

gravitation with a scalar potential proved to be another instance of his habit. Within a few months Einstein had endorsed (if not initiated) a most interesting escape.

12. NORDSTRÖM'S SECOND THEORY

On July 24, 1913, Nordström submitted another version of his gravitation theory to *Annalen der Physik*, (Nordström 1913b). This version of the theory finally took proper notice of what Einstein had presented as obvious in his *Entwurf* paper. The only possible scalar that can represent gravitational source density is the trace of the stress-energy tensor and this choice, in conjunction with Laue's work on complete static systems, enables satisfaction of the requirement of equality of inertial and gravitational mass. Moreover the version of the requirement satisfied is an Einsteinian version in which the quantity of gravitational source is proportional to the total energy. This differs from the version embodied in Nordström's equation (16) in which the motion of a body in free fall is merely independent of its rest mass. Finally the theory offered an ingenious escape from Einstein's *Entwurf* objection. It turned out that the objection failed if one assumed that the proper length of a body would vary with the gravitational potential. This new version of this theory is sufficiently changed that it is now customarily known as Nordström's second theory.

There is room for interesting speculation on the circumstances under which Nordström came to modify his theory. In the introduction (p. 533) he thanked Laue and Einstein for identifying the correct gravitational source density. As we shall see, at two places in the paper (p. 544, 554), he also attributed arguments and results directly to Einstein without citation. Since I know of no place in which Einstein published these results, it seems a reasonable conjecture that Nordström learned of them either by correspondence or personal contact. That it was personal contact during a visit to Zurich at this time is strongly suggested by the penultimate line of the paper, which, in standard *Annalen der Physik* style, gives a place and date. It reads "Zurich, July 1913."[53] Since Nordström does thank both Laue and Einstein directly and in that order and since the wording of Einstein's *Entwurf* suggests a personal communication directly from Laue, we might conjecture also that there was similar direct contact between Laue and Nordström. Laue was also in Zurich at this time at the University of Zurich.

One cannot help but sense a somewhat sheepish tone in the introduction to (Nordström 1913b, 533), when he announced that this earlier presentation (Nordström 1913a) was "not completely unique" and that "the rest density of matter was defined in a fairly arbitrary way." In effect he was conceding that he had bungled the basic idea of his earlier (Nordström 1913a), that one had to take notice of the mechanics of stressed bodies in defining the gravitational source density and that this ought to be

53 An entry in Ehrenfest's Diary ("I", NeLR, Ehrenfest Archive, Scientific Correspondence, ENB: 4–15) reveals a visit to Zurich by Nordström in late June. See (CPAE 4, 294–301), "Einstein on Gravitation and Relativity: the Collaboration with Marcel Grossmann."

done in a way that preserved the equality of inertial and gravitational mass. Laue and Einstein were now telling him how he ought to have written that paper.

12.1 The Identification of the Gravitational Source

The first task of Nordström (1913b) was to incorporate the new source density into his theory and this was tackled in its first section. The final result would be to define this density in terms of what he called the "elastic-material tensor" $T_{\mu\nu}$, which corresponded to the sum of Nordström's (1913a) material tensor and elastic stress tensor as given above in (31). Following Einstein and Laue, he ended up selecting $1/c^2$ times the negative[54] trace D of $T_{\mu\nu}$ as his source density ν

$$\nu = \frac{1}{c^2}D = -\frac{1}{c^2}(T_{xx} + T_{yy} + T_{zz} + T_{uu}). \tag{38}$$

However unlike Laue and Einstein, that selection came at the conclusion of a fairly lengthy derivation. Nordström would show that the requirement of equality of inertial and gravitational mass in the case of a complete static system would force this choice of source density.

To begin, Nordström chose essentially the same field equation for the potential Φ and gravitational force density equations as in (Nordström 1913a):

$$\frac{\partial^2\Phi}{\partial x^2} + \frac{\partial^2\Phi}{\partial y^2} + \frac{\partial^2\Phi}{\partial z^2} + \frac{\partial^2\Phi}{\partial u^2} = g(\Phi)\nu, \tag{12''}$$

$$\mathfrak{K}_\mu^g = -g(\Phi)\nu\frac{\partial\Phi}{\partial x_\mu}. \tag{18'}$$

Here ν remained the as yet undetermined gravitational source density. The important innovation was that the gravitation factor g was now allowed to vary as a function of the potential Φ. In an attempt to bring some continuity to the development of (Nordström 1913b) from (1913a), he recalled that in the former paper (p. 873, 878) he had foreshadowed the possibility that g might be a function of the inner constitution of bodies. Indeed the paper had closed with the speculation that such a dependence might enable the molecular motions of a falling body to influence its acceleration of fall, presumably as part of a possible escape from Einstein's original objection to his theory.

As it happened, however, the Φ dependence of g was introduced for an entirely different purpose in (Nordström 1913b). Einstein's version of the equality of inertial and gravitational mass required that the *total* energy of a system would be the measure of its gravitational source strength. This total energy would include the energy of the gravitational field itself. This requirement, familiar to us from Einstein's treat-

---

54 Since he retained his standard coordinate system of $x, y, z, u = ict$, the negative sign is needed to preserve the positive sign of ν.

ment of general relativity, leads to non-linearity of field equations. In Nordström's theory, this non-linearity would be expressed as a Φ dependence of g.

Nordström now turned to complete static systems, whose properties would yield not just the relation (38) but also a quite specific function for $g(\Phi)$. At this point Nordström seemed able to give Laue a little taste of his own medicine. As I pointed out above, Laue's 1911 definition of a complete static system had excluded such systems as bodies rotating uniformly about their axis of symmetry. Nordström now made the obvious extension, defining what he called a "complete *stationary*" system. Curiously he made no explicit statement that his was a more general concept. Readers simply had to guess that his replacement of Laue's "static" by his "stationary" was no accident. Or perhaps they had to wait until Laue's (1917, 273) own concession that Nordström was first to point out the extension.

Nordström's complete stationary system had the following defining characteristics: it was a system of finite bodies for which a "justified"[55] reference system existed in which the gravitational field was static, that is, $\partial\Phi/\partial t = 0$. In particular in the relevant reference system, instead of Laue' s condition (28) which required the vanishing everywhere of the momentum density $\mathfrak{g}$, Nordström required merely that the *total* momentum $\mathfrak{G}$ vanished,

$$\mathfrak{G} = \int \mathfrak{g}\, dv = 0$$

where v is the volume of the body in its rest frame. The two illustrations Nordström gave—surely not coincidentally—were exactly two systems that Laue's earlier definition did not admit: a body rotating about its axis of symmetry and a fluid in stationary flow. Of course the first example was one of great importance to Nordström. It was precisely the example discussed in the final paragraphs of both (Nordström 1912 and 1913a). That such a system would fall more slowly than a non-rotating system was the substance of Einstein's original objection. Now able to apply the machinery of Laue's complete static systems to this example, Nordström could try to show that these rotating bodies did not fall slower in the new theory.

Nordström proceeded to identify the three stress-energy tensors which could contribute to the total energy of a complete stationary system. They were the "elastic-material tensor" $T_{\mu\nu}$, mentioned above; the "electromagnetic tensor" $L_{\mu\nu}$, which we would otherwise know as the stress-energy tensor of the electromagnetic field; and finally the "gravitation tensor" $G_{\mu\nu}$. This last tensor was the stress-energy tensor of the gravitational field itself. It had been identified routinely in (Nordström (1912, 1128; 1913a, 875). It was given by[56]

$$G_{\mu\nu} = \frac{\partial\Phi}{\partial x_\mu}\frac{\partial\Phi}{\partial x_\nu} - \frac{1}{2}\delta_{\mu\nu}\frac{\partial\Phi}{\partial x_\lambda}\frac{\partial\Phi}{\partial x_\lambda} \tag{39}$$

---

55   In this context I read this to mean "inertial".

where Φ is the gravitational potential. Invoking Laue's basic result (29), which was also used by Einstein for the same end, Nordström could represent the total energy E_0 of a complete stationary system in its rest frame as the integral over all space of the sum of the traces of these three tensors

$$E_0 = -\int T + G + L \ dv \ .$$

However we have $L = 0$ and have written $T = -D$. Finally the integral of the trace G could be written in greatly simplified fashion if one assumed special properties for the complete stationary system. In particular, the gravitational potential Φ must approach the limiting constant value Φ_a at spatial infinity. From this and an application of Gauss' theorem, Nordström inferred that[57]

$$\int G \ dv = \int (\Phi - \Phi_a) g(\Phi) v \ dv \ . \tag{40}$$

Combining and noting that the inertial rest mass m of the system is E_0/c^2, we have

$$m = \frac{E_0}{c^2} = \frac{1}{c^2} \int \{ D - (\Phi - \Phi_a) g(\Phi) v \} dv \ .$$

However we also have the total gravitational mass M_g of the system is

$$M_g = \int g(\Phi) v dv \ . \tag{41}$$

Nordström was now finally able to invoke what he calls "Einstein's law of equivalence," the equality of inertial and gravitational mass. Presumably viewing the completely stationary system from a great distance, one sees that it is a system with inertial mass m lying within a potential Φ_a so that we must be able to write

---

56   I continue to compress Nordström's notation. He did not use the Kronecker delta $\delta_{\mu\nu}$ and wrote individual expressions for G_{xx}, G_{xy}, etc. The derivation of (39) is brief and entirely standard. Writing ∂_μ for $\partial/\partial x_\mu$ we have, from substituting (12") into (18')

$$\mathfrak{R}^g_\mu = -g(\Phi) v \partial_\mu \Phi = -(\partial_\nu \partial_\nu \Phi) \partial_\mu \Phi = -\left(\partial_\nu (\partial_\mu \Phi \partial_\nu \Phi) - \frac{1}{2} \partial_\mu (\partial_\lambda \Phi \partial_\lambda \Phi) \right)$$

$$= -\partial_\nu \left(\partial_\mu \Phi \partial_\nu \Phi - \frac{1}{2} \delta_{\mu\nu} (\partial_\lambda \Phi \partial_\lambda \Phi) \right) = -\partial_\nu G_{\mu\nu} \ .$$

57   Nordström's derivation of (40) seems to require the additional assumption that v is non-zero only in some finite part of space. For a completely stationary system, the field equation reduces to the Newtonian $\partial_i \partial_i \Phi = g(\Phi) v$. If v satisfies this additional assumption one can recover from it Nordström's result (his equation (5)), that $|\nabla \Phi| = |\partial_i \Phi| = \frac{1}{4\pi r^2} \int g(\Phi) v dv$, at large distances r from the system. This result seems to be needed to complete the derivation of (40).

$$M_g = g(\Phi_a)m .$$

These last three equations form an integral equation which can only be satisfied identically if

$$D = g(\Phi)v\left\{\Phi - \Phi_a + \frac{c^2}{g(\Phi_a)}\right\} .$$

Finally Nordström required that this relation between D and v be independent of Φ_a, from which the two major result of the analysis followed: first, $g(\Phi)$ is given by

$$g(\Phi) = \frac{c^2}{A + \Phi} \tag{42}$$

where A is a universal constant; second, he recovered the anticipated identity of the source density v

$$v = \frac{1}{c^2}D = -\frac{1}{c^2}(T_{xx} + T_{yy} + T_{zz} + T_{uu}) . \tag{38}$$

The constant A in equation (42) is taken by Nordström to be an arbitrary additive gauge factor, corresponding to the freedom in Newtonian gravitation theory of setting an arbitrary zero point for the gravitational potential. However, in contrast with the Newtonian case, there is a natural gauge of $A = 0$ in which the equations are greatly simplified. Writing the potential that corresponds to the choice of $A = 0$ as Φ', the expression for $g(\Phi')$ is

$$g(\Phi') = \frac{c^2}{\Phi'} \tag{43}$$

and in this gauge one recovers a beautifully simple expression for the relationship between the total rest energy E_0, inertial rest mass m and gravitational mass M_g of a completely stationary system

$$E_0 = mc^2 = M_g\Phi'_a . \tag{44}$$

In particular, it contains exactly the Newtonian result that the energy of a system with gravitational mass M_g in a gravitational field with potential Φ'_a is $M_g\Phi'_a$. This was an improvement over Nordström's first theory were the closest corresponding result was (15), a dependence of mass on an exponential function of the potential.

12.2 Dependence of Lengths and Times on the Gravitational Potential

Satisfactory as these results were, Nordström had not yet answered the objection of Einstein's *Entwurf* paper. Indeed it is nowhere directly mentioned in Nordström's paper, although Nordström (1913b, 544) does cite the relevant part of the *Entwurf* paper to acknowledge Einstein's priority concerning the expression for the gravitational source density in equation (38). However at least the second, third and fourth parts of Nordström (1913b) are devoted implicitly to escaping from Einstein's objection. Nordström there developed a model of a spherical electron within his theory and showed that its size must vary inversely with the gravitational potential Φ'. It only becomes apparent towards the middle of Section 3 that this result must hold for the dimensions of all bodies and that this general result provides the escape from Einstein's objection. The general result is demonstrated by an argument attributed without any citation to Einstein. Since I know of no place where this argument was published by Einstein and since we know that Nordström was visiting Einstein at the time of submission of his paper, it is a reasonable supposition that he had the argument directly from Einstein in person. Since the general result appears only in the context of this argument, it is a plausible conjecture that the result, as well as its proof, is due to Einstein. Of course the successful recourse to an unusual kinematical effect of this type is almost uniquely characteristic of Einstein's work.

Einstein's argument takes the *Entwurf* objection and reduces it to its barest essentials. The violation of energy conservation inferred there depended solely on the behavior of the massless members of the systems that were oriented transverse to the direction of the field in which they moved. The transverse members lowered were stressed and thus were endowed with a gravitational mass so that work was recovered. The transverse members elevated were unstressed so that they had no gravitational mass and no work was required to elevate them. The outcome was a net gain in energy. Einstein's new argument considered this effect in a greatly simplified physical system. Instead of using radiation pressure to stress the transverse members, Einstein now just imagined a single transverse member—a non-deformable rod—tensioned between vertical rails. The gravitational mass of the rod is increased by the presence of these stresses. Since the rod expands on falling into the gravitational field, the rails must diverge and work must be done by the forces that maintain the tension in the rod. It turns out that this work done exactly matches the work released by the fall of the extra gravitational mass of the rod due to its stressed state. The outcome is that no net work is released. In recounting Einstein's argument, Nordström makes no mention of Einstein's *Entwurf* objection. This is puzzling. It is hard to imagine that he wished to avoid publicly correcting Einstein when the history of the whole theory had been a fruitful sequence of objections and correction and this correction was endorsed and even possibly invented by Einstein. In any case it should have been clear to a contemporary reader who understood the mechanism of Einstein's objection that the objection would be blocked by an analogous expansion of the systems as they fell into the gravitational field. Certainly Einstein (1913, 1253), a few months later, reported that his *Entwurf* objection failed because of the tacit assumption of the constancy of dimensions of the systems as they move to regions of different potential.[58]

Einstein's argument actually establishes that the requirement of energy conservation for such cycles necessitates a presumed isotropic expansion of linear dimensions to be in inverse proportion to the gravitational potential Φ'. Nordström's report of the argument reads:

> Herr Einstein has proved that the dependence in the theory developed here of the length dimensions of a body on the gravitational potential must be a general property of matter. He has shown that otherwise it would be possible to construct an apparatus with which one could pump energy out of the gravitational field. In Einstein's example one considers a non-deformable rod that can be tensioned moveably between two vertical rails. One could let the rod fall stressed, then relax it and raise it again. The rod has a greater weight when stressed than unstressed, and therefore it would provide greater work than would be consumed in raising the unstressed rod. However because of the lengthening of the rod in falling, the rails must diverge and the excess work in falling will be consumed again as the work of the tensioning forces on the ends of the rod.
>
> Let S be the total stress (stress times cross-sectional area) of the rod and l its length. Because of the stress, the gravitational mass of the rod is increased by[59]
>
> $$\frac{g(\Phi)}{c^2} Sl = \frac{1}{\Phi'} Sl \,.$$
>
> In falling [an infinitesimal distance in which the potential changes by $d\Phi'$ and the length of the rod by dl], this gravitational mass provides the extra work
>
> $$-\frac{1}{\Phi'} Sl d\Phi'$$
>
> However at the same time at the ends of the rod the work is lost [to forces stressing the rod]. Setting equal these two expressions provides
>
> $$-\frac{1}{\Phi'} d\Phi' = \frac{1}{l} dl$$
>
> which yields on integration
>
> $$l\Phi' = \text{const.} \qquad\qquad [(45)]$$
>
> This, however, corresponds with [Nordström's] equation (25a) [the potential dependence of the radius of the electron].[60]

This result (45) was just one of a series of dependencies of basic physical quantities on gravitational potential. In preparation in Section 2 for his analysis of the electron,

---

58 He gives no further analysis. It is clear, however, that as the mirrored box is lowered into the field and expands, work would be done by the radiation pressure on its walls. It will be clear from the ensuing analysis that this work would reduce the total energy of the radiation by exactly the work released by the lowering of the gravitational mass of the system. Thus when the radiation is elevated in the mirrored shaft the radiation energy recovered would be diminished by exactly the amount needed to preserve conservation of energy in the entire cycle.

59 [JDN] To see this, align the x axis of the rest frame with the rod. The only non-vanishing component of $T_{\mu\nu}$ is T_{xx} which is the stress (per unit area) in the rod. Therefore, from (38) and (41), the gravitational mass

$$M_g = \int g(\Phi) v \, dv = -\frac{g(\Phi)}{c^2} Sl \,.$$

Nordström had already demonstrated that the inertial mass m of a complete stationary system varied in proportion to the external gravitational potential Φ':

$$\frac{m}{\Phi'} = \text{const.} \qquad (46)$$

whereas the gravitational mass of the system M_g was independent of Φ'. The proof considered the special case of a complete stationary system for which the external field Φ'_a would be uniform over some sphere at sufficient distance from the system and directed perpendicular to the sphere. We might note that a complete stationary system of finite size within a constant potential field would have this property. He then imagined that the external field Φ'_a is altered by a slow displacement of yet more distant masses. He could read directly from the expression for the stress energy tensor of the gravitational field what was the resulting energy flow through the sphere enclosing the system and from this infer the alteration in total energy and therefore mass of the system. The result (46) followed immediately and from it the constancy of M_g through (44).[61] Calling on (45), (46) and other specific results in his analysis of the electron, Nordström was able to infer dependencies on gravitational potential for the stress S in the electron's surface, the gravitational source density v and the stress tensor p_{ab}:

$$\frac{S}{\Phi'^3} = \text{const.} \quad \frac{v}{\Phi'^4} = \text{const.} \quad \frac{P_{ab}}{\Phi'^4} = \text{const.}$$

Finally, after these results had been established, the entire Section 5 was given to establishing that the time of a process T would depend on the potential Φ' according to

$$T\Phi' = \text{const.} \qquad (47)$$

In particular it followed from this result that the wave lengths of spectral lines depends on the gravitational potential. Nordström reported that the wavelength of light from the Sun's surface would be increased by one part in two million. He continued to report that "The same—possibly even observable—shift is given by several other new theories of gravitation." (p. 549) While he gave no citation, modern readers need hardly be told that all of Einstein's theories of gravitation from this period give this effect, including his 1907–1912 scalar theory of static fields, his *Entwurf* form of

---

60   A footnote here considers a complication that need not concern us. It reads:
      "If the rod is deformable, in stressing it, some work will be expended and the rest energy of the rod will be correspondingly increased. Thereby the weight also experiences an increase, which provides the added work dA in falling. However, since in falling the rest energy diminishes, the work recovered in relaxing the rod is smaller than that consumed at the stressing and the difference amounts to exactly dA."

61   On pp. 545–46, he showed that, for a complete stationary system, the external potential Φ'_a varied in direct proportion to the potential Φ' at some arbitrary point within the system, so that Φ' and Φ'_a could be used interchangeably in expressing the proportionalities of the form of (45), (46) and (47).

general relativity and the final 1915 generally covariant version of the theory. The very numerical value that Nordström reported—one part in two million—was first reported in Einstein's earliest publication on gravitation, (Einstein 1907a, 459).

Nordström devoted some effort to the proof of (47). He noted that it followed immediately from (45) and the constancy of the velocity of light for the time taken by a light signal traversing a a rod. Anxious to show that it held for other systems, he considered a small mass orbiting another larger mass M_1 in a circular orbit of radius r within an external potential field Φ'_a. The analysis proved very simple since the speed of the small mass, its inertial mass and the potential along its trajectory were all constant with time. He showed that its orbital period T satisfied

$$\frac{M_1 c^2}{4\pi r \Phi'_a} = \frac{4\pi^2 r^2}{T^2}. \tag{48}$$

As the potential Φ'_a varies, $r\Phi'_a$ remains constant according to (45). Therefore the equality requires that T vary in direct proportion to r from which it follows that T satisfies (47). Again he showed the same effect for the period of a simple harmonic oscillator of small amplitude.

12.3 Applications of Nordström's Second Theory: The Spherical Electron and Free Fall

So far we have seen the content of Nordström's second theory and how he established its coherence. The paper also contained two interesting applications of the theory. The first was an analysis of a spherical electron given in his Section 3. It turned out to yield an especially pretty illustration of the result that a gravitational mass is associated with a stress. For the *entire* gravitational mass of Nordström's electron proves to be due to an internal stress. The electron was modelled as a massless shell carrying a charge distributed on its surface. (See Appendix for details.) The shell must be stressed to prevent mechanical disintegration of the electron due to repulsive forces between parts of the charge distribution. The electric field does not contribute to the electron's gravitational mass since the trace of its stress-energy tensor vanishes. Since the shell itself is massless it also does not contribute to the gravitational mass when unstressed. However, when stressed in reaction to the repulsive electric forces, it acquires a gravitational mass which comprises the entire gravitational mass of the electron.

Nordström's model of the electron was not self contained in the sense that it required only known theories of electricity and gravitation. Like other theories of the electron at this time, it had to posit that the stability of the electron depended on the presence of a stress bearing shell whose properties were largely unknown. While one might hope that the attractive forces of gravitation would replace this stabilizing shell, that was not the case in Nordström's electron. Rather he was superimposing the effects of gravitation on a standard model of the electron.[62]

The second illustration was an analysis in his final Section 7 of the motion in free fall of a complete stationary system. In particular, Nordström was concerned to determine just how close this motion was to the corresponding motion of a point mass. The results were not entirely satisfactory. He was able to show that complete stationary systems fell like point masses only for the case of a homogeneous external gravitational field, that is, one whose potential was a linear function of all four coordinates (x, y, z, u). He showed that a complete stationary system of mass m, falling with four velocity $\mathfrak{B}_\mu$ in a homogeneous external field Φ_a, obeys equations of motion

$$-g(\Phi'_a)m\frac{\partial\Phi_a}{\partial x_\mu} = \frac{d}{d\tau}m\mathfrak{B}_\mu \qquad (49)$$

which corresponded to the equations (13) for a point mass in his first theory, excepting the added factor $g(\Phi'_a)$. Allowing that the mass m varies inversely with Φ'_a through equation (46), it follows that explicit mention of the mass m can be eliminated from these equations of motion which become[63]

$$-c^2\frac{\partial}{\partial x_\mu}\ln\Phi'_a = \frac{d}{d\tau}\mathfrak{B}_\mu - \mathfrak{B}_\mu\frac{d}{d\tau}\ln\Phi_a .$$

Nordström's concern was clearly still Einstein's original objection to this first theory recounted above in Section 7. A body rotating about its axis of symmetry could form a complete stationary system. He could now conclude that such a body would fall exactly as if it had no rotation, contrary, as he noted, to the result of his earlier theory. Also, he concluded without further discussion that molecular motions would have no influence on free fall. However, the vertical acceleration of free fall would continue to be slowed by its initial velocity according to (19) of his first theory.

We might observe that stresses would play a key role in the cases of the rotating body and the kinetic gas. The rotating body would be stressed to balance centrifugal forces and the walls containing a kinetic gas of molecules would be stressed by the forces of the gas pressure. These stresses would add to the gravitational mass of the spinning body and the contained gas allowing them to fall independently of their internal motions. No such compensating stresses would be present in the case of a point mass or a complete stationary system projected horizontally, so they would fall slower due to their horizontal velocity.

Thus, while Nordström's theory finally satisfied the requirement of equality of inertial and gravitational mass in Einstein's sense, it still did not satisfy the requirement that all bodies fall alike in a gravitational field. This Einstein (1911, §1) called "Galileo's principle," elsewhere (1913, 1251) citing it as the fact of experience supporting the equality of inertial and gravitational mass. Galileo's principle held only under rather restricted conditions: the system must be in vertical fall and in a homo-

---

62 We can see the Nordström had no real hope of eliminating this shell with gravitational attraction. For the electric field by itself generates no gravitational field in his theory. Another element must be present in the structure of the electron if gravitational forces are to arise.

63 I have corrected Nordström's incorrect "+" to "−" on the right-hand side.

geneous field.[64] At least, however, he could report that Einstein had extended the result to systems that were not complete stationary systems. He had shown that the *average* acceleration of an elastically oscillating system accorded with (19). Since this last result is nowhere reported in Einstein's publications, we must assume that he had it directly from Einstein.

Finally, in the course of his exposition, Nordström could note that the mass dependence on Φ' of relation (46) now replaces the corresponding condition (15) of his first theory. The new variable factor of $g(\Phi') = c^2/\Phi'$ in the in the equation of motion (49) causes (14) to be replaced by $\dfrac{mc^2}{\Phi'}\dfrac{d\Phi'}{d\tau} = c^2\dfrac{dm}{d\tau}$, which integrates to yield (46).

13. EINSTEIN FINALLY APPROVES:
THE VIENNA LECTURE OF SEPTEMBER 1913

In September 1913, Einstein attended the 85th Congress of the German Natural Scientists and Physicians. There he spoke on the subject of the current state of the problem of gravitation, giving a presentation of his new *Entwurf* theory and engaging in fairly sharp dispute in discussion. A text for this lecture with ensuing discussion was published in the December issue of *Physikalische Zeitschrift* (Einstein 1913). Einstein made clear (p. 1250) his preference for Nordström's theory over other gravitation theories, including Abraham's and Mie's. Nordström's latest version of his gravitation theory was the only competitor to Einstein's own new *Entwurf* theory satisfying four requirements that could be asked of such gravitation theories:

   1. "Satisfaction of the conservation law of momentum and energy;"

   2. "Equality of inertial and gravitational masses of closed systems;"

   3. Reduction to special relativity as a limiting case;

   4. Independence of observable natural laws from the absolute value of the gravitational potential.

What Einstein did not say was that the satisfaction of 1. and 2. by Nordström's theory was due in significant measure to Einstein's pressure on Nordström and Einstein's own suggestions.

Einstein devoted a sizeable part of his lecture to Nordström's theory, giving a self-contained exposition of it in his Section 3. That exposition was a beautiful illustration of Einstein's ability to reduce the complex to its barest essentials and beyond. He simplified Nordström's development in many ways, most notably:

---

64   One might think that this would give Einstein grounds for rejecting the competing Nordström theory in favor of his own *Entwurf* theory. At the appropriate place, however, Einstein (1913, 1254) did *not* attack Nordström on these grounds. Perhaps that was for the better since it eventually turned out that general relativity fared no better. In general relativity, for example, a rotating body falls differently, in general, from a non-rotating body. See (Papapetrou 1951; Corinaldesi and Papapetrou 1951).

- Einstein selected the natural gauge (43) for the potential Φ', writing the resulting potential without the prime as φ.
- Einstein eradicated the implicit potential dependence of the mass m in (46), using a new mass m which did not vary with potential. This meant that Einstein's m coincided with the gravitational mass, not the inertial mass of a body.

To begin, Einstein used as the starting point the "Hamiltonian" equation of motion (36) which he had first recommended in Section 7 of his *Entwurf* paper. Using coordinates $(x_1, x_2, x_3, x_4) = (x, y, z, ict)$, he wrote this equation of motion of a mass m as

$$\delta\left\{\int H\, dt\right\} = 0 \qquad (50)$$

where

$$H = -m\varphi\frac{d\tau}{dt} = m\varphi\sqrt{c^2 - q^2}\ .$$

Here q is the coordinate three-speed and τ is the Minkowski interval given by

$$d\tau^2 = -dx^2 - dy^2 - dz^2 + c^2 dt^2\ . \qquad (51)$$

Since Einstein varied the three spatial coordinates of the particle trajectory x_i, the resulting equation of motion governed the three-velocity $\dot{x}_i = \dfrac{dx_i}{dt}$

$$\frac{d}{dt}\left\{m\varphi\frac{\dot{x}_i}{\sqrt{c^2 - q^2}}\right\} - m\frac{\partial\varphi}{\partial x_i}\sqrt{c^2 - q^2} = 0\ .$$

It also followed that the momentum (increased by a multiplicative factor c) I_i and the conserved energy E were given by

$$I_i = m\varphi\frac{\dot{x}_i}{\sqrt{c^2 - q^2}}\ , \quad E = m\varphi\frac{c^2}{\sqrt{c^2 - q^2}}\ .$$

In particular, one could read directly from these formulae that the inertial mass of a body of mass m at rest is given by $m\varphi/c^2$ and that its energy is $m\varphi$.

Einstein then introduced the notion that had rescued Nordström's theory from his own recent attack: directly measured lengths and times might not coincide with those given by the Minkowski line element (51). He called the former quantities "natural" and indicated them with a subscript 0. He called the latter "coordinate" quantities. The magnitude of the effect was represented by a factor ω which would be a function of φ and was defined by

$$d\tau_0 = \omega d\tau\ . \qquad (52)$$

Allowing for the dependence of energy on φ and the effects of the factor ω, Einstein developed an expression for the stress-energy tensor $T_{\mu\nu}$ of "flowing, incoherent matter"—we would now say "pressureless dust"—in terms of its natural mass density ρ_0 and the corresponding gravitational force density k_μ:

$$T_{\mu\nu} = \rho_0 c \varphi \omega^3 \frac{dx_\mu dx_\nu}{dt \ dt} , \quad k_\mu = -\rho_0 c \varphi \omega^3 \frac{\partial \varphi}{\partial x_\mu} .$$

The two quantities were related by the familiar conservation law

$$\frac{\partial T_{\mu\nu}}{\partial x_\nu} = k_\mu .$$

The next task was to re-express this conservation law in terms of the trace $T_{\sigma\sigma}$ of the stress-energy tensor. Mentioning Laue's work, Einstein remarked that this quantity was the only choice for the quantity measuring the gravitational source density. For the special case of incoherent matter, $T_{\sigma\sigma} = -\rho_0 c \varphi \omega^3$, so that the conservation law took on a form independent of the special quantities involved in the case of incoherent matter flow

$$\frac{\partial T_{\mu\nu}}{\partial x_\nu} = T_{\sigma\sigma} \frac{1}{\varphi} \frac{\partial \varphi}{\partial x_\mu} , \tag{53}$$

Einstein announced what was really an assumption: this form of the law governed arbitrary types of matter as well.

   This general form of the conservation law allowed Einstein to display the satisfaction by the theory of the second requirement he had listed. That was the equality of inertial and gravitational masses of closed systems. His purpose in including the additional words "closed systems" now became clear. In effect he meant by them Laue's complete static systems. His demonstration of the satisfaction of this result was admirably brief but damnably imprecise, compared to the careful attention Nordström had lavished on the same point. Einstein simply assumed that he had a system over whose spatial extension there was little variation in the φ term $\frac{1}{\varphi} \frac{\partial \varphi}{\partial x_\mu} = \frac{\partial \log \varphi}{\partial x_\mu}$

on the right-hand side of (53). An integration of (53) over the spatial volume v of such a system revealed that the four-force acting on the body is

$$\frac{\partial \log \varphi}{\partial x_\mu} \int T_{\sigma\sigma} dv = \frac{\partial \log \varphi}{\partial x_\mu} \int T_{44} dv ,$$

where the terms in T_{11}, T_{22}, and T_{33} were eliminated by Laue's basic result (29). Since $\int T_{44} dv$ is the negative of the total energy of the system, Einstein felt justified to conclude: "Thereby is proven that the weight of a closed system is determined by its total measure [of energy]." Einstein's readers might well doubt this conclusion and suspect that the case of constant φ considered was a special case that may not be representative of the general case. Fortunately such readers could consult (Nordström 1913b) for a more precise treatment.

In his lecture, Einstein was seeking to give an exposition of both Nordström's and his new theory of gravitation and reasons for deciding between them. Thus we might anticipate that he had to cut corners somewhere. And that place turned out to be the singular novelty of Nordström's theory in 1913, the potential dependence of lengths and times. His introduction of this effect and concomitant retraction of his *Entwurf* objection was so brief that only someone who had followed the story closely and read the report of Einstein's argument in (Nordström 1913b) could follow it. Virtually all he had to say lay in a short paragraph (p. 1253):

> Further, equation [(53)] allows us to determine the function [ω] of φ left undetermined from the physical assumption that no work can be gained from a static gravitational field through a cyclical process. In §7 of my jointly published work on gravitation with Herr Grossmann I generated a contradiction between the scalar theory and the fundamental law mentioned. But I was there proceeding from the tacit assumption that ω = const[ant]. The contradiction is resolved, however, as is easy to show, if one sets[65]
>
> $$ l = \frac{l_0}{\omega} = \frac{\text{const.}}{\varphi} $$
>
> or
>
> $$ \omega = \text{const.} \cdot \varphi . \qquad\qquad [(54)] $$
>
> We will give yet a second substantiation for this stipulation later.

That second substantiation followed shortly, immediately after Einstein had given the field equation of Nordström's theory. He considered two clocks. The first was a "light clock," a rod of length l_0 with mirrors at either end and a light signal propagating in a vacuum and reflected between them. The second was a "gravitation clock," two gravitationally bound masses orbiting about one another at constant distance l_0. He gave no explicit analysis of these clocks. His only remark on their behavior was that their relative speed is independent of the absolute value of the gravitational potential, in accord with the fourth of the requirements he had laid out earlier for gravitation theories. This, he concluded, "is an indirect confirmation of the expression for ω given in equation [(54)]."

Einstein's readers would have had to fill in quite a few details here. Clearly the dependence of l_0 on the potential would cause the period of the light clock also to vary according to (47). But readers would also need to know of the analysis of the gravitation clock given by (Nordström 1913b) which led to (48) above and the same dependence on potential for the clock's period. Thus the dependence of both periods is the same so that the relative rate of the two clocks remains the same as the external potential changes. Had this result been otherwise, the fourth requirement would have been violated. That it was not presumably displays the coherence of the theory and thereby provides the "*indirect* confirmation." Curiously Einstein seems not to be

---

65   *l* is defined earlier as the length of a body. This retraction is also mentioned more briefly (p. 261) in the addendum to the later printing of (Einstein and Grossmann 1913) in the *Zeitschrift*.

making the obvious point that his equations (52) and (54) together yield the same potential dependence for periodic processes as follows from the behavior of these two clocks—or perhaps he deemed that point too obvious to mention.

The final component of the theory was its field equation. Recalling that "Laue's scalar" $T_{\sigma\sigma}$ must enter into this equation, Einstein simply announced it to be

$$-\kappa T_{\sigma\sigma} = \varphi \frac{\partial^2}{\partial x_\tau^2} \varphi . \tag{55}$$

It became apparent that the additional factor of φ on the right-hand side was included to ensure compatibility with the conservation of energy and momentum.[66] To display this compatibility he noted that stress-energy tensor $t_{\mu\nu}$ of the gravitational field is

$$t_{\mu\nu} = \frac{1}{\kappa} \left\{ \frac{\partial \varphi}{\partial x_\mu} \frac{\partial \varphi}{\partial x_\nu} - \frac{1}{2} \delta_{\mu\nu} \left(\frac{\partial \varphi}{\partial x_\tau} \right)^2 \right\} .$$

This tensor satisfies the equalities

$$T_{\sigma\sigma} \frac{1}{\varphi} \frac{\partial \varphi}{\partial x_\mu} = -\frac{1}{\kappa} \frac{\partial \varphi}{\partial x_\mu} \frac{\partial^2 \varphi}{\partial x_\tau^2} = -\frac{\partial t_{\mu\nu}}{\partial x_\nu}$$

The first depends on substitution of $T_{\sigma\sigma}$ by the field equation and the second holds identically. Substituting into the conservation law (53) yields an expression for the joint conservation of gravitational and non-gravitational energy momentum,[67]

$$\frac{\partial}{\partial x_\nu} (T_{\mu\nu} + t_{\mu\nu}) = 0 .$$

All that remained for Einstein was to give his reasons for not accepting Nordström's theory. In our time, of course, the theory is deemed an empirical failure because it does not predict any deflection of a light ray by a gravitational field and does not explain the anomalous motion of Mercury. However in late 1913, there had been no celebrated eclipse expeditions and Einstein's own *Entwurf* theory also did not explain the anomalous motion of Mercury. Thus Einstein's sole objection to the theory was not decisive, although we should not underestimate its importance to Einstein.

---

66 Although Einstein does not make this point, it is helpful to divide both sides by φ and look upon $T_{\sigma\sigma}/\varphi$ as the gravitational source density. The trace $T_{\sigma\sigma}$ represents the mass-energy density and division by φ cancels out this density's φ dependence to return the gravitational mass density.

67 As Michel Janssen has repeatedly emphasized to me, Einstein's analysis is a minor variant of the method he described and used to generate the field equations of his *Entwurf* theory. Had Einstein begun with the identity mentioned, the expression for $t_{\mu\nu}$ and the conservation law (53), a reversal of the steps of Einstein's argument would generate the field equation. For further discussion of Einstein's method, see (Norton 1995).

According to Nordström's theory, the inertia of a body with mass m was $m\varphi/c^2$. Therefore, as the gravitational field in the neighborhood of the body was intensified by, for example, bringing other masses closer, the inertia of the body would actually decrease. This was incompatible with Einstein's idea of the "relativity of inertia" according to which the inertia of a body was caused by the remaining bodies of the universe, the precursor of what he later called "Mach's Principle." This deficiency enabled Einstein to ask after the possibility of extending the principle of relativity to accelerated motion, to see the real significance of the equality of inertial and gravitational mass in his principle of equivalence (which was not satisfied by Nordström's theory) and to develop his *Entwurf* theory.

14. EINSTEIN AND FOKKER: GRAVITATION IN NORDSTRÖM'S THEORY AS SPACETIME CURVATURE

It was clear by the time of Einstein's Vienna lecture that Nordström's most conservative of approaches to gravitation had led to a something more than a conservative Lorentz covariant theory of gravitation, for it had become a theory with kinematical effects very similar to those of Einstein's general theory of relativity. Gravitational fields would slow clocks and alter the lengths of rods. All that remained was the task of showing just how close Nordström's theory had come to Einstein's theory. This task was carried out by Einstein in collaboration with a student of Lorentz', Adriaan D. Fokker, who visited Einstein in Zurich in the winter semester of 1913–1914 (Pais 1982, 487). Their joint (Einstein and Fokker 1914), submitted on February 19, 1914, was devoted to establishing essentially one result, namely, in modern language, Nordström's theory was actually the theory of a spacetime that was only conformal to a Minkowski spacetime with the gravitational potential the conformal factor, so that the presence of a gravitational field coincided with deviations of the spacetime from flatness. That, of course, was not how Einstein and Fokker described the result. Their purpose, as they explained in the title and introduction of the paper, was to apply the new mathematical methods of Einstein's *Entwurf* theory to Nordström's theory. These methods were the "absolute differential calculus" of Ricci and Levi-Civita (1901). They enabled a dramatic simplification of Nordström's theory. It will be convenient here to summarize the content of the theory from this new perspective as residing in three basic assumptions:

I. Spacetime admits preferred coordinate systems $(x_1, x_2, x_3, x_4) = (x, y, z, ct)$ in which the spacetime interval is given by

$$ds^2 = \Phi^2(dx^2 + dy^2 + dz^2 - c^2dt^2) \tag{56}$$

and in which the trajectory of point masses in free fall is given by

$$\delta\int ds = 0 \, .$$

That such a characterization of the spacetime of Nordström's theory is possible is implicit in Einstein's Vienna lecture. In fact, once one knows the proportionality of ω and φ, the characterization can be read without calculation from Einstein's expression (52) for the natural proper time and the equation of motion (50). Einstein and Fokker emphasized that the preferred coordinate systems are ones in which the postulate of the constancy of the velocity of light obtains. For, along a light beam $ds^2 = 0$, so that

$$dx^2 + dy^2 + dz^2 - c^2dt^2 = 0 .$$

We see here in simplest form the failure of the theory to yield a deflection of a light beam in a gravitational field. This failure is already evident, of course, from the fact that a light beam has no gravitational mass since the trace of its stress-energy tensor vanishes.

II. The conservation of gravitational and non-gravitational energy momentum is given by the requirement of the vanishing of the covariant divergence of the stress-energy tensor $T_{\mu\nu}$ for non-gravitational matter. At this time, Einstein preferred to write this condition as[68]

$$\sum_\nu \frac{\partial \mathfrak{T}_{\sigma\nu}}{\partial x_\nu} = \frac{1}{2} \sum_{\mu\nu\tau} \frac{\partial g_{\mu\nu}}{\partial x_\sigma} \gamma_{\mu\tau} \mathfrak{T}_{\tau\nu} ,$$

since they could interpret the term on the right-hand side as representing the gravitational force density.

Noting, as Einstein and Fokker did on pp. 322–23, that the $T_{\mu\nu}$ of the Vienna lecture corresponds to the tensor density $\mathfrak{T}_{\mu\nu}$ of the new development, they evaluated this conservation law in the preferred coordinate systems of I. It yielded the form of the conservation law (53) of the Vienna lecture.

Finally Einstein and Fokker turned to the field equation which was to have the form

$$\Gamma = k\mathfrak{T}$$

where κ is a constant. The quantity $\mathfrak{T}$ had to be a scalar representing material processes. In the light of the earlier discussion, we know there was only one viable choice, the trace of the stress-energy tensor $T = \dfrac{1}{\sqrt{-g}} \sum_\tau \mathfrak{T}_{\tau\tau}$. For the quantity Γ, which must be constructed from the metric tensor and its derivatives, they reported that the researches of mathematicians allowed only one quantity to be considered, the

---

68   Here Einstein had not yet begun to use modern notational conventions. Summation over repeated indices is not implied. All indices are written as subscript so that $\gamma_{\mu\nu}$ is the fully contravariant form of the metric, which we would now write as $g^{\mu\nu}$. $\mathfrak{T}_{\mu\nu}$ is the mixed tensor density which we could now write as $\sqrt{-g}T^\nu_\mu$.

full contraction of the Riemann-Christoffel tensor (ik, lm) of the fourth rank, where they allowed i, k, l and m to vary over 1, 2, 3 and 4. This assumed that the second derivative of $g_{\mu\nu}$ enters linearly into the equation. Therefore we have:

III. The gravitational field satisfies the field equation which asserts the proportionality of the fully contracted Riemann-Christoffel tensor and the trace of the stress energy tensor

$$\sum_{iklm} \gamma_{ik}\gamma_{lm}(ik, lm) = \kappa \frac{1}{\sqrt{-g}} \sum_{\tau} \mathfrak{T}_{\tau\tau} .$$

Evaluation of this field equation in the preferred coordinate systems of I. yields the field equation (55) of the Vienna lecture.

Einstein and Fokker were clearly and justifiably very pleased at the ease with which the methods of the *Entwurf* theory had allowed generation of Nordström's theory. In the paper's introduction they had promised to show that (p. 321)

... one arrives at Nordström's theory instead of the Einstein-Grossmann theory if one makes the single assumption that it is possible to choose preferred reference systems in such a way that the principle of the constancy of the velocity of light obtains.

Their concluding remarks shine with the glow of their success when they boast that (p. 328)

... one can arrive at Nordström's theory from the foundation of the principle of the constancy of the velocity of light through purely formal considerations, i.e. without assistance of further physical hypotheses. Therefore it seems to us that this theory earns preference over all other gravitation theories that retain this principle. From the physical stand point, this is all the more the case, as this theory achieves strict satisfaction of the equality of inertial and gravitational mass.

Of course Einstein retained his objection that Nordström's theory violates the requirement of the relativity of inertia.[69] The new formulation gives us vivid demonstration of this failure: the disposition of the preferred coordinate systems of I. will be entirely unaffected by the distribution of matter in spacetime. Einstein must then surely have been unaware that it would prove possible to give a generally covariant formulation of Nordström's theory on the basis of Weyl's work (Weyl 1918). The requirement that the preferred coordinate systems of I. exist could be replaced by the generally covariant requirement of the vanishing of the conformal curvature tensor. This formal trick, however, does not alter the theory's violation of the relativity of inertia and the presence of preferred coordinate systems in it.

---

69   As we know from lecture notes taken by a student, Walter Dallenbach, (EA 4 008, 41-42), Einstein in his teaching at the ETH in Zurich at this time included the claim that one arrives at the Nordström theory merely by assuming there are specialized coordinate system in which the speed of light is constant. There he remarks that this theory violates the relativity of inertia.

There remained a great irony in Einstein and Fokker's paper, which their readers would discover within two short years. While the existence of preferred coordinate systems was held against the Nordström theory, Einstein's own *Entwurf* theory was not itself generally covariant and would not be until November 1915, when Einstein would disclose the modern field equations to the Prussian Academy. Einstein and Grossmann (1913) had settled upon gravitational field equations which were not generally covariant. We now know that the generally covariant field equations of the completed general theory of relativity can be derived by means of the Riemann-Christoffel tensor through an argument very similar to the one used to arrive at the generally covariant form of the field equation of the Nordström theory. Einstein and Grossmann had considered and rejected this possibility in §4.2 of Grossmann's part of their joint paper. The obvious ease with which consideration of the Riemann-Christoffel tensor led to the field equation of Nordström's theory clearly gave Einstein an occasion to rethink that rejection. For Einstein and Fokker's paper concluded with the tantalizing remark that the reasons given in Grossmann's §4 of their joint paper against such a connection did not withstand further examination. Whatever doubt this raised in Einstein's mind seem to have subsided by March 1914, at which time he reported in a letter to this confidant Michele Besso that the "general theory of invariants functioned only as a hindrance" in construction of his system (Speziali 1972, 53).

Thus the conservative path struck by Nordström and Einstein led not just to the connection between gravitation and spacetime curvature but to the first successful field equation which set an expression in the Riemann-Christoffel curvature tensor proportional to one in the stress-energy tensor of matter.

15. WHAT EINSTEIN KNEW IN 1912

Einstein and Fokker's characterization in 1914 of the Nordström theory gives us a convenient vantage point from which to view Einstein's theory of 1912 for static gravitational fields. In particular we can see clearly that this theory already contained many of the components that would be assembled to form Nordström's theory. Indeed we shall see that Einstein's theory came very close to Nordström's theory. However we shall also see that a vital component was missing—the use of the stress-energy tensor and Laue's work on complete static systems. This component enables a scalar Lorentz covariant theory of gravitation to satisfy some version of the requirement of the equality of inertial and gravitational mass. We must already suspect that Einstein was unaware of this possibility prior to his August 1912 move to Zurich for his July 1912 response to Abraham (Einstein 1912d), quoted in Section 4 above, purports to show that no Lorentz covariant theory of gravitation could satisfy this requirement.

Einstein (1912a, 1912b) was the fullest development of a relativistic theory of static gravitational fields based on the principle of equivalence and in which the gravitational potential was the speed of light c. By Einstein's own account the following year (Einstein and Grossmann 1913, I, §1, §2), the theory was actually a theory of a spacetime with the line element

$$ds^2 = -dx^2 - dy^2 - dz^2 + c^2 dt^2 \qquad (57)$$

where c is now a function of x, y and z and behaves as a gravitational potential. Einstein (1912a, 360) offered the field equation

$$\nabla^2 c = kc\sigma \qquad (58)$$

where k is a constant and σ the rest density of matter.[70] What Einstein did not mention in his *Entwurf* reformulation of the 1912 theory was that this field equation corresponded to the generally covariant field equation

$$R = \frac{k}{2}T$$

where R is the fully contracted Riemann-Christoffel tensor and T the trace of the stress–energy tensor, in the case of an unstressed, static matter distribution. This is exactly the field equation of Nordström's theory!

This field equation (58) had an extremely short life, for in (Einstein 1912b, §4), a paper submitted to *Annalen der Physik* on March 23, 1912, just a month after February 26, when he had submitted (Einstein 1912a), he revealed the disaster that had befallen his theory and would lead him to retract this field equation. Within the theory the force density $\mathfrak{F}$ on a matter distribution σ at rest is

$$\mathfrak{F} = -\sigma \operatorname{grad} c \, .$$

Einstein conjoined this innocuous result with the field equation (58) and applied it to a system of masses at rest held together in a rigid massless frame within a space in which c approached a constant value at spatial infinity. He concluded that the total gravitational force on the frame

$$\int \mathfrak{F} d\tau = -\int \sigma \operatorname{grad} c \, d\tau = -\frac{1}{k} \int \frac{\nabla^2 c}{c} \operatorname{grad} c \, d\tau$$

in general does not vanish. That is, the resultant of the gravitational forces exerted by the bodies on one another does not vanish. Therefore the system will set itself into motion, a violation of the equality of action and reaction, as Einstein pointed out. In effect the difficulty lay in the theory's failure to admit a gravitational field stress tensor, for the gravitational force density $\mathfrak{F}$ is equal to the divergence of this tensor. Were the tensor to be definable in Einstein's theory, that fact alone, through a standard application of Gauss' theorem, would make the net resultant force on the system vanish.[71]

---

70 The factor of c on the right-hand side of this otherwise entirely classical equation is introduced in order to leave c undetermined by a multiplicative gauge factor rather than an additive one.

Einstein then proceeded to consider a number of escapes from this disaster. The second and third escapes involved modifications to the force law and the field equation. The former failed but the latter proved workable. Einstein augmented the source density σ of (58) with a term in c:

$$\nabla^2 c = k\left\{c\sigma + \frac{1}{2k}\frac{\text{grad}^2 c}{c}\right\}.$$

The extra term was constructed to allow the formation of a gravitational field stress tensor and the conclusion that there would be no net force on the system of masses. Einstein was especially pleased to find that this extra term proved to represent the gravitational field energy density so that the source term of the field equation was now the total energy density of the system, gravitational and non-gravitational.[72]

For our purposes what is most interesting is the first escape that Einstein considered and rejected. Mentioning vaguely "results of the old theory of relativity," he considered the possibility that the stressed frame of the system might have a gravitational mass. That possibility was dismissed however with an argument that is surprising to those familiar with his work of the following year: that possibility would violate the equality of inertial and gravitational mass! Einstein considered a box with mirrored walls containing radiation of energy E. He concluded from his theory that, if the box were sufficiently small, the radiation would exert a net force on the walls of the box of $-E\,\text{grad}\,c$. He continued (Einstein 1912b, 453):

> This sum of forces must be equal to the resultant of forces which the gravitational field
> exerts on the whole system (box together with radiation), if the box is massless and if the
> circumstance that the box walls are subject to stresses as a result of the radiation pressure
> does not have the consequence that the gravitational field acts on the box walls. Were the

---

71 Writing $t_{im} = \frac{1}{kc}\left(\partial_i c\,\partial_m c - \frac{1}{2}\delta_{im}(\partial_n c\,\partial_n c)\right)$ for the quantity that comes closest to the stress tensor,

we have the following in place of the standard derivation of the stress tensor (analogous to the derivation of (39)). Substituting field equation (58) into the expression for $\mathfrak{F}$, we recover:

$$\mathfrak{F}_i = -\sigma\partial_i c = -\frac{1}{kc}\partial_m c\,\partial_m c\,\partial_i c = -\partial_m t_{im} - \frac{1}{2k}\frac{\partial_m c\,\partial_m c}{c^2}\partial_i c.$$

The first term of the final sum is a divergence which would vanish by Gauss' theorem when integrated over the space containing the masses of the frame, leaving no net force. The problem comes from the second term, which is present only because of the factor of c on the source side of the field equation (58). In this integration it will not vanish in general, leaving the residual force on the masses. The need to eliminate this second term also dictates the precise form of the modification to the field equation that Einstein ultimately adopted. When the field equation source σ was augmented to become

$\sigma + \frac{1}{2k}\frac{\partial_m c\,\partial_m c}{c^2}$, this second term no longer arose in the above expression for $\mathfrak{F}_i$.

72 However Einstein was disturbed to find that the new field equation only allowed his principle of equivalence to apply to infinitesimally small parts of space. See (Norton 1985, §4.2, §4.3).

latter the case, then the resultant of the forces exerted by the gravitational field on the box (together with its contents) would be different from the value $-E \operatorname{grad} c$, *i.e.* the gravitational mass of the system would be different from E.

Einstein could not have written this were he aware of the relevant properties of "Laue's scalar" T. As Einstein himself showed the following year, the use of T as the gravitational source density in exactly this example of radiation enclosed in a mirrored cavity allowed one to infer *both* that the walls of the cavity acquired a gravitational mass because of their stressed state and that the gravitational mass of the entire system was given by its total energy. We must then take Einstein at his word and conclude that he learned of these properties of T from Laue. Presumably this means after his move to Zurich in August 1912 where Laue also was, and after completion of his work on his scalar theory of static gravitational fields in 1912.

Had Einstein been aware of these results earlier in 1912, they would probably not have pleased him in the long run. To begin, he did believe at the time of writing the *Entwurf* paper that the selection of T as the gravitational source density in a scalar theory of gravitation led to a contradiction with the conservation of energy. Had he seen past this to its resolution in the gravitational potential dependence of lengths he would have arrived at a most remarkable outcome: his theory of 1912 would have become exactly Nordström's final theory! As we saw above, his first field equation of 1912 was already equivalent to Nordström's final field equation in covariant terms. His equation of motion for a mass point was already the geodesic equation for a spacetime with the line element (57). This line element already entailed a dependence of times on the gravitational potential. The consistent use of Laue's scalar T as a source density would finally have led to a similar dependence for spatial length so that the line element (57) would be replaced by Nordström's (56). Since the expressed purpose of Einstein's 1912 theory was to extend the principle of relativity, this out come would not have been a happy one for Einstein. For his path would have led him to a theory which entailed the existence of coordinate systems in which the speed of light was globally constant. That is, the theory had resurrected the special coordinate systems of special relativity.

16. THE FALL OF NORDSTRÖM'S THEORY OF GRAVITATION

Revealing as Einstein and Fokker's formulation of the theory had been, Nordström himself clearly did not see it as figuring in the future development of his theory. Rather, Nordström embedded his 1913 formulation of his gravitation theory in his rather short lived attempts to generate a unified theory of electricity and gravitation within a five dimensional spacetime (Nordström 1914c, 1914d, 1915). Other work on the theory in this period was devoted to developing a clearer picture of the behavior of bodies in free fall and planetary motion according to the theory. Behacker (1913) had computed this behavior for Nordström's first theory and (Nordström 1914a) performed the same service for his second theory. In both cases the behavior demanded by the theories was judged to be in complete agreement with experience.

Nordström also had to defend his theory from an attack by Gustav Mie. Mie had made painfully clear in the discussion following Einstein's Vienna lecture of 1913 (published in *Physikalische Zeitschrift*, *14*, 1262–66) that he was outraged over Einstein's failure even to discuss Mie's own theory of gravitation in the lecture. Einstein explained that this omission derived from the failure of Mie's theory to satisfy the requirement of the equality of inertial and gravitational mass. Mie counterattacked with a two part assault (Mie 1914) on Einstein's theory. In an appendix (§10) Mie turned his fire upon Nordström's theory, claiming that it violated the principle of energy conservation. Nordström's (1914b) response was that Mie had erroneously inferred the contradiction within Nordström's theory by improperly importing a result from Mie's own theory into the derivation. Laue (1917, 310–13) pointed to errors on both sides of this dispute.

However it was not Mie's theory that led to the demise of Nordström's theory. Rather it was the rising fortunes of Einstein's general theory of relativity. Einstein completed the theory in a series of papers submitted to the Prussian Academy in November 1915. Within a few years, with the success of Eddington's eclipse expedition, Einstein had become a celebrity and his theory of gravitation eclipsed all others. One of the papers from that November 1915 (Einstein 1915) reported the bewitching success of the new theory in explaining the anomalous motion of Mercury. This success set new standards of empirical adequacy for gravitation theories. Prior to this paper, the pronouncements of a gravitation theory on the minutia of planetary orbits were not deemed the ultimate test of a new theory of gravitation. Einstein's own *Entwurf* theory failed to account for the anomalous motion of Mercury. Yet this failure is not mentioned in Einstein's publications from this period and one cannot even tell from these publications whether he was then aware of it. Thus the treatment in (Nordström 1914a) of the empirical adequacy of his theory to observed planetary motions was entirely appropriate by the standards of 1914. He showed that his theory predicted a very slow retardation of the major axis of a planet's elliptical orbit. Computing this effect for the Earth's motion he found it to be 0.0065 seconds of arc per year, which could be dismissed as "very small in relation to the astronomical perturbations [due to other planets]" (p.1109) Thus he could proceed to the overall conclusion (p. 1109) that

> ... the laws derived for [free] fall and planetary motion are in the *best* agreement with experience [my emphasis]

Standards had changed so much by the time of Laue's (1917) review article on the Nordström theory that even motions much smaller than the planetary perturbations were decisive in the evaluation of a gravitation theory. Einstein's celebrated 43 seconds of arc per century advance of Mercury's perihelion is less than a tenth of the perihelion motion due to perturbations from the other planets. Laue (p. 305) derived a formula for the predicted retardation—not advance—of a planet's perihelion. Without even bothering to substitute values into the formula he lamented

> Therefore the perihelion moves opposite to the sense of rotation of the orbit. In the case
> of Mercury, the impossibility of explaining its perihelion motion with this calculation
> lies already in this difference of sign concerning the perihelion motion.

Through this period, Nordström's theory had its sympathizers and the most notable of these was Laue himself.[73] He clearly retained this sympathy when he wrote the lengthy review article, (Laue 1917). Einstein's theory had become so influential by this time that Laue introduced the review with over four pages of discussion of Einstein's theory (pp. 266–70). That discussion conceded that Einstein's theory had attracted the most adherents of any relativistic gravitation theory. It also contained almost two pages of continuous and direct quotation from Einstein himself, as well as discussion of the epistemological and empirical foundations of Einstein's theory. His discussion was not the most up-to-date, for he reported Einstein's *Entwurf* 0.84 seconds of arc deflection for a ray of starlight grazing the Sun, rather than the figure of 1.7 of the final theory of 1915. All this drove to the conclusion that there were no decisive grounds for accepting Einstein's theory and provided Laue with the opportunity to review a gravitation theory based on special relativity, Nordström's theory, which he felt had received less attention than it deserved.

The fall of Nordström's theory was complete by 1921. By this time even Laue had defected. In that year he published a second volume on general relativity to accompany his text on special relativity (Laue 1921). On p.17, he gave a kind appraisal of the virtues and vices of his old love, Nordström's theory. However he was firm in his concluding the superiority of Einstein's theory because of the failure of Nordström's theory to yield any gravitational light deflection—a defect, he urged, that must trouble any Lorentz covariant gravitation theory. Laue never lost his affection for the theory and years later took the occasion of Einstein's 70th birthday to recall the virtues of Nordström's theory (Laue 1949). The theory's obituary appeared in Pauli's encyclopedic distillation of all that was worth knowing in relativity theory (Pauli 1921, 144). He pronounced authoritatively

> The theory solves in a logically quite unexceptionable way the problem sketched out
> above, of how to bring the Poisson equation and the equation of motion of a particle into
> a Lorentz-covariant form. Also, the energy-momentum law and the theorem of the equal-
> ity of inertial and gravitational mass are satisfied. If, in spite of this, Nordström's theory
> is not acceptable, this is due, in the first place, to the fact that it does not satisfy the prin-
> ciple of *general* relativity (or at least not in a simple and natural way ...). Secondly, it is
> in contradiction with experiment: it does not predict the bending of light rays and gives
> the displacement of the perihelion of Mercury with the wrong sign. (It is in agreement
> with Einstein's theory with regard to the red shift.)

He thereby rehearsed generations of physicists to come in the received view of Nordström's theory and relieved them of the need to investigate its content any further.

---

73 In a letter of October 10, 1915, to Wien, Mie had identified Laue as an adherent of Nordström's the-
    ory, explaining it through Laue's supposed failure to read anything else! I am grateful to John Stachel
    for this information.

17. CONCLUSION

The advent of the general theory of relativity was so entirely the work of just one person— Albert Einstein—that we cannot but wonder how long it would have taken without him for the connection between gravitation and spacetime curvature to be discovered. What would have happened if there were no Einstein? Few doubt that a theory much like special relativity would have emerged one way or another from the researches of Lorentz, Poincaré and others. But where would the problem of relativizing gravitation have led? The saga told here shows how even the most conservative approach to relativizing gravitation theory still did lead out of Minkowski spacetime to connect gravitation to a curved spacetime. Unfortunately we still cannot know if this conclusion would have been drawn rapidly without Einstein's contribution. For what led Nordström to the gravitational field dependence of lengths and times was a very Einsteinian insistence on just the right version of the equality of inertial and gravitational mass. Unceasingly in Nordström's ear was the persistent and uncompromising voice of Einstein himself demanding that Nordström see the most distant consequences of his own theory.

APPENDIX: NORDSTRÖM'S MODEL OF THE ELECTRON

Nordström's (1913b) development of his second theory contains (§3) a model of the electron which accounts for the effect of gravitation. The electron is modelled as a massless spherical shell of radius a carrying charge e distributed uniformly over its surface.[74] Three types of matter are present: an electric charge and its field; the shell stressed to balance the repulsive electric forces between different parts of the charge distribution; and the gravitational field generated by all three types of matter. See Fig. 4. Taking each in turn, we have

---

74   "Rational" units of charge are used, which means, in effect, that the electrostatic field equation is $\Delta \Psi = -\rho$, for charge density ρ.

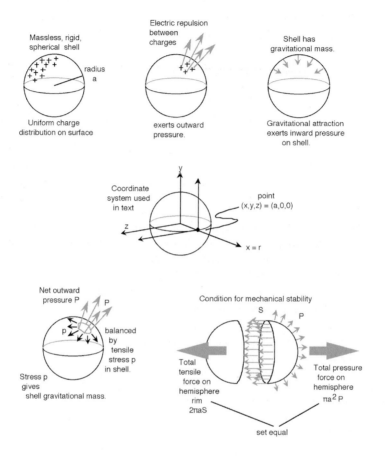

Figure 4: Nordström's Model of the Electron

Electric Charge and its Field

Using familiar results of electrostatics in the rational system of units, the electric charge e generates an electric potential Ψ at radius r from the center of the shell, for the case of $r \geq a$, which satisfies

$$\Psi = \frac{e}{4\pi r}, \quad \frac{\partial \Psi}{\partial r} = -\frac{e}{4\pi r^2}.$$

The latter value is all that is required to compute the Maxwell stress tensor at an arbitrary point on the shell which is representative of all its points due the rotational symmetry of the shell. We choose convenient coordinates $(x_1, x_2, x_3) = (x, y, z)$ for this point. We set the origin at the center of the shell, align the x axis with a radial arm and consider a point on the surface of the shell at which $(x, y, z) = (a, 0, 0)$. Writing ∂_i for $(\partial/\partial x_i)$, we have that the Maxwell stress tensor is[75]

$$L_{ik} = -\left(\partial_i \Psi \partial_k \Psi - \frac{1}{2}\delta_{ik}(\delta_m \Psi \partial_m \Psi)\right)$$

$$= -\begin{bmatrix} \dfrac{e^2}{32\pi^2 r^4} & 0 & 0 \\[2ex] 0 & -\dfrac{e^2}{32\pi^2 r^4} & 0 \\[2ex] 0 & 0 & -\dfrac{e^2}{32\pi^2 r^4} \end{bmatrix}.$$

We read directly from the coefficients of this tensor that the charges of the shell (at position $r = a$) are subject to an outwardly directed pressure of magnitude $e^2/32\pi a^4$ which seeks to cause the shell to explode radially outwards. That is, these charges are subject to a net electric force density given by the negative divergence of this stress tensor, $\partial_k L_{ik}$. With $r = a$, this force density is of magnitude $4e^2/32\pi^2 a^5$ directed radially outward.

Gravitational Field

The stresses in the shell will generate a gravitational field. For the moment, we shall write the total gravitational mass as M_g and note that it must be distributed uniformly over the shell. Since the source gravitational mass is all located in the shell over which the gravitational potential is constant, the field equation and stress tensor of the gravitational field reduce to the analogous equations of electrostatics, excepting a sign change. Thus the gravitational potential for $r \geq a$ satisfies

$$\Phi = -\frac{M_g}{4\pi r} + \Phi_a, \qquad \frac{\partial \Phi}{\partial r} = \frac{M_g}{4\pi r^2},$$

where Φ_a is the external gravitational potential. Choosing the same point and coordinate system as in the analysis of the electric field, we find that the gravitational field stress tensor, as given by the spatial parts of the gravitational stress-energy tensor (39) is

---

75   The nonstandard minus sign follows the convention Nordström used in paper of requiring that (force density) =—(divergence of stress tensor). See (Nordström 1913b, 535, eq. 7).

$$G_{ik} = \begin{bmatrix} \dfrac{M_g^2}{32\pi^2 r^4} & 0 & 0 \\[2ex] 0 & -\dfrac{M_g^2}{32\pi^2 r^4} & 0 \\[2ex] 0 & 0 & -\dfrac{M_g^2}{32\pi^2 r^4} \end{bmatrix}.$$

This reveals an inwardly directed pressure $M_g^2/(32\pi^2 a^4)$ which seeks to implode the shell. That is, the shell is subject to a net gravitational force density given by the negative divergence of this stress tensor, $\partial_k T_{ik}$. With $r = a$, this force density is of magnitude $4M_g^2/32\pi^2 a^5$ directed radially inwards.

Stressed Shell

The combined effect of both electric and gravitational forces is a net outward pressure on the shell of magnitude

$$P = \frac{e^2 - M_g^2}{32\pi^2 a^4}.\tag{59}$$

Mechanical stability is maintained by a tensile stress p in the shell. At the point considered above in the same coordinate systems, this stress will correspond to a stress tensor T_{ik} given by

$$T_{ik} = \begin{bmatrix} 0 & 0 & 0 \\ 0 & p & 0 \\ 0 & 0 & p \end{bmatrix},$$

where p will have a negative value. If this tensile stress is integrated across the thickness of the shell, we recover the tensile force S per unit length active in the shell

$$S = -\int p \, ds.$$

The condition for mechanical stability is[76]

$$2S = aP.\tag{60}$$

---

76 This standard result from the theory of statics can be derived most easily, as Nordström points out, by considering the pressure forces due to P acting on a hemisphere of the shell. A simple integration shows this force is $\pi a^2 P$. This force must be balanced exactly by the tensile force S along the rim of the hemisphere. That rim is of length $2\pi a$, so the total force is $2\pi a S$. Setting $2\pi a S = \pi a^2 P$ entails the result claimed.

Computation of gravitational mass M_g and inertial mass m of the electron

The as yet undetermined gravitational mass M_g of the electron is now recovered by combining the results for the three forms of matter. The source density v is determined by the stress-energy tensor $T_{\mu\nu}$ through equation (38). By assumption, there is no energy associated with the tensile stress in the shell in its rest frame. Thus in a rest frame $T_{uu} = 0$. The spatial components of $T_{\mu\nu}$ are given by the stress tensor T_{ik} above. Therefore

$$v = -\frac{1}{c^2}(T_{xx} + T_{yy} + T_{zz} + T_{uu}) = -\frac{1}{c^2}2p .$$

We can now recover the gravitational mass M_g from (41) by integrating over the shell

$$M_g = \int g(\Phi)v\,dv = \frac{g(\Phi)}{c^2}8\pi a^2 S .$$

We now substitute S in this expression with the condition (60) for mechanical stability and thence for P with the condition (59). By means of (42), we can also express $g(\Phi)$ in terms of $g(\Phi_a) = g_a$ using

$$g(\Phi) = \frac{g_a}{1 + \frac{g_a}{c^2}(\Phi - \Phi_a)} = \frac{g_a}{1 - \frac{g_a}{c^2}\frac{M_g}{4\pi a}} .$$

After some algebraic manipulation, we recover an implicit expression for M_g

$$M_g = \frac{g_a}{c^2}\frac{e^2 + M_g^2}{8\pi a} .$$

Since this gravitational mass M_g of this complete stationary system resides in an external potential Φ_a, the total mass of the system satisfies $M_g = g_a m$, so that we have for the rest mass m and rest energy E_0 of the electron

$$m = \frac{E_0}{c^2} = \frac{e^2 + M_g^2}{8\pi c^2 a} . \tag{61}$$

As Nordström points out of this final result of §3 of his paper is an extremely satisfactory one. The total energy of his electron is made up solely of the sum of an electric component $e^2/8\pi a$ and a gravitational component $M_g^2/8\pi a$. These two components agree exactly with the corresponding classical values. This agreement is not a foregone conclusion since the gravitational mass of the electron arises in an entirely non-classical way: it derives from the fact that the electron shell is stressed. Presumably this agreement justifies Nordström's closing remark in his §3, "Thus the expression found for m contains a verification of the theory."

In his §4, Nordström proceeded to use his expression (61) for the mass m of an electron to introduce the dependence of length on gravitational potential. In accordance with (46), derived in his §2, the mass m must vary in proportion to the external field Φ'_a in the appropriate gauge. However it was not clear how one could recover this same variability from the quantities in the expression (61) for m. He had found in §2 that M_g is independent of the gravitational potential and he asserted that the same held for e according to the basic equations of electrodynamics. Thus he concluded that the radius a of the electron must vary with gravitational potential according to (45). He then turned to Einstein's more general argument for (45).

ACKNOWLEDGEMENTS

This paper was first published in *Archive for History of Exact Sciences* (Norton 1992b). I am grateful to the Albert Einstein Archives, Hebrew University of Jerusalem, the copyright holder, for their kind permission to quote from Einstein's unpublished writings. I am grateful also to Michel Janssen and Jürgen Renn for helpful discussion.

REFERENCES

Abraham, Max. 1909. "Zur elektromagnetischen Mechanik." *Physikalische Zeitschrift*, 10: 737–41.
———. 1910. "Die Bewegungsgleichungen eines Massenteilchens in der Relativtheorie." *Physikalische Zeitschrift*, 11: 527–31.
———. 1912a. "Zur Theorie der Gravitation." *Physikalische Zeitschrift*, 13: 1–4. (English translation in this volume.)
———. 1912b. "Das Elementargesetz der Gravitation." *Physikalische Zeitschrift*, 13: 4–5.
———. 1912c. "Relativität und Gravitation. Erwiderung auf einer Bemerkung des Hrn. A. Einstein." *Annalen der Physik*, 38: 1056–58.
———. 1912d. "Nochmals Relativität und Gravitation. Bemerkung zu A. Einsteins Erwiderung." *Annalen der Physik*, 39: 444–48.
———. 1912e. "Eine neue Gravitationstheorie." *Archiv der Mathematik und Physik* (3), XX: 193–209. (English translation in this volume.)
Behacker, M. 1913. "Der freie Fall und die Planetenbewegung in Nordströms Gravitationstheorie." *Physikalische Zeitschrift*, 14: 989–992.
Corinaldesi, E. and A. Papapetrou, A. 1951. "Spinning Test Particles in General Relativity II." *Proceedings of the Royal Society, London*, A209: 259–68.
CPAE 2. 1989. John Stachel, David C. Cassidy, Jürgen Renn, and Robert Schulmann (eds.), *The Collected Papers of Albert Einstein*. Vol. 2. *The Swiss Years: Writings, 1900–1909*. Princeton: Princeton University Press.
CPAE 3. 1993. Martin J. Klein, A. J. Kox, Jürgen Renn, and Robert Schulmann (eds.), *The Collected Papers of Albert Einstein*. Vol. 3. *The Swiss Years: Writings, 1909–1911*. Princeton: Princeton University Press.
CPAE 4. 1995. Martin J. Klein, A. J. Kox, Jürgen Renn, and Robert Schulmann (eds.), *The Collected Papers of Albert Einstein*. Vol. 4. *The Swiss Years: Writings, 1912–1914*. Princeton: Princeton University Press.
CPAE 6. 1996. A. J. Kox, Martin J. Klein, and Robert Schulmann (eds.), *The Collected Papers of Albert Einstein*. Vol. 6. *The Berlin Years: Writings, 1914–1917*. Princeton: Princeton University Press.
Ehrenfest, Paul. 1907. "Die Translation deformierbarer Elektronen und der Flächensatz." *Annalen der Physik*, 23: 204–205.
Einstein, Albert. 1907a. "Über das Relativitätsprinzip und die aus demselben gezogenen Folgerungen." *Jahrbuch der Radioaktivität und Elektronik*, (1907) 4: 411–462; (1908) 5: 98–99, (CPAE 2, Doc. 47).
———. 1907b. "Bemerkungen zu der Notiz von Hrn Paul Ehrenfest: 'Die Translation deformierbarer Elektronen und der Flächensatz'." *Annalen der Physik*, 23: 206–208, (CPAE 2, Doc. 44).

————. 1907c. "Über die vom Relativitätsprinzip gefordete Trägheit der Energie." *Annalen der Physik*, 23: 371–384.

————. 1911. "Über den Einfluss der Schwerkraft auf die Ausbreitung des Lichtes." *Annalen der Physik*, 35: 898–908, (CPAE 3, Doc. 23).

————. 1912a. "Lichtgeschwindigkeit und Statik des Gravitationsfeldes." *Annalen der Physik*, 38: 355–69, (CPAE 4, Doc. 3).

————. 1912b. "Zur Theorie des Statischen Gravitationsfeldes." *Annalen der Physik*, 38: 443–58, (CPAE 4, Doc. 4).

————. 1912c. "Gibt es eine Gravitationswirkung die der elektrodynamischen Induktionswirkung analog ist?" *Vierteljahrsschrift für gerichtliche Medizin und öffentliches Sanitätswesen*, 44: 37–40, (CPAE 4, Doc. 7).

————. 1912d. "Relativität und Gravitation. Erwiderung auf eine Bemerkung von M. Abraham." *Annalen der Physik*, 38: 1059–64, (CPAE 4, Doc. 8).

————. 1912e. "Bemerkung zu Abrahams vorangehender Auseinandersetzung 'Nochmals Relativität und Gravitation'." *Annalen der Physik*, 39: 704, (CPAE 4, Doc. 9).

————. 1913. "Zum gegenwärtigen Stande des Gravitationsproblems." *Physikalische Zeitschrift*, 14: 1249–1262, (CPAE 4, Doc. 17). (English translation in this volume.)

————. 1915. "Erklärung der Perihelbewegung des Merkur aus der allgemeinen Relativitätstheorie." *Königlich Preussische Akademie der Wissenschaften* (Berlin). *Sitzungsberichte*. 1915: 831–39, (CPAE 6, Doc. 24).

————. 1933. "Notes on the Origin of the General Theory of Relativity." In *Ideas and Opinions*, 285–290. Translated by Sonja Bargmann. New York: Crown, 1954.

————. 1949. *Autobiographical Notes*. Open Court, 1979.

Einstein, Albert and Adriaan D. Fokker. 1914. "Die Nordströmsche Gravitationstheorie vom Standpunkt des absoluten Differentialkalküls." *Annalen der Physik*, 44: 321–28, (CPAE 4, Doc. 28).

Einstein, Albert and Marcel Grossmann. 1913. *Entwurf einer verallgemeinerten Relativitätstheorie und einer Theorie der Gravitation*. Leipzig: B.G.Teubner (separatum); with addendum by Einstein in *Zeitschrift für Mathematik und Physik*, 63(1913): 225–61, (CPAE 4, Doc. 13).

Holton, Gerald. 1975. "Finding Favor with the Angel of the Lord. Notes towards the Psychobiographical Study of Scientific Genius." In Yehuda Elkana (ed.), *The Interaction between Science and Philosophy*. Humanities Press.

Isaksson, Eva. 1985. "Der finnische Physiker Gunnar Nordström und sein Beitrag zur Entstehung der allgemeinen Relativitätstheorie Albert Einsteins." *NTM-Schriftennr. Gesch. Naturwiss., Technik, Med.* Leipzig, 22, 1: 29–52.

Janssen, Michel. 1995. "A Comparison between Lorentz's Ether Theory and Einstein's Special Theory of Relativity in the Light of the Experiments of Trouton and Noble." PhD dissertation, University of Pittsburgh.

Laue, Max. 1911a. "Zur Dynamik der Relativitätstheorie." *Annalen der Physik*, 35: 524–542.

————. 1911b. *Das Relativitätsprinzip*. Braunschweig: Friedrich Vieweg und Sohn.

————. 1911c. "Ein Beispiel zur Dynamik der Relativitätstheorie." *Verhandlungen der Deutschen Physikalischen Gesellschaft*, 1911: 513–518.

————. 1912. "Zur Theorie des Versuches von Trouton und Noble." *Annalen der Physik*, 38: 370–84.

————. 1917. "Die Nordströmsche Gravitationstheorie." *Jahrbuch der Radioaktivität und Elektronik*, 14: 263–313.

————. 1921. *Die Relativitätstheorie*. Vol. 2: *Die allgemeine Relativitätstheorie und Einsteins Lehre von der Schwerkraft*. Braunschweig: Friedrich Vieweg und Sohn.

————. 1949. "Zu Albert Einsteins 70-tem Geburtstag." *Rev. Mod. Phys.*, 21: 348–49.

Lewis, Gilbert N. and Richard C. Tolman. 1909. "The Principle of Relativity, and Non-Newtonian Mechanics." *Philosophical Magazine*, 18: 510–523.

Liu, Chuang. 1991. *Relativistic Thermodynamics: Its History and Foundation*. Ph.D. Dissertation, University of Pittsburgh.

Maxwell, James C. 1864. "A dynamical theory of the electromagnetic field." In W. D. Niven (ed.), *The Scientific Papers of James Clerk Maxwell*. Cambridge University Press, 1890. Reprinted 1965, New York: Dover, 526–597.

Mie, Gustav. 1914. "Bemerkungen zu der Einsteinschen Gravitationstheorie. I and II." *Physikalische Zeitschrift*, 14: 115–122; 169–176.

Minkowski, Hermann. 1908. "Die Grundgleichungen für die elektromagnetischen Vorgänge in bewegten Körpern." *Königlichen Gesellschaft der Wissenschaften zu Göttingen, Mathematisch-Physikalische Klasse, Nachrichten* 1908: 53–111; In *Gesammelte Abhandlung von Hermann Minkowski* Vol.2, Leipzig, 1911, 352–404. Reprinted New York: Chelsea. Page citations from this edition. (English translation of the appendix "Mechanics and the Relativity Postulate" in this volume.)

———. 1909. "Raum und Zeit." *Physikalische Zeitschrift*, 10: 104–111. In *Gesammelte Abhandlung von Hermann Minkowski* Vol.2, Leipzig, 1911, 431–444. Reprinted New York: Chelsea (page citations from this edition). Translated as "*Space and Time*," in H.A.Lorentz et al., *Principle of Relativity*, 1923, 75–91. Reprinted in 1952, New York: Dover.

Misner, Charles W., Kip S. Thorne, and John A. Wheeler. 1973. *Gravitation*. San Francisco: Freeman.

Nordström, Gunnar. 1909. "Zur Elektrodynamik Minkowskis." *Physikalische Zeitschrift*, 10: 681–87.

———. 1910. "Zur elektromagnetischen Mechanik." *Physikalische Zeitschrift*, 11: 440–45.

———. 1911. "Zur Relativitätsmechanik deformierbar Körper." *Physikalische Zeitschrift*, 12: 854–57.

———. 1912. "Relativitätsprinzip und Gravitation." *Physikalische Zeitschrift*, 13: 1126–29. (English translation in this volume.)

———. 1913a. "Träge und schwere Masse in der Relativitätsmechanik." *Annalen der Physik*, 40: 856–78. (English translation in this volume.)

———. 1913b. "Zur Theorie der Gravitation vom Standpunkt des Relativitätsprinzips." *Annalen der Physik*, 42: 533–54. (English translation in this volume.)

———. 1914a. "Die Fallgesetze und Planetenbewegungen in der Relativitätstheorie." *Annalen der Physik*, 43: 1101–10.

———. 1914b. "Über den Energiesatz in der Gravitationstheorie." *Physikalische Zeitschrift*, 14: 375–80.

———. 1914c. "Über die Möglichkeit, das elektromagnetische Feld und das Gravitationsfeld zu vereinigen." *Physikalische Zeitschrift*, 15: 504–506.

———. 1914d. "Zur Elektrizitäts- und Gravitationstheorie." *Ofversigt af Finska Vetenskaps-Societetens Förhandlingar*, 57, (1914–1915), Afd. A, N:o.4: 1–15.

———. 1915. "Über eine Mögliche Grundlage einer Theorie der Materie," *Ofversigt af Finska Vetenskaps-Societetens Förhandlingar*, 57, (1914–1915), Afd. A, N:o.28: 1–21.

Norton, John D. 1984. "How Einstein found his Field Equations: 1912–1915." *Historical Studies in the Physical Sciences*, 14: 253–316. Reprinted 1989 in Don Howard and John Stachel (eds.), *Einstein and the History of General Relativity (Einstein Studies* vol. 1). Boston/Basel/Berlin: Birkhäuser, 101–159.

———. 1985. "What was Einstein's Principle of Equivalence?" *Studies in History and Philosophy of Science*, 16, 203–246. Reprinted 1989 in Don Howard and John Stachel (eds.), *Einstein and the History of General Relativity. (Einstein Studies* vol. 1.) Boston/Basel/Berlin: Birkhäuser, 3–47. (Page citations from the former.)

———. 1992a. "The Physical Content of General Covariance." In J. Eisenstaedt and A. Kox (eds.), *Studies in the History of General Relativity. (Einstein Studies*, vol. 3.) Boston/Basel/Berlin: Birkhäuser.

———. 1992b. "Einstein, Nordström and the Early Demise of Scalar, Lorentz Covariant Theories of Gravitation." *Archive for History of Exact Sciences* 45: 17–94.

———. 1993. "Einstein and Nordström: Some Lesser-Known Thought Experiments in Gravitation." In J. Earman, M. Janssen and J. D. Norton (eds.), *The Attraction of Gravitation: New Studies in the History of General Relativity. (Einstein Studies*, vol. 5.) Boston/Basel/Berlin: Birkhäuser, 3–29.

———. 1995. "Eliminative Induction as a Method of Discovery: How Einstein Discovered General Relativity." In J. Leplin (ed.), *The Creation of Ideas in Physics: Studies for a Methodology of Theory Construction*. Dordrecht: Kluwer, 29–69.

Pais, Abraham. 1982. *Subtle is the Lord ...: The Science and the Life of Albert Einstein*. Oxford: Clarendon.

Papapetrou, Achilles. 1951. "Spinning Test Particle in General Relativity I." *Proceedings of the Royal Society, London*, A209, 248–58.

Pauli, Wolfgang. 1921. "Relativitätstheorie." In *Encyklopädie der mathematischen Wissenschaften, mit Einschluss an ihrer Anwendung*. Vol. 5, *Physik*, Part 2. Arnold Sommerfeld (ed.). Leipzig: B.G. Teubner, 1904–1922, 539–775. [Issued November 15, 1921.] English translation *Theory of Relativity*. With supplementary notes by the author. G. Field. Translated in 1958, London: Pergamon. (Citations from the Pergamon edition.)

Planck, Max. 1908. "Zur Dynamik bewegter Körper." *Annalen der Physik*, 26: 1–34.

Poincaré, Henri. 1905. "Sur la Dynamique de l' Électron." *Comptes Rendus des Séances de l' Academie des Sciences*, 140: 1504–1508.

———. 1906. "Sur la Dynamique de l' Électron," *Rendiconti del Circolo Matematico di Palermo*, 21: 129–75.

Ricci, Gregorio and Tullio Levi-Civita. 1901. "Méthodes de Calcul Différentiel Absolu et leurs Application." *Math. Ann.*, 54: 125–201. Reprinted 1954 in T. Levi-Civita, *Opere Matematiche*, vol. 1, Bologna, 479–559.

Rohrlich, Fritz. 1960. "Self-Energy and Stability of the Classical Electron." *American Journal of Physics*, 28: 639–43.

Sommerfeld, Arnold. 1910. "Zur Relativitätstheorie I. Vierdimensionale Vektoralgebra." *Annalen der Physik*, 32: 749–776; "Zur Relativitätstheorie II. Vierdimensionale Vektoranalysis." *Annalen der Physik*, 33: 649–89.

Speziali, Pierre (ed.). 1972. *Albert Einstein-Michele Besso: Correspondance 1903–1955*. Paris: Hermann.

Tolman, Richard C. 1934. *Relativity, Thermodynamics and Cosmology.* Oxford: Oxford University Press; Dover reprint, 1987.

Trouton, Frederick T. and Henry R. Noble. 1903. "The Mechanical Forces Acting on a Charged Condensor moving through Space." *Philosophical Transactions of the Royal Society of London*, 202: 165–181.

Weyl, Hermann. 1918. "Reine Infinitesimal Geometrie." *Mathematische Zeitschrift*, 2: 384–411.

Wheeler, John A. 1979. "Einstein's Last Lecture." In G. E. Tauber (ed.), *Albert Einstein's Theory of General Relativity.* New York: Crown, 187–190.

GUNNAR NORDSTRÖM

THE PRINCIPLE OF RELATIVITY AND GRAVITATION

Originally published as "Relativitätsprinzip und Gravitation" in Physikalische Zeitschrift 13, 1912, 23, pp. 1126–1129. Received October 23, 1912. Author's date: Helsingfors, October 20, 1912.

Einstein's hypothesis that the speed of light c depends upon the gravitational potential[1] leads to considerable difficulties for the principle of relativity, as the discussion between Einstein and Abraham shows us.[2] Hence, one is led to ask if it would not be possible to replace Einstein's hypothesis with a different one, which leaves c constant and still adapts the theory of gravitation to the principle of relativity in such a way that gravitational and inertial mass are equal.[3] I believe that I have found such a hypothesis, and I will present it in the following.

Let x, y, z, u be the four coordinates, with

$$u = ict.$$

Like Abraham,[4] I set

$$\frac{\partial^2 \Phi}{\partial x^2} + \frac{\partial^2 \Phi}{\partial y^2} + \frac{\partial^2 \Phi}{\partial z^2} + \frac{\partial^2 \Phi}{\partial u^2} = 4\pi f \gamma, \tag{1}$$

designating the rest density of matter by γ and the gravitational potential by Φ. Φ as well as γ are four-dimensional quantities; f is the gravitational constant. In a gravitational field we have a four-vector

---

1   A. Einstein, *Ann. d. Phys.* **35**, 898, 1911

2   See *Ann. d. Phys.* **38**, 355, 1056, 1059; **39**, 444, 1912

3   By the equality of inertial and gravitational mass, I do not mean, however, that every inertial phenomenon is caused by the inertial and gravitational mass. For elastically stressed bodies, according to Laue (see below), one obtains a momentum that cannot at all be traced back to a mass. I will return to these questions in a future communication.

4   M. Abraham, this journal, **13**, 1, 1912.

Jürgen Renn (ed.). *The Genesis of General Relativity,* Vol. 3
Gravitation in the Twilight of Classical Physics: Between Mechanics, Field Theory, and Astronomy.
© 2007 Springer.

$$\mathfrak{F}_x = -\frac{\partial \Phi}{\partial x}, \qquad \mathfrak{F}_y = -\frac{\partial \Phi}{\partial y},$$
$$\mathfrak{F}_z = -\frac{\partial \Phi}{\partial z}, \qquad \mathfrak{F}_u = -\frac{\partial \Phi}{\partial u}, \qquad (2)$$

which must be the cause of the acceleration of a mass point located in the field. However, if one considers the four-vector $\mathfrak{F}$ as the *accelerating force* [*bewegende Kraft*] acting on an unchanging unit mass, then the constancy of the speed of light cannot be maintained. In this case, namely, $\mathfrak{F}$ would be equal to the four-dimensional acceleration vector of a mass point and could not remain perpendicular to the velocity vector $\mathfrak{a}$ for arbitrary directions of motion, as demanded by the constancy of the speed of light.[5]

Keeping the speed of light constant, one can nevertheless still eliminate the difficulty in two ways. Either one takes not $\mathfrak{F}$ itself but only its component perpendicular to the velocity vector as the *accelerating force*,[6] or one takes the mass of a mass point to be not constant but dependent on the gravitational potential. On each of these two assumptions, the four-vectors $\mathfrak{F}$ and $\mathfrak{a}$ do not remain parallel: in the first case due to an extra force [*Zusatzkraft*] added to $\mathfrak{F}$, in the second case due to the variability of the mass. As we shall see, the two methods lead to the same laws for the motion of a mass point, but they correspond to two different interpretations of the concept of force.

In accordance with the position of most researchers in the domain of relativity theory, I will first use the second method. Thus, we treat

$$m\mathfrak{F}_x = -m\frac{\partial \Phi}{\partial x} \quad etc.$$

as the components of an *accelerating* force acting on a mass point, but view the rest mass of that point as variable. If the components of the velocity vector are $\mathfrak{a}_x, \mathfrak{a}_y, \mathfrak{a}_z, \mathfrak{a}_u$, and τ is the proper time, the equations of motion of the mass point are

$$-m\frac{\partial \Phi}{\partial x} = \frac{d}{d\tau}(m\mathfrak{a}_x) = m\frac{d\mathfrak{a}_x}{d\tau} + \mathfrak{a}_x\frac{dm}{d\tau},$$
$$-m\frac{\partial \Phi}{\partial y} = \frac{d}{d\tau}(m\mathfrak{a}_y) = m\frac{d\mathfrak{a}_y}{d\tau} + \mathfrak{a}_y\frac{dm}{d\tau},$$
$$-m\frac{\partial \Phi}{\partial z} = \frac{d}{d\tau}(m\mathfrak{a}_z) = m\frac{d\mathfrak{a}_z}{d\tau} + \mathfrak{a}_z\frac{dm}{d\tau}, \qquad (3)$$
$$-m\frac{\partial \Phi}{\partial u} = \frac{d}{d\tau}(m\mathfrak{a}_u) = m\frac{d\mathfrak{a}_u}{d\tau} + \mathfrak{a}_u\frac{dm}{d\tau}.$$

---

5   M. Abraham, loc. cit., eq. (5).
6   Minkowski treats the electrodynamic force in a similar way. Compare *Gött. Nachr.*, 1908, p. 98, eq. (98).

| We multiply the equations in turn by $\mathfrak{a}_x$, $\mathfrak{a}_y$, $\mathfrak{a}_z$, $\mathfrak{a}_u$ and add them. Since [1127]

$$\frac{\partial\Phi}{\partial x}\mathfrak{a}_x + \frac{\partial\Phi}{\partial y}\mathfrak{a}_y + \frac{\partial\Phi}{\partial z}\mathfrak{a}_z + \frac{\partial\Phi}{\partial u}\mathfrak{a}_u = \frac{\partial\Phi}{\partial\tau},$$

$$\mathfrak{a}_x\frac{d\mathfrak{a}_x}{d\tau} + \mathfrak{a}_y\frac{d\mathfrak{a}_y}{d\tau} + \mathfrak{a}_z\frac{d\mathfrak{a}_z}{d\tau} + \mathfrak{a}_u\frac{d\mathfrak{a}_u}{d\tau} = 0,$$

$$\mathfrak{a}_x^{\,2} + \mathfrak{a}_y^{\,2} + \mathfrak{a}_z^{\,2} + \mathfrak{a}_u^{\,2} = -c^2,$$

we obtain

$$-m\frac{d\Phi}{d\tau} = -c^2\frac{dm}{d\tau},$$

$$\frac{1}{m}\frac{dm}{d\tau} = \frac{1}{c^2}\frac{d\Phi}{d\tau}. \tag{4}$$

Integration yields

$$\log m = \frac{1}{c^2}\Phi + \text{const},$$

or

$$m = m_0 e^{\frac{\Phi}{c^2}}. \tag{5}$$

This equation shows that the mass m depends on the gravitational potential according to a simple law.

Using (4) the equations of motion (3) can also be written in the following form:

$$\left.\begin{aligned}
-\frac{\partial\Phi}{\partial x} &= \frac{d\mathfrak{a}_x}{d\tau} + \frac{\mathfrak{a}_x}{c^2}\frac{d\Phi}{d\tau}, \\[4pt]
-\frac{\partial\Phi}{\partial y} &= \frac{d\mathfrak{a}_y}{d\tau} + \frac{\mathfrak{a}_y}{c^2}\frac{d\Phi}{d\tau}, \\[4pt]
-\frac{\partial\Phi}{\partial z} &= \frac{d\mathfrak{a}_z}{d\tau} + \frac{\mathfrak{a}_z}{c^2}\frac{d\Phi}{d\tau}, \\[4pt]
-\frac{\partial\Phi}{\partial u} &= \frac{d\mathfrak{a}_u}{d\tau} + \frac{\mathfrak{a}_u}{c^2}\frac{d\Phi}{d\tau}.
\end{aligned}\right\} \tag{6}$$

As one can see, the mass m drops out of these equations. The laws according to which a mass point moves in a gravitational field are thus completely independent of the mass of the point.

The considerations so far are based on the assumption that $m\mathfrak{F}$ is the accelerating force. Now, for the moment, we wish to assume that the component of $m\mathfrak{F}$ perpendicular to the velocity vector $\mathfrak{a}$, rather than $m\mathfrak{F}$ itself, is the accelerating force. This part of $m\mathfrak{F}$ is a four-vector having an x-component[7]

$$m\mathfrak{F}_x + m\frac{\mathfrak{a}_x}{c^2}\{\mathfrak{F}_x\mathfrak{a}_x + \mathfrak{F}_y\mathfrak{a}_y + \mathfrak{F}_z\mathfrak{a}_z + \mathfrak{F}_u\mathfrak{a}_u\}.$$

The second term is the x-component of an extra force added to $m\mathfrak{F}$. According to (2) the expression can be changed to

$$-m\left\{\frac{\partial\Phi}{\partial x} + \frac{\mathfrak{a}_x}{c^2}\frac{\partial\Phi}{d\tau}\right\}.$$

Since we now view m as constant, the first of the equations of motion of a mass point is

$$-m\left\{\frac{\partial\Phi}{\partial x} + \frac{\mathfrak{a}_x}{c^2}\frac{\partial\Psi}{d\tau}\right\} = m\frac{d\mathfrak{a}_x}{d\tau}.$$

But this is just the first of the equations of motion (6).

From the two alternative assumptions we obtain precisely the same laws describing the motion of a mass point in a gravitational field, only the force and the mass are conceptualized differently in the two cases. The latter way of thinking corresponds to Minkowski's original, the way treated first corresponds to that held by Laue and Abraham.[8]

So far we have considered an isolated point mass. Now we would like to investigate the motion of arbitrary bodies in a gravitational field and develop the law of conservation of energy for this process. We assume only that the mass of each particle of the bodies actually is something real, so that we can speak of the rest density γ of the spacetime points. This is certainly the case when no tangential stresses are present in the body.[9] The rest density γ is of course a function of the four coordinates

$$\gamma = \gamma(x, y, z, u).$$

We again view mass as variable and accept the concept of force equation (3) is based on. Then the components of the force exerted by gravitation on a *unit volume* of matter are[10]

$$\left.\begin{array}{ll} \mathfrak{K}_x = -\gamma\dfrac{\partial\Phi}{\partial x}, & \mathfrak{K}_y = -\gamma\dfrac{\partial\Phi}{\partial y}, \\[2ex] \mathfrak{K}_z = -\gamma\dfrac{\partial\Phi}{\partial z}, & \mathfrak{K}_u = -\gamma\dfrac{\partial\Phi}{\partial u}. \end{array}\right\} \tag{7}$$

---

7   Cf. H. Minkowski, loc. cit.

8   Cf. the discussion between Abraham and the author, this journal **10**, 681, 737, 1909; **11**, 440, 527, 1910. I now take the position then taken up by Abraham.

9   Cf. M. Laue, *Das Relativitätsprinzip*, Braunschweig 1911, p. 151 f.; G. Nordström, this journal **12**, 854, 1911.

10 If the four-vector $\mathfrak{K}$ is taken to be the *accelerating* force, then it should be designated as the "*accelerating force per unit rest volume.*"

For the sake of generality, we assume that besides gravitation an "external" force $\mathfrak{K}'$ with components

$$\mathfrak{K}_x', \quad \mathfrak{K}_y', \quad \mathfrak{K}_z', \quad \mathfrak{K}_u'$$

acts on the unit volume of matter. We can then write the equations of motion of matter in the following general form[11]

$$
\left.
\begin{aligned}
-\gamma\frac{\partial\Phi}{\partial x} + \mathfrak{K}_x' &= \frac{\partial}{\partial x}\gamma\,\mathfrak{a}_x{}^2 + \frac{\partial}{\partial y}\gamma\,\mathfrak{a}_x\mathfrak{a}_y + \frac{\partial}{\partial z}\gamma\,\mathfrak{a}_x\mathfrak{a}_z + \frac{\partial}{\partial u}\gamma\,\mathfrak{a}_x\mathfrak{a}_u, \\
&\text{—} \quad \text{—} \quad \text{—} \quad \text{—} \quad \text{—} \quad \text{—} \quad \text{—} \\
&\text{—} \quad \text{—} \quad \text{—} \quad \text{—} \quad \text{—} \quad \text{—} \quad \text{—} \\
-\gamma\frac{\partial\Phi}{\partial u} + \mathfrak{K}_u' &= \frac{\partial}{\partial x}\gamma\,\mathfrak{a}_u\mathfrak{a}_x + \frac{\partial}{\partial y}\gamma\,\mathfrak{a}_u\mathfrak{a}_y + \frac{\partial}{\partial z}\gamma\,\mathfrak{a}_u\mathfrak{a}_z + \frac{\partial}{\partial u}\gamma\,\mathfrak{a}_u{}^2.
\end{aligned}
\right\}
\tag{8}
$$

[1128]

If we wish to introduce the ordinary three-dimensional velocity v and the ordinary mass density ρ, we have to set

$$\mathfrak{a}_x = \frac{v_x}{\sqrt{1-q^2}}, \quad \cdots\cdots, \quad \mathfrak{a}_u = \frac{ic}{\sqrt{1-q^2}},$$

$$\gamma = \rho\sqrt{1-q^2},$$

where $q = v/c$ has been substituted for reasons of simplicity. We multiply the last of the equations (8) by $-ic$ and insert the expressions above into its right-hand side. Continuing to use the notation of three-dimensional vector analysis, the equation becomes

$$\gamma\frac{\partial\Phi}{\partial t} - ic\mathfrak{K}_u' = c^2\,\mathrm{div}\,\frac{\rho v}{\sqrt{1-q^2}} + c^2\frac{\partial}{\partial t}\frac{\rho}{\sqrt{1-q^2}}. \tag{9}$$

We wish to transform the first term. Equation (1) yields

$$4\pi f\gamma = \mathrm{div}\,\nabla\Phi - \frac{1}{c^2}\frac{\partial^2\Phi}{\partial t^2},$$

and thus

---

11 G. Nordström, this journal **11**, 441, eq. (4'), 1910.

$$\gamma \frac{\partial \Phi}{\partial t} = \frac{1}{4\pi f}\left\{ \frac{\partial \Phi}{\partial t}\mathrm{div}\,\nabla\Phi - \frac{1}{c^2}\frac{\partial \Phi}{\partial t}\frac{\partial}{\partial t}\!\left(\frac{\partial \Phi}{\partial t}\right)\right\}$$

$$= \frac{1}{4\pi f}\left\{ \mathrm{div}\!\left(\frac{\partial \Phi}{\partial t}\nabla\Phi\right) - \nabla\Phi\cdot\frac{\partial}{\partial t}\nabla\Phi - \frac{1}{2c^2}\frac{\partial}{\partial t}\!\left(\frac{\partial \Phi}{\partial t}\right)^2 \right\},$$

$$\gamma \frac{\partial \Phi}{\partial t} = \frac{1}{4\pi f}\mathrm{div}\!\left(\frac{\partial \Phi}{\partial t}\nabla\Phi\right) - \frac{1}{8\pi f}\frac{\partial}{\partial t}\left\{(\nabla\Phi)^2 + \frac{1}{c^2}\!\left(\frac{\partial \Phi}{\partial t}\right)^2\right\}.$$

We insert this expression into equation (9) and obtain the following equation, which expresses the law of conservation of energy:

$$\begin{aligned}
-ic\mathfrak{K}_u{}' = {}& -\frac{1}{4\pi f}\mathrm{div}\!\left(\frac{\partial \Phi}{\partial t}\nabla\Phi\right) + \frac{1}{8\pi f}\frac{\partial}{\partial t}\left\{(\nabla\Phi)^2 + \frac{1}{c^2}\!\left(\frac{\partial \Phi}{\partial t}\right)^2\right\} \\
& + c^2\mathrm{div}\frac{\rho v}{\sqrt{1-\mathfrak{q}^2}} + c^2\frac{\partial}{\partial t}\frac{\rho}{\sqrt{1-\mathfrak{q}^2}}.
\end{aligned} \tag{10}$$

The quantity $-ic\mathfrak{K}_u{}'$ represents the energy influx caused by the external force $\mathfrak{K}'$ per unit volume and per unit time. Of the terms on the right-hand side, the first two terms relate to the gravitational field, the last two to the matter of the bodies. We set

$$\mathfrak{S}^g = -\frac{1}{4\pi f}\frac{\partial \Phi}{\partial t}\nabla\Phi. \tag{11}$$

$$\psi^g = \frac{1}{8\pi f}\left\{(\nabla\Phi)^2 + \frac{1}{c^2}\!\left(\frac{\partial \Phi}{\partial t}\right)^2\right\}. \tag{12}$$

$$\mathfrak{S}^m = \frac{c^2\rho v}{\sqrt{1-\mathfrak{q}^2}}, \tag{13}$$

$$\psi^m = \frac{c^2\rho}{\sqrt{1-\mathfrak{q}^2}}. \tag{14}$$

ψ^g is the *energy density* of the gravitational field, $\mathfrak{S}^g$ is the *energy flux* of this field, ψ^m and $\mathfrak{S}^m$ are the energy density and the convective energy flux of matter. For these quantities, we have found the expressions (13) and (14) already known earlier.[12]

We note that according to (12) the energy density of the field is always positive.

Finally, the energy equation is written as

---

12   G. Nordström, loc. cit., eqs. (11) and (12); M. Laue, loc. cit., § 24.

$$-ic\Re_u{}' = \text{div}(\mathfrak{S}^g + \mathfrak{S}^m) + \frac{\partial}{\partial t}(\psi^g + \psi^m). \tag{15}$$

We see that the law of conservation of energy is satisfied.

The quantities $\mathfrak{S}^g$ and ψ^g depend on a four-dimensional tensor, which also yields fictitious stresses for the gravitational force $\Re$. This tensor is precisely the same as that which Abraham obtained using different assumptions.[13] The ten components of the gravitation tensor are

$$
\left.
\begin{aligned}
X_x &= \frac{1}{4\pi f}\left\{-\left(\frac{\partial\Phi}{\partial x}\right)^2 + \Psi\right\}, \\
&\text{\textemdash\ \textemdash\ \textemdash\ \textemdash\ \textemdash\ \textemdash\ \textemdash} \\
&\text{\textemdash\ \textemdash\ \textemdash\ \textemdash\ \textemdash\ \textemdash\ \textemdash} \\
U_u &= \frac{1}{4\pi f}\left\{-\left(\frac{\partial\Phi}{\partial u}\right)^2 + \Psi\right\}, \\
X_y &= Y_x = -\frac{1}{4\pi f}\frac{\partial\Phi}{\partial x}\frac{\partial\Phi}{\partial y}, \\
&\text{\textemdash\ \textemdash\ \textemdash\ \textemdash\ \textemdash\ \textemdash\ \textemdash} \\
&\text{\textemdash\ \textemdash\ \textemdash\ \textemdash\ \textemdash\ \textemdash\ \textemdash} \\
&\text{\textemdash\ \textemdash\ \textemdash\ \textemdash\ \textemdash\ \textemdash\ \textemdash} \\
&\text{\textemdash\ \textemdash\ \textemdash\ \textemdash\ \textemdash\ \textemdash\ \textemdash} \\
Z_u &= U_z = -\frac{1}{4\pi f}\frac{\partial\Phi}{\partial z}\frac{\partial\Phi}{\partial u},
\end{aligned}
\right\} \tag{16}
$$

where Ψ is the following four-dimensional scalar |

$$\Psi = \frac{1}{2}\left\{\left(\frac{\partial\Phi}{\partial x}\right)^2 + \left(\frac{\partial\Phi}{\partial y}\right)^2 + \left(\frac{\partial\Phi}{\partial z}\right)^2 + \left(\frac{\partial\Phi}{\partial u}\right)^2\right\}. \tag{16a}$$ [1129]

It can be easily shown that in fact

$$-\gamma\frac{\partial\Phi}{\partial x} = \frac{\partial X_x}{\partial x} + \frac{\partial X_y}{\partial y} + \frac{\partial X_z}{\partial z} + \frac{\partial X_u}{\partial u}, \quad \text{etc.}$$

$$\mathfrak{S}_x{}^g = icU_x, \qquad \mathfrak{S}_y{}^g = icU_y, \qquad \mathfrak{S}_z{}^g = icU_z, \qquad \psi^g = U_u.$$

13   M. Abraham, loc. cit., p. 3.

Because the gravitation tensor is symmetric, the momentum density is equal to the energy flux divided by c^2.

Equation (4), which expresses the variability of the mass of a mass point, can be easily generalized to extended masses. For this purpose, we have to treat the system of equations (8) in the same way as we treated the system of equations (3) earlier. We multiply the equations (8) in turn by $\mathfrak{a}_x$, $\mathfrak{a}_y$, $\mathfrak{a}_z$, $\mathfrak{a}_u$ and add them. If no causes other than the gravitational field lead to a variability of mass, the external force $\mathfrak{R}'$ is perpendicular to $\mathfrak{a}$, and after some rearranging one obtains

$$\frac{\partial}{\partial x}\gamma\,\mathfrak{a}_x + \frac{\partial}{\partial y}\gamma\,\mathfrak{a}_y + \frac{\partial}{\partial z}\gamma\,\mathfrak{a}_z + \frac{\partial}{\partial u}\gamma\,\mathfrak{a}_u = \frac{\gamma}{c^2}\frac{d\Phi}{d\tau},\tag{17}$$

or

$$\operatorname{div}\rho\upsilon + \frac{\partial\rho}{\partial t} = \frac{\rho}{c^2}\left\{\upsilon\nabla\Phi + \frac{\partial\Phi}{\partial t}\right\},\tag{18}$$

or still

$$\frac{d}{dt}(\rho\,dv) = \frac{\rho\,dv}{c^2}\frac{d\Phi}{dt}\tag{18a}$$

(dv is the volume element). These three equivalent equations express in general the law of the dependence of mass on the gravitational field.

Equation (1) can be integrated in a well-known manner. One obtains the well-known expression for the retarded potential

$$\Phi(x_0, y_0, z_0, t) = -f\cdot\int\frac{dx\;dy\;dz}{r}\gamma_{\,t-\frac{r}{c}} + \text{const},\tag{19}$$

where $r = \sqrt{(x-x_0)^2 + (y-y_0)^2 + (z-z_0)^2}$.

The integration is over three-dimensional space, and γ is evaluated at the time $t-\frac{r}{c}$.

From equations (5) and (19) it becomes apparent that point masses cannot really exist because within such a mass point $\Phi = -\infty$, and hence the mass would be zero. If a body contracts, its mass decreases and with vanishing volume its mass would also vanish. As far as I can see, these consequences of the theory do not lead to contradictions.

Obviously, the theory developed here has much in common with the one which Abraham presented in this journal **13**, 1, 1912, but later refuted.[14] The theory developed here, however, is free from all the maladies which are brought about by the variability of the speed of light in the theories of Einstein and Abraham.

Addendum to proofs: From a letter from Herr Prof. Dr. A. Einstein I learn that earlier he had already concerned himself with the possibility I used above for treating gravitational phenomena in a simple way. However, he became convinced that the

---

14   M. Abraham, this journal **13**, 793, 1912.

consequences of such a theory cannot correspond with reality. In a simple example he shows that, according to this theory, a rotating system in a gravitational field will acquire a smaller acceleration than a non-rotating system.

I do not find this result dubious in itself, for the difference is too small to yield a contradiction with experience. Of course, the result under discussion shows that my theory is not compatible with Einstein's hypothesis of equivalence, according to which an unaccelerated reference system in a homogeneous gravitational field is equivalent to an accelerated reference system in a gravitation free space.

In this circumstance, however, I do not see a sufficient reason to reject the theory. For, even though Einstein's hypothesis is extraordinarily ingenious, on the other hand it still provides great difficulties. Therefore other attempts at treating gravitation are also desirable and I want to provide a contribution to them with my communication.

GUNNAR NORDSTRÖM

INERTIAL AND GRAVITATIONAL MASS
IN RELATIVISTIC MECHANICS

Originally published as "Träge und schwere Masse in der Relativitätsmechanik" in Annalen der Physik, 40, 1913, pp. 856–878. Received 21. January 1913. Author's date: Helsingfors, January 1913.

In several recent papers in the field of relativistic mechanics, the concept of the mass of bodies plays a very subordinate role. The reason is easy to understand. As Laue[1] and Herglotz[2] have shown, one can develop the entire mechanics of extended bodies without exploiting the concept of inertial mass in any way. Thus the concept of mass is not absolutely essential for mechanics, and on the other hand, if one considers bodies subject to arbitrary elastic stresses, this concept is also not sufficient to describe all inertial phenomena of matter.

But the question of the mass of matter is nevertheless of considerable importance for the theory of relativity, especially for the assessment of the way in which the theory of gravitation is to be integrated into the theory of relativity. In any case inertia and gravity [*Schwere*] of matter must stand in close relation to each other, and it would be easiest to account for this unity of essence [*Wesenseinheit*] via the mass underlying these two phenomena. One would attempt to retain such a concept of mass, even though it is known that according to relativity theory there exist inertial phenomena which cannot be traced back to mass in any way. In such cases, one must make use of a specially defined momentum, which depends, for example, upon the state of elastic stress of a body I rather than upon its mass.   [857]

In the present paper, I will treat the relativistic mechanics of deformable bodies in such a way that the possibility of generally maintaining the concept of mass is clearly emphasized. On this occasion, I will also investigate the influence of the heat conduction on mechanical processes. Finally, I will consider gravitation by also ascribing gravity to the inertial mass.

---

1   M. Laue, *Das Relativitätsprinzip*, Braunschweig 1911, VII; *Ann. d. Phys.*, 35, p. 524, 1911.
2   G. Herglotz, *Ann. d. Phys.*, 36, p. 493, 1911.

Jürgen Renn (ed.). *The Genesis of General Relativity*, Vol. 3
Gravitation in the Twilight of Classical Physics: Between Mechanics, Field Theory, and Astronomy.
© 2007 Springer.

1. THE FOUNDATIONS OF THE RELATIVISTIC MECHANICS OF
DEFORMABLE BODIES

We consider a body in an arbitrary state of motion and arbitrary state of stress. In addition to the elastic forces, a spatially distributed ponderomotive force $\mathfrak{K}$ of any kind may act on the bodies. $\mathfrak{K}$ is a four-vector which is to be designated the "external" ponderomotive force per unit volume, or the "external" accelerating force [*bewegende Kraft*] per unit of rest volume.[3]

According to Laue,[4] there is a symmetric four-dimensional tensor T, whose components give the spatial stresses as well as the mechanical energy-momentum density. Accordingly, we can write the equations of motion of the body in the following form:

$$\mathfrak{K}_x = \frac{\partial T_{xx}}{\partial x} + \frac{\partial T_{xy}}{\partial y} + \frac{\partial T_{xz}}{\partial z} + \frac{\partial T_{xu}}{\partial u},$$

$$\mathfrak{K}_y = \frac{\partial T_{yx}}{\partial x} + \frac{\partial T_{yy}}{\partial y} + \frac{\partial T_{yz}}{\partial z} + \frac{\partial T_{yu}}{\partial u},$$

$$\mathfrak{K}_z = \frac{\partial T_{zx}}{\partial x} + \frac{\partial T_{zy}}{\partial y} + \frac{\partial T_{zz}}{\partial z} + \frac{\partial T_{zu}}{\partial u}, \qquad (1)$$

$$\mathfrak{K}_u = \frac{\partial T_{ux}}{\partial x} + \frac{\partial T_{uy}}{\partial y} + \frac{\partial T_{uz}}{\partial z} + \frac{\partial T_{uu}}{\partial u},$$

where $x, y, z, u = ict$ are the four coordinates; the speed of light c is supposed to be a universal constant. |

[858]   We want to assign to each spacetime point of matter a certain *rest-mass density* v. This quantity is to be a four-dimensional scalar, but otherwise for the time being we leave it completely undetermined, so that we still have the freedom to further specify the concept of mass. From the rest density, the usual mass density μ is determined by the equation

$$v = \mu \sqrt{1 - \frac{v^2}{c^2}}, \qquad (2)$$

where v represents the (three-dimensional) velocity of the point in question. For simplicity we set

$$\mathfrak{q} = \frac{v}{c},$$

and therefore have

$$v = \mu \sqrt{1 - \mathfrak{q}^2}.$$

---

3   H. Minkowski, *Gött. Nachr.*, 1908, p. 107 and 108 [excerpts from this article are contained in this volume]; compare also eqs. (6) and (9) below.

4   M. Laue, *Das Relativiätsprinzip*, p. 149.

Now we take the four-dimensional tensor $\boldsymbol{T}$ to be the sum of two such tensors by setting

$$
\left.
\begin{aligned}
\boldsymbol{T}_{xx} &= \boldsymbol{p}_{xx} + \nu\mathfrak{B}_x{}^2, \\
&\cdot\ \cdot\ \cdot\ \cdot\ \cdot\ \cdot\ \cdot \\
&\cdot\ \cdot\ \cdot\ \cdot\ \cdot\ \cdot\ \cdot \\
\boldsymbol{T}_{uu} &= \boldsymbol{p}_{uu} + \nu\mathfrak{B}_u{}^2, \\
\boldsymbol{T}_{xy} &= \boldsymbol{p}_{xy} + \nu\mathfrak{B}_x\mathfrak{B}_y, \\
&\cdot\ \cdot\ \cdot\ \cdot\ \cdot\ \cdot\ \cdot \\
&\cdot\ \cdot\ \cdot\ \cdot\ \cdot\ \cdot\ \cdot \\
&\cdot\ \cdot\ \cdot\ \cdot\ \cdot\ \cdot\ \cdot \\
&\cdot\ \cdot\ \cdot\ \cdot\ \cdot\ \cdot\ \cdot \\
\boldsymbol{T}_{zu} &= \boldsymbol{p}_{zu} + \nu\mathfrak{B}_z\mathfrak{B}_u,
\end{aligned}
\right\}
\tag{3}
$$

where $\mathfrak{B}$ represents the four-dimensional velocity vector, which, as is well known, is related to the velocity v by the equations

$$
\mathfrak{B}_x = \frac{v_x}{\sqrt{1-\mathfrak{q}^2}}, \quad \ldots \quad \mathfrak{B}_u = \frac{ic}{\sqrt{1-\mathfrak{q}^2}}.
\tag{4}
$$

We call the four-dimensional tensor $\boldsymbol{p}$ introduced in eq. (3)[1] the *elastic stress tensor*. Like $\boldsymbol{T}$, it is symmetric, since $\boldsymbol{p}_{xy} = \boldsymbol{p}_{yx}$ etc. The second part of the tensor $\boldsymbol{T}$ can be called the *material* tensor. |

We set

$$
\left.
\begin{aligned}
\mathfrak{K}_x{}^e &= -\frac{\partial \boldsymbol{p}_{xx}}{\partial x} - \frac{\partial \boldsymbol{p}_{xy}}{\partial y} - \frac{\partial \boldsymbol{p}_{xz}}{\partial z} - \frac{\partial \boldsymbol{p}_{xu}}{\partial u}, \\
&\cdot\ \cdot\ \cdot\ \cdot\ \cdot\ \cdot\ \cdot\ \cdot\ \cdot\ \cdot\ \cdot\ \cdot \\
&\cdot\ \cdot\ \cdot\ \cdot\ \cdot\ \cdot\ \cdot\ \cdot\ \cdot\ \cdot\ \cdot\ \cdot \\
\mathfrak{K}_u{}^e &= -\frac{\partial \boldsymbol{p}_{ux}}{\partial x} - \frac{\partial \boldsymbol{p}_{uy}}{\partial y} - \frac{\partial \boldsymbol{p}_{uz}}{\partial z} - \frac{\partial \boldsymbol{p}_{uu}}{\partial u},
\end{aligned}
\right\}
\tag{5}
$$

[859]

and call the vector $\mathfrak{K}^e$ the *elastic* ponderomotive force. Our equations of motion (1) are now

$$
\left.
\begin{aligned}
\mathfrak{K}_x + \mathfrak{K}_x^{\;e} &= \frac{\partial}{\partial x}\nu\mathfrak{B}_x^{\;2} + \frac{\partial}{\partial y}\nu\mathfrak{B}_x\mathfrak{B}_y + \frac{\partial}{\partial z}\nu\mathfrak{B}_x\mathfrak{B}_z + \frac{\partial}{\partial u}\nu\mathfrak{B}_x\mathfrak{B}_u, \\[4pt]
\mathfrak{K}_y + \mathfrak{K}_y^{\;e} &= \frac{\partial}{\partial x}\nu\mathfrak{B}_y\mathfrak{B}_x + \frac{\partial}{\partial y}\nu\mathfrak{B}_y^{\;2} + \frac{\partial}{\partial z}\nu\mathfrak{B}_y\mathfrak{B}_z + \frac{\partial}{\partial u}\nu\mathfrak{B}_y\mathfrak{B}_u, \\[4pt]
\mathfrak{K}_z + \mathfrak{K}_z^{\;e} &= \frac{\partial}{\partial x}\nu\mathfrak{B}_z\mathfrak{B}_x + \frac{\partial}{\partial y}\nu\mathfrak{B}_z\mathfrak{B}_y + \frac{\partial}{\partial z}\nu\mathfrak{B}_z^{\;2} + \frac{\partial}{\partial u}\nu\mathfrak{B}_z\mathfrak{B}_u, \\[4pt]
\mathfrak{K}_u + \mathfrak{K}_u^{\;e} &= \frac{\partial}{\partial x}\nu\mathfrak{B}_u\mathfrak{B}_x + \frac{\partial}{\partial y}\nu\mathfrak{B}_u\mathfrak{B}_y + \frac{\partial}{\partial z}\nu\mathfrak{B}_u\mathfrak{B}_z + \frac{\partial}{\partial u}\nu\mathfrak{B}_u^{\;2}.
\end{aligned}
\right\}
\tag{6}
$$

In order to understand the meaning of the right-hand sides, we transform them. We denote the volume of a material particle of the body by dv, and the rest volume of the same by dv_0, where

$$
dv_0 = \frac{dv}{\sqrt{1-\mathfrak{q}^2}}.
\tag{7}
$$

Furthermore, if τ denotes the proper time

$$
d\tau = dt\sqrt{1-\mathfrak{q}^2},
\tag{8}
$$

then by introducing $\mathfrak{v}$ and using a well known formula[5] one obtains |

[860]

$$
\left.
\begin{aligned}
&\frac{\partial}{\partial x}\nu\mathfrak{B}_x^{\;2} + \frac{\partial}{\partial y}\nu\mathfrak{B}_x\mathfrak{B}_y + \frac{\partial}{\partial z}\nu\mathfrak{B}_x\mathfrak{B}_z + \frac{\partial}{\partial u}\nu\mathfrak{B}_x\mathfrak{B}_u \\[4pt]
&= \frac{1}{dv_0}\frac{d}{d\tau}(\nu\mathfrak{B}_x dv_0) = \frac{1}{dv}\frac{d}{dt}\left\{\frac{\mu\mathfrak{v}_x}{\sqrt{1-\mathfrak{q}^2}}dv\right\}.
\end{aligned}
\right\}
\tag{9}
$$

Changing the index x to y, z, u respectively, one obtains the corresponding equations. Inserting these expressions into (6), one obtains the equations of motion in a similar form, appropriate for a material point.

As usual, the first three of the equations of motion are supposed to express the law of conservation of momentum, the fourth that of conservation of energy. In order to study the first law more closely, we set

$$
\mathfrak{g}_x^{\;e} = -\frac{i}{c}\mathbf{P}_{xu}, \qquad \mathfrak{g}_y^{\;e} = -\frac{i}{c}\mathbf{P}_{yu}, \qquad \mathfrak{g}_z^{\;e} = -\frac{i}{c}\mathbf{P}_{zu},
\tag{10}
$$

---

5                           $\dfrac{d}{dt}\displaystyle\int \varphi\,dv = \int\left\{\operatorname{div}\ \varphi\mathfrak{v} + \frac{\partial\varphi}{\partial t}\right\}dv,$

where φ is an arbitrary function of x, y, z, t and the integration on the left extends over any particular part of the matter. The vector symbols in this essay are those explained in Abraham's *Theorie der Elektrizität*, Vol. I; they always refer to *three-dimensional* vectors.

and call the three-dimensional vector $\mathfrak{g}^e$ the *elastic momentum density*. The vector $\mathfrak{g}^m$, constructed in a similar way from the material tensor, with the components

$$\left(\mathfrak{g}_x{}^m = -\frac{i}{c}v\mathfrak{B}_x\mathfrak{B}_u\right) \quad \text{etc.,}$$

shall be called the *material momentum density*. One finds

$$\mathfrak{g}^m = \frac{\mu v}{\sqrt{1-\mathfrak{q}^2}}. \tag{11}$$

We furthermore introduce the *relative stresses* $\boldsymbol{t}$ [6] through the following equation

$$\left.\begin{aligned}
\boldsymbol{t}_{xx} &= \boldsymbol{p}_{xx} + \frac{i}{c}\boldsymbol{p}_{xu}v_x, \\[2mm]
\boldsymbol{t}_{xy} &= \boldsymbol{p}_{xy} + \frac{i}{c}\boldsymbol{p}_{xu}v_y, \\[2mm]
&\text{etc.,}
\end{aligned}\right\} \tag{12}$$

or by (10),

$$\left.\begin{aligned}
\boldsymbol{t}_{xx} &= \boldsymbol{p}_{xx} - \mathfrak{g}_x{}^e v_x, \\[2mm]
\boldsymbol{t}_{xy} &= \boldsymbol{p}_{xy} - \mathfrak{g}_x{}^e v_y, \\[2mm]
&\text{etc.,}
\end{aligned}\right\} \tag{12a}$$

| The relative stresses form a three-dimensional asymmetric tensor. Clearly, the calcu- [861] lation of these stresses is similar to that of the pressure on a moving surface in electrodynamics. Furthermore, it should be noted that writing $\boldsymbol{T}$ instead of $\boldsymbol{p}$ in (12) yields the same relative stresses, because the second (material) tensor into which we partitioned $\boldsymbol{T}$ contributes zero relative stress. For this reason, the relative stresses defined by eq. (12) are identical with those introduced by Laue (loc. cit.).

We can now transform the expressions for the spatial components of $\mathfrak{R}^e$. We obtain from (10) and (12a)

$$\begin{aligned}
\mathfrak{R}_x{}^e = &-\left\{\frac{\partial \boldsymbol{t}_{xx}}{\partial x} + \frac{\partial \boldsymbol{t}_{xy}}{\partial y} + \frac{\partial \boldsymbol{t}_{xz}}{\partial z}\right\} \\[2mm]
&-\left\{\frac{\partial}{\partial x}\mathfrak{g}_x{}^e v_x + \frac{\partial}{\partial y}\mathfrak{g}_x{}^e v_y + \frac{\partial}{\partial z}\mathfrak{g}_x{}^e v_z + \frac{\partial}{\partial t}\mathfrak{g}_x{}^e\right\},
\end{aligned} \tag{13}$$

and the corresponding expressions for $\mathfrak{R}_y{}^e$ and $\mathfrak{R}_z{}^e$.

---

6   M. Abraham, "Zur Elektrodynamik bewegter Körper," *Rend. Circ. Matem. Palermo*, eq. (10) 1909; M. Laue, loc. cit. p. 151.

We multiply eq. (13) by dv and integrate over a (three-dimensional) space v filled with mass. The integral of the expression in the first bracket can be converted into a surface integral by Gauss's theorem. Also, applying the formula in footnote 5 [p. 859 in the original] to the final bracketed expression, we obtain

$$\int \Re_x^e dv = -\int \{ t_{xx} df_x + t_{xy} df_y + t_{xz} df_z \} - \frac{d}{dt} \int g_x^e dv. \tag{14}$$

Here, df_x, df_y, df_z are the components of an area element of the surface bounding the region v under consideration, with df taken as a vector in the direction of the external normal. The symbol d/dt denotes the temporal change in a bounded region of the matter.

Corresponding expressions hold for the remaining spatial axis directions, and it is clear that the elastic force is partially determined by the relative elastic stresses acting as area forces [*Flächenkräfte*], and partially by the change of the elastic momentum.

According to (6), (9) and (14), we can now write the first of the equations of motion in the following integral form: |

$$\int \Re_x dv - \int \{ t_{xx} df_x + t_{xy} df_y + t_{xz} df_z \} - \frac{d}{dt} \int g_x^{\ e} dv$$

$$= \frac{d}{dt} \int \frac{\mu v_x}{\sqrt{1 - q^2}} dv = \frac{d}{dt} \int g_x^{\ m} dv. \tag{15}$$

These equations and the two analogous ones for the remaining axis directions express the law of conservation of momentum for a bounded region of the matter.

The asymmetry of the relative stress tensor implies that the elastic forces in general apply a torque[7] to each part of the body. According to the theory of elasticity, the torque acting on the unit volume about an axis parallel to the x-axis is

$$t_{yz} - t_{zy} = v_y g_x^{\ e} - v_z g_y^{\ e}.$$

Hence, expressed vectorially one has for the torque n per unit volume

$$\mathfrak{n} = [v g^e]. \tag{16}$$

Thus this torque must always appear when the momentum density has a component perpendicular to the velocity. The torque is thus necessary also to retain the uniform translatory motion of the elastically stressed body. As is well known, this is a significant difference between classical and relativistic mechanics, the reason for which becomes clear when establishing the area law.[8] However, we do not want to discuss this here.

---

7   M. Laue, loc. cit., p. 168.
8   M. Laue, *Ann. d. Phys.*, 35, p. 536, 1911.

Whereas the first three equations of motion express the law of conservation of momentum, the last equation expresses the law of conservation of energy. We set

$$\mathfrak{S}^e = c^2 \mathfrak{g}^e, \tag{17}$$

$$\mathfrak{S}^m = c^2 \mathfrak{g}^m = \frac{c^2 \mu \mathfrak{v}}{\sqrt{1-q^2}}, \tag{18}$$

hence

$$\mathfrak{S}_x^e = -ic\boldsymbol{p}_{xu} \qquad \text{etc.,} \tag{17a}$$

$$\mathfrak{S}_x^m = -icv\mathfrak{V}_x\mathfrak{V}_u \qquad \text{etc.,} \tag{18a}$$

I and furthermore [863]

$$\psi^e = -\boldsymbol{p}_{uu}, \tag{19}$$

$$\psi^m = -v\mathfrak{V}_u^2 = \frac{c^2\mu}{\sqrt{1-q^2}}. \tag{20}$$

Equation (5) yields an expression for $ic\mathfrak{K}_u^e$, which is, written in vector form,

$$ic\mathfrak{K}_u^e = \operatorname{div}\mathfrak{S}^e + \frac{\partial\psi^e}{\partial t}. \tag{21}$$

We can now write the last of the equations of motion (6), multiplied by $-ic$, as

$$-ic\mathfrak{K}_u = \operatorname{div}(\mathfrak{S}^e + \mathfrak{S}^m) + \frac{\partial}{\partial t}(\psi^e + \psi^m). \tag{22}$$

This is the equation expressing the conservation of energy. We recognize that the vectors $\mathfrak{S}^e$ and $\mathfrak{S}^m$ express the *elastic* and the *material energy flux* respectively and that the quantities ψ^e and ψ^m are the corresponding *energy densities*. The quantity $-ic\mathfrak{K}_u$ gives the energy influx per unit volume and time produced by the action of the external force $\mathfrak{K}$. The meaning of the right-hand side is obvious. By integrating over an arbitrary volume and applying Gauss's theorem, one obtains the law of conservation of energy, expressed for a spatial region, fixed in the (x, y, z) reference frame employed.

2. DETAILED INVESTIGATION OF THE ELASTIC STATE VARIABLES

In order to gain a clear conception of the elastic variables, we transform the stress tensor $\boldsymbol{p}$ to rest at the spacetime point under consideration. Then the components of the tensor take the form

$$
\left.\begin{array}{llll}
\overset{0}{\boldsymbol{p}}_{xx} & \overset{0}{\boldsymbol{p}}_{xy} & \overset{0}{\boldsymbol{p}}_{xz} & 0, \\[4pt]
\overset{0}{\boldsymbol{p}}_{yx} & \overset{0}{\boldsymbol{p}}_{yy} & \overset{0}{\boldsymbol{p}}_{yz} & 0, \\[4pt]
\overset{0}{\boldsymbol{p}}_{zx} & \overset{0}{\boldsymbol{p}}_{zy} & \overset{0}{\boldsymbol{p}}_{zz} & 0, \\[4pt]
0 & 0 & 0 & \overset{0}{\boldsymbol{p}}_{uu};
\end{array}\right\}
\tag{23}
$$

because in the case of rest the elastic state of stress can produce no energy flux and thus also no momentum.[9] |

[864]　　We note incidentally that for the case of rest the usual laws of the theory of elasticity apply. We can thus relate the six spatial stress components $\overset{0}{\boldsymbol{p}}_{xx}$, $\overset{0}{\boldsymbol{p}}_{xy}$, ... to the deformation quantities at rest.[10] However, we do not want to discuss this in more detail.

One easily sees that $\overset{0}{\boldsymbol{p}}_{uu}$ must be a four-dimensional scalar. It is

$$
\begin{aligned}
-c^2 \overset{0}{\boldsymbol{p}}_{uu} = {} & \boldsymbol{p}_{xx}\mathfrak{B}_x{}^2 + \boldsymbol{p}_{yy}\mathfrak{B}_y{}^2 + \boldsymbol{p}_{zz}\mathfrak{B}_z{}^2 + \boldsymbol{p}_{uu}\mathfrak{B}_u{}^2 \\
& + 2\boldsymbol{p}_{xy}\mathfrak{B}_x\mathfrak{B}_y + 2\boldsymbol{p}_{xz}\mathfrak{B}_x\mathfrak{B}_z + \dots ,
\end{aligned}
$$

since the right-hand side is invariant under Lorentz transformations, and upon transformation to rest, nine of the ten terms disappear, whereby one obtains the identity $-c^2 \overset{0}{\boldsymbol{p}}_{uu} = -c^2 \overset{0}{\boldsymbol{p}}_{uu}$.

Furthermore, by (19) and (20)

$$
\Psi = -\overset{0}{\boldsymbol{p}}_{uu} + c^2 \nu
\tag{24}
$$

is the *rest-energy density* of the matter, which is also a four-dimensional scalar. Since ν is still undetermined, one could define this quantity in such a way that $\overset{0}{\boldsymbol{p}}_{uu} = 0$. However, for the time being we do not want to impose such a condition.

By means of this transformation to rest, one also recognizes the validity of the following system of equations:

$$
\left.\begin{array}{l}
\boldsymbol{p}_{xx}\mathfrak{B}_x + \boldsymbol{p}_{xy}\mathfrak{B}_y + \boldsymbol{p}_{xz}\mathfrak{B}_z + \boldsymbol{p}_{xu}\mathfrak{B}_u = \overset{0}{\boldsymbol{p}}_{uu}\mathfrak{B}_x, \\[4pt]
\boldsymbol{p}_{yx}\mathfrak{B}_x + \boldsymbol{p}_{yy}\mathfrak{B}_y + \boldsymbol{p}_{yz}\mathfrak{B}_z + \boldsymbol{p}_{yu}\mathfrak{B}_u = \overset{0}{\boldsymbol{p}}_{uu}\mathfrak{B}_y, \\[4pt]
\boldsymbol{p}_{zx}\mathfrak{B}_x + \boldsymbol{p}_{zy}\mathfrak{B}_y + \boldsymbol{p}_{zz}\mathfrak{B}_z + \boldsymbol{p}_{zu}\mathfrak{B}_u = \overset{0}{\boldsymbol{p}}_{uu}\mathfrak{B}_z, \\[4pt]
\boldsymbol{p}_{ux}\mathfrak{B}_x + \boldsymbol{p}_{uy}\mathfrak{B}_y + \boldsymbol{p}_{uz}\mathfrak{B}_z + \boldsymbol{p}_{uu}\mathfrak{B}_u = \overset{0}{\boldsymbol{p}}_{uu}\mathfrak{B}_u.
\end{array}\right\}
\tag{25}
$$

9　If heat flow occurs, its influence is to be included in the action of the external force $\mathfrak{R}$. Compare section 5 below.

10　Cf. Herglotz, loc. cit.

From the first three of these equations we obtain the expressions for the components of the elastic energy flux and the momentum density. Namely, using eqs. (4) we find that

$$-ic\boldsymbol{p}_{ux} = -\overset{0}{\boldsymbol{p}}_{uu}\mathfrak{v}_x + \boldsymbol{p}_{xx}\mathfrak{v}_x + \boldsymbol{p}_{xy}\mathfrak{v}_y + \boldsymbol{p}_{xz}\mathfrak{v}_z,$$

hence

$$\left.\begin{aligned}
\mathfrak{S}_x^{\,e} &= c^2\mathfrak{g}_x^{\,e} = -\overset{0}{\boldsymbol{p}}_{uu}\mathfrak{v}_x + \boldsymbol{p}_{xx}\mathfrak{v}_x + \boldsymbol{p}_{xy}\mathfrak{v}_y + \boldsymbol{p}_{xz}\mathfrak{v}_z, \\
\mathfrak{S}_y^{\,e} &= c^2\mathfrak{g}_y^{\,e} = -\overset{0}{\boldsymbol{p}}_{uu}\mathfrak{v}_y + \boldsymbol{p}_{yx}\mathfrak{v}_x + \boldsymbol{p}_{yy}\mathfrak{v}_y + \boldsymbol{p}_{yz}\mathfrak{v}_z, \\
\mathfrak{S}_z^{\,e} &= c^2\mathfrak{g}_z^{\,e} = -\overset{0}{\boldsymbol{p}}_{uu}\mathfrak{v}_z + \boldsymbol{p}_{zx}\mathfrak{v}_x + \boldsymbol{p}_{zy}\mathfrak{v}_y + \boldsymbol{p}_{zz}\mathfrak{v}_z.
\end{aligned}\right\} \qquad (26)$$

ǀ We can also express these vector components in terms of the relative stresses, by using (12a) to eliminate $\boldsymbol{p}$. We obtain   [865]

$$\mathfrak{S}_x^{\,e}(1 - \mathfrak{q}^2) = -\overset{0}{\boldsymbol{p}}_{uu}\mathfrak{v}_x + \boldsymbol{t}_{xx}\mathfrak{v}_x + \boldsymbol{t}_{xy}\mathfrak{v}_y + \boldsymbol{t}_{xz}\mathfrak{v}_z, \qquad (26a)$$

and the corresponding expressions for the two remaining components.

The following expression for the elastic energy density arises from the last of eqs. (25):

$$\psi^e = -\overset{0}{\boldsymbol{p}}_{uu} + \mathfrak{g}^e\mathfrak{v}. \qquad (27)$$

3. THE CHANGES OF THE MASS AND OF THE REST-ENERGY

In order to obtain the law describing the variability of mass, we multiply the eqs. (6) in turn by $\mathfrak{B}_x, \mathfrak{B}_y, \mathfrak{B}_z, \mathfrak{B}_u$ and add them, where we take into account that

$$\mathfrak{B}_x\left\{\frac{\partial}{\partial x}v\mathfrak{B}_x^{\,2} + \frac{\partial}{\partial y}v\mathfrak{B}_x\mathfrak{B}_y + \frac{\partial}{\partial z}v\mathfrak{B}_x\mathfrak{B}_z + \frac{\partial}{\partial u}v\mathfrak{B}_x\mathfrak{B}_u\right\}$$

$$= \mathfrak{B}_x^{\,2}\left\{\frac{\partial}{\partial x}v\mathfrak{B}_x + \frac{\partial}{\partial y}v\mathfrak{B}_y + \frac{\partial}{\partial z}v\mathfrak{B}_z + \frac{\partial}{\partial u}v\mathfrak{B}_u\right\}$$

$$+ \frac{1}{2}v\left\{\mathfrak{B}_x\frac{\partial\mathfrak{B}_x^{\,2}}{\partial x} + \mathfrak{B}_y\frac{\partial\mathfrak{B}_x^{\,2}}{\partial y} + \mathfrak{B}_z\frac{\partial\mathfrak{B}_x^{\,2}}{\partial z} + \mathfrak{B}_u\frac{\partial\mathfrak{B}_x^{\,2}}{\partial u}\right\} \quad \text{etc.}$$

Since furthermore, according to the principles of the theory of relativity,

$$\mathfrak{B}_x^{\,2} + \mathfrak{B}_y^{\,2} + \mathfrak{B}_z^{\,2} + \mathfrak{B}_u^{\,2} = -c^2, \qquad (28)$$

we obtain

$$\left.\begin{aligned}
\mathfrak{B}_x(\mathfrak{K}_x + \mathfrak{K}_x{}^e) + \mathfrak{B}_y(\mathfrak{K}_y + \mathfrak{K}_y{}^e) + \mathfrak{B}_z(\mathfrak{K}_z + \mathfrak{K}_z{}^e) + \mathfrak{B}_u(\mathfrak{K}_u + \mathfrak{K}_u{}^e) \\
= -c^2\left\{\frac{\partial}{\partial x}\nu\mathfrak{B}_x + \frac{\partial}{\partial y}\nu\mathfrak{B}_y + \frac{\partial}{\partial z}\nu\mathfrak{B}_z + \frac{\partial}{\partial u}\nu\mathfrak{B}_u\right\}.
\end{aligned}\right\} \tag{29}$$

This equation gives the variability of mass with respect to time, because if dv_0 designates the rest volume of a material particle, then (compare eq. (9))

$$\frac{\partial}{\partial x}\nu\mathfrak{B}_x + \frac{\partial}{\partial y}\nu\mathfrak{B}_y + \frac{\partial}{\partial z}\nu\mathfrak{B}_z + \frac{\partial}{\partial u}\nu\mathfrak{B}_u = \frac{1}{d\tau_0}\frac{d}{d\tau}(\nu dv_0), \tag{30}$$

where $\nu dv_0 = \mu dv$ is the mass of the particle.

Therefore, if the sum of the external and elastic forces are perpendicular to the velocity vector $\mathfrak{B}$, the mass of the matter is constant in time, but otherwise it is not. I

[866]   Further, we want to develop a formula for the elastic force, and for this purpose differentiate the equations (25) with respect to x, y, z, u and add. Simply converting terms, taking (5) into consideration, we thus obtain

$$\left.\begin{aligned}
\mathfrak{B}_x\mathfrak{K}_x{}^e + \mathfrak{B}_y\mathfrak{K}_y{}^e &+ \mathfrak{B}_z\mathfrak{K}_z{}^e + \mathfrak{B}_u\mathfrak{K}_u{}^e \\
= \; \boldsymbol{P}_{xx}\frac{\partial\mathfrak{B}_x}{\partial x} &+ \boldsymbol{P}_{yy}\frac{\partial\mathfrak{B}_y}{\partial y} + \boldsymbol{P}_{zz}\frac{\partial\mathfrak{B}_z}{\partial z} + \boldsymbol{P}_{uu}\frac{\partial\mathfrak{B}_u}{\partial u} \\
+ \boldsymbol{P}_{xy}&\left\{\frac{\partial\mathfrak{B}_x}{\partial y} + \frac{\partial\mathfrak{B}_y}{\partial x}\right\} + \boldsymbol{P}_{xz}\left\{\frac{\partial\mathfrak{B}_x}{\partial z} + \frac{\partial\mathfrak{B}_z}{\partial x}\right\} \\
+ \boldsymbol{P}_{xu}&\{..\} + \boldsymbol{P}_{yz}\{..\} + \boldsymbol{P}_{yu}\{..\} + \boldsymbol{P}_{zu}\{..\} \\
- \left\{\frac{\partial}{\partial x}\boldsymbol{P}_{uu}^0\mathfrak{B}_x\right. &+ \frac{\partial}{\partial y}\boldsymbol{P}_{uu}^0\mathfrak{B}_y + \frac{\partial}{\partial z}\boldsymbol{P}_{uu}^0\mathfrak{B}_z + \left.\frac{\partial}{\partial u}\boldsymbol{P}_{uu}^0\mathfrak{B}_u\right\}.
\end{aligned}\right\} \tag{31}$$

Subtracting (29) from (31), taking (24) into account, one obtains

$$
\left.
\begin{aligned}
&\frac{\partial}{\partial x}\Psi\mathfrak{B}_x + \frac{\partial}{\partial y}\Psi\mathfrak{B}_y + \frac{\partial}{\partial z}\Psi\mathfrak{B}_z + \frac{\partial}{\partial u}\Psi\mathfrak{B}_u \\
&= -\{\mathfrak{B}_x\mathfrak{R}_x + \mathfrak{B}_y\mathfrak{R}_y + \mathfrak{B}_z\mathfrak{R}_z + \mathfrak{B}_u\mathfrak{R}_u\} \\
&\quad - \boldsymbol{p}_{xx}\frac{\partial\mathfrak{B}_x}{\partial x} - \boldsymbol{p}_{yy}\frac{\partial\mathfrak{B}_y}{\partial y} - \boldsymbol{p}_{zz}\frac{\partial\mathfrak{B}_z}{\partial z} - \boldsymbol{p}_{uu}\frac{\partial\mathfrak{B}_u}{\partial u} \\
&\quad - \boldsymbol{p}_{xy}\left\{\frac{\partial\mathfrak{B}_x}{\partial y} + \frac{\partial\mathfrak{B}_y}{\partial x}\right\} - \boldsymbol{p}_{xz}\left\{\frac{\partial\mathfrak{B}_x}{\partial z} + \frac{\partial\mathfrak{B}_z}{\partial x}\right\} \\
&\quad - \boldsymbol{p}_{xu}\{..\} - \boldsymbol{p}_{yz}\{..\} - \boldsymbol{p}_{yu}\{..\} - \boldsymbol{p}_{zu}\{..\}.
\end{aligned}
\right\}
\tag{32}
$$

This equation expresses the law of conservation of energy for a rest volume carried along by the matter, in contrast to eq. (22), which refers to a unit volume fixed in the spatial coordinate system used. Equation (32) is completely symmetric with respect to x, y, z, u; that it really expresses the law of conservation of energy is easily seen by transforming to rest.

4. THE DEFINITION OF INERTIAL MASS

Until now, we have taken the rest-mass density to be a completely arbitrary function of the four coordinates of the spacetime points of matter. Now we want to end this indeterminacy, I and for that reason we will focus on various possibilities for the moment. [867]

In eq. (24),

$$
\Psi = -\boldsymbol{p}_{uu}^0 + c^2\nu
$$

the rest-energy density Ψ is a defined quantity, one of the quantities $\boldsymbol{p}_{uu}^0$ and ν, however, can be freely specified. We demand that $\boldsymbol{p}_{uu}^0$ becomes zero if no elastic stresses exist in the body under consideration, because the tensor $\boldsymbol{p}$ should represent the elastic state of stress and only that state. Therefore, if all spatial components of $\boldsymbol{p}$ are zero when transformed to the state of rest (schema (23)), then $\boldsymbol{p}_{uu}^0$ should also be zero. But this can be achieved in various ways.

If one restricts attention to bodies in which there is a normal pressure from all sides, one can easily define the rest density ν in such a way that (if no heat conduction takes place) the total inertia of the body will be determined by its mass. To do so, one only has to set

$$
\boldsymbol{p}_{xx}^0 = \boldsymbol{p}_{yy}^0 = \boldsymbol{p}_{zz}^0 = \boldsymbol{p}_{uu}^0,
$$
$$
0 = \boldsymbol{p}_{xy}^0 = \boldsymbol{p}_{xz}^0 = \boldsymbol{p}_{xu}^0 = \dots,
$$

in the schema of (23),[11] from which ν is determined according to (24).

Then the stress tensor p has degenerated into a scalar and the elastic momentum density $\mathfrak{g}^e$ becomes equal to zero, independent of the state of motion.

However, as far as I am aware, this conception of mass cannot be naturally extended to the general case of bodies in which (relative) tangential stresses also exist. For the general case, it appears to me that the simplest and most expedient definition lies in the stipulation

$$c^2 \mathbf{v} = \Psi. \tag{33}$$

The rest-mass density is thus set proportional to the rest-energy density. Then, according to (24)

$$\overset{0}{\mathbf{p}}_{uu} = 0, \tag{34}$$

[868] | and several of our previous equations become simpler as a result.

Of course, the factual content of relativistic mechanics remains completely unaffected by the way in which we define inertial mass. Our definition gains more than a merely formal meaning only later, in section §6, when we also attribute weight to the inertial mass.

Having now fixed the concept of mass through the definition (33), we have to note that each moving and elastically stressed body possesses a momentum $\mathfrak{g}^e$ that is determined not by the mass but by the state of elastic stress of the body. According to (26) and (26a) we obtain

$$\left.\begin{aligned}
\mathfrak{g}_x^{\,e} &= \frac{1}{c^2}(\mathbf{p}_{xx}\mathfrak{v}_x + \mathbf{p}_{xy}\mathfrak{v}_y + \mathbf{p}_{xz}\mathfrak{v}_z) \\[1mm]
&= \frac{1}{c^2(1-\mathfrak{q}^2)}(\mathbf{t}_{xx}\mathfrak{v}_x + \mathbf{t}_{xy}\mathfrak{v}_y + \mathbf{t}_{xz}\mathfrak{v}_z) \quad \text{etc.}
\end{aligned}\right\} \tag{35}$$

This momentum also appears when there is a normal pressure from all sides in the body under consideration. In this case, $\mathfrak{g}^e$ can be derived from a *virtual* inertial mass which is added to that defined by eq. (33).

5. THE INFLUENCE OF HEAT CONDUCTION

All the equations developed above are also valid if heat conduction takes place in the bodies considered, because the effects of the heat conduction can be attributed to a ponderomotive force $\mathfrak{K}^w$ appearing in the heat conduction field, and this force can be included in the external force $\mathfrak{K}$. Of the heat conduction force $\mathfrak{K}^w$ the energy component $\mathfrak{K}_u^{\,w}$ plays the essential role. According to our fundamental assumptions, like all ponderomotive forces, $\mathfrak{K}^w$ should also be derivable from a symmetric four-dimensional tensor. We denote the heat conductivity tensor by w, and thus have

---

11   G. Nordström, *Physik. Zeitschr.*, 12. p. 854, 1911; M. Laue, *Das Relativitätsprinzip*, p. 151.

$$\mathfrak{R}_x^{\,w} = -\frac{\partial \mathbf{w}_{xx}}{\partial x} - \frac{\partial \mathbf{w}_{xy}}{\partial y} - \frac{\partial \mathbf{w}_{xz}}{\partial z} - \frac{\partial \mathbf{w}_{xu}}{\partial u},$$

$$\cdot \ \cdot \ \cdot \ \cdot \ \cdot \ \cdot \ \cdot \ \cdot \ \cdot \ \cdot \ \cdot \ \cdot \ \cdot \ \cdot$$

$$\cdot \ \cdot \ \cdot \ \cdot \ \cdot \ \cdot \ \cdot \ \cdot \ \cdot \ \cdot \ \cdot \ \cdot \ \cdot$$

$$\mathfrak{R}_u^{\,w} = -\frac{\partial \mathbf{w}_{ux}}{\partial x} - \frac{\partial \mathbf{w}_{uy}}{\partial y} - \frac{\partial \mathbf{w}_{uz}}{\partial z} - \frac{\partial \mathbf{w}_{uu}}{\partial u},$$

$$\left. \begin{array}{c} \\ \\ \\ \\ \end{array} \right\}$$

(36) I

where, e. g.,

[869]

$$\mathbf{w}_{xu} = \mathbf{w}_{ux} \quad \text{etc.}$$

For the case of rest, the tensor $\mathbf{w}$ takes the following schema:

$$\left. \begin{array}{cccc} 0 & 0 & 0 & \overset{0}{\mathbf{w}}_{xu}, \\[4pt] 0 & 0 & 0 & \overset{0}{\mathbf{w}}_{yu}, \\[4pt] 0 & 0 & 0 & \overset{0}{\mathbf{w}}_{zu}, \\[4pt] \overset{0}{\mathbf{w}}_{ux} & \overset{0}{\mathbf{w}}_{uy} & \overset{0}{\mathbf{w}}_{uz} & 0, \end{array} \right\}$$

(37)

because in the state of rest, all the spatial stresses are given by the tensor $\mathbf{p}$ (schema 23) and the total energy density of the matter is given by $\Psi = c^2\nu$. Therefore, the real components of $\mathbf{w}$ must be zero for the state of rest.

If we take heat conduction into account, then we have three four-dimensional tensors pertaining to matter: the thermal conductivity tensor, the elastic tensor, and the material tensor. For the case of rest, all three can be combined in the following common schema:

$$\left. \begin{array}{ccc|c} \overset{0}{\mathbf{p}}_{xx} & \overset{0}{\mathbf{p}}_{xy} & \overset{0}{\mathbf{p}}_{xz} & \overset{0}{\mathbf{w}}_{xu} \\[4pt] \overset{0}{\mathbf{p}}_{yx} & \overset{0}{\mathbf{p}}_{yy} & \overset{0}{\mathbf{p}}_{yz} & \overset{0}{\mathbf{w}}_{yu} \\[4pt] \overset{0}{\mathbf{p}}_{zx} & \overset{0}{\mathbf{p}}_{zy} & \overset{0}{\mathbf{p}}_{zz} & \overset{0}{\mathbf{w}}_{zu} \\[4pt] \hline \overset{0}{\mathbf{w}}_{ux} & \overset{0}{\mathbf{w}}_{uy} & \overset{0}{\mathbf{w}}_{uz} & -c^2\nu \end{array} \right\}.$$

(38)

Because we made the stipulation (33), the decomposition of the total tensor into the three parts is unambiguous.

We can introduce a four-vector $\mathfrak{W}$ by the system of equations

$$\left. \begin{aligned} w_{xx}\mathfrak{B}_x + w_{xy}\mathfrak{B}_y + w_{xz}\mathfrak{B}_z + w_{xu}\mathfrak{B}_u &= -\mathfrak{W}_x, \\ \cdots \cdots \cdots \cdots \cdots \cdots \cdots \cdots \cdots \cdots \cdots \\ \cdots \cdots \cdots \cdots \cdots \cdots \cdots \cdots \cdots \cdots \cdots \\ w_{ux}\mathfrak{B}_x + w_{uy}\mathfrak{B}_y + w_{uz}\mathfrak{B}_z + w_{uu}\mathfrak{B}_u &= -\mathfrak{W}_u. \end{aligned} \right\} \tag{39}$$

As we are going to show, this vector is to be designated the *rest heat flow*. The law of energy conservation for heat conduction (compare equation 21) is expressed by the equation

$$ic\mathfrak{R}_u^w = -ic\left\{ \frac{\partial w_{ux}}{\partial x} + \frac{\partial w_{uy}}{\partial y} + \frac{\partial w_{uz}}{\partial z} + \frac{\partial w_{uu}}{\partial u} \right\},$$

[870]　| and for the case of rest one obtains from (39)

$$ic w_{xu}^0 = -\mathfrak{W}_x^0,$$

$$ic w_{yu}^0 = -\mathfrak{W}_y^0,$$

$$ic w_{zu}^0 = -\mathfrak{W}_z^0,$$

$$0 = -\mathfrak{W}_u^0,$$

from which the asserted meaning of the vector $\mathfrak{W}$ becomes clear.

From the last equations one also sees that the four-vector $\mathfrak{W}$ is orthogonal to the velocity vector $\mathfrak{B}$, so that

$$\mathfrak{W}_x\mathfrak{B}_x + \mathfrak{W}_y\mathfrak{B}_y + \mathfrak{W}_z\mathfrak{B}_z + \mathfrak{W}_u\mathfrak{B}_u = 0, \tag{40}$$

since the left hand side of this equation is invariant under Lorentz transformations, and it equals zero when transformed to rest.

The tensor w can also be expressed as the "tensor product"[12] of the two four-vectors $\mathfrak{W}$ and $\mathfrak{B}$. Transforming to the state of rest, one finds the following expressions for the components of w

---

12　Cf. W. Voigt, *Gött. Nachr.*, p. 500, 1904.

$$
\left.\begin{aligned}
\mathbf{W}_{xx} &= \frac{2}{c^2}\mathfrak{W}_x\mathfrak{B}_x, \\[2mm]
\mathbf{W}_{uu} &= \frac{2}{c^2}\mathfrak{W}_u\mathfrak{B}_u, \\[2mm]
\mathbf{W}_{xy} &= \frac{1}{c^2}\{\mathfrak{W}_x\mathfrak{B}_y + \mathfrak{W}_y\mathfrak{B}_x\}, \\[2mm]
\mathbf{W}_{xu} &= \frac{1}{c^2}\{\mathfrak{W}_x\mathfrak{B}_u + \mathfrak{W}_u\mathfrak{B}_x\},
\end{aligned}\right\}
\tag{41}
$$

etc.

For the energy density ψ^w and for the energy flux $\mathfrak{S}^w$ of the heat conduction field one has of course

$$
\psi^w = -\mathbf{W}_{uu}, \tag{42}
$$

$$
\mathfrak{S}^w_x = -ic\mathbf{W}_{xu}, \qquad \mathfrak{S}^w_y = -ic\mathbf{W}_{yu}, \qquad \mathfrak{S}^w_z = -ic\mathbf{W}_{zu}. \tag{43}
$$

These quantities can also be expressed using the vector $\mathfrak{W}$. First, we find from (40)

$$
-ic\mathfrak{W}_u = \mathfrak{W}v, \tag{44}
$$

| where the right-hand side is the scalar product of two three-dimensional vectors. [871] Furthermore, from (41) we obtain

$$
\psi^w = \frac{2}{c^2\sqrt{1-q^2}}\mathfrak{W}v, \tag{42a}
$$

$$
\left.\begin{aligned}
\mathfrak{S}^w &= \frac{1}{\sqrt{1-q^2}}\left\{\mathfrak{W} + \frac{v}{c^2}(\mathfrak{W}v)\right\} \\[3mm]
&= \frac{\mathfrak{W}}{\sqrt{1-q^2}} + v\frac{\psi^w}{2}.
\end{aligned}\right\}
\tag{43a}
$$

The energy flux $\mathfrak{S}^w$ corresponds, of course, to the momentum density

$$
\mathfrak{g}^w = \frac{1}{c^2}\mathfrak{S}^w.
$$

We can write the last of the equations (36), multiplied by ic, as

$$
ic\mathfrak{R}_u^{\,w} = \operatorname{div}\mathfrak{S}^w + \frac{\partial\psi^w}{\partial t}, \tag{45}
$$

which can be inserted into the energy eq. (22). If $\mathfrak{K}^w$ is the only "external" force acting, naturally one has to set $\mathfrak{K} = \mathfrak{K}^w$ in all the equations of the previous sections, and thus specifically in (22) $\mathfrak{K}_u = \mathfrak{K}_u{}^w$.

We further wish to develop a few formulas for $\mathfrak{K}^w$. Inserting the expressions (41) into the system of equations (36), we obtain after simple transformations[13]

$$
\begin{aligned}
c^2 \mathfrak{K}_x{}^w = &-\frac{d\mathfrak{B}_x}{d\tau} - \mathfrak{B}_x \left\{ \frac{\partial \mathfrak{B}_x}{\partial x} + \frac{\partial \mathfrak{B}_y}{\partial y} + \frac{\partial \mathfrak{B}_z}{\partial z} + \frac{\partial \mathfrak{B}_u}{\partial u} \right\} \\[2mm]
&- \left\{ \mathfrak{B}_x \frac{\partial \mathfrak{B}_x}{\partial x} + \mathfrak{B}_y \frac{\partial \mathfrak{B}_x}{\partial y} + \mathfrak{B}_z \frac{\partial \mathfrak{B}_x}{\partial z} + \mathfrak{B}_u \frac{\partial \mathfrak{B}_x}{\partial u} \right\} \\[2mm]
&- \mathfrak{B}_x \left\{ \frac{\partial \mathfrak{B}_x}{\partial x} + \frac{\partial \mathfrak{B}_y}{\partial y} + \frac{\partial \mathfrak{B}_z}{\partial z} + \frac{\partial \mathfrak{B}_u}{\partial u} \right\},
\end{aligned}
\tag{46}
$$

and corresponding expressions for the remaining components of $\mathfrak{K}^w$. Multiplying these expressions by $\mathfrak{B}_x$, $\mathfrak{B}_y$, $\mathfrak{B}_z$, $\mathfrak{B}_u$ I and adding them, we obtain further, in light of (40) and (28),

[872]

$$
\begin{aligned}
\mathfrak{B}_x \mathfrak{K}_x{}^w &+ \mathfrak{B}_y \mathfrak{K}_y{}^w + \mathfrak{B}_z \mathfrak{K}_z{}^w + \mathfrak{B}_u \mathfrak{K}_u{}^w \\[2mm]
&= \frac{\partial \mathfrak{B}_x}{\partial x} + \frac{\partial \mathfrak{B}_y}{\partial y} + \frac{\partial \mathfrak{B}_z}{\partial z} + \frac{\partial \mathfrak{B}_u}{\partial u} \\[2mm]
&+ \frac{1}{c^2} \left\{ \mathfrak{B}_x \frac{d\mathfrak{B}_x}{d\tau} + \mathfrak{B}_y \frac{d\mathfrak{B}_y}{d\tau} + \mathfrak{B}_z \frac{d\mathfrak{B}_z}{d\tau} + \mathfrak{B}_u \frac{d\mathfrak{B}_u}{d\tau} \right\}.
\end{aligned}
\tag{47}
$$

This equation makes it possible to take heat conduction into account in the formulae (29) and (32).

---

13   If φ is an arbitrary function of the four coordinates, then

$$
\frac{d\varphi}{d\tau} = \mathfrak{B}_x \frac{\partial \varphi}{\partial x} + \mathfrak{B}_y \frac{\partial \varphi}{\partial y} + \mathfrak{B}_z \frac{\partial \varphi}{\partial z} + \mathfrak{B}_u \frac{\partial \varphi}{\partial u}.
$$

6. GRAVITATION

Various approaches to treat the phenomena of gravitation from the standpoint of relativity theory have been attempted. The theories of Einstein[14] and Abraham[15] are especially noteworthy. According to these two theories, however, the speed of light c would depend on the gravitational field rather than being constant, and this would require at least a complete revolution of the foundations of the present theory of relativity.

However, through a modification of the theory of Abraham one can, as I have shown elsewhere,[16] maintain the constancy of the speed of light, and develop a theory of gravitation which is compatible with the theory of relativity in its present form. Since I want to generalize this theory in one respect, its foundations are briefly recounted here.

I introduce the gravitational potential Φ and set, using rational units,

$$\frac{\partial^2 \Phi}{\partial x^2} + \frac{\partial^2 \Phi}{\partial y^2} + \frac{\partial^2 \Phi}{\partial z^2} + \frac{\partial^2 \Phi}{\partial u^2} = g\nu. \tag{48}$$

Here, ν is the rest density of matter as defined in eq. (33). The gravitational potential Φ and the quantity g are also four-dimensional scalars; we call g the *gravitational factor*. |

The gravitational field exerts forces on the bodies present within the field. For the [873] ponderomotive gravitational force $\Re^g$ per unit volume, I set

$$\Re_x{}^g = -g\nu\frac{\partial \Phi}{\partial x}, \quad \Re_y{}^g = -g\nu\frac{\partial \Phi}{\partial y}, \quad \Re_z{}^g = -g\nu\frac{\partial \Phi}{\partial z}, \quad \Re_u{}^g = -g\nu\frac{\partial \Phi}{\partial u}. \tag{49}$$

The equations (48) and (49), along with the principle of constant c,

$$c = \text{a universal constant}, \tag{50}$$

constitute the complete basis of my theory of gravitation. These equations also determine the rational units of Φ and g. For the time being, we consider the gravitational factor g to be a universal constant, but here I would like to remark that since g occurs only as a factor of ν, nothing prevents us from assuming that g depends upon the internal state of matter.

The fundamental equations (48), (49) and (50) demand that the mass of a material particle depends on the gravitational potential at its location. In order to obtain the law of this dependence, accordingly we consider the motion of a mass point of mass m in an arbitrary gravitational field. We assume that no forces except gravitation act on the mass point. Then we can write the equations of motion of the mass point in the following way (compare eq. (6) and (9))

---

14   A. Einstein, *Ann. d. Phys.*, 35, p. 898, 1911.
15   M. Abraham, *Physik. Zeitschr.*, 13, p. 1, 1912 [in this volume].
16   G. Nordström, *Physik. Zeitschr.*, 13, p. 1126, 1912 [in this volume].

$$-gm\frac{\partial\Phi}{\partial x} = m\frac{d\mathfrak{B}_x}{d\tau} + \mathfrak{B}_x\frac{dm}{d\tau},$$

$$-gm\frac{\partial\Phi}{\partial y} = m\frac{d\mathfrak{B}_y}{d\tau} + \mathfrak{B}_y\frac{dm}{d\tau},$$

$$-gm\frac{\partial\Phi}{\partial z} = m\frac{d\mathfrak{B}_z}{d\tau} + \mathfrak{B}_z\frac{dm}{d\tau},$$ (51)

$$-gm\frac{\partial\Phi}{\partial u} = m\frac{d\mathfrak{B}_u}{d\tau} + \mathfrak{B}_u\frac{dm}{d\tau}.$$

We multiply the equations in turn by $\mathfrak{B}_x, \mathfrak{B}_y, \mathfrak{B}_z, \mathfrak{B}_u$ and add them. Taking (28) into account, due to

$$\frac{d\Phi}{d\tau} = \mathfrak{B}_x\frac{\partial\Phi}{\partial x} + \mathfrak{B}_y\frac{\partial\Phi}{\partial y} + \mathfrak{B}_z\frac{\partial\Phi}{\partial z} + \mathfrak{B}_u\frac{\partial\Phi}{\partial u},$$

we obtain

$$-gm\frac{d\Phi}{d\tau} = -c^2\frac{dm}{d\tau},$$

[874] | or

$$\frac{1}{m}\frac{dm}{d\tau} = \frac{g}{c^2}\frac{d\Phi}{d\tau}.$$ (52)

If g is assumed to be constant, integration yields

$$m = m_0 e^{\frac{g}{c^2}\Phi},$$ (53)

and this equation gives the dependence of the mass on the gravitational potential.

The equations of motion can also be written in the following form, from (52):

$$-g\frac{\partial\Phi}{\partial x} = \frac{d\mathfrak{B}_x}{d\tau} + \frac{g}{c^2}\mathfrak{B}_x\frac{d\Phi}{d\tau},$$

$$-g\frac{\partial\Phi}{\partial y} = \frac{d\mathfrak{B}_y}{d\tau} + \frac{g}{c^2}\mathfrak{B}_y\frac{d\Phi}{d\tau},$$

$$-g\frac{\partial\Phi}{\partial z} = \frac{d\mathfrak{B}_z}{d\tau} + \frac{g}{c^2}\mathfrak{B}_z\frac{d\Phi}{d\tau},$$ (54)

$$-g\frac{\partial\Phi}{\partial u} = \frac{d\mathfrak{B}_u}{d\tau} + \frac{g}{c^2}\mathfrak{B}_u\frac{d\Phi}{d\tau},$$

whereby the mass m cancels out of the equations of motion.

The reason for the variability of the mass m is that the gravitational force $\mathfrak{K}^g$ is not orthogonal to the velocity vector $\mathfrak{V}$ (compare p. 507 [p. 865 in original]). Multiplying the equations (49) by $\mathfrak{V}_x$, $\mathfrak{V}_y$, $\mathfrak{V}_z$, $\mathfrak{V}_u$ and adding them, we obtain

$$\mathfrak{V}_x\mathfrak{K}_x{}^g + \mathfrak{V}_y\mathfrak{K}_y{}^g + \mathfrak{V}_z\mathfrak{K}_z{}^g + \mathfrak{V}_u\mathfrak{K}_u{}^g = -g\nu\frac{d\Phi}{d\tau}. \tag{55}$$

We can insert this expression into the eq. (29) for the change of the mass; the gravitational force $\mathfrak{K}^g$ is, of course, part of the "external" force $\mathfrak{K}$.

The gravitational force $\mathfrak{K}^g$ is derived from a symmetric four-dimensional tensor G, in that

$$\left. \begin{aligned} \mathfrak{K}_x{}^g &= -\frac{\partial G_{xx}}{\partial x} - \frac{\partial G_{xy}}{\partial y} - \frac{\partial G_{xz}}{\partial z} - \frac{\partial G_{xu}}{\partial u}, \\ &\quad \cdots \cdots \cdots \cdots \cdots \\ &\quad \cdots \cdots \cdots \cdots \cdots \\ \mathfrak{K}_u{}^g &= -\frac{\partial G_{ux}}{\partial x} - \frac{\partial G_{uy}}{\partial y} - \frac{\partial G_{uz}}{\partial z} - \frac{\partial G_{uu}}{\partial u}. \end{aligned} \right\} \tag{56}$$

One obtains equations of this form by inserting the expression (48) for $g\nu$ into (49), and then performing a further | transformation. Then one also finds the following [875] expressions for the tensor components:[17]

$$\left. \begin{aligned} G_{xx} &= \frac{1}{2}\left\{ \left(\frac{\partial\Phi}{\partial x}\right)^2 - \left(\frac{\partial\Phi}{\partial y}\right)^2 - \left(\frac{\partial\Phi}{\partial z}\right)^2 - \left(\frac{\partial\Phi}{\partial u}\right)^2 \right\}, \\ &\quad \cdots \cdots \cdots \cdots \cdots \cdots \\ &\quad \cdots \cdots \cdots \cdots \cdots \cdots \\ G_{uu} &= \frac{1}{2}\left\{ -\left(\frac{\partial\Phi}{\partial x}\right)^2 - \left(\frac{\partial\Phi}{\partial y}\right)^2 - \left(\frac{\partial\Phi}{\partial z}\right)^2 + \left(\frac{\partial\Phi}{\partial u}\right)^2 \right\}, \\ G_{xy} &= \frac{\partial\Phi}{\partial x}\frac{\partial\Phi}{\partial y}, \\ &\quad \cdots \cdots \cdots \\ &\quad \cdots \cdots \cdots \\ &\quad \cdots \cdots \cdots \\ G_{zu} &= \frac{\partial\Phi}{\partial z}\frac{\partial\Phi}{\partial u}. \end{aligned} \right\} \tag{57}$$

---

17   Abraham obtains precisely the same expression in his theory mentioned above; M. Abraham, loc. cit. p. 3.

These quantities give the fictitious gravitational stresses (pressure taken as positive), as well as momentum density, energy flux and energy density in the gravitational field. For the energy flux $\mathfrak{S}^g$ and for the momentum density $\mathfrak{g}^g$, one has

$$\mathfrak{S}_x{}^g = c^2 \mathfrak{g}_x{}^g = -ic\mathbf{G}_{xu} \qquad \text{etc.,}$$

and for the energy density ψ^g

$$\psi^g = -\mathbf{G}_{uu}.$$

Hence, according to (57), in the notation of vector analysis

$$\mathfrak{S}^g = c^2 \mathfrak{g}^g = -\frac{\partial \Phi}{\partial t} \nabla \Phi, \tag{58}$$

$$\psi^g = \frac{1}{2}\left\{ (\nabla\Phi)^2 + \frac{i}{c^2}\left(\frac{\partial \Phi}{\partial t}\right)^2 \right\}. \tag{59}$$

One sees that ψ^g is always positive.

The last of the equations (56), multiplied by ic, is now

$$ic\mathfrak{K}_u{}^g = \operatorname{div}\mathfrak{S}^g + \frac{\partial \psi^g}{\partial t}, \tag{60}$$

[876] which is the equation expressing the law of conservation of energy for the gravitational field. For regions outside of the material bodies, we have, I of course, $\mathfrak{K}_u{}^g = 0$. For regions within the bodies, eq. (60) is to be combined with eq. (22).

———

Equation (48) can obviously be viewed as a four-dimensional Poisson equation, and its integration can be performed accordingly.[18] However, the form of the eq. (48) also shows that one can calculate Φ according to the well-known formula for the retarded potential. Taking into account the possibility that g might be variable, one has

$$\left.\begin{aligned}
\Phi(x_0, y_0, z_0, t_0) &= -\frac{1}{4\pi}\int \frac{dx\,dy\,dz}{r}(g\mathsf{v})_{x, y, z, t} + \text{const.} \\[2mm]
&= -\frac{1}{4\pi}\int \frac{dx\,dy\,dz}{r}(g\mu\sqrt{1 - \mathsf{q}^2})_{x, y, z, t} + \text{const.,}
\end{aligned}\right\} \tag{61}$$

where

18   M. Abraham, *Physik. Zeitschr.* 13, p. 5, 1912; A. Sommerfeld, *Ann. d. Phys.*, 33, p. 665, 1910.

$$r = \sqrt{(x - x_0)^2 + (y - y_0)^2 + (z - z_0)^2},$$
$$t = t_0 - \frac{r}{c}.$$
(61a)

The integration is to be carried out over three-dimensional space.

7. FREE-FALL MOTION

We first wish to establish an equation for the motion of a mass point in an arbitrary *static* gravitational field. On this occasion we should make two comments. First, our theory does not allow real point-like masses, because at such a point by (61) we would have $\Phi = -\infty$, and hence, by (53), the mass of the point would be zero. Thus a "mass point" must always have a certain extension. Second, it should be noted that in order to allow one to treat the field as static, the particle moving in the field must be constituted such that its own field is weak in comparison to the external field, even in its immediate vicinity. |

In the static field one has

[877]

$$\frac{\partial \Phi}{\partial t} = 0.$$

We multiply the first three of the equations (54) by v_x, v_y, v_z and add them. On the left hand side we obtain $-gv\nabla\Phi$. Furthermore, one has generally

$$\frac{d\mathfrak{B}_x}{d\tau} = \frac{1}{1 - \mathfrak{q}^2}\frac{dv_x}{dt} + \frac{v_x}{c^2(1 - \mathfrak{q}^2)^2}v\frac{dv}{dt} \qquad \text{etc.,}$$
(62)

and hence

$$v_x\frac{d\mathfrak{B}_x}{d\tau} + v_y\frac{d\mathfrak{B}_y}{d\tau} + v_z\frac{d\mathfrak{B}_z}{d\tau} = \frac{1}{(1 - \mathfrak{q}^2)^2}v\frac{dv}{dt}.$$

Since furthermore in our case

$$\frac{d\Phi}{d\tau} = \frac{1}{\sqrt{1 - \mathfrak{q}^2}}v\nabla\Phi,$$

we obtain

$$-gv\nabla\Phi = \frac{i}{(1 - \mathfrak{q}^2)^2}v\frac{dv}{dt} + g\frac{\mathfrak{q}^2}{1 - \mathfrak{q}^2}v\nabla\Phi,$$

and finally

$$-gv\nabla\Phi = \frac{1}{1 - \mathfrak{q}^2}v\frac{dv}{dt}.$$
(63)

Now we wish to assume more specifically that the gravitational field is homogeneous and parallel to the z-axis, and hence that

$$\frac{\partial \Phi}{\partial z} = \text{const.}, \qquad \frac{\partial \Phi}{\partial x} = \frac{\partial \Phi}{\partial y} = \frac{\partial \Phi}{\partial u} = 0,$$

and investigate the motion of a mass point in this field. The third of the equations (54) gives

$$-g\frac{\partial \Phi}{\partial z} = \frac{1}{1-\mathfrak{q}^2}\frac{dv_z}{dt} + \frac{v_z}{c^2(1-\mathfrak{q}^2)^2}v\frac{dv}{dt} + g\frac{v_z{}^2}{c^2(1-\mathfrak{q}^2)}\frac{\partial \Phi}{\partial z};$$

taking (63) into account we find that the last two terms cancel one another and we obtain

$$\frac{dv_z}{dt} = -(1-\mathfrak{q}^2)g\frac{\partial \Phi}{\partial z}.$$

The first of the equations (54) yields in a similar manner

$$0 = \frac{1}{1-\mathfrak{q}^2}\frac{dv_x}{dt} + \frac{v_x}{c^2(1-\mathfrak{q}^2)^2}v\frac{dv}{dt} + g\frac{v_x v_z}{c^2(1-\mathfrak{q}^2)}\frac{\partial \Phi}{\partial z}.$$

Here too, the last two terms cancel one another, and the equation becomes
[878] $dv_x/dt = 0$. Since the same must be true for dv_y/dt, ∣ for a mass-point in a homogeneous gravitational field we obtain the equations of motion

$$\left.\begin{aligned} \frac{dv_z}{dt} &= -\left(1-\frac{v^2}{c^2}\right)g\frac{\partial \Phi}{\partial z}, \\ \frac{dv_x}{dt} &= \frac{dv_y}{dt} = 0. \end{aligned}\right\} \tag{64}$$

These equations state the following: *The velocity component perpendicular to the field direction is uniform. Gravitational acceleration becomes smaller as the velocity increases, but this is independent of the direction of the velocity. A body projected horizontally falls slower than one without initial velocity falling vertically.* One also sees that a rotating body must fall slower than a non-rotating one. Of course, for attainable rotational speeds the difference is much too small to be amenable to observation.

These results raise the question of whether the molecular motions within a falling body also have an influence on the gravitational acceleration. At least one cannot deny the possibility that this is the case. The theory of gravitation is then simply to be modified by considering the gravitational factor g as dependent on the molecular motions within the body rather than as a constant. For this reason we have left this possibility open in the foregoing treatment. In this context, it should be pointed out that also the mass density of a body depends upon the molecular motions, since the rest-energy density, which determines ν according to eq. (33), is influenced by these motions.

However, those questions of the theory of gravitation which are related to the atomic structure of matter lie beyond the scope of this essay.

EDITORIAL NOTE

[1] In the original, Nordström mistakenly refers to eq. (4) rather than eq. (3).

GUNNAR NORDSTRÖM

ON THE THEORY OF GRAVITATION FROM THE STANDPOINT OF THE PRINCIPLE OF RELATIVITY

Originally published as "Zur Theorie der Gravitation vom Standpunkt des Relativitätsprinzips" in Annalen der Physik, 42, 1913, pp. 533–554. Received July 24, 1913. Author's date: Zurich, July 1913.

In the present communication, I wish to develop further several aspects of the theory of gravitation whose fundamentals I published in two previous essays and discuss it.[1] The theory presented in the last essay is not completely unambiguous. First—as emphasized on p. 509 [p. 867 in the original]—the rest density of matter was defined in a fairly arbitrary way; though a different definition of the concept of mass would not change the general laws of mechanics, it would modify the laws of gravitation. Second, in the theory of gravitation, the possibility has been left open that the gravitational factor g is not a constant, but could depend on various circumstances. One can think of this scalar quantity as being dependent on the internal state of the object as well as on the gravitational potential at the location in question. A dependence of the gravitational factor on the state of stress of the body is equivalent to a change in the definition of mass, but a dependence on the gravitational potential will have a deeper significance for the theory.

1. DEFINITE FORMULATION OF THE THEORY

All the aforementioned ambiguities of the theory can be removed by a very plausible stipulation which I owe to Mr. Laue and Mr. Einstein. Mr. Laue has shown that one can maintain Einstein's theorem of equivalence—though not in its full scope—by defining | the rest density of matter in an appropriate manner,[2] namely by means of the sum [p. 534]

$$T_{xx} + T_{yy} + T_{zz} + T_{uu} = -D \tag{1}$$

---

1 G. Nordström, *Physik. Zeitschr.*, 13, p. 1126, 1912; *Ann. d. Phys.*, 40, p. 856, 1913 [both in this volume]. The present communication is a continuation of the latter, and the symbol loc. cit. in the text refers to the same.

2 See A. Einstein, "Entwurf einer verallgemeinerten Relativitätstheorie und einer Theorie der Gravitation." *Zeitschr. f. Math. Phys.*, 62, p. 21, 1913.

Jürgen Renn (ed.). *The Genesis of General Relativity*, Vol. 3
Gravitation in the Twilight of Classical Physics: Between Mechanics, Field Theory, and Astronomy.
© 2007 Springer.

of the diagonal components of the tensor T, which represents the state of the matter. T is the dynamical tensor introduced by Laue,[3] and it is equal to the sum of the two tensors which I earlier called the elastic stress tensor and the material tensor.[4] Following Einstein, we will call the invariant D defined by eq. (1) Laue's scalar, and we will find that when divided by c^2 it represents the rest density.

Furthermore, it will turn out that Einstein's theorem of equivalence demands a very particular dependence of the gravitational factor g on the gravitational potential Φ; we put

$$g = g(\Phi).$$

If we furthermore denote the rest density of matter by ν, the fundamental equations for the gravitational field are[5]

$$\frac{\partial^2\Phi}{\partial x^2} + \frac{\partial^2\Phi}{\partial y^2} + \frac{\partial^2\Phi}{\partial z^2} + \frac{\partial^2\Phi}{\partial u^2} = g(\Phi)\nu, \tag{2}$$

$$\left.\begin{aligned}
\Re_x^g &= -g(\Phi)\nu\frac{\partial\Phi}{\partial x}, & \Re_y^g &= -g(\Phi)\nu\frac{\partial\Phi}{\partial y},\\
\Re_z^g &= -g(\Phi)\nu\frac{\partial\Phi}{\partial z}, & \Re_u^g &= -g(\Phi)\nu\frac{\partial\Phi}{\partial u}.
\end{aligned}\right\} \tag{3}$$

Equation (2) determines the gravitational field produced by a given distribution of masses. The system of eqs. (3) determines the ponderomotive force $\Re^g$, which the field exerts on matter.

The task now is to define the rest density ν and to determine the function $g(\Phi)$ in such a way that the theorem of equivalence is valid in the widest possible sense.

[p. 535] For this purpose, we consider a system of finite bodies such that an appropriate reference system exists, | in which the gravitational field is static so that we have everywhere $\partial\Phi/\partial t = 0$. However, bodies rotating about their symmetry axis and stationary flows of fluids may occur. In any case, in the reference system under consideration, the total momentum is equal to zero,[6]

$$\mathfrak{B} = \int g\,dv = 0,$$

and the system as a whole is at rest. A system which satisfies these conditions is to be called a complete stationary system.

3   M. Laue, *Das Relativitätsprinzip*, 2nd. Ed., p. 182.
4   G. Nordström, *Ann. d. Phys.*, loc. cit., p. 858.
5   G. Nordström, loc. cit., eqs. (48) and (49).
6   We must exclude heat transport, since otherwise the total momentum is not zero. Besides, heat transport would change with time the energy distribution, and thus also the mass distribution, and therefore make the gravitational field time-dependent.

As the state does not change with time, eq. (2) yields, according to the usual potential theory

$$\Phi = -\frac{1}{4\pi} \int \frac{dv}{r} g(\Phi)v + \Phi_a, \tag{4}$$

where the integral extends over all of xyz-space. Φ_a is the value of Φ at infinity and has its origin in other systems of masses, which we assume to be far away. For large distances r, one has

$$|\nabla\Phi| = \frac{1}{4\pi r^2} \int g(\Phi)vdv, \tag{5}$$

and the direction of $\nabla\Phi$ is away from the system of masses.

In the most general case, we have in the system three different world-tensors, which give the spatial stresses, the energy flux, and the energy-momentum density: the elastic-material tensor $\mathbf{T}$, the gravitation tensor $\mathbf{G}$ and the electromagnetic tensor $\mathbf{L}$. For the components of the gravitation tensor, the eqs. (57) apply, loc. cit.

$$\left.\begin{aligned}
\mathbf{G}_{xx} &= \frac{1}{2}\left\{\left(\frac{\partial\Phi}{\partial x}\right)^2 - \left(\frac{\partial\Phi}{\partial y}\right)^2 - \left(\frac{\partial\Phi}{\partial z}\right)^2 - \left(\frac{\partial\Phi}{\partial u}\right)^2\right\}, \\
\mathbf{G}_{xy} &= \frac{\partial\Phi}{\partial x}\frac{\partial\Phi}{\partial y} \qquad \text{etc.,}
\end{aligned}\right\} \tag{6}$$

and we have

$$\mathfrak{R}_x^g = -\left\{\frac{\partial\mathbf{G}_{xx}}{\partial x} + \frac{\partial\mathbf{G}_{xy}}{\partial y} + \frac{\partial\mathbf{G}_{xz}}{\partial z} + \frac{\partial\mathbf{G}_{xu}}{\partial u}\right\} \qquad \text{etc.} \tag{7}$$

| We want to form the sum of the diagonal components for the *total tensor* [p. 536] $\mathbf{T} + \mathbf{G} + \mathbf{L}$ and to integrate over all of three-dimensional space in our system of reference. Since $\partial\Phi/\partial t = 0$ the trace of the gravitation tensor is equal to $-(\nabla\Phi)^2$; the trace of $\mathbf{T}$ has been denoted by $-D$, and the trace of the electromagnetic tensor is equal to zero. Hence, we form the integral

$$-\int\{D + (\nabla\Phi)^2\}dv$$

extended over all of space. But since, according to a theorem of Laue[7] we have

$$\int\{\mathbf{T}_{xx} + \mathbf{G}_{xx} + \mathbf{L}_{xx}\}dv = 0$$

and two corresponding equations for the yy- and zz-components hold, we obtain

---

7   M. Laue, loc. cit., p. 209.

$$-\int\{D+(\nabla\Phi)^2\}dv = \int\{T_{uu}+G_{uu}+L_{uu}\}dv = -E_0,$$

where E_0 denotes the energy of the whole system in the frame of reference used (the rest energy).

Since $\partial^2\Phi/\partial u^2 = 0$, eq. (2) yields

$$\mathrm{div}\nabla\Phi = g(\Phi)\mathrm{v},$$
$$\mathrm{div}\Phi\nabla\Phi = \Phi g(\Phi)\mathrm{v} + (\nabla\Phi)^2.$$

The integration of $(\nabla\Phi)^2$ over a ball of infinitely large radius, taking (5) into account, yields

$$\int(\nabla\Phi)^2 dv = -\int(\Phi-\Phi_a)g(\Phi)\mathrm{v}dv.$$

Therefore, for E_0 one obtains

$$E_0 = \int Ddv - \int(\Phi-\Phi_0)g(\Phi)\mathrm{v}dv. \tag{8}$$

Since the total momentum in the system of reference under consideration is zero, in a different system of reference, in which our system of masses is moving with the speed v, one has the following expressions for the energy E and for the momentum $\mathfrak{G}$,[8]

$$E = \frac{E_0}{\sqrt{1-q^2}}, \qquad \mathfrak{G} = \frac{E_0\cdot\mathrm{v}}{c^2\sqrt{1-q^2}},$$

where we have set $q = \mathrm{v}/c$. The inertial mass of the system is thus:

$$m = \frac{E_0}{c^2} = \frac{1}{c^2}\int\{D-(\Phi-\Phi_a)g(\Phi)\mathrm{v}\}dv \tag{9}$$

From eqs. (4), (5) and (3), one sees that the quantity |

[p. 537]

$$M_g = \int g(\Phi)\mathrm{v}dv \tag{10}$$

determines the gravitational effects that the system exerts and experiences. M_g will be called the *gravitational mass* of the system. Einstein's equivalence theorem implies that for various systems M_g and m are proportional to each other. Then M_g/m can only be a function of Φ_a, and this function can be none but $g(\Phi_a)$, so that one has

---

8   M. Laue, loc. cit., p. 209.

$$g(\Phi_a)m = \int g(\Phi)v\,dv. \tag{11}$$

Now the task is to define the rest density v and to determine the function $g(\Phi)$ by means of this stipulation. We equate the expressions obtained for m from (9) and (11) and obtain

$$\frac{1}{c^2}\int D\,dv = \frac{1}{c^2}\int g(\Phi)v\left\{\Phi - \Phi_a + \frac{c^2}{g(\Phi_a)}\right\}dv.$$

In order to satisfy this equation identically, we set

$$D = g(\Phi)v\left\{\Phi - \Phi_a + \frac{c^2}{g(\Phi_a)}\right\}. \tag{12}$$

Since furthermore, the potential Φ_a of the external field must drop out of the equation, we set

$$\frac{c^2}{g(\Phi)} - \Phi = A, \qquad g(\Phi) = \frac{c^2}{A + \Phi}, \tag{13}$$

where A signifies a universal constant.

From (12), one now obtains

$$v = \frac{1}{c^2}D = -\frac{1}{c^2}(T_{xx} + T_{yy} + T_{zz} + T_{uu}), \tag{14}$$

whereby the rest density of the matter is defined.[9]

It is to be noted that the value of the constant A is unknown, since the absolute value of the potential Φ cannot be calculated at any point. Denoting the gravitational potential at a point accessible to investigation by Φ, I and that part of Φ which arises [p. 538] from masses external to our solar system by Φ_0, then only the difference $\Phi - \Phi_0$ can be determined by means of any kind of observations concerning the quantities Φ and Φ_0. Let $g(\Phi_0)$ be denoted by g_0. If we eliminate the quantity A from the equations

$$g(\Phi) = \frac{c^2}{A + \Phi} \quad \text{and} \quad g_0 = \frac{c^2}{A + \Phi_0},$$

we obtain for the function $g(\Phi)$:

---

9   In addition it should be noted that one can also express v by means of the relative stress t and the rest energy density Ψ of matter. One finds

$$v = \frac{1}{c^2}\{\Psi - t_{xx} - t_{yy} - t_{zz}\}.$$

$$g(\Phi) = \frac{g_0}{1 + \dfrac{g_0}{c^2}(\Phi - \Phi_0)}. \tag{15}$$

Only experimentally accessible quantities appear in this equation. Naturally, one could also fix the initial potential Φ_0 in a different way, since for two arbitrary values of Φ one has

$$g(\Phi_2) = \frac{g(\Phi_1)}{1 + \dfrac{g(\Phi_1)}{c^2}(\Phi_2 - \Phi_1)}. \tag{15a}$$

The eqs. (14) and (15) for v and g uniquely determine the theory of gravitation. If we set

$$\Phi' = \Phi + A = \frac{c^2}{g_0} + \Phi - \Phi_0, \tag{16}$$

then the new gravitational potential Φ' is not afflicted with the ambiguity of Φ. We obtain from (13) and (11)

$$g(\Phi) = \frac{c^2}{\Phi'}, \tag{17}$$

$$M_g = \frac{mc^2}{\Phi_a'} = \frac{E_0}{\Phi_a'}, \tag{18}$$

and if one inserts Φ' into the fundamental eqs. (2), (3), they become

$$\Phi'\left\{\frac{\partial^2\Phi'}{\partial x^2} + \frac{\partial^2\Phi'}{\partial y^2} + \frac{\partial^2\Phi'}{\partial z^2} + \frac{\partial^2\Phi'}{\partial u^2}\right\} = c^2 v, \tag{19}$$

$$\left.\begin{array}{ll} \Re_x^g = -c^2 v\dfrac{\partial}{\partial x}\ln\Phi', & \Re_y^g = -c^2 v\dfrac{\partial}{\partial y}\ln\Phi', \\ \Re_z^g = -c^2 v\dfrac{\partial}{\partial z}\ln\Phi', & \Re_u^g = -c^2 v\dfrac{\partial}{\partial u}\ln\Phi'. \end{array}\right\} \tag{20}$$

If they are written in this way, no universal constant corresponding to the gravitational constant appears in the fundamental equations.[10] In eq. (19) one can, to a certain approximation, take Φ' in front of the brackets as a constant equal to c^2/g_0. By integration one then obtains the usual formula for the retarded potential.

[p. 539]

---

10 But in §6 it is to be shown that such a universal constant plays a role in the definition of the fundamental units.

2. DEPENDENCE OF A BODY'S MASS ON THE GRAVITATIONAL POTENTIAL

We wish to prove that the inertial mass of a system depends on the properties of the gravitational potential Φ_a of the external field outlined in §1. We examine the state of affairs from the system of reference in which the bodies produce a static gravitational field and construct about the bodies a spherical surface of very large radius r. At the points on this surface, $\nabla\Phi$ is directed vertically outwards and has, according to (5), the magnitude

$$|\nabla\Phi| = \frac{M_g}{4\pi r^2}.$$

We imagine that the gravitational potential Φ_a of the external field is produced by masses which lie very far away from our system of bodies and outside the spherical surface. For the time being, Φ_a is spatially and temporally constant inside this surface. Then we imagine Φ_a being changed by $d\Phi_a$ due to a slow displacement of the distant masses. This change engenders a certain flow of energy through the spherical surface, which we want to calculate. The energy flux $\mathfrak{S}^g$ in the gravitational field is according to eq. (58), loc. cit., as well as by (6)

$$\mathfrak{S}^g = -\frac{\partial\Phi}{\partial t}\nabla\Phi.$$

By integrating over the spherical surface, we find that the change $d\Phi_a$ of the external potential results in energy transport through the spherical surface to the interior, having the magnitude

$$4\pi r^2 |\nabla\Phi| d\Phi_a = M_g d\Phi_a.$$

Hence, the amount by which the rest energy E_0 of our system has been increased is:

$$dE_0 = M_g d\Phi_a.$$

I If we insert here $E_0 = c^2 m$ from (9) and $M_g = g(\Phi_a)m$ from (11), we obtain the [p. 540] equation

$$\frac{1}{m}dm = \frac{g(\Phi_a)}{c^2}d\Phi_a, \qquad (a)$$

which agrees with the eq. (52), loc. cit., found by a different method. According to (17), we have further

$$d\ln m = d\ln\Phi_a',$$

and obtain finally through integration

$$\frac{m}{\Phi_a'} = \text{const.} \qquad (21)$$

The inertial mass of a body is thus directly proportional to the gravitational potential Φ_a' of the external field. According to (16), we can also write this dependence in the following form

$$m = m_0 \left\{ 1 + \frac{g_0}{c_2}(\Phi_a - \Phi_0) \right\},\qquad (21a)$$

where m_0 is the inertial mass associated with the gravitational potential Φ_0.[11]
    According to (21) and (18) we have

$$M_g = \int g(\Phi)\nu d\upsilon = \text{const.}\qquad (22)$$

Thus, in contrast to the inertial mass, the gravitational mass is a characteristic constant for each body that does not depend upon the external gravitational potential.

3. INERTIAL AND GRAVITATIONAL MASS OF A SPHERICAL ELECTRON

As an example of the theory, we want to establish the formulas for the inertial and gravitational mass of a spherical electron with uniform surface charge. Let the electric charge of the electron expressed in rational units be e, and the radius a. In order to prevent the unlimited expansion of the electron as a result of the force of repulsion between equal electric charges, certain elastic stresses must act within the electron. Most conveniently, we assume that these stresses are concentrated on the surface of
[p. 541]   the electron as well. I According to eq. (14), the elastic stresses give the electron a mass, which also produces gravitational effects. The gravitational field is superimposed on the electric field and both fields act back on the electron. The electron is at rest. We have to think of the surface of the electron as an infinitely thin shell, in which the elastic tensor $\boldsymbol{T}$ is different from zero. We assume that the component T_{uu}^{0} equals zero, so that, for the case of rest, the tensor $\boldsymbol{T}$ does not contribute to the energy of the electron. Hence, the tensor $\boldsymbol{T}$ reduces (for the case of rest) to a spatial stress tensor, which for reasons of symmetry must have one principal axis in the direction of the radius. Since we should have a tensile stress [*Zugspannung*] parallel to the shell only, the principal component of $\boldsymbol{T}$ in the direction of the radius equals zero; the two other principal components are equal to each other and will be called p. The trace of $\boldsymbol{T}$ is thus $2p$ and the rest density ν becomes

$$\nu = -\frac{1}{c^2}2p.$$

The tensile stress S in the surface is equal to the line integral

---

11   The eqs. (21) and (21a) take the place of (53), loc. cit., which presupposes a constant g and thus has now lost its validity.

$$S = -\int pds$$

integrated across the shell.

We now form the integral $\int v dv$ extended over the whole shell and obtain

$$\int v dv = -\frac{2}{c^2}4\pi a^2\int pds = \frac{1}{c^2}8\pi a^2 S.$$

Since for reasons of symmetry, $g(\Phi)$ must have the same value at all points of the surface, we obtain for the gravitational mass M_g of the electron

$$M_g = \int g(\Phi)v dv = \frac{g(\Phi)}{c^2}8\pi a^2 S.$$

On the surface of the electron, the gravitational potential has the value

$$-\frac{M_g}{4\pi a} + \Phi_a,$$

where Φ_a is the potential of the external field (not produced by the electron). Hence, according to (15a), we can substitute the expression |

$$g(\Phi) = \frac{g_a}{1 - \frac{g_a}{c^2}\frac{M_g}{4\pi a}}, \tag{23}$$

[p. 542]

for $g(\Phi)$, and thus obtain the equation

$$M_g = \frac{g_a}{c^2}\frac{8\pi a^2 S}{1 - \frac{g_a}{c^2}\frac{M_g}{4\pi a}}. \tag{a}$$

The task is now to calculate S from the forces the electric field and the gravitational field exert on the surface of the electron. On each element of the electron's surface, the electric field exerts a force perpendicular to and directed outwards from the surface, whose magnitude per unit of surface area is given by the Maxwellian stresses. One finds for this outward-directed force per unit area the expression

$$\frac{e^2}{32\pi^2 a^4}.$$

In a similar manner, the gravitational field exerts a force on each surface element of the electron's surface, which is perpendicular but directed inwards. Outside of the electron one has

$$\Phi = -\frac{M_g}{4\pi r} + \Phi_a, \qquad \frac{d\Phi}{dr} = \frac{M_g}{4\pi r^2}.$$

The fictitious gravitational stress on the surface is perpendicular to the surface, and by (6) it has the magnitude

$$\frac{1}{2}\left(\frac{d\Phi}{dr}\right)^2 = \frac{M_g^{\,2}}{32\pi^2 a^4}.$$

This force is exerted by the gravitational field on the electron's surface per unit of surface area. The combined force which the two fields exert on a unit of surface area is thus

$$P = \frac{e^2 - M_g^{\,2}}{32\pi^2 a^4}, \tag{b}$$

where positive is outward-directed. The force P and the elastic stress S in the electron's surface should now maintain equilibrium. It is easy to find that the condition for equilibrium is

$$2S = aP. \tag{c}$$

[p. 543] To derive this relation, one can for example think of the spherical electron surface as divided into two equal halves. I The normal force P attempts to drive the two halves apart with a total force of $\pi a^2 P$, whereas the stress S holds the two halves together with a total force of $2\pi a S$. By equating the two expressions for the force, one obtains the relation (c).

The eqs. (a), (b), (c) yield

$$M_g = \frac{g_a}{c^2}\frac{e^2 - M_g^{\,2}}{8\pi a} \cdot \frac{1}{1 - \dfrac{g_a}{c^2}\dfrac{M_g}{4\pi a}}.$$

We thus have a quadratic equation for M_g:

$$M_g - \frac{g_a M_g^{\,2}}{c^2 8\pi a} = \frac{g_a}{c^2}\frac{c^2}{8\pi a}.$$

From this we obtain

$$M_g = \frac{g_a}{c^2} \cdot \frac{e^2 + M_g^{\,2}}{8\pi a}, \tag{23}$$

and since according to (11) $M_g = g_a \cdot m$, we obtain for the inertial mass m of the electron the expression

$$m = \frac{e^2 + M_g{}^2}{8\pi c^2 a} = \frac{E_0}{c^2}. \tag{24}$$

The rest energy E_0 is composed of two parts: the energy of the electric field and that of the gravitational field. These two parts of E_0 are

$$\frac{e^2}{8\pi a} \quad \text{and} \quad \frac{M_g{}^2}{8\pi a}.$$

Thus the expression found for m contains a confirmation of the theory.

4. DEPENDENCE OF THE DIMENSIONS OF LENGTH ON THE GRAVITATIONAL POTENTIAL

An important conclusion can furthermore be drawn from the expression (24) for the inertial mass of an electron. If the gravitational potential Φ_a of the external field (not stemming from the electron) is changed, then m changes according to eq. (21). However, by (22), M_g remains constant, and the same applies for e, according to the fundamental equations of electrodynamics. Therefore, the radius a must vary inversely with m, and one obtains from (21)

$$a\Phi_a{}' = \text{const.} \tag{25}$$

| Since on the electron surface   [p. 544]

$$\Phi' = \Phi_a{}' - \frac{M_g}{4\pi a} = \Phi_a{}' \cdot \text{const.},$$

one also has

$$a\Phi' = \text{const.} \tag{25a}$$

The elastic tension S in the electron surface varies also with Φ_a. From (a), p. 531 [p. 542 in the original], we find the law for this if we take into account that according to (17) and (25) g_a/a remains constant. We see that $a^3 S$ must be constant, and that therefore

$$\frac{S}{\Phi_a{}'^3} = \text{const.}$$

Mr. Einstein has proved that the dependence in the theory developed here of the length dimensions of a body on the gravitational potential must be a general property of matter. He has shown that otherwise it would be possible to construct an apparatus with which one could pump energy out of the gravitational field. In Einstein's example one considers a non-deformable rod that can be constrained to move between two vertical rails. One could let the rod fall while stressed, then remove the stress and raise it back up. The rod has a greater weight when stressed than when unstressed, and therefore it would do more work than would be consumed in raising the unstressed rod. However, because the rod lengthens while falling, the rails must diverge, and the excess work from the fall will be consumed again as the work done by the tensioning forces on the ends of the rod.

Let S be the total stress (stress times cross-sectional area) of the rod and l its length. Because of the stress, the gravitational mass of the rod is increased by

$$\frac{g(\Phi)}{c^2}Sl = \frac{1}{\Phi'}Sl.$$

In falling, this gravitational mass provides the extra work

$$-\frac{1}{\Phi'}sld\Phi'.$$

[p. 545] | However at the same time at the ends of the rod the work

$$Sdl$$

is lost. Equating these two expressions yields

$$-\frac{1}{\Phi'}d\Phi' = \frac{1}{l}dl,$$

which, on integration, gives

$$l\Phi' = \text{const.}$$

But this corresponds precisely to eq. (25a).[12]

The result found for the stressed rod, as well as other examples, shows that the eqs. (25) and (25a) possess a general validity for a material body's dimensions of length. Of course it is the real gravitational potential Φ' existing at a point, and not that of the external field, which influences the length; however, we can easily see that we may also generally use the potential Φ_a' of the external field when calculating changes in length, since Φ' and Φ_a' are proportional to one another. For a system of the kind considered in §1 at rest, we have from (4) and (17)

$$\Phi' = \Phi_a' - \frac{c^2}{4\pi}\int\frac{vdv}{\Phi'r}.\tag{a}$$

According to the results found earlier, the gravitational mass of the system does not change when the gravitational potential Φ_a' of the external field is changed. Therefore, upon such a change we have

---

12   If the rod is deformable, some work will be expended in stressing it, and the rest energy of the rod will be correspondingly increased. In this way too, the weight experiences an increase, which provides an added work dA in falling. However, since in falling the rest energy diminishes, the work recovered in relaxing the rod is smaller than that consumed in stressing and the difference amounts to exactly dA.

$$c^2 \int \frac{v \, dv}{\Phi'} = \text{const.}$$

The quantities v and dv do change in a certain way with Φ', but in such a way that $v \, dv / \Phi'$ remains constant for each particular element of the system. If on the left in eq. (a) Φ' denotes the potential at a certain point of the material system, I the integral [p. 546] on the right varies inversely with the length r. We obtain

$$\Phi_a' = \Phi'(1 + \text{const.}),$$

that is, at each point of the system, Φ' changes in proportion to Φ_a'.

For these reasons, the dependence of a body's linear dimensions l (at rest) on the gravitational potential is given generally by the two equivalent equations

$$l\Phi' = \text{const.}, \qquad l\Phi_a' = \text{const.}, \tag{26}$$

and further, corresponding to (21a)

$$l = \frac{l_0}{1 + \dfrac{g_0}{c^2}(\Phi_a - \Phi_0)}. \tag{26a}$$

For the volume dv of a particle of a body transformed to rest, one has, of course, from (26)

$$dv\Phi'^3 = \text{const.}$$

Since above we found $v \, dv / \Phi' = \text{const.}$, it follows that

$$\frac{v}{\Phi'^4} = \text{const.} \tag{27}$$

Since according to (14) $-c^2 v$ is the sum of the diagonal components of the tensor $\mathbf{T}$, the components of $\mathbf{T}$, and thus in particular the elastic stresses $\mathbf{p}_{ab}$, depend on Φ' in the same way as v:

$$\frac{\mathbf{p}_{ab}}{\Phi'^4} = \text{const.} \tag{28}$$

The result found earlier (p. 533) [p. 544 in the original] for the electron's surface tension S is thus in agreement with the above since S is a stress times a length.

5. THE DEPENDENCE OF A PROCESS'S TIME DEVELOPMENT
ON THE GRAVITATIONAL POTENTIAL

The dependence of a body's linear dimensions on Φ' raises the question of whether a physical process's time development is also influenced by the gravitational potential.

[p. 547] For a simple case we can answer the question | without difficulty. Due to the constancy of the speed of light, it is clear that the time during which a light signal propagates from one end of a rod to the other grows in the same proportion as the rod lengthens. This time is thus inversely proportional to the gravitational potential.

Another process that can be treated without difficulty is circular motion under the influence of the gravitational attraction of a central body. Let a mass point with gravitational mass M_2 move in a circular orbit around another mass point with gravitational mass M_1, with M_2 much smaller than M_1. Let M_1 be so large in relation to M_2 that we may consider the former as being at rest. In §7 we will further investigate the question of when a body may be viewed as a mass point, and derive the laws of its motion. Here, we only need to know that for a mass point the equations of motion (51), loc. cit. apply, where Φ denotes the gravitational potential of the field not produced by the mass point itself. This potential is

$$\Phi = \Phi_a - \frac{M_1}{4\pi r},$$

where Φ_a is the constant external potential and r is the distance from the mass M_1. Because Φ has the same value at all points on the circular orbit, the inertial mass of the moving point remains unchanged, and the equations of motion (51), loc. cit. yield:

$$-g(\Phi)\frac{\partial\Phi}{\partial x} = \frac{d\mathfrak{B}_x}{d\tau}, \text{ etc.}$$

According to (17), we can also write the equation as

$$-\frac{\partial\Phi}{\partial x} = \frac{\Phi' d\mathfrak{B}_x}{c_2 \, d\tau}$$

etc. Of course,

$$|\nabla\Phi| = \frac{M_1}{4\pi r^2}.$$

Furthermore, since the moving mass point has no tangential acceleration, we have

$$\frac{d\mathfrak{B}_x}{d\tau} = \frac{1}{1-q^2}\frac{dv_x}{dt}$$

etc. (e.g., compare eq. (62), loc. cit.). For the absolute value of the acceleration, |

$$\left|\frac{dv}{dt}\right| = \frac{v^2}{r}$$

[p. 548]

applies, where r is the radius of the circular orbit. By using all of these equations we obtain the following equation of motion

$$\frac{M_1}{4\pi r^2} = \frac{\Phi_a' - \dfrac{M_1}{4\pi r}}{c^2(1-\mathfrak{q}^2)} \cdot \frac{v^2}{r}.$$

A transformation yields

$$\frac{M_1}{4\pi r} = \Phi_a'\frac{v^2}{c^2}$$

or, if we introduce the period of rotation T,

$$\frac{M_1 c^2}{4\pi r \Phi_a'} = \frac{4\pi^2 r^2}{T^2}. \tag{a}$$

This equation connects the three quantities r, T and Φ_a' with one another. Setting $c^2/\Phi' = g$ according to (17), we obtain precisely the equation which classical mechanics would give.

We now imagine the two mass points M_1 and M_2 and also the measuring rods, with which we measure length, transferred to another location with a different external gravitational potential Φ_a'. Then all lengths have changed in inverse proportion to Φ_a', and if we wish to re-establish the previous process, we measure the distance r such that $r \cdot \Phi_a'$ has the same value in both cases. Therefore, according to (a), r/T also has the same value, which means that the time of revolution changes in proportion to the bodies' linear dimensions. Therefore according to (26) one has,

$$T\Phi_a' = \text{const.} \tag{29}$$

We further want to investigate the behavior of the period of oscillation of a material point, oscillating about a fixed equilibrium position as a result of an elastic (or a "quasi-elastic") force. For a sufficiently small amplitude of oscillation we can use the usual harmonic oscillator equation:

$$m\frac{d^2x}{dt^2} = -ax,$$

where m is the inertial mass of the material point, x the | displacement from equilibrium, and a an elastic constant. For the period of oscillation T one obtains the well known expression (the easiest way to obtain this is by substituting $x = Ce^{2\pi i(t/T)}$)

[p. 549]

$$T^2 = 4\pi^2\frac{m}{a}.$$

Upon a change of the gravitational potential one has from (21)

$$\frac{m}{\Phi'} = \text{const.}$$

Since ax is the elastic force, it is to be treated like a pull on a stretched string. Thus upon a change of Φ', it behaves like a stress times an area (the cross-section of the string), and according to (28) and (26)

$$\frac{ax}{\Phi'^2} = \text{const.}$$

Since x is a length, one has

$$\frac{a}{\Phi'^3} = \text{const.}$$

Hence for the period of oscillation

$$T\Phi' = \text{const.}$$

precisely in agreement with the two earlier results. It may be supposed that the course of all physical processes is influenced in a corresponding manner.

From the last example, it follows that the wavelength of a spectral line depends upon the gravitational potential. A numerical calculation shows that the wavelengths on the surface of the Sun must be greater by about one part in two million than those of terrestrial light sources. Several other recent theories of gravitation also give the same — perhaps even observable — displacement.

6. REMARKS ON THE DEFINITION OF FUNDAMENTAL UNITS

From the dependence of the linear dimensions and masses of the bodies as well as of the time development of phenomena on the gravitational potential, it follows that in [p. 550] defining the fundamental units, the gravitational potential has to be taken | into account. By a centimeter, we thus understand the length of a reference rod at a certain temperature and at a certain gravitational potential. For the latter, one takes of course the potential present on the surface of the Earth. The same applies for the definition of the unit of time and the unit of inertial mass.

If the units of length and time have been established for a location with a certain gravitational potential, from this location one can in principle measure all lengths and times in the world by means of a telescope and the exchange of light signals, because light signals propagate in straight lines with a constant speed c. Hence, no transport of measuring rods and clocks from one place to another is necessary to compare lengths and times at different locations.

The gravitational potential Φ_0, in terms of which the fundamental units have been defined, is to be considered a universal constant, and the same applies to $g(\Phi_0)$. Thus the universal constant of the theory of gravitation developed here does

not appear in the field eqs. (19), (20), but plays a part in the definition of the fundamental units.

7. THE EQUATIONS OF MOTION OF A BODY WHICH MAY BE TREATED AS A MASS POINT

In order to obtain the equations of motion of a material point, and additionally to gain
clear insight into the conditions under which a body may be viewed as a material
point, we consider a body that has the properties of the complete stationary system
discussed in §1, moving freely in an external homogeneous gravitational field. We
have

$$\Phi = \Phi_1 + \Phi_2,$$

where

$$\Phi_1 = -\frac{1}{4\pi}\int\frac{dv}{r}\{g(\Phi)v\}_{t-\frac{r}{c}}.$$

The gravitational potential Φ_2 of the external field does not need to be constant in
time as long as $\partial\Phi_2/\partial t$ is constant. I We thus have   [p. 551]

$$\Phi_2 = ax + by + fz + hu,$$

where the coefficients $a...h$ are constant in space and time.

According to the general foundations of relativistic mechanics, one has (compare
§ 1)

$$\left.\begin{array}{l}\frac{\partial}{\partial x}(G_{xx} + T_{xx} + L_{xx}) + \frac{\partial}{\partial y}(G_{xy} + T_{xy} + L_{xy}) \\[2mm] + \frac{\partial}{\partial z}(G_{xz} + T_{xz} + L_{xz}) + \frac{\partial}{\partial u}(G_{xu} + T_{xu} + L_{xu}) = 0,\end{array}\right\} \quad (a)$$

and another three equations obtained by interchanging the first index x with y, z, u.
We integrate the expression (a) over all of xyz-space, and for the time being deal
with the gravitation tensor G separately. According to (7) and (3) we have:

$$\left.\begin{array}{l}\int\{\frac{\partial G_{xx}}{\partial x} + \frac{\partial G_{xy}}{\partial y} + \frac{\partial G_{xz}}{\partial z} + \frac{\partial G_{xu}}{\partial u}\}dv \\[3mm] = \int g(\Phi)v\frac{\partial\Phi_1}{\partial x}dv + \frac{\partial\Phi_2}{\partial x}\int g(\Phi)vdv.\end{array}\right\} \quad (b)$$

We wish to transform the first integral on the right. Since the second derivatives of
Φ_2 are all zero, in the eq. (2) on the left we can set Φ_1 instead of Φ. If we insert the
expression so obtained for $g(\Phi)v$, we obtain, after a transformation similar to the
one that leads from eq. (3) to (7),

$$\int g(\Phi) v \frac{\partial \Phi_1}{\partial x} dv = \int \left\{ \frac{\partial G_{xx}^{1}}{\partial x} + \frac{\partial G_{xy}^{1}}{\partial y} + \frac{\partial G_{xz}^{1}}{\partial z} \right\} dv + \int \frac{\partial G_{xu}^{1}}{\partial u} dv.$$

Here, G_{xx}^{1} etc. are the expressions one obtains by writing Φ_1 instead of Φ in the eqs. (6). We can transform the first integral on the right in the last equation into a surface integral over an infinitely large spherical surface using Gauss' theorem, and this integral becomes zero because expressions corresponding to eq. (5) apply for the first derivatives of Φ_1. Therefore, we obtain

$$\int g(\Phi) v \frac{\partial \Phi_1}{\partial x} dv = \frac{d}{du} \int \frac{\partial \Phi_1}{\partial x} \frac{\partial \Phi_1}{\partial u} dv. \tag{c}$$

[p. 552] I Upon integration of (a) the two remaining world tensors T and L give the result

$$\frac{d}{du} \int \{ T_{xu} + L_{xu} \} dv,$$

since the remaining terms can also be transformed into a surface integral that becomes zero. Therefore the integration of (a) over all of space yields the following result, taking (b) and (c) into account,

$$-\frac{\partial \Phi_2}{\partial x} \int g(\Phi) v dv = \frac{d}{du} \int \int \left\{ \frac{\partial \Phi_1}{\partial x} \frac{\partial \Phi_1}{\partial u} + T_{xu} + L_{xu} \right\} dv. \tag{d}$$

Of course

$$\mathfrak{G}_x = -\frac{i}{c} \int \int \left\{ \frac{\partial \Phi_1}{\partial x} \frac{\partial \Phi_1}{\partial u} + T_{xu} + L_{xu} \right\} dv$$

is the x-component of the total momentum of the moving body. If the motion is *quasi-stationary*,[13] which we must now assume, we can calculate $\mathfrak{G}_x$ using the same formula that applies to uniform motion of the body, that is (compare p. 536)

$$\mathfrak{G}_x = \frac{m v_x}{\sqrt{1 - \mathfrak{q}^2}} = m \mathfrak{B}_x.$$

m is the inertial mass of the body. If the body rotates or if stationary motions occur in its interior, then the three-dimensional velocity v and the four-dimensional velocity vector $\mathfrak{B}$ relate to the body as a whole, according to §1, and therefore give the center of mass's change of position. We must assume further that the body is of such moder-

---

13   Compare M. Abraham, *Theorie der Elektrizität II*. Leipzig 1905, p. 183.

ate dimensions that one can view $g(\Phi_2)$ as spatially constant within the body, which is practically always the case. We then have from (11)

$$g(\Phi_2)m = \int g(\Phi)v dv_0 = \frac{1}{\sqrt{1-\mathfrak{q}^2}}\int g(\Phi)v dv,$$

because eq. (11) relates to a system of reference in which the velocity v is momentarily zero, and for I a volume element dv_0 in this system of reference we have   [p. 553]

$$dv_0 = \frac{dv}{\sqrt{1-\mathfrak{q}^2}}.$$

Since furthermore

$$d\tau = dt\sqrt{1-\mathfrak{q}^2} = \frac{du}{ic}\sqrt{1-\mathfrak{q}^2},$$

where τ denotes the proper time of the body, we finally obtain from (d)

$$-g(\Phi_2)m\frac{\partial\Phi_2}{\partial x} = \frac{d}{d\tau}m\mathfrak{B}_x.$$

This is the first of the equations of motion for the body; naturally one obtains the remaining three by exchanging y, z, u for x.

We have assumed that the external field is homogeneous. If this is not the case, then the equations are only valid to the degree of accuracy to which $\nabla\Phi_2$ and $\partial\Phi_2/\partial t$ are spatially constant within the body. (The field outside of the body can not, of course, act on the body itself.) To this degree of accuracy, we can consider the body as a material point, and thus the following equations of motion apply to it, according to the above considerations:

$$\left.\begin{array}{c}-g(\Phi_a)m\dfrac{\partial\Phi_a}{\partial x} = \dfrac{d}{d\tau}m\mathfrak{B}_x, \\ \cdots\cdots\cdots\cdots \\ \cdots\cdots\cdots\cdots \\ -g(\Phi_a)m\dfrac{\partial\Phi_a}{\partial u} = \dfrac{d}{d\tau}m\mathfrak{B}_u,\end{array}\right\} \quad\quad (30)$$

where the gravitational potential of the external field is denoted by Φ_a. This system of equations are equivalent to the eqs. (51), loc. cit.

If we multiply the eqs. (30) by $\mathfrak{B}_x, \mathfrak{B}_y, \mathfrak{B}_z, \mathfrak{B}_u$ and add them, we find the law for the change of the inertial mass in exactly the same manner as on p. 516, loc. cit. [p. 873 in the original], and we arrive at the formulas already derived in §2. Using these formulas, we can bring the equations of motion (30) into the following form: I

[p. 554]

$$
\left.\begin{array}{l}
-c^2 \dfrac{\partial}{\partial x}\ln\Phi_a{}' = \dfrac{d\mathfrak{B}_x}{d\tau} + \mathfrak{B}_x \dfrac{d}{d\tau}\ln\Phi_a{}', \\[2ex]
-c^2 \dfrac{\partial}{\partial y}\ln\Phi_a{}' = \dfrac{d\mathfrak{B}_y}{d\tau} + \mathfrak{B}_y \dfrac{d}{d\tau}\ln\Phi_a{}', \\[2ex]
-c^2 \dfrac{\partial}{\partial z}\ln\Phi_a{}' = \dfrac{d\mathfrak{B}_z}{d\tau} + \mathfrak{B}_z \dfrac{d}{d\tau}\ln\Phi_a{}', \\[2ex]
-c^2 \dfrac{\partial}{\partial u}\ln\Phi_a{}' = \dfrac{d\mathfrak{B}_u}{d\tau} + \mathfrak{B}_u \dfrac{d}{d\tau}\ln\Phi_a{}'.
\end{array}\right\} \tag{31}
$$

One sees from this that the motion of a body in a gravitational field is completely independent of the constitution of the body, if only it may be treated as a material point.

In a static, homogeneous field in particular, every quasi-stationarily moving body that satisfies the conditions in §1 may be treated as a material point, and thus all such bodies fall in the same manner. In the case of free fall, eqs. (64), loc. cit. apply; it is only to be noted that g depends on Φ, and that Φ denotes the potential of the external field. Since a body rotating about its axis of symmetry satisfies the conditions in §1, it must fall exactly like a non-rotating body. Therefore the assertion about rotating bodies made on p. 520, loc. cit. [p. 878 in the original] does not apply in our present theory; furthermore, molecular motions have no influence on the falling motion. In contrast, a body thrown horizontally falls slower than one which does not have an initial velocity, as demanded by eqs. (64), loc. cit.

Systems in an external field that do not satisfy the conditions in §1 generally move approximately according to the equations of motion (30), (31). For example, Mr. Einstein has shown that according to the theory developed here, an elastically oscillating system's gravitational acceleration must change with the phase of the oscillation, but that the mean acceleration is given by (64), loc. cit.

ALBERT EINSTEIN

ON THE PRESENT STATE OF THE
PROBLEM OF GRAVITATION

Originally published as "Zum gegenwärtigen Stande des Gravitationsproblems" in Physikalische Zeitschrift 14 (1913), pp. 1249–1262. Reprinted in "The Collected Papers of Albert Einstein," Vol. 4, Doc. 17: An English translation is given in its companion volume.

1. GENERAL FORMULATION OF THE PROBLEM

The first domain of physical phenomena where a successful theoretical elucidation was achieved was that of the general attraction of masses. The laws of weight and of the motions of celestial bodies were reduced by Newton to a simple law of motion for a mass point and to a law of interaction for two gravitating mass points. These laws have proved to hold so exactly that, from an empirical point of view, there is no decisive reason to doubt their strict validity. If, despite this, one can scarcely find a physicist today who believes in the exact validity of these laws, this is due to the transformative influence of the development of our knowledge of electromagnetic processes over the last few decades.

Before Maxwell, electromagnetic processes were attributed to elementary laws built as closely as possible on the model of Newton's force law. According to these laws, electrical masses, magnetic masses, current elements, and so on, are supposed to exert actions-at-a-distance on each other, which require no time for propagation through space. Then 25 years ago, Hertz showed with his brilliant experimental investigation of the propagation of electrical force that electrical effects require time for their propagation. By doing so he contributed to the victory of Maxwell's theory, which replaced unmediated action-at-a-distance with partial differential equations. Following this demonstration of the untenability of action-at-a-distance theory in the area of electrodynamics, confidence in the correctness of Newton's action-at-a-distance gravitational theory was also | shaken. The conviction that Newton's law of [1250] gravitation encompasses as little of the totality of gravitational phenomena as Coulomb's law of electrostatics and magnetostatics captures of the totality of electromagnetic phenomena had to come to light. Newton's law previously sufficed for calculating the motions of the celestial bodies due to the small velocities and accelerations of those motions. In fact, it is easy to demonstrate that the motion of celestial bodies determined by electrical forces acting on electrical charges they bear would not reveal Maxwell's laws of electrodynamics to us if their velocities and accelerations were of the same order of magnitude as in the motions of the celestial bodies with which we are familiar. One would be able to describe such motions with great accuracy on the basis of Coulomb's law.

Even though confidence in the comprehensiveness of Newton's action-at-a-distance law was thus shaken, there were still no direct reasons to force an extension of Newton's theory. However, today there is such a direct reason for those who adhere to the correctness of relativity theory. According to relativity theory, in nature there is no means that would permit us to send signals with a velocity greater than that of light. Yet on the other hand, it is clear that if Newton's law were strictly valid, we would be able to use gravitation to send instantaneous signals from a place A to a distant place B; since the motion of a gravitating mass at A would lead to simultaneous changes of the gravitational field, $B-$ in contradiction to relativity theory.

The theory of relativity not only forces us to modify Newton's theory, but fortunately it also strongly constrains the possibilities for such a modification. If this were not the case, the attempt to generalize Newton's theory would be a hopeless undertaking. To see this clearly, one need only imagine being in the following analogous situation: suppose that of all electromagnetic phenomena, only those of electrostatics are known experimentally. Yet one knows that electrical effects cannot propagate with superluminal velocity. Who would have been able to develop Maxwell's theory of electromagnetic processes on the basis of these data? Our knowledge of gravitation corresponds precisely to this hypothetical case: we only know the interaction between masses at rest, and probably only in the first approximation. Relativity theory limits the bewildering manifold of possible generalizations of the theory, because according to it in every system of equations the time coordinate appears in the same manner as the three spatial coordinate, up to a difference in sign. This formal insight of Minkowski's, which is here only roughly foreshadowed, has been a tool of utmost importance in the search for equations compatible with relativity theory.

2. PLAUSIBLE PHYSICAL HYPOTHESES CONCERNING THE GRAVITATIONAL FIELD

In what follows we shall specify several general postulates, which can be employed by a gravitational theory, although it need not employ all of them:

1. Satisfaction of the laws of energy and momentum conservation.

2. Equality of the *inertial* and the *gravitational* mass for isolated systems.

3. Validity of the theory of relativity (in the restricted sense); i.e., the systems of equations are covariant with respect to linear orthogonal substitutions (generalized Lorentz transformations).

4. The observable laws of nature do not depend on the absolute magnitude of the gravitational potential (or gravitational potentials). Physically, this means the following: The embodiment of relations between observable quantities that one can

determine in a laboratory is not changed if I bring the whole laboratory into a region with a different (spatially and temporally constant) gravitational potential.

We note the following regarding these postulates. All theorists agree with one another that postulate 1 must be upheld. There is not such a general consensus regarding adherence to postulate 3. Thus, M. Abraham has developed a gravitational theory that does not comply with postulate 3. I could subscribe to this point of view, if Abraham's system were covariant with respect to transformations that turn into linear orthogonal transformations in regions of constant gravitational potential, but this does not appear to be the case with Abraham's theory. I Therefore this theory does not [1251] contain relativity theory, as previously developed without connection with gravitation, as a special case. All of the arguments that have been put forward in favor of relativity theory in its current form militate against such a theory. In my opinion, it is absolutely necessary to hold fast to postulate 3 as long as there are no compelling reasons against doing so; the moment we give up this postulate, the manifold of possibilities will become impossible to survey.

Postulate 2 calls for a more precise examination, and, in my opinion, we must hold on to it unconditionally until there is proof to the contrary. The postulate is initially based on the fact of experience that all bodies fall with the same acceleration in a gravitational field; we will have direct our attention to this important point again later on. Here it should only be said that the equality (proportionality) of gravitational and inertial mass was proved with great accuracy by Eötvös's investigation,[1] which is of highest significance to us; he proved this proportionality by establishing experimentally that the resultant of weight and of the centrifugal force due to the Earth's rotation is independent of the nature of the material (the relative difference between the two masses is $<10^{-7}$). In combination with one of the main results of the ordinary relativity theory, postulate 2 leads to a consequence that can already be drawn at this point. According to relativity theory, the inertial mass of a closed system (treated as a whole) is determined by its energy. From postulate 2, the same must also hold for *gravitational* mass. Therefore, if the state of the system undergoes an arbitrary change without altering its total energy, then the gravitational action-at-a-distance does not change, even if a part of the system's energy is converted into gravitational energy. The gravitational mass of a system is fixed by its total energy, including gravitational energy.

Finally, postulate 4 arguably cannot be grounded on experience. It is only justified by our confidence in the simplicity of the laws of nature, and we cannot have as much right to depend on it as we do with the three axioms named above.

I am fully aware that the postulates 2–4 resemble a scientific profession of faith more than a firm foundation. I am also far from claiming that the two generalizations of Newton's theory presented in the following are the only ones possible, but I dare say that given the current state of our knowledge they must be the *most natural* ones.

---

1   B. Eötvös, *Mathem. und naturw. Ber. aus Ungarn* 8, 1890; Supplement 15: 688, 1891.

3. NORDSTRÖM'S THEORY OF GRAVITATION

According to the familiar relativity theory in connection with gravitational theory, an isolated material point moves uniformly in a straight line in accord with Hamilton's equation

$$\delta \left\{ \int d\tau \right\} = 0, \tag{1}$$

where we have set, in the usual way,

$$\left. \begin{aligned} d\tau &= \sqrt{- dx_1^{\,2} - dx_2^{\,2} - dx_3^{\,2} - dx_4^{\,2}} = \\ &= \sqrt{c^2 dt^2 - dx^2 - dy^2 - dz^2} = dt\sqrt{c^2 - q^2}. \end{aligned} \right\} \tag{2}$$

We can also write equation (1) as

$$\left. \begin{aligned} \delta &\left\{ \int H dt \right\} = 0 \\ &\text{where} \\ H &= -m\frac{d\tau}{dt} = -m\sqrt{c^2 - q^2} \end{aligned} \right\} \tag{1a}$$

is the Lagrangian function of the moving point, and m is a constant characteristic of it, its "mass." The momentum (I_x, I_y, I_z) and the energy E of the point follows directly from this, as Planck has shown, in the familiar way.[2]

$$I_x = \frac{\partial H}{\partial \dot{x}} = m\frac{\dot{x}}{\sqrt{c^2 - q^2}}$$

$$E = \frac{\partial H}{\partial x}\dot{x} + \frac{\partial H}{\partial y}\dot{y} + \frac{\delta H}{\delta z}\dot{z} - H = m\frac{c^2}{\sqrt{c^2 - q^2}}.$$

From here it is easy to arrive at Nordström's theory if we make the following assumptions. The covariance of the equation with respect to linear orthogonal substitutions still stands, as is the case in the familiar relativity theory. The gravitational field can be described as a scalar. The motion of a material point in the gravitational field can be represented with an equation of Hamiltonian form. In that case one obtains the following equation for the motion for a mass point:[3] |

---

2   These expressions differ from the customary ones only by the constant factor $1/c$.
3   Taking into consideration the fact that the Hamiltonian integral must be an invariant.

$$\delta \left\{ \int \varphi d\tau \right\} = 0, \tag{1'} \quad [1252]$$

such that (2) with constant c remains valid and φ is the scalar fixed by the gravitational field. For the propagation of this light ray we have $dt = 0$, and so $q = c$; i.e., the speed of light propagation is equal to the constant c. The light rays are not bent by the gravitational field.

In the place of equations (1a) we have

$$
\left.
\begin{aligned}
&\delta \left\{ \int H d\tau \right\} = 0, \\[1em]
&\text{whereupon} \\[1em]
&H = -m\varphi \frac{d\tau}{dt} = -m\varphi \sqrt{c^2 - q^2}.
\end{aligned}
\right\} \tag{1a'}
$$

The Lagrangian equations of motion read:

$$\frac{d}{dt} \left\{ m\varphi \frac{\dot{x}}{\sqrt{c^2 - q^2}} \right\} + m \frac{\partial \varphi}{\partial x} \sqrt{c^2 - q^2} = 0 \text{ etc. }^{[1]}$$

From this it follows that the expressions for the impulse, energy, and the force $\mathfrak{K}$ exerted by the gravitational field at a point are:

$$
\left.
\begin{aligned}
I_x &= m\varphi \frac{\dot{x}}{\sqrt{c^2 - q^2}} \text{ etc.} \\[1em]
E &= m\varphi \frac{c^2}{\sqrt{c^2 - q^2}} \\[1em]
\mathfrak{K}_x &= -m \frac{\partial \varphi}{\partial x} \sqrt{c^2 - q^2} \text{ etc.}
\end{aligned}
\right\} \tag{2a}
$$

Thus m is a constant characteristic of the mass point, independent of φ and q. The expression for $\mathfrak{K}$ shows that φ plays the role of the gravitational potential. Furthermore, the expressions for I_x and E show that according to Nordström's theory the inertia of a mass point is determined by the product $m\varphi$; the smaller φ is, i.e., the more mass we pile up in the region of the mass point, the smaller the inertial resistance the body exerts in response to a change in its velocity becomes. This is one of the most important physical consequences of the scalar theory of gravitation, to which we must return later.

In this theory, as well as in the theory explained below, the coordinate differences do not have as simple a physical meaning as they do in the usual relativity theory. Let us consider a given moveable unit measuring rod and a moveable clock, which ticks such that in a vacuum light traverses a distance equal to one unit measuring rod[4] during one unit of time, as measured by the clock. We will call the four-dimensional interval between two infinitely close spacetime points, which can be measured with these measuring tools in the same way as in the usual relativity theory, the "natural" four-dimensional interval $d\tau_0$ of the spacetime point. By definition this is an invariant, and hence in the case of the usual relativity theory it is equal to $d\tau$. We call the latter quantity the "coordinate interval," in contrast to the natural interval and according to its definition, or also briefly as the "interval" of the spacetime point. In our case it is possible, however, that the natural interval $d\tau_0$ differs from the coordinate interval $d\tau$ by a factor that is a function of φ. Thus we set

$$d\tau_0 = \omega d\tau. \tag{3}$$

We can further speak of the natural length l_0 and the natural volume V_0 of a body. These are the length and volume, respectively, that are measured using comoving unit measuring rods. The lengths l and volumes V measured in coordinates also play a role. It follows that the relation between the coordinate volume V and the natural volume V_0 is:

$$\frac{1}{V} = \frac{\omega^3 c\, dt}{V_0\, d\tau} = \frac{\omega^3 c}{V_0 \sqrt{c^2 - q^2}}. \tag{4}$$

In addition, by a unit mass we understand the mass of water enclosed in a natural volume of magnitude unity. The mass of a body is the ratio of its inertia to that of a unit mass, which is thus a scalar. We understand the natural density ρ_0 to be relative to the density of water or the mass in a natural volume with magnitude 1; ρ_0 is thus a scalar by definition.

We can derive further consequences from the results obtained above if we pass from material points to the continuum. We achieve this by treating the material point as a continuum of coordinate volume V and natural volume V_0. One multiplies the expressions for I_x, E and $\mathfrak{R}_x$ given above in (2a) by $1/V$, using (4), so that one obtains the impulse i_x etc., the energy η, and the pondermotive force $\mathfrak{k}_x$ etc., per unit volume for an incoherent mass stream. Taking the relation

$$\rho_0 = \frac{m}{V_0}$$

[1253] into account, one obtains |

---

4 We will make the assumption that this is achievable at all locations and at all times; this is a special case of postulate (4).

$$ici_x = \frac{icI_x}{V} = \rho_0 c\varphi\omega^3 \frac{dx_1}{d\tau}\frac{dx_4}{d\tau} \left.\begin{array}{c} \\ \\ \\ \\ \\ \\ \end{array}\right\}$$

$$-\eta = -\frac{E}{V} = \rho_0 c\varphi\omega^3 \frac{dx_4}{d\tau}\frac{dx_4}{d\tau} \qquad\qquad (2b)$$

$$\mathfrak{k}_x = \frac{\mathfrak{K}_x}{V} = -\rho_0 c\omega^3 \frac{\partial\varphi}{\partial x}.$$

In the first equations i denotes the unit imaginary number. We now recall the expressions for the law of energy-momentum conservation in relativity theory. If X_x and etc. are the generalized pressure-stresses, and f_x etc. are the components of the energy flux density, then the quantities

$$\begin{array}{cccc} X_x & X_y & X_z & ici_x \\ Y_x & Y_y & Y_z & ici_y \\ Z_x & Z_y & Z_z & ici_z \\ \frac{i}{c}f_x & \frac{i}{c}f_y & \frac{i}{c}f_z & -\eta \end{array}$$

form a symmetric tensor, that we will write $T_{\mu\nu}$ (μ and ν are indices running from 1 to 4). Furthermore, denoting the work transferred by external forces to the material per unit volume with 1, then

$$\mathfrak{k}_x, \mathfrak{k}_y, \mathfrak{k}_z, \frac{i}{c}1$$

is a four vector, with its components referred to by k_μ. The law of energy-momentum conservation is then expressed by the equation

$$\sum_\nu \frac{\partial T_{\mu\nu}}{\partial x_\nu} = k_\mu \qquad (\mu = 1 \text{ to } \mu = 4). \qquad (5)$$

As equations (2b) illustrate, this schema can find direct application in our case of an incoherent flow of matter in a gravitational field, insofar as one sets

$$T_{\mu\nu} = \rho_0 c\varphi\omega^3 \frac{dx_\mu}{dt}\frac{dx_\nu}{d\tau} \left.\begin{array}{c} \\ \\ \\ \\ \end{array}\right\}$$

$$k_\mu = -\rho_0 c\omega^3 \frac{\partial\varphi}{\partial x_\mu}. \qquad\qquad (5a)$$

So far we have treated only the question of how the gravitational field acts on matter, but not the question of by which law, in turn, the matter determines the gravitational field. According to Nordström's theory, the latter is given by a scalar φ; thus, what enters into the differential equation for φ we are seeking must also be a scalar associated with the field generating process. This scalar can only be the scalar

$$\sum_\sigma T_{\sigma\sigma}$$

whose existence and meaning was notably highlighted by von Laue. Setting up this scalar for the case of a case of an incoherent mass stream, we obtain with the help of (5a)

$$\sum_\sigma T_{\sigma\sigma} = -\rho_0 c \varphi \omega^3$$

$$k_\mu = \sum_\sigma T_{\sigma\sigma} \cdot \frac{1}{\varphi} \frac{\partial \varphi}{\partial x_\mu}.$$

Thus instead of (5) we have

$$\sum_\nu \frac{\partial T_{\mu\nu}}{\partial x_\nu} = \sum_\sigma T_{\sigma\sigma} \cdot \frac{1}{\varphi} \frac{\partial \varphi}{\partial x_\mu}. \tag{5b}$$

This equation is particularly important in that there is nothing in it to remind us of the case of an incoherent mass stream discussed so far. According to Nordström's theory, equation (5b) expresses the energy balance of an arbitrary material process, if the stress-energy tensor corresponding to this process is substituted for $T_{\mu\nu}$.

From equation (5b) it follows that Nordström's theory satisfies postulate 2. If one were to observe a system on such a small scale that one could regard the $\partial \lg\varphi / \partial x_\mu$ as clearly constant for the spatial extent of the system, then one obtains for the total force exerted on the system by the gravitational field in the X-direction:

$$\frac{\partial \lg\varphi}{\partial x_\mu} \int \sum T_{\sigma\sigma} dv = \frac{\partial \lg\varphi}{\partial x_\mu} \int T_{44} dv = -\frac{\partial \lg\varphi}{\partial x_\mu} \int \eta \, dv,$$

where dv is the three-dimensional volume element. This reformulation is based on Laue's theorem, that for a closed system

$$\int \Sigma T_{11} dv = \int \Sigma T_{22} dv = \int \Sigma T_{33} dv = 0.$$

This proves that what determines the gravity of a closed system is its total quantity.

Equation (5b) further allows us to determine the function φ, which has been left undetermined so far, on the basis of the physical assumption that no work can be extracted from a static gravitational field via cyclic processes. In section 7 of my paper on gravitation published jointly with Mr. Grossmann, I obtained a contradiction between a scalar theory and this basic principle, based, however, on the tacit assumption that $\omega = $ const. But it is easy to show that the contradiction vanishes if one sets

$$l = \frac{l_0}{\omega} = \frac{\text{const.}}{\varphi}$$

or

$$\omega = \text{const.} \cdot \varphi. \tag{6}$$

I Later we will give yet another justification for this stipulation.   [1254]

Now it is simple to establish the general equation for the gravitational field, which is to be regarded as a generalization of Poisson's equation for the gravitational field. That is, one has to set Laue's scalar equal to a scalar differential expression of the quantity φ such that the conservation laws hold for the material process and the gravitational field taken together. One achieves this by setting,

$$-\kappa \Sigma T_{\sigma\sigma} = \varphi \Box \varphi, \tag{7}$$

where κ is a universal constant (the gravitational constant), and $\Box$ denotes the operator

$$\sum_\tau \frac{\partial^2}{\partial x_\tau^2} \quad (\tau \text{ from 1 to 4}).$$

The fact that the conservation laws are satisfied follows from the equations (5b) and (7), by virtue of the identity which follows from (7)

$$\sum T_{\sigma\sigma} \frac{1}{\varphi} \frac{\partial \varphi}{\partial x_\mu} = -\frac{1}{\kappa} \frac{\partial \varphi}{\partial x_\mu} \sum \frac{\partial^2 \varphi}{\partial x_\sigma^2} = -\sum \frac{\partial t_{\mu\nu}}{\partial x_\nu},$$

where we set

$$t_{\mu\nu} = \frac{1}{\kappa} \left\{ \frac{\partial \varphi}{\partial x_\mu} \frac{\partial \varphi}{\partial x_\nu} - \frac{1}{2} \delta_{\mu\nu} \sum \frac{\partial \varphi^2}{\partial x_\tau} \right\}. \tag{8}$$

$\delta_{\mu\nu}$ denotes 1 respectively 0, depending on whether $\mu = \nu$ or $\mu \neq \nu$. The component of the stress-energy tensor of the gravitational field is $t_{\mu\nu}$; then it follows from the penultimate equation and (5b) that

$$\sum_\nu \frac{\partial}{\partial x_\nu} (T_{\mu\nu} + t_{\mu\nu}) = 0. \tag{9}$$

Thus postulate 1 is satisfied. It can also be shown that, in accord with postulate 2, the number of gravitational lines emanating from a closed stationary system to infinity depends on the total energy of the system.

Furthermore, the following is in conformity with postulate 4. If one places two mirrors at the ends of a natural length l_0 facing each other, and allows a light ray to go back and forth between them in a vacuum, then this system represents a clock (light clock). If two masses m_1 and m_2 circle each other at the natural distance l_0 under the influence of their gravitational interaction, then this system also represents a clock (gravitational clock). With the help of the equations derived above, one can easily show that the relative rate of these two clocks, supposing that they are found in the same gravitational potential, is independent of the absolute value of the potential. This is an indirect confirmation of the expression given for ω in equation (6).

In conclusion we can say that Nordström's scalar theory, which holds firmly onto the postulate of the constancy of the speed of light, satisfies all the requirements for a theory of gravitation that can be imposed on the basis of current experience. Only one unsatisfactory circumstance remains, namely that according to this theory the inertia of bodies seems to not be *caused* by other bodies, even though it is *influenced* by them, because the inertia of a body is greater the farther other bodies are from it.

4. IS THE ATTEMPT TO EXTEND RELATIVITY THEORY JUSTIFIED?[5]

If we wish to show a neophyte the extent to which the formulation of relativity theory is empirically justifiable, we can point out the following to him. For a person located in a railway car travelling uniformly in a straight line with its windows covered it is not even possible to decide what direction and at what speed the car travels; if we abstract from the inevitable shaking of the car, it is not even possible to decide whether the car is moving or not. Expressed abstractly: the laws of events described with respect to the system moving uniformly with respect to the original coordinate system (the Earth's surface) are the same as with respect to the original coordinate system (the Earth's surface). We call this proposition the relativity principle for uniform motion.

Yet one might be apt to add: it is surely different if the railway car moves non-uniformly; if the car changes its velocity, the passenger gets a jolt through which he detects the acceleration of the car. Speaking abstractly, there is no relativity principle for nonuniform motion. But concluding in this way is not irreproachable, because it is, after all, not certain whether the occupant of the car must necessarily ascribe the jolt he felt to the acceleration of the wagon. From the following example one sees that this conclusion is premature.

Two physicists, A and B, wake from a drug-induced slumber and discover that [1255] they are in a closed box with opaque walls, equipped with | all of their instruments. The have no idea how the box is situated, and how and whether it is moving. Now they determine that all bodies that they bring to the middle of the box and release fall in the same direction—let's say downward—with the same acceleration γ. What can the physicists conclude from this? A concludes that the box sits still on a celestial body, and that the downward direction must be towards the center of the celestial body, if it is taken to be spherical. But B adopts the point of view that the box could be moving with constant acceleration upward with the acceleration γ; due to an externally applied force, and there need not be a celestial body nearby. Is there a criterion that the two physicists could use to determine who is correct? We do not know of any such criterion, but we also do not know whether there is no criterion. *In any*

---

5   Cf. A. Einstein, *Ann. d. Phys.* (4) 35: 898, 1911.

case, Eötvös's exact experimental result regarding the equality of inertial and gravitational mass supports the view that there is no such criterion. One sees that, in this regard, Eötvös's experiment plays a similar role to that of Michelson's experiment with respect to the physical verifiability of *uniform* motion.

If it is really in principle impossible for the two physicists to decide which of the two views is correct, then *acceleration* has as little absolute physical meaning as *velocity.*[6] The same reference system can be taken to be accelerating or non-accelerating with equal justice, but then, according to the view chosen, one must postulate the presence of a gravitational field that determines the motion of freely moving bodies with respect to the reference system together with the possible acceleration of the system.

The circumstance that bodies behave in exactly the same in what is, according to our view, a nonaccelerated reference system in the presence of a gravitational field, as in an accelerating reference system, forces us to seek an extension of the principle of relativity to the case of accelerating reference systems.

From a mathematical standpoint, this amounts to demanding covariance of the equations expressing laws of nature not only under linear orthogonal substitutions, but also with respect to other, in particular non-linear, transformation; for only the non-linear substitutions correspond to a transformation between relatively *accelerated* systems. But then we face the difficulty that our scant empirical knowledge of the gravitational field permits no reliable deduction of the substitutions for which the covariance of the equations must be demanded. In an investigation undertaken with my friend Grossmann,[7] it turned out that it is possible and expedient to initially demand covariance with respect to *arbitrary* substitutions.

One further comment before proceeding to dispel a natural misunderstanding. An adherent of current relativity theory has some right to call the velocity of a point mass "apparent." In fact he can choose a coordinate system, such that the velocity is zero at the instant in question. But if he is dealing with a system of points whose mass points have different velocities, he cannot introduce a reference system such that all of the velocities of the mass points vanish with respect to it. Analogously, a physicist sharing our point of view can call the gravitational field "apparent," for by a suitable choice of the state of acceleration he can achieve the result that there is no gravitational field present at a given spacetime point. But it is clear that for extended gravitational fields this elimination of the gravitational field by a transformation cannot be achieved, in general. For example, it would not be possible to make the Earth's gravitational field vanish by choosing an appropriate reference system.

---

6   This point of view will be modified in section 6; but for the time being we will stick with it firmly.

7   *Entwurf einer verallgemeinerten Relativitätstheorie und einer Theorie der Gravitation*, Leipzig: B. G. Teubner, 1913.

5. CHARACTERIZATION OF THE GRAVITATIONAL FIELD;
ITS EFFECT ON PHYSICAL PROCESSES

Since we are uncertain about the class of admissible spacetime substitutions, the most natural thing, as already mentioned above, is to consider arbitrary substitutions of the spacetime variables x, y, z, t, which we can more conveniently write as x_1, x_2, x_3, x_4. It turns out to be pointless to introduce an imaginary time coordinate in the case of the generalization considered below.

[1256]   First we consider a spacetime region, in which there is no gravitational field in an appropriately chosen coordinate system. | We are then faced with the case that is familiar from the usual relativity theory. A free mass point moves uniformly and in a straight line according to the equation

$$\delta\left\{\int \sqrt{-dx^2 - dy^2 - dz^2 + c^2 dt^2}\right\} = 0.$$

Introducing new coordinates x_1, x_2, x_3, x_4 through an arbitrary substitution, it then follows that the motion of the point relative to the new system obeys the equation

$$\left.\begin{aligned} &\delta\left\{\int ds\right\} = 0 \\ &\text{where we set} \\ &ds^2 = \sum_{\mu\nu} g_{\mu\nu} dx_\mu dx_\nu. \end{aligned}\right\} \tag{1b}$$

From this we can also assume that

$$\left.\begin{aligned} &\delta\left\{\int H dt\right\} = 0 \\ &\text{where we set} \\ &H = -m\frac{ds}{dt}. \end{aligned}\right\} \tag{1b'}$$

H is the Hamiltonian function.

In the new system the quantities $g_{\mu\nu}$, determine the motion of the mass point, which according to the general observations of the foregoing section can be conceived of as the components of the gravitational field, as long as we treat the new system as "at rest." In general each gravitational field is defined by the ten components $g_{\mu\nu}$, which are functions of x_1, x_2, x_3, x_4. The motion of material points will always be determined by equations of the given form. Given its physical meaning, the element ds must be an invariant with respect to all substitutions. Through this the transformation laws for the components $g_{\mu\nu}$ is established if the coordinate transformation is given. ds is the only invariant associated with the four-dimensional line element (dx_1, dx_2, dx_3, dx_4). We call it the value or magnitude of the line element. If there is no gravitational field, then given a suitable choice of variables the system of $g_{\mu\nu}$'s reduces to the system

$$
\begin{array}{cccc}
-1 & 0 & 0 & 0 \\
0 & -1 & 0 & 0 \\
0 & 0 & -1 & 0 \\
0 & 0 & 0 & c^2.
\end{array}
$$

Thus we have come back to the case of the usual relativity theory.

The following equation determines the law for the velocity of light:

$$ds = 0.$$

With this one recognizes that in general the velocity of light depends not only on the spacetime point but also on the direction. The reason why we do not notice anything like this is that in the region of spacetime accessible to us the $g_{\mu\nu}$ are almost constant, and we can choose the reference system such that, up to small deviations, the $g_{\mu\nu}$ will have the constant values given above.

We can speak here of the natural length of a four-dimensional element exactly as in Nordström's theory. This is the element's length as measured by a moveable unit measuring rod and moveable clock. By definition this natural length is a scalar, and must therefore be equal to the magnitude ds up to a constant, which we set to 1. This gives the relation between coordinate differentials, on one hand, and measurable lengths and times, on the other; since they have this dependence on the quantities $g_{\mu\nu}$, the coordinates by themselves have no physical meaning. The stipulations regarding mass and natural density remain applicable without modification

Now we can set up the Lagrangian equations of motion for a material point, just as in our analysis of Nordström's theory, starting with equations (1b) and (1b'). From them we borrow the expressions for the momentum I and the energy E of a mass point, and the force $\Re$ exerted by a gravitational field on the mass point. Just as above, we can derive the corresponding expressions for the unit volume, and we obtain

$$\left.\begin{aligned}
i_x &= -\rho_0\sqrt{-g}\sum_{vs}g_{1v}\frac{dx_v dx_4}{dx_s dx_s}\\[4pt]
-\eta &= -\rho_0\sqrt{-g}\sum_{vs}g_{4v}\frac{dx_v dx_4}{dx_s dx_s}\\[4pt]
\mathfrak{k}_x &= -\frac{1}{2}\rho_0\sqrt{-g}\sum_{vs}\frac{dg_{\mu v}}{dx_1}\frac{dx_\mu dx_v}{dx_s dx_s}.
\end{aligned}\right\} \tag{2c}$$

From this we obtain, as above, the law of energy-momentum conservation for the incoherent mass stream:

$$\sum_{\mu v}\frac{\partial}{\partial x_v}(\sqrt{-g}\,g_{\sigma\mu}\Theta_{\mu v})-\frac{1}{2}\sum_{\mu v}\sqrt{-g}\frac{\partial g_{\mu v}}{\partial x_\sigma}\Theta_{\mu v}=0 \qquad (\sigma=1,2,3,4)$$

$$\Theta_{\mu v}=\rho_0\frac{dx_\mu dx_v}{ds\ ds}. \tag{5b}$$

Here g denotes the determinant of the $g_{\mu v}$. The first three equations of (5b) express the law of momentum conservation, and the last states the law of energy conservation. We can give this system of equations a somewhat more perspicuous form if we introduce the quantities |

$$\left.\begin{aligned}
&\mathfrak{T}_{\sigma v}=\sqrt{-g}\,g_{\sigma\mu}\Theta_{\mu v}\\
&\text{It follows that}\\[4pt]
&\sum_v\frac{\partial\mathfrak{T}_{\sigma v}}{\partial x_v}=\frac{1}{2}\sum_{\mu v\tau}\frac{\partial g_{\mu v}}{\partial x_\sigma}\gamma_{\mu\tau}\mathfrak{T}_{\tau v},
\end{aligned}\right\} \tag{5c}$$

[1257]

where $\gamma_{\mu\tau}$ denotes the subdeterminant of $g_{\mu v}$ divided by g. The physical meaning of the quantities $T_{\sigma v}$ emerges from the following schema:

$$\begin{array}{cccc}
\mathfrak{T}_{11} & \mathfrak{T}_{12} & \mathfrak{T}_{13} & \mathfrak{T}_{14}\\
\mathfrak{T}_{21} & \mathfrak{T}_{22} & \mathfrak{T}_{23} & \mathfrak{T}_{24}\\
\mathfrak{T}_{31} & \mathfrak{T}_{32} & \mathfrak{T}_{33} & \mathfrak{T}_{34}\\
\mathfrak{T}_{41} & \mathfrak{T}_{42} & \mathfrak{T}_{43} & \mathfrak{T}_{44}
\end{array}
=
\begin{array}{cccc}
X_x & Y_y & Z_z & i_x\\
Y_x & Y_y & Y_z & i_y\\
Z_x & Z_y & Z_z & i_z\\
f_x & f_y & f_z & \eta,
\end{array}$$

where the relations given on the right-hand side have the same meaning as in section 3. The right-hand side of (5c) expresses the momentum $(\delta=1-3)$ and energy $(\delta=4)$ given off by the gravitational field per unit volume and time.

The equations (5b) and (5c), without a doubt, have a meaning that extends far beyond the case of incoherent mass streams we have considered; they probably express the energy-momentum balance between a physical process and the gravitational field in general. But for each particular physical domain the quantities $\Theta_{\mu\nu}$ and $\mathfrak{T}_{\mu\nu}$ must be expressed in a specific manner.

6. COMMENTS ON THE MATHEMATICAL METHOD

In the theory we have just sketched the conventional theory of vectors and tensors cannot be applied, since according to it Σdx_{ν}^2 is not an invariant. The fundamental invariant, which we have called the magnitude of the line element, is rather

$$ds^2 = \Sigma g_{\mu\nu} dx_\mu dx_\nu.$$

The theory of covariance of a four-dimensional manifold defined by its line element has already been developed under the name "absolute differential calculus," by Ricci and Levi-Civita[8] in particular, who based their work primarily on a fundamental paper by Christoffel.[9] One can find a concise account of the most important theorems in the part of our work cited above penned by Mr. Grossmann.

In this theory one distinguishes several kinds of tensors, namely covariant, contravariant, and mixed, which are governed by algebraic rules similar to the well-known case characterized by the Euclidean line element. Differential operators that, when applied to tensors, produce tensors again have also been worked out, so that one can specify algebraic and differential relations for the general line element corresponding to those of the conventional theory of vectors and tensors.

It should be noted that dx_ν is the νth component of a contravariant tensor of the first rank (i.e., with one index). $g_{\mu\nu}$ and $\gamma_{\mu\nu}$, respectively, are components of a covariant and a contravariant tensor of the second rank, which we call the "fundamental tensor" based on its significance for the line element. $\Theta_{\mu\nu}$ is a second rank contravariant tensor, and $\frac{1}{\sqrt{-g}}\mathfrak{T}_{\sigma\nu}$ is a second rank mixed tensor.

Equation (5b) expresses the vanishing of the "divergence" of the tensor $\Theta_{\mu\nu}$. From this it follows that equation (5b) is covariant with respect to arbitrary substitutions, which naturally must also be demanded from a physical point of view.

By replacing the equations of relativity theory with the corresponding equations by means of the absolute differential calculus, one obtains a system of equations that account for the effect of the gravitational field on the domain of phenomena under consideration. This problem has already been solved by Köttler for the case of electromagnetic processes in a vacuum.[10]

From what has been said it follows that the question of the influence of the gravitational field on physical processes has been satisfactorily solved in principle, and in

---

8   Ricci and Levi-Civita, "Méthodes de calcul différentiel absolu et leurs applications," *Math. Ann.* 54: 125, 1900.

9   Christoffel, "Über Transformation der homogenen Differentialausdrücke zweiten Ranges," *Journ. f. Math.* 70: 46, 1869.

10   Köttler, "Über die Raumzeitlinien der Minkowskischen Welt," *Wien. Ber.* 121, 1912.

such a way that the equations are covariant under arbitrary substitutions. With that the spacetime variables are reduced to intrinsically meaningless, auxiliary variables that can be chosen arbitrarily. The whole problem of gravitation would thus be satisfactorily solved if one could find equations *covariant under arbitrary substitutions* that are satisfied by the quantities $g_{\mu\nu}$ fixed by the gravitational field itself. However, we have not succeeded in solving the problem in this manner.[11] We have obtained a solution by instead subsequently restricting the reference system. One is led to this

[1258] method naturally by the following considerations. It is clear that for any | material process by itself (i.e., without its gravitational field) the conservation theorems for momentum and energy cannot be satisfied. This situation corresponds to the appearance of the term on the right-hand side of (5c). On the other hand, we certainly must demand that the conservation theorems are satisfied for the material process and the gravitational field together. From this it follows that we must demand the existence of an expression $t_{\sigma\nu}$ for the stress, momentum, and energy flux and energy density of the gravitational field that, together with the corresponding quantity $\mathfrak{T}_{\sigma\nu}$ for the material process, fulfills the relation

$$\sum_\nu \frac{\partial(\mathfrak{T}_{\sigma\nu} + t_{\sigma\nu})}{\partial x_\nu} = 0.$$

If $t_{\sigma\nu}$ should have the same character as $\mathfrak{T}_{\sigma\nu}$, according to the theory of invariants, then the left-hand side of this equation cannot be covariant under arbitrary transformations; it is probably so only with respect to arbitrary *linear* transformations.

Therefore by demanding the validity of the conservation theorems, we restrict the reference systems to a great extent, and thereby relinquish the construction of gravitational equations in generally covariant form.

Thus, here is where the limit of applicability of the arguments given in section 4 lies. If one begins with a reference system with respect to which the conservations laws in the form given above hold and introduces a new reference system through an acceleration transformation, then with respect to the latter the conservation theorems are no longer satisfied. Nevertheless, I believe that the equations derived on the basis of the considerations in section 1 do not lose their footing because of this. On the one hand, it is certainly possible to describe the processes with respect to arbitrary reference systems; on the other hand, I do not see how the specialization of the reference system introduced here could bring about the specialization of the equations.

---

11 A short time ago I found a proof to the effect that such a generally covariant solution cannot exist at all.

7. SYSTEM OF EQUATIONS FOR THE GRAVITATIONAL FIELD

The sought-after system of equations should be a generalization of Poisson's equation

$$\Delta\varphi = 4\pi k\rho.$$

Since in our theory the gravitational field is determined by the 10 quantities $g_{\mu\nu}$ in place of φ, we will obtain 10 equations in place of this *one*. By the same token, $\Theta_{\mu\nu}$ appears on the right-hand side of the field equations as the field source instead of ρ, so that the sought-after equation will be of the form

$$\Gamma_{\mu\nu} = \kappa\Theta_{\mu\nu}.$$

$\Gamma_{\mu\nu}$ is a differential expression built up from the quantities $g_{\mu\nu}$, from which we know that it must be covariant with respect to linear transformations. I further assume that $\Gamma_{\mu\nu}$ does not contain anything higher than second derivatives. Furthermore, the conservation theorem necessitates the following: if we replace the second term of (5b) $\Theta_{\mu\nu}$ with $(1/\kappa)\Gamma_{\mu\nu}$, then we must allow this term to be transformed such that, like the first term of (5b), it can be written as a sum of derivatives. So far as I can see, these considerations gave me a unique way of identifying the $\Gamma_{\mu\nu}$ and hence the sought-after equations. These read:

$$\Delta_{\mu\nu}(\gamma) = \kappa(\Theta_{\mu\nu} + \vartheta_{\mu\nu}), \tag{7a}$$

where we set

$$\Delta_{\mu\nu}(\gamma) = \sum_{\alpha\beta}\frac{1}{\sqrt{-g}}\frac{\partial}{\partial x_\alpha}\left(\gamma_{\alpha\beta}\sqrt{-g}\frac{\partial\gamma_{\mu\nu}}{\partial x_\beta}\right) - \sum_{\alpha\beta\tau\rho}\gamma_{\alpha\beta}g_{\tau\rho}\frac{\partial\gamma_{\mu\tau}}{\partial x_\alpha}\frac{\partial\gamma_{\nu\rho}}{\partial x_\beta}$$

and

$$-2\kappa\vartheta_{\mu\nu} = \sum_{\alpha\beta\tau\rho}\left(\gamma_{\alpha\mu}\gamma_{\beta\nu}\frac{\partial\gamma_{\tau\rho}}{\partial x_\alpha}\frac{\partial\gamma_{\tau\rho}}{\partial x_\beta} - \frac{1}{2}\gamma_{\mu\nu}\gamma_{\alpha\beta}\frac{\partial g_{\tau\rho}}{\partial x_\alpha}\frac{\partial\gamma_{\tau\rho}}{\partial x_\beta}\right). \text{[2]}$$

The energy-momentum equation for material process and the gravitational field together assume the form

$$\sum\frac{\partial}{\partial x_\nu}\{\sqrt{-g}\,g_{\sigma\mu}(\Theta_{\mu\nu} + \vartheta_{\mu\nu})\} = 0. \tag{9a}$$

From (9a) one sees that $\vartheta_{\mu\nu}$ plays the same role for the gravitational field that $\Theta_{\mu\nu}$ plays for material processes. $\vartheta_{\mu\nu}$ is a covariant tensor with respect to linear transformations, and we will call it the stress-energy tensor of the gravitational field. In accord with postulate 2, $\vartheta_{\mu\nu}$ appears like $\Theta_{\mu\nu}$, as a field-generating cause.

The equations become simpler when one introduces the stress components themselves in the equations:

$$\mathfrak{T}_{\sigma\nu} = \sqrt{-g}\,g_{\sigma\mu}\Theta_{\mu\nu}$$

and

$$t_{\sigma\nu} = \sqrt{-g}\,g_{\sigma\mu}\vartheta_{\mu\nu}.$$

The equations then take the form: |

[1259]

$$\sum_{\alpha\beta\mu}\frac{\partial}{\partial x_\alpha}\left(\sqrt{-g}\,\gamma_{\alpha\beta}g_{\sigma\mu}\frac{\partial\gamma_{\mu\nu}}{\partial x_\beta}\right) = \kappa(\mathfrak{T}_{\sigma\nu} + t_{\sigma\nu}) \tag{7b}$$

$$-2\kappa t_{\sigma\nu} = \sqrt{-g}\left(\sum_{\beta\tau\rho}\gamma_{\beta\nu}\frac{\partial g_{\tau\rho}}{\partial x_\sigma}\frac{\partial\gamma_{\tau\rho}}{\partial x_\beta} - \frac{1}{2}\sum_{\alpha\beta\tau\rho}\partial_{\sigma\nu}\gamma_{\alpha\beta}\frac{\partial g_{\tau\rho}}{\partial x_\alpha}\frac{\partial\gamma_{\tau\rho}}{\partial x_\beta}\right).$$

Then the conservation theorem assumes the form

$$\sum_{\nu}\frac{\partial}{\partial x_\nu}(\mathfrak{T}_{\sigma\nu} + t_{\sigma\nu}) = 0. \tag{9b}$$

Equation (7b) allows us to conclude that the equations obtained above satisfy postulate 2.[12]

8. THE NEWTONIAN GRAVITATIONAL FIELD

The gravitational equations we have established are certainly very complicated. But several important consequences can be easily derived from them based on the following considerations. If the usual relativity theory in its familiar form were exactly correct, the components of $g_{\mu\nu}$ respectively $\gamma_{\mu\nu}$ would be given by the following tables:

Table of the $g_{\mu\nu}$				Table of the $\gamma_{\mu\nu}$			
−1	0	0	0	−1	0	0	0
0	−1	0	0	0	−1	0	0
0	0	−1	0	0	0	−1	0
0	0	0	c^2	0	0	0	$\frac{1}{c^2}$

The gravitational field equations do not allow that the components of the fundamental tensor could actually have these values in a finite region, if some physical process occurs in it. However, it appears that the departures of the tensor components from the given constant values can be taken to be very small quantities for the region

---

12 Because from equation (7b) one can see, for example, that the quantities $t_{\sigma\nu}$ of the gravitational field, which play the same role for this field that the quantities $\mathfrak{T}_{\sigma\nu}$ do for the material process, have the same field-inducing effect as the quantities $\mathfrak{T}_{\sigma\nu}$, in conformity with postulate (2).

of the world accessible to us. We obtain a far-reaching approximation if we take these deviations, which we will write with $g^*_{\mu\nu}$ and respectively $\gamma^*_{\mu\nu}$, along with their derivatives into consideration only when they enter linearly, and disregard all terms in which two such quantities are multiplied together. Then the equations (7a) and (7b) assume the form:

$$\Box g^*_{\mu\nu} = \frac{\partial^2 g^*_{\mu\nu}}{\partial x^2} + \frac{\partial^2 g^*_{\mu\nu}}{\partial y^2} + \frac{\partial^2 g^*_{\mu\nu}}{\partial z^2} - \frac{1}{c^2}\frac{\partial^2 g^*_{\mu\nu}}{\partial t^2} = \kappa T_{\mu\nu}, \tag{7c}$$

where the $T_{\mu\nu}$ gives an incoherent mass flow according to the schema

$$\left.\begin{array}{cccc}
\dfrac{\rho_0}{c^2-q^2}\dot{x}\dot{x} & \dfrac{\rho_0}{c^2-q^2}\dot{x}\dot{y} & .. & -\dfrac{\rho_0 c^2}{c^2-q^2}\dot{x} \\[3mm]
\dfrac{\rho_0}{c^2-q^2}\dot{y}\dot{x} & .. & .. & -\dfrac{\rho_0 c^2}{c^2-q^2}\dot{y} \\[3mm]
 & & & -\dfrac{\rho_0 c^2}{c^2-q^2}\dot{z} \\[3mm]
.. & .. & .. & \\[3mm]
-\dfrac{\rho_0 c^2}{c^2-q^2}\dot{x} & .. & .. & \dfrac{\rho_0 c^4}{c^2-q^2}.
\end{array}\right\} \tag{8}$$

We obtain the Newtonian system insofar as we introduce the following approximations:

1. Only the mass flow is regarded as the field source.
2. The influence of the velocity of the field-generating masses is neglected, and hence the field is treated as static.
3. The velocity and acceleration components in the equations of motion of a material point are treated as small quantities, and only quantities of the lowest order are retained.

Finally, we also have to assume that at infinity the $g^*_{\mu\nu}$ vanish.

It then follows from (7c) and (8) that, if we write the Laplacian operator as Δ,

$$\left.\begin{array}{l}
\Delta g^*_{\mu\nu} = 0 \qquad (\text{unless } \mu = \nu = 4) \\[3mm]
\Delta g^*_{44} = \kappa c^2 \rho_0.
\end{array}\right\} \tag{7d}$$

From this, as is well known, it follows that

$$\left.\begin{array}{l}
g^*_{\mu\nu} = 0 \qquad (\text{except in the case where } \mu = \nu = 4) \\[3mm]
g^*_{44} = \dfrac{\kappa c^2}{4\pi}\displaystyle\int \frac{\rho_0 dv}{r},
\end{array}\right\} \tag{10}$$

where the integration extends over three-dimensional space, and r is the distance from dv to the origin. It follows from (1b) and (1b'), taking the approximation postulated above taken into account, that

$$\ddot{x} = -\frac{1}{2}\frac{\partial g_{44}^*}{\partial x}.$$ (1c)

Equations (9) and (1c) contain Newton's gravitational theory, where our constant κ is connected to the usual gravitational constant K by the relation

$$K = \frac{\kappa c^2}{8\pi},$$ (11)

from which it follows that

$$K = 6,7 \cdot 10^{-8} \qquad \kappa = 1,88 \cdot 10^{-27}. \,|$$

[1260]   In the approximation considered here, for the "natural" four-dimensional volume element ds, we have

$$ds = \sqrt{-dx^2 - dy^2 - dz^2 + g_{44}dt^2},$$

whereby

$$g_{44} = c^2\left(1 - \frac{\kappa}{4\pi}\int\frac{\rho_0 dv}{r}\right).$$

One can recognize that the coordinate length is identical to the natural length ($dt = 0$); hence measuring rods undergo no distortion in a "Newtonian" gravitational field. By contrast, the rate of a clock depends upon the gravitational potential. For ds/dt gives a measure of the clock's rate, if one sets $dx = dy = dz = 0$. One obtains:

$$\frac{ds}{dt} = \sqrt{g_{44}} = \text{const.}\left(1 - \frac{\kappa}{8\pi}\int\frac{\rho_0 dv}{r}\right).$$

Thus, the greater the mass placed in its vicinity, the slower the clock ticks.[13] It is interesting that the theory has this result in common with Nordström's theory.

For the propagation of light ($ds = 0$) one obtains the velocity

$$\mathfrak{L} = \left|\sqrt{\frac{dx^2 + dy^2 + dz^2}{dt^2}}\right|_{ds=0} = \sqrt{g_{44}} = c\left(1 - \frac{\kappa}{8\pi}\int\frac{\rho_0 dv}{r}\right).$$

Thus according to the foregoing theory, and in contradiction with Nordström's theory, light rays are bent by the gravitational field. This is the only consequence of the theory find so far that is accessible to experience.

---

13 According to postulate (4), this result holds for the rate of any process whatsoever.

Without continuing this consideration of the use of approximations in field calculations, we will give the exact equations of motion for a point in the field considered here. From the general equations of motion (1b') we obtain

$$\frac{d}{dt}\left\{-m\sum_{\nu}g_{\sigma\nu}\frac{dx_{\nu}}{ds}\right\} = -\frac{1}{2}m\sum_{\mu\nu}\frac{\partial g_{\mu\nu}}{\partial x_{\sigma}}\frac{dx_{\mu}}{ds}\frac{dx_{\nu}}{dt}.$$
(1b")

For the special case of the Newtonian field this yields

$$\frac{d}{dt}\left\{m\frac{\dot{x}}{\sqrt{g_{44}-q^2}}\right\} = -\frac{1}{2}m\frac{\dfrac{\partial g_{44}}{\partial x}}{\sqrt{g_{44}-q^2}}.$$
(1c')

9. ON THE RELATIVITY OF INERTIA

From (1c') it follows that the momentum I and the energy E of a mass point moving slowly in the Newtonian gravitational field are given by the equations:

$$I_x = m\left(1 + \frac{\kappa}{8\pi}\int\frac{\rho_0 dv}{r}\right)\frac{x}{c} \text{ etc.}$$

and

$$E = mc\left(1 - \frac{\kappa}{8\pi}\int\frac{\rho_0 dv}{r}\right) + \frac{1}{2}\frac{m}{c}\left(1 + \frac{\kappa}{8\pi}\int\frac{\rho_0 dv}{r}\right)q^2.$$
(12)

Thus, although the energy of a point mass at rest decreases with the accumulation of masses in its vicinity, as the first term of the expression for E shows, the same accumulation leads to an *increase* of the inertia of the point mass under consideration. This result is of great theoretical interest. For if the inertia of a body can *increase* due to the piling up of mass in its vicinity, then we have no choice but to regard the inertia of a point as being caused by the presence of the other masses.Thus, inertia appears to be *caused* by a kind of interaction between the point mass to be accelerated and all of the other point masses.

This result appears quite satisfactory if one reflects on the following. It makes no sense to speak of the motion, and hence also the acceleration, of a body A by itself. One can only speak of the motion or acceleration of a body A relative to other bodies B, C, etc. Whatever holds true kinematically regarding acceleration must also hold true for the inertial resistance with which bodies oppose acceleration; it is to be expected a priori, if not exactly necessarily,[14] that inertial resistance is nothing but a resistance of the designated body A to relative acceleration with respect to the totality of all other bodies B, C, etc. It is well known that E. Mach first defended this point of view, with perfect acuity and clarity, in his history of mechanics, so that here I can simply refer to his arguments. Let me also refer to a witty brochure by the Viennese mathematician W. Hoffman that independently argues for the same position. I will

call the I conception sketched here the "hypothesis of the relativity of inertia."

To avoid misunderstandings, let me say once more that, like Mach, I do not think that the relativity of inertia is a logical necessity. But a theory which grants the relativity of inertia is more satisfactory than our current theory, because in the latter theory, inertial systems are introduced which, on the one hand, have a state of motion that does not depend on the states of observable objects, and thus is not caused by anything accessible to observation, but, on the other hand, are supposed to determine the behavior of material points.

The concept of the relativity of inertia requires, however, not only that the inertia of a mass A increases when masses at rest pile up in its surroundings, but also that this increase of inertial resistance will not take place if the masses BC ... are accelerated with the mass A. One can express this point as follows: the acceleration of the masses BC ... must induce an accelerative force on A that is in the same direction as the acceleration. With this one can see that this accelerating force must overcompensate for the increase of inertia produced by the mere presence of BC ..., for according to the relation between the inertia and energy of systems, the system ABC ... as a whole must have less inertia the smaller its gravitational energy.

In order to see that our theory fulfills this requirement, we must take into account the terms on the right-hand side of the system of equations (7c), which are proportional to the first power of the velocity of the field-producing masses. We then obtain the following instead of the system of equations (7d):

$$
\left.
\begin{aligned}
\Box g^*_{\mu\nu} &= 0 \qquad (\text{if } \mu \neq 4 \text{ and } \nu \neq 4) \\
\Box g^*_{14} &= -\kappa\rho_0\dot{x} \\
\Box g^*_{24} &= -\kappa\rho_0\dot{y} \\
\Box g^*_{34} &= -\kappa\rho_0\dot{z} \\
\Box g^*_{44} &= -\kappa c^2\rho_0 .
\end{aligned}
\right\}
\tag{7e}
$$

The equations of motion of the material point (1b″) differ from (1c′), in that now g_{14}, also differ from zero. They read in full:

$$
\left(\frac{d}{dt}\left\{ m\left(\frac{dx}{ds} - g_{14}\frac{dt}{ds}\right)\right\}\right) = -\frac{1}{2}m\left(2\frac{\partial g_{14}}{\partial x}\frac{dx}{ds} + 2\frac{\partial g_{24}}{\partial x}\frac{dy}{ds} + 2\frac{\partial g_{34}}{\partial x}\frac{dz}{ds} + \frac{\partial g_{44}}{\partial x}\frac{dt}{ds}\right)\right) \quad \text{etc.}
$$

---

14 One typically avoids the consequences of such arguments by introducing reference systems (inertial systems) with respect to which freely moving mass points are in rectilinear uniform motion. What is unsatisfactory is that it remains unexplained how the inertial systems can be privileged with respect to other systems.

For a slowly moving point one can write these equation as follows in the usual three-dimensional vector notation:

$$\ddot{r} = -\frac{1}{2}\operatorname{grad} g_{44} + \dot{g} - [\dot{r}, o]. \tag{1d}$$

Here

$\qquad$ r $\;=\;$ radius vector of the mass point

$\qquad \dot{r} \;=\; \dfrac{dr}{dt}$ etc.

$\qquad$ g $\;=\;$ a vector with the components $g_{14}, g_{24}, g_{34},$

$\qquad$ o $\;=\;$ curl g

If we denote the velocity of the field-producing masses (comp. $\dot{x}, y, \dot{z}$) with v, then we can write (7e) in more concisely:

$$\left.\begin{aligned} \Box g &= -\kappa\rho_0 v \\ \Box g_{44}^{*} &= \kappa c^2 \rho_0. \end{aligned}\right\} \tag{7e'}$$

The equations (7e') and (1d) show how slowly moving masses influence each other according to the new gravitational theory. To a great extent, the equations correspond to those in electrodynamics, g_{44} corresponds to the scalar potential of electrical masses up to the sign and up to the circumstance that the factor $1/2$ appears in the first term of the right-hand side of (1d). g corresponds to the vector potential of electric currents; the second term on the right-hand side of (1d), which corresponds to an electric field strength resulting from a temporal change of the vector potential, yields precisely those induction effects, directed like the acceleration, that we must expect based on our ideas regarding the inertia of energy. The vector o corresponds to the magnetic field strength (curl of the vector potential), so the last term in (1d) corresponds to the Lorentz force.

It should further be remembered that a term of the form $(\dot{r}, o)$ occurs in the theory of relative motion in mechanics, and is known as the Coriolis force. One can show from (7e') that a field with vector o, exists on the inside of a rotating spherical shell, which leads to the result that the plane of oscillation of a pendulum set up inside the spherical shell does not stay fixed in space, but rather, due to the sphere's rotation, must take part in a precessional motion in the same direction as this rotation. This result is also to be expected from the meaning of the concept of the relativity of inertia, and has long been anticipated. It is noteworthy that the theory also agrees with the above conception with regard to this point, but unfortunately I the expected [1262] effect is so slight that we cannot hope to confirm it via terrestrial experiments or astronomy.

10. CONCLUDING REMARKS

In the foregoing discussion we have sketched the most natural paths that a gravitational theory can follow. One either stands by the usual relativity theory, i.e., one assumes that the equations expressing laws of nature remain covariant only under linear orthogonal substitutions. Then one can develop a scalar theory of gravitation (Nordström's theory), which is fairly simple and sufficiently satisfies the most important requirements to be imposed on a gravitational theory, although it does not include the relativity of inertia as a consequence. Or one augments the relativity theory in the way sketched above. One certainly attains equations of considerable complexity, but, in exchange, the sought after equations follow from the basic principles with the help of surprisingly few hypotheses, and the conception of the relativity of inertia is satisfied.

Whether the first or the second way corresponds essentially to nature must be decided by observations of stars appearing near the Sun during solar eclipses. Hopefully the solar eclipse of 1914 will already resolve this important decision.

EDITORIAL NOTES

[1] In the original, the following equation was mistakenly given the equation number (2), which appeared already on p. 546 [p. 1251 in the original].

[2] In the original, the first $\gamma_{\tau\rho}$ and $\gamma_{\alpha\beta}$ are misprinted as $y_{\tau\rho}$ and $\gamma_{\beta\nu}$ respectively.

FROM HERETICAL MECHANICS
TO A NEW THEORY OF RELATIVITY

JULIAN B. BARBOUR

EINSTEIN AND MACH'S PRINCIPLE

INTRODUCTION

Einstein's attempt to realize Machian ideas in the construction of general relativity was undoubtedly a very major stimulus to the creation of that theory. Indeed, the very name of the theory derives from Einstein's conviction that a theory which does justice to Mach's critique of Newton's notion of absolute space must be generally relativistic, or covariant with respect to the most extensive possible transformations of the spacetime coordinates.

The extent to which general relativity is actually Machian is, however, the subject of great controversy. During the last six months, I have been examining closely all of Einstein's papers that concern the special and general theory of relativity together with a substantial proportion of his correspondence related to relativity. There were several things that I wished to establish: 1) What precisely was the defect (or defects) in the Newtonian scheme that Einstein sought to rectify in his general theory of relativity? 2) How did Einstein propose to rectify the perceived defect(s)? 3) What relation does Einstein's work on his Machian ideas bear to the other ideas and work of his predecessors and contemporaries on the problem of absolute and relative motion? 4) Finally, to what extent did general relativity solve that great and ancient problem of the connection between and status of absolute and relative motion?

In this paper, which addresses the first three issues and gives my main conclusions (which are being presented in more detail together with my attempt at an answer to the fourth question in a forthcoming book (Barbour, in preparation)), I begin by reviewing the most important contributions to the discussion of absolute and relative motion made by Einstein's predecessors and contemporaries. As we shall see, this work identified certain key problems and went some way to providing the solutions to them. In particular, in 1902 Poincaré (1902; 1905, 75–78 and 118) provided a very valuable criterion for when a theory could be said to be Machian. Moreover, Mach (1883, 1960), Hofmann (1904), and Reissner (1914, 1915) made definite proposals of non-relativistic models of particle mechanics that meet this criterion. The examination of Einstein's entire relativity opus shows that this work made virtually no impact on him. Moreover, there is rather strong evidence which indicates a surprising lack of awareness on Einstein's part of the central problem with which the absolute-relative debate is concerned—*the problem of defining velocity*, i.e., change of position (and, more generally, *change* of any kind). For reasons that can be at least partly under-

Jürgen Renn (ed.). *The Genesis of General Relativity,* Vol. 3
Gravitation in the Twilight of Classical Physics: Between Mechanics, Field Theory, and Astronomy.
© 2007 Springer.

stood, Einstein saw this as a relatively trivial matter and regarded *acceleration* as more problematic.

In fact, Einstein associated with Mach's name two specific problems.

The first may be called the **absolute-space problem**, but it could equally well be called the problem of the *distinguished frames of reference*. Einstein initially presented it as the great mystery of why there seem to exist distinguished frames of reference for the expression of the laws of nature, though later he often spoke of the unacceptability of there being a thing (absolute space) that could influence the behavior of matter without itself being affected by matter.

The second may be called the **inertial-mass problem**. This problem was first mentioned explicitly by Einstein in 1912, when he asserted that Mach had sought to explain the *inertial mass* of bodies through a kind of interaction with all the masses of the universe.

In the years up to the definitive formulation of general relativity in 1915 and a little beyond, Einstein repeatedly mentioned these two problems. However, in 1918, following a critique by Kretschmann (Kretschmann 1917), Einstein (Einstein 1918a) said that he had not hitherto distinguished properly between these two problems (and between the means by which he proposed to resolve them). He then gave a formal definition of what he called *Mach's Principle*, which took the form of the requirement that all the local inertial properties of matter should be completely determined by the distribution of mass-energy throughout the universe. He said that this was "a generalization of Mach's requirement that inertia should be derived from an interaction of bodies." At the same time, Einstein gave a definition of the relativity principle that took from it all the specific empirical content it had previously seemed to possess in Einstein's work and transformed it into a very general necessary condition on the very possibility of stating any laws of nature: "The laws of nature are merely statements about spacetime coincidences; they therefore find their only natural expression in generally covariant equations."

Towards the end of his life, Einstein admitted (not very publicly but explicitly in a letter to Felix Pirani)[1] that his 1918 formulation of Mach's Principle made no sense mathematically and from the physical point of view had been made obsolete by the development of physical notions that had displaced material bodies from the pre-eminence they had possessed in Newtonian theory. However, to the end of his life he retained the 1918 formulation of the relativity principle, which he admitted carried little real physical content. However, he asserted that in conjunction with a requirement of simplicity it possessed great heuristic value, namely that, in a choice between rival theories, preference should be given to those theories that, when expressed in generally covariant form, took a simple and harmonious form.

This faith in *simplicity* as a criterion for selecting physical theories is extremely characteristic of Einstein and gives expression to his deep faith in the ultimate rationality of physics. It is, however, a notoriously slippery criterion. It is also a fact that

1   Einstein to Felix Pirani, 1954 (EA 17-447).

when, in the years up to and including 1916, Einstein said that a satisfactory theory of gravity and inertia must be generally covariant he undoubtedly thought that this requirement had a deep physical significance going far beyond the bland 1918 formulation of the relativity principle.

Mach made the comment that the creators of great theories are seldom the best people to present those theories in a logically concise and consistent form. In this book devoted to alternative strategies that could have been adopted (and in some cases were) to the development of relativity theory, I hope that the following attempt to establish what Einstein was trying to do, actually did, and might have done will help to cast light on the extremely tangled story of the creation of one of the wonders of theoretical physics: the general theory of relativity. In particular, I hope this paper will complement the articles by Jürgen Renn and John Norton[2] (both of which I found very useful in my own work) by looking at Einstein's work closely from the perspective of the specific problem of absolute *vs* relative motion. John Norton has done a splendid technical and conceptual job in comparing Einstein's approach with the more conventional 'Lorentz-invariant field theoretical' approach (to use Norton's useful anachronism) that virtually all his contemporaries adopted to the finding of a relativistic field theory of gravitation. Jürgen Renn, for his part, has emphasized the vital importance of Einstein's more wide-ranging approach and the inclusion of epistemological problems from the foundations of mechanics in the set of issues to be resolved in a satisfactory theory of gravitation. He brings out the value of Einstein's philosophical and integrative outlook. Examination of Einstein's work from the specific absolute *vs* relative perspective brings to light some further issues and aspects of Einstein's work that are not so readily revealed in their approaches.

I hope and believe that nearly all the articles in this book will have not only historical and philosophical interest but also serve a useful purpose for current research. It is widely agreed that the greatest current problem that has to be solved in theoretical physics is that of the relationship between quantum theory and the general theory of relativity. It is my conviction (Barbour 1994, 1995, in preparation) that general relativity is deeply Machian in a sense that unfortunately Einstein never managed to pinpoint accurately and that precisely this very Machian nature of general relativity is the main cause of the difficulties that stand in the way of its quantization. I therefore hope that the present article will have not only historical relevance but also help to clarify some central issues of current research.

In this article, it will not be possible to give a comprehensive account. I aim merely to identify some of the most important issues and ask the reader to consult my forthcoming monograph for a more detailed account. See also the *Notes Added in Proof* at the end of this article.

---

2   See *The Third Way to General Relativity* and *Einstein, Nordström, and the Early Demise of Scalar, Lorentz Covariant Theories of Gravitation* (both in this volume).

1. THE ORIGIN AND EARLY HISTORY OF
THE ABSOLUTE *VS* RELATIVE DEBATE

The whole absolute *vs* relative debate arose from Descartes's claim in his *Principles of Philosophy* (1644) that *motion is relative* (Barbour 1989). Descartes argued that position can only be defined relative to definite reference bodies. Since there is evidently no criterion for choosing certain reference bodies in preference to others, Descartes argued that there can be no unique definition of motion—a given body has as many different motions as there are reference bodies (which, in general, will, of course, be moving relative to each other) with which it can be compared.

Despite this rather cogent argument, Descartes then proceeded, in a manifest *non sequitur*, to formulate definite laws of motion, the first two of which were identical in their content to the law that Newton subsequently adopted as his first law: Any body free of disturbing forces will either remain at rest or move in a straight line with uniform speed. It is evident that such a law presupposes a definite frame of reference—a reference space—and an independent time (an external clock) if it is to make any sense. About this mysterious reference space Descartes said not a word.

We know from Newton's tract *De Gravitatione* (Hall and Hall 1962), written around 1670 but published only in 1962, that Newton was intensely aware of the flagrant contradiction between Descartes's espousal of relativism and the vortex theory, on the one hand, and his anticipation and formulation of the law of inertia, on the other. In a world in which all matter is in ceaseless relative motion (as it is in accordance with Cartesian vortex theory or the atomistic theories so prevalent in the 17th century), Cartesian relativism seems to make it utterly impossible to define a definite motion; in particular, it would appear to be impossible to say that any given body is moving in a straight line. Commenting sarcastically on Descartes's law, Newton said: "That the absurdity of this position may be disclosed in full measure, I say that thence it follows that a moving body has no determinate velocity and no definite line in which it moves." This may truly be called the *fundamental problem of motion*: If all motion is relative and everything in the universe is in motion, how can one ever set up a determinate theory of motion?

The entire story of the absolute *vs* relative debate flows from this dilemma that Newton posed so clearly in around 1670. For completeness, one should also add the temporal part of the story: Motion can never be measured by *time* in the abstract but only by a definite comparison motion. For scientific purposes, the comparison motion was for millennia the rotation of the Earth, though more recently a global network of atomic clocks has been introduced as the official standard of time. Thus, statements in physics involving time are really statements about physical clocks, for which a theory based on first principles is needed (given the fundamental importance of time).

Having formed the deep conviction that no sensible mathematically well-defined dynamics could be based upon Cartesian relativism, Newton insisted on the introduction of a rigidly fixed absolute space and a uniformly flowing external absolute time as the kinematic framework for the definition of motion. However, he was still very conscious of the cogency of Descartes's relativism and in the famous Scholium in the

Principia on absolute and relative motion admitted freely the need to show how absolute motions, which cannot be observed directly ("because the parts of that immovable space, in which those motions are performed, do by no means come under the observation of the senses"), could be deduced from the observed relative motions. This task may be appropriately called the *Scholium problem*: Given observed relative motions, find the corresponding absolute motions. Although Newton actually claimed at the end of the Scholium that he wrote the *Principia* specifically in order to show how that problem is to be solved, he never spelled out the solution explicitly and in the Scholium merely advanced some first qualitative arguments designed to show that absolute space must exist. Even less effort was made to demonstrate the existence of absolute time.

Despite eloquent criticism of the notions of absolute space and time by Newton's contemporaries Huygens, Leibniz, and Berkeley, the absolute *vs* relative problem remained effectively in a state of limbo for very nearly 200 years until it was taken up again by the mathematician Carl Neumann in 1870 (Neumann 1870) and by Ernst Mach in 1872 (Mach 1872, 25; 1911) at the end of an extended essay on the conservation of energy and then again in his famous book on mechanics in 1883 (Mach 1883, 1960). Parallel but less influential work was done in Britain (Scotland to be precise) by William Thomson (later Lord Kelvin) and Tait (Thomson and Tait 1867, §§208ff.; Tait 1883) and also Lord Kelvin's brother James Thomson (Thomson 1883). The interventions of Neumann and Mach brought two issues to the fore.

The first was essentially the Scholium problem: under the assumption that Newton's scheme is in essence correct, how can one make correct epistemological sense of his notions of absolute space and time? Important and significant contributions to the resolution of this problem were made by Neumann (Neumann 1870), Tait (in an unfortunately little noted elegant piece of work (Tait 1883)), Ludwig Lange (Lange 1884, 1885, 1886), the logician Frege (Frege 1891), and above all Poincaré (Poincaré 1898 and 1902; 1905, 75–78 and 118). This work will be considered in Sec. 3.

The second issue brought to the fore was Mach's proposal, made already in 1872 and then repeated (though not quite so clearly or unambiguously as one might wish) in his 1883 *Mechanik* and all its subsequent editions, to the effect that Newton's mechanics might actually be *physically incorrect* and should be replaced by a dynamics of a different form in which only relative separations of bodies occur. The physical cogency of this proposal was made much more impressive by Mach's ability to counter Newton's bucket argument from the undoubted existence of centrifugal force to the need for an absolute space to explain it. Mach observed that the distant masses of the universe rather than some absolute space could be the ultimate origin of the centrifugal forces and that if this were the case local material bodies, such as the wall of Newton's bucket, could be expected to have only a minuscule and unobservable effect.

2. DIRECT ATTEMPTS TO IMPLEMENT MACH'S PROPOSAL
AND THEIR LACK OF IMPACT ON EINSTEIN

Although they have attracted very little notice, attempts at a direct implementation of Mach's proposal were made throughout the twentieth century. The first such attempts were made early enough for them to have influenced Einstein in his work on general relativity. In this section, this work and its very marginal impact on Einstein will be considered.

A proposal for a new, non-Newtonian mechanics was already advanced by Mach, in a very tentative and mathematically rather unsatisfactory form, in the *Mechanik* in 1883.[3] His ideas were advanced in several interesting ways by the Friedlaender brothers in a rather obscure booklet published in 1896 (Friedlaender and Friedlaender 1896). In a simple and beautiful example,[4] Benedict Friedlaender showed how distant rotating masses (the 'stars' as seen from someone rotating with Newton's bucket) could very well generate centrifugal forces away from the axis of rotation and thus make absolute space unnecessary. In his contribution to this volume, Renn discusses the various interesting points and also anticipations of Einstein's later work that can be found in the Friedlaenders' booklet.

A rather general way of generating (nonrelativistic) relational theories of the kind envisaged by Mach was found by a certain Wenzel Hofmann of Vienna, who in 1904 (Hofmann 1904) published an even more obscure booklet[5] than the Friedlaenders' which would surely have been lost forever had it not been for fleeting references to it by Mach in the 5th and 6th editions of the *Mechanik* and by Einstein in 1913 (Einstein 1913a). In modern terms, the essence of Hofmann's proposal was to replace the Newtonian kinetic energy T, which occurs in the Lagrange function $T - V$ of the classical mechanics of n point particles and consists of a sum over individual masses of the form

$$1/2 \sum m_i \dot{r}_i \cdot \dot{r}_i, i = 1, \ldots, n, \tag{1}$$

where m_i is the mass of particle i, r_i is its position vector in absolute space, and the dot denotes the time derivative, by a sum over all pairs of the n particles of the form

$$\sum_{i < j} m_i m_j f(r_{ij}) \dot{r}_{ij}^2, \tag{2}$$

where r_{ij} is the (Euclidean) separation of particles i and j, $f(r_{ij})$ is some function of this separation, and the dot has the same meaning as in (1).

Hofmann was able to show qualitatively that in a realistic cosmological model, in which there are many stars distributed more or less uniformly over a large area,

---

3   See (Mach 1960, §VI.7, 286–7) and the discussion of this section by Norton (who questions whether it is a proposal for a new mechanics) and myself in (Barbour and Pfister 1995).
4   Translated in part in (Barbour and Pfister 1995).
5   Mach's proposal reduced essentially to the special case $f = 1$ of Hofmann's general proposal (2).

masses such as those in the solar system would behave in accordance with laws that approximated quite well Newton's laws but in an effective space determined explicitly by the matter distribution in the universe.

Hofmann's idea has since been independently rediscovered many times. The first person to do that was Reissner in 1914 and 1915 (Reissner 1914, 1915), when he chose the particular form $1/r_{ij}$ for $f(r_{ij})$ in (2). This choice is physically plausible and has some remarkably interesting consequences as was shown in part by Reissner himself and also Schrödinger (Schrödinger 1925) in a very beautiful paper at least partly inspired by Reissner's work.

More recently, Bertotti and I (Barbour and Bertotti 1977, 1982) considered a very general framework for constructing relational theories of this kind, including a relational treatment of time. The basic idea is taken straight from Mach. One assumes that dynamics must be formulated for the universe as a whole[6] and, in a variational formulation, insists that only the relative quantities r_{ij} and their rates of change may appear on the Lagrangian that describes the dynamics of the universe. Time is treated relationally by insisting that all changes are measured, not by comparison with some abstract external time t but always by comparison with other actual changes in the universe. This has the effect that Newton's abstract time is replaced by an appropriate average of the totality of changes in the universe.

It turns out that within this large class of possible Machian theories there exist at least two distinct subclasses. One is essentially the class discovered by Hofmann, but it has the disadvantage that it leads to an effective inertial mass that is anisotropic in the presence of nearby accumulations of mass. Schrödinger, in particular, was well aware of this anisotropy and knew that it could lead to an experimental refutation of such theories. He attempted to investigate the effect of the Galaxy and found it to be just below the then existing observational accuracy. He was however using a much too low value for the mass of the Galaxy, and modern data rule out such a theory completely. Such theories are therefore of interest mainly as examples of what Machian theories might look like. In contrast, in the theories of the second class, which Bertotti and I base on a notion called the intrinsic derivative (or *best matching*), mass anisotropy is completely absent. Indeed, one can construct intrinsic models of Machian mechanics that in their locally (but not globally) observable consequences are completely indistinguishable from Newtonian mechanics. I shall return to this briefly at the end of the next section.

The fact that the basic idea of relational mechanics was rediscovered many times[7] indicates that it is a very natural and direct way of realizing Mach's ideas and thereby eliminating absolute motions (and with them absolute space and time) from the foundations of physics. Given Einstein's passionate desire to implement Mach's ideas, it

---

6   This is implicit in the proposal of Mach and is made explicit by the appearance of the crucial summation in Hofmann's expression (2).

7   Apart from Hofmann, Reissner, and Schrödinger in the early part of this century, at least five other people besides Bertotti and myself hit on the same basic idea in the period 1960–1990, as noted in the articles by myself and Assis in (Barbour and Pfister 1995).

has always seemed to me most surprising that the basic idea—the insistence that only relative quantities should appear in the laws of nature—never seems to have been considered seriously by Einstein. All of Einstein's work on relativity—from 1905 right through to his death in 1955—has a quite different 'flavour.' In fact, it is quite difficult to find evidence that Einstein was even aware of the possibility.

Unless more evidence comes to light in the as yet unpublished correspondence, the only really clear statement of Einstein which does show that he was aware of what might be done along these lines comes from a paper published in 1918 (Einstein 1918b) with the title "Dialogue on objections to the theory of relativity," which includes the following:

> We want to distinguish more clearly between quantities that belong to a physical system as such (are independent of the choice of the coordinate system) and quantities that depend on the coordinate system. Ones initial reaction would be to require that physics should introduce in its laws only the quantities of the first kind. However, it has been found that this approach cannot be realized in practice, as the development of classical mechanics has already clearly shown. One could, for example, think—and this was actually attempted—of introducing in the laws of classical mechanics only the distances of material points from each other instead of coordinates; *a priori* one could expect that in this manner the aim of the theory of relativity should be most readily achieved. However, the scientific development has not confirmed this conjecture. It cannot dispense with coordinate systems and must therefore make use in the coordinates of quantities that cannot be regarded as the results of definable measurements

In the absence of definite references, it is impossible to know for sure whose work Einstein had in mind with his "this was actually attempted" but it is plausible to suppose that he was referring to Mach's original proposal of 1883, Hofmann's 1904 booklet, which he had mentioned briefly in 1913 (Einstein 1913a), describing it as "ingenious," and also perhaps Reissner's two papers.[8] It must also be said that, if he was thinking of the work of Hofmann and Reissner, Einstein had clearly failed to grasp what had been achieved in that work. Both authors had in fact succeeded in finding a genuine alternative to Newtonian inertia governed by absolute space. Moreover, the alleged difficulty to which Einstein refers, that of dispensing with coordinate systems, is simply nonexistent. Both Hofmann and Reissner *did* dispense with coordinate systems in the formulation of their proposed law and worked directly with "only the distances of material points from each other instead of coordinates."

Since these last cited words of Einstein do perfectly encapsulate what Mach had advocated, and since also Einstein repeatedly expressed the greatest admiration for Mach's critique of Newtonian mechanics, his remarks in 1918 present something of a

---

8   No correspondence from Einstein to Reissner survives. There is one letter from Reissner to Einstein in the Einstein Archives. It dates from 1915 but concerns Reissner's work on general relativity. Reissner makes no mention of his Machian papers. In September 1925, Einstein (Einstein to Schrödinger, September 26, 1925 (EA 22-003) thanked Schrödinger for sending him a copy of his 1925 paper on the relativity principle. Einstein merely said it was "interesting." Had the work of Hofmann and Reissner truly made any impact on him, one might have expected Einstein to point out to Schrödinger that his work had been anticipated by them.

puzzle, as I noted a little earlier: Why did Einstein take so little interest in a serious and direct attempt to implement Mach's proposal? To this query one may add the observation that Einstein's frequent references to Mach in his papers in the period 1912 to 1923 seldom reflect accurately what Mach actually said and sometimes even represent a serious distortion. The most serious distortion concerns a straight confusion between two quite distinct meanings of the word *inertia*. It is worth saying something about this.

Both in Mach's time and now, the word inertia meant two things: first, as expressed in Newton's first law, the law of inertia, namely the tendency of a body to continue in rest or in uniform motion in a straight line unless acted upon by some force; second, the quantitative measure of resistance to acceleration as expressed by the presence of m, the *inertial mass*, in Newton's second law $F = ma$. Mach (Mach 1872, 25; 1883) pointed out that Newton had failed to give a meaningful definition of inertial mass and proceeded to supply one himself. He believed that his definition removed all difficulty surrounding the use of the concept of inertial mass in Newtonian dynamics. In contrast, he felt that Newton's formulation of the law of inertia was very seriously deficient and probably incapable of being given adequate expression without some actual change in its physical content. Mach insisted that genuine content must be given to expressions like "uniform motion in a straight line": uniform with respect to what and straight with respect to what? He considered it absolutely impermissible to invoke invisible time and space to answer these questions, and his discussion of these issues takes us straight back to the problems with which Newton grappled in *De Gravitatione*.

Very careful examination of *all* of Einstein's numerous comments on issues related to Mach have led me to a very surprising conclusion. Einstein *never once* even mentioned this problem—the fundamental problem of motion—at the heart of dynamics. He seems to have been more or less completely blind to its existence. He very often used the word inertia but never once made the distinction between the two meanings of it. When he was most explicit about Mach and inertia, he incorrectly attributed to Mach the idea that the *inertial mass* should arise in some manner from a kind of interaction of all the bodies in the universe (Einstein 1912, 1917). Now it is true that the m_i's that appear in Hofmann's proposal (2) are best interpreted as inertial *charges*. In the theory to which (2) and other similar proposals give rise, one then obtains effective *inertial masses*, which are indeed determined by interaction with all the bodies in the universe. This was clearly demonstrated by both Reissner and Schrödinger, but it was already qualitatively clear to Hofmann.

Einstein may very well have had a correct intuitive appreciation that some such effect could come out of a Machian theory of motion, but his repeated assertions that this was what Mach had called for are unfortunate on several counts: 1) They are historically inaccurate. 2) The effect arises in a certain class of Machian theories—the class considered by Hofmann, Reissner and Schrödinger—but not in another, which Bertotti and I discovered (Barbour and Bertotti 1982). This second class of theories is impeccably Machian and actually includes general relativity as a special and remarkably interesting example (Barbour 1995, see also the *Notes Added in Proof*). 3) Einstein's concentration on the inertial mass deflects attention away from the true and profound problem that underlies the absolute *vs* relative debate: How are time and motion to be defined?

This is the fundamental question that, very surprisingly, Einstein never addressed directly. In the final section of this paper, I shall try to establish why this was so. However, before then, in the following section, I want to complete the review of the work of Einstein's predecessors and contemporaries. As noted earlier, the critique of Neumann and Mach raised two issues: 1) Can Newtonian theory be recast in an epistemologically satisfactory manner without change of its essential physical content? 2) Can Newtonian theory be replaced by a physically different theory based on Machian ideas?

This section has essentially considered the answer to the second question. In the next section, we shall consider the answer to the first.

3. THE EPISTEMOLOGICAL WORK OF NEUMANN, LANGE, AND POINCARÉ AND ITS IMPACT ON EINSTEIN

In his habilitation lecture of 1870, Neumann posed a general problem and provided a partial solution to a small part of it. The general problem was this: As formulated by Newton, the laws of mechanics simply cannot be tested because absolute space and time are invisible and inaccessible to experimentalists. The question then was: Is it nevertheless possible to make epistemological sense of Newton's laws by identifying operational surrogates of absolute space and time?

To begin to make progress in this direction, Neumann assumed that particles moving freely of all forces (force-free particles) exist and could be identified as such and that also by some means absolute space (or a suitable surrogate of it) could be observed directly. If the second assumption is satisfied, one can then observe the motion of some chosen force-free particle. Neumann pointed out that, in the absence of an external clock, it is meaningless to say that such a particle is moving uniformly (though, if absolute space has been 'made visible', one can verify that it is moving in a straight line). However, what one can do is observe further force-free particles and see how they behave relative to the original particle, which is taken as a reference body. One can use the distance traversed by this reference body as a measure of time (inertial clock) and see if, relative to this inertial clock, a second force-free body moves uniformly. In this way, Neumann was able to give genuine operational content to the part of Newton's first law which asserts the uniformity of the motion of a force-free body. However, Neumann admitted that he was unable to solve the problem of making absolute space 'visible.'

This problem was taken up by the youthful Ludwig Lange (he was only 21) in 1884. He proceeded very much in the spirit of Neumann and assumed the existence of force-free particles that could be identified as such. His basic idea was to use *three* such particles to define a spatial frame of reference. Just as in the case of Neumann's inertial clock, for which it is meaningless to say that the clock itself is moving uniformly, Lange noted that it would be meaningless to say that his three reference bodies are moving rectilinearly. Instead, they *define* a frame of reference, with respect to which one can then verify that other bodies are moving rectilinearly. Moreover, using any one of the three chosen reference bodies as a Neumann inertial clock, one can simultaneously verify that further bodies are moving uniformly as well as rectilinearly.

Lange's actual construction of the spatial frame of reference using three force-free bodies is in fact rather awkward and clumsy, so I shall not attempt to describe it here, especially since I shall shortly describe a much neater construction due to Tait (Tait 1883). However, it is worth emphasizing the crucial point of the construction, which Lange was the first to recognize clearly and for which he deserves great credit. It will be recalled that Newton criticized Cartesian relativism because it made the motion of a considered body dependent on the choice of the reference bodies used to determine its motion. Since the choice of reference bodies is entirely arbitrary, it would appear that motion itself cannot be defined in any unique way. However, the situation is radically altered if one insists that the reference bodies—no matter which are chosen—*are themselves moving in accordance with Newton's laws*. This is the crucial stipulation that takes the seemingly fatal arbitrariness out of a relational definition of motion. Once this basic fact has been recognized, precise definitions merely reduce to a working out of details.

One severe problem with the Neumann-Lange approach—Lange never succeeded in overcoming it—was that of recognizing when bodies are free of forces. The construction depends crucially on the existence of unambiguously identifiable force-free bodies. This raises *two* problems: 1) How can one tell if a body is free of forces? 2) What can one do if nature fails to provide *any* force-free bodies? In fact, this is exactly the case with gravity, to which all bodies are subject. These serious difficulties were pointed out clearly by the logician Frege (Frege 1891) in an otherwise positive review of Lange's work. Frege correctly emphasized that the axioms of dynamics form a closed system and can only be tested in their totality. Since forces are an integral part of dynamics, their existence must be taken into account in the foundations of any method used to determine the distinguished frames of reference that play such an important role in Newtonian dynamics.

As it happens, the requirement that Frege raised was (in its essentials) met in three studies that unfortunately received very little attention. The first was actually the work of Tait in 1883 that I already mentioned. The other two were published in 1898 and 1902 by Poincaré.

Tait did not solve the problem of finding the dynamical frame of reference in full generality in the case when no force-free bodies are available. He did, however, give a solution to the problem for purely inertial motion that yields the Newtonian frames of reference given purely relative data and simultaneously confirms that all the considered bodies are actually free of all forces.

Tait solved the following problem, which had been posed by James Thomson (Thomson 1883). Suppose that at certain unknown instants of time we are given all the relative separations r_{ij} between a set of n point particles. Thus, we are, as it were, given 'snapshots' of the relative configurations of the particles. Using these snapshots and nothing else, can we verify if there exist a frame of reference and a measure of time, both of which must be deduced from the snapshots, in which all the particles are moving in accordance with Newton's first law?

To solve this problem, Tait supposed that the answer is yes. I shall consider the solution he gave for the case of three particles, since it fully illustrates the underlying principle. If all the particles are moving in accordance with Newton's first law, then one can certainly always choose the frame of reference in such a way that one of the particles is permanently at rest at the origin of the frame. If we exclude the special cases in which there are collisions of the particles, then if we consider some second particle there must exist a time at which it passes the first one at a distance a of closest approach. We can then choose x and y axes of the frame of reference in such a way that at $t = 0$ this second particle is at the point $(a, 0, 0)$ and at time t is at the point $(a, t, 0)$. Thus, we choose the unit of time such that particle 2 has unit velocity. It becomes a Neumann inertial clock. The spatiotemporal framework is then uniquely defined (up to reflections). At $t = 0$, the third particle will have some initial position (x_3, y_3, z_3) and initial velocity $(\dot{x}_3, \dot{y}_3, \dot{z}_3)$ Thus, this three-body problem will have *seven* essential unknowns. The problem of inertial motion is more or less trivial and one can find an analytical solution for the observable separations r_{ij} in terms of these seven unknowns. Given observed values of r_{ij}, these can be compared with the analytical solution and the seven unknowns determined.

As Tait noted, the most interesting point concerns the number of snapshots needed to find the seven unknowns. Each snapshot yields three independent data—the three sides of the triangle—but each snapshot is taken at an unknown time, so that only *two* useful data are supplied with each. It is thus clear that to determine the spatiotemporal framework and test whether all three particles are moving inertially in accordance with Newton's first law one needs at least four snapshots, since they give eight data, from which the seven unknowns can be determined and one verification made of the conjecture. Each extra snapshot yields a further two verifications.

Several important points emerge from Tait's analysis. First, contrary to a very widespread opinion engendered by Lange's work, three particles are already sufficient to establish the spatiotemporal framework and to test whether Newton's first law is satisfied. Lange, and many of his followers, believed three particles were needed to define the framework and that only a fourth would permit a nontrivial verification of Newton's law. Second, attention should be drawn to the central importance of the complete configurations of the three particles, which, in a sense, *define* the instants of time, and to the fact that both time and the spatial reference frame are best and mostly effectively determined together from the raw observational data—the relative separations. Third, knowledge of the spatial frame of reference is a vital prerequisite for determination of all quantities of primary concern in dynamics, above all time, which in the Tait procedure is read off from distance traversed in the spatial inertial frame of reference, and velocity and momentum, both of which can only be found once the complete spatiotemporal framework has been determined.

Two further points should be made here. In analytical mechanics, great emphasis is placed on the possibility of representing dynamics in completely arbitrary frames of reference. However, this does not alter the fact that somehow or other the primary dynamical quantities such as momentum and energy must be found in an inertial spatiotemporal framework. It is only then that a transformation to an arbitrary framework can be performed. Many people, even experts, are quite unaware of this fact. The second remark concerns the definition of a clock. It is widely believed that the essential basis of a clock is a strictly periodic process, the 'ticks' of which measure time. This belief is wrong on *two* scores. First, the Neumann-Lange-Tait procedure shows that linear distance traversed in an inertial frame of reference by a force-free particle is a perfectly good measure of time. Thus, a periodic process is not needed. Second, the inertial frame of reference and distance traversed in it are (in mechanics at least) always the ultimate source of a scientifically meaningful definition of time. Ironically, a pendulum clock, the rate of which depends upon the strength of the gravitational field in which it is set up, is not really a good clock, since its rate is not exclusively determined by its local inertial frame of reference. Thus, a pendulum clock goes faster near sea level than on the top of a mountain, but (as Einstein's general theory of relativity established) clocks that measure proper time go slower at sea level. This highlights the salient point: A clock, to function properly, must 'lock onto' or 'tap' processes directly and exclusively governed by the local inertial frame of reference.

We still have to consider the realistic general case in which no force-free particles are available at all. How is the inertial spatiotemporal framework to be determined in that case? As preparation to the answer to this question, it is worth noting that in the case of Tait's problem in the general case of n point particles, the number of unknowns to be determined is $1 + 6(n - 2) = 6n - 11$ (giving our 7 for $n = 3$). On the other hand, each snapshot of n particles yields $3n - 6$ independent mutual separations or $3n - 7$ useful bits of information (since the time of the snapshot is unknown). Thus, two snapshots can only yield $6n - 14$ data, while $6n - 11$ are needed to determine the inertial spatiotemporal framework and, from it, the dynamically relevant quantities. Two snapshots are therefore never enough information but, if n is large, three are comfortably more than enough. The reason why two snapshots always fail to yield enough information is that, in Newtonian terms, they contain no data at all on the change of the *orientation* of the system of n particles as a whole in absolute space.

This fundamental fact was made the point of departure of a very interesting analysis of the problem of absolute *vs* relative motion made by Poincaré in his *La Science et l'Hypothèse* in 1902 (Poincaré 1902; 1905, 75–78 and 118). Before considering this, it is worth mentioning that unfortunately Poincaré never, so far as I know, published a single comprehensive study of the problem of determining the complete spatiotemporal framework of dynamics from observable relative quantities. He considered the temporal and spatial problems separately (the former in his "*Mèsure du temps*" in (Poincaré 1898) and the latter in 1902). Both studies were rather qualitative in nature, and both attracted much less attention than they might otherwise have done on account of the creation in 1905 of the special theory of relativity. This then attracted most of the serious attention of scientists concerned with foundational problems and also introduced a host of new issues. This was unfortunate, since a solid authoritative study by Poincaré, of which he was undoubtedly capable, would have become an important landmark in the absolute *vs* relative debate. As it is, his work has very largely passed unnoticed (in part, at least, because Einstein did not notice it, as we shall see).

In his *La Science et l'Hypothèse*, Poincaré asked what if anything was 'wrong' with Newton's use of absolute rather than relative quantities in the foundations of dynamics. Instead of asking the epistemological question—how do we find the absolute quantities given the relative quantities?—Poincaré posed a very interesting question, which was this: If, in the case of the n- body problem of celestial dynamics, one has access to only *relational* initial data (which will be the mutual separations r_{ij} of bodies and their various derivatives $\dot{r}_{ij}, \ddot{r}_{ij}, \ldots$, with respect to the time t (Poincaré assumed t known for the purposes of his discussion)), what *initial data* must be specified if one is to be able to predict the observable future evolution of the system uniquely? Since the ability to predict the future is the acid test of dynamical theory, Poincaré's question could not be better designed to cast much needed light on the role of absolute and relative quantities in dynamics.

Poincaré then noted that if, like the relationists, one believed the relative quantities were truly fundamental and all that counted, one might then suppose that (given known masses of the bodies and under the assumption that they were moving in accordance with Newton's laws, including the law of universal gravitation) knowledge of the r_{ij} at one instant together with the rates of change of these r_{ij}, i.e., the $\dot{r}_{ij}$'s, would be sufficient to determine the future uniquely. However, he then drew attention to the fact with which we are already familiar from Tait's analysis of the inertial case, namely, that even in that simplest of cases two snapshots are not sufficient to determine the absolute quantities, which, as Poincaré pointed out, are needed to make dynamical calculations. (The initial-value problem of celestial mechanics is well posed if, in addition to the masses and specification of the law of interaction, one is given initial positions and initial velocities in *absolute space*.) The situation is no different if interactions occur. In Poincaré's view, this failure of the initial-value problem if one is given only relative quantities is the clearest indication that dynamics involves something more than just relations of bodies among themselves—and that 'something more' is what Newton called absolute space.

It is important to realize, as Poincaré was careful to emphasize, that it is perfectly possible to express the entire content of Newtonian mechanics in purely relational terms. However, the resulting equations, unlike Newton's equations, which contain at the highest *second* derivatives with respect to the time, must contain at least some *third* derivatives. Although he did not explicitly mention him by name, Poincaré almost certainly had in mind here Lagrange's famous study of the three-body problem of celestial mechanics made in 1772 (Lagrange 1772). Lagrange (1772) had assumed the validity of Newton's equations in absolute space and, in an outstanding piece of work, had then proceeded to find equations that govern the variation in time of the *sides of the triangle* formed by the three particles, i.e., precisely the r_{ij}'s for this problem. Lagrange had found three equations, each containing the r_{ij}'s and their derivatives symmetrically and all containing first, second, and *third* derivatives of the r_{ij} with respect to the time. He was also able to show that two of the equations could be integrated once, giving two equations of the form

$$F_1(r_{ij}, \dot{r}_{ij}, \ddot{r}_{ij}) = E, \tag{3}$$

$$F_2(r_{ij}, \dot{r}_{ij}, \ddot{r}_{ij}) = M^2, \tag{4}$$

where E is the total energy of the system and M^2 is the square of the total angular momentum of the system (both in the center-of-mass system). These equations show very graphically that whereas the fundamental dynamical quantities such as energy and angular momentum are functions of the coordinates and their *first* time derivatives *in absolute space*, the expressions for the same quantities in relative quantities also necessarily contain the *second* derivatives.

Poincaré considered this a decidedly mysterious and unsatisfying feature of Newtonian mechanics and felt that it was the only thing one could fault in the Newtonian scheme. He felt, repugnant though this state of affairs was to a philosophically minded person, that one still had to accept it as a fact. He was however prepared to speculate as to how things might be in an ideal world, and this led him to a very interesting speculation as to the form that the relativity principle might have taken.

He noted that the ordinary Galilean relativity principle of classical mechanics had very interesting consequences for the initial data that had to be specified in mechanics. An n-particle system requires formally the specification of $3n$ initial positions and $3n$ initial velocities in absolute space. However, because of the fundamental symmetries of classical mechanics, it is sufficient to specify these quantities with respect to the center of mass of the system. This reduces the number of data that need to be given by 6. In addition, the initial orientation of the system in absolute space has no physical significance, so three more data are redundant. However, essentially that is as far as the reduction to relative quantities can go. It remains crucially important to know at the initial instant how the orientation of the system as a whole in absolute space is changing. This cannot be obtained from purely relative quantities and is the reason why third derivatives of the r_{ij} occur in one of Lagrange's equations.

Such considerations then led Poincaré to comment that "for the mind to be fully satisfied" the law of relativity would have to be formulated in such a way that the initial-value problem of dynamics would hold for a *completely relational* specification of the initial data. One should not be left with the curious absolute-relative mixture just described.

This analysis and suggestion of Poincaré are both extremely valuable. They show that the problem with Newtonian dynamics is not that it cannot be cast into relational form—Lagrange's work is the clearest demonstration of the incorrectness of that belief (which is actually quite widely held).[9] The problem is that when Newtonian theory is recast in a relational (or generally covariant) form it turns out to be *less predictive* than one would like it to be. In addition, Poincaré's analysis also shows what a Machian theory, expressed solely in relative quantities as Mach required, must achieve if it is to represent any improvement on Newtonian theory: It must be able to predict the future uniquely given only r_{ij} and $\dot{r}_{ij}$ at an initial instant. Mach's critique of Newtonian mechanics was unfortunately couched in rather vague terms and the same goes for his proposal for a relational alternative. Poincaré's analysis provides a most welcome clarification and sharpening of the issues involved.

It should be mentioned that all the Machian models of the Hofmann-Reissner-Schrödinger type together with the alternative (intrinsic) type considered by Bertotti and myself meet the requirement of the relativity principle in the stronger form as formulated by Poincaré. It is also the case that the special set of *Newtonian* solutions of an *n*-body universe for which the total angular momentum in the center-of-mass system vanishes are described by equations of a form different from those that hold in the general case. In this special case, the constants E and M^2 disappear from the right-hand sides of Eqs. (3) and (4) and the third derivative also disappears from Lagrange's third equation. Therefore, the corresponding set of equations for this special case satisfy Poincaré's requirement. Indeed, it is a very interesting fact that when Newton's equations are expressed in a generally covariant form (as Lagrange in effect did, using quantities completely independent of all coordinate systems), the complete set of possible solutions breaks up into distinct classes corresponding to the general case with both $E, M^2 \neq 0$ and the various special cases with either one or both of E and M^2 equal to zero. The most interesting special case

$$E = 0, M^2 = 0 \qquad (5)$$

arises very naturally from the intrinsic Machian dynamics developed by Bertotti and myself and referred to in the previous section.

In fact, such a situation was foreseen to quite an extent by Poincaré, who pointed out that, when one is considering the complete universe, it is appropriate to consider

---

9   Lagrange's work does in fact represent the complete solution (for the three-body case) of the problem that Newton posed in the Scholium: Given relative observations, how can one find the absolute quantities? First, Lagrange found equations that govern the evolution of the sides of the triangle. Second, he showed how, once these equations for the sides of the triangle had been solved, one could find the position of the triangle in absolute space (the position of its center of mass and—a much greater problem—its orientation) by quadrature (i.e., by straightforward integration of functions known from the solution of the problem for the sides). A good account of all this is given by Dziobek (Dziobek 1888, 1892). It is somewhat ironic that Lagrange was evidently much more interested in practical problems of celestial mechanics than Newton's Scholium problem and did his work at a time when absolute space had ceased to be a problematic issue. Its importance for the Scholium problem was not noted and escaped Neumann, Lange, and Mach. It is truly a great pity that Poincaré did not flesh out his very perceptive remarks in *La Science et l'Hypothèse* and draw explicit attention to Lagrange's work and its bearing on the Scholium problem.

these various different cases as actually corresponding to fundamentally different dynamical laws of the universe. An important point to note is that if an n- body universe as a whole does satisfy the condition (5) isolated subsystems of it can still perfectly well have nonvanishing values of their energy and angular momentum. They would then appear to be governed by perfectly standard Newtonian dynamics, even though the universe as a whole is governed by a more powerful and more predictive dynamics. This is the reason why Bertotti and I were able to recover Newtonian behavior exactly for local observations. It may also be mentioned that the formalism of intrinsic dynamics is completely general and is not restricted to nonrelativistic mechanics. Unlike the Hofmann-Reissner-Schrödinger approach, it can readily be applied to field theory and even to dynamic geometry. Indeed, it turns out that general relativity is itself of the general type of intrinsic theories, and this is the reason why Bertotti and I have concluded that it is actually perfectly Machian (Barbour 1995).

Let me now go back to Poincaré's earlier paper of 1898 on the topic of time. This paper has received significantly more attention than the analysis of the absolute *vs* relative question in *La Science et l'Hypothèse*, but its Machian implications have nevertheless been completely missed.

Poincaré noted that in recent years there had been considerable discussion of the problem of measuring time. What does it mean to say that a second today has the same *duration* as a second tomorrow? What criterion is to be used to choose the unit of time and identify clocks? Poincaré noted that these questions had become especially topical and acute for the astronomers, who had been finding anomalies in the observed motion of the Moon, one possible explanation of which could be irregularities in the rotation rate of the Earth. (This has since been confirmed. It is due to tidal effects of the Moon.) Since for millennia the rotation of the Earth had constituted the sole reliable clock for use in astronomy, this placed the astronomers in a serious quandary.

Poincaré then proceeded to outline the solution to which the astronomers were moving. Their point of departure was that Newtonian theory was in fact correct, namely, that there did exist a frame of reference and time for which Newton's laws were correct. The entire problem consisted of finding the invisible frame of reference and time from things that could actually be observed. The only material on which they could work was the motions of the bodies making up the solar system. Fortunately, this could, on account of the immense distance of the stars, be treated as an effectively isolated dynamical system. However, in contrast to the gedanken experiments considered by Lange and Tait, the bodies of the solar system were certainly not free of forces, since they all interacted with one another through universal gravitation. The astronomers were therefore confronted with the task that Frege a few years earlier had said needed to be solved by Lange.

The solution proposed by the astronomers, and endorsed in principle by Poincaré, was to seek a frame of reference and time in such a way that the observed motions did indeed accord with Newton's laws when referred to the obtained frame of reference and time. This is a rather obvious generalization of the method initiated by Neumann, Lange, and Tait, but, of course, entailed much greater mathematical difficulties on account of the need to take into account interactions. Fortunately for the astronomers, they did not have to start completely from scratch, since excellent approximations to the conjectured Newtonian frame of reference and time already existed.

A very significant difference of this astronomical procedure from the Tait-Lange procedure is that in the latter time and the frame of reference can in principle be found from just *three* bodies, but the astronomical procedure entails consideration of *all* the dynamically significant bodies in the solar system. If accuracy adequate for astronomical purposes is to be achieved, it is in principle necessary to take into account even relatively small asteroids. This means that effectively the only clock available to the astronomers is the complete solar system.

About forty years after Poincaré wrote his 1898 paper, the astronomers did indeed go over to such a definition of time (which by then had to take into account small relativistic corrections as well). It was initially called *Newtonian* time, but is now known as *ephemeris time* (Clemence 1957). A rather beautiful feature of ephemeris time is that it is actually a weighted average of all the dynamically significant motions of the bodies in the solar system in its center-of-mass inertial frame. Were the solar system to consist of a system of point particles, the expression for the infinitesimal increment of ephemeris time would be given as follows. Let the position of particle $i, i = 1, ..., N$, at one instant of time be given by x_i and at a slightly later instant by $x + dx_i$, the positions being measured in the inertial frame of reference. Then the increment dt of ephemeris time is given by

$$dt = \frac{k\sqrt{\sum m_i dx_i \cdot dx_i}}{\sqrt{E - V}},$$

where E is the total energy of the system, V is the instantaneous potential energy, and k is a constant.

Note also that but for the fortunate fact that the solar system is almost perfectly isolated an accurate determination of time would require the summation in the above expression to be extended to the complete universe. Ultimately, the only reliable clock is the complete universe!

I have gone into this detail about ephemeris time (the theory of which was outlined rather more sketchily by Poincaré in his 1898 paper) because, first, it rectifies the shortcoming of Newton's treatment in the Scholium, and, second, it has passed almost without notice for over a century. This remarkable state of affairs has arisen because a quite different aspect of time—the problem of defining simultaneity at spatially separated points—came to dominate discussions once Einstein had created the special theory of relativity.

As it happens, Poincaré also mentioned this problem of simultaneity in his 1898 paper and noted that in some respects it was a more immediate problem than that of defining duration but that hitherto it had hardly been noted. It is on account of this remarkable early anticipation of the key problem of special relativity that Poincaré's

1898 paper is mentioned relatively often today, but I am not aware of any discussion of the duration problem even though it is certainly very fundamental.

The reason for this lack of notice is, I suspect, to be traced to the immense influence of Einstein, and this is an appropriate point at which to consider how his own work on special and, more particularly, general relativity relates to the topics discussed in this and the previous section. At the end of the previous section, I noted that Einstein seems to have had not much accurate knowledge of the work done by Hofmann and Reissner and to have taken little interest in it. The same comment is true of the epistemological work reported in this section. So far as I can judge from his published papers and the correspondence I have examined, *none* of the work described in this section made any significant impact on him. In the remainder of this section, I shall substantiate this claim; in the following section, I shall try to establish why Einstein seems to have been remarkable insensitive to what might be called the classical issues in the absolute *vs* relative debate.

Let me start with the topic last discussed—the definition of duration and a clock. To the best of my knowledge, this question was never once discussed by Einstein (in striking contrast to his numerous discussions of the definition of simultaneity). Throughout his entire work on relativity, Einstein simply assumed, as a phenomenological fact, that clocks (like rods to measure distance) exist and can be used to measure the fundamental interval ds of relativity theory.

Already in the 1920s (Einstein 1923) and then again in the *Autobiographical Notes* (Einstein 1949) written towards the end of his life, Einstein noted that his consistently phenomenological treatment of rods and clocks, which made it necessary to introduce them formally as separate entities in the framework of his theory, was a logical defect of the theory that ought to be eliminated. Rods and clocks should be constructed explicitly from the truly fundamental physical quantities in the theory— preferably fields alone, but, if particles could not be eliminated as fundamental entities, then from fields and particles together.

From the way Einstein wrote about this, I get the strong impression that he did not think anything particularly interesting would come out of this exercise. However, I think it can be argued that he was actually insensitive to a fundamental issue. This is reflected in the fact that he invariably described a clock as being realized through some strictly periodic process. However, this immediately begs the question that Neumann set out to answer with his inertial clock: How can one say of a single motion that it is uniform? I have not seen anything in Einstein's writings which shows an awareness of the fact that a measure of time can be extracted only from the totality of the motions within a dynamically isolated system and that, if it is to give true readings, a clock must somehow 'lock onto' and reflect the inertial spatiotemporal framework. I shall return to this.

A similar rather perfunctory attitude characterizes Einstein's references to the determination of inertial frames of reference. In his published papers, he never once referred to the procedures of Lange or Tait or drew attention to the difficulties that Newton 250 years earlier had already recognized so clearly. Generally, he simply says that an inertial frame of reference is one in which a force-free particle moves rectilinearly and uniformly, giving no indication at all how such a frame of reference is to be found. Very characteristic of his approach is the following passage written in the early 1920s (Einstein 1923):

> In classical mechanics, an inertial system and time are best determined together by means of a suitable formulation of the law of inertia: It is possible to establish a time and give the coordinate system a state of motion (inertial system) such that relative to it material points not subject to the action of forces do not undergo acceleration.

A little later, Einstein noted that such a definition had a logical weakness "since we have no criterion to establish whether a material point is free of forces or not; therefore the concept of an 'inertial system' remains to a certain degree problematic" This passage (with its incorrect conclusion) suggests to me that Einstein never gave much serious thought to the issue of the determination of inertial frames of reference.

Confirmation that this is the case can be found in some remarkably interesting late correspondence between Einstein and his old friend Max von Laue. Among the leading relativists, von Laue is the only one who mentions the work of Lange. In 1948 (von Laue 1948), he wrote an appreciation of Lange and his work, in which he stated: "Ludwig Lange progressed so far in the solution of the problem of the physical frame of reference, which Copernicus, Kepler, and Newton did not completely solve, that only Einstein's theory of relativity added something new." In 1951, he published a new edition of his book on the theory of relativity (von Laue 1955), which opens with the definition of the inertial time scale and inertial system as given by Lange, calling it a great achievement. Not surprisingly, he sent Einstein a copy of the new edition. In response,[10] Einstein commented:

> I was surprised that you find Lange's treatment of the inertial system significant. It merely says that there exists a coordinate system (with time) in which 'uninfluenced' material points move rectilinearly and uniformly. This is Newton's '*absolute* space.' It is not absolute because no transformations exist that conserve the law of inertia but because it must be prescribed in order to give the concept of acceleration a clear meaning.

In the same letter, Einstein remarked: "Provided one considers action-at-a-distance forces that decrease with r sufficiently rapidly, the word 'uninfluenced' has a direct meaning." This comment implies, like the one made in the 1920s, that inertial frames of reference can only be determined if force-free bodies are available. As we have noted earlier, this is simply not true, though unfortunately the correct state of affairs had never been clearly stated in the literature (see footnote 9). However, I am convinced that had Einstein really made a serious effect to find out the truth he would certainly have succeeded. What we must try to establish (in the next section) is why he was insensitive to the issue.

To conclude this section, it is worth mentioning a connection between Einstein's lack of concern about the definition of the inertial frame of reference and his belief

---

10   Einstein to Max von Laue, 17 January 1952 (EA 16–168).

that a Machian theory of motion should provide some kind of cosmic derivation of inertial mass (rather than a cosmic derivation of the law of inertia). It is a very striking fact that the expressions 'relativity of position' and especially 'relativity of velocity' (the truly fundamental problem of the absolute *vs* relative question) hardly ever occur in Einstein's writings, whereas he frequently mentions the relativity of acceleration. In fact, almost the only case in which relativity of velocity occurs is in the following passage (Einstein 1913b), in which Einstein is discussing his first attempt at a general theory of relativity undertaken with Grossmann in 1913:

> The theory sketched here overcomes an epistemological defect that attaches not only to the original theory of relativity, but also to Galilean mechanics, and that was especially stressed by E. Mach. It is obvious that one cannot ascribe an absolute meaning to the concept of acceleration of a material point, no more so than one can ascribe it to the concept of velocity. Acceleration can only be defined as relative acceleration of a point with respect to other bodies. This circumstance makes it seem senseless to simply ascribe to a body a resistance to an acceleration (inertial resistance of the body in the sense of classical mechanics); instead, it will have to be demanded that the occurrence of an inertial resistance be linked to the relative acceleration of the body under consideration with respect to other bodies. It must be demanded that the inertial resistance of a body could be increased by having unaccelerated inertial masses arranged in its vicinity; and this increase of the inertial resistance must disappear again if these masses accelerate along with the body.

Einstein then proceeds to claim that the 1913 theory does indeed contain an effect of the desired kind.

The above passage is remarkable on two scores. First, there is the already noted incorrect claim that Mach was concerned about the definition of inertial resistance. Second, Einstein states that both velocity and acceleration are relative and presents this as a major problem. However, he never once in his papers attempted to show how general relativity attacked the fundamental kinematic problem of the relativity of velocity. The idea that inertial resistance, like acceleration, must be relative, is expressed very prominently in Einstein's writings from 1912 through to about 1922. However, Einstein never once attempted to show how such an idea (and still less the even more fundamental relativity of motion alluded to above) was implemented in the basic kinematic and dynamic structure of the theory he was constructing.

This is in very striking contrast to the epistemological work of Neumann, Tait, Lange, and Poincaré and the manifestly relational proposals of Hofmann and Reissner. All of these authors attacked the relativity of motion head on. What are the reasons for Einstein's conspicuous failure to follow their example?

4. EINSTEIN'S PRIORITIES WHEN CREATING GENERAL RELATIVITY

Let me now attempt to begin to answer the question with which the previous section ended by considering the evidence that can be gleaned from Einstein's early papers and correspondence. It is quite clear that by the time he had left school and commenced university studies Einstein had set himself a supremely ambitious task. He was going to attack and make an extremely serious attempt to solve the great topical problems of physics. In later years, he may have liked to cultivate the image of a somewhat indolent student, but a very different picture emerges from his correspondence. There were certain fundamental issues that he followed avidly, above all anything related to Maxwellian field theory and also anything that could provide evidence for the existence of atoms. These were the burning topics of the time, and he followed them closely.

It seems to me that with regard to the absolute *vs* relative debate, the situation was somewhat different. There is no doubt that it was a topic of genuine widespread interest; Poincaré's inclusion of it in *La Science et l'Hypothèse* is clear evidence of that. However, it was a topic with relatively few (but by no means none at all) opportunities for decisive experimental tests;[11] both Mach and Poincaré tended to treat the topic in a rather passive manner, drawing attention to problems but without proposing an energetic programme for their resolution. For an ambitious young man like Einstein, with a strong awareness of the importance of experiment and clearly determined to make a name for himself as quickly as he could, the problems of electromagnetism and atomism must surely have appeared to offer far better prospects. This could well explain why Einstein's imagination was clearly caught, through his reading of Mach's *Mechanik* around 1898 (CPAE 1), by the great issue of absolute space without this leading him on to a more detailed consideration of the details. Whatever the reason, in the period 1898 to 1905 (and, indeed, up to the end of his life) Einstein had the opportunity to go into the details and really come to grips with the central problems of defining time, clocks, and motion. He did not or, at least, not directly (except, of course, with regard to simultaneity).

There are, I believe, at least three clearly identifiable reasons for Einstein's *indirect* attack on the problem of absolute space. All three are important and interrelated and already played a decisive role in his creation of the *special* theory of relativity.

The first, and surely the most important, is that the principle of Galilean relativity suggested to Einstein an indirect but extremely effective way of making absolute space redundant in physics. He saw the success of special relativity as an important first step in that direction and then attempted, with great consistency, to generalize the relativity principle to the maximum extent possible. He believed that this would make absolute space completely redundant as a concept in physics.

The second reason for Einstein's indirect strategy is to be found in the phenomenological concept of the rigid body and the important work done by Helmholtz on the empirical foundations of geometry. The phenomenological rigid body played a vitally important role in both special and general relativity but, as we shall see, made it extremely difficult to address directly the relativity of motion in any obviously Machian manner.

The third reason for Einstein's indirect approach may seem somewhat surprising at the first glance—it was Planck's discovery of the quantum of action in 1900. We shall see that this discovery greatly diminished Einstein's confidence in the possibil-

---

11   For a discussion of early experiments, see (Norton 1995).

ity of finding quickly any explicit and detailed dynamical equations that could be taken to describe the behavior of particles and fields at the fundamental microscopic level. Instead, he consciously sought general principles such as those established in phenomenological thermodynamics by means of which he could obtain constraints on the behavior of matter. This strengthened his faith in the value of the relativity principle and his indirect approach to implementation of Mach's ideas. It also persuaded him that it would be useless to attempt to construct a microscopic theory of rods and clocks.

Let me now expand on these three points in more detail.

It seems to me entirely possible that an overall strategy for eliminating absolute space from physics started to take shape in Einstein's mind very soon after he had read Mach's *Mechanik* around 1898. The basic idea arose from consideration of a problem that Mach had not considered at all: electrodynamics. Much of the later development of relativity theory is clearly prefigured in a comment of Einstein to his future wife in a letter written in August 1899 (CPAE 1):

> I am more and more convinced that the electrodynamics of moving bodies, as presented today, is not correct, and that it should be possible to present it in a simpler way. The introduction of the term "aether" into the theories of electricity led to the notion of a medium of whose motion one can speak without being able, I believe, to associate a physical meaning with this statement.

This train of thought then led on to the clear formulation in 1905 of the relativity principle, in accordance with which uniform motion relative to the supposed aether is completely undetectable. As Einstein (Einstein 1905) famously remarked, this then meant that "the introduction of a 'luminiferous aether' will prove to be superfluous". Moreover, by the end of the 19th century, the aether had more or less come to be identified with absolute space, a rigid substrate that besides being the carrier of electromagnetic excitations also served as the ultimate standard of rest for all bodies in the universe. In his famous 1895 paper on electrodynamics with which Einstein was certainly familiar, Lorentz said of the aether (Lorentz 1895, 4): "When for brevity I say that the aether is at rest this means merely that no part of this medium is displaced relative to any other part and that all observable motions of the heavenly bodies are relative motions with respect to the aether."[12]

Having banished the aether from the foundations of physics, Einstein felt that he had made an important first step on the way to the complete elimination of the notion of absolute space. Einstein felt that a thing could only be said to exist if it had observable effects. The 1905 relativity principle showed that *uniform* motion relative to the putative aether (or absolute space) had no observable consequences. If the relativity principle could be extended further, to all accelerated motions, then all residual arguments for the existence of absolute space would be eliminated. Einstein's 1933 Gibson lecture (Einstein 1933) suggests rather strongly that this train of thought had

---

12 It is worth noting that this is a remarkably naive concept of motion compared with the subtlety of Lange's construction.

taken shape in Einstein's mind already by 1905, but that at that stage he was unable to take the idea any further. It was only in autumn 1907 (Einstein 1907) that the potential of what he later called the equivalence principle struck him; for it suggested that the restricted principle of relativity could be extended from uniform motions to *uniformly accelerated* motions as well. This then opened up the prospect of extension of the relativity principle even further—to all motions whatsoever.

This logic is spelled out very clearly in the Gibson lecture, from which the following quotation is taken:

> After the special theory of relativity had shown the equivalence for formulating the laws of nature of all so-called inertial systems (1905) the question of whether a more general equivalence of coordinate systems existed was an obvious one. In other words, if one can only attach a relative meaning to the concept of velocity, should one nevertheless maintain the concept of acceleration as an absolute one? From the purely kinematic point of view the relativity of any and every sort of motion was indubitable; from the physical point of view, however, the inertial system seemed to have a special importance which made the use of other moving systems of coordinates appear artificial.

> I was, of course, familiar with Mach's idea that inertia might not represent a resistance to acceleration as such, so much as a resistance to acceleration relative to the mass of all the other bodies in the world. This idea fascinated me; but it did not provide a basis for a new theory.

Note how Einstein insists that the idea of a more general equivalence of coordinate systems "was an obvious one". It certainly was not so to his contemporaries. If there is one aspect of Einstein's work on gravitation that most clearly distinguished him from them all, it was his insistence on the need to generalize the relativity principle and on the equivalence principle as the means to do so. All of the truly original steps which eventually led Einstein to the general theory of relativity sprang from this conviction. It is certainly the case that Mach's vehement opposition to Newton's absolute space as a nonexistent monstrosity was completely shared by Einstein and served as the main stimulus to the creation of general relativity.

However, it is important to note that the two men disliked absolute space for rather different reasons. Mach tended very much to concentrate on the things in the world that could be directly observed—bodies—and on the relationships between them, which were expressed in the first place by the mutual separations between them. This gut instinct is expressed very clearly in Mach's famous comment (Mach 1960): "The world is not *twice* given, with an earth at rest and an earth in motion, but only *once*, with its *relative* motions, along determinable." Given Einstein's great enthusiasm for Mach, it is remarkably difficult to find evidence which shows unambiguously that Einstein understood what Mach really wanted to do: base mechanics solely on the relative separations of bodies. As I have already noted, many of Einstein's remarks about Mach actually represent a distortion of the older man's thought. The passage from 1918 quoted earlier is a clear expression of what Mach wanted to do ("introducing in the laws of classical mechanics only distances of material points from each other"), but there is no direct attribution to Mach. The solitary direct attribution I have found is in a very late letter to Pirani,[13] in which Einstein says [my translation]:

There is much talk of Mach's Principle. It is, however, not easy to associate a clear notion with it. Mach's stimulus was this. It is unbearable [unerträglich] that space (or the inertial system) influences all ponderable things by determining the inertial behaviour without the ponderable things exerting a determining reaction back on space. Mach rediscovered what Leibniz and Huygens had correctly faulted in Newton's theory. He sought to eliminate this evil by attempting to abolish space and replace it by the relative inertia of the ponderable bodies with respect to each other. Space should be replaced by the distances between the bodies taken in pairs (with these distances as independent concepts). This evidently did not work, quite apart from the fact that the time with its absolute nature remained.

Several comments can be made about this opening paragraph of Einstein's letter.

First, there is Einstein's admission that it is not easy to associate a clear notion with Mach's Principle. This, however, is what the criterion of Poincaré considered earlier does do.

Second, the idea that something should not be able to influence another thing without suffering a back reaction on itself is not, so far as I know, to be found anywhere in Mach's writings. It is, however, an idea that Einstein himself frequently advanced from around 1914, mainly I think as a result of his work on Nordström's theory, in which the propagation of light is governed by an absolute structure and is not subject to the influence of gravitation. For example, in 1914 (Einstein 1914) he wrote: "It seems to me unbelievable that the course of any process (e.g., that of the propagation of light in a vacuum) could be conceived of as independent of all other events in the world."

Next, it should be noted that even the account of what Mach proposed is not strictly correct, since Mach did not propose to eliminate the relations of Euclidean space and regard the distances between bodies as completely independent. It should be noted that for n bodies in Euclidean space there are $n(n-1)/2$ mutual separations, of which only $3n-6$ are independent (for $n \geq 3$). In his proposal for a new law of inertia, Mach did not include any suggestion that this basic fact of three-dimensional Euclidean geometry should be relaxed. However, in a very early paper he had speculated (Mach 1872, 25; 1911, 51–53) that such a relaxation might occur in the interior of atoms and play an important role in the formation of spectral lines. He later explicitly withdrew (Mach 1911, 94) this theoretical speculation, which hinged on a putative representation of atoms and molecules in Euclidean spaces of more than three dimensions. Einstein read Mach's booklet on the *Conservation of Energy*, where the idea is discussed, in 1909,[14] so it is possible that in his old age he muddled it up with Mach's proposals for inertia.

Finally, we note in Einstein's "this evidently did not work" an echo of the comment in 1918 that the proposal to found mechanics solely on relative separations had not proved feasible. However, the papers of Hofmann, Reissner, and Schrödinger had shown the approach to be perfectly feasible. With the possible exception of Reiss-

13   Einstein to Felix Pirani, 1954 (EA 17–447).
14   As we know from a letter Einstein wrote to Mach in August 1909 (CPAE 5E, Doc. 174).

ner's work, Einstein had read these papers. It therefore seems that they made very little impression on him; he was certainly confused about their content, since all demonstrated the Machian approach *was* feasible. Had Einstein from the beginning shared Mach's gut instinct that only relative separations count and that the central problem was to reflect this in the foundations of mechanics, then surely he might have been expected to have taken more notice of what had been achieved.

The fact that he did not suggests that Einstein's objection to absolute space had a somewhat different psychological origin. For this conclusion, there is much evidence. Rather than consider objects in space, Einstein was evidently wont to contemplate the notion of empty space by itself. Evidence for this can be found, for example, in (Einstein 1921). Given the perfect uniformity of space, Einstein then found it an affront to the principle of sufficient reason that such a featureless thing should contain within it *distinguished frames of reference* for the formulation of the laws of nature. Numerous arguments on such lines can be found in Einstein's papers from 1913 up to the *Autobiographical Notes*. They always invoke the point mentioned in the 1933 Gibson lecture—that "from the purely kinematic point of view the relativity of any and every sort of motion [in space] was undubitable." Thus, the only way to create a theory perfectly in accord with the principle of sufficient reason was through generalization of the relativity principle to the absolutely greatest extent possible.

The difference between Mach and Einstein can be summarized very simply: Mach wanted to eliminate coordinate systems entirely, Einstein wanted to show that all coordinate systems were equally valid. Given the tremendous success of the special theory of relativity, which established the equivalence of all coordinate systems in uniform motion relative to each other, and the promise offered by the equivalence principle for extension to accelerated motion, it is very easy to see why Einstein became so totally committed to his approach and took virtually no notice of the alternative.

The question of whether one (and then which one) or both of these two approaches are valid is very complex. It is a subject that I cannot follow further in this paper, in which I have set myself the more modest task of identifying some characteristic differences between the approaches of Mach (and his contemporaries) and Einstein, finding the reasons for Einstein's choices, and placing his work in the perspective of other work on the absolute *vs* relative debate. Let me just say that, in my opinion, Mach's approach (augmented by Poincaré's analysis) is deeper and more consistent than Einstein's but that nevertheless Einstein's theory, when properly analyzed as a dynamical theory, does perfectly implement Mach's ideas. However, the reason for this has more to do with deep intrinsic properties of the absolute differential calculus, which Einstein took over 'ready made' from the mathematicians, than with Einstein's covariance arguments. All this will be spelled out in my forthcoming monograph (Barbour, in preparation). See also my *Notes Added in Proof* at the end of this article.

Because it ties in very well with Einstein's conception of space that we have just been considering, let me now turn to the role played by the rigid body and Helmholtz's work on the empirical foundation of geometry (Helmholtz 1868) in Einstein's development of both special and general relativity. Helmholtz's study was made about a decade after Riemann's famous habilitation lecture of 1854 (Riemann 1867). Initially Helmholtz was unaware of Riemann's work, which was not published until 1867, and found that he had largely rediscovered already known results. However, there was one respect in which Helmholtz went significantly beyond Riemann. This concerned the hypothesis that Riemann had made for the form of the line element.

Riemann had assumed, more or less on grounds of simplicity and to match Pythagoras's theorem, that the fundamental line element ds of his generalized geometry should be the square of a *quadratic* form in the coordinate differences. He noted, however, that *a priori* one could not rule out, say, taking the fourth root of a quartic form. In contrast, Helmholtz considered the empirical realization of geometry by rigid bodies and congruence relations between them. If such bodies are to be brought to congruence, they must satisfy certain conditions of mobility and remain congruent in different positions and different orientations. Their congruence must also be independent of the paths by which they are brought to congruence. Helmholtz was able to show that if these conditions are to be met then the quadratic form of the line element adopted by Riemann as a simplicity hypothesis is indeed uniquely distinguished. This established a very beautiful connection between empirical geometry based on physical measuring rods and a particular mathematical formalism. Helmholtz concluded his important paper with the following words:

> the whole possibility of the system of our space measurements ... depends on the existence of natural bodies that correspond sufficiently closely to the concept of rigid bodies that we have set up. The fact that congruence is independent of position, of the direction of the objects brought to congruence, and of the way in which they have been brought to each other—that is the basis of the measurability of space.

The influence of Helmholtz's study is manifest throughout Einstein's entire relativity opus. In a newspaper article published in 1926, Einstein (Einstein 1926) described the practical geometry of the experimental physicist in which "rigid bodies with marks made on them realize, provided certain precautions are taken, the geometrical concept of interval" and said [my italics]:

> Then the geometrical "interval" corresponds to a definite object of nature, and thus all the propositions of geometry acquire the nature of assertions about real bodies. This point of view was particularly clearly expressed by Helmholtz; *one may add that without this viewpoint it would have been practically impossible to arrive at the theory of relativity.*

The Helmholtzian conception was crucial for two reasons in particular: It provided a definite framework in which Einstein could comprehend length contraction and simultaneously gave a method for *position determination*. It also gave Einstein a way of 'making space visible' that perhaps made him less concerned than Mach about the problems of position determination. Taken together, these factors led Einstein to a method of position determination that appears to be decidedly un-Machian.

Indeed, a complete method of position coordination appeared already in the famous Kinematical Part of his 1905 paper (Einstein 1905). Einstein opens that section as follows: "Let us take a system of coordinates in which the equations of Newtonian mechanics hold good." Thus, he simply *presupposes* the outcome of a Tait-Lange procedure for finding an inertial frame of reference. He then continues:

> If a material point is at rest relatively to this system of coordinates, its position can be defined relatively thereto by the employment of rigid standards of measurement and the methods of Euclidean geometry, and can be expressed in Cartesian coordinates.

The standards of measurement (Helmholtz's rigid bodies) serve two supremely important purposes. First, they can be imagined to "fill" the whole space of the inertial system; since one can also suppose that the bodies carry marks permanently scratched on them, other bodies can be unambiguously located by means of these marks. Space has been made visible. Then, second, comes the really great convenience of such rigid bodies—intervals defined by the marks on them satisfy the congruence conditions required in Helmholtz's phenomenological foundation of geometry. Thus, the coordinates can be associated with the marks on the rigid bodies in such a way that they simultaneously *give physical distances directly*.

Convenient as all this is, it still does not contain anything that goes beyond Helmholtz's scheme. However, the scheme turned out to be wonderfully adapted to the exigencies of relativity theory and length contraction. Here the important thing is the underlying conception Einstein had of what might be called the true physical nature of rigid bodies (and therefore of measuring rods). He certainly did not think of them as ultimate elements incapable of further explication. On the contrary, Einstein was a convinced atomist and he conceived of a measuring rod as being made up of a definite number of atoms governed by quite definite laws of nature. Provided external circumstances (pressure, temperature, etc.) remained the same, such a system of atoms could be expected to 'settle into' a unique equilibrium configuration. Two such systems constituted by identical collections of atoms would settle into the same configuration and therefore be congruent to each other. Thus, Helmholtz's phenomenological foundation of geometry would have a theoretical underpinning in atomism and the laws governing it.

A vital part in this overall picture was played by the notion of an inertial system coupled with Einstein's formulation of the (restricted) relativity principle, in accordance with which the laws of physics must have the identical form in all inertial systems obtained from each other by a uniform translational motion. Coupled with Einstein's (long tacit but later explicit (Einstein 1923)) atomistic conception, the relativity principle ensured that the identical phenomenological Helmholtzian geometry must be realized in each inertial system. However, it left open the connection between the geometries (and chronometry, which I have not considered here) in different inertial systems. The Helmholtzian scheme had *just* enough flexibility to allow and accommodate those marvellous *bombes surprises* of relativity: length contraction and time dilation. Moreover, the underlying atomistic conception meant that one could still talk about 'the same measuring rod' in two different inertial systems. One merely had to suppose two rods constituted of the same atoms and subject to the same external conditions. Then in their respective inertial systems they would settle into identical configurations, and comparison of these configurations between inertial systems would become epistemologically valid. One would be talking about the 'same things.'

Another extraordinary convenience of this whole conception was that it could be generalized almost unchanged to the framework of the general theory of relativity. It was merely necessary to repeat the entire exercise, not globally, but 'in the small.' Very important here was the presumed existence of local approximations to inertial systems in the neighborhood of every point of spacetime; for the distinguished 'equilibrium' configurations into which rods could be assumed to 'settle' exist only in an inertial system.

Einstein's attitude to his phenomenological treatment of rods and clocks is summarized very clearly in his *Autobiographical Notes*:[15]

> One is struck [by the fact] that the theory (except for the four-dimensional space) introduces two kinds of physical things, i.e., (1) measuring rods and clocks, (2) all other things, e.g., the electro-magnetic field, the material point, etc. This, in a certain sense, is inconsistent; strictly speaking measuring rods and clocks would have to be represented as solutions of the basic equations (objects consisting of moving atomic configurations), not, as it were, as theoretically self-sufficient entities. However, the procedure justifies itself because it was clear from the very beginning that the postulates of the theory are not strong enough to deduce from them sufficiently complete equations for physical events sufficiently free from arbitrariness, in order to base upon such a foundation a theory of measuring rods and clocks. If one did not wish to forego a physical interpretation of the coordinates in general (something which, in itself, would be possible), it was better to permit such inconsistency—with the obligation, however, of eliminating it at a later stage of the theory. But one must not legalize the mentioned sin so far as to imagine that intervals are physical entities of a special type, intrinsically different from other physical variables ("reducing physics to geometry," etc.).

Before commenting on this passage and its bearing on the central issues of the absolute *vs* relative question, let us consider the third factor that I identified as shaping Einstein's overall strategy in the creation of both relativity theories: the quantum. In the above passage, Einstein merely says "it was clear from the very beginning that the postulates of the theory are not strong enough to deduce ... a theory of measuring rods and clocks." However, while this statement is obviously true it is at the same time something of an inversion of the actual historical development. The fact is that Einstein *deliberately*, already in the period 1904/5, chose to develop his ideas on the basis of very general postulates. His reasons for doing so are very well known and were explained by Einstein himself in the *Autobiographical Notes*.

The truth is that Planck's discovery of the quantum of action in 1900 made a tremendous impression on Einstein and quickly persuaded him that some very strange things indeed must be happening at the microscopic level. In particular, he was certain that Maxwell's equations (for which he had the very greatest respect) could not be valid in their totality for microscopic phenomena. This comes out graphically in a letter that Einstein wrote to von Laue in January 1951:[16]

---

15 He had already made very similar comments in (Einstein 1923).
16 Einstein to Max von Laue, January 1951 (EA 16-154).

If one goes through your collection of the verifications of the special theory of relativity, one gets the impression that Maxwell's theory is secure enough to be grasped. But in 1905 I already knew for certain that it yields false fluctuations of the radiation pressure and thus an incorrect Brownian motion of a mirror in a Planck radiation cavity.

It is well known that Einstein's complete conviction that Maxwell's theory could be at best partly right was a decisive factor in his selection, as a postulate in the foundations of his special theory of relativity, of the one minimal part of Maxwellian theory in which he did retain confidence: the light propagation postulate.

More generally, it led him at the same time to formulate consciously the idea of a theory based on principles that wide experience had demonstrated had universal validity. The classic paradigms of such principles were the denials of the possibility of construction of perpetual motion machines of the first and second kind, which played such a fruitful role in the phenomenological thermodynamics that had been created around the middle of the 19th century. The great strength of such theories was that they enabled one to make many important predictions without attempting to find a detailed theory at a fundamental microscopic level. Such a theory Einstein called a *constructive* theory, in contrast to a theory based on principles. In a very clear account of the distinction between the two kinds of theory that he included in a piece that he wrote for *The Times* (Einstein 1919), Einstein said that the theory of relativity was one of the latter kind, which he called *fundamental*.

Let me now conclude by considering some of the consequences of these three aspects of Einstein's overall strategy—the programme to eliminate absolute space by generalization of the relativity principle, the use of Helmholtzian rigid bodies to define position, and the eschewal of constructive theories. Both together and separately, they had the consequence that virtually all the issues that one might have expected to feature prominently in a frontal attack on the absolute *vs* relative question—and that was certainly a very major part of Einstein's undertaking—were actually missing as explicit elements in Einstein's work. It is almost a case of Hamlet without the Prince of Denmark.

One of the biggest problems with Einstein's approach is that distinguished frames of reference figured crucially in his work, but he never explicitly considered their status and origin. For example, in his work on special relativity he would invariably start by assuming the existence of inertial frames of reference and then postulate the existence of laws of nature that had to be expressed relative to these frames of reference. Once this step had been taken, the relativity principle could come into play—the laws of nature must take the same form in all frames of reference related by Lorentz transformations. It is however legitimate to ask what determines the frames of reference: Are there laws of nature that determine them? The whole logic of Einstein's approach made it impossible for him to pose, let alone answer, this question, since the frames of reference had to be there before he could formulate the laws of nature. Einstein bequeathed us an unresolved vicious circle at the heart of his theory.

It should not be thought that the transition to general relativity, in which all frames of reference are purportedly allowed, eliminates this problem. The fact is that the *local* existence of distinguished frames of reference (locally inertial frames) is an absolute prerequisite of the theory, since it is only when such frames exist that Einstein's phenomenological treatment of rods and clocks can be taken over from special relativity. But Einstein never once seems to have considered seriously how the local frames of reference and his rods and clocks could arise from first principles.

Of course, we know that he was at least partly aware of this issue, since he more than once said that one should elaborate a theory of rods and clocks. However, I suspect that Einstein did not quite appreciate the true nature of the problem, which is already evident from Neumann's theory of the inertial clock. This showed clearly that there is a *twofold* problem. First, one must have access to an inertial system; second, one must track some physical object or process whose behavior stands in a known one-to-one relationship to the framework defined by the inertial system. In the simplest case of Neumann's inertial clock, this is done directly by the motion of a force-free particle. Of these two problems, the first is by far the most difficult; indeed, the second problem is effectively solved together with the first, the very posing of which is a decidedly subtle matter (as the long and painful elaboration of the problem demonstrates).

If we now examine Einstein's relatively terse comments about rods and clocks, rather strong evidence emerges that his understanding of the issue never really advanced beyond the level of Neumann's inertial clock defined in a *known* inertial system. For example, in the same *Autobiographical Notes* from which the earlier quotation about rods and clocks was taken, Einstein refers to a clock as an 'in itself determined periodic process realized by a system of sufficiently small spatial extension' and then shortly afterwards comments:

> The presupposition of the existence (in principle) of (ideal, viz, perfect) measuring rods and clocks is not independent of each other; since a light signal, which is reflected back and forth between the ends of a rigid rod, constitutes an ideal clock, provided that the postulate of the constancy of the light-velocity in vacuum does not lead to contradictions.

From this it is clear that Einstein already presupposed knowledge of the positions of objects constituting his model clocks in an inertial frame of reference—for if the rigid rod is not moving strictly inertially, Einstein's ideal light clock is useless. All the evidence I have examined is consistent with my conclusion that Einstein never grasped this fact and that he did not properly understand the problem posed by determination of the inertial frames of reference. His disparaging remarks to von Laue about Lange's work are strong support for this view.

This is not deny the correctness of Einstein's supposition that the quantum problems made it premature to try and make truly realistic microscopic models of rods and clocks. But what Einstein had in mind was the theory of such things in *a known inertial frame of reference*, whereas the more fundamental problem concerned the origin of the frame. And to address that problem he did not really need such advanced physics. The quantum bogey gave Einstein a valid excuse for not constructing microscopic theories of actual physical clocks, but may have misled him by seeming to locate the problem in the wrong place.

Mention should also be made here of the rather vague way in which Einstein formulated the relativity principle. He invariably simply said that the laws of nature, the form of which he deliberately left vague, must take the same form in all frames of reference. He never attempted anything like the subtly refined formulation proposed by Poincaré based on identification of the kind of initial data needed to predict the future uniquely. A strength of Poincaré's approach is that it avoids the vicious circle inherent in Einstein's approach whereby the status and origin of the necessary distinguished frames (the local existence of which is still needed in general relativity despite the general covariance of that theory) is left open. In Poincaré's approach, the question of the distinguished frames of reference does not arise, since he formulated the initial-value problem deliberately in such a way that they do not enter into it at all. For some reason or other, Einstein never seems to have thought of general relativity as a dynamical theory and Poincaré's comments seem to have made no impression on him.

Finally, we must consider the effect of Einstein's Helmholtzian method of position determination. It was this above all that precluded any directly Machian implementation of relativity of position and velocity after the manner of Hofmann and Reissner. For them, like Mach, position and velocity were determined by the set of distances to all other bodies in the universe. But Einstein was completely and irrevocably tied to local position determination by means of Helmholtzian rigid bodies that 'filled' the space of inertial systems, which Mach insisted must be understood as arising from relations to other matter in the universe, whereas Einstein simply took them as given. Thus, Einstein's technique was doubly un-Machian. In the 1918 paper quoted in Sec. 2, Einstein said that "the historical development" had shown that it was not possible to "dispense with coordinate systems." For 'historical development' we must here understand the foundations of Einstein's own work: coordination by Helmholtzian rigid bodies and relativity transformations between such coordinates. Note also that in the passage cited earlier from his 1913 paper Einstein said:

> It is obvious that one cannot ascribe an absolute meaning to the concept of acceleration of a material point, no more so than one can ascribe it to the concept of velocity. Acceleration can only be defined as relative acceleration of a point with respect to other bodies.

In this passage, Einstein is quite clearly saying that position and velocity are defined relative to other bodies in exactly the same sense as did Mach (and all the other relationists). Yet he did not give any indication how that requirement was implemented in his own theory. He merely pinned his hopes on a *resistance* to acceleration induced by distant matter. These hopes came to nothing—and meanwhile the Machian issues were never directly addressed.

However, general relativity is an immensely rich and sophisticated theory, and the same can be said of the veritable odyssey of its discovery by Einstein. One can find ironies, serendipity, and utter brilliance throughout the entire saga. Just because Einstein's chosen route to the creation of general relativity did not directly address certain central issues of the absolute *vs* relative debate, this does not necessarily mean that his wonderful theory fails to solve them. After all, Newton posed some critical problems in the Scholium at the time he wrote the *Principia*, but some two hundred years passed before they were more or less completely resolved within the framework of Newtonian theory. As already indicated, I am convinced that the problems considered in this paper, which has considered alternative issues that Einstein might have addressed (and his contemporaries, above all Poincaré, did address), are actually resolved within the heart of general relativity by virtue of the exquisite mathematics on which it is based. However, that is quite a long story too and will have to be considered elsewhere (Barbour, in preparation).

NOTES ADDED IN PROOF

Since writing this article, I have continued to research and write about Mach's principle and related issues. This activity has generated much material that relates to the topics discussed in this article and envisaged for (Barbour, in preparation). First, there is my book *The End of Time* (Barbour 1999a), which considers the quantum cosmological implications of a relational treatment of time and motion. Second, the three review papers (Barbour 1997, 1999b, 2001) complement the present paper. Third, collaboration with Edward Anderson, Brendan Foster, Bryan Kelleher and Niall Ó Murchadha has resulted in the publication of about ten research papers, of which I mention (Barbour, Foster and Ó Murchadha 2002, Anderson and Barbour 2002, Barbour 2003a, Anderson, Barbour, Foster and Ó Murchadha 2003, Barbour 2003b).

These research papers take very much further the approach to Mach's principle initiated Hofmann, Reissner and Schrödinger as modified in (Barbour and Bertotti 1982) through the use of best matching (Sec. 2) to avoid anisotropy of inertial mass. The relational treatment of time, implemented through reparametrization invariance, also plays a central role. If one assumes that space is Riemannian and that all interactions are local, the two principles of *best matching* and *reparametrization invariance* lead almost uniquely to Einstein's general theory of relativity. Quite unexpectedly, they also enforce the existence of a universal lightcone and gauge theory. One starts with the notion of Riemannian space (*not* pseudo-Riemannian spacetime) and three-dimensional fields (scalar, spinor and vector) defined on it. Then implementation of the Machian principles of the relativity of motion (through best matching) and time (through reparametrization) creates a four-dimensional spacetime with all the basic features of modern physics. I believe that my claim that general relativity is perfectly Machian (as regards the relativity of time and motion) is strongly vindicated.[17]

Another issue that we have investigated is relativity of scale (Barbour 2003a, Anderson, Barbour, Foster and Ó Murchadha 2003). The same intuitive convictions that lead one to require relativity of time and motion suggest that physics ought to be scale invariant too. In the two cited papers, we have extended the notion of best matching to scale transformations. We have shown that general relativity is almost scale invariant but not quite perfectly so. Specifically, in the case of a spatially closed universe one can change the (spatial) scale arbitrarily at all points. However, this must be done subject to the solitary requirement that these local transformations do

---

17   See, especially (Barbour, Foster and Ó Murchadha 2002).

not change the spatial volume of the universe. This remarkably weak restriction is, in fact, what permits expansion of the universe to be a meaningful concept. Modern cosmology depends crucially on this single residual 'Machian defect.' Work on this topic and our general approach, which we call the 3-space approach, is continuing. My personal feeling is that we are close to a definitive formulation of the principles and main conclusions of the 3-space approach. I may at last be in the position to complete (Barbour, in preparation)! In fact, I put aside work on it because I felt that it would not be complete without a proper Machian treatment of the relativity of scale.

ACKNOWLEDGMENTS

I am grateful to Domenico Giulini for drawing my attention to Frege's paper and to the Albert Einstein Archives at the Hebrew University of Jerusalem for permission to quote from the Einstein correspondence.

REFERENCES

Anderson, Edward and Julian B. Barbour. 2002. "Interacting Vector Fields in Relativity without Relativity." *Classical and Quantum Gravity* 19: 3249–3261 (gr-qc 0201092).

Anderson, Edward, Julian B. Barbour, Brendan Z. Foster, and Niall Ó Murchadha. 2003. "Scale-Invariant Gravity: Geometrodynamics." *Classical and Quantum Gravity* 20: 3217–3248 (gr-qc 0211022).

Barbour, Julian B. 1989. *Absolute or Relative Motion*, Vol. 1. *The Discovery of Dynamics*. Cambridge: Cambridge University Press. Reprinted 2001 as the paperback *The Discovery of Dynamics*. New York: Oxford University Press.

———. 1994. "The Timelessness of Quantum Gravity. I. The Evidence from the Classical Theory. II. The Appearance of Dynamics in Static Configurations." *Classical and Quantum Gravity* 11: 2853–2897.

———. 1995. "General relativity as a Perfectly Machian Theory." In (Barbour and Pfister 1995).

———. 1997. "Nows Are All We Need." In H. Atmanspacher and E. Ruhnau (eds.), *Time, Temporality, Now*. Berlin: Springer-Verlag.

———. 1999a. *The End of Time*. London: Weidenfeld & Nicholson: New York: Oxford University Press (2000).

———. 1999b. "The Development of Machian Themes in the Twentieth Century." In J. Butterfield (ed.), *The Arguments of Time*. Oxford: Oxford University Press.

———. 2001. "On General Covariance and Best Matching." In C. Callender and N. Huggett (eds.), *Physics Meets Philosophy at the Planck Scale*. Cambridge: Cambridge University Press.

———. 2003a. "Scale-Invariant Gravity: Particle Dynamics." *Classical and Quantum Gravity* 20: 1543-1570 (gr-qc 0211021).

———. 2003b. "Dynamics of Pure Shape, Relativity and the Problem of Time." In H.-T. Elze (ed.), *Decoherence and Entropy in Complex Systems. Lecture Notes in Physics*, Vol. 663. Berlin: Springer-Verlag (gr-qc 0308089).

———. (in preparation). *Absolute or Relative Motion?* Vol. 2: *The Frame of the World*. New York: Oxford University Press.

Barbour, Julian B. and Bruno Bertotti. 1977. "Gravity and Inertia in A Machian Framework." *Nuovo Cimento* 38B: 1–27.

———. 1982. "Mach's Principle and the Structure of Dynamical Theories." *Proceedings of the Royal Society of London*, Series A 382: 295–306.

Barbour, Julian B., Brendan Z. Foster and Niall Ó Murchadha. 2002. "Relativity without Relativity." *Classical and Quantum Gravity* 19: 3217–3248 (gr-qc 0012089).

Barbour, Julian B. and Herbert Pfister (eds.). 1995. *Mach's Principle: From Newton's Bucket to Quantum Gravity. Einstein Studies*, Vol. 6. Boston: Birkhäuser.

Clemence, G.M. 1957. "Astronomical Time." *Reviews of Modern Physics* 29, 2–8.

CPAE 1: John Stachel, David C. Cassidy, Robert Schulmann, and Jürgen Renn (eds.), *The Collected Papers of Albert Einstein*. Vol. 1. *The Early Years, 1879–1902*. Princeton: Princeton University Press, 1987.

CPAE 2. 1989. John Stachel, David C. Cassidy, Jürgen Renn, and Robert Schulmann (eds.), *The Collected Papers of Albert Einstein*. Vol. 2. *The Swiss Years: Writings, 1900–1909*. Princeton: Princeton University Press.

CPAE 4. 1995. Martin J. Klein, A. J. Kox, Jürgen Renn, and Robert Schulmann (eds.), *The Collected Papers of Albert Einstein*. Vol. 4. *The Swiss Years: Writings, 1912–1914*. Princeton: Princeton University Press.

CPAE 5E: *The Collected Papers of Albert Einstein*. Vol. 5. *The Swiss Years: Correspondence, 1902–1914*. English edition translated by Anna Beck, consultant Don Howard. Princeton: Princeton University Press, 1995.

CPAE 6. 1996. A. J. Kox, Martin J. Klein, and Robert Schulmann (eds.), *The Collected Papers of Albert Einstein*. Vol. 6. *The Berlin Years: Writings, 1914–1917*. Princeton: Princeton University Press.

CPAE 7. 2002. Michel Janssen, Robert Schulmann, József Illy, Christoph Lehner, and Diana Kormos Buchwald (eds.), *The Collected Papers of Albert Einstein*. Vol. 7. *The Berlin Years: Writings, 1918–1921*. Princeton: Princeton University Press.

Dziobek, Otto. 1888. *Die mathematischen Theorien der Planeten-Bewegungen*. Leipzig: J. A. Barth.

———. 1892. *Mathematical Theories of Planetary Motions*. Register Publishing Company. Reprinted by Dover in 1962.

Einstein, Albert. 1905. "Zur Elektrodynamik bewegter Körper." *Annalen der Physik* 17, 891–921, (CPAE 2, Doc. 23). English translation in (Lorentz et al. 1923).

———. 1907. "Über das Relativitätsprinzip und die aus demselben gezogenen Folgerungen." V. §17. *Jahrbuch der Radioaktivität und Elektronik* 4: 411–462, (CPAE 2, Doc. 47).

———. 1912. "Gibt es eine Gravitationswirkung, die der elektrodynamischen Induktionswirkung analog ist?" *Vierteljahrschrift für gerichtliche Medizin und öffentliches Sanitätswesen* 44: 37–40, (CPAE 4, Doc. 7).

———. 1913a. "Zum gegenwärtigen Stande des Gravitationsproblems." *Physikalische Zeitschrift* 14: 1249–1266, (CPAE 4, Doc. 17). (English translation in this volume.)

———. 1913b. "Physikalische Grundlagen einer Gravitationstheorie." *Vierteljahrsschrift der Naturforschenden Gesellschaft Zürich* 58: 284–290, (CPAE 4, Doc. 16).

———. 1914. "Die formale Grundlage der allgemeinen Relativitätstheorie." *Sitzungsberichte der Preussischen Akademie der Wissenschaften*, 1030–1085, Part 2, (CPAE 6, Doc. 9).

———. 1917. "Kosmologische Betrachtungen zur allgemeinen Relativitätstheorie." *Sitzungsberichte der Preussischen Akademie der Wissenschaften*: 142–145, (CPAE 6, Doc. 43).

———. 1918a. "Prinzipielles zur allgemeinen Relativitätstheorie." *Annalen der Physik* 55: 241–244, (CPAE 7, Doc. 4).

———. 1918b. "Dialog über Einwände gegen die Relativitätstheorie." *Die Naturwissenschaften* 6: No. 48, 697–702, (CPAE 7, Doc. 13).

———. 1919. "What is the Theory of Relativity?" *The Times*, 29 November 1919. Republished in (Einstein 1954).

———. 1921. "Geometrie und Erfahrung." *Sitzungsberichte der Preussischen Akademie der Wissenschaften* Vol. 1, 123–130.

———. 1923. "Grundgedanken und Probleme der Relativitätstheorie." In *Nobelstiftelsen, Les Prix Nobel en 1921–1922*. Stockholm: Imprimerie Royale.

———. 1926. "Nichteuklidsche Geometrie in der Physik." *Neue Rundschau*, January.

———. 1933. "Notes on the Origin of the General Theory of Relativity." In *Albert Einstein. Ideas and Opinions*, 285–290. Translated by Sonja Bargmann. New York: Crown, 1954.

———. 1949. "Autobiographical Notes." In P.A. Schilpp (ed.), *Albert Einstein, Philosopher-Scientist*. Evanston, Illinois: The Library of Living Philosophers, Inc., Illinois, 29.

———. 1954. *Ideas and Opinions*. New York: Crown Publishers.

Frege, Gustav. 1891. "Über das Trägheitsgesetz." *Zeitschrift für Philosophie und philosophische Kritik* 98, 145–161.

Friedlaender, Benedict and Immanuel Friedlaender. 1896. *Absolute oder relative Bewegung?* Berlin: Leonhard Simion. (English translation in this volume.)

Hall, Alfred R. and Marie B. Hall. 1962. *Unpublished Scientific Papers of Isaac Newton*. Cambridge: Cambridge University Press.

Helmholtz, Hermann L. F. 1968. "Über die Tatsachen, die der Geometrie zum Grunde liegen." *Nachrichten von der Königlichen Gesellschaft der Wissenschaften zu Göttingen*, No. 9, June 3rd.

Hofmann, W. 1904. *Kritische Beleuchtung der beiden Grundbegriffe der Mechanik: Bewegung und Trägheit und daraus gezogene Folgerungen betreffs der Achsendrehung der Erde und des Foucault'schen Pendelversuchs*. Wien und Leipzig: M. Kuppitsch.

Kretschmann, Erich. 1917. "Über den physikalischen Sinn der Relativitätspostulate, A. Einsteins neue und seine ursprüngliche Relativitätstheorie." *Annalen der Physik* 53: 575–614.

Lagrange, Joseph-Louis. 1772. "Essai sur le problème des trois corps." Republished in: *Oeuvres de Lagrange*, Vol. 6, Paris: Gauthier-Villars, 229 (1873).

Lange, Ludwig. 1884. "Über die wissenschaftliche Fassung des Galilei'schen Beharrungsgesetz." *Philosophische Studien* 2: 266–297.

———. 1885. "Über das Beharrungsgesetz." *Berichte der Königlichen Sächsischen Gesellschaft der Wissenschaften, Math.-Physik.* Klasse 333–351.

———. 1886. *Die geschichtliche Entwicklung des Bewegungsbegriffs und ihr voraussichtliches Endergebnis. Ein Beitrag zur historischen Kritik der mechanischen Prinzipien.* Leipzig: W. Engelmann.

Laue, Max von. 1948. "Dr. Ludwig Lange. 1863–1936. (Ein zu unrecht Vergessener.)" *Die Naturwissenschaften* 35, 193.

———. 1955. *Die Relativitätstheorie*, Vol. 1. *Die Spezielle Relativitätstheorie.* Braunschweig: Vieweg.

Lorentz, Hendrik Anton. 1895. *Versuch einer Theorie der electrischen und optischen Erscheinungen in bewegten Körpern.* Leiden: Brill.

Lorentz, Hendrik Antoon et al. 1923. *The Principle of Relativity.* London: Methuen.

Mach, Ernst. 1872. *Die Geschichte und die Wurzel des Satzes von der Erhaltung der Arbeit.* Prague: Calve.

———. 1883. *Die Mechanik in ihrer Entwickelung. Historisch-kritisch dargestellt.* Leipzig: F.A. Brockhaus.

———. 1911. *History and Root of the Principle of the Conservation of Energy.* Chicago: Open Court.

———. 1960. *The Science of Mechanics. A Critical and Historical Account of Its Development.* LaSalle: Open Court.

Neumann, Carl. 1870. *Ueber die Prinzipien der Galilei-Newtonschen Theorie.* Leipzig: Teubner.

Norton, John. 1995. "Mach's Principle before Einstein." In (Barbour and Pfister 1995).

Poincaré, Henri. 1898. "La Mesure du Temps." Republished in Poincaré, Henri (1905). *La Valeur de la Science.*

———. 1902. *La Science et l'Hypothèse.* Paris: Flammarion.

———. 1905. *Science and Hypothesis.* London: Walter Scott Publ. Co.

Reissner, Hans. 1914. "Über die Relativität der Beschleunigungen in der Mechanik." *Physikalische Zeitschrift* 15: 371–75.

———. 1915. "Über eine Möglichkeit die Gravitation als unmittelbare Folge der Relativität der Trägheit abzuleiten." *Physikalische Zeitschrift* 16: 179–85.

Riemann, Bernhard. 1867. "Über die Hypothesen, welche der Geometrie zu grunde liegen." *Abhandlungen der Königlichen Gesellschaft der Wissenschaften zu Göttingen* 13.

Tait. Peter G. 1883. "Note on Reference Frames." *Proceedings of the Royal Society of Edinburgh,* Session 1883–84: 743–745.

Thomson, James. 1883. "On the law of inertia; the principle of chronometry; and the principle of absolute clinural rest, and of absolute rotation." In *Proceedings of the Royal Society of Edinburgh,* Session 1883–84, 568 and 730.

Thomson, William and Peter G. Tait. 1867. *Elements of Natural Philosophy.* Cambridge: Cambridge University Press.

ALBERT EINSTEIN

ON THE RELATIVITY PROBLEM

Originally published as "Zum Relativitätsproblem" in Scientia 15, 1914, pp. 337–348. Reprinted in "The Collected Papers of Albert Einstein," Vol. 4, Doc. 31: An English translation is given in its companion volume.

After two eminent specialists have presented their objections to relativity theory in this journal, it must be not undesirable for the readers if an adherent of this new theoretical direction expounds his view. This shall be done as concisely as possible in the following.

Currently we have to distinguish two theoretical systems, both of which fall under the name "relativity theory." The first of these, which we will call "relativity theory in the narrower sense," is based on a considerable body of experience and is accepted by the majority of theoretical physicists to be one of the simplest theoretical expressions of these experiences. The second, which we will call "relativity theory in the broader sense," is as yet by no means established on the basis of physical experience. The majority of my colleagues regard this second system either sceptically or dismissively. It should be said immediately that one can certainly be an adherent of the relativity theory in the narrow sense without admitting the validity of relativity theory in the wider sense. For that reason we will discuss the two theories separately.

I. RELATIVITY THEORY IN THE NARROWER SENSE

It is well known that the equations of the mechanics established by Galileo and Newton are not valid with respect to an arbitrarily moving coordinate system, if one adheres to the requirement that the description of motion admits only central forces satisfying the | law of equality of action and reaction. But if the motion is referred to [338] a coordinate system K, such that Newton's equations are valid in the indicated sense, then that coordinate system is not the *only one* with respect to which those laws of mechanics hold. Rather, all of the coordinate systems K' with arbitrary spatial orientation that have uniform translational motion with respect to K have the property that relative to them the laws of motion hold. We call the assumption of the equal value of all these coordinate systems K, K', etc. for the formulation of the laws of motion, actually for the general laws of physics, the "relativity principle" (in the narrow sense).

As long as one believed that classical mechanics lies at the foundation of the theoretical representation of all processes, one could not doubt the validity of this relativity principle. But even abstaining from that, it is difficult from an empirical standpoint
to doubt the validity of that principle. In fact if it did not hold, then the processes of
nature referred to a reference system at rest with respect to the Earth would appear to
be influenced by the motion (velocity) of the Earth's yearly orbital motion around the
Sun; the terrestrial space of observations would have to behave physically in an anisotropic manner due to the existence of this motion. But despite the most arduous
searching, physicists have never observed such an apparent anisotropy.

The relativity principle is hence as old as mechanics, and no one could ever have
questioned its validity from an empirical standpoint. That it has been nonetheless
doubted, and is again doubted today, is due to the fact that it seemed to be incompatible with Maxwell-Lorentz electrodynamics. Whoever is in a position to judge this
theory, in light of its inclusiveness, the small number of its fundamental assumptions,
and its successful representation of phenomena in the domain of electrodynamics and
optics, will find it difficult to dispel the impression that the main features of this theory are as definitively established as are the equations of mechanics. It has also not
been accomplished to set another theory against this one that could even tolerably
compete with it. |

[339]   It is easy to specify wherein lies the apparent contradiction between Maxwell-
Lorentz electrodynamics and the relativity principle. Suppose that the equations of
that theory hold relative to the coordinate system K. This means that every light ray
propagates in vacuo with a definite velocity c, with respect to K, which is independent of direction and of the state of motion of the light source; this proposition will be
called the "principle of the constancy of the speed of light" in the following. Now if
one such light ray were to be observed by an observer moving relative to K, then the
propagation speed of this light ray, as estimated from the standpoint of this observer,
in general seems to be different than c. For example, if the light ray propagates in the
direction of the positive x- axis of K with speed c, and our observer moves in the
same direction with the temporally constant speed v, then one would believe that
one can immediately conclude that the light ray's propagation speed must be $c - v$
according to the moving observer. Relative to the observer, that is, relative to a coordinate system K' moving with the same velocity, the principle of the constancy of the
speed of light does not appear to hold. Hence, here is an apparent contradiction with
the principle of relativity.

However, an exact analysis of the physical content of our spatial and temporal
determinations leads to the well-known result that the implied contradiction is only
apparent, since it depends on both of the following arbitrary assumptions:

1. The assertion that whether two events occurring in different places occur simulta-
    neously has content independently of the choice of a reference system.

2. The spatial distance between the places in which two simultaneous events occur is
    independent of the choice of a reference system.

Given that the Maxwell-Lorentz theory as well as the relativity principle are empirically supported to such a large degree, one must therefore decide to drop both the aforementioned arbitrary assumptions, the apparent evidence for which rests solely on the facts that light gives us information about distant events *apparently instantaneously*, and that the objects | we deal with in daily life have velocities that [340] are small compared to the velocity of light c.

By abandoning these arbitrary assumptions, one achieves compatibility between the principle of the constancy of the speed of light, which results from Maxwell-Lorentz electrodynamics, and the relativity principle. One can retain the assumption that one and the same light ray propagates with velocity c relative to all reference systems K' in uniform translational motion with respect to a system K, rather than only relative to K. One only has to choose the transformation equations, which exist between the spacetime coordinates (x, y, z, t) with respect to K and those (x', y', z', t') with respect to K', in an appropriate way; one will thereby be led to the system of transformation equations called the "Lorentz transformation." This Lorentz transformation supercedes the corresponding transformations that until the development of relativity theory were regarded as the only conceivable ones, which, however, were based on the assumptions (1) and (2) given above.

The heuristic value of relativity theory consists in the fact that it provides a constraint that all of the systems of equations that express general laws of nature must satisfy. All such systems of equations must be constructed such that with the application of a Lorentz transformation they go into a system of equations of the same form (covariance with respect to the Lorentz transformations). Minkowski presented a simple mathematical schema to which equation systems must be reducible if they are to behave covariantly with respect to Lorentz transformations; thereby he achieved the advantage, that for the accommodation of the system of equations with the constraint mentioned above it is certainly not necessary to in fact carry out a Lorentz transformation on those systems.

From what has been said it clearly follows that relativity theory by no means gives us a tool for deducing previously unknown laws of nature from nothing. It only provides an always applicable criterion that constrains the possibilities; in this respect | it is comparable to the law of energy conservation or the second law of thermody- [341] namics.

It follows from a close examination of the most general laws of theoretical physics that Newtonian mechanics must be modified to satisfy the criterion of relativity theory. These altered mechanical equations have proven to be applicable to cathode rays and β- rays (motion of free electrical particles). Moreover, the implementation of relativity theory has lead to neither a logical contradiction nor a conflict with empirical results.

Only one result of relativity theory will be given here in particular, because it is of importance for the following analyses. According to Newtonian mechanics the inertia of a system constituted by a collection of material points (that is, the inertial resistance against acceleration of the system's center of gravity) is independent of the state of the system. By contrast, according to relativity theory the inertia of an isolated system (floating in a vacuum) depends upon the state of the system, such that the inertia increases with the energy content of the system. Thus according to relativity theory, it is ultimately *energy* that inertia can be attributed to. The energy, rather than the inertial mass of a material point, is what we have ascribed indestructibility to; hence the theorem of the conservation of mass is incorporated into the theorem of the conservation of energy.

It was remarked above that it would be a great mistake to regard relativity theory as a universal method that allows one to develop an unequivocally appropriate theory for a domain of phenomena regardless of how little this has been explored empirically. Relativity theory only *reduces* by a significant amount the empirical conclusions necessary for the development of a theory. There is only one domain of fundamental importance where we have such poor empirical knowledge that this knowledge, in combination with relativity theory, is not sufficient, by a wide margin, for a clear determination of the general theory. This is the domain of gravitational phenomena. Here we can only reach our goal by complementing what is empirically [342] known with physical hypotheses in order to complete the basis | of the theory. The following considerations are firstly to show how one arrives at what are, in my opinion, the most natural such hypotheses.

When we speak of a body's *mass*, we associate with this word two definitions that are logically completely independent. By *mass* we understand, first, a constant inherent to a body, which measures its resistance to acceleration ("inertial mass"), and second, the constant of a body that determines the magnitude of the force that it experiences in a gravitational field ("gravitational mass").

It is in no way self-evident *a priori* that the inertial and gravitational mass of a body must agree; we are simply *accustomed* to assume their agreement. The belief in this agreement stems from the empirical fact that the acceleration that various bodies experience in a gravitational field is independent of their material constitution. Eötvös has shown that, in any case, inertial and gravitational mass agree with very great precision, in that, through his experiments with a torsion balance, he ruled out the existence a relative deviation of the two masses from each other on the order of magnitude of 10^{-8}.[1]

Enormous quantities of energy in the form of heat are discharged into the environment by radioactive processes. According to the result regarding the inertia of energy presented above, the decay products generated by the reaction taken together must have a smaller inertial mass than that of the material existing before the radioactive decay. This change of inertial mass is, for the kind of reaction with known heat effect,

---

1 Eötvös' experimental method is based on the following. The Earth's gravity and the centrifugal force influence a body found on the Earth's surface. For the first the body's gravitational mass, and for the second the inertial mass, is the determining factor. If the two were not identical, then the direction of the resultant of the two (the apparent weight) would depend on the material of the body. With his torsion balance experiments, Eötvös proved with great precision the non-existence of such a dependence.

of the order 10^{-4}. If the gravitational mass did not change simultaneously with the inertial mass of the system, then the inertial mass would have to differ from the gravitational mass for various elements far more than Eötvös's | experiments would allow. [343] Langevin was the first to call attention to this important point.

From what has been said the identity of the inertial and gravitational mass of closed systems (at rest) follows with great probability; I think that based on the present state of empirical knowledge we should adhere to the assumption of this identity unconditionally. We have thereby attained one of the most important physical demands that, from my point of view, must be imposed on any gravitational theory.

This demand involves a far-reaching constraint on gravitational theories, which one recognizes especially in conjunction with the theorem of the inertia of energy. All energy corresponds to inertial mass, and all inertial mass corresponds to gravitational mass; the gravitational mass of a closed system must therefore be determined by its energy. The energy of its gravitational field also belongs to the energy of a closed system; hence the gravitational field energy itself contributes to the system's gravitational mass rather than only its inertial mass.

Abraham and Mie have proposed gravitational theories. Abraham's theory contradicts the relativity principle, and Mie's theory contradicts the demand of the equality of the inertial and gravitational mass of a closed system. According to the latter theory, were a body to be heated the *inertial* mass *grows* in proportion to the energy gain, but not the gravitational mass; the latter would actually decrease for a gas with rising temperature.[2]

By way of contrast, a gravitational theory recently proposed by Nordström complies with both the relativity principle and the requirement of the gravity of energy of closed systems, with one restriction to be indicated in the following. Abraham's claim to the contrary made in a paper appearing in this journal is not | correct. In fact, I [344] believe that a cogent argument against Nordström's theory cannot be drawn from experience.

According to Nordström's theory the principle of the gravity of energy of closed systems at rest holds as a statistical principle. The gravitational mass of a closed system (with the whole system at rest) is in general an oscillating quantity, whose temporal average is determined by the total energy of the system. As a consequence of the oscillatory character of mass, such a system must emit standing longitudinal gravitational waves. Yet the energy loss expected according to the theory is so small that it must escape our notice.

After a more detailed study of Nordström's theory, everyone will have to admit that this theory, when regarded from an empirical standpoint, is an unobjectionable integration of gravitation into the framework of relativity theory (in the narrower

---

2   Due to their smallness, these effects are certainly not accessible to experiments. But it seems to me that there is much to be said for taking the connection between inertial and gravitational mass to be warranted in principle, regardless of what forms of energy are taken into account. According to Mie, one can account for the fact that the equality of inertial and gravitational mass holds for radioactive transformations only with assumptions regarding the special nature of energy in the interior of atoms.

sense). Even though I am of the opinion that we cannot be satisfied with this solution, my reasons for this have an epistemological character that will be described in the following.

II. RELATIVITY THEORY IN THE BROADER SENSE

Classical mechanics, as well as relativity theory in the narrow sense briefly described above, suffer from a fundamental defect, which no one can deny, that is accessible to epistemological arguments. The weaknesses of our physical world picture to be discussed below were already uncovered with full clarity by E. Mach in his deeply penetrating investigations of the foundations of Newtonian mechanics, so that what I will assert in this respect can have no claim to novelty. I will explain the essential point with an example, which is chosen to be quite elementary, in order to allow what is essential to stand out.

[345]   Two masses float in space at a great distance from all celestial bodies. Suppose that these two are close enough together that they can exert an influence on each other. Now an observer watches the motion of both bodies, such that he continually looks along the direction of the line l connecting the two masses toward the sphere of fixed stars. He will observe that the line of sight traces out a closed line on the visible sphere of fixed stars, which does not change its position with respect to the visible sphere of fixed stars. If the observer has any natural intelligence, but has learned neither geometry nor mechanics, he would conclude: "My masses carry out a motion, which is at least in part causally determined by the fixed stars. The law by which masses in my surroundings move is co-determined by the fixed stars." A man who has been schooled in the sciences would smile at the simple-mindedness of our observer and say to him: "The motion of your masses has nothing to do with the heaven of fixed stars; it is rather fully determined by the laws of mechanics entirely independently of the remaining masses. There is a space R in which these laws hold. These laws are such that the masses remain continually in a plane in this space. However, the system of fixed stars cannot rotate in this space, because otherwise it would be disrupted by powerful centrifugal forces. Thus it necessarily must be at rest (at least almost!), if it is to exist permanently; this is the reason that the plane in which your masses move always goes through the same fixed stars." But our intrepid observer would say: "You may be incomparably learned. But just as I could never be brought to believe in ghosts, so I cannot believe in this gigantic thing that you speak of and call space. I can neither see something like that nor conceive of it. Or should I think of your space R as a subtle net of bodies that the remaining things are all referred to? Then I can imagine a second such net R' in addition to R, that is moving in an arbitrary manner relative to R (for example, rotating). Do your equations also hold at the same time with respect to R'?" The learned man denies this with certainty. In reply to which the ignoramus: "But how do the masses know which "space" R, R', etc. with respect to which they should move according to your equations, how do they recognize the space or spaces they orient themselves with respect to?" Now our

learned man is in a quite embarrassing position. He certainly insists that there must be such privileged spaces, but he knows no reason to give for why such spaces could be distinguished from | other spaces. The ignoramus's reply: "Then I will take, for the [346] time being, your privileged spaces as an idle fabrication, and stay with my conception, that the sphere of fixed stars co-determines the mechanical behavior of my test masses."

I will explain the violation of the most elementary postulate of epistemology of which our physics is guilty in yet another way. One would try in vain to explain what one understands by the simple acceleration of a body. One would only succeed in defining *relative* acceleration of bodies with respect to each other. However, having said that, we base our mechanics on the premise that a force (cause) is necessary for the generation of a body's acceleration, ignoring the fact that we cannot explain what it is that we understand by "acceleration," exactly because only *relative* accelerations can be an object of perception.

The dubious aspect of proceeding in this vein is very nicely illustrated by a comparison, which I owe to my friend Besso. Suppose we think back to an earlier time, when it was assumed that the surface of the Earth must be approximately *flat*. Imagine that the following conception exists among the learned. In the world there is a physically distinguished direction, the vertical. All objects fall in this direction if they are not supported. Because of this the surface of the Earth is essentially perpendicular to this direction, and this is why it tends towards the form of a plane. While in this case, the mistake lies in privileging one direction over all others without good reason (fictitious cause), rather than simply regarding the Earth as the cause of falling, the mistake in our physics lies in the introduction without good reason of privileged reference systems as fictitious causes; both cases are characterized by forgoing the establishment of a sufficient reason.

Since relativity theory in the narrower sense, rather than only classical mechanics, exhibits the fundamental deficiency explained above, I set myself the goal of generalizing relativity theory in such a way that this imperfection will be avoided. First of all, I recognized that gravitation in general | must be assigned a fundamental role in [347] any such theory. Then from what was explained earlier it already follows that every physical process must also produce a gravitational field, because of the quantity of energy corresponding to it. On the other hand, the empirical fact that all bodies fall with the same speed in a gravitational field suggests the idea that physical processes happen in a gravitational field exactly as they do relative to an accelerated reference system (equivalence hypothesis). In taking this idea as a foundation, I came to the conclusion that the velocity of light is not to be regarded as independent of the gravitational potential. Thus the principle of the constancy of the speed of light is incompatible with the equivalence hypothesis; that is why relativity theory in the narrower sense cannot be made consistent with the equivalence hypothesis. This led me to take relativity theory in the narrow sense to be applicable only in regions within which no noticeable differences in the gravitational potential occur. Relativity theory (in the

narrower sense) has to be replaced by a general theory, which contains the former as
a limiting case.

The path leading to this theory can only be very incompletely described in
words.[3] The equations of motion for material points in a gravitational field that follow from the equivalence hypothesis can be easily written in a form such that these
laws are completely independent of the choice of variables determining place and
time. By leaving the choice of these variables as *a priori* completely arbitrary, and
thus not privileging any spacetime system, one averts the epistemological objection
discussed above. A quantity

$$ds^2 = \sum_{\mu\nu} g_{\mu\nu} dx_\mu dx_\nu,$$

appears in that law of motion, and it is invariant, i.e., it is a quantity that is independent of the choice of reference system (i.e., of the choice of four spacetime coordinates). The quantities $g_{\mu\nu}$ are functions of $x_1 \ldots x_4$ and represent the gravitational
[348]   field. |

With the help of the absolute differential calculus, which has been developed by
Ricci and Levi-Civita based on Christoffel's mathematical investigations, one can
succeed, based on the existence of the invariant above, in replacing the well-known
systems of equations of physics with equivalent systems (when all $g_{\mu\nu}$ are constant),
which are valid independent of a choice of the spacetime coordinate system x_ν. All
such systems of equations include the quantities $g_{\mu\nu}$, i.e., the quantities that determine the gravitational field. Thus the latter influence all physical processes.

Conversely, physical processes must also determine the gravitational field, i.e.,
the quantities $g_{\mu\nu}$. One arrives at the differential equations that determine these
quantities by means of the hypothesis that the conservation of momentum and energy
must hold for material events and the gravitational field taken together. This hypothesis subsequently constrains the choice of spacetime variables x, without thereby
evoking again the epistemological doubts analyzed above. Because according to this
generalized relativity theory there are no longer privileged spaces with peculiar physical qualities. The quantities $g_{\mu\nu}$ control the course of all processes, which for their
part are determined by the physical events in all the rest of the universe.

The principle of the inertia and gravitation of energy is completely satisfied in this
theory. Furthermore, the equations of motion for gravitational masses are such that it
is, as one must demand based on the considerations above, acceleration with respect
to other bodies rather than absolute acceleration (acceleration with respect to
"space") that appears as that which is decisive for the appearance of inertial resistance.

Relativity theory in the broader sense signifies a further development of the earlier relativity theory, rather than an abandonment of it, that seems necessary to me for
the epistemological reasons I cited.

---

3   Cf. A. Einstein and M. Grossmann, *Zeitschrift f. Math. & Physik* 62 (1914): p. 225.

ALBERT EINSTEIN

ETHER AND THE THEORY OF RELATIVITY

Originally published in 1920 as "Äther und Relativitätstheorie" at Springer, Berlin, on the basis of an address delivered on May 5th, 1920, at the University of Leyden. The English version reproduced here first appeared in 1922 in Albert Einstein: "Sidelights of Relativity" at Methuen, London, pp. 3–24. The page numbers given here refer to the latter edition.

How does it come about that alongside of the idea of ponderable matter, which is [3] derived by abstraction from everyday life, the physicists set the idea of the existence of another kind of matter, the ether? The explanation is probably to be sought in those phenomena which have given rise to the theory of action at a distance, and in the properties of light which have led to the undulatory theory. Let us devote a little while to the consideration of these two subjects.

Outside of physics we know nothing of action at a distance. When we try to connect cause and effect in the experiences which natural objects afford us, it seems at first as if there were no other I mutual actions than those of immediate contact, e.g. [4] the communication of motion by impact, push and pull, heating or inducing combustion by means of a flame, etc. It is true that even in everyday experience weight, which is in a sense action at a distance, plays a very important part. But since in daily experience the weight of bodies meets us as something constant, something not linked to any cause which is variable in time or place, we do not in everyday life speculate as to the cause of gravity, and therefore do not become conscious of its character as action at a distance. It was Newton's theory of gravitation that first assigned a cause for gravity by interpreting it as action at a distance, proceeding from masses. Newton's theory is probably the greatest stride ever made in the effort towards the causal nexus of natural phenomena. And yet this theory evoked a lively sense of discomfort among Newton's contemporaries, because it seemed to be in conflict with the principle springing from the rest of experience, that there can be reciprocal I action only through contact, and not through immediate action at a distance. [5]

It is only with reluctance that man's desire for knowledge endures a dualism of this kind. How was unity to be preserved in his comprehension of the forces of nature? Either by trying to look upon contact forces as being themselves distant forces which admittedly are observable only at a very small distance—and this was the road which Newton's followers, who were entirely under the spell of his doctrine,

Jürgen Renn (ed.). *The Genesis of General Relativity*, Vol. 3
Gravitation in the Twilight of Classical Physics: Between Mechanics, Field Theory, and Astronomy.
© 2007 Springer.

mostly preferred to take; or by assuming that the Newtonian action at a distance is only apparently immediate action at a distance, but in truth is conveyed by a medium permeating space, whether by movements or by elastic deformation of this medium. Thus the endeavour toward a unified view of the nature of forces leads to the hypothesis of an ether. This hypothesis, to be sure, did not at first bring with it any advance in the theory of gravitation or in physics generally, so that it became customary to
[6] treat Newton's law of force as an axiom | not further reducible. But the ether hypothesis was bound always to play some part in physical science, even if at first only a latent part.

When in the first half of the nineteenth century the far-reaching similarity was revealed which subsists between the properties of light and those of elastic waves in ponderable bodies, the ether hypothesis found fresh support. It appeared beyond question that light must be interpreted as a vibratory process in an elastic, inert medium filling up universal space. It also seemed to be a necessary consequence of the fact that light is capable of polarisation that this medium, the ether, must be of the nature of a solid body, because transverse waves are not possible in a fluid, but only in a solid. Thus the physicists were bound to arrive at the theory of the "quasi-rigid" luminiferous ether, the parts of which can carry out no movements relatively to one another except the small movements of deformation which correspond to light-waves.

[7] This theory—also called the theory of | the stationary luminiferous ether—moreover found a strong support in an experiment which is also of fundamental importance in the special theory of relativity, the experiment of Fizeau, from which one was obliged to infer that the luminiferous ether does not take part in the movements of bodies. The phenomenon of aberration also favoured the theory of the quasi-rigid ether.

The development of the theory of electricity along the path opened up by Maxwell and Lorentz gave the development of our ideas concerning the ether quite a peculiar and unexpected turn. For Maxwell himself the ether indeed still had properties which were purely mechanical, although of a much more complicated kind than the mechanical properties of tangible solid bodies. But neither Maxwell nor his followers succeeded in elaborating a mechanical model for the ether which might furnish a satisfactory mechanical interpretation of Maxwell's laws of the electromagnetic field. The laws were clear and simple, the mechanical
[8] interpretations | clumsy and contradictory. Almost imperceptibly the theoretical physicists adapted themselves to a situation which, from the standpoint of their mechanical programme, was very depressing. They were particularly influenced by the electrodynamical investigations of Heinrich Hertz. For whereas they previously had required of a conclusive theory that it should content itself with the fundamental concepts which belong exclusively to mechanics (e.g. densities, velocities, deformations, stresses) they gradually accustomed themselves to admitting electric and magnetic force as fundamental concepts side by side with those of mechanics, without requiring a mechanical interpretation for them. Thus the purely mechanical view of nature was gradually abandoned. But this change led to a fundamental dualism which in the

long-run was insupportable. A way of escape was now sought in the reverse direction, by reducing the principles of mechanics to those of electricity, and this especially as confidence in the strict validity of the equations of Newton's I mechanics was shaken [9] by the experiments with β- rays and rapid cathode rays.

This dualism still confronts us in unextenuated form in the theory of Hertz, where matter appears not only as the bearer of velocities, kinetic energy, and mechanical pressures, but also as the bearer of electromagnetic fields. Since such fields also occur in vacuo—i.e. in free ether—the ether also appears as bearer of electromagnetic fields. The ether appears indistinguishable in its functions from ordinary matter. Within matter it takes part in the motion of matter and in empty space it has every-where a velocity; so that the ether has a definitely assigned velocity throughout the whole of space. There is no fundamental difference between Hertz's ether and pon-derable matter (which in part subsists in the ether).

The Hertz theory suffered not only from the defect of ascribing to matter and ether, on the one hand mechanical states, and on the other hand electrical states, which do not stand in any conceivable I relation to each other; it was also at variance [10] with the result of Fizeau's important experiment on the velocity of the propagation of light in moving fluids, and with other established experimental results.

Such was the state of things when H. A. Lorentz entered upon the scene. He brought theory into harmony with experience by means of a wonderful simplification of theoretical principles. He achieved this, the most important advance in the theory of electricity since Maxwell, by taking from ether its mechanical, and from matter its electromagnetic qualities. As in empty space, so too in the interior of material bodies, the ether, and not matter viewed atomistically, was exclusively the seat of electro-magnetic fields. According to Lorentz the elementary particles of matter alone are capable of carrying out movements; their electromagnetic activity is entirely confined to the carrying of electric charges. Thus Lorentz succeeded in reducing all electro-magnetic happenings to Maxwell's equations for free space.

As to the mechanical nature of the I Lorentzian ether, it may be said of it, in a [11] somewhat playful spirit, that immobility is the only mechanical property of which it has not been deprived by H. A. Lorentz. It may be added that the whole change in the conception of the ether which the special theory of relativity brought about, consisted in taking away from the ether its last mechanical quality, namely, its immobility. How this is to be understood will forthwith be expounded.

The spacetime theory and the kinematics of the special theory of relativity were modelled on the Maxwell-Lorentz theory of the electromagnetic field. This theory therefore satisfies the conditions of the special theory of relativity, but when viewed from the latter it acquires a novel aspect. For if K be a system of co-ordinates rela-tively to which the Lorentzian ether is at rest, the Maxwell-Lorentz equations are valid primarily with reference to K. But by the special theory of relativity the same equations without any change of meaning also hold in relation to any new system of co-ordinates I K' which is moving in uniform translation relatively to K. Now comes [12] the anxious question:—Why must I in the theory distinguish the K system above all

K' systems, which are physically equivalent to it in all respects, by assuming that the ether is at rest relatively to the K system? For the theoretician such an asymmetry in the theoretical structure, with no corresponding asymmetry in the system of experience, is intolerable. If we assume the ether to be at rest relatively to K, but in motion relatively to K', the physical equivalence of K and K' seems to me from the logical standpoint, not indeed downright incorrect, but nevertheless inacceptable.

The next position which it was possible to take up in face of this state of things appeared to be the following. The ether does not exist at all. The electromagnetic fields are not states of a medium, and are not bound down to any bearer, but they are independent realities which are not reducible to anything else, exactly like the atoms

[13] of ponderable matter. This | conception suggests itself the more readily as, according to Lorentz's theory, electromagnetic radiation, like ponderable matter, brings impulse and energy with it, and as, according to the special theory of relativity, both matter and radiation are but special forms of distributed energy, ponderable mass losing its isolation and appearing as a special form of energy.

More careful reflection teaches us, however, that the special theory of relativity does not compel us to deny ether. We may assume the existence of an ether; only we must give up ascribing a definite state of motion to it, i.e. we must by abstraction take from it the last mechanical characteristic which Lorentz had still left it. We shall see later that this point of view, the conceivability of which I shall at once endeavour to make more intelligible by a somewhat halting comparison, is justified by the results of the general theory of relativity.

Think of waves on the surface of water. Here we can describe two entirely differ-

[14] ent things. Either we may observe how | the undulatory surface forming the boundary between water and air alters in the course of time; or else—with the help of small floats, for instance—we can observe how the position of the separate particles of water alters in the course of time. If the existence of such floats for tracking the motion of the particles of a fluid were a fundamental impossibility in physics—if, in fact, nothing else whatever were observable than the shape of the space occupied by the water as it varies in time, we should have no ground for the assumption that water consists of movable particles. But all the same we could characterize it as a medium.

We have something like this in the electromagnetic field. For we may picture the field to ourselves as consisting of lines of force. If we wish to interpret these lines of force to ourselves as something material in the ordinary sense, we are tempted to interpret the dynamic processes as motions of these lines of force, such that each sep-

[15] arate line of | force is tracked through the course of time. It is well known, however, that this way of regarding the electromagnetic field leads to contradictions.

Generalizing we must say this:—There may be supposed to be extended physical objects to which the idea of motion cannot be applied. They may not be thought of as consisting of particles which allow themselves to be separately tracked through time. In Minkowski's idiom this is expressed as follows:—Not every extended conformation in the four-dimensional world can be regarded as composed of worldthreads. The special theory of relativity forbids us to assume the ether to consist of particles observable through time, but the hypothesis of ether in itself is not in conflict with the special theory of relativity. Only we must be on our guard against ascribing a state of motion to the ether.

Certainly, from the standpoint of the special theory of relativity, the ether hypothesis appears at first to be an empty hypothesis. In the equations of the | electromag- [16] netic field there occur, in addition to the densities of the electric charge, only the intensities of the field. The career of electromagnetic processes in vacuo appears to be completely determined by these equations, uninfluenced by other physical quantities. The electromagnetic fields appear as ultimate, irreducible realities, and at first it seems superfluous to postulate a homogeneous, isotropic ether-medium, and to envisage electromagnetic fields as states of this medium.

But on the other hand there is a weighty argument to be adduced in favour of the ether hypothesis. To deny the ether is ultimately to assume that empty space has no physical qualities whatever. The fundamental facts of mechanics do not harmonize with this view. For the mechanical behavior of a corporeal system hovering freely in empty space depends not only on relative positions (distances) and relative velocities, but also on its state of rotation, which physically may be taken as a characteristic not appertaining to the system in itself. In order | to be able to look upon the rotation of [17] the system, at least formally, as something real, Newton objectivizes space. Since he classes his absolute space together with real things, for him rotation relative to an absolute space is also something real. Newton might no less well have called his absolute space "ether"; what is essential is merely that besides observable objects, another thing, which is not perceptible, must be looked upon as real, to enable acceleration or rotation to be looked upon as something real.

It is true that Mach tried to avoid having to accept as real something which is not observable by endeavouring to substitute in mechanics a mean acceleration with reference to the totality of the masses in the universe in place of an acceleration with reference to absolute space. But inertial resistance opposed to relative acceleration of distant masses presupposes action at a distance; and as the modern physicist does not believe that he may accept this action at a distance, he comes back once more, if he | follows Mach, to the ether, which has to serve as medium for the effects of inertia. [18] But this conception of the ether to which we are led by Mach's way of thinking differs essentially from the ether as conceived by Newton, by Fresnel, and by Lorentz. Mach's ether not only *conditions* the behavior of inert masses, but *is also conditioned* in its state by them.

Mach's idea finds its full development in the ether of the general theory of relativity. According to this theory the metrical qualities of the continuum of spacetime differ in the environment of different points of spacetime, and are partly conditioned by the matter existing outside of the territory under consideration. This spacetime variability of the reciprocal relations of the standards of space and time, or, perhaps, the recognition of the fact that empty space in its physical relation is neither homogeneous nor isotropic, compelling us to describe its state by ten functions (the gravitation potentials $g_{\mu\nu}$), has, I think, finally disposed of the view that | space is [19] physically empty. But therewith the conception of the ether has again acquired an intelligible content, although this content differs widely from that of the ether of the mechanical undulatory theory of light. The ether of the general theory of relativity is a medium which is itself devoid of *all* mechanical and kinematical qualities, but helps to determine mechanical (and electromagnetic) events.

What is fundamentally new in the ether of the general theory of relativity as opposed to the ether of Lorentz consists in this, that the state of the former is at every place determined by connections with the matter and the state of the ether in neighboring places, which are amenable to law in the form of differential equations; whereas the state of the Lorentzian ether in the absence of electromagnetic fields is conditioned by nothing outside itself, and is everywhere the same. The ether of the general theory of relativity is transmuted conceptually into the ether of Lorentz if we [20] substitute constants for the functions of space which describe the I former, disregarding the causes which condition its state. Thus we may also say, I think, that the ether of the general theory of relativity is the outcome of the Lorentzian ether, through relativation.

As to the part which the new ether is to play in the physics of the future we are not yet clear. We know that it determines the metrical relations in the spacetime continuum, e.g. the configurative possibilities of solid bodies as well as the gravitational fields; but we do not know whether it has an essential share in the structure of the electrical elementary particles constituting matter. Nor do we know whether it is only in the proximity of ponderable masses that its structure differs essentially from that of the Lorentzian ether; whether the geometry of spaces of cosmic extent is approximately Euclidean. But we can assert by reason of the relativistic equations of gravitation that there must be a departure from Euclidean relations, with spaces of cosmic order of magnitude, if there exists a positive mean density, no matter how small, of [21] the matter in the universe. I In this case the universe must of necessity be spatially unbounded and of finite magnitude, its magnitude being determined by the value of that mean density.

If we consider the gravitational field and the electromagnetic field from the standpoint of the ether hypothesis, we find a remarkable difference between the two. There can be no space nor any part of space without gravitational potentials; for these confer upon space its metrical qualities, without which it cannot be imagined at all. The existence of the gravitational field is inseparably bound up with the existence of space. On the other hand a part of space may very well be imagined without an electromagnetic field; thus in contrast with the gravitational field, the electromagnetic field seems to be only secondarily linked to the ether, the formal nature of the electromagnetic field being as yet in no way determined by that of gravitational ether. From the present state of theory it looks as if the electromagnetic field, as opposed to the [22] gravitational field, rests upon an entirely new formal *motif*, as though I nature might just as well have endowed the gravitational ether with fields of quite another type, for example, with fields of a scalar potential, instead of fields of the electromagnetic type.

Since according to our present conceptions the elementary particles of matter are also, in their essence, nothing else than condensations of the electromagnetic field, our present view of the universe presents two realities which are completely separated from each other conceptually, although connected causally, namely, gravitational ether and electromagnetic field, or—as they might also be called—space and matter.

Of course it would be a great advance if we could succeed in comprehending the gravitational field and the electromagnetic field together as one unified conformation. Then for the first time the epoch of theoretical physics founded by Faraday and Maxwell would reach a satisfactory conclusion. The contrast between ether and matter would fade away, and, through the general theory of relativity, the whole of I physics [23] would become a complete system of thought, like geometry, kinematics, and the theory of gravitation. An exceedingly ingenious attempt in this direction has been made by the mathematician H. Weyl; but I do not believe that his theory will hold its ground in relation to reality. Further, in contemplating the immediate future of theoretical physics we ought not unconditionally to reject the possibility that the facts comprised in the quantum theory may set bounds to the field theory beyond which it cannot pass.

Recapitulating, we may say that according to the general theory of relativity space is endowed with physical qualities; in this sense, therefore, there exists an ether. According to the general theory of relativity space without ether is unthinkable; for in such space there not only would be no propagation of light, but also no possibility of existence for standards of space and time (measuring-rods and clocks), nor therefore any spacetime intervals in the physical sense. But this ether may not be thought of as endowed with the quality characteristic I of ponderable media, as consisting of parts [24] which may be tracked through time. The idea of motion may not be applied to it.